HANDBOOK OF FLUOROPOLYMER SCIENCE AND TECHNOLOGY

HANDBOOK OF FLUOROPOLYMER SCIENCE AND TECHNOLOGY

Edited by

DR. DENNIS W. SMITH JR.

Department of Chemistry
University of Texas at Dallas
Richardson, TX

DR. SCOTT T. IACONO

Department of Chemistry
Chemistry Research Center
United States Air Force Academy
Colorado Springs, CO

DR. SURESH S. IYER

3M Corporate R&D
St. Paul, MN

WILEY

Published by John Wiley & Sons, Inc., Hoboken, New Jersey.
Published simultaneously in Canada.

Library of Congress Cataloging-in-Publication Data:

Handbook of fluoropolymer science and technology / edited by Dr. Dennis W. Smith, Jr., Department of Chemistry, the University of Texas at Dallas, Richardson, TX, Dr. Scott T. Iacono, Chemistry Research Center, United States Air Force Academy, Colorado Springs, CO, Dr. Suresh S. Iyer, 3M Corporate R & D, St. Paul, MN, Dr. Suresh S. Iyer, 3M Corporate R & D, St. Paul, MN.

pages cm

Includes bibliographical references and index.

ISBN 978-0-470-07993-5 (cloth)

1. Fluoropolymers. I. Smith, Dennis W., Jr., editor of compilation. II. Iacono, Scott T., editor of compilation. III. Iyer, Suresh S., editor of compilation.

QD383.F48H36 2014

547′.84–dc23

2014002866

MIX
Paper from responsible sources
FSC® C013604
FSC www.fsc.org

CONTENTS

FOREWORD

This book is long overdue in view of substantial progress in fluoropolymers during more than a decade. Individual articles have been written by contributing authors but the combination of integrated highlights in each chapter and comprehensive coverage of topics will be welcomed by a multitude of scientists in this field.

To me, the need for this handbook was perhaps forshadowed by the ACS symposium: *Fluorine-Containing Polymeric Materials* held in Boston in 1990 where the room was filled to overflowing, much to the surprise of the organizers. This led to a follow-on symposium and eventually to a biennial, international, weeklong workshop on fluoropolymers which is well attended.

The span of this text has earned the designation as a handbook in that it covers comprehensively the subject of fluoropolymers from synthesis to properties to copolymers, to composites to processing to applications, even to decomposition, all sorely needed and wisely included herein.

The inclusion of chapters by the leaders in their fields speaks to the quality of this consolidated work. These authors, many of them, have seen the science and technology develop over the past two decades or more and have, therefore, a valuable overview of where we have been in the fluoropolymer endeavor and where future efforts might lead.

PATRICK E. CASSIDY

Distinguished Professor Emeritus
Texas State University

IN MEMORIAM

Dr. Patrick E. Cassidy
1937–2013

Patrick E. Cassidy, born November 8, 1937, died on December 6, 2013, after a lengthy battle with cancer. He is survived by his wife, Jeanne, and three children. Pat was born and raised in East Moline, Illinois. Pat received a B.S. with Honors in Chemistry from the University of Illinois and M.S. and Ph.D. degrees from the University of Iowa. He was a postdoctoral fellow at the University of Arizona with Professor C. S. Marvel. Most of us in the fluoropolymer community are familiar with his work at Texas State University, San Marcos, where his career spanned 41 years, including 15 years as Associate Vice President of Academic Affairs. During his tenure at the university, he founded and directed several institutes and centers, including the Polymer Research Group. Pat published and presented internationally more than 400 papers. Pat was well known internationally, had done extensive work on fluorous polyimides, and had investigated gas separation characteristics of these polymers. He retired as Distinguished Professor Emeritus.

His awards included, among others, being named a Fellow of the ACS Polymer Chemistry Division and also Fellow of the ACS Polymer Materials Science and Engineering Division. He received the Southwest American Chemical Society Award and the Texas State University President's Award for Excellence in Research. During his tenure, he was awarded nine patents, mostly concerning fluoropolymer materials.

Pat was active in the ACS Polymer Chemistry Division and often attended the Executive Committee dinner on Sunday evening. At the last meeting in New Orleans, I enjoyed an extended conversation with Pat and we reminisced about long careers—both his and mine. It is a special, appropriate, and poignant feature of this book to have a Preface authored by Patrick Cassidy. Pat will be long remembered by his friends and associates for his ready smile, his thoughtful ways, and of course his expertise in fluoropolymers.

KENNETH J. WYNNE

Commonwealth Professor
Virginia Commonwealth University
January 2014

PREFACE

The emergence of organofluorine chemistry was prompted by the synthesis of chlorofluorocarbons, which were highly sought after for their desirable thermodynamic properties as refrigerants. A joint research venture between GM and DuPont led by Roy Plunkett in the 1930s produced an abundant line of commercialized refrigerants and has also led to the chance discovery of a material that, in part, has changed much of the modern world. This thermoplastic material was poly(tetrafluoroethylene) (PTFE) and trademarked Teflon®, which was quickly transitioned to the consumer as a household brand name. Initially commissioned for the Manhattan Project in the late 1940s, Teflon® was being produced at a rate of 900 tons per year and has maintained a current growth rate above the US GDP. Now, more than 80 years later, considerable advances in the ability to tailor fluoropolymer properties continue to forge a wide range of technologies in order to tune optical, electronic, processing, environmental stability, and surface properties.

The concept of this book began after a noticeable publication gap was apparent for topical, consolidated reviews on fluorine-containing polymers since *Topics in Applied Chemistry: Fluoropolymers 1 & 2* (Editors: Hougham, Cassidy, Johns, and Davison, Kluwer Academic/Plenum Publishers, 1999) and *Modern Fluoropolymers: High Performance Polymers for Diverse Applications* (Editor: Schiers, John Wiley & Sons, 1997). Other notable book publications on fluorine-containing materials have emerged with emphasis in specialized developments to include: a systematic account of commercial fluoropolymers for industrial practice in *Technology of Fluoropolymers* (Drobny, 2nd ed., CRC Press, 2004), tailoring fluorine-based polymers or emerging applications in *Well-Architectured Fluoropolymers: Synthesis, Properties and Applications* (Ameduri and Boutevin, Elsevier, 2004), and more recently a reoccurring *ACS Symposium Series: Advances in Fluorine Containing Polymers* (Editors:

Smith, Iacono, Boday, and Kettwich, Oxford University Press, 2012). Since 2000, a growing fluoropolymer community of academic and industry membership continued to meet at the biannual Fluoropolymer Workshop sponsored by the American Chemical Society, Division of Polymer Chemistry. Over this time, it was readily apparent that fluoropolymer research is vibrant as ever and continues to morph in order to provide solutions for today's ever-changing advanced material challenges among the commercial, defense, and civilian aerospace corridors.

The purpose of this book is to present a timely assembly of review topics of new discoveries from leading international laboratories including academia, government, and industrial institutions. Emphasis is placed on emerging technologies in energy, optics, space exploration, fuel cells, microelectronics, gas separation membranes, biomedical instrumentation, and much more. In addition, the book covers the current environmental concerns associated with fluoropolymers as well as relevant regulations and potential growth opportunities. The book is targeted for practicing scientists and engineers who are engaged in research and development of fluoropolymer synthesis, characterization, and materials processing. In addition, this handbook serves as an academic instructional tool for advanced undergraduate or graduate-level special topics courses in chemistry, chemical engineering, or materials science and engineering.

Dennis W. Smith Jr.
Scott T. Iacono
Suresh S. Iyer

September 2013

CONTRIBUTORS

Olumide I. Adebolu, Department of Chemistry, University of Connecticut, Storrs, CT, USA

Anupriya Agrawal, Department of Chemistry, Clemson University, Clemson, SC, USA

Harry R. Allcock, Department of Chemistry, Pennsylvania State University, University Park, PA, USA

Bruno Ameduri, Institute Charles Gerhardt of Montpellier, France

Nasim Anousheh, Department of Chemistry, Université de Sherbrooke, Sherbrooke, Québec, Canada

Marco Apostolo, Solvay Specialty Polymers, R&D Center, Bollate (MI), Italy

Vincenzo Arcella, Solvay Specialty Polymers, R&D Center, Bollate (MI), Italy

Dipak Aryal, Department of Chemistry, Clemson University, Clemson, SC, USA

Alexandru D. Asandei, Institute of Materials Science, Polymer Program and Department of Chemistry, University of Connecticut, Storrs, CT, USA

Jessi Baughman, Department of Chemistry, University of Akron, Akron, OH, USA

Paul Brothers, Experimental Station, E.I. du Pont de Nemours and Co., Wilmington, DE, USA

Kennard Brunson, Department of Chemical and Life Science Engineering, Virginia Commonwealth University, Richmond, VA, USA

Tao Cai, NUS Graduate School for Integrative Science and Engineering, National University of Singapore, Kent Ridge, Singapore

Patrick E. Cassidy, Department of Chemistry and Biochemistry, Texas State University, San Marcos, TX, USA

Asima Chakravorty, Department of Chemical and Life Science Engineering, Virginia Commonwealth University, Richmond, VA, USA

Gregory Chapman, Washington Laboratory, E.I. du Pont de Nemours and Co., Washington, WV, USA

T.C. Mike Chung, Department of Materials Science and Engineering, Pennsylvania State University, University Park, PA, USA

James L. Clark Jr., Office of Research Safety, Clemson University, Clemson, SC, USA

Stephen Creager, Department of Chemistry and the Advanced Materials Research Laboratory, Clemson University, Anderson, SC, USA

Philip Crouse, Department of Chemical Engineering, University of Pretoria, Pretoria, South Africa

Darryl D. DesMarteau, Department of Chemistry and the Advanced Materials Research Laboratory, Clemson University, Anderson, SC, USA

Mathias Destarac, Department of Chemistry, University Paul Sabatier of Toulouse, Toulouse, France

Kimberly Farnsworth, Washington Laboratory, E.I. du Pont de Nemours and Co., Washington, WV, USA

Richard E. Fernandez, Rf Consulting, 122 Loudwater Drive, Anderson, SC USA and Department of Chemistry and the Advanced Materials Research Laboratory, Clemson University, Anderson, SC, USA

Tomoko Fujiwara, Department of Chemical and Life Science Engineering, Virginia Commonwealth University, Richmond, VA, USA

Chun Gao, Department of Chemistry, University of Akron, Akron, OH, USA

Etienne Girard, Department of Chemistry, University Paul Sabatier of Toulouse, Toulouse, France

Karen K. Gleason, Department of Chemical Engineering, Massachusetts Institute of Technology, Cambridge, MA, USA

Gary S. Grest, CINT, Sandia National Laboratories, Albuquerque, NM, USA

Daniel A. Hercules, Department of Chemistry and the Advanced Materials Research Laboratory, Clemson University, Anderson, SC, USA

Yuji Higaki, Institute for Materials Chemistry and Engineering, Kyushu University, Fukuoka, Japan

Klaus Hintzer, 3M/Dyneon GmbH, Burgkirchen, Germany

Scott T. Iacono, Department of Chemistry and Chemistry Research Center, United States Air Force Academy, Colorado Springs, CO, USA

Ryohei Ishige, Institute for Materials Chemistry and Engineering, Kyushu University, Fukuoka, Japan

Dieter Jehnichen, Leibniz-Institut für Polymerforschung Dresden e.V.

Jianyong Jin, School of Chemical Sciences, University of Auckland, Auckland, New Zealand

En-Tang Kang, Department of Chemical and Biomolecular Engineering, National University of Singapore, Kent Ridge, Singapore

Kotaro Koike, Keio Photonics Research Institute, Keio University, Kanagawa, Japan

Yasuhiro Koike, Keio Photonics Research Institute, Keio University, Kanagawa, Japan

Pinar Kurt, Department of Chemical and Life Science Engineering, Virginia Commonwealth University, Richmond, VA, USA

Linlin Li, Department of Chemistry, University of Akron, Akron, OH, USA

Xiaohong Li, College of Chemistry, Chemical Engineering and Materials Science, Soochow University, Suzhou, Jiangsu, China

Joseph Mabry, Air Force Research Laboratory, Rocket Propulsion Division, Edwards Air Force Base, CA, USA

Umit Makal, Department of Chemical and Life Science Engineering, Virginia Commonwealth University, Richmond, VA, USA

Jean-Daniel Marty, Department of Chemistry, University Paul Sabatier of Toulouse, Toulouse, France

Elizabeth F. McCord, Experimental Station, E.I. du Pont de Nemours and Co., Wilmington, DE, USA

Luca Merlo, Solvay Specialty Polymers, R&D Center, Bollate (MI), Italy

František Mikeš, Polymer Research Institute, NYU Polytechnic School of Engineering, Brooklyn, NY, USA

Richard Morgan, Washington Laboratory, E.I. du Pont de Nemours and Co., Washington, WV, USA

Oliver Mulamba, Mechanical Engineering Department, Texas Tech University, Lubbock, TX, USA

Koon-Gee Neoh, Department of Chemical and Biomolecular Engineering, National University of Singapore, Kent Ridge, Singapore

Dennis E. Ohman, Department of Microbiology and Immunology, Virginia Commonwealth University, McGuire Veterans Affairs Medical Center, Richmond, VA, USA

Yoshiyuki Okamoto, Polymer Research Institute, NYU Polytechnic School of Engineering, Brooklyn, NY, USA

Michael Owen, Michigan Molecular Institute, Midland, MI, USA

Michelle Pantoya, Mechanical Engineering Department, Texas Tech University, Lubbock, TX, USA

Dvora Perahia, Department of Chemistry, Clemson University, Clemson, SC, USA

Riccardo Pieri, Solvay Specialty Polymers, R&D Center, Bollate (MI), Italy

François Porzio, Department of Chemistry, Université de Sherbrooke, Sherbrooke, Québec, Canada

Doris Pospiech, Leibniz-Institut für Polymerforschung Dresden e.V.

Gerard Puts, Department of Chemical Engineering, University of Pretoria, Pretoria, South Africa

Mustapha Raihane, Faculty of Sciences and Techniques, Marrakech, Morocco

Sean Ramirez, Air Force Research Laboratory, Rocket Propulsion Division, Edwards Air Force Base, CA, USA

Peter L. Rinaldi, Department of Chemistry, University of Akron, Akron, OH, USA

Hideo Sawada, Department of Frontier Materials Chemistry, Graduate School of Science and Technology, Hirosaki University, Hirosaki, Japan

Werner Schwertfeger, 3M/Dyneon GmbH, Burgkirchen, Germany

Iqbal Sharif, Department of Chemistry and the Advanced Materials Research Laboratory, Clemson University, Anderson, SC, USA

Christopher P. Simpson, Institute of Materials Science, Polymer Program, University of Connecticut, Storrs, CT, USA

Dennis W. Smith Jr., Department of Chemistry and the Alan G. MacDiarmid NanoTech Institute, University of Texas at Dallas, Richardson, TX, USA

Armand Soldera, Department of Chemistry, Université de Sherbrooke, Sherbrooke, Québec, Canada

Atsushi Takahara, Institute for Materials Chemistry and Engineering, International Institute for Carbon-Neutral Energy Research (WPI-I2CNER), Japan Science and Technology Agency ERATO Takahara Soft Interfaces Project, Kyushu University, Fukuoka, Japan

Joseph S. Thrasher, Department of Chemistry and the Advanced Materials Research Laboratory, Clemson University, Anderson, SC, USA

Paolo Toniolo, Solvay Specialty Polymers, R&D Center, Bollate (MI), Italy

Francesco Triulzi, Solvay Specialty Polymers, R&D Center, Bollate (MI), Italy

Eric B. Twum, Department of Chemistry, University of Akron, Akron, OH, USA

Lynn Wood, Department of Microbiology and Immunology, Virginia Commonwealth University, McGuire Veterans Affairs Medical Center, Richmond, VA, USA

Kenneth J. Wynne, Department of Chemical and Life Science Engineering, Virginia Commonwealth University, Richmond, VA, USA

Faith J. Wyzgoski, Department of Chemistry, Ohio State University—Mansfield, Mansfield, OH, USA

Jose L. Yagüe, Department of Chemical Engineering, Massachusetts Institute of Technology, Cambridge, MA, USA

ABOUT THE EDITORS

Dennis W. Smith Jr. joined The Dow Chemical Company Central Research Laboratory as Sr. Research Chemist in 1993, working primarily on the synthesis and characterization of high performance thermosets for thin film microelectronics applications. He then joined Clemson in 1998 and was promoted to Professor of Chemistry in 2006, and in 2010, he joined the University of Texas at Dallas as Robert A. Welch Professor of Chemistry and was elected as Fellow of the American Chemical Society. His research interests include synthesis, mechanism, structure–property relationships, and applications of polymeric materials and composites. Smith received his BS from Missouri State University and his PhD from the University of Florida under the guidance of Prof. Ken Wagener on the scope and mechanism of acyclic diene metathesis (ADMET) polymerization.

Scott T. Iacono was a research chemist at the Air Force Research Laboratory at Edwards Air Force Base, CA, synthesizing inorganic–organic hybrid fluoropolymers for solid rocket motor propulsion components. In 2010, he joined the Department of Chemistry at the United States Air Force Academy and has since been promoted to associate professor and concurrently serves as the Chemistry Research Center Director. His current research efforts focus on organofluorine methodologies for monomer development for fluorosiloxane polymers, metalized fluoropolymer composites, and fulvene-based conjugated polymers for light emissive applications. He has authored collectively over 60 journal papers, conference papers, and contributed book chapters. Iacono completed his BS at Louisiana State University and PhD studies with emphasis organic and polymer chemistry at Clemson University.

Dr. Suresh Iyer is currently a Technical Manager at the 3M Safety & Graphics business lab, 3M India Ltd. His group provides design, synthesis and evaluation of polymers for new technology and products at 3M. Suresh has a Ph.D. in Polymer Chemistry from National Chemical Laboratory, Pune (India) and pursued research in academic institutions before joining 3M in 2006. At 3M he continued his research on high performance polymers, functional coatings and low surface energy materials. He also successfully completed the Lean Six Sigma Black Belt Training from 3M in 2011. Suresh has over 25 technical publications and 40 patents/applications. He was a member of organizing committee of FLUOROPOLYMER 2012 and Chair Elect – 2014 for Minnesota local section of the American Chemical Society.

1

FLUORINATED POLYPHOSPHAZENES

Harry R. Allcock
The Pennsylvania State University

1.1 BACKGROUND

Polyphosphazenes (Structure 1.1) are a broad class of macromolecules with a backbone of alternating phosphorus and nitrogen atoms and with two side groups (R) attached to each phosphorus atom.

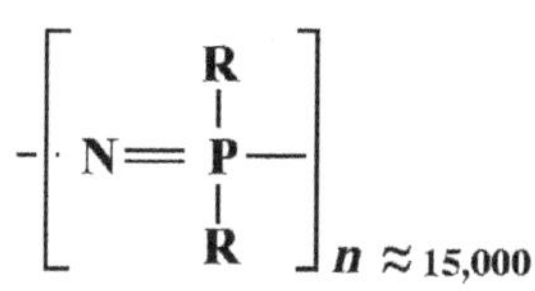

STRUCTURE 1.1

The skeletal architecture may be linear, branched, star, or dendritic, or it may be part of a di- or triblock copolymer in conjunction with organic macromolecules or poly(organosiloxanes) (silicones). However, it is in the wide variety of side groups that this system differs from many other polymer platforms. More than 250 different organic, organometallic, or inorganic side groups have been utilized in single-substituent arrays or in di-, tri-, or higher mixed-substituent patterns. Thus, hundreds of different polyphosphazenes are known with a corresponding diversity of properties

Handbook of Fluoropolymer Science and Technology, First Edition. Edited by Dennis W. Smith Jr., Scott T. Iacono, and Suresh S. Iyer.

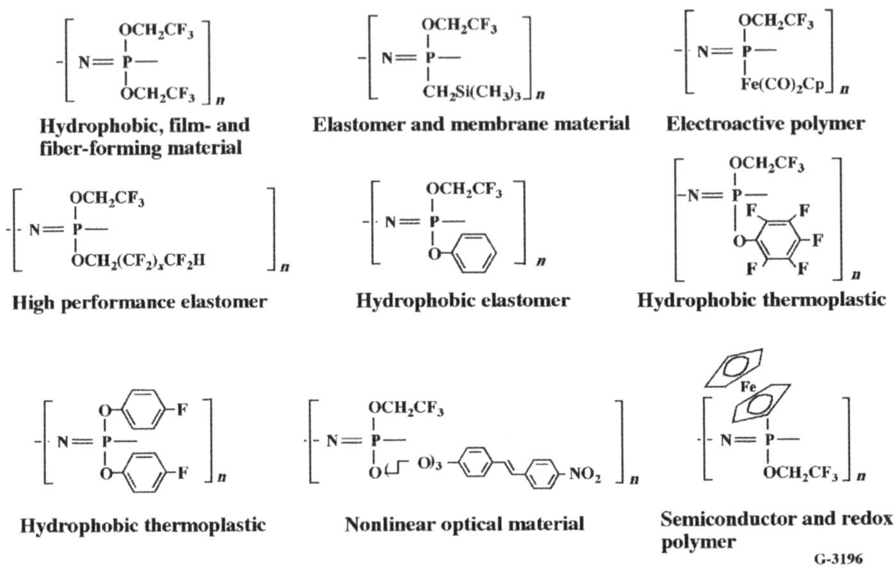

FIGURE 1.1 Examples of fluorinated organophosphazene polymers.

and potential uses [1]. These can be divided into different "families" of polymers such as inert biomaterials, bioerodible polymers, optical materials, membranes, ionic conductors, and so on. One of the most important and most interesting families consists of polyphosphazenes that bear fluorinated organic side groups. Examples of polymers within this group are shown in Figure 1.1.

1.2 SYNTHESIS METHODS AND PROPERTY DEVELOPMENT

A number of different access routes have been developed to poly (organophosphazenes) [1]. We have focused on a two-stage sequence that involves first the preparation of a linear polymeric reaction intermediate, poly(dichlorophosphazene), $(NPCl_2)_n$, followed in a second step by replacement of the chlorine atoms in this polymer by organic side groups (Figure 1.2) [2–4]. The reactive intermediate is accessible either by a ring-opening polymerization of a cyclic trimer, $(NPCl_2)_3$, or via a living cationic condensation polymerization of a phosphoranimine (Figure 1.2) [5–15]. Another route to poly(dichlorophosphazene) is via the condensation reactions of $Cl_3P{=}N\text{-}POCl_2$ [16], a method that yields lower molecular weight polymers than the ring-opening route. Replacement of the chlorine atoms in $(NPCl_2)_n$ is accomplished by reactions with nucleophiles such as alkoxides, aryloxides, amines, or organometallic reagents.

This is a very different protocol than that exists for most classical polymers, where the side groups destined for the final polymer must be in place on the monomer

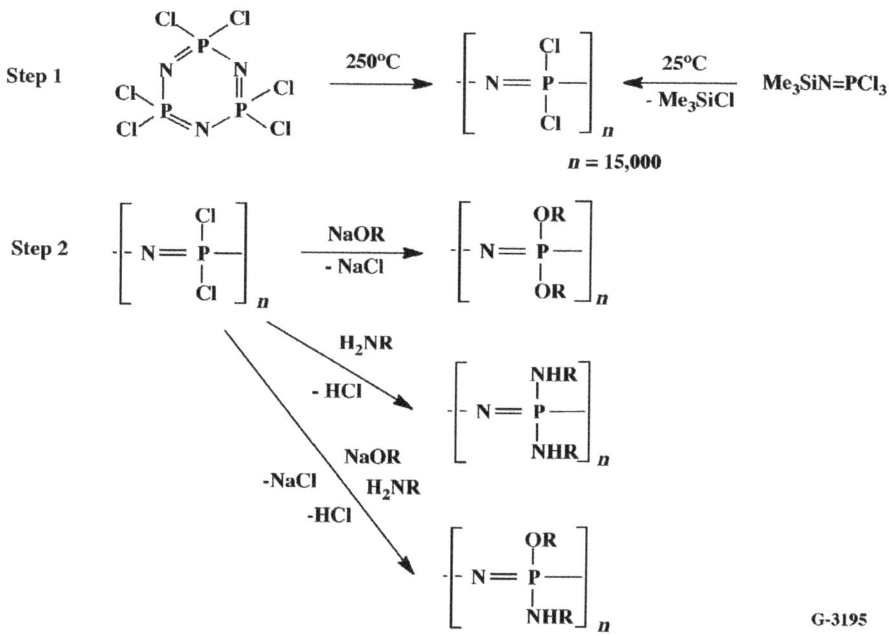

FIGURE 1.2 Two-stage synthesis of poly(organophosphazenes).

before polymerization [17]. Modification of the side groups in conventional macromolecules after polymerization is restricted to simple reactions such as hydrolysis of esters or partial sulfonation. It is the high reactivity of poly(dichlorophosphazene) that allows the broad diversity of structure and properties that are a characteristic of poly(organophosphazenes). Other valuable methods have been developed to prepare poly(organophosphazenes) that involve the condensation reactions of organic-substituted phosphoranimines [18–21], but the range of side groups used in that process is more restricted than in the macromolecular substitution method, and the molecular weights tend to be lower.

Use of these synthetic techniques has led to the development of numerous different classes of phosphazene materials, many of which contain fluorine, but others that do not [1]. For example, a versatile class of hydrogel polyphosphazenes and ion conductive materials possesses nonfluorinated oligoethyleneoxy side chains. Nonfluorinated aryloxy substituents give fire-retardant polymers. Amino acid ester side groups or nonfluorinated alkoxy groups generate bioerodible properties that have been developed extensively for tissue engineering applications. Nevertheless, the presence of fluorine in the side group structure has led to some of the most intriguing developments, and this is the focus of the rest of this article. Using the two-step synthesis protocol, molecular diversity is accomplished in several different ways.

Method 1. Different nucleophiles give polymers with different side groups and diverse properties [1–4]. For example, oligoethyleneoxy side groups give water-soluble, water-stable polymers [22]. Aryloxy side units generate hydrophobic, water-insoluble polymers. Amino acid ester side groups or oligopeptide units linked to the polymer skeleton through the amino terminus generate bioerodible characteristics [23]. Fluoroalkoxy or fluoroaryloxy side groups generate hydrophobic, water- and radiation-stable polymers [24].

Method 2. A second method for structural and property tuning involves the introduction of two or more different side groups along the same polymer chain. For example, amphiphilic character is accessible by the use of fluoroalkoxy groups and oligoethyleneoxy side chains, with the exact properties being controlled by the ratios of the two. Two different fluoroalkoxy side groups on the same chain have a striking effect on the polymer morphology. A polyphosphazene with only trifluoroethoxy side groups is a film- or fiber-forming microcrystalline material, similar to poly(tetrafluoroethylene) in surface properties but, unlike Teflon, soluble in ordinary organic solvents such as acetone or methylethylketone. By contrast, the related polymer with both trifluoroethoxy and longer chain telomer fluoroalkoxy groups is an amorphous elastomer, prized for its low glass transition temperature (approximately −60°C), solvent and oil resistance, and impact-absorbing character.

For polyphosphazene molecules that bear two or more different side groups, serious questions exist about the ratios and pattern of distribution of these groups along the polymer chains. These two factors have a significant influence on the properties of the final polymer. For example, a random distribution often precludes crystallinity and favors elastomeric character. Properties that result from a regular distribution may depend on whether the substituents are geminal or nongeminal, cis or trans, or if block structures are present (Figures 1.3 and 1.4). ^{31}P NMR spectra can sometimes provide clues about the distribution.

Note that the *order* of introduction of the side groups often plays a major role in controlling the distribution pattern and the properties. The distribution of the first substituent will control the positioning of the second substituent and, in turn, the precise combination will control the properties of the fully substituted polymer. *Simultaneous* addition of the two nucleophiles often yields results that depend on the electronic and steric characteristics of the reagents and also on the reaction conditions.

Method 3. The third opportunity for structural diversity arises through organic side group exchange chemistry (Figure 1.5). Thus, specific polyphosphazenes in solution can be modified by the replacement of one type of organic side group by exposure to another nucleophile. Fluoroalkoxide A will replace fluoroalkoxy group B. Aryloxy side groups with electron-withdrawing substituents may be replaced by trifluoroethoxy groups. Trichloroethoxy side groups can also be exchanged for trifluoroethoxy side units [25]. This is a useful way to fine-tune polymer properties such as solubility or morphology. It can also be used as a method for cross-linking polyphosphazene chains by the use of a difunctional nucleophile that can exchange with some of the organic side groups. Moreover, it provides an excellent method for modification of the *surface* of a solid polyphosphazene without affecting the composition of the interior [26,27].

Method 4. Finally, the preparation of polymers with multifunctional organic side groups raises a synthetic challenge. For example, if the objective is to produce a

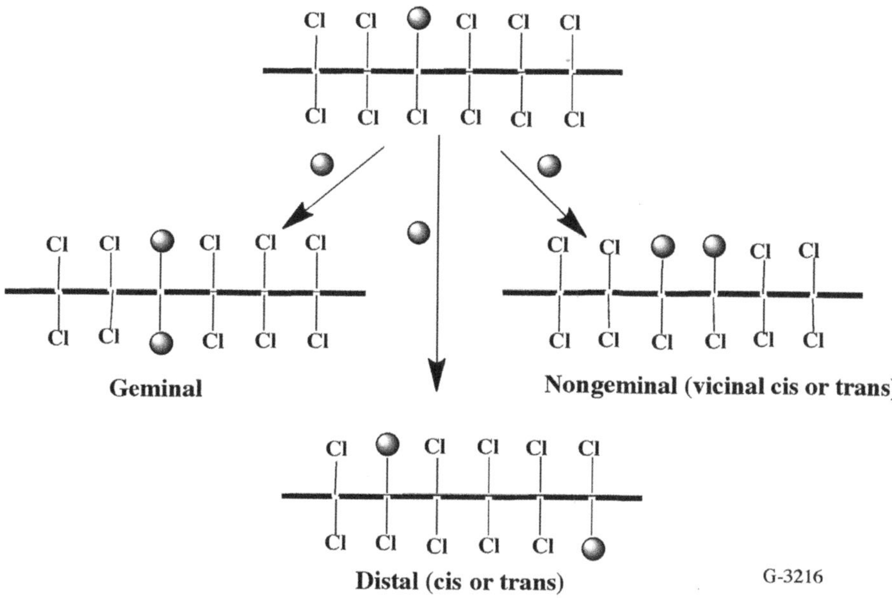

FIGURE 1.3 Initial possible steps after the replacement of the first chlorine atom per chain by the same organic group (spheres).

polymer with the side groups connected to the main chain by the reaction of a hydroxyl group with the chlorophosphazene, but the incoming nucleophile also bears an amino unit, a carboxylic acid group, or a second hydroxy group, the di- or tri-functional nucleophile will cross-link the chains, precipitate the polymer before all the chlorine atoms are replaced and prevent complete halogen replacement. Hence, a first step must be the protection of the noncoupling functional groups, and a deprotection of these units once chlorine replacement along the polymer chain is complete. A simple example, is the use of the sodium salt of an ester of p-hydroxybenzoic acid

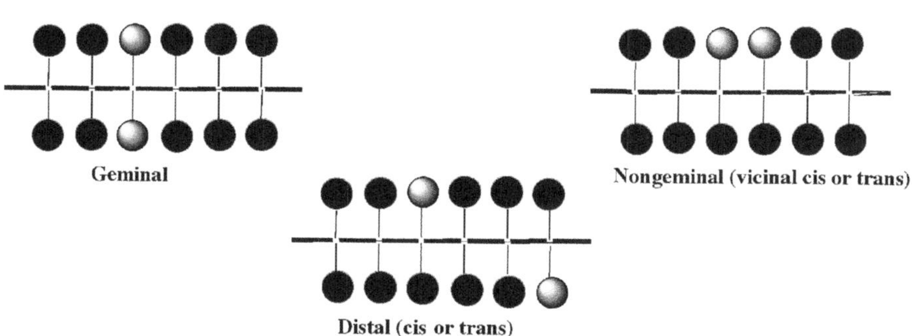

FIGURE 1.4 Simplified representation of the side group disposition following the introduction of a second type of side group (black spheres) after the distribution of the first side groups has been established.

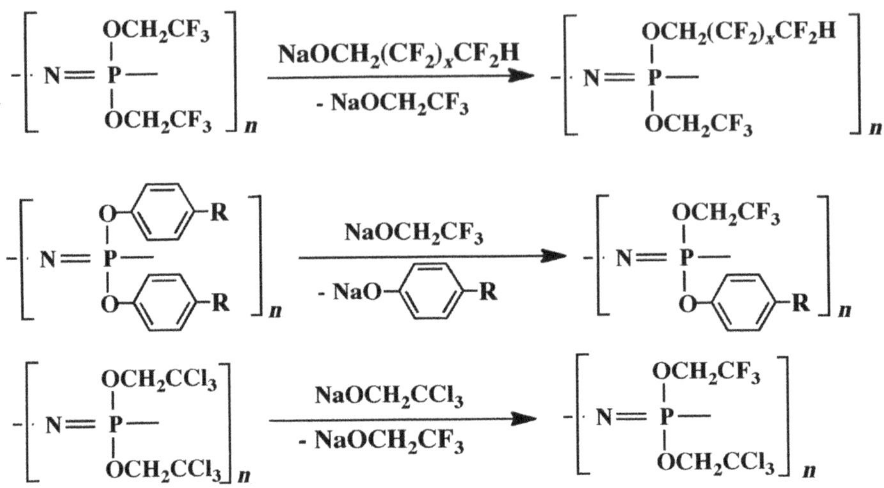

FIGURE 1.5 Substituent exchange as an alternative route to the preparation of mixed-substituent polymers.

as the nucleophile to link the side group to the chain via the hydroxyl group, followed by hydrolysis of the ester function to generate the free carboxylic acid or carboxylate moiety (Figure 1.6). We have developed more complex protection–deprotection procedures to utilize numerous multifunctional biological side groups [23, 28].

In addition to the above methods, mixed side group systems have been synthesized through block copolymers with other polyphosphazenes, or with organic polymers or organosilicon polymers, and through the assembly of stars, or dendrimers, all by use of the living cationic polymerization route (Figure 1.2). These options further widen the opportunities for property development. As an example, amphiphilic diblock copolymers with one block bearing hydrophobic side groups and the second block being hydrophilic have been studied for micelle formation.

There is one additional experimental issue that is crucial for all this synthesis work. This is a consequence of the hydrolytic sensitivity of chlorophosphazenes. Both the cyclic trimer, $(NPCl_2)_3$, and the high polymer, $(NPCl_2)_n$, react with atmospheric moisture to produce P—OH units that can trigger ring or chain cleavage reactions or may undergo intermolecular condensation reactions to cross-link phosphazene molecules via P—O—P linkages. In the high polymer, this is a major source of unwanted cross-linking that can also restrict the extent of chlorine replacement by organic nucleophiles. Reports of difficulties with the preparation of polyphosphazenes by inexperienced investigators can almost always be traced to failure to purify commercial $(NPCl_2)_3$ or to protect the trimer or the polymer from atmospheric moisture before chlorine replacement. Moreover, dry reagents and solvents must be

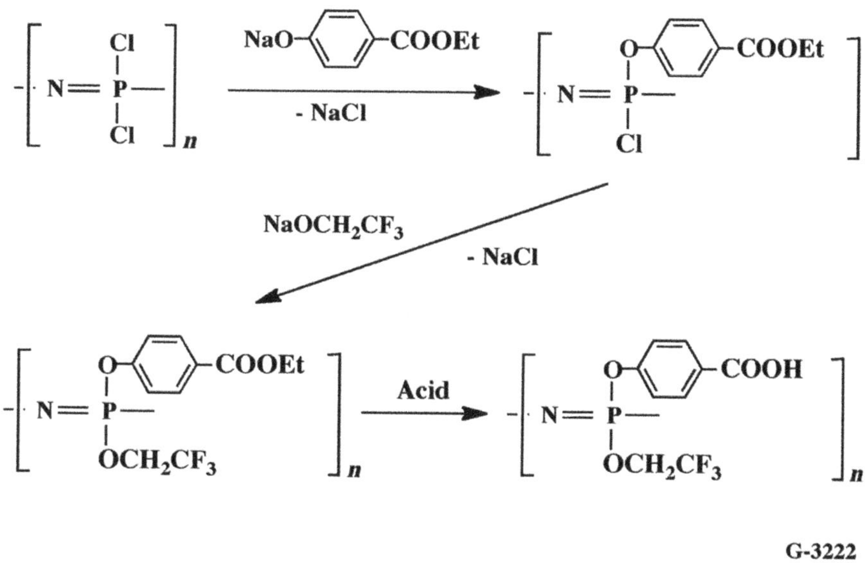

FIGURE 1.6 A variety of protection–deprotection strategies have been developed to prevent cross-linking when a polyphosphazene reacts with di- or tri-functional reagents. The simplest of these involves deprotection by the hydrolysis of an ester function.

used at every stage of the syntheses until all the chlorine atoms have been replaced by organic groups.

1.3 THE ROLE OF FLUORINE IN POLYPHOSPHAZENE SYNTHESIS CHEMISTRY

Polyphosphazenes with fluoro-organic side groups have played a major role in the development of this field. This is a consequence of the unique properties imparted to polyphosphazenes by fluorinated organic groups, but it also reflects the role played by fluorine in the synthesis chemistry as well. There are four ways in which fluorine plays an important role in the synthesis of polyphosphazenes.

1.3.1 Facilitation of Chlorine Replacement by Fluorinated Nucleophiles

In this first influence, fluorine as a component of various alkoxide, aryloxide, or amino nucleophiles accelerates the chlorine replacement reactions on which the syntheses of most poly(organophosphazenes) depend. This role is attributed to the strong electron-withdrawal associated with carbon–fluorine bonds, an effect that favors dissociation of, for example, the sodium–oxygen bond in sodium alkoxides or aryloxides and thus accelerates the substitution process. This effect supplements the driving force brought about by the precipitation of sodium chloride from organic media.

1.3.2 Enhancement of the Hydrolytic Stability and Resistance to Other Reagents When Fluorine Is Present in the Organic Side Groups

Carbon–fluorine bonds in the side groups withdraw electrons from the skeletal nitrogen atoms, thus retarding side reactions in which reagents such as water, acids, or bases attack and cleave the skeletal linkages. Fluorine in the organic side groups also imparts other useful properties such as hydrophobicity, low glass transition temperatures, resistance to oil, and general chemical resistance. These attributes become evident when, for example, poly[diethoxyphosphazene], $[NP(OEt)_2]_n$, is compared with poly[bis(trifluoroethoxy)phosphazene], $[NP(OCH_2CF_3)_2]_n$. The nonfluorinated ethoxy derivative is hydrolytically unstable whereas the trifluoroethoxy counterpart is resistant to hydrolysis.

1.3.3 Influence by the Structure of the Fluorinated Alkoxide Nucleophile

Although polyphosphazenes with trifluoroethoxy side groups are among the most interesting and useful polymers, deviations from this structure lead to synthetic challenges. Thus, the preparation of sodium fluoroalkoxides by the treatment of $HOCH_2(CF_2)_xCF_2H$ with metallic sodium or sodium hydride is a facile reaction provided two criteria are met. First, mild conditions must be ensured to avoid elimination of HF from the $-CF_2H$ terminus [29]. However, the formation of terminal unsaturation can be utilized for cross-linking processes. Second, fluoroalkoxy nucleophiles with two adjacent methylene units, such as $-O(CH_2)_2(CF_2)_xCF_3$ are especially prone to HF elimination in the presence of sodium metal or sodium hydride [30] and this ultimately leads to the formation of colored, insoluble products rather than chlorine replacement. Both small molecule model reactions and polymer syntheses have been used to study these side reactions.

1.3.4 Fluoroaryloxy Side Groups

Replacement of chlorine atoms in poly(dichlorophosphazene) occurs readily with fluorine-substituted aryloxides such as 4-fluorophenoxide, 4-trifluoromethylphenoxide, or 3,5-bis(trifluoromethyl)phenoxide; although, for steric reasons the preparation of mixed-substituent polymers with both fluoroaryloxy and trifluoroethoxy groups is easier than the introduction of single-substituent fluoroaryloxy derivatives [31, 32]. However, attempts to synthesize polymers that bear *only* pentafluorophenoxy groups have proved problematic due to the high crystallinity and insolubility of the polymer as it approaches 100% chlorine replacement. On the other hand, mixed-substituent polymers with both pentafluorophenoxy and trifluoroethoxy substituents are relatively easy to produce. Like the other reactions, these fluoro-aryloxide substitutions can be modeled fairly well by the reactions of the cyclic trimer, $(NPCl_2)_3$ before expending effort on the polymer syntheses.

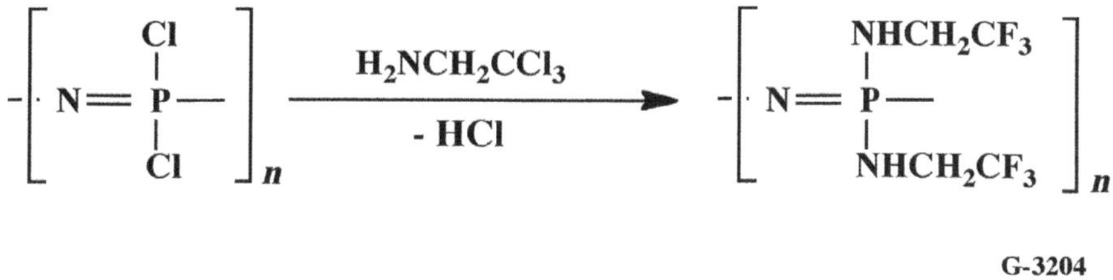

FIGURE 1.7 Fluoroalkylamino groups can be linked to the polyphosphazene chain. These reactions are facilitated by the presence of nonfluorinated tertiary amines as hydrochloride acceptors.

1.3.5 Fluorinated Alkylamino Side Groups

Although fluorinated alkoxy and aryloxy polyphosphazenes have been investigated in some detail, fluorinated *amino*phosphazene polymers have received less attention. A possible reason is the lower reactivity of these amines compared to the fluoroalkoxides. However, an example of this class has been investigated recently with the synthesis of 2,2,2-trifluoroethylamino derivatives (Figure 1.7) [33].

1.3.6 Poly(difluorophosphazene) and Derivatives as Alternatives to Poly(dichlorophosphazene)

Reactions of organometallic reagents with phosphazenes are facilitated if the macromolecular intermediate bears P—F rather than P—Cl bonds. Reactions of poly(dichlorophosphazene) with organometallic reagents such as Grignard, organolithium or organocopper compounds are often accompanied by coordination of the organometallic species to the lone pair electrons of the skeletal nitrogen atoms, a process that can eventually lead to insolubilization or phosphorus–nitrogen bond cleavage.

However, when phosphorus–fluorine rather than phosphorus–chlorine bonds are present in the macromolecular intermediate, the chemistry favors halogen replacement by organometallic reagents [34]. The nitrogen lone pair electron density is reduced by the direct electron-withdrawal by the proximal P–F bonds, thus favoring fluorine replacement over nitrogen coordination. The use of poly(difluorophosphazene) as a macromolecular intermediate is complicated by its insolubility in nearly all solvents [35]. However, polymerization of monophenylpentafluorocyclotriphosphazene, $N_3P_3F_5Ph$ [36], yields a linear polymer with one phenyl group on every third phosphorus atom [37]. This polymer is sufficiently soluble in common organic solvents that it can be used for the replacement of the fluorine atoms using organometallic reagents. Polymers with phenyl, trimethylsilyl, and trifluoroethoxy side groups have been prepared by this route (Figure 1.8) [38].

An alternative source of a fluorophosphazene macromolecular intermediate is via the living cationic polymerization of $Me_3SiN{=}PF_2Ph$, a counterpart of the reaction shown in Figure 1.2 [39]. The overall polymerization reaction is shown in Figure 1.9.

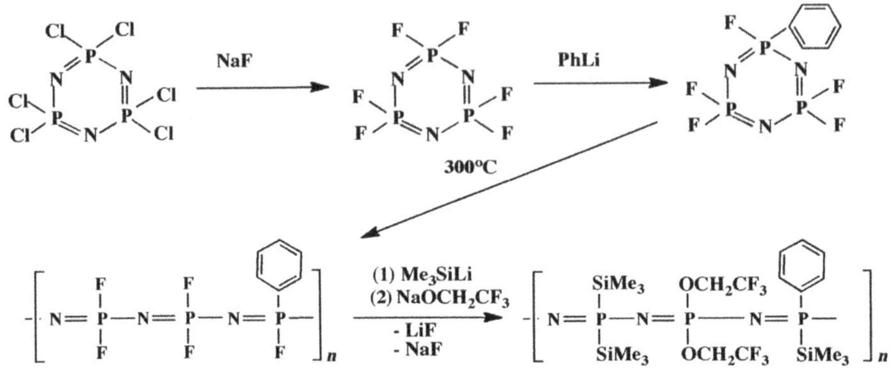

FIGURE 1.8 The presence of phenyl groups along a poly(fluorophosphazene) chain generates solubility in common organic solvents, and this allows the fluorine atoms to be replaced by the use of organometallic reagents.

1.3.7 Block Copolymers and Micelles

Access to the room temperature cationic condensation polymerization, shown in Figure 1.2, has provided a method for precise control of the chain length of poly(dichlorophosphazene) and synthesis of block copolymers via reactions of the living chain termini. One aspect involves the formation of block copolymers with two phosphazene blocks that bear different side groups [39]. A polyphosphazene chain can also be coupled to organic polymers or poly(organosiloxanes) via the use of terminal reactivity in both polymers [40]. Some examples are shown in Figures 1.10 and 1.11. In diblock copolymers, if one block bears hydrophilic side groups and the other block contains fluorinated hydrophobic groups (or *vice versa*) then access is gained to micelles. We have explored micelle formation and its consequences in some detail [41–43], especially when the hydrophobic blocks bear fluoroalkoxy side groups. These micelles are of interest as nanoscale carriers for hydrophobic guest molecules in aqueous media.

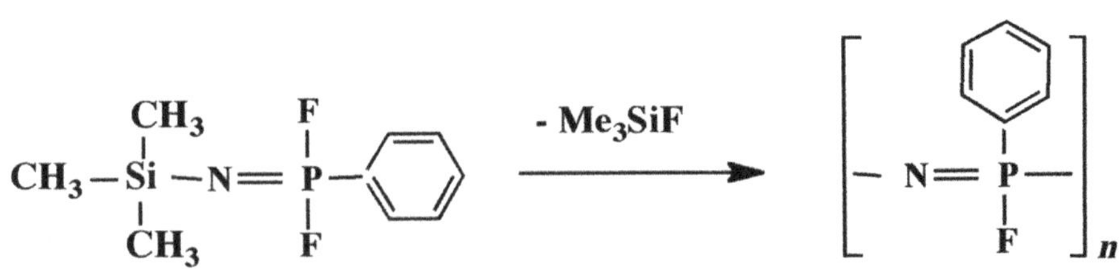

G-3221

FIGURE 1.9 Preparation of a polyphosphazene with one fluorine and one phenyl group on every phosphorus atom using the living cationic polymerization technique.

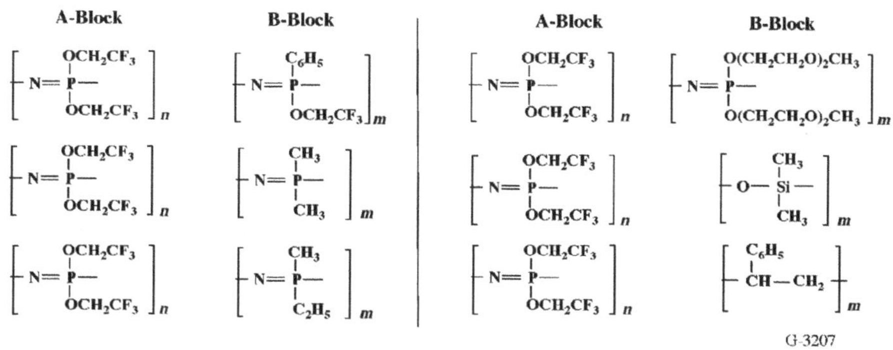

FIGURE 1.10 Diblock polymers containing polyphosphazene units, produced via the living cationic polymerization technique.

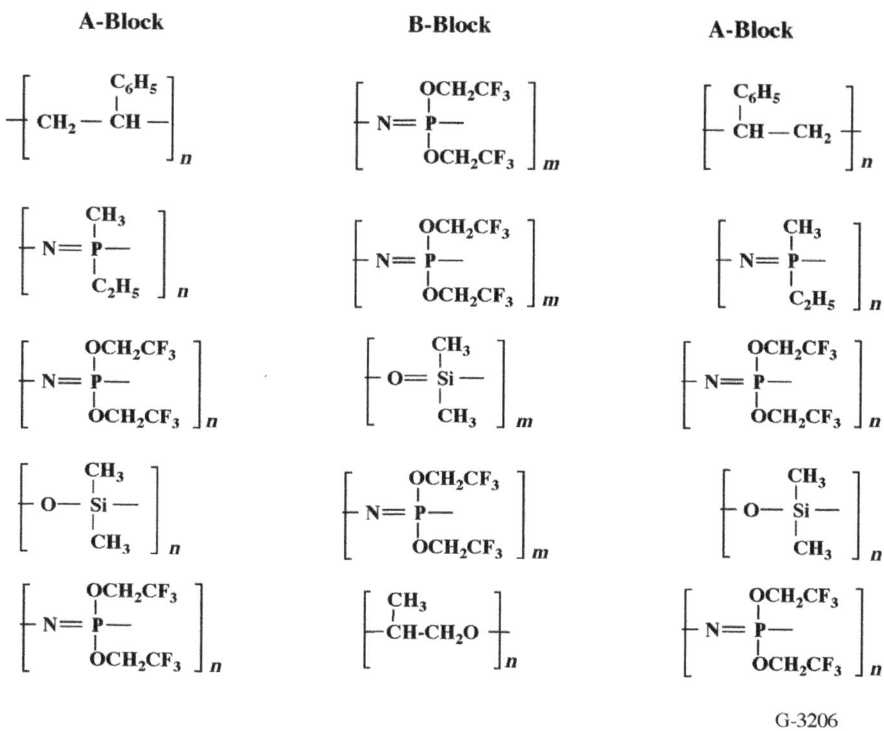

FIGURE 1.11 Triblock copolymers containing polyphosphazene units, accessed via the living cationic polymerization method.

1.4 PROPERTIES OF FLUORINATED POLYPHOSPHAZENES

Fluorinated poly(organophosphazenes) have distinctly different properties from their nonfluorinated counterparts. This is due to the stabilizing effect of the fluorinated side groups on the backbone, the hydrophobicity or super-hydrophobicity of films or fibers, and the general biomedical stability and compatibility of polymers with fluorocarbon units in the side group structure. The following examples illustrate some of the main properties.

1.4.1 Solubility as a Function of Side Group Structure

The solubility of a polymer in different solvents holds the key to its use in many fabrication techniques such as solution fiber spinning, electrospinning, film fabrication, and spin coating. The lack of solubility in hydrocarbon fuels and lubricants and aggressive hydraulic fluids is important for many engineering applications. Fluoroalkoxy groups linked to a polyphosphazene chain control solubility in organic solvents in the following ways. Trifluoroethoxy side groups confer solubility in solvents such as tetrahydrofuran (THF), acetone, or methylethylketone. Co-substituent side groups of the type $-OCH_2(CF_2)_xCF_2H$ permit solubility in the same solvents but with the solubility decreasing as the number of CF_2 groups increases. With the exception of trifluoroethoxy side groups, fluoroalkoxy units of type $OCH_2(CF_2)_xCF_3$ with CF_3 terminal units yield polymers that are less soluble in the same solvents. Fluorinated *aryloxy* side groups give rise to polymers that are less soluble than their alkoxy counterparts in nearly all solvents.

1.4.2 Solid State Properties

A comparison of polyphosphazenes that bear fluorinated or nonfluorinated side groups reveals that the nonfluorinated species are usually amorphous whereas the fluorinated examples are microcrystalline. For example, the polyphosphazene with ethoxy side groups is an amorphous gum. Its counterpart with 2,2,2-trifluoroethoxy side groups is a microcrystalline film- and fiber-former. However, polymers with two or more fluorinated side chains of different length have lower crystallinity or are amorphous presumably due to the lack of molecular order. The presence of fluorine in the side chains increases the glass transition temperature compared to nonfluorinated analogs but, even so, T_g values as low as $-66°C$ have been recorded for trifluoroethoxy-substituted derivatives. This value can be compared with the $-84°C$ T_g for ethoxy-substituted polyphosphazenes. The presence of crystallinity in the trifluoroethoxy-substituted species can be attributed to the polarity generated by the fluorine atoms and possibly to a lower torsional mobility of the fluorinated groups compared to ethoxy substituents.

The crystallinity of the fluorinated derivatives (particularly the trifluoroethoxy-single-substituent polymer) is responsible for the formation of flexible films, fibers, and nanofibers by solution casting, solution–precipitation spinning, or by electrospinning. Orientation of films and fibers by stretching can yield highly crystalline, tough, and strong materials. As mentioned above, mixed-substituent phosphazene polymers tend to be elastomers. This applies to polymers with two different fluoroalkoxy side groups or to those with both fluoroalkoxy and nonfluorinated alkoxy or aryloxy side units, although the exact properties depend on the ratios of the two (or more) different substituents. A recent example is the elastomer that bears both trifluoroethoxy and trichloroethoxy side groups [33].

1.4.3 Surface Properties

It is no surprise that polyphosphazenes with fluorinated alkoxy or aryloxy side groups are hydrophobic. Contact angles to water in the range of 100° are found routinely [32, 43]. Superhydrophobicity, with contact angles in the range of 160° are generated by nanofiber mats produced by electrospinning (Figure 1.12) [43]. The surface hydrophobicity can be converted to hydrophilicity if the interface is treated with, for example, a hydrophilic nucleophile such as the sodium salt of p-hydroxybenzoic acid. Fluoroalkoxyphosphazene polymer films also respond well to environmental plasma processing. For example, films of poly[bis(trifluoroethoxy)phosphazene] can be patterned via masking to generate films with hydrophilic images by the use of

FIGURE 1.12 Poly[bis(trifluoroethoxy)phosphazene] surfaces. Top left: a solvent cast film. Top right: solvent-electrospun ~80 nm nanofibers. Bottom left: water droplet on film surface with advancing contact angle of 104°. Bottom right: water droplet on nanofiber mat with advancing contact angle of 155°.

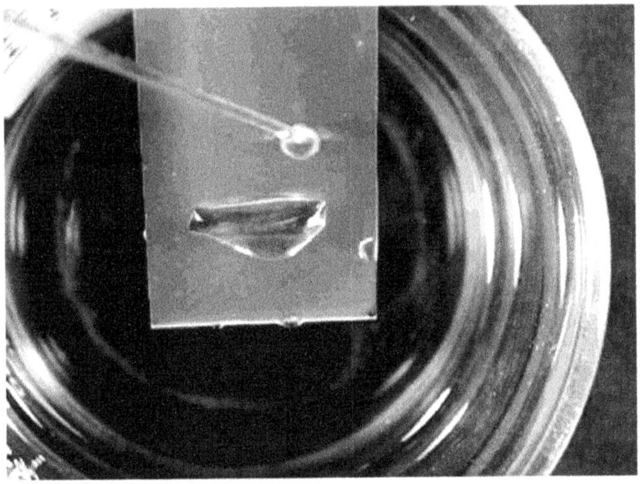

FIGURE 1.13 Film of poly[bis(trifluoroethoxy)phosphazene] masked and exposed to an environmental plasma to convert a region of the surface interface to hydrophilic.

oxygen, nitrogen, or ammonia plasmas (Figure 1.13) [44]. The same polymer surface exposed to a CF_4 plasma generates a superhydrophobic interface with a contact angle to water of 150°.

1.4.4 Thermal and Thermo-Oxidative Stability

The thermal stability of all polymers can be understood in terms of three different factors—resistance to fragmentation, depolymerization, and combustion. The upper thermal stability limits of polymers are set by the onset temperature of thermal cleavage of the backbone and the bonds in the side groups. This theoretical limit is rarely reached for fluorinated alkoxy or aryloxy polyphosphazenes due to their propensity to depolymerize to cyclic small molecules at elevated temperatures. Thus, poly[bis(trifluoroethoxy)phosphazene] depolymerizes to a mixture of cyclic trimeric, tetrameric, and higher ring systems with the same repeating structure as the high polymer at temperatures above 300°C. This is a consequence of an entropy-driven process in which the ΔS term in the Gibbs equation exceeds the ΔH term at a "ceiling temperature," so that many small molecules constitute a more stable system than a few high polymer molecules. This is the same reason why poly(organosiloxanes) (silicones) depolymerize at moderately high temperatures. However, the resistance to depolymerization of polymers can often be enhanced by cross-linking, and this is possible for polyphosphazenes. In general, the smaller the dimensions of the side groups, the higher will be the ceiling temperature.

Resistance to combustion is an aspect where nearly all polyphosphazenes are superior to their all-organic counterparts or poly(organosiloxanes). Small molecule phosphorus compounds have long been employed as fire-retardant additives for other polymers because of the ability of phosphorus to interrupt the chain reactions that are responsible for fire and also their ability to generate an intumescent char that isolates the polymer from oxygen. Nitrogen, along with phosphorus, is especially useful in this regard. Thus, the backbone of polyphosphazenes is an inherent fire retardant which enhances the combustion resistance of these polymers. This effect is present in both the fluorinated and nonfluorinated species and also in the aryloxy counterparts.

1.5 EXISTING AND EMERGING APPLICATIONS FOR FLUORINATED POLYPHOSPHAZENES

The fundamental science of polyphosphazenes has far outstripped the commercial development of this field. To a large extent this is a consequence of the novelty in the industrial preparation of the starting materials and the unusual synthesis procedures compared to the main methods used for the manufacture of conventional organic polymers. Nevertheless, commercial interest in these polymers continues to grow, an interest that is driven by the recognition that unique combinations of properties are accessible. A large part of the recent research in our program has involved the development of an understanding of structure–property relationships in this field as a prelude to identifying potential applications. Recent structure–property work for nonfluorinated polyphosphazenes has been focused on bioabsorbable biomedical applications, optical materials, fire-resistant polymers, membranes, gels, and polymer electrolytes. However, the fluorinated derivatives are in several cases closer to widespread utilization, especially in the following fields mentioned here.

1.5.1 Fibers and Nanofibers

As mentioned earlier, poly[bis(trifluoroethoxy)phosphazene] is a fiber-forming, microcrystalline polymer that can be fabricated into microfibers by solution extrusion from THF into a nonsolvent such as heptane. Alternatively, it can be electrospun to nanofibers from a solution in THF. The structure–property relationships for this polymer are fairly well understood, and these provide a basis for expanding the understanding to polymers with other side groups.

1.5.2 Impact-Absorbing Elastomers

As described above, the presence of two different fluoroalkoxy side groups generates low temperature elasticity (*Tg* approximately −60°C) (Figure 1.14) coupled with hydrophobicity, solvent and oil resistance, and resistance to combustion. These properties were the reason for the development of so-called PNF elastomers for military applications in the Cold War era [45–48]. Interest in these polymers for biomedical applications continues at the present time, with applications in dentistry [49] and cardiovascular devices being prominent. Research carried out in our program in recent years has provided tools for expanding the range of properties and potential uses for these elastomers and for optimizing the synthesis procedures.

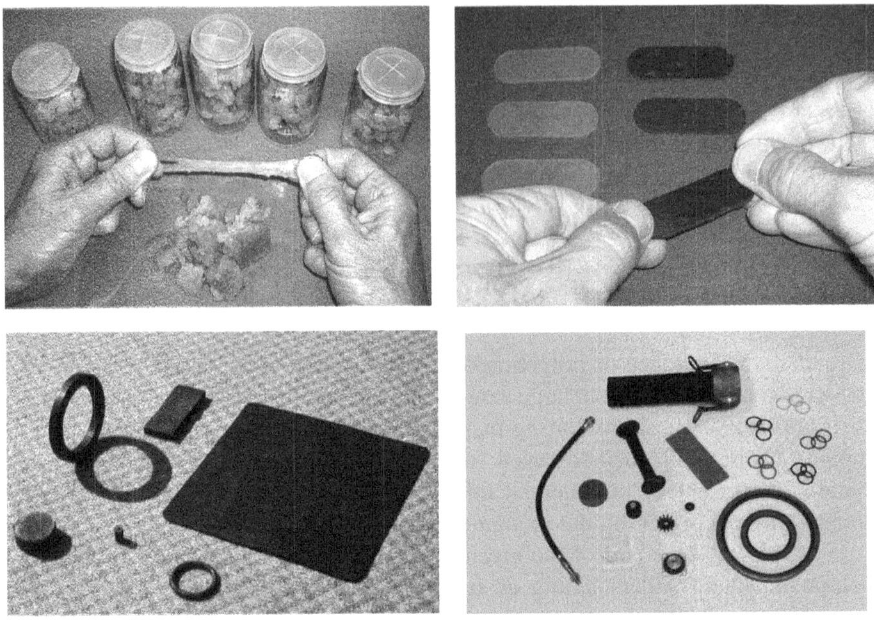

FIGURE 1.14 PNF-type elastomers. Top left: raw gum-stock after synthesis and purification. Top right: the same polymer after cross-linking, with and without carbon reinforcement. Bottom left: fabricated seals and O-rings. Courtesy of Ethyl Corporation and used with permission. Bottom right: aerospace components. Courtesy of Firestone Tire Company and used with permission.

1.5.3 Surface Coatings

The high hydrophobicity of soluble fluorinated polyphosphazenes has drawn attention to their possible use as surface coatings in applications where insoluble fluoropolymers like Teflon cannot be used. This also applies to PNF-type elastomers that, in the uncross-linked state, are highly adhesive. One application that has shown promise is as ice-phobic coatings on helicopter rotor blades. Another reported application is the use of trifluoroethoxy-substituted polyphosphazenes for coating biomedical devices.

1.5.4 Radiation Resistant Polymers

Poor resistance to high energy radiation is a characteristic of many classical organic polymers. The polyphosphazene backbone provides enhanced resistance to both free-radical damage and ionization decomposition when exposed to wavelengths from the mid-ultraviolet through to the X-ray and γ-ray regions. The energy transmission window of the backbone extends throughout the visible wavelengths and to the 220 nm region in the ultraviolet, and the lack of absorptions within this window avoids the sensitivity to photochemical decomposition. Specific transparency windows also exist in the infrared region. In this sense, the skeleton has characteristics more typical of inorganic compounds than carbon-containing backbones. Thus, the organic side groups are the most sensitive sites to photo-decomposition in phosphazene polymers, and this can be exploited for cross-linking purposes. Nevertheless, preliminary evidence suggests that fluorinated side groups are among the least vulnerable substituents, and such polymers show impressive resistance to visible and ultraviolet radiation. Polymers with both fluorinated and nonfluorinated side groups allow tailoring of the properties to favor radiation cross-linking without causing general decomposition.

1.5.5 Membranes

The ease of changing side groups and the film-forming properties of many polyphosphazenes have attracted attention for their possible use as membranes for gas- and liquid-phase separations. Their promise for these applications was recognized by studies in the 1990s [50] and by later engineering evaluations [51].

1.5.6 Micelles

Amphiphilic block copolymers are well-known starting points for the preparation of micelles—either organized structures in the solid state or individual spherical or worm-like micelles suspended in a liquid medium. In polyphosphazene science, it is relatively easy to prepare diblock copolymers. As discussed above, both blocks can be phosphazenes or one can be a phosphazene and the other an organic polymer block or a poly(organosiloxane). Provided one block is hydrophilic and the other hydrophobic, micellar characteristics (Figure 1.15) are accessible. One example

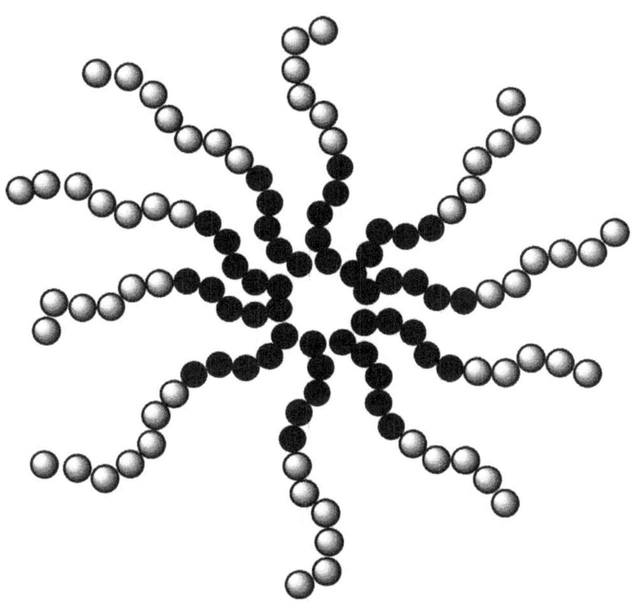

FIGURE 1.15 Schematic illustration of a micelle generated in water. The black circles represent hydrophobic repeating units; the lighter circles, the hydrophilic components.

prepared recently in our program that readily generates spherical micelles in aqueous media is an amphiphilic block copolymer constructed from a hydrophobic poly[bis(trifluoroethoxy)phosphazene] and a hydrophilic poly[dimethylamino)ethyl methacrylate] block (Structure 1.2) [52]. The fluorinated hydrophobic core can be used to "solubilize" hydrophobic drug molecules or nanoparticles in aqueous media.

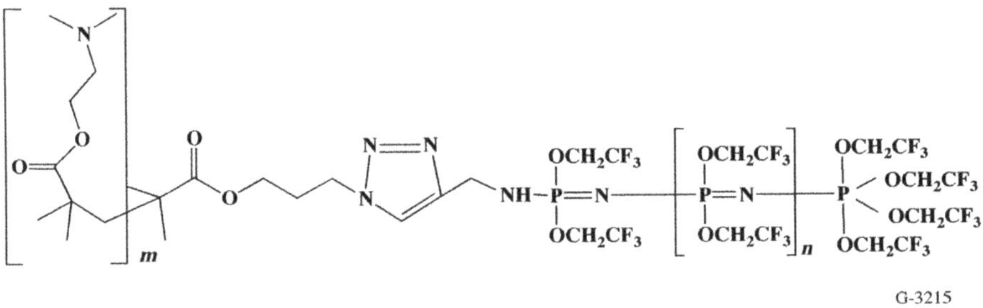

STRUCTURE 1.2

1.5.7 Uses for Fluorophosphazene Oligomers

Although this review is about fluorophosphazene high polymers, it is worthwhile to note that a few of the cyclic oligomers have been developed for commercial uses. For example, liquid cyclic trimers with both trifluoroethoxy and aryloxy side groups on the same ring were developed as advanced hydraulic fluids, presumably because of their thermal stability and their general inertness in contact with gaskets and O-rings. Similar compounds are reported to be lubricants for use in machinery and hard disk drives [53]. In principle, there are many other potential applications for the oligomeric derivatives now that the cyclic trimer, $(NPCl_2)_3$, is readily available on a large scale.

Discussion about the applications of any evolving polymer system inevitably raises inquiries about the feasibility of scale-up of the syntheses to the manufacturing level. For polyphosphazenes, this question has intrigued many observers. Scale-up of the fluorinated PNF-type polyphosphazene elastomers has already been demonstrated by several organizations including the Firestone Tire and Rubber Company (now Bridgestone), Ethyl Corporation, the US military, and by smaller companies in the United States, as well as organizations in the former Soviet Union. Perhaps the greatest impediment to wider commercialization is the difference in chemistry compared to that used in conventional petrochemical polymer manufacturing. In this sense, the novel chemistry that was needed for the commercial development of the poly(organosiloxanes) (silicones) following World War II is an example of what can be accomplished if the incentive is strong enough. Thus, the technical requirements for the commercial development of polyphosphazenes are high quality inorganic–organic chemistry, with meticulous purification of starting materials, and protection of starting materials and intermediates from atmospheric moisture. Polyphosphazenes may never be as widely available as their conventional fluorocarbon counterparts or even silicone polymers, but they have properties that make them ideal for many high performance applications in engineering and medicine for which there are no current alternatives.

REFERENCES

1. Allcock, H. R. *Chemistry and Applications of Polyphosphazenes*; Wiley-Interscience: Hoboken, NJ, 2003, pp 725.
2. Allcock, H. R.; Kugel, R. L. *J. Am. Chem. Soc.* **1965**, *87*, 4216.
3. Allcock, H. R.; Kugel, R. L.; Valan, K. J. *Inorg. Chem.* **1966**, *5*, 1709.
4. Allcock, H. R.; Kugel, R. L. *Inorg. Chem.* **1966**, *5*, 1716.
5. Honeyman, C. H.; Manners, I.; Morrissey, C. T.; Allcock, H. R. *J. Am. Chem. Soc.* **1995**, *117*, 7035.
6. Allcock, H. R.; Crane, C. A.; Morrissey, C. T.; Nelson, J. M.; Reeves, S. D.; Honeyman, C. H.; Manners, I. *Macromolecules* **1996**, *29*, 7740.
7. Allcock, H. R.; Nelson, J. M.; Reeves, J. M.; Honeyman, C. H.; Manners, I. *Macromolecules* **1997**, *30*, 50.
8. Nelson, J. M.; Allcock, H. R. *Macromolecules* **1997**, *30*, 1854.
9. Allcock, H. R.; Reeves, S. D.; Nelson, J. M.; Crane, C. A.; Manners, I. *Macromolecules* **1997**, *30*, 2213.
10. Nelson, J. M.; Allcock, H. R.; Manners, I. *Macromolecules* **1997**, *30*, 3191.
11. Allcock, H. R.; Crane, C. A.; Morrissey, C. T.; Olshavsky, M. A. *Inorg. Chem.* **1999**, *38*, 280.
12. Allcock, H. R.; Nelson, J. M.; Prange, R.; Crane, C. A.; de Denus, C. R. *Macromolecules* **1999**, *32*, 5736.
13. Prange, R.; Allcock, H. R. *Macromolecules* **1999**, *32*, 6390.
14. Allcock, H. R.; Reeves, S. D.; Nelson, J. M.; Manners, I. *Macromolecules* **2000**, *33*, 3999.
15. Allcock, H. R.; Reeves, S. D.; de Denus, C. R.; Crane, C. A. *Macromolecules* **2001**, *34*, 748.
16. De Jaeger, R.; Potin, P. In *Synthesis and Characterization of Polyphosphazenes*; Gleria, M., De Jaeger, R., Eds.; Nova Science: New York, 2004; Chapter 2, pp 25–48.
17. Allcock, H. R.; Lampe, F. W.; Mark, J. E. *Contemporary Polymer Chemistry*, 3rd. ed.; Prentice-Hall: Upper Saddle River, NJ, 2003, p 814.
18. Neilson, R. H.; Wisian-Neilson, P. *J. Am. Chem. Soc.* **1980**, *102*, 2848.
19. Neilson, R. H.; Wisian-Neilson, P. *Chem. Rev.* **1988**, *88*, 541.
20. Wisian-Neilson, P. In *Phosphazenes: A Worldwide Insight*; De Jaeger, R., Gleria, M., Eds.; Nova Science: Hauppauge, New York, 2002, Chapter 5.
21. Montague, R. A.; Matyjaszewski, K. *J. Am. Chem. Soc.* **1990**, *112*, 6721.
22. Allcock, H. R.; Austin, P. E.; Neenan, T. X.; Sisko, J. T.; Blonsky, P. M.; Shriver, D. F. *Macromolecules* **1986**, *19*, 1508.
23. Allcock, H. R.; Morozowich, N. L. *R.S.C. Polym. Chem.* **2012**, *3*, 578.
24. Allcock, H. R.; Steely, L. B.; Singh, A. *Polym. Int.* **2006**, *55*, 621.
25. Liu, X.; Breon, J.; Chen, C.; Allcock, H. R. *Inorg. Chem.* **2012**, *51*, 11910.

26. Books, J. T.; Klobucar, W. D.; Kolich, C. H, US Patent 4,945,139, 1998.
27. Allcock, H. R.; Fitzpatrick, R. J. *Chem. Mater.* **1991**, *3*, 450.
28. Morozowich, N. L.; Nichol, J. L.; Allcock, H. R. *Chem. Mater.* **2012**, *24*, 3500.
29. Allcock, H. R.; Suszko, P. R.; Evans, T. L. *Organometallics* **1982**, *1*, 1443.
30. Tian, Z.; Liu, X.; Manseri, A.; Ameduri, B.; Allcock, H. R. *Polymer Eng. Sci.* published online Aug. 19, **2013**.
31. Hess, A. R.; Allcock, H. R. **2012** unpublished results.
32. King, A.; Presnall, D.; Steeley, L. B.; Allcock, H. R.; Wynne, K. J. *Polymer* **2013**, *54*, 1123–1129.
33. Chen, C.; Liu, X.; Tian, Z.; Allcock, H. R. *Macromolecules* **2012**, *45*, 9085–9091.
34. Allcock, H. R.; Lavin, K. D.; Riding, G. H.; Suszko, P. R.; Whittle, R. R. *J. Am. Chem. Soc.* **1984**, *106*, 2337.
35. Allcock, H. R.; Patterson, D. B.; Evans, T. L. *Macromolecules* **1979**, *12*, 172.
36. Allen, C. W.; Moeller, T. *Inorg. Chem.* **1968**, *11*, 2177.
37. Allcock, H. R.; McDonnell, G. S.; Desorcie, J. L. *Macromolecules* **1990**, *23*, 3873.
38. Allcock, H. R.; Coggio, W. D. *Macromolecules* **1993**, *26*, 764–771.
39. Nelson, J. M.; Allcock, H. R.; Manners, I. *Macromolecules* **1997**, *30*, 3191.
40. Nelson, J. M.; Primrose, A. P.; Hartle, T. J.; Allcock, H. R.; Manners, I. *Macromolecules* **1998**, *31*, 947.
41. Chang, Y.; Lee, S. C.; Kim, C.; Reeves, S. D.; Allcock, H. R. *Macromolecules* **2001**, *34*, 269.
42. Chang, Y.; Bender, J. D.; Phelps, M. V. B.; Allcock, H. R. *Biomacromolecules* **2002**, *3*, 1364.
43. Allcock, H. R.; Steeley, L. B.; Singh, A.; Hindenlang, M. D. *J. Adhesion. Sci. Technol.* **2009**, *23*, 435.
44. Steely, L. B.; Allcock, H. R.; Kim, S. H.; Kim, J. H.; Kang, B. K. *Mater. Res. Symp.* **2007**, 951E, 0951E-12-41.
45. Rose, S. H. *J. Polym. Sci. (B)* **1968**, *6*, 837.
46. Tate, D. P. *J. Polym. Sci. Polym. Symp.* **1974**, *48*, 33.
47. Penton, H. R. *ACS Symp. Ser.* **1988**, *360*, 277.
48. Singler, R. E.; Schneider, N. S.; Hagnauer, G. L. *Polym. Eng. Sci.* **1975**, *15*, 321.
49. Gettleman, L. Uses of Polyphosphazene in Dentistry. In *Applicative Aspects of Poly(organophosphazenes)*; Gleria, M., De Jaeger, Eds.; Nova Science: New York, 2004; Chapter 2, pp. 33–47.
50. Nagai, K.; Freeman, B. D.; Cannon, A. M.; Allcock, H. R. *J. Membrane Sci.* **2000**, *172*, 167.
51. Peterson, E. S.; Stone, M. L.; Cummings, D. G.; McCaffrey, R. R. *Sep. Sci. Technol.* **1993**, *28*, 271.
52. Tian, Z.; Liu, X.; Chen, C.; Allcock, H. R. *Macromolecules* **2012**, *45*, 2502.
53. Singler, R. E.; Bieberich, M. J. In *Synthetic Lubricants and High Performance Functional Fluids*; Shubkin, R. L., Ed.; Dekker: New York, 1992, p. 217.

2

$Mn_2(CO)_{10}$-VISIBLE LIGHT PHOTOMEDIATED, CONTROLLED RADICAL POLYMERIZATION OF MAIN CHAIN FLUORINATED MONOMERS AND SYNTHESIS OF BLOCK COPOLYMERS THEREOF

Alexandru D. Asandei, Olumide I. Adebolu, and Christopher P. Simpson
University of Connecticut

2.1 INTRODUCTION

Fluoropolymers constitute a fundamental class of specialty materials endowed with wide morphological versatility, high thermal, chemical, aging, and weather resistance, as well as low surface energy, dielectric constant, flammability, refractive index, and moisture absorption. As such, their applications range from paints and coatings, pipe liners, transmission fluids, O-rings for extreme temperatures, fuel cell membranes, antifouling layers, etc., to optical fibers and high power capacitors [1]. Therefore, their precise synthesis is very relevant.

However, while controlled/living radical polymerizations (C/LRPs) have lately undergone remarkable developments [2–6], and atom transfer, nitroxide or addition–fragmentation methods [2] have proven quite successful for acrylates or styrenes,

Handbook of Fluoropolymer Science and Technology, First Edition. Edited by Dennis W. Smith Jr., Scott T. Iacono, and Suresh S. Iyer.

their applicability in the CRP of *side-chain* fluorinated monomers such as pentafluoro styrene ($CH_2{=}CH(C_6F_5)$), fluorinated (meth)acrylates ($CH_2{=}CHCOO(CH_2)_nR_F$) or α-trifluoromethyl acrylates ($CH_2{=}C(CF_3)COOR_{alk}$) remains in its incipient stages [1]. Most importantly, their application in the CRP of the industrially significant, very reactive, gaseous, and thus experimentally difficult *main chain* fluorinated alkene monomers (FM), such as vinylidene fluoride (VDF, $CH_2{=}CF_2$), hexafluoropropene ($CF_2{=}CF(CF_3)$), tetrafluoroethylene ($CF_2{=}CF_2$), trifluorochloroethylene ($CF_2{=}CFCl$), still awaits demonstration.

Hence, due to the current lack of suitable CRP chemistry for these FMs and for the synthesis of well-defined complex architectures thereof (blocks, graft, hyperbranched, stars, etc.), the study and understanding of their self-assembly and of the properties and applications thereby derived, lag significantly behind those of the corresponding materials based on conventional monomers (styrene, acrylates, dienes, etc.). Thus, the development of FM-CRP and the synthesis of elaborate FM polymer designs is a worthy endeavor [7, 8]. Still, such polymerizations are very challenging on laboratory scale, as all FMs are gases ($b_p^{VDF} = -83°C$) and typical telo/polymerizations are carried out at $T > 100$–$150°C$ [1], in high pressure metal reactors. Moreover, in additional contrast with most monomers, the regioselectivity defects in VDF propagation leads to the formation of both $-CF_2-CH_2-Y$ and $-CH_2-CF_2-Y$ chain ends. Thus, regardless of the Y group of the reversible termination method in any potential LRP mechanism, the C—Y bond in $-CF_2-CH_2-Y$ will always be stronger than the one in $-CH_2-CF_2-Y$, thereby rendering the corresponding chain ends *dead*, not dormant, as far as reversible activation is concerned. Such chain ends will thus accumulate, broaden the PDI, and eventually lead to loss of control. Indeed, to the best of our knowledge there is no experimental evidence of the reversible activation of the "bad" chain ends in PVDF.

To date, the most successful approach to FM-CRP [1] remains the iodine degenerative transfer (IDT) [3, 9–14] (IDT: $P_n^\bullet + P_m{-}I \rightleftharpoons P_n{-}I + P_m^\bullet$) [1, 15, 16], an outcome of the earlier research on the high temperature (100–230°C) free radical VDF telomerization [17, 18] with polyhalides [19], especially (per)fluorinated iodo chain transfer (CT) agents [16] such as $CF_3{-}I$ [20–23], $CF_3{-}CF_2{-}I$ [24], $CF_3{-}(CF_2)_3{-}I$ [25, 26], $CF_3{-}(CF_2)_5{-}I$ [21, 27–29], $(CF_3)_2CF{-}I$ [26], $Cl{-}CF_2{-}CFCl{-}I$ [23], $I{-}(CF_2)_{4-6}{-}I$ [10, 29, 30], but also with less active $HCF_2CF_2CH_2{-}I$ [27, 28], $C_6F_{13}CH_2CF_2{-}I$ [28, 31], and even CH_2I_2 [32], $R_F{-}CH_2{-}CH_2{-}I$ [33], or $CH_3{-}I$ [34]. In fact, IDT may be one of the oldest CRP methods, as a linear dependence of M_n on polymer yield was already demonstrated] in the early 1980s, and was the first industrial implemented CRP protocol [10–14].

In-depth modeling [30, 35] and kinetic [20, 27, 28] investigations have later revealed the importance of the structure and reactivity of the chain transfer agent (CTA) (I > Br > Cl ∼ H and difunctional better than monofunctional) [10, 16], as well as that of the contributions of side reactions (transfer to monomer and polymer, etc.) and monomer addition mode (1,2- vs. 2,1-) to the quality of the polymerization livingness. The synthesis of the iodo CTAs as well as some derivatization of the PVDF-I chain ends is available [16, 36].

However, VDF–IDT–CRPs always require a free radical source (*e.g.*, ^{t}butyl peroxide (TBPO)), as *direct* metal catalyzed initiation from perfluoroiodides or any other halides is not available. This is a serious drawback with respect to the precise synthesis of block or graft copolymers based on FMs, as such systems would inevitably lead to mixtures of homo- and copolymers with the current technology. Therefore, the availability of an initiation method directly from halides, most likely mediated by transition metal catalysis, would be highly valuable.

Unfortunately, while VDF polymerization can proceed at ambient temperature [37], only very low VDF telomers (DP <1–3) may be obtained, and only at high temperature (>100°C) from transition metal salts and polyhalides by redox catalysis [1, 26, 38]. Moreover, while the 1:1 adduct formation by thermal or metal (Cu [39], Zn [40], Pd [41], $SnCl_2/CH_3COOAg$ [42], Cp_2TiCl [43], PPh_3 [44], AIBN [45, 46], $(NH_4)_2S_2O_8$/HCOONa [47]) catalyzed addition of R_F–I derived perfluoroalkyl radicals to alkenes or of the R_F–Mt–I (Zn [48, 49], Zn/Cu [50]) organometallics to carbonyls and amides is available, the corresponding metal catalyzed addition of such electrophilic radicals to electrophilic fluorinated alkene substrates (FMs) at T <100°C, and especially at ambient temperatures is conspicuously absent from the literature. Thus, to the best of our knowledge, prior to our recent work [51,52], there were no reports on metal-mediated FM or VDF polymerizations, let alone VDF–CRP. Consequently, the ability to carry out such reactions under mild conditions would be of great synthetic use, particularly in view of the great demand for perfluoroalkylation and trifluoromethylation reactions in organic chemistry [52] as well as in the initiation of the CRP of FMs and the synthesis of block, graft, and star FM copolymers.

Accordingly, and by contrast to styrene or acrylate CRPs which can be conveniently sampled from the side arm of a Schlenk tube on, for example, a 1 g scale, kinetic studies of VDF polymerizations involve many one-data-point experiments. This is very time-consuming and expensive due to the typical lab unavailability of a large number of costly metal reactors, which moreover still require hundreds of grams of monomer. Thus, development of methods that enable small-scale (1 g) polymerizations to proceed at room temperature (rt) in inexpensive, low pressure glass tubes would be highly desirable, as they could be adapted for fast screening of a wide range of catalysts and reaction conditions, and take advantage of photochemistry.

2.2 VDF PHOTOPOLYMERIZATION CATALYST SELECTION

Toward the above goal, in a large number of preliminary experiments we initially attempted the VDF polymerization at moderate temperatures (25–60°C) in low pressure glass tubes *via* conventional LRP methods (e.g., Cu-ATRP [53], nitroxides, RAFT, Cp_2TiCl [54, 55]). However, all polymerizations were unsuccessful, most likely due to the inability of the particular systems to generate radicals reactive enough to add to VDF, or to reactivate the PVDF-Y chain ends at or around rt, or due to PVDF *vs.* catalyst solvent incompatibility [54]. Indeed, while, for example,

CuX/L abstracts halides from activated substrates (esters, benzyl, etc.), it hardly does so from perfluorohalides ($k_{(CH3)2C(COOEt)-Br}/k_{C8F17-Br} > 10^2$) [56], and thus barely activates the C—I bond in —CF_2—CF_2—I, let alone in the "good" —CH_2—CF_2—I, and especially in the "bad" —CF_2—CH_2—I PVDF chain end, and explains the failure of Cu-chemistry to provide PVDF-block copolymers. Moreover, activating amine Cu ligands are impractical with R_F–I initiators due to the formation of ammonium salts or charge transfer complexes [57]. In addition, potential metal-mediated radical processes involving C—Mt bond formation will be plagued by β-H and β-F eliminations [54]. As such, as none of the conventional R–CF_2–I activations or LRP protocols tested proved effective, we decided to investigate alternative means of radical generation and trapping [58, 59].

Interestingly, high power UV-photoinduced *telomerizations* are available for VDF [19, 60, 61]. Moreover, we have also shown [54, 62–65] that the free radical polymerization (FRP) of VDF occurs readily under UV irradiation at rt in the presence of TBPO. However, UV irradiation degrades most organometallic complexes, as well as RAFT reagents. To the best of our knowledge, prior to our work, there were [51, 52] no reports on the VDF polymerization using regular *visible* light. Thus, to maintain an ambient reaction temperature and a low pressure polymerization in the glass tubes, we decided to investigate photopolymerizations mediated by commercially available, low wattage (<30 W), spiral, compact, white light fluorescent bulbs.

Since VDF is a very reactive monomer, successful rt initiators should provide highly reactive, destabilized alkyl, semifluorinated, or perfluoroalkyl radicals. Thus, the visible light generated $L_nMt^\bullet$ metal radical species should be a very good halide abstractor, and, for possible catalyzed CRP, L_nMt–X should also be a very good halide donor.

As such, similarly to the $ClCp_2Ti$–$TiCp_2Cl$ dimer chemistry we previously described for the CRP of styrene [66] and isoprene [67], but which was unsuccessful for VDF [54], prototypical examples [58, 59] of radically photolyzable transition metal complexes are based on dimers of the $(CO)_nMt$–$Mt(CO)_n$ type. Out of many known examples, the following $Re(CO)_5^\bullet > Mn(CO)_5^\bullet > CpW(CO)_3^\bullet >$ $CpMo(CO)_3^\bullet > CpFe(CO)_2^\bullet > Co(CO)_4^\bullet$ order is available in terms of their ability to abstract halides [68]. Although $Re(CO)_5^\bullet$ (λ_{max} = 535–550 nm) abstracts Cl from CCl_4 ~65 times faster than $Mn(CO)_5^\bullet$ [68, 69], the stronger Re–Re [70] bond dissociation energy (BDE) and its higher cost make the relatively inexpensive $Mn_2(CO)_{10}$ [71] dimer [72] the most popular reagent in the series [59]. At rt, in the dark the dimer is stable, ($K_{eq} < 2.4 \times 10^{-19}$) [73], but as the Mn–Mn linkage is relatively weak (20–40 kcal/mol) [59, 74, 75], and further decreased by extra ligands [76], $Mn_2(CO)_{10}$ undergoes both thermal (~90°C) [77] and facile rt photo homolysis. Indeed, while UV irradiation leads to CO loss, forming $Mn_2(CO)_9$ and $Mn(CO)_5^\bullet$, near-UV and visible longer wavelength (λ = 366–400 nm, $\lambda_{max}{}^{Mn_2(CO)_{10}}$ = 324 nm) provide the $Mn(CO)_5^\bullet$ 17e$^-$ paramagnetic metalloradical ($\lambda_{max}{}^{Mn(CO)_5}$ = 780–830 nm) [78] with good quantum efficiency [79, 80].

$Mn(CO)_5^\bullet$ cleanly effects both hydride and halide abstraction from good RH donors such as Bu_3SnH [81], and respectively from RX halides with moderate BDEs (<310 kJ/mol) [82] such as CCl_4 [83], I_2 [83], polyhalides [82], allyl and benzyl halides [82, 83], CH_3I [84], and other inactivated alkyl iodides, sometimes in the presence of $InCl_3$ [75, 82]. Phosphine ligands (PPh_3) depress the rates of such reactions [85]. Interestingly, and by contrast to other radical reagents, due to the CO ligand sterics, $Mn(CO)_5^{\bullet}$ reacts faster with primary than secondary or tertiary halides [82]. Yet, while Mn-alkyls photolyze, they are not effective in radical reactions [75]. In addition, prior to our work [51–55], there were no examples of the use $Mn_2(CO)_{10}$ for the radical activation of perfluoroalkyl halides, including trifluoromethylation reactions with CF_3–I or CF_3–SO_2–I.

Interestingly, early reports proposed photo or thermal $Mn_2(CO)_{10}$ or $Re_2(CO)_{10}$-mediated polymerizations initiated via either H abstraction from the monomer [86], addition of the metaloradical to tetrafluoroethylene to initiate its polymerization [87, 88], MMA [87, 89, 90] block copolymerization [91], or addition to 1,2-disubstituted acetyls and alkenes [92]. However, while preliminary polymerizations with $Mn_2(CO)_{10}$ or $Re_2(CO)_{10}$ and CCl_4 [93] or grafting from N-halogenated polyamides [94] were carried out in the 1960s, $Mn_2(CO)_{10}$ was only recently employed in the thermal (60–90°C) [95] FRP of MMA or in the PMMA chain extensions from –CCl_3 groups on SiO_2 [96], chitosan [97], or from polystyrene-Br [98–100].

Very recently, $Mn_2(CO)_{10}$ was, however, used as a photo-coinitiator for activated alkyl iodides (IDT) [101–103], or RAFT reagents [104] in controlled radical photo(co)polymerizations of VAc, acrylates, styrene, and alkenes, where $Mn(CO)_5^{\bullet}$ irreversibly activated iodine terminated chains [101], but the *in situ* generated $Mn(CO)_5$–I [83] was not involved in the IDT.

Therefore, as $Mn_2(CO)_{10}$ was never employed in polymerizations of main chain fluorinated monomers, or with inactivated alkyl halides or with perfluoroalkyl iodides, we decided to assess its scope and limitations as photo-coinitiator and to demonstrate that such L_nMt–MtL_n photolyzable transition metal complexes afford the initiation of VDF polymerization *directly* from a variety of regular and (per)fluorinated alkyl halides (Cl, Br, I) even at rt, thus opening up novel synthetic avenues for the photomediated synthesis of block and graft copolymers based on FMs. Second, we also set to kinetically explore such polymerization and investigate the possibility of $Mn_2(CO)_{10}$-mediated IDT–VDF–CRP. Third, we aimed to demonstrate the first examples of the synthesis of well-defined PVDF-block copolymers.

2.3 EFFECT OF REACTION PARAMETERS AND MECHANISTIC CONSIDERATIONS

2.3.1 Effect of Temperature and Light

While visible light could be invoked in the photodissociation of RX, control experiments revealed no polymerization in the dark, or with illumination, but in the absence of $Mn_2(CO)_{10}$ [51]. Although polymerizations could easily be carried out anywhere from T = 0°C to T = 100°C, we selected T = 40°C for all experiments, as a reasonable compromise between rate and a safe pressure inside the tube. In fact, simply lowering the bottom part of the light bulb inside the oil bath helped maintain such temperature, with minimal additional heating from the hot plate [51].

2.3.2 Solvent Effect

Typical VDF laboratory polymerizations are carried out in acetonitrile (ACN) [18] that is a nonsolvent and a weak CTA. However, there is very little data [105–107] on the solvent effect in VDF polymerizations, let alone photopolymerizations. As such, as $Mn_2(CO)_{10}$ is soluble in and inert toward most typical organic solvents (except for CCl_4, $CHCl_3$), we decided to scan [51] a wide variety of polymerization media, noting that while good monomer/polymer solubilization and a photosensitizing solvent effect would be desirable, minimization of solvent CT outweighs solubility considerations. As such, the trends discussed below are the combined effect of the solvent CT transfer and polymer swelling ability, where the best solvents (alkyl carbonates) display minimum transfer and highest swelling. While we also carried out a similar study using Cp_2TiCl_2, a solvent compatible with both $Cp_2TiCl^{\bullet}$ and PVDF could not be found [54].

Previous investigations of PVDF solution properties [108–113] have indicated that the best solvents are polar aprotic and H-bonding capable, and that HH sequences decrease solubility [108]. A common structural motif is a favorable H bonding interaction of the polar Y═O bond (Y = C, S, P) of the solvent, with the mildly acidic Hs of $-CF_2-CH_2-CF_2-$. Accordingly, 2-pentanone, propionitrile, benzonitrile, nitrobenzene, $(MeO)_3P$, $(NEt_2)SO_2$, sulfolane, and methylpyridines were classified as bad solvents. $(BuO)_3P{=}O$ only swells PVDF, whereas γ-butyrolactone, propylenecarbonate, cyclohexanone, CH_3COPh, and $Ph_2C{=}O$ solutions gel on annealing at rt. Short chain phosphates ($(RO)_3P{=}O$; R = Me, Et, and $(CH_3)(MeO)_2P{=}O$) are good solvents, while tertiary amides ($HCONR_2$; R = Me (DMF), Et, DMAC, N-formylmorpholine, NMP) and DMSO are very good solvents. Finally, cyclopentanone, N-alkyl cyclic urethanes as well as linear and cyclic ureas ($(Me_2N)_2C{=}O$, $[-N(CH_3)-(CH_2)_{2\text{-}3}-N(CH_3)-]C{=}O$, DMEU and DMPU) dissolve PVDF even at rt, with $[(CH_3)_2N]_3P{=}O$ (HMPA) and $(CH_3)[(CH_3)_2N]_2P{=}O$ considered excellent solvents.

The solvent effect in $Mn_2(CO)_{10}$ polymerizations was explored with over 40 solvents [51] and perfluorobutyl iodide (PFBI), easily activated by $Mn(CO)_5^{\bullet}$ (*vide infra*) was used as initiator. The comparisons were made under the same conditions ($VDF/PFBI/Mn_2(CO)_{10}$ = 25/1/0.2; VDF/solvent = 1/3 wt/v) and revealed that in accordance with the principles outlined above, no polymerization occurred in anisole, α,α,α-trifluorotoluene, diglyme, diethylene glycol monoethyl ether, ethyl ether, THF, dioxane, *o*-cresol, isopropanol, trifluoroacetic anhydride, tetramethylurea, sulfolane, benzonitrile, cyclopentanone even after 1–3 days. Poor reaction rates ($k_p^{app} < 1 - 9 \times 10^{-3}\ h^{-1}$) were observed in HMPA, 2-butanone, methyl pentanone, DMF, DMSO, PEO, DMAc, δ-valerolactone, ^tBuOH, and EtOAc. Slightly faster polymerizations ($k_p^{app} = 1 - 1.7 \times 10^{-2}\ h^{-1}$) were seen in CH_2Cl_2, acetic anhydride, γ-butyrolactone, 1,2-dichloroethane, TMP, MeOH, pentafluorobutane, and HFBz, whereas β-butyrolactone, ε-caprolactone, ACN, and H_2O displayed medium rates ($k_p^{app} = 2.4 - 2.9 \times 10^{-2}$ h^{-1}).

While fast polymerizations occurred in diethyl-, di^{t}Bu- and propylene carbonate ($k_p^{app} = 3 - 3.3 \times 10^{-2}$ h^{-1}), the fastest rates were provided by ethylene carbonate, HFIPA ($k_p^{app} = 5 - 7 \times 10^{-2}$ h^{-1}) and especially by dimethyl carbonate (DMC, $k_p^{app} = 1 \times 10^{-1}$ h^{-1}). The conversion, polymerization rate (k_p^{app}), M_n, and PDI were largely insensitive to the amount of DMC, ($V_{tube} = 35$ mL, 1 g VDF, 1–12 mL DMC), whereas the rate significantly increased with the amount of VDF (*i.e.*, pressure and monomer concentration in solution, with 1–4 g VDF, 3–12 mL DMC). These are typical features of heterogeneous polymerizations of gaseous monomers [114], reminiscent of precipitation/dispersion polymerization of PVDF in $scCO_2$ with [29] and without [115] iodine CTAs. However, while fluorination enhances polymer solubilization, $scCO_2$ is a very poor PVDF solvent [116], even at high temperatures and pressures.

Thus, unfortunately, many of the good PVDF solvents (DMF, DMAc, ureas, etc.) act as strong CTAs and lead to no or very low conversions. However, remarkable trends are observed with carbonates, and especially with DMC, a green solvent [117]. Indeed, although DMC does not dissolve PVDF at rt, and similarly to ACN affords a heterogeneous polymerization, it provides by far the fastest reaction rates, at least thrice those obtained in ACN. Moreover, DMC consistently outperformed ACN not only for PFBI, but for all initiators tested including in polymerizations from poor initiators such as CF_3—$(CF_2)_2$—CO—Cl or Cl—CF_2—$(CF_2)_6$—CF_2—Cl, which occurred *only* in DMC. Possible explanations for this trend include a photosensitizing effect or better monomer/polymer solubilization.

DMC is stable to pyrolysis and photolysis up to 350°C [118], but gas phase irradiation with high power Hg UV lamps induces some acetone promoted [119] decomposition at T > 150–200°C [120], while methyl and methoxy radicals form in the presence of nitric oxide [121]. Liquid phase DMC UV-photolysis with H_2O_2 [122], NO, NO_2 [123] or ^{t}butylperoxide [124], additives occurs with H abstraction and formation of CH_3—O—CO—O—$CH_2^{\bullet}$, and solid state ^{60}Co γ-irradiation at 77°K produces methyl and methoxycarbonyl radicals [125, 126]. However, to the best of our knowledge, there are no reports on the photolysis or a potential photosensitizing effect of DMC under ambient visible light. Moreover, control experiments revealed that no photopolymerization occurs in DMC with VDF alone, VDF/$Mn_2(CO)_{10}$, or VDF/PFBI even after prolonged visible light exposure. Thus, while it may act as a weak CTA, DMC does not generate radicals under visible light. Accordingly, the better polymerization rates afforded by DMC are not a photo effect, but a consequence of its better VDF solubilization and PVDF swelling, which also enables better monomer diffusion to the propagating center. These results are consistent with studies on the solvent diffusion coefficient, and on PVDF swelling by carbonate solvents used as electrolytes for Li-ion batteries [127] containing PVDF microporous membranes [128–131], which revealed the same DMC ≫ PC ~ EC > DEC trend, a PVDF/DMC swelling ratio = 1.2 at 40°C, and possible dissolution of PVDF by DMC at high temperatures (T > 80°C).

2.3.3 Polymerization Mechanism and Initiator Evaluation

The proposed reaction mechanism is outlined in Scheme 2.1, with the setup previously described [51]. Initiation begins with the reversible photodissociation of $Mn_2(CO)_{10}$ (Equation 2.1). Subsequent irreversible [101, 102] halide abstraction from R–X (and later PVDF-X), driven by the formation of high BDE Mn–X [74, 75, 132a], X = Cl, Br, I, (Equation 2.2) affords $Mn(CO)_5$–X and $R^\bullet$, which, if reactive enough, adds to VDF, thus initiating polymerization (Equation 2.3). As VDF is asymmetric, both 1,2- and 2,1-units, (Equation 2.4, head to tail HT, ~95% [20, 106] and respectively head to head, HH) occur in FRP (Equation 2.4). $Mn(CO)_5^\bullet$ remains slowly but continuously photogenerated from the dimer throughout the polymerization, and, by abstracting X from R–X and PVDF-X (especially PVDF–CH_2–CF_2–I), maintains a steady state radical concentration by compensating for termination reactions.

2.3.3.1 Effect of Initiator Chain Transfer Constant on the Polymerization Mechanism If the amount of $Mn_2(CO)_{10}$ is less than that of the initiator, depending on the value of its CT constant, ($CT_{RX}^{PVDF^*} = k_{transfer}^{PVDF^* toRX}/k_{propagation}^{VDF}$), excess RX may act as a CTA *vs.* the propagating chains (Equations 2.5 and 2.6). Higher values (CT > 1) and X = I are required for controlled DT polymerizations [16], where such initiators are consumed very early in the process to provide iodide terminated polymeric CTAs

Initiation and Propagation

$$Mn_2(CO)_{10} \underset{}{\overset{h\nu}{\rightleftharpoons}} 2\dot{M}n(CO)_5 \quad (2.1)$$

$$R{-}X + \dot{M}n(CO)_5 \not\rightleftharpoons R^\bullet + Mn(CO)_5X \quad (2.2)$$

$$R^\bullet \xrightarrow{k_i,\ VDF} R{-}VDF^\bullet \xrightarrow{nVDF, k_p} P^\bullet_{n+1} \quad (2.3)$$

$$P_m{-}CH_2{-}\dot{C}F_2 \underset{k_p}{\xrightarrow{VDF}} P_m{-}CH_2{-}CF_2{-}CH_2{-}\dot{C}F_2 + P_m{-}CH_2{-}CF_2{-}CF_2{-}\dot{C}H_2 \quad (2.4)$$

Chain Transfer to Initiator

$$P_n{-}CH_2{-}\dot{C}F_2 + R{-}X \xrightarrow{k_{1tr}} P_n{-}CH_2{-}CF_2{-}X + R^\bullet \quad (2.5)$$

$$P_m{-}CF_2{-}\dot{C}H_2 + R{-}X \xrightarrow{k_{2tr}} P_m{-}CF_2{-}CH_2{-}X + R^\bullet \quad (2.6)$$

Iodine Degenerative Transfer

$$P_m{-}CH_2{-}\dot{C}F_2 + P_n{-}CH_2{-}CF_2{-}I \overset{K_{ex}=1}{\rightleftharpoons} P_m{-}CH_2{-}CF_2{-}I + P_n{-}CH_2{-}\dot{C}F_2 \quad (2.7)$$

$$P_m{-}CH_2{-}\dot{C}F_2 + P_n{-}CF_2{-}CH_2{-}I \overset{K_{ex}<<1}{\rightleftharpoons} P_m{-}CH_2{-}CF_2{-}I + P_n{-}CF_2{-}\dot{C}H_2 \quad (2.8)$$

$$P_m{-}CF_2{-}\dot{C}H_2 + P_n{-}CF_2{-}CH_2{-}I \not\rightleftharpoons P_m{-}CF_2{-}CH_2{-}I + P_n{-}CF_2{-}\dot{C}H_2 \quad (2.9)$$

Bimolecular termination

$$P_m{-}CH_2{-}\dot{C}F_2 + P_n{-}CH_2{-}\dot{C}F_2 \longrightarrow P_m{-}CH_2{-}CF_2{-}CF_2{-}CH_2{-}P_n \quad (2.10)$$

H Transfer

$$P_m{-}CF_2{-}\dot{C}H_2 + S{-}H \longrightarrow P_m{-}CF_2{-}CH_2H \quad (2.11)$$

$$P_n{-}CH_2{-}\dot{C}F_2 + S{-}H \longrightarrow P_n{-}CH_2{-}CF_2H \quad (2.12)$$

SCHEME 2.1 Mechanism of $Mn_2(CO)_{10}$-mediated VDF–FRP and VDF–IDT–CRP.

[28, 51]. For PVDF, two such chain ends (*i.e.*, P_n—CH_2—CF_2—I and P_m—CF_2—CH_2—I) are thus obtained [16, 27, 28, 31, 51].

Unfortunately, the $CT_{RX}^{PVDF^*}$ values of the initiators used herein or for related structures are only available in a few cases ($CT_{CCl3Br} = 35$, $CT_{CCl4} = 0.25$, $CT_{CHCl3} = 0.06$ at 141°C [133], $CT_{C6F13-I} = 0.8$, $CT_{C6F13-Br} = 0.08$, $CT_{C6F13-H} = 0.0002$ in $scCO_2$ at 120°C [29]). Interestingly, perfluoroiodides and PVDF–CH_2–CF_2–I have similarly large CT values ($CT_{C6F13-I} = 7.9$, $CT_{C6F13-CH2-CF2-I} = 7.4$), whereas the "bad" PVDF–CF_2–CH_2–I 2,1-chain end is 25 times less active ($CT_{HCF2CF2CH2-I} = 0.3$ at 75°C) [28]. Thus, the PVDF–CF_2–$CH_2^\bullet$ propagating chain is correspondingly more reactive than PVDF–CH_2–$CF_2^\bullet$. Obviously, the above CT data are apparent values, mol fraction-averaged over the two unequal reactivity growing chains, where clearly, $CT_{RX}^{VDF-CH2^*} > CT_{RX}^{VDF-CF2^*}$.

The possibility of initiation via H abstraction [81] was first tested from the —CF_2–H group of 1H,1H,7H-dodecaheptafluoro acrylate, the C—H group of $(CF_3)_2CH$—OH, as well as from PVDF—CH_2—CF_2—H chain ends, but no polymer was obtained, indicating that although such protons are more acidic, they are not abstracted by $Mn(CO)_5^\bullet$ at rt. A wide variety of over 40 halide structures never previously reported in conjunction with $Mn_2(CO)_{10}$ (except for CH_3—I [134], CCl_4, $RCCl_3$ [93, 97], $(CH_3)_2C(COOEt)Br$ [101, 102]), were subsequently evaluated [51].

However, no initiation was observed [51] from I_2, ${}^tBu$—I, CH_3—SO_2Cl, CH_3O—Ph—SO_2Cl, CH_2Cl_2, CH_2I_2, $CHCl_2$—$CHCl_2$, $CHBr_3$, CHI_3, CBr_4, CH_2═CH—CH_2—Cl/Br/I, Ph—CH_2—Cl/Br/I, Ph—CH(CH_3)—Br, Ph(CH_2—Br/I$)_2$, CH_3—CH(CN)—Br, CH_2(CN)—I, $(CH_3)_2C(COOEt)$—Br/I, I—Ph—O—CH_3, and NBS under a wide variety of conditions. As $Mn(CO)_5^\bullet$ has a very high halide affinity [59], abstraction is available in all cases. Therefore, the lack of initiation results from the corresponding radicals being more stable than the propagating PVDF$^\bullet$ radical, and thus failing to add at moderate temperatures, and being consumed by recombination or transfer.

By contrast, very reactive alkyl, polyhalide, as well as semi- and perfluorinated halide analogs of the above initiators such as $CHCl_3$, CCl_4, CCl_3—CCl_3, $CF_3(CF_2)_2CO$—Cl, CF_3—SO_2—Cl, Cl—CF_2—CClF—Cl, Cl—$(CF_2)_8$—Cl, —(CF_2—CFCl$)_n$—, CCl_3—Br, EtOOC—CF_2—Br, Br—$(CH_2)_{10}$—Br, Br—CF_2—CH_2—CF_2—Br, Br—$(CF_2)_4$—Br, CH_3—I, $CH_3(CH_2)_5$—I, I—$(CH_2)_{10}$—I, C_6F_5—CF_2—I, H—CF_2—CF_2—CH_2—I, EtOOC—CF_2—I, Cl—CF_2—CFCl—I, CF_3—I, CF_3CF_2—I, $(CF_3)_2CF$—I, $(CF_3)_3C$—I, $CF_3(CF_2)_3$—I, and I—$(CF_2)_{4,6}$—I, all led to polymer formation [51]. The NMR demonstration of these initiations is outlined in Figure 2.1.

Remarkably, for VDF as well as for CF_2═CFCl, CF_2═CCl_2, CF_2═CFBr, CH_2═CFH, and for VDF random copolymers with CF_2═CF(CF_3) and CF_2═CF(OCF_3) [51], initiation occurred not only from polyhalides and all R_F–I structures (which also provide IDT and elimination of HH defects), but even from semifluorinated models of the bad PVDF chain end (H—CF_2—CF_2—CH_2—I), and especially from simple inactivated alkyl iodides (CH_3—I). This indicates the feasibility of $Mn_2(CO)_{10}$-mediated block or graft VDF copolymerization *directly* from the halides above or their congeners regardless of their DT capability, when anchored on polymeric chains, surfaces, etc.

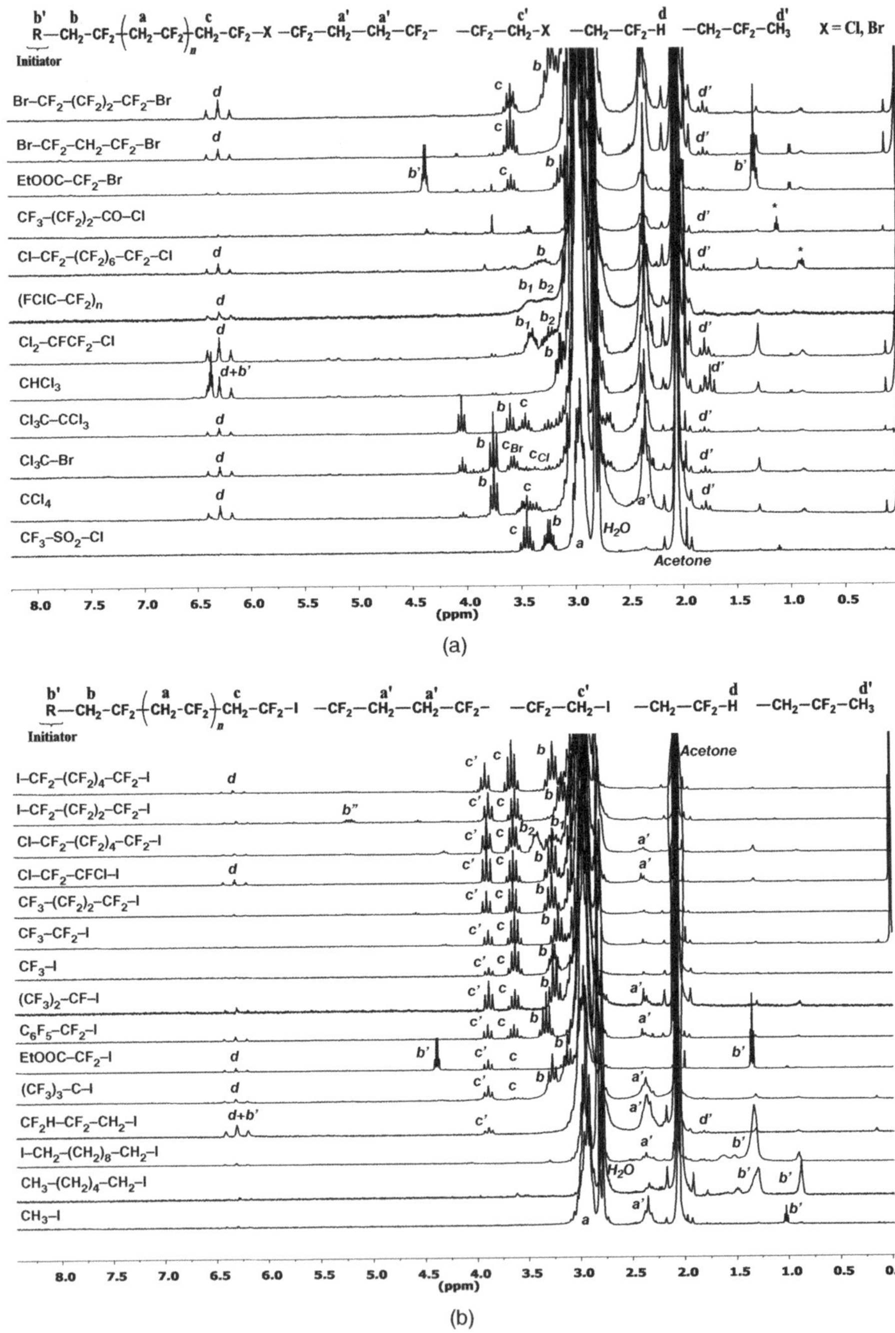

FIGURE 2.1 (a) ^{1}H-NMR spectra of $Mn_2(CO)_{10}$-photoinitiated PVDF from Cl and Br substrates. (b) ^{1}H-NMR spectra of $Mn_2(CO)_{10}$-photoinitiated PVDF from iodine substrates. Reprinted with permission from Reference 51. Copyright 2012 American Chemical Society.

Subsequent to initiation, the polymerization outcome is controlled by the combined effect of the relative order of $CT_{RX}^{PVDF^*}$ values (*i.e.*, (C—X BDE in RX) [18, 29] (Equations 2.5 and 2.6)), the reactivity of fluorinated radicals (i.e., more branched, more electrophilic) [135], and the preferential activation of primary *vs.* secondary or tertiary halides [82] by $Mn(CO)_5^{\bullet}$. Accordingly, the initiators fall into three classes which require different amounts of $Mn_2(CO)_{10}$ for activation, and VDF undergoes FRP for R—X (X = Cl, Br, I) and IDT–CRP for R_F—I. These distinctions can easily be followed by inspecting the features of the halide chain ends in the corresponding PVDF NMRs (Figure 2.1).

First are the R—X species that provide only initiation and no chain transfer. Thus, initiators with strong R—X bonds (R—I, $CHCl_3$, and R_F—Cl, *i.e.*, Cl—CFCl—CF_2—Cl, Cl—(CF_2—CFCl)$_{3-6}$—Cl, $CHCl_3$, CF_3—$(CF_2)_2$—CO—Cl, Cl—CF_2—$(CF_2)_6$—CF_2—Cl, CH_3—I, CH_3—$(CH_2)_4$—CH_2—I, and I—CH_2—$(CH_2)_8$—CH_2—I), do not undergo noticeable CT with PVDF$^{\bullet}$, demand stoichiometric $Mn_2(CO)_{10}$ activation and afford PVDF with no halide chain ends, and where the HH propagation is clearly noticed. The lack of polymer halide termini, stemming from the very low initiator CT value ($CT_{R-X} \ll 1$) and the larger amount of $Mn_2(CO)_{10}$ required, leads to VDF–FRP. Indeed, only $Mn(CO)_5^{\bullet}$, and not the propagating chain can abstract the halide from RX.

In the second group, substrates with weak R—X bonds (CF_3SO_2—Cl, R—CCl_3, R_F—X, X = Br, I; *i.e.*, CF_3—SO_2—Cl, CCl_4, CCl_3Br, CCl_3—CCl_3, Br—CF_2—CH_2—CF_2—Br, Br—CF_2—CF_2—CF_2—CF_2—Br, EtOOC—CF_2—Br), do undergo CT (Equations 2.5 and 2.6), require reduced (10%) amounts of $Mn_2(CO)_{10}$, and thus, at least one or both halide functionalized PVDF—X chain ends (—CH_2—CF_2—X and —CF_2—CH_2—X, X = Cl, Br, I, Figure 2.1) are observed. Moreover, if CT_{RX} is high (e.g., CF_3—SO_2—Cl), HH defects may even be apparently suppressed. Here, the initiator is an efficient CTA, but the resulting PVDF halide chain ends are less reactive ($CT_{RX} > 1 > CT_{PVDF-CF2-X} \gg CT_{PVDF-CH2-X}$). Thus, as the Cl— and Br—DT exchange is inexistent, only VDF–FRP may ensue. Since none of the chain ends are reversibly activated, there will be no accumulation of PVDF–CH_2—X, as both chain ends will abstract the halide from the initiator to provide the typical ~10/1 ratio of "good"/"bad" chain ends, similar to HT *vs.* HH propagation, unless the CT_{R-X} is so low, that only the more reactive PVDF—$CH_2^{\bullet}$ can abstract. Moreover, as $Mn(CO)_5^{\bullet}$ could activate PVDF—X chains (X = Cl, Br) throughout the polymerization, a molecular weight increase may occur, albeit in a poorly controlled manner.

Finally, the best initiators for controlled VDF–IDT photopolymerizations and for high functionality PVDF–I for subsequent chain end derivatization or block copolymer synthesis are based on the semi and perfluorinated initiators. Thus, while good Cl and Br CTAs can at best provide efficient telomerizations [18], uncatalyzed halide DT–CRP occurs only for iodine [15–19].

Here, while poorer results are initially observed with H—CF_2—CF_2—CH_2—I or $(CF_3)_3C$—I where due to sluggish IDT ($CT_{RI} \sim CT_{PVDF-CH2-I} < 1$), and respectively $Mn(CO)_5^{\bullet}$ *vs.* R_FI sterics issues, the PVDF—CH_2—I chain ends and the HH units are still present, activated perfluoroalkyliodides such as CF_3—CF_2—I ~ $(CF_3)_2CF$—I

< C_6F_5—CF_2—I, EtOOC—CF_2—I < Cl—CF_2—CFCl—I < CF_3—$(CF_2)_2$—CF_2—I < CF_3—I, < I—$(CF_2)_{4,6}$—I provide not only *both* types of iodine chain ends but also suppression of HH propagation defects, and of termination by H transfer (—CF_2—H and —CH_2—H), to below 1%. Most likely, their CT_{RF-I} is comparable or better than that of C_6F_{13}—I ($CT_{C6F13-I} = 7.9$) [28], thus enabling very efficient competition with both propagation and termination. As such, these high CT R_F—I initiators suitable for IDT–CRPs [16] are converted early in the process into macromolecular PVDF–I CTAs [28], where the terminal P_m–CF_2–CH_2—I [16, 20–30] 2-1 unit is about 25 times less reactive toward IDT than the isomeric P_n—CH_2—CF_2—I 1,2-unit [28].

Once all the R_F—I initiator is consumed via CT, no *new* PVDF–I chains are generated, and the only productive, thermodynamically neutral, uncatalyzed, *reversible* IDT, equilibrium ($K_{equil\,(ex1)} = 1$) between equally reactive, propagating and dormant P_n—CH_2—$CF_2^{\bullet}$ and P_m—CH_2—CF_2—I terminal 1,2-units (Equation 2.7), operates. Here, the exchange constant, $C_{ex,1} = k_{1,exchange}^{PVDF-CF2^*,PVDF-CF2-I}/k_{propagation,12-addition}^{PVDF-CF2^*,VDF}$ is $\gg 1$ and thus exchange is favored over propagation and termination. However, due to the much stronger —CH_2—I bond, the cross-IDT between the 1,2- and 2,1-units (Equation 2.8) is shifted toward the irreversible buildup of P_n—CF_2—CH_2—I chain ends, whereas the IDT of the 2,1- terminal units is virtually inexistent (Equation 2.9) [27, 28].

Nonetheless, Equations 2.7ab, enable IDT–CRP, as demonstrated [51] (Figure 2.2) by the linear dependence of M_n on conversion and moderate PDI values, which indicate that $Mn_2(CO)_{10}$ supports a photo-CRP over a wide range of molecular weights, and represents *the first example of metal-mediated VDF–CRP and the first example of visible light-mediated VDF–CRP.*

Mechanistically, however, with the exception of iodide chain end reactivation by $Mn(CO)_5^{\bullet}$ to compensate for termination, this remains a conventional IDT. Thus, while IDT catalysis would lead to PDI decrease [3, 9], control experiments [51] reveal that, consistent with PVAc-IDT [101, 102], the photochemically inactive [136a] $Mn(CO)_5$–I is incapable of donating I. Although R_F–$Mn(CO)_5$ ($R_F = CH_2F$, CF_2H) [136b] derivatives exist, organometallic CRP mediation with PVDF–$Mn(CO)_5$ via reversible C–Mn bond homolysis is discounted in view of the observed –I, not –H or –$Mn(CO)_5$ chain ends, of the successful CRP with *catalytic,* not stoichiometric $Mn_2(CO)_{10}$ *vs.* R_F–I, and of the relative order of the R_F–$Mn(CO)_5$ (34) [74] < $(CO)_5Mn$–$Mn(CO)_5$ (38) [132b] < R_F–I (48) [137] < I–$Mn(CO)_5$ (54) [132] BDEs (kcal/mol), consistent with the photoinstability of Mn alkyls [132b].

2.3.3.2 *Effect of Mono vs. Difunctional R_F–I and I–R_F–I Initiators* I–R_F–I initiators are particularly suitable for FM-CRPs, as bidirectional growth from difunctional propagating species [10], in conjunction with initiator or chain end halide activation by the continuously photogenerated $Mn(CO)_5^{\bullet}$ [138], (Equation 2.2) compensate for termination by radical coupling or transfer [10] and helps maintain a steady state radical concentration [51]. Indeed, termination is not as pronounced for the I–R_F–I-initiated polymerizations, where dimerization of two propagating I–PVDF$^{\bullet}$ radicals provides I–$(PVDF)_2$–I, or where H transfer affords I–PVDF–H. Thus, iodine chain end functionality is still retained, and can be reactivated by $Mn(CO)_5^{\bullet}$.

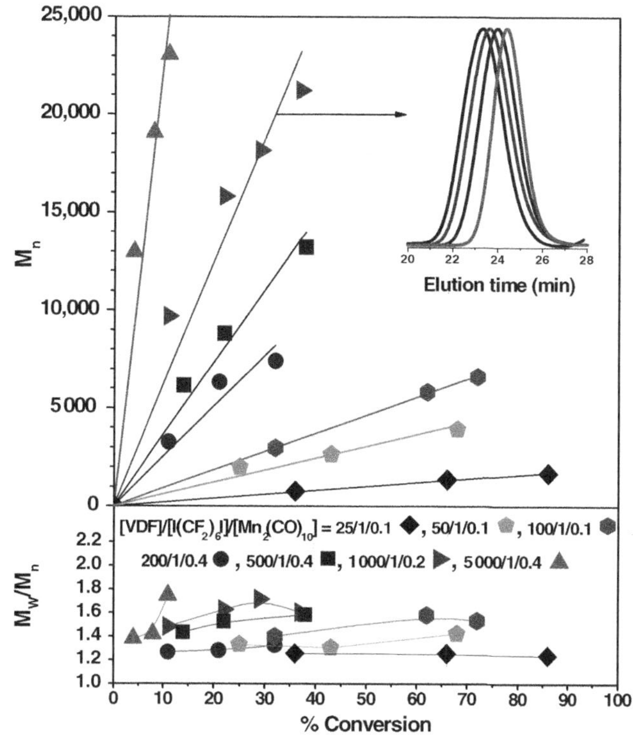

FIGURE 2.2 Dependence of M_n and M_w/M_n on conversion in $Mn_2(CO)_{10}$-photomediated VDF–IDT–CRP. Reprinted with permission from Reference 51. Copyright 2012 American Chemical Society.

Indeed, even if two radicals are consumed, and one propagating chain end becomes endcapped as $-CF_2-CH_2-I$ or terminated by H transfer, the remaining $-CH_2-CF_2-I$ chain end may still get reversibly activated and propagate for some time and even undergo additional dimerizations. Thus, the probability of having $-CH_2-H$ or $-CF_2-H$ end groups is greatly reduced and the lifetime of the chains is increased. This dramatically improves the livingness of the polymerization, narrows the PDI values, and almost doubles the iodine chain end functionality.

2.3.3.3 Dependence of Halide Chain Ends and HH Units on Conversion As demonstrated by NMR, in IDT, HH defects (δ = 2.3–2.4 ppm) are dramatically suppressed (Figure 2.1b.), being intercepted as $P_m-CF_2-CH_2-I$, and thus show no conversion dependence [51]. Obviously, IDT cannot control the regioselectivity of propagation, and the 2,1-addition is just *apparently* suppressed via CT to R_F-I. In reality, as $\sim$PVDF–$CF_2-CH_2^\bullet$ is more reactive than $\sim$PVDF–$CH_2-CF_2^\bullet$ ($k_{p,21} > k_{p,12}$), the diminishing of HH or of the terminal $-CH_2-H$ and $-CF_2-H$ units, is due to the faster CT to the reactive R_F-I initiators by comparison to propagation

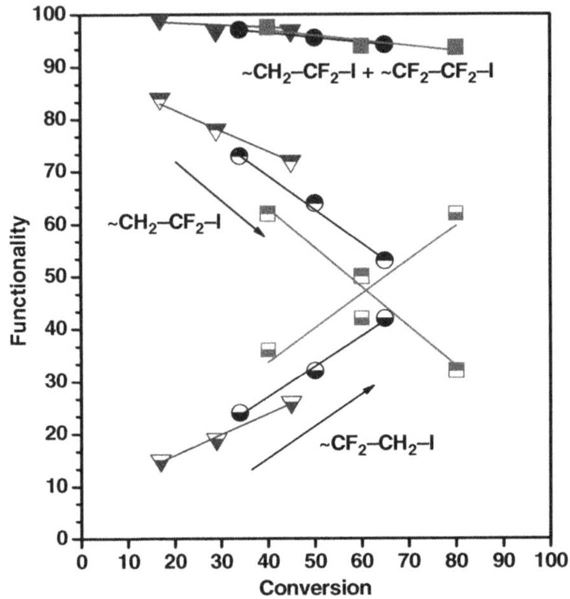

FIGURE 2.3 Dependence of the iodide chain end functionality on conversion in the $Mn_2(CO)_{10}$-mediated VDF–IDT–CRP. $[VDF]/[I(CF_2)_6I]/[Mn_2(CO)_{10}]$ = 50/1/0.1 (▼), 50/1/0.2 (●), 50/1/0.4 (■). Filled symbols = total functionality, top filled = $\sim CH_2-CF_2-I$, bottom filled = $\sim CF_2-CH_2-I$. Reprinted with permission from Reference 51. Copyright 2012 American Chemical Society.

($CT_{RX}^{PVDF-CH2^*} > CT_{RX}^{PVDF-CF2^*} > 1$), dimerization or H abstraction. Accordingly, once the 2,1-propagating unit is intercepted by the I of R_F-I, it is no longer able to propagate, transfer, or terminate, and instead, the potential HH unit is observed as the inactive $PVDF-CF_2-CH_2-I$. Nonetheless, its eventual reactivation if using excess $Mn(CO)_5^{\bullet}$, or much slower by IDT with $\sim PVDF-CF_2-CH_2^{\bullet}$, could still provide an HH unit.

Since the $PVDF-CH_2-I$ termini are much less reactive, and thus not involved in IDT, they will accumulate (Figure 2.3) and will belong to a lower molecular weight population than the corresponding *dormant* $PVDF-CF_2-I$, which can still propagate. Similarly to VAc-IDT [101, 139], their accumulation will broaden the PDI, while their reactivation will require stronger halide abstractors than for the 1,2-$PVDF-CF_2-I$ unit [51]. Unfortunately, this also implies that the activator forms an even stronger bond with the halide. Indeed, while such reversible IDT *catalysis* with $Mn(CO)_5-I$ would have prevented accumulation of $P_n-CF_2-CH_2-I$, this was not the case. As such, the $P_n-CF_2-CH_2-I$ chain ends are pretty much unresponsive as far as conventional, free radical initiated IDT or other metal mediated or organic chain end derivatizations are concerned [27]. Moreover, their concentration increases continuously with conversion, leaving a very bleak prospect for block copolymerizations. Nonetheless, as explained above, the *total* (—CF_2—CH_2—I + —CF_2—CH_2—I) iodine functionality remains >95% in the case of difunctional initiators. This is a reasonable value, on condition that *both* chain ends can be reactivated, as is the case with stoichiometric $Mn(CO)_5^\bullet$, below.

2.4 SYNTHESIS OF WELL-DEFINED BLOCK COPOLYMERS FROM PVDF–I AND I–PVDF–I CHAIN ENDS

Earlier tryouts at the synthesis of pseudo PVDF blocks included either attempting VDF initiation from macromolecular R_F—I in the presence of free radical initiators [31, 140, 141] (which inherently produces PVDF homopolymer), or *assuming* that the chain ends of PVDF—X [141–144] were reactive enough to permit the radical initiation of another monomer (Cu/ATRP [142–144] or thermal IDT [141]), or polycondensations with polysulfone *bis*-nucleophiles [143]. Moreover, the dependence of the chain end functionality on conversion was never studied and acknowledged. Thus, with one exception [141], no details on the PVDF–X halide chain ends were provided, or the fact that mixtures are actually produced, was recognized and understood. In reality, CuX/L hardly activates perfluoroalkylhalides ($k_{abstr}^{(CH3)2C(COOEt)-Br}/k_{abstr}^{C8F17-Br} \sim 10^2$) [145], and thus, would barely initiate from —CF_2CF_2—I, let alone from —CH_2CF_2—I, and especially from the unreactive —CF_2—CH_2—I chain end. Similarly, radical ethyleneation [36, 146, 147] or azidation [148] are again available only for the —CH_2—CF_2—I chain end, and only at high temperature (150–200°C) or under microwave irradiation. Finally, F strongly deactivates nucleophilic substitutions [36].

In retrospect, due to the failure of the respective chemistries to activate the stronger and dominant —CF_2—CH_2—X termini, (as such experiments used high conversion PVDF—I samples, thus containing >80% of the inactive PVDF—CH_2—I), it is clear that all previous endeavors were futile and fundamentally incomplete, and that all so-called "blocks" were in fact *always* inseparable, ill-defined mixtures of PVDF—CH_2—I with PVDF-block copolymers [31, 140–144]. Likewise, as the dead PVDF—CH_2—I termini accumulate, they are always of *lower* molecular weight than the dormant/propagating PVDF—CF_2—I chain ends. Conceivably, selective precipitation/fractionation may enable enrichment in the PVDF—CF_2—I chain end [28], but this would be inefficient and difficult. Thus, the clean synthesis of "pure," well-defined PVDF-block copolymers requires *complete* activation and/or derivatization of both types of PVDF chain ends, and especially of ~CF_2CH_2–X.

As such, while the concentration of active —CH_2—CF_2—I decreases and that of unreactive —CF_2—CH_2—I increases with conversion, (Figure 2.3) [51], the *total* (—CH_2—CF_2—I + —CF_2—CH_2—I) iodine functionality remains at least 95%, even at larger levels of $Mn_2(CO)_{10}$ [51]. This is adequate for block copolymer synthesis, if *both* halide chain ends can be activated, and this is where the high $Mn(CO)_5^\bullet$ halide affinity [59] comes into play. Indeed, as seen above, $Mn(CO)_5^\bullet$ was able to activate not only the ~CF_2—I based initiators, but even the CH_3—I, CH_3—$(CH_2)_5$—I, as well as H—CF_2—CF_2—CH_2—I models of the reverse PVDF—CF_2—CH_2—I 2,1-chain end addition. Initiation from these inactivated, primary alkyl iodides is a remarkable result and indicates that $Mn_2(CO)_{10}$ can easily activate even strong C—I bonds. Yet, the progressively more reactive ~CF_2—CH_2—I and ~CH_2—CF_2—I PVDF iodine chain are even easier to activate than regular alkyl iodides. As such, $Mn(CO)_5^{\bullet}$ affords the clean and *quantitative* activation of *both* —CH_2—CF_2—I and —CF_2—CH_2—I chain ends. As such, regardless of polymerization conversion and thus their ratio in PVDF—I, they *both* represent viable substrates for the $Mn(CO)_5^{\bullet}$-based initiation, and enable the synthesis of well-defined PVDF-block copolymers with any radically polymerizable vinyl monomer. Therefore, $Mn_2(CO)_{10}$ not only mediates IDT, but more importantly, and by contrast to other metal systems (e.g., Cu–ATRP), activates all chain ends for the initiation of a block copolymerization.

Selected examples of the ^{1}H-NMR characterization of such PVDF blocks are presented in Figure 2.4 [51]. Besides acetone and water (δ = 2.05 and 2.84 ppm), the HT —CF_2—[CH_2—CF_2]$_n$—CH_2-, (**a**), and HH, —CF_2—CH_2—CH_2—CF_2— (**a′**) PVDF linkages [36, 149] are seen at δ = 2.8–3.1 ppm and δ = 2.3–2.4 ppm. Resonance **b** (δ = 3.25 ppm) confirms the R_F—CH_2—CF_2— connectivity with the first polymer unit, while the 1,2-CH_2—CF_2—I (**c**) and 2,1-CF_2—CH_2—I (**c′**), iodine chain ends are

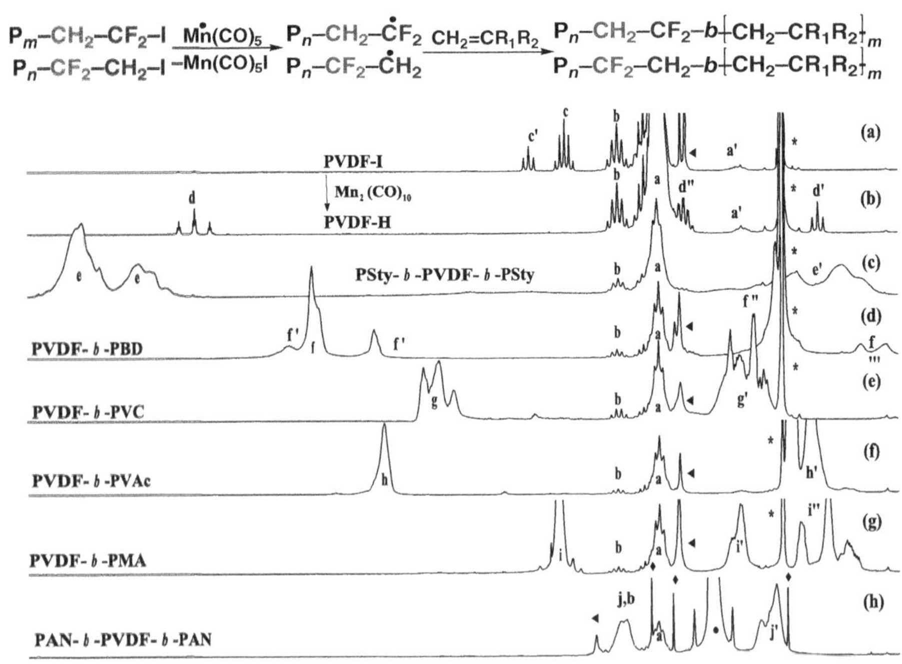

FIGURE 2.4 500 MHz 1H-NMR spectra of PVDF–I, PVDF–H, and various PVDF-block copolymers. All in d_6-acetone except PAN in d_6-DMSO. ◀ = H_2O, * = acetone, ♦ = DMAC, ● = DMSO. Reprinted with permission from Reference 51. Copyright 2012 American Chemical Society.

seen [27] at $\delta = 3.62$ ppm and $\delta = 3.87$ ppm. Trace termination by H transfer to PVDF$^{\bullet}$ (Equations 2.11 and 2.12), (i.e., $-CH_2-CF_2-H$ and $-CF_2-CH_3$, peaks **d**, **d′**) is seen at $\delta = 6.30$ ppm and $\delta = 1.80$ ppm [149].

Upon treatment of PVDF–I with stoichiometric $Mn(CO)_5^{\bullet}$, as demonstrated by the disappearance of the ***c*** and ***c′*** peaks, *complete* radical activation of *both* iodide chain ends occurs, and the resulting PVDF$^{\bullet}$ radicals are deactivated by H abstraction from solvent, to afford the corresponding PVDF–H chain ends, seen as a dramatic increase in the **d** and **d′** peaks, and a more resolved $-CH_2-CF_2-CH_2-CF_2-H$ **d″**, $\delta = 2.77$ ppm [51, 149].

Thus, performing the activation in the presence of radically polymerizable alkenes leads to the first examples of well-defined AB or ABA-type PVDF-block copolymers with styrene (**e**, **e′**), butadiene (**f**, **f′**, **f″**,**f‴**), vinyl chloride (**g**, **g′**), vinyl acetate (**h**, **h′**), methyl acrylate (**i**, **i′**, **i″**), and acrylonitrile (**j**, **j′**), initiated from *both* the PVDF halide chain ends. While here $Mn_2(CO)_{10}$ simply performs irreversible halide activation, and there is no IDT, control of the block copolymerization can be envisioned by other CRP methods.

2.5 CONCLUSIONS

We have demonstrated for the first time, that metal-catalyzed initiation of the polymerization of main chain fluorinated gaseous monomers such as VDF, can easily be accomplished at 0–100°C, including at rt upto 40°C, *directly* from a wide variety of alkyl, semifluorinated and perfluoroalkyl halides (Cl, Br, I), using a visible light, $Mn_2(CO)_{10}$ photomediated protocol, carried out in low pressure glass tubes, and which is especially successful in dimethyl carbonate.

The use of perfluorinated alkyl iodides, particularly the difunctional $I-R_F-I$, enables the IDT mechanism, leading to CRPs, with very high (>95%) total iodide chain end functionality and <1% HH defects.

While the concentration of the active $-CH_2-CF_2-I$ decreases and that of unreactive $-CF_2-CH_2-I$ PVDF chain ends increases with conversion, their subsequent quantitative activation with $Mn_2(CO)_{10}$ affords the first examples of well-defined PVDF-block copolymers with a variety of other classes of monomers, irrespective of the $-CH_2-CF_2-I/-CF_2-CH_2-I$ ratio in the PVDF–I sample.

The direct halide VDF initiation, VDF–IDT–CRP, and the quantitative iodide chain ends activation open up previously unavailable strategies for the photomediated synthesis of pure and well-defined, architecturally complex fluoromaterials. As such, main chain fluorinated monomers can be block copolymerized or grafted directly from any substrates featuring suitable halide initiators, while on the other hand, the polymerization of other monomers can be quantitatively initiated from the halide chain ends of such fluoropolymers, without the formation of mixtures. With multifunctional initiators, star and hyperbranched fluoropolymers can also be envisioned. Lastly, the $R_F-I/Mn_2(CO)_{10}$ protocol is also applicable in radical trifluoromethylation and perfluoroalkylation reactions which are in great demand in organic chemistry [52].

REFERENCES

1. Ameduri, B. *Macromolecules* **2010**, *43*, 10163–10184.
2. Lena, F.; Matyjaszewski, K. *Prog. Polym. Sci.* **2010**, *35*, 959–1021.
3. Matyjaszewski, K.; Davis, T. P. *Handbook of Radical Polymerization*; John Wiley & Sons: New York, 2002.
4. Braunecker, W. A.; Matyjaszewski, K. *Prog. Polym. Sci.* **2007**, *32*, 93–146.
5. Goto, A.; Fukuda, T. *Progr. Polym. Sci.* **2004**, *29*, 329.
6. Fukuda, T. *J. Polym. Sci.: Part A: Polym. Chem.* **2004**, *42*, 4743–4755.
7. Ameduri, B. *Chem. Rev.* **2009**, *109*, 6632–6686.
8. Hansen, N. M. L.; Jankova, K.; Hvilsted, S. *Eur. Polym. J.* **2007**, *43*, 255–293.
9. Fukuda, T.; Goto, A.; Tsujii, Y. Kinetics of Living Radical Polymerization. In *Handbook of Radical Polymerization*; Matyjaszewski, K., Davis, T. P., Eds.; John Wiley & Sons: New York, 2002; pp 407–462.
10. Oka, M.; Tatemoto, M. Vinylidene Fluoride-Hexafluoropropylene Copolymer Having Terminal Iodines. In *Contemporary Topics in Polymer Science*; Bailey, W. J., Tsuruta, T., Eds.; Plenun Press: New York, 1984; Vol. 4, pp 763–781.
11. Tatemoto, M. *Int. Poly. Sc. Tech.* 1985, *12*, 85–98.
12. Tatemoto, M. In *Polymeric Materials Encyclopedia*; Salamone, J. C., Ed.; CRC: Boca Raton, FL, 1996; Vol. 5, pp 3847–3862.
13. Tatemoto, M.; Shimizu, T. Thermoplastic Elastomers. In *Modern Fluoropolymers*; Scheirs, J., Ed., John Wiley & Sons: New York, 1997; pp 565–576.
14. Tatemoto, M.; Nakagawa, T. US Patent 4,158,678, 1979.
15. Gaynor, S. G.; Wang, J. S.; Matyjaszewski, K. *Macromolecules* **1995**, *28*, 8051–8056.
16. David, G.; Boyer, C.; Tonnar, J.; Ameduri, B.; Lacroix-Desmazes, P.; Boutevin, B. *Chem. Rev.* **2006**, *106*, 3936–3962.
17. Boutevin, B. *J. Polym. Sci.: Part A: Polym. Chem.* **2000**, *38*, 3235–3243.
18. Ameduri, B.; Boutevin, B. *Top. Curr. Chem.* **1997**, *192*, 165–233.
19. Ameduri, B.; Boutevin, B. *Well Architectured Fluoropolymers: Synthesis, Properties and Applications*; Elsevier: Amsterdam, 2004; pp 1–99.
20. Ameduri, B.; Ladavière, C.; Delolme, F.; Boutevin, B. *Macromolecules* **2004**, *37*, 7602–7612.
21. Montefusco, F.; Bongiovanni, R.; Priola, A.; Ameduri, B. *Macromolecules* **2004**, *37*, 9804–9813.
22. Cape, J. N.; Greig, A. C.; Tedder, J. M.; Walton, J. C. *J. Chem. Soc., Faraday Trans. 1: Phys. Chem. Condens. Phases* **1975**, *71*, 592.
23. Hauptschein, M.; Braid, M.; Lawlor, F. E. *J. Am. Chem. Soc.* **1958**, *80*, 846–851.
24. Soueni, A. E.; Tedder, J. M.; Walton, J. C. *J. Fluor. Chem.* **1978**, *11*, 407–417.
25. Combes, J. R.; Guan, Z.; DeSimone, J. M. *Macromolecules* **1994**, *27*, 865–886.
26. Balague, J.; Ameduri, B.; Boutevin, B.; Caporiccio, G. *J. Fluorine Chem.* **1995**, *70*, 215–223.
27. Boyer, C.; Valade, D.; Sauguet, L.; Ameduri, B.; Boutevin, B. *Macromolecules* **2005**, *38*, 10353–10362.

28. Boyer, C.; David, V.; Lacroix-Desmazes, P.; Ameduri, B.; Boutevin, B. *J. Polym. Sci.: Part A: Polym. Chem.* **2006**, *44*, 5763–5777.
29. Imran-ul-haq, M.; Förster, N.; Vukicevic, R.; Herrmann, K.; Siegmann, R.; Beuermann, S. *ACS Symp. Ser.* **2009**, *1024*, 233–243.
30. Apostolo, M.; Arcella, V.; Storti, G.; Morbidelli, M. *Macromolecules* **2002**, *35*, 6154–6166.
31. Valade, D.; Boyer, C.; Ameduri, B.; Boutevin, B. *Macromolecules* **2006**, *39*, 8639–8651.
32. MacMurray, N.; Tedder, J. M.; Vertommen, L. L. T.; Walton, J. C. *J. Chem. Soc., Perkin Trans.* **1976**, *2*, 63–69.
33. Hung, M. H. US Patent 5,231,154, 1993.
34. Low, H. C.; Tedder, J. M.; Walton, J. C. *J. Chem. Soc., Faraday Trans. 1: Phys. Chem. Condens. Phases* **1976**, *72*, 1707.
35. Apostolo, M.; Arcella, V.; Storti, G.; Morbidelli, M. *Macromolecules* **1999**, *32*, 989.
36. Ameduri, B.; Boutevin, B. *J. Fluorine. Chem.* **1999**, *100*, 97–116.
37. Zhang, Z. C.; Chung, T. C. *Macromolecules* **2006**, *39*, 5187–5189.
38. Balague, J.; Ameduri, B.; Boutevin, B.; Caporiccio, G. *J. Fluorine Chem.* **2000**, *102*, 253–268.
39. Nguyen, B. V.; Yang, Z. Y.; Burton, D. J. *J. Org. Chem.* **1998**, *63*, 2887–2891.
40. Li, A. R.; Chen, Q. Y. *J. Fluorine Chem.* **1997**, *81*(2), 99–101.
41. Chen, M. Y.; Yang, Z. Y.; Zhao, C. X.; Qiu, Z. M. *J. Chem. Soc. Perkin Trans. I.* **1988**, *3*, 563.
42. Metzerg, J. O.; Linker, U. *Liebigs Ann. Chem.* **1992**, *3*, 209–216.
43. Hu, C. M.; Qiu, Y. L. *J. Chem. Soc. Perkin Trans. 1.* **1992**, *13*, 1569–1572.
44. Huang, W. Y.; Zhang, H. Z.; Lu, L. *J. Fluorine Chem.* **1990**, *50*(1), 133–140
45. Leung, L.; Linclau, B. *J. Fluorine Chem.* **2008**, *129*, 986–990.
46. Fang, X.; Yang, X.; Yang, X.; Mao, S.; Wang, Z.; Chena, G.; Wua, F. *Tetrahedron.* **2007**, *63*, 10684–10692.
47. Hu, C. M.; Qing, F. L. *J. Org. Chem.* **1991**, *56*, 6348–6351.
48. Yang, Z. Y.; Burton, D. J. *J. Org. Chem.* **1991**, *56*, 1037–1041.
49. Tsunoi, S.; Ryu, I.; Fukushima, H.; Tanaka, M.; Komatsu, M.; Noburu, S. *Synlett.* **1999**, *5*, 1249–1252.
50. Benefice-Malouet, S.; Blancou, H.; Commeyras, A. *J. Fluorine Chem.* **1993**, *63*(3), 217–226.
51. Asandei, A. D.; Adebolu, O. I.; Simpson, C. P. *J. Am. Chem. Soc.* **2012**, *134*, 6080–6083.
52. Asandei, A. D.; Adebolu, O. I.; Simpson, C. P.; Kim, J. S. **Angew. Chem. Int. Ed. 2013**, *52*, 10027–10030.
53. Asandei, A. D.; Simpson, C. P.; Adebolu, O.; Chen, Y. *Polym. Prepr.* **2011**, *52*(2), 728–729.
54. Asandei, A. D.; Adebolu, O. I.; Simpson, C. P. *Acs. Symp. Ser.* **2012**, *1106*, 47–63.
55. Asandei, A. D.; Simpson, C. P.; Adebolu, O.; Chen, Y. *Polym. Prepr.* **2011**, *52*(2), 554–555.
56. Sauguet, L.; Boyer, C.; Ameduri, B.; Boutevin, B. *Macromolecules* **2006**, *39*, 9087–9101.
57. Messina, M. T.; Metrangolo, P.; Resnati, G. *ACS Symp. Ser.* **1999**, *746*, 239–254.

58. Rowlands, G. J. *Tetrahedron* **2009**, *65*, 8603–8655.
59. Gilbert, B. C.; Parsons, A. F. *J. Chem. Soc., Perkin Trans. 2.* **2002**, *3*, 367–387.
60. Haszeldine, R. N.; Steele, B. R. *J. Chem. Soc.* **1954**, 923.
61. Saint-Loup, R.; Ameduri, B. *J. Fluorine Chem.* **2002**, *116*, 27–34.
62. Asandei, A. D.; Chen, Y. *Polym. Mater: Sci. Eng.* **2008**, *98*, 346–347.
63. Asandei, A. D.; Chen, Y. *Polym. Prepr.* **2007**, *48*(2), 452–453.
64. Asandei, A. D.; Chen, Y. *Polym. Mater: Sci. Eng.* **2007**, *97*, 270–271.
65. Asandei, A. D.; Chen, Y. *Polym. Prepr.* **2005**, *46*(2), 633.
66. Asandei, A. D.; Moran, I. W. *J. Am. Chem. Soc.* **2004**, *126*, 15932–15933.
67. Asandei, A. D.; Simpson, C. P.; Yu, H. S.; Adebolu, O. I.; Saha, G.; Chen, Y. *ACS Symp. Ser.* **2009**, *1024*, 149–166.
68. Abrahamson, H. B.; Wrighton, M. S. *J. Am. Chem. Soc.* **1977**, *99*, 5510–5512.
69. Meckstroth, W. K.; Walters, R. T.; Waltz, W. L.; Wojcicki, A.; Dorfman, L. M. *J. Am. Chem. Soc.* **1982**, *104*, 1842–1846.
70. Fawcett, J. P.; Poe, A.; Sharma, K. R. *J. Chem. Soc., Dalton Trans.: Inorg. Chem.* **1979**, *12*, 1886–1890.
71. Brimm, E. O.; Lynch, M. A., Jr.; Sesny, W. J. *J. Am. Chem. Soc.*, **1954**, *76*(14), 3831–3835.
72. Treichel, P. M. Manganese carbonyls and manganese carbonyl halides. In *Comprehensive Organometallic Chemistry II Vol. 6, Manganese Group*; Abel, E. W., Stone, F. G. A., Wilkinson, G., Casey, C. P., Eds.; Pergamon, 1995; pp 1–19.
73. Tenhaeff, S. C.; Covert, K. J.; Castellani, M. P.; Grunkemeier, J.; Kunz, C.; Weakley, T. J. R.; Koenig, T.; Tyler, D. R. *Organometallics* 1993, *12*, 5000–5004.
74. Martinho Simoes, J. A.; Beauchamp, J. A. *Chem. Rev.* **1990**, *90*, 629–688.
75. Friestad, G. K.; Qin, J. *J. Am. Chem. Soc.* **2001**, *123*, 9922–9923.
76. Jackson, R. A.; Poe, A. *Inorg. Chem.* **1978**, *17*, 997–1003.
77. Fawcett, J. P.; Poe, A.; Sharma, K. R. *J. Am. Chem. Soc.* **1976**, *98*, 1401.
78. Rothberg, L. J.; Cooper, J. N.; Peters, K. S.; Vaida, V. *J. Am. Chem. Soc.* **1982**, *104*, 3536–3537.
79. (a) Robert, H. C.; Michael, P., Eds. *Comprehensive Organometallic Chemistry III, Volume 5*; Elsevier: Providence, RI, 2007; pp 440–746. (b) Zhang, J. Z.; Harris, C. B. *J. Chem. Phys.* **1991**, *95*(6), 4024–4032.
80. Sarakha, M. Ferraudi, G. *Inorg. Chem.* **1999**, *38*, 4605–4607.
81. Sullivan, R. J.; Brown, T. L. *J. Am. Chem. Soc.* **1991**, *113*, 9155–9161.
82. Gilbert, B. C.; Kalz, W.; Lindsay, C. I.; McGrail, P. T.; Parsons, A. F.; Whittaker, D. T. E. *J. Chem. Soc., Perkin Trans. 1.* **2000**, 1187–1194.
83. Gilbert, B. C.; Lindsay, C. I.; McGrail, P. T.; Parsons, A. F.; Whittaker, D. T. E. *Synth. Commun.* **1999**, *29*(15), 2711–2718.
84. Hallock, S. A.; Wojcicki, A. *J. Organomet. Chem.* **1979**, *182*(4), 521–535.
85. Herrick, R. S.; Herrinton, T. R.; Walker, H. W.; Brown, T. L. *Organometallics* **1985**, *4*, 42–45.
86. Bamford, C. H.; Mahmud, M. U. *J. Chem. Soc., Chem. Commun.* **1972**, *13*, 762–763.
87. Bamford, C. H.; Mullik, S. U. *Polymer* **1973**, *14*(1), 38–39.

88. Bamford, C. H.; Mullik, S. U. *Polymer.* **1976**, *17*(3), 225–230.
89. Aliwi, S. M.; Bamford, C. H.; Mullik, S. U. *J. Polym. Sci., Polym. Symp.* **1975**, *50*, 33–50.
90. Bamford, C. H.; Mullik, S. U. *J. Chem. Soc., Faraday Trans. 1.* **1975**, *71*(3), 625–636.
91. Bamford, C. H.; Mullik, S. U. *Polymer* **1976**, *17*, 94–95.
92. Bamford, C. H.; Mullik, S. U. *J. Chem. Soc., Faraday Trans. 1.* **1976**, *72*(2), 368–375.
93. Bamford, C. H.; Denyer, R. *Nature* **1968**, *217*, 59–60.
94. Bamford, C. H.; Duncan, F. J.; Reynolds, R. J. W.; Seddon, J. D. *J. Polym. Sci., Part C: Polym. Symp.* **1968**, *23*, 419–432.
95. Bamford, C. H.; Dyson, R. W.; Eastmond, G. C. *J. Polym. Sci., Part C.* **1967**, *16*, 2425.
96. Jiang, M.; Wang, S.; Jin X. *J. Mater. Sci. Lett.* **1990**, *9*, 1239–1240.
97. Jenkins, D. W.; Hudson, S. M. *Macromolecules* **2002**, *35*, 3413.
98. Eastmond, G. C.; Parr, K. J.; Woot, J. *Polymer* **1988**, *29*(5), 950–957.
99. Eastmond, G. C.; Richardson, J. E. *Macromolecules* **1991**, *24*, 3189–3200.
100. Iskin, B.; Yilmaz, G.; Yagci, Y. *Macromol. Chem. Phys.* **2013**, *214*, 94–98.
101. Koumura, K.; Satoh, K.; Kamigaito, M. *Macromolecules* **2008**, *41*(20), 7359–7367.
102. Koumura, K.; Satoh, K.; Kamigaito, M. *J. Polym. Sci: Part A: Polym. Chem.* **2009**, *47*, 1343–1353.
103. Koumura, K.; Satoh, K.; Kamigaito, M. *Macromolecules* **2009**, *42*, 2497–2504.
104. Kimura, K.; Satoh, K, Kamigaito, M. *Polym. J.* **2009**, *41*, 595–603.
105. Doll, W. W.; Lando, J. B. *J. Appl. Polym. Sci.* **1970**, *14*(7), 1767–1773.
106. Russo, S.; Behari, K.; Chengji, S.; Pianca, M.; Barchiesi, E.; Moggi, G. *Polymer* **1993**, *34*(22), 4777–4781.
107. Duc, M.; Ameduri, B.; Boutevin, B.; Kharroubi, M.; Sage, J.-M. *Macromol. Chem. Phys.* **1998**, *199*, 1271–1289.
108. Galin, M.; Maslinkot, L. *Macromolecules* **1985**, *18*, 2192–2196.
109. Galin, J. C.; Lutringer, G.; Galin, M. *J. Appl. Polym. Sci.* **1989**, *37*, 487–498.
110. Lutringer, G.; Meurer, B.; Weill, G. *Polymer* **1991**, *32*, 884–891.
111. Kutringer, G.; Weill, G. *Polymer* **1991**, *32*, 877–883.
112. Bottino, A.; Campanelli, G.; Munari, S.; Turturro, A. *J. Polym. Sci.: Part B: Polym. Phys.* **1988**, *26*, 785–794.
113. Lee, W. H. Cyclic Carbonates. In *Chemistry of Nonaqueous Solvents Vol. 4*; Lagowski, J. J., Ed.; Academic Press: New York, 1976; p 167.
114. Odian G. *Principles of Polymerization*, 4th ed.; John Wiley & Sons, 2004; pp 292–298.
115. Ahmeda, T. S.; DeSimone, J. M.; Roberts, G. W. *Chem. Eng. Sci.* **2010**, *65*, 651–659.
116. DiNoia, T. P.; Conway, S. E.; Lim, J. S.; Mchugh, M. A. *J. Polym. Sci., Part B: Polym. Phys.* **2000**, *38*, 2832–2840.
117. Tundo, P.; Selva, M. *Acc. Chem. Res.* **2002**, *35*, 706–716.
118. Gordon, A. S.; Norris, W. P. *J. Phys. Chem.* **1965**, *69*(9), 3013–3017.
119. Thynne, J. C. J.; Gray, P. *Trans. Faraday Soc.* **1962**, *58*, 2403–2409.
120. Wijnen, M. H. J. *J. Chem. Phys.* **1961**, *34*, 1465–1466.
121. Yee Quee, M. J.; Thynne, J. C. J. *Trans. Faraday Soc.* **1966**, *62*, 3154–3161.
122. Zeldes, H.; Livingston, R. *J. Magn. Reson.* **1976**, *21*(1), 109–113.

123. Bilde, M.; Møgelberg, T. E.; Sehested, J.; Nielsen, O. J.; Wallington, T. J.; Hurley, M. D.; Japar, S. M.; Dill M., Orkin, V. L.; Buckley, T. J.; Huie, R. E.; Kurylo, M. J. *J. Phys. Chem. A.* **1997**, *101*, 3514–3525.
124. Griller D.; Roberts, B. P. *J. Chem. Soc. Perkin Trans. II.* **1972**, *6*, 747–751.
125. McRae, J. A.; Symons, M. C. R. *J. Chem. Soc. (B), Phys. Org.* **1968**, 428–430.
126. Hudson, R. L.; Williams, F. *J. Phys. Chem.* **1981**, *85*(5), 510–511.
127. Freunberger, S. A.; Chen, Y.; Peng, Z.; Griffin, J. M.; Hardwick, L. J.; Barde, F.; Novak, P.; Bruce, P. G. *J. Am. Chem. Soc.* **2011**, *133*, 8040–8047.
128. Burchill, M. T. Diffusion Effects of Carbonate Solvents in PVDF and PVDF Copolymers. In *Progress in Batteries and Battery Materials;* 1998, Vol. 17, pp 144–156, ITE-JEC Press.
129. Saunier, J.; Alloin, F.; Sanchez, J. Y.; Barriere, B. *J. Polym. Sci.: Part B: Polym. Phys.* **2004**, *42*, 532–543.
130. Saunier, J.; Alloin, F.; Sanchez, J. Y.; Barriere, B. *J. Polym. Sci.: Part B: Polym. Phys.* **2004**, *42*, 544–552.
131. Saunier, J.; Alloin, F.; Sanchez, J. Y.; Maniguet, L. *J. Polym. Sci.: Part B: Polym. Phys.* **2004**, *42*, 2308–2317.
132. (a) Drago, R. S.; Wong, N. M.; Ferris, D. C. *J. Am. Chem. Soc.* **1992**, *114*, 91–98. (b) Friestad, G. K; Marie, J. C.; Suh, Y. S.; Qin, J. J. *Org. Chem.* **2006**, *71*, 7016.
133. Duc, M.; Ameduri, B.; Ghislain, D.; Boutevin, B. *J. Fluorine Chem.* **2007**, *128*, 144–149.
134. Hallock, S. A.; Wojcicki, A. *J. Organomet. Chem.* **1979**, *182*, 521–535.
135. Dolbier, W. R., Jr. *Top. Curr. Chem.* **1997**, *192*, 97–163.
136. (a) Pan, X.; Philbin, C. E.; Castellani, M. P.; Tyler, D. R. *Inorg. Chem.* **1988**, *27*, 671–676. (b) Martinho Simoes, J. A.; Beauchamp, J. L. *Chem. Rev.* **1990**, *90*, 629–688.
137. Okafo, E. N.; Whittle, E. *Int. J. Chem. Kinet.* **1975**, *7*, 287–300.
138. Sarakha, M.; Ferraudi, G. *Inorg. Chem.* **1999**, *38*, 4605.
139. Koumura, K.; Satoh, K.; Kamigaito, M. *Macromolecules* **2006**, *39*, 4054–4061.
140. Gelin, M.; Ameduri, B. *J. Polym. Sci.: Part A. Polym. Chem.* **2003**, *41*, 160–171.
141. Laruelle, G.; Nicol, E.; Ameduri, B.; Tassin, J. F, Ajellal, N. *J. Polym. Sci. Part A: Polym. Chem.* **2011**, *49*, 3960–3969.
142. Ming, E.; Tsang, W.; Shi, Z.; Holdcroft, S. *Macromolecules* **2011**, *44*, 8845–8857.
143. Yang, Y.; Shi, Z.; Holdcroft, S. *Macromolecules* **2004**, *37*, 1678–1681.
144. Jol, S. M.; Lee, W. S.; Ahn, B.S.; Park, K Y.; Kim, K. A.; Paeng, I. R. *Polym. Bull.* **2000**, *44*, 1–8.
145. Sauguet, L.; Boyer, C.; Ameduri, B.; Boutevin, B. *Macromolecules* **2006**, *39*, 9087–9101.
146. Manseri, A.; Ameduri, B.; Boutevin, B.; Kotora, M.; Caporiccio, G. *J. Fluorine Chem.* **1997**, *81*, 103–113.
147. Balague, J.; Ameduri, B.; Boutevin, B.; Caporiccio, G. *J. Fluorine Chem.* **2000**, *102*, 253–268.
148. Vukicevic, R.; Beuermann, S. *Macromolecules* **2011**, *44*, 2597–2603.
149. Wormald, P.; Ameduri, B.; Harris, R. K.; Hazendonk, P. *Polymer* **2008**, *49*, 3629–3638.

3

INTERFACIAL RESPONSE OF SEMIFLUORINATED MULTI-BLOCK COPOLYMERS

Anupriya Agrawal, Dipak Aryal, and Dvora Perahia
Clemson University

Gary S. Grest
Sandia National Laboratories

3.1 INTRODUCTION

Soft materials have a remarkable ability to respond to stimuli in a variety of ways [1–19]. From conformational changes and migration of molecules to induced chemical reactions, response of soft materials has enabled a large variety of new technologies. The impact of responsiveness transcends across disciplines from biotechnology such as drug release and polymeric actuators for muscle regeneration, to smart coatings for selective transport and adhesion. Dynamic surfaces that respond to external stimuli such as changes of temperature, pressure, pH, and presence of specific substances, polymers often respond by changing their interfacial conformation and composition. As a result, the interfacial characteristics change. Properties such as adhesion, affinity to chemicals, and permeability, as well as optical characteristics such as refractive indices, are modified. This interfacial response may result in perpetual or reversible transformations. Among soft materials, polymers offer a rich

Handbook of Fluoropolymer Science and Technology, First Edition. Edited by Dennis W. Smith Jr., Scott T. Iacono, and Suresh S. Iyer.

variety of dynamic modes that enable responsiveness. The ultimate, responsive polymeric interface should be durable, however sufficiently dynamic to adapt. Despite the immense advances in responsive materials, transforming the idea of responsiveness of soft materials to large-scale dynamic coatings requires the development of materials that can form stable layers over extended areas while retaining conformational mobility in the polymer [1].

Building on the basic premise that tailoring multiple blocks with significantly different response into one polymer introduces a unique polymeric manifold that can be simultaneously responsive and stable, we introduce a molecular dynamics study of model semifluorinated polymers [10, 13, 17]. Semifluorinated polymers are exciting materials where the properties of the fluorinated segments play several contradictory roles. Fluorinated polymers are known to exhibit high chemical and temperature stability, whereas at the same time fluorinated segments imbedded in protonated polymers drive responsiveness due to their low interfacial energies and inherent segregation from the protonated segments [15–25]. Striking the balance between the dynamics and stability opens a new paradigm in the design of responsive but mechanically stable polymeric interfaces.

The technological significance of responsive and self-healing macromolecular interfaces has driven a large effort to tailor polymeric synthetic manifolds capable of reorganization [1–25]. These include grafted polymer layers and polymeric layers impregnated with nanoparticles, as well as the use of block copolymers, where each block exhibits unique characteristics. The conformational changes and surface rearrangements vary with the specific polymeric systems. For example, a collapsed to a swollen conformation transition (or vice versa) takes place in end-grafted polymers as the polymer layer is exposed to solvents. This transition may result in changes of surface adhesion. In thin films, mitigation of alignment and ordering at the interface upon temperature changes not only affect the adhesion of such an interface, but also drive the healing of mechanical damage. The ability to respond requires a driving force for a change and often a sufficient mobility to rearrange. To transform responsive polymeric surfaces to a technologically viable entity, one has to form durable interfaces that retain their mechanical integrity, together with the chemical stability as the film responds. For large-scale surfaces, effective manufacturing processes are required, along with the response.

Among the most promising pathways to design robust though responsive polymeric coatings that would transcend from a laboratory scale to industrial applications are thin films of structured polymers. These macromolecules consist of highly segregated block copolymers with tailored properties, including a component that is able to easily rearrange as a response to stimuli while the other component retains mechanical stability. The highly segregated nature of the segments is a crucial feature. It allows block copolymers with relatively short segments to form well-defined, structured phases without losing overall mobility [17].

With an effort to determine the condition in which semifluorinated polymer surfaces become responsive, we have probed several families of semifluorinated polymers both experimentally [26–29] and computationally [30–32]. Our studies have shown that semifluorinated polymers form compounded interface where both fluorinated segments and protonated ones reside at the interface. Small temperature changes and solvent presence changes the interfacial composition of fluorine and in turn change the interfacial characteristics. These studies have also shown that the field of parameters that influence the migration to the interface is large and varies significantly with the specific structure.

Here we present a new insight into the interfacial behavior of semifluorinated polymers and its origin as derived from molecular dynamics simulations of a model polymer that consists of semifluorinated multi-blocks of alternating protonated and fluorinated linear carbon chains of varying segment length while the total number of monomers remains constant. This study revealed for the first time the effects of topology of the polymers and rigidity of the blocks on the interfacial characteristics of semifluorinated polymers. Examples of films formed by polymers with an identical number of carbons, whose topology is varied by varying the dimension of fluorinated and protonated segments, are shown in Figure 3.1. Qualitatively, short segments

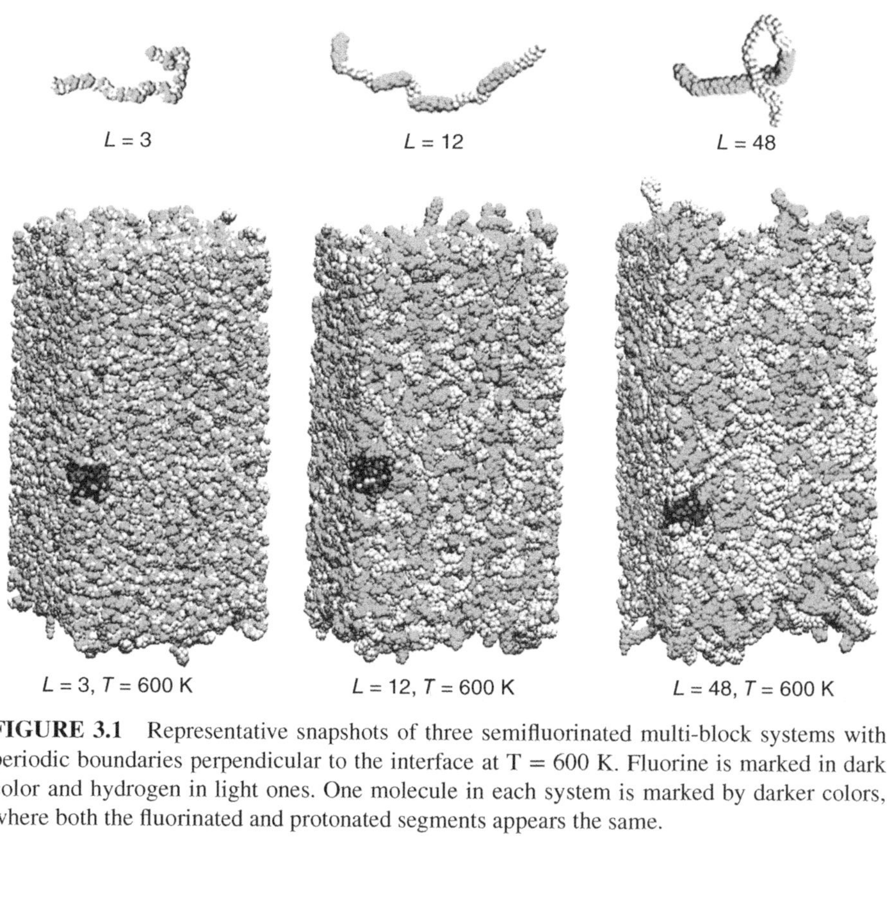

FIGURE 3.1 Representative snapshots of three semifluorinated multi-block systems with periodic boundaries perpendicular to the interface at T = 600 K. Fluorine is marked in dark color and hydrogen in light ones. One molecule in each system is marked by darker colors, where both the fluorinated and protonated segments appears the same.

form fully miscible systems where larger segment phase segregate, impacting the interfacial characteristics.

Our previous molecular dynamics simulations have surprisingly determined that the interface of short diblock copolymers that consist of fluorinated and protonated segments exhibits structured interfaces, with nano domains of protonated and fluorinated segments at the surface [29–32]. The composition of the H and F segments slightly varies as a function of the distance from the surface. These computational results are consistent with our early experimental studies of PFCB-based polymers that have shown a fluorine-rich air interface [27]. The computational studies have shown that the surface tension depends on the degree of fluorine at the interface. These studies have also shown that small changes of fluorine content at the interface impact significantly the surface tension of the films.

The strong segregation between fluorinated and protonated segments, the rigidity versus low interacting energies of the fluorinated segments, and the unique interfacial affinity of fluorine distinguish the semifluorinated from other block copolymers and offer driving forces for rearrangements. While the chemical structure of the fluorine and phase structure induced due to the high segregation contributes significantly to the stability, the dynamics at the interfacial region pertaining to the segmental motion and interfacial alignment changes the overall characteristics of the interface. The present study probes a model system using atomistic molecular dynamics simulations. Computational studies are ideal to probe small changes in the interfacial characteristics that are able to change the properties of the surface.

3.2 SIMULATION MODEL AND METHODOLOGY

The polymers were simulated using all-atom optimized potentials (OPLS-AA) developed by Jorgensen and coworkers [33–36]. Nonbonded interactions consisted of a sum of the standard 12–6 Lennard–Jones (LJ) and electrostatic potentials. Berthelot geometric mixing rules $\varepsilon_{ij} = (\varepsilon_i * \varepsilon_j)^{1/2}$ and $\sigma_{ij} = (\sigma_i * \sigma_j)^{1/2}$, where ε_{ij} and σ_{ij} are the LJ units of energy and distance between atoms i and j, respectively were used for all LJ interactions between unlike atoms, except for a 25% reduction in the H—F LJ interaction energy as discussed by Song et al. [37] and Escobedo and Chen [38] and used in our previous work [31,32]. This reduction was found necessary to accurately describe mixtures of protonated and perfluorinated alkanes. Nonbonded interactions were calculated for all atomic pairs on different molecules as well as between atoms on the same molecule separated by three or more bonds. This interaction was reduced by a factor of two for atoms separated by three bonds. All LJ interactions were cut off at $r_c = 1.6$ nm to minimize the size of the tail correction to the surface tension due to cutting off the LJ interaction at r_c [30]. The long-range electrostatic interactions were calculated using the particle–particle particle-mesh (PPPM) algorithm [39]. The precision of the PPPM algorithm was set to one part in 10^4 and a switching radius of 1.6 nm between the real space and k-space contributions to the PPPM was used. Four distinct atom types were used to describe SFA within this framework, fluorinated carbon, protonated carbon, fluorine, and hydrogen. The OPLS-AA potential also includes harmonic bond and angle terms and torsional interactions. The full set of OPLS-AA parameters needed to completely describe the SFA as given by Pierce et al. [31].

Simulated systems contained 400 linear semifluorinated chains in each sample, that consist of alternating blocks of protonated and fluorinated linear carbon chains: $H(C_nH_{2n}C_nF_{2n})_mF$, with $2n \times m = 96$ and $n = 3, 6, 12, 24$, and 48. The fully protonated system $C_{96}H_{194}$ and the pure perfluorinated system $C_{96}F_{194}$ were also studied and they serve as a reference. We will define here L = n for all multi-block copolymers; $L = 0$ for fully protonated chains, and $L = 96$ for fully fluorinated samples. The number of molecules and box dimensions were chosen to allow a reasonably thick film in which the mass density profiles approached a plateau at the center of the film, thus approximating the bulk. The construction of systems began with the building of a bulk melt sample for the various systems and equilibrating the systems at constant pressure with periodic boundaries in all three dimensions at 600 K. Following equilibration, the bulk samples were transformed into one with two liquid/vapor interfaces by a two-step process. In the first step, all molecules that crossed the z-periodic boundary were reconnected by shifting the location of the atoms crossing this boundary L_z. In the second step, L_z was extended to 60 nm, six times the length in the other two periodic directions $L_{\|} = L_x = L_y = 10$ nm. As the thickness of the films is ~20 nm, this choice of L_z effectively eliminated any interaction between the upper and lower interface of the sample across the z-periodic boundary. Examples of films for $L = 6$, 12, and 48 at $T = 600$ K are given in Figure 3.1. All the interface simulations were carried out at constant volume.

All simulations were performed using the LAMMPS classical MD package [40]. The equations of motion were integrated using a velocity Verlet algorithm with a 1 fs time step. A Langevin thermostat with a 1 ps damping time was used to regulate the system temperature. The simulations were run for a minimum of 20 ns with most run for 30 ns. Results for density and surface tension were averaged over the last part of the run after discarding the first 10 ns.

The liquid–vapor surface tension γ_p was calculated using the Kirkwood–Buff formula which in the geometry used in our system is given by [41]

$$\gamma_p = \frac{L_Z}{2}\left[< p_z > - \frac{< p_x > + < p_y >}{2} \right] \tag{3.1}$$

where $<p_z>$ and ($<p_x> + <p_y>/2$) are the ensemble averaged normal and tangential pressure components, respectively. The factor of 1/2 outside the bracket takes into account the presence of two interfaces in the system. The surface tension calculated using this formula does not take into account the full nonbonded potential because interactions within only a cutoff sphere are considered. To estimate the total surface tension γ we include a tail correction term γ_{tail} that represents contribution of the potential to the surface tension for distances longer than the cutoff r_c. The tail correction is required only for the Lennard–Jones interaction because the long-range electrostatic interactions were handled by the PPPM method [42]. The tail correction was determined from [42–44]

$$\gamma_{\text{tail}} = \frac{\pi}{2}\int_{-\infty}^{\infty}\int_{-1}^{1}\int_{r_C}^{\infty} r^3 \frac{dU(r)}{dr} g(r)(1-3s^2)(\rho(z)\rho(z-sr)-(\rho_G(z))^2)dr\,ds\,dz \tag{3.2}$$

where $U(r)$ is the dispersion (r^{-6}) portion of the LJ pair-wise potential and $g(r)$ is the radial distribution function. For the large cutoff r_c =1.6 nm that was used here, $g(r) = 1$ for $r > r_c$. The density $\rho(z)$ is the measured density profile and $\rho_G(z)$ is the Gibbs dividing surface given by

$$\rho_G(z) = \rho_c + \frac{\Delta\rho}{2} s\,g\,n(z) \tag{3.3}$$

where $\rho_c = (\rho_L + \rho_V)/2$ is the average density of the liquid and vapor phases and $\Delta\rho = \rho_L - \rho_V$ is the difference between the average densities of the two phases. The density profiles were fit to an error function $\text{erf}(z/\sqrt{2}\Delta)$ for fitting the tail correction to the surface tension [42, 45, 46]. It is important to note that the measured interfacial widths depend on inherent factors that control the interface, together with capillary waves that depend logarithmically on the interfacial area. Here we present results for the interfacial width Δ for interfacial area $A = 100$ nm^2.

3.3 RESULTS AND DISCUSSION

As interfacial interactions often affect the properties of the polymer, we have probed the entire sample including the center of the film, which we refer to here as bulk, and the interfacial regions. The overall density of the copolymers was extracted as a function of temperature and compared with those of protonated and fluorinated polymers of the same length. The density decreases with decreasing temperature for all block lengths and is only slightly dependent on block length as shown in Figure 3.2. The measured density appears to be close to the average of values for the pure protonated and fluorinated polymers. The limited dependence of the density on the topology suggests that while the size of the fluorinated and protonated domains does change with block size, the overall configuration of the molecules and the degree of stretching are only slightly affected. With increasing temperature however, the density decreases linearly for all block lengths as well as for the neat polymers. The values of the densities are within the range of those found for semifluorinated diblock copolymers.

In contrast to the liquid densities that exhibit hardly any dependence on block length, the root mean square of the end-to-end distance $\langle R^2 \rangle^{1/2}$ depends on L as shown in Figure 3.3. The measurements separated the bulk and interfacial regions. The end-to-end distance of the bulk protonated chain is ~44 Å and the interfacial values are

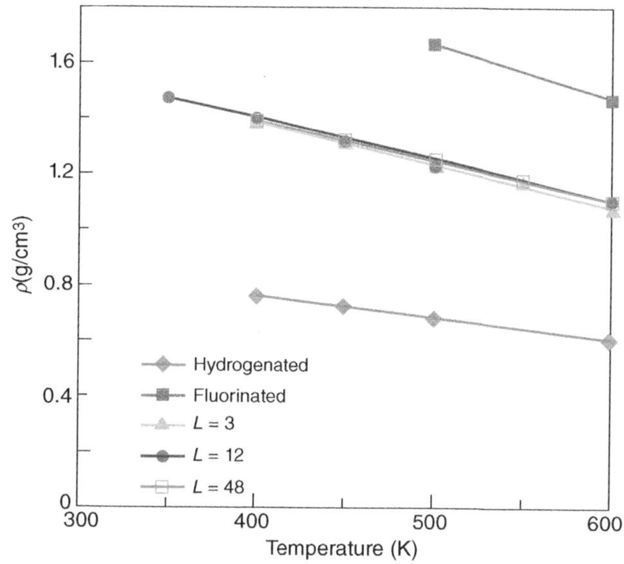

FIGURE 3.2 Bulk density of the multi-block copolymer as a function of temperature at the indicated block length.

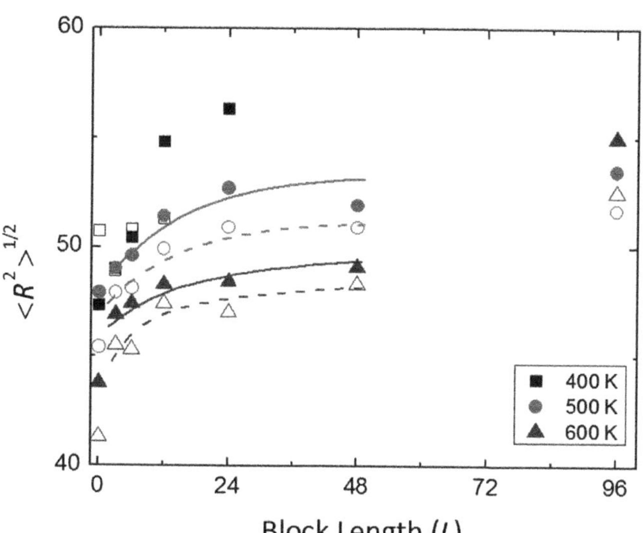

FIGURE 3.3 The root mean square end-to-end distance $<R^2>^{1/2}$ of the polymer chains as a function of the block length at the indicated temperatures. The bold symbols correspond to the bulk region and open symbols to interfacial region. $L = 0$ corresponds to the fully protonated system and $L = 96$ corresponds to fully fluorinated one.

a bit lower, while for the perfluorinated polymer it is ~55 Å with similar interfacial trends. The size of all polymers becomes smaller with increasing temperature as they become more flexible. The temperature dependence is much more pronounced for the fully protonated chains. Surprisingly, with increasing L the end-to-end distance increases and levels of at the dimensions of the fully fluorinated segment at $L = 24$. We extract separately the average dimensions for the fluorinated and protonated segments for $L = 6$, 24, and 48. At 600 K for $L = 6$ the dimensions of the individual fluorinated and protonated segments are determined by the inherent chemical structure of fully stretched carbon chains (~6 Å). For block length $L = 24$, the end-to-end distance of the protonated block is 19 Å, and the fluorinated block is 22 Å. Further increasing the block length to $L = 48$, some flexibility is observed and the over dimensions are dominated by the fluorinated segment whose $\langle R^2 \rangle^{1/2} = 36$ Å.

With increasing temperature, the chains become more flexible and the dimensions of the chains decrease. In comparison with the overall dimensions of the perfluorinated polymer, the dimensions of the semi fluorinated polymers are dominated by the overall content of the fluorinated segment mediated by a small stretching at low L. As the dimensions of the protonated segments are increased, the polymer becomes more flexible and can fold. Overall the distribution of the blocks along the polymer backbone in the bulk does not impact the overall densities. For chains in the interfacial region, the end-to-end distance $\langle R^2 \rangle^{1/2}$ is reduced by several Å on average from its value in the bulk. This is consistent with an interfacial conformation in which the chains are either in a mushroom configuration or lie along the interface.

The interfacial region is visually depicted in Figure 3.4, where a top view of the samples is shown for $L = 3$, 12, and 48 at 400 and 600 K. With decreasing temperature the chains become more extended enhancing the packing of the molecules. These top projections also reveal that for small L the interface is rather homogenous with approximately the same fraction of fluorinated and protonated segments. With increasing L surface segregation is increased and fluorinated and protonated domains are clearly observed. With further increase in L the interface becomes dominated by fluorine. This point will be further demonstrated through close analysis of the interfacial topology.

To correlate the polymer topology with interfacial morphology and the properties of the interfaces formed by semifluorinated polymers we measured the liquid–vapor surface tension γ. The results are described in Figure 3.5. The surface tension values for the multi-blocks assumes values between those of the fully protonated and the perfluorinated polymers. The surface tension decreases with temperature for all systems as shown in Figure 3.5a. In contrast to the behavior of bulk densities, the surface tension strongly depends on the block dimension as shown in Figure 3.5b. With increasing length the surface tension increases and then decreases. Similar to the overall dimension of the polymers, at low block length, the rigidity of the segments dominates the interface and as the block length is increased, the flexibility of the protonated segments manifests itself in determining both the dimensions and the interfacial energies. These results clearly show that while the overall ratios of H to F in the polymers remain constant, the interfacial characteristics depend on the morphology: that is, it depends on the distribution of the segments within the chain.

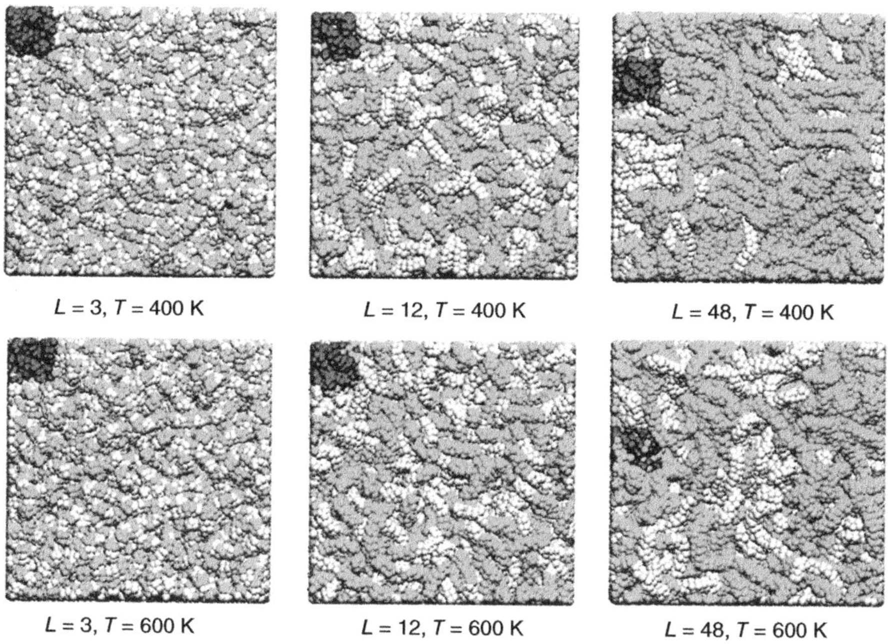

FIGURE 3.4 Top view of three polymer interface for block lengths $L = 3$, 12, and 48 for $T = 400$ and 600 K. Fluorine is marked in dark and hydrogen in light colors. One molecule at the interface is marked by black for resolving individualize molecule.

The density profiles across the films for $L = 3$, 12, and 48 for both the protonated blocks and fluorinated ones are presented in Figure 3.6. At low L values, the density across the film is homogenous with some roughness at the interface. With increasing L segregation takes place as is denoted by the oscillations across the film. At larger segments a clear excess of fluorine is observed at the vapor interface. Both the degree

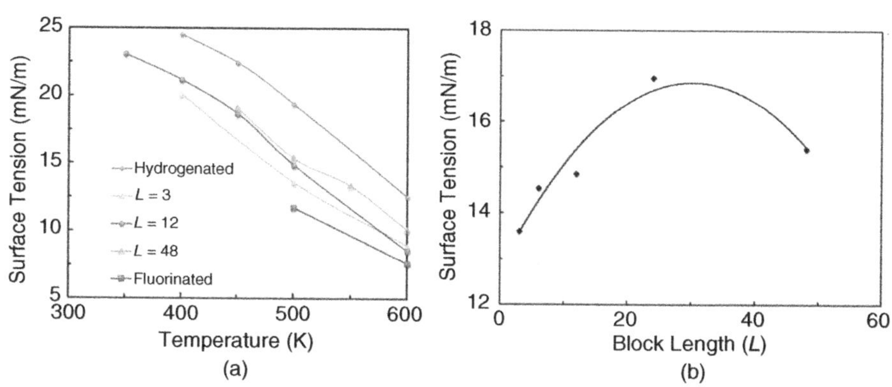

FIGURE 3.5 (a) Surface tension as a function of temperatures for the indicated block lengths. (b) Surface tension as a function of block length at 500°C.

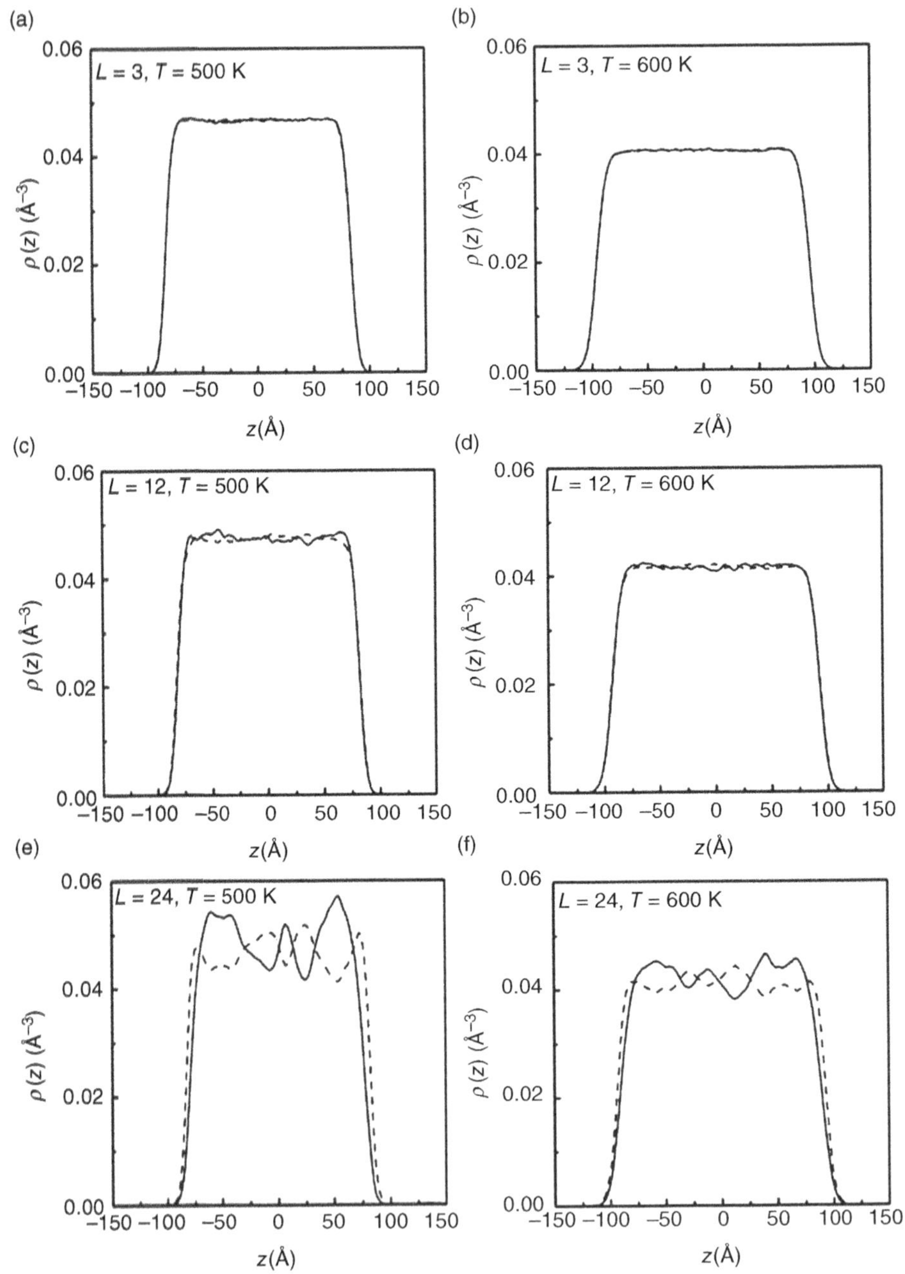

FIGURE 3.6 Polymer profiles at the indicated temperatures and L values. The solid lines correspond to the fluorinated domain's density across the film and the dashed line to the protonated ones. The center of the film is at $z = 0$.

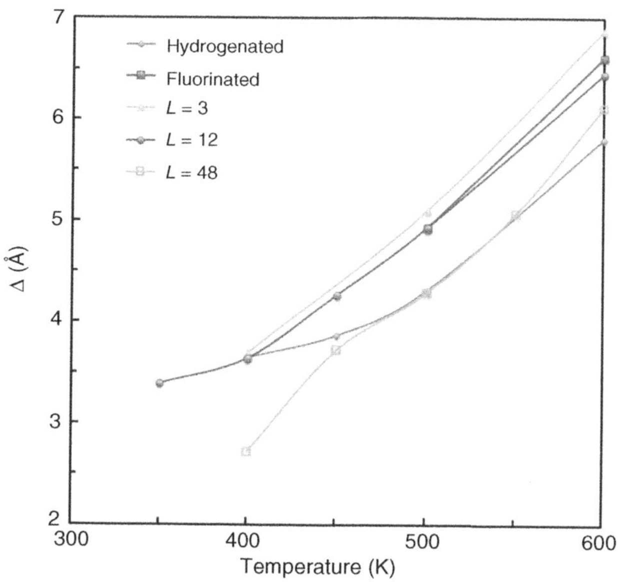

FIGURE 3.7 Interfacial width Δ as a function of temperatures for the indicated block lengths for interfacial area $A = 100$ nm^2.

of segregation and the excess of fluorine at the interface increase with increasing temperature.

The interfacial width Δ is shown in Figure 3.7. The interfacial width increases with increasing temperature. While the inherent roughness decreases with temperature, thermal fluctuations are enhanced. It is interesting to note that the actual values for $L = 3$, 12, and the fully fluorinated chains are similar whereas the values for $L = 48$ and the protonated ones are similar. This is again a demonstration where the topology of the polymer impacts the behavior.

3.4 CONCLUSIONS

Our studies have shown that the topology of semifluorinated copolymers impacts their bulk dimensions and interfacial characteristics. While keeping the overall number of protonated and fluorinated segments in a copolymer constant, the study followed the bulk and surface behavior of a polymer film. We found that while the bulk density is hardly affected by the block dimension, the end-to-end distance of the entire polymer changes with L. Polymer chains that reside at the interface are marginally smaller than those in the bulk. The size of the polymer chains first increases with increasing block length and then levels off. At this point the dimensions are essentially determined by the dimensions of the fluorinated blocks. The degree of segregation of fluorinated and protonated domains increases with increasing the size of the blocks, both in the bulk and at the interfaces. The degree of fluorination at the interface is enhanced with increasing the size of the block and the interfacial width decreases. The surface tension however peaks at the dimensions were the end-to-end distance levels off. The interfacial characteristics of semifluorinated polymers are impacted not only by the degree of fluorination but also by the distribution of the fluorine along the polymer backbone.

ACKNOWLEDGMENTS

We thank NSF 206-2007419 and DOE DE-FG02-12ER46843 for partial support of this work. This work was performed, in part, at the Center for Integrated Nanotechnologies, a US Department of Energy, and Office of Basic Energy Sciences user facility. Sandia National Laboratories is a multi-program laboratory managed and operated by Sandia Corporation, a wholly owned subsidiary of Lockheed Martin Corporation, for the US Department of Energy's National Nuclear Security Administration under Contract No. DE-AC04-94AL85000. We acknowledge Flint Pierce for constructing some copolymers at the initial stage of the study. We thank Mesfin Tsige for his code for analyzing the tail corrections to the surface tension.

REFERENCES

1. Stuart, M. A. C.; Huck, W. T. S.; Genzer, J.; Muller, M.; Ober, C.; Stamm, M.; Sukhorukov, G. B.; Szleifer, I.; Tsukruk, V. V.; Urban, M.; Winnik, F.; Zauscher, S.; Luzinov, I.; Minko, S. *Nature Mater.* **2010**, *9*, 101–113.
2. Urban, M. W. *Nature Chem.* **2012**, *4*, 80–82.
3. Aumsuwan, N.; Ye, S.-H.; Wagner, W. R.; Urban, M. W. *Langmuir* **2011**, *27*, 11106–11110.
4. Gopishetty, V.; Tokarev, I.; Minko, S., *J. Mater. Chem.* **2012**, *22*, 19482–19487.
5. Attivi, D.; Wehrle, P.; Ubrich, N.; Damge, C.; Hoffman, M.; Maincent, P. *Drug Dev. Ind. Pharm.* **2005**, *31*, 179–189.
6. White, S. R.; Caruso, M. M.; Moore, J. S. *MRS Bull.* **2008**, *33*, 766–769.
7. Biesalski, M. A.; Knaebel, A.; Tu, R.; Tirrell, M. *Biomaterials* **2006**, *27*, 1259–1269.
8. Nyilas, E.; Morton, W. A.; Cumming, R. D.; Lederman, D. M.; Chiu, T. H.; Baier, R. E. *J. Biomed. Mater. Res.* **1977**, *11*, 51–68.
9. Zhang, X. B.; Pint, C. L.; Lee, M. H.; Schubert, B. E.; Jamshidi, A.; Takei, K.; Ko, H.; Gillies, A.; Bardhan, R.; Urban, J. J.; Wu, M.; Fearing, R.; Javey, A. *Nano Lett.* **2011**, *11*, 3239–3244.
10. Estillore, N. C.; Advincula, R. C. *Macromol. Chem. Phys.* **2011**, *212*, 1552–1566.
11. Brittain, W. J.; Boyes, S. G.; Granville, A. M.; Baum, M.; Mirous, B. K.; Akgun, B.; Zhao, B.; Blickle, C.; Foster, M. D. *Adv. Polym. Sci.* **2006**, *198*, 125–147.
12. Russell, T. P. *Science* **2002**, *297*, 964–967.

13. Kulkarni, M. M.; Yager, K. G.; Sharma, A.; Karim, A. *Macromolecules* **2012**, *45*, 4303–4314.

14. Uhlmann, P.; Ionov, L.; Houbenov, N.; Nitschke, M.; Grundke, K.; Motornov, M.; Minko, S.; Stamm, M. *Prog. Org. Coat.* **2006**, *55*, 168–174.

15. Granville, A. M.; Boyes, S. G.; Akgun, B.; Foster, M. D.; Brittain, W. J. *Macromolecules* **2005**, *38*, 3263–3270.

16. Xu, H.; Norsten, T. B.; Uzun, O.; Jeoung, E.; Rotello, V. M. *Chem. Commun.* **2005**, 41, 5157–5159.

17. Arnold, M. E.; Nagai, K.; Spontak, R. J.; Freeman, B. D.; Leroux, D.; Betts, D. E.; DeSimone, J. M.; DiGiano, F. A.; Stebbins, C. K.; Linton, R. W. *Macromolecules* **2002**, *35*, 3697–3707.

18. You, F. X.; Paik, M. Y.; Hackel, M.; Kador, L.; Kropp, D.; Schmidt, H. W.; Ober, C. K. *Adv. Funct. Mater.* **2006**, *16*, 1577–1581.

19. Makal, U.; Wynne, K. J. *Langmuir* **2005**, *21*, 3742–3745.

20. Makal, U.; Uslu, N.; Wynne, K. J. *Langmuir* **2007**, *23*, 209–216.

21. Fujiwara, M.; Satoh, K.; Kondo, S.; Ujiie, S. *Macromolecules* **2006**, *39*, 5836–5842.

22. Russell, T. P.; Rabolt, J. F.; Twieg, R. J.; Siemens, R. L.; Farmer, B. L. *Macromolecules* **1986**, *19*, 1135–1143.

23. Li, X. F.; Andruzzi, L.; Chiellini, E.; Galli, G.; Ober, C. K.; Hexemer, A.; Kramer, E. J.; Fischer, D. A. *Macromolecules* **2002**, *35*, 8078–8087.

24. Sivaniah, E.; Genzer, J.; Fredrickson, G. H.; Kramer, E. J.; Xiang, M.; Li, X.; Ober, C.; Magonov, S. *Langmuir* **2001**, *17*, 4342–4346.

25. El Abed, A. I.; Ionov, R.; Daoud, M.; Abillon, O. *Phys. Rev. E* **2004**, *70*, 1607.

26. Traiphol, R.; Shah, H.; Smith, D. W.; Perahia, D. *Macromolecules* **2001**, *34*, 3954–3961.

27. Traiphol, R.; Smith, D. W.; Perahia, D. *J. Polym. Sci. Part B—Polym. Phys.* **2002**, *40*, 2817–2824.

28. Jin, J. Y.; Smith, D. W.; Glasser, S.; Perahia, D.; Foulger, S. H.; Ballato, J.; Kang, S. W.; Kumar, S. *Macromolecules* **2006**, *39*, 4646–4649.

29. Perahia, D.; Ratnaweera, D. R.; Shrestha, U. M.; Iacono, S. T.; Smith, D. W.; Mabry, J.; Majewski, J. *ACS Symp. Ser.* **2012**, *1106*, 81–93.

30. Tsige, M.; Grest, G. S. *J. Phys. Chem. C* **2008**, *112*, 5029–5035.

31. Pierce, F.; Tsige, M.; Borodin, O.; Perahia, D.; Grest, G. S. *J. Chem. Phys.* **2008**, *128*, 214903-1–214903-14.

32. Pierce, F.; Tsige, M.; Perahia, D.; Grest, G. S. *J. Phys. Chem. B* **2008**, *112*, 16012–16020.

33. Jorgensen, W. L.; Madura, J. D.; Swenson, C. J. *J. Am. Chem. Soc.* **1984**, *106*, 6638–6646.

34. Jorgensen, W. L.; Maxwell, D. S.; Tirado-Rives, J. *J. Am. Chem. Soc.* **1996**, *118*, 11225–11236.

35. Watkins, E. K.; Jorgensen, W. L. *J. Phys. Chem. A* **2001**, *105*, 4118–4125.

36. Padua, A. A. H. *J. Phys. Chem. A* **2002**, *106*, 10116–10123.

37. Song, W.; Rossky, P. J.; Maroncelli, M. *J. Chem. Phys.* **2003**, *119*, 9145–9163.

38. Escobedo, F. A.; Chen, Z. *J. Chem. Phys.* **2004**, *121*, 11463–11473.

39. Hockney, R. W.; Eastwood, J. W. *Computer Simulation Using Particles*; Adam Hilger-IOP: Bristol, UK, 1988.
40. Plimpton, S. *J. Comp. Phys.* **1995**, *117*, 1–19.
41. Kirkwood, J. G.; Buff, F. P. *J. Chem. Phys.* **1949**, *17*, 338–343.
42. Ismail, A. E.; Grest, G. S.; Stevens, M. J. *J. Chem. Phys.* **2006**, *125*, 014702.
43. Chapela, G. A.; Saville, G.; Thompson, S. M.; Rowlinson, J. S. *J. Chem. Soc., Faraday Trans. 2.* **1977**, *73*, 1133–1144.
44. Blokhuis, E. M.; Bedeaux, D.; Holcomb, C. D.; Zollweg, J. A. *Mol. Phys.* **1995**, *85*, 665–669.
45. Huang, J. S.; Webb, W. W. *J. Chem. Phys.* **1969**, *50*, 3677–3693.
46. Beysens, D.; Robert, M. *J. Chem. Phys.* **1987**, *87*, 3056–3061.

4

FLUOROPOLYMER NANOCOMPOSITES

HIDEO SAWADA
Hirosaki University

4.1 INTRODUCTION AND OVERVIEW

A considerable interest has been devoted to block copolymers containing longer perfluoroalkyl groups which exhibit low surface energy and self-assembled polymeric micellar aggregates in aqueous and organic media: these adsorption and aggregation properties are not displayed by corresponding random perfluoroalkylated polymers [1–23]. In these fluorinated block copolymers, partially fluoroalkylated polymers, especially fluoroalkyl end-capped oligomers [R_F—$(M)_n$—R_F; R_F = fluoroalkyl groups; M = radical polymerizable monomers] are attractive functional materials because they exhibit various unique properties such as high solubility, surface active properties, biological activities, and nanometer size-controlled molecular aggregates, which cannot be achieved by the corresponding nonfluorinated, randomly or block-type fluoroalkylated polymers, and low molecular weight fluorinated surfactants [24–30]. Thus, from the developmental viewpoint of new fluorinated polymeric materials, it is in particular interest to develop fluoroalkyl end-capped oligomeric nanocomposites-encapsulated a variety of inorganic particles possessing numerous unique characteristics imparted by not only fluorine but also inorganic particles.

Fluoroalkyl end-capped oligomers can form nanometer size-controlled self-assembled oligomeric aggregates through the aggregations of end-capped fluoroalkyl groups (see Figure 4.1) [31–33]. Fluoroalkyl end-capped oligomeric aggregates can

Handbook of Fluoropolymer Science and Technology, First Edition. Edited by Dennis W. Smith Jr., Scott T. Iacono, and Suresh S. Iyer.

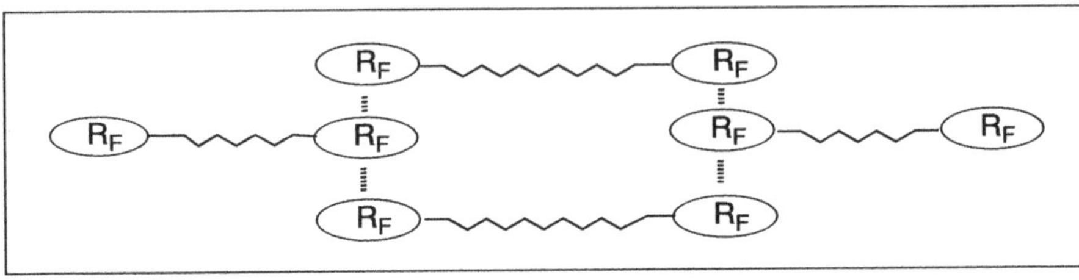

FIGURE 4.1 Schematic model for the formation of fluoroalkyl end-capped oligomeric aggregates with the aggregation of end-capped fluoroalkyl groups.

also interact with guest molecules to afford fluorinated aggregates/guest molecules nanocomposites; although the corresponding nonfluorinated oligomers cannot form such molecular aggregates to interact with guest molecules [34]. Especially, it is suggested that silica nanoparticles should act as a guest molecules in fluorinated oligomeric aggregate cores to give new fluorinated oligomer-coated silica nanoparticles (fluorinated oligomers/silica nanocomposites) under the *in situ* preparative conditions as shown in Figure 4.2 [35]. On the other hand, in general, organic polymer/inorganic composites such as polystyrene-grafted silica nanoparticles can be prepared by the copolymerization of silica nanoparticles containing methacryoyloxypropyl groups with styrene monomer catalyzed by potassium persulfate as a radical initiator (see Scheme 4.1) [36]. Therefore, fluoroalkyl end-capped oligomers have high potential for the preparation of novel fluorinated oligomers/guest molecules nanocomposites.

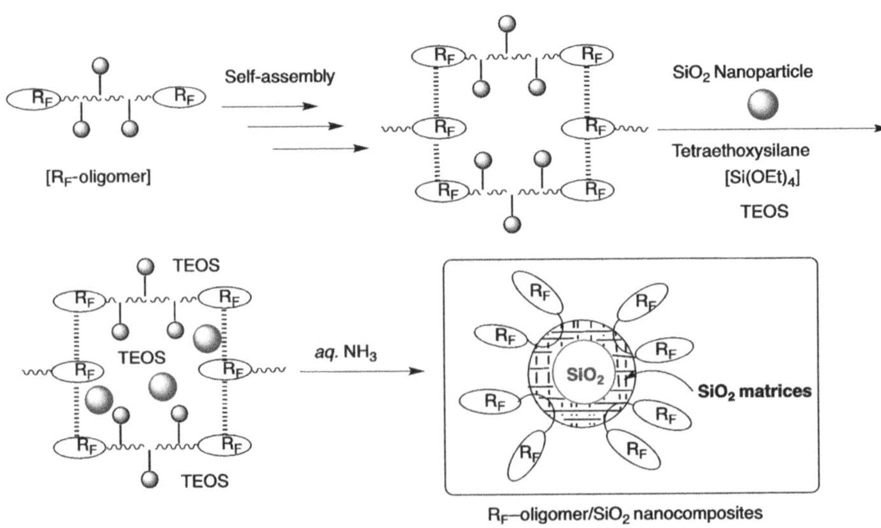

FIGURE 4.2 Preparation of fluoroalkyl end-capped oligomers/SiO_2 nanocomposites.

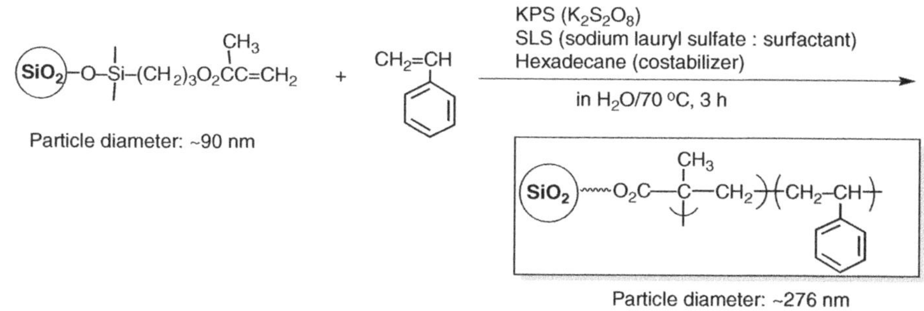

SCHEME 4.1 Preparation of polystyrene-grafted silica nanoparticles.

In this chapter, we would like to review the recent development of preparation and applications of fluoropolymer nanocomposites by the use of fluoroalkyl end-capped oligomers as key intermediates.

4.2 PREPARATION OF FLUOROALKYL END-CAPPED OLIGOMERS

Fluoroalkanoyl peroxides [R_F—C(=O)—O—O—(O=)C—R_F; R_F = fluoroalkyl groups] decompose homolytically through the three bond radical fissions to afford $R_F^{\bullet}$ radicals, although the corresponding alkanoyl peroxides [R—C(=O)—O—O—(O=)C—R; R = alkyl groups] decompose homolytically through the stepwise radical fissions to afford R—C(=O)—O$^{\bullet}$ radicals [24]. This suggests that fluoroalkanoyl peroxide is a convenient tool for the direct introduction of the corresponding fluoroalkyl groups into organic molecules via a radical process. In fact, as shown in Scheme 4.2, fluoroalkanoyl peroxides reacted with acrylic acid monomer to give two fluoroalkyl end-capped acrylic acid oligomers in good isolated yields [37]. Acrylic acid oligomers with two fluoroalkyl end-groups were obtained via primary radical termination or radical chain transfer to the peroxides [37]. In this oligomerization process, primary radical termination of electron poor $R_F^{\bullet}$ radicals toward poly(acrylic acid) propagation radical [~CH_2—CH(COOH)$^{\bullet}$] should be predominately occurred to produce R_F end-capped oligomers due to the effective interaction between HOMO of propagation radical and SOMO of $R_F^{\bullet}$ radical, differing from the production of usual macromolecules by the use of nonfluorinated R—(C=O)—O$^{\bullet}$ radical or R$^{\bullet}$ radical [38].

In contrast, it is well known that styrene is a useful radical polymerizable monomer, and diacyl peroxide such as benzoyl peroxide and alkanoyl peroxide are widely used as conventional radical initiators. Since the interaction between the LUMO of diacyl peroxide (e.g., butyryl peroxide) and the HOMO of styrene become extremely weak compared to that of perfluorobutyryl peroxide (see Figure 4.3), the interaction of the SOMO of acyloxy radicals, which produce in the radical decomposition of diacyl peroxides, should interact with the HOMO of styrene to afford the radical polymerization of styrene. In contrast, styrene reacted with fluoroalkanoyl peroxides to afford not the radical polymerizable products but 1:1 adducts [PhCH(OCOR$_F$)CH$_2$R$_F$] via

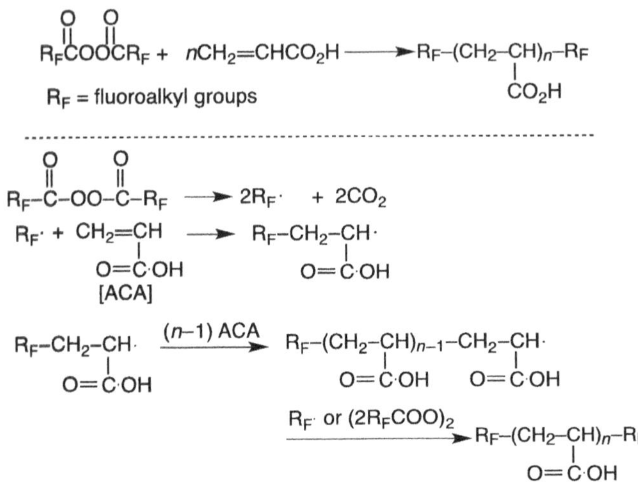

SCHEME 4.2 Synthesis of two fluoroalkyl end-capped acrylic acid oligomer by using fluoroalkanoyl peroxide.

a single electron transfer from the HOMO of styrene to the LUMO of peroxide (see Scheme 4.3) [39], due to the extremely lower LUMO energy level of peroxide compared to that of the corresponding nonfluorinated peroxide [40,41]. Thus, we are unable to synthesis fluoroalkyl end-capped styrene oligomers by using fluoroalkanoyl peroxides. However, very recently, we have found that head-to-head type styrene and substituted styrene dimers bearing two fluoroalkyl end-groups can been efficiently

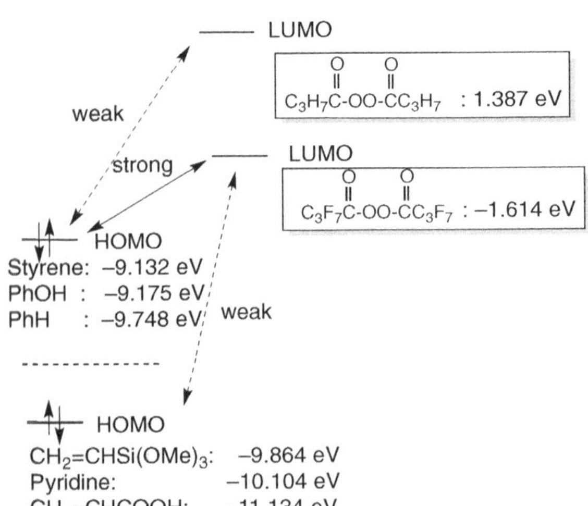

FIGURE 4.3 LUMO energy levels of perfluorobutyryl peroxide and butyryl peroxide, together with the HOMO–LUMO interactions.

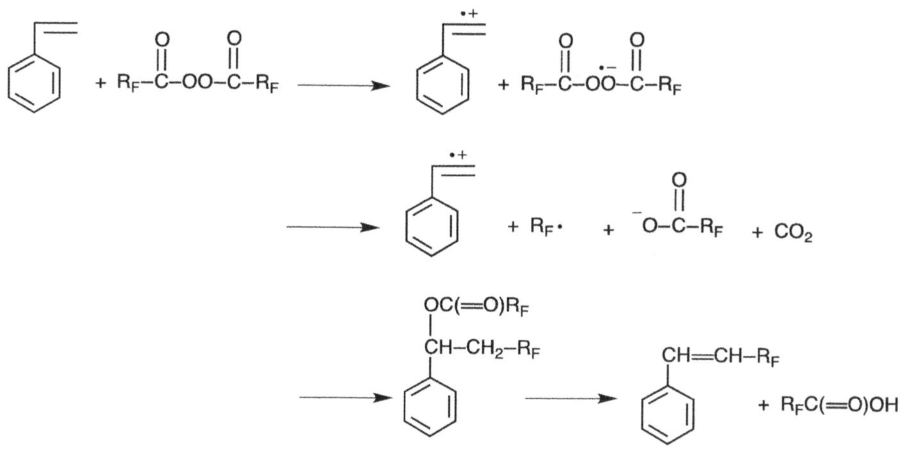

SCHEME 4.3 Reaction of perfluorobutyryl peroxide with styrene.

synthesized by a simple reaction of perfluoroalkyl iodide with styrene under radical conditions as a mixture of *meso* and *racemic* forms as shown in Scheme 4.4 [42]. The *meso* form obtained from the mixture by recrystallization gave a crystal suitable for X-ray diffraction study and the crystal structure was found to be based on π-stacking of benzene rings and aggregation of fluoroalkyl chains (see Figure 4.4) [42]. Especially, *meso*-styrene dimers bearing two fluoroalkyl end-groups can form the nanometer size-controlled self-assemblies through the intermolecular π-stacking of benzene rings and aggregation of end-capped fluoroalkyl groups in methanol [42].

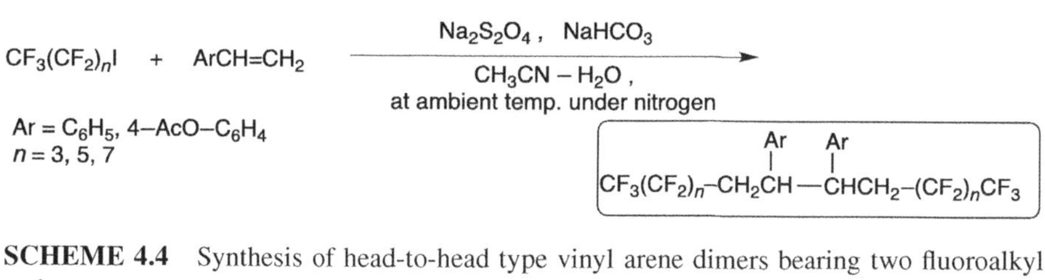

SCHEME 4.4 Synthesis of head-to-head type vinyl arene dimers bearing two fluoroalkyl end-groups.

4.3 PREPARATION AND THERMAL STABILITY OF FLUOROALKYL END-CAPPED OLIGOMERS/SILICA NANOCOMPOSITES

Fluoroalkyl end-capped *N*-(1,1-dimethyl-3-oxobutyl)acrylamide oligomer/silica nanocomposites [R_F–$(DOBAA)_n$–R_F/SiO_2] were prepared by the sol-gel reactions of the corresponding oligomer with tetraethoxysilane (TEOS) under alkaline conditions (see Scheme 4.5) [43–45].

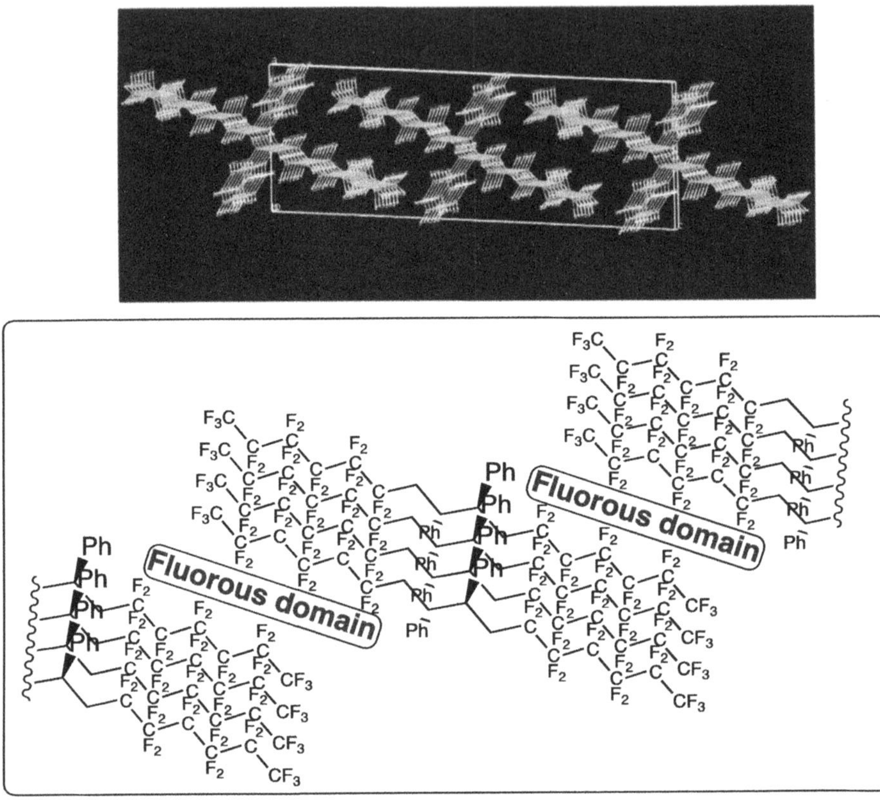

FIGURE 4.4 X-ray analysis of self-assembled *meso*-perfluorohexylated styrene dimer $[C_6F_{13}—CH_2CHPh—CHPh—CH_2—C_6F_{13}]$ based on π-stacking of benzene rings and aggregation of fluoroalkyl chains: fluorous domains are constructed by self-assembly of fluoroalkyl chains.

R_F—$(DOBAA)_n$—R_F/SiO_2 nanocomposite, which was prepared under acidic conditions, showed a clear weight loss (25%) at 800°C, whose value is quite similar to that (24%) of elementary analyses of fluorine in this composite [43–45]. However, R_F—$(DOBAA)_n$—R_F/SiO_2 nanocomposite, which was prepared under alkaline conditions, exhibited no weight loss characteristic even after calcination at 800°C, although the content of R_F—$(DOBAA)_n$—R_F oligomer in this nanocomposite was estimated to be 23% by elementary analyses of fluorine (see Figure 4.5) [46].

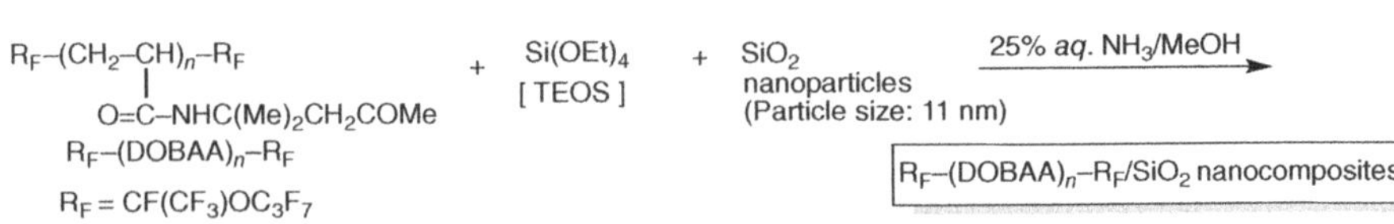

SCHEME 4.5 Preparation of fluoroalkyl end-capped *N*-(1,1-dimethyl-3-oxobutylacrylamide) oligomers $[R_F$—$(DOBAA)_n$—$R_F]$/silica nanocomposites.

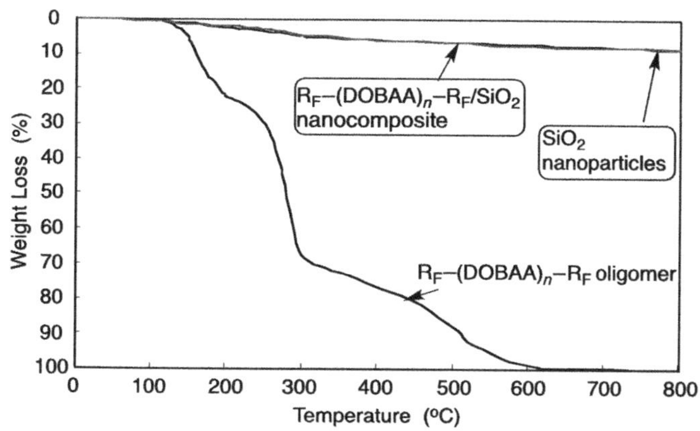

FIGURE 4.5 Thermogravimetric analyses of R_F—(DOBAA)$_n$—R_F/SiO_2 nanocomposite [R_F=$CF(CF_3)OC_3F_7$].

These findings suggest that the nanocomposite reaction of R_F—(DOBAA)$_n$—R_F oligomer with silica nanoparticles should proceed smoothly under alkaline conditions to afford not only the expected R_F—(DOBAA)$_n$—R_F/silica nanocomposite but also ammonium hexafluorosilicate as shown in the plausible reaction mechanism (see Scheme 4.6) [46].

The formation of ammonium hexafluorosilicate during the composite reactions can afford a nonflammable characteristic toward R_F—(DOBAA)$_n$—R_F oligomer [46,47]. That is, R_F—(DOBAA)$_n$—R_F oligomer in nanocomposite should be encapsulated

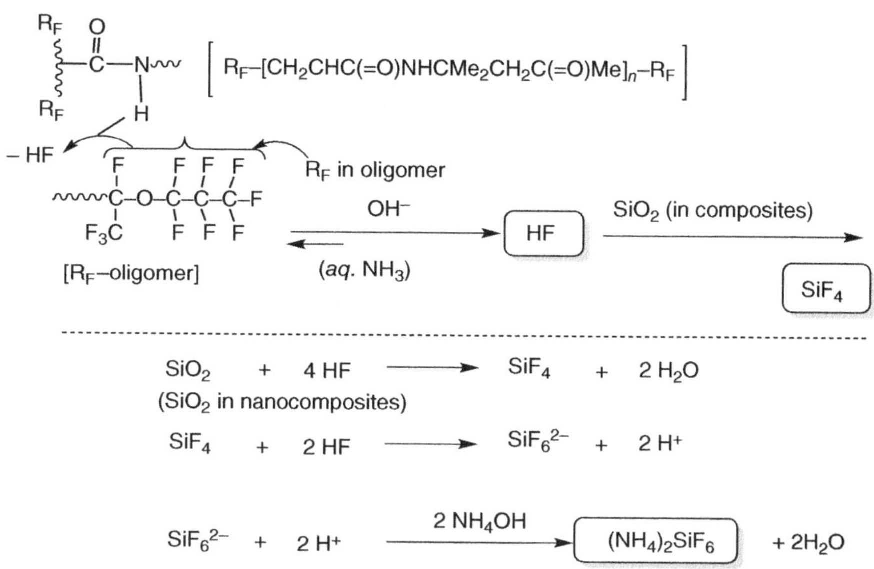

SCHEME 4.6 Plausible reaction mechanism for the formation of ammonium hexafluorosilicate.

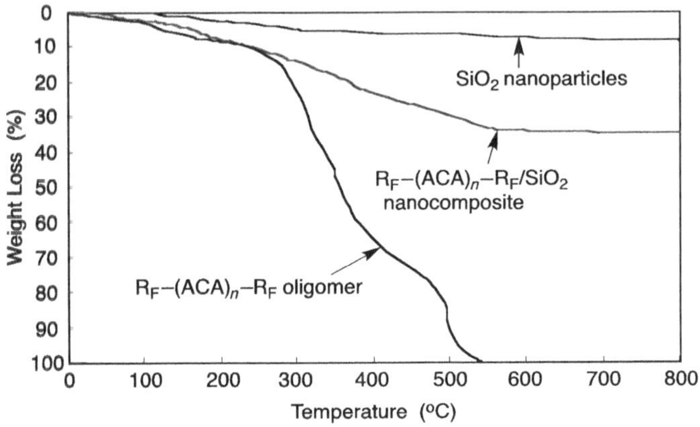

FIGURE 4.6 Thermogravimetric analyses of R_F—$(ACA)_n$—R_F/SiO_2 nanocomposite [R_F=$CF(CF_3)OC_3F_7$].

quite effectively into the nanometer size-controlled silica gel matrices (see Figure 4.2) through the molecular-level synergistical combination, which is due to not only the strong interaction between fluorine in oligomer and silicon in silica gel nanocomposite but also the effective interaction between ammonium hexafluorosilicate and R_F—$(DOBBA)_n$—R_F oligomer in silica gel matrices, to afford a nonflammable characteristic for oligomer [48]. No dehydrofluorination in the nanocomposite reaction under acidic conditions affords a usual flammable behavior for R_F—$(DOBBA)_n$—R_F oligomer [46, 47].

Fluoroalkyl end-capped acrylic acid oligomer [R_F—$(CH_2CHCOOH)_n$—R_F/SiO_2] [R_F—$(ACA)_n$—R_F]/silica nanocomposites decompose completely at 800°C to afford the weight loss in proportion to the content of R_F—$(ACA)_n$—R_F oligomer in composites due to the nonformation of ammonium hexafluorosilicate during the calcination process (see Figure 4.6) [43].

Fluoroalkyl end-capped acrylic acid oligomer [R_F—$(ACA)_n$—R_F] can also react with (TEOS) and silica nanoparticles in the presence of low molecular weight aromatic compounds [Ar—H] such as cetylpyridinium chloride (CPC) and bisphenol AF under alkaline conditions to afford R_F—$(ACA)_n$—R_F/SiO_2 nanocomposites-encapsulated Ar—H in excellent to moderate isolated yields (see Scheme 4.7) [49].

These fluorinated silica nanocomposites-encapsulated Ar—H can exhibit no weight loss behavior corresponding to the contents of Ar—H even after calcination at 800°C, although fluoroalkyl end-capped acrylic acid oligomer in the nanocomposites decomposed completely under similar conditions [49].

Traditional organic polymers such as poly(vinyl chloride), poly(ethylene), poly(propylene), and poly(styrene) are well-known to decompose completely around 400–500°C; however, perfluorinated polymers such as poly(tetrafluoroethylene) (PTFE) decompose at higher temperatures (around 600°C) due to the bond-strengthening effect of fluorine for C—C and C—F bonds in highly fluorinated compounds [1, 2, 4–6]. Hybridization of PTFE with silica gels is of particular interest

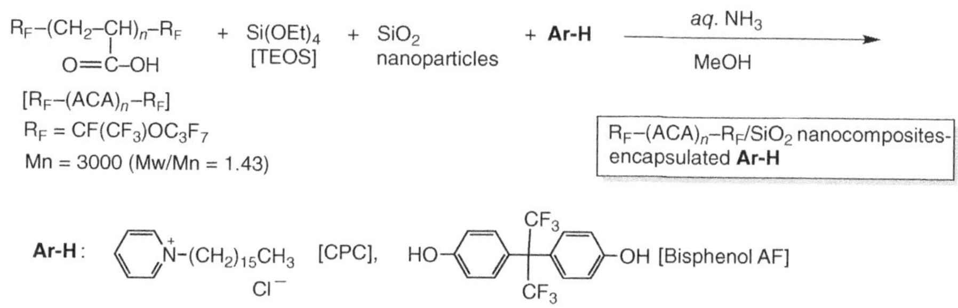

SCHEME 4.7 Preparation of R_F—(ACA)$_n$—R_F/SiO_2 nanocomposites-encapsulated Ar—H.

from the developmental viewpoint of novel fluorinated functional polymeric materials. Chen et al. have already prepared PTFE/SiO_2 hybrids; however, parent PTFE in the silica gel hybrids decomposes completely around 700°C [8]. On the other hand, the molecular weights of fluoroalkyl end-capped oligomers are in general within 10,000, and their thermal stability is extremely poor compared to that of usual perfluorinated polymers or perfluoroalkylated polymers [37, 38]. However, as mentioned above, unexpectedly, R_F—(DOBAA)$_n$—R_F oligomer can be used as novel fluorinated functional materials possessing a perfectly nonflammable characteristic at 800°C by the composite reactions of the corresponding oligomers with silica nanoparticles under alkaline conditions [45, 46]. Therefore, our fluoroalkyl end-capped oligomers are attractive functional fluorinated polymers from the development of new fluorinated heat-resistant materials.

Novel cross-linked fluoroalkyl end-capped trimethoxyvinylsilane oligomeric nanoparticles [R_F—(VM-SiO_2)$_n$—R_F]-encapsulated 1,1′-bi(2-naphthol) (BINOL) were prepared by the sol-gel reaction of fluoroalkyl end-capped trimethoxyvinylsilane oligomer in the presence of BINOL under alkaline conditions (see Scheme 4.8) [50].

R_F—(VM-SiO_2)$_n$—R_F/BINOL nanocomposites thus obtained were found to afford a clear weight loss at 800°C, which corresponds to the content of fluorinated oligomer in the composite [50]. However, unexpectedly, R_F—(VM—SiO_2)$_n$—R_F/BINOL nanocomposites were found to exhibit no weight loss corresponding to the presence of BINOL (the content of BINOL in the composites: ~8%) in the composite at 800°C under atmospheric conditions (see Figure 4.7) [50].

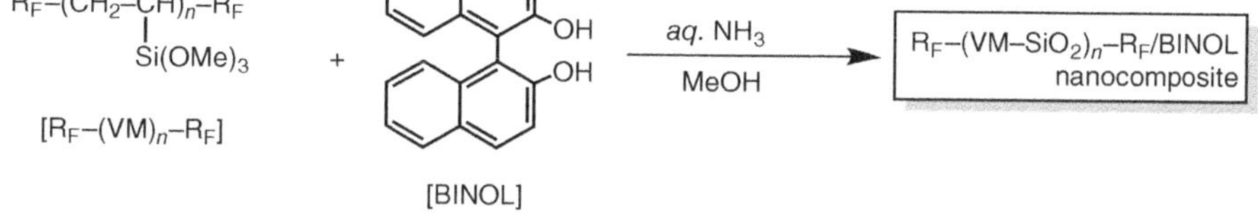

SCHEME 4.8 Preparation of fluoroalkyl end-capped trimethoxyvinylsilane oligomeric nanoparticles [R_F—(VM-SiO_2)$_n$—R_F]—encapsulated 1,1′-bi(2-naphthol) (BINOL).

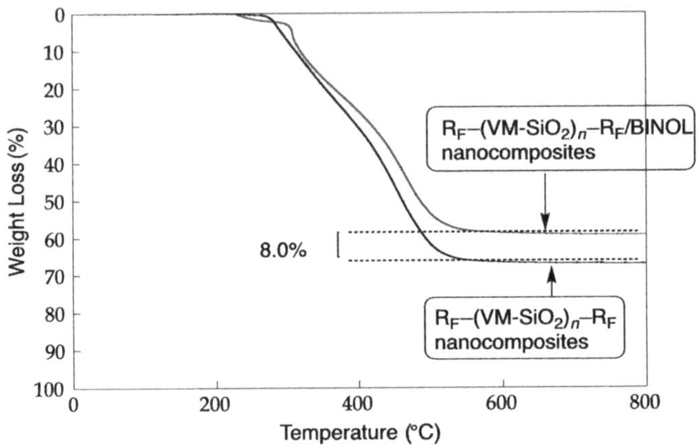

FIGURE 4.7 Thermogravimetric analyses of R_F—(VM—SiO_2)$_n$—R_F/BINOL nanocomposite, and R_F—(VM—SiO_2)$_n$—R_F nanocomposite.

Fluoroalkyl end-capped vinyltrimethoxysilane oligomeric nanocomposites having biphenylene units [R_F—(VM—SiO_2)$_n$—R_F/**Ar—SiO$_2$**] were prepared by the sol-gel reaction of the corresponding oligomer [R_F—(VM)$_n$—R_F] with 4,4′-bis(triethoxysilyl)-1,1′-biphenyl [**Ar—Si(OEt)$_3$**] under alkaline conditions (see Scheme 4.9) [51]. The shape of R_F—(VM—SiO_2)$_n$—R_F/**Ar—SiO$_2$** nanocomposite particles is morphologically controlled cubic particles, although the shape of fluoroalkyl end-capped oligomeric nanocomposites [R_F—(VM—SiO_2)$_n$—R_F] and biphenylene units-containing silica composites [**Ar—SiO$_2$**] is spherically fine particles, respectively (see Figure 4.8) [51].

R_F—(VM-SiO_2)$_n$—R_F/**Ar—SiO$_2$** nanocomposites were applied to the surface modification of poly(methyl methacrylate) (PMMA) to exhibit not only a good oleophobicity imparted by fluorine but also a fluorescent emission ability on the surface [51]. Methanol sol solutions of R_F—(VM—SiO_2)$_n$—R_F/**Ar—SiO$_2$** nanocomposites were effective for the surface modification of glass through the dipping technique to exhibit good oleophobicity with superhydrophobicity on the modified glass surface [51]. On the other hand, 1,2-dichloroethane sol solutions enabled R_F—(VM—SiO_2)$_n$—R_F/**Ar—SiO$_2$** nanocomposites to exhibit both superhydrophobic (water contact angle value: 180°) and superoleophobic (dodecane contact angle value: 98°) characteristics on the modified surface through dipping the glass in these sol solutions (see Figure 4.9) [51].

4.4 PREPARATION AND PROPERTIES OF FLUOROALKYL END-CAPPED OLIGOMERS/TITANIUM OXIDE NANOCOMPOSITES

Titanium oxide (TiO_2) is not only the most important white pigment in the plastic industry mainly because of the high refractive index of TiO_2 nanoparticles, but also

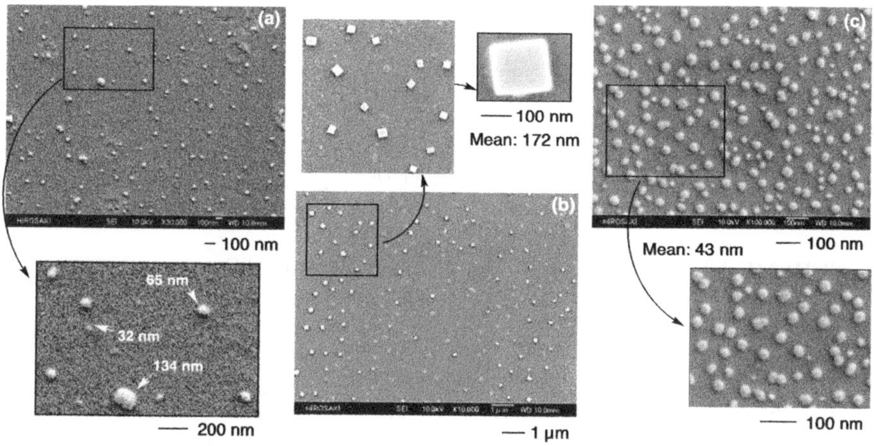

FIGURE 4.8 FE-SEM (field-emission scanning electron microscopy) images of $\mathbf{Ar{-}SiO_2}$ nanoparticles (a), $R_F{-}(VM{-}SiO_2)_n{-}R_F/\mathbf{Ar{-}SiO_2}$ nanocomposites (b), and $R_F{-}(VM{-}SiO_2)_n{-}R_F$ nanoparticles (c) in methanol solutions.

has extensive applications in a wide variety of fields such as photocatalyst, chemical sensor, and energy conversions, since titanium oxide (TiO_2) is highly photoactive, biologically and chemically inert, nontoxic, and relatively inexpensive [52–67]. Therefore, it is in particular interest to prepare organic polymers/TiO_2 nanocomposites based on blends of organic polymers and titanium oxide dispersed at a molecular level for the aim of the development of new functional materials possessing

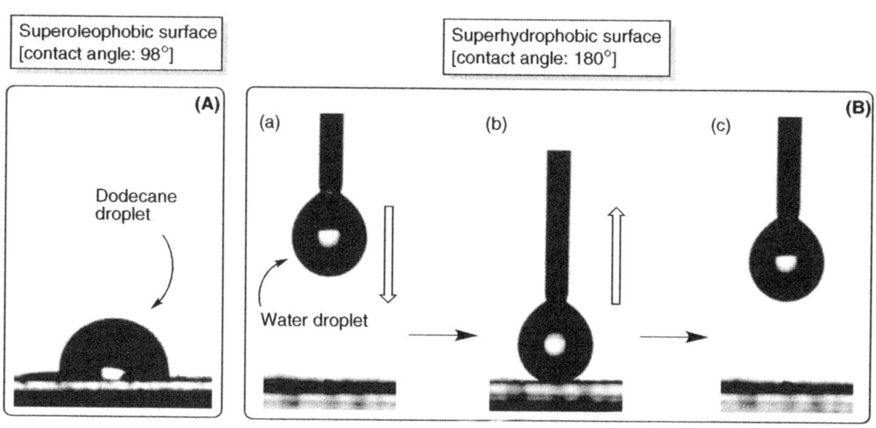

FIGURE 4.9 CCD (charge-coupled device) camera images of the dodecane droplet (A) and the water droplets (B) on the modified glass surface treated with $R_F{-}(VM{-}SiO_2)_n{-}R_F/\mathbf{Ar{-}SiO_2}$ nanocomposites: (a) water droplet which adhered to the needle tip (process before adhesion of the water droplet on the modified glass surface); (b) water droplet on the modified glass surface; (c) pull-up process of the needle from the modified glass surface.

SCHEME 4.9 Preparation of R_F—$(VM—SiO_2)_n$—R_F/**Ar—SiO$_2$** nanocomposites.

synergistic characteristics imparted by both organic polymers and titanium oxide. However, in general, the traditional organic polymer/TiO_2 composites have a poor dispersibility and stability in a variety of solvents due to the relatively lower surface activity of the parent organic polymers [64]. Thus, from the developmental viewpoint of organic polymer/titanium oxide nanocomposites possessing a good dispersibility and stability, it is of particular interest to use fluorinated polymers possessing a high surface active characteristic for the preparation of these titanium oxide nanocomposites [68]. In fact, we have found that fluoroalkyl end-capped vinyltrimethoxysilane oligomer [R_F—$(VM)_n$—R_F] suffered the sol-gel reaction under alkaline conditions in the presence of anatase titanium oxide nanoparticles (*an*-TiO_2) in tetrahydrofuran to afford the corresponding fluorinated oligomer/anatase titanium oxide nanocomposites [R_F—$(VM—SiO_2)_n$—R_F/*an*-TiO_2] as shown in Scheme 4.10 [69].

These nanocomposites were applied to the surface modification of glass to exhibit not only a completely superhydrophobic characteristic with a nonwetting property against water droplets but also a good oleophobicity on their surface [69]. Especially, the wettability for water can be switched between superhydrophobicity and superhydrophilicity by alternation of ultraviolet (UV) irradiation and dark storage with keeping a good oleophobicity on the modified surface treated with these anatase titanium oxide nanocomposite [69].

It is well-known that anatase transforms irreversibly to rutile at elevated temperatures [70]. Normally, the phase transformation temperature in air from anatase to rutile titanium oxide is between 600°C and 750°C [71]. From an applicable viewpoint of anatase titanium oxide into a variety of fields, it is deeply desirable to develop thermally stable anatase titanium oxide composites without the phase transformation to rutile under such higher temperature conditions. We have very recently found that fluoroalkyl end-capped vinyltrimethoxysilane oligomer/anatase titanium oxide nanocomposites [R_F—$(VM—SiO_2)_n$—R_F/*an*-TiO_2] can keep completely its

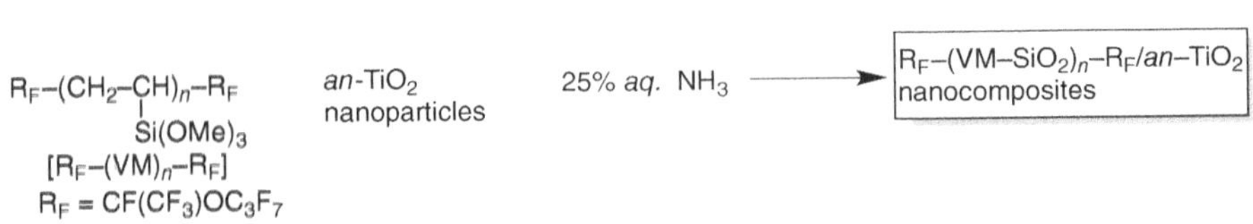

SCHEME 4.10 Preparation of R_F—$(VM—SiO_2)_n$—R_F/*an*-TiO_2 nanocomposites.

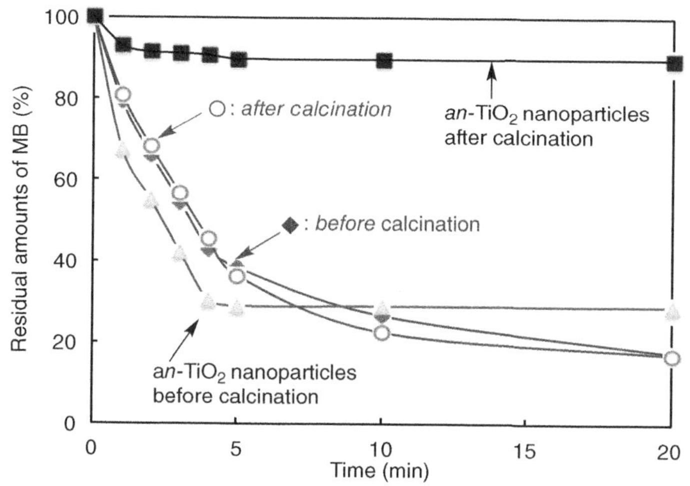

FIGURE 4.10 Relationship between the residual amounts of methylene blue (MB) and photoirradiation (λ_{max}: 365 nm) time in the presence of R_F—$(VM—SiO_2)_n$—R_F/*an*-TiO_2 nanocomposites (or original *an*-TiO_2 nanoparticles) before and after calcination at 1000°C: Concentration of MB: 2.5 mg/dm^3; concentration of nanocomposites (or original *an*-TiO_2 nanoparticles before and after calcination at 1000°C): 25 mg/dm^3.

structure without phase transformation to rutile even after calcination at 1000°C, and R_F—$(VM—SiO_2)_n$—R_F/*an*-TiO_2 nanocomposite before and after calcination at 1000°C exhibited the similar photocatalytic activity for the decolorization of methylene blue under UV light irradiation as shown in Figure 4.10 [72].

Polyaniline (PAn)/inorganic composites have been considered as new class of materials due to their improved properties compared with those of pure conducting polymers and inorganic materials [73–77]. For example, the combination of electrical conductivity of PAn and UV sensitivity of anatase TiO_2 are expected to find applications in electrochromic devices, nonlinear optical system, and photochemical devises [73]. From this point of view, it is in particular interest to prepare fluoroalkyl end-capped oligomers/PAn/titanium oxide nanocomposites. In fact, fluoroalkyl end-capped acrylic acid oligomer [R_F—$(ACA)_n$—R_F]/, 2-methacryloyloxyethane sulfonic acid oligomer [R_F—$(MES)_n$—R_F]/, 2-acrylamido-2-methylpropanesulfonic acid

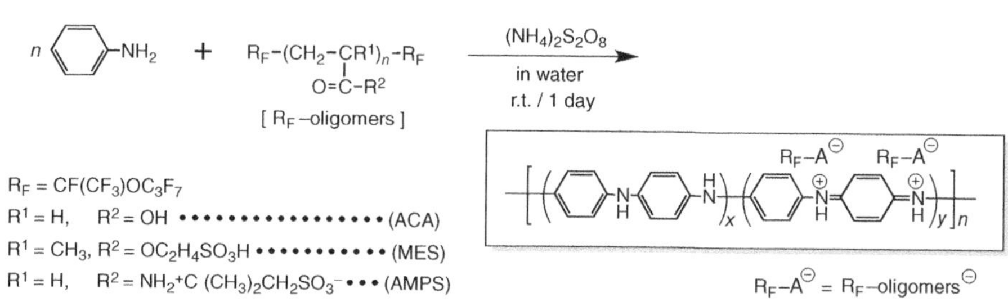

SCHEME 4.11 Preparation of fluoroalkyl end-capped oligomers/PAn nanocomposites.

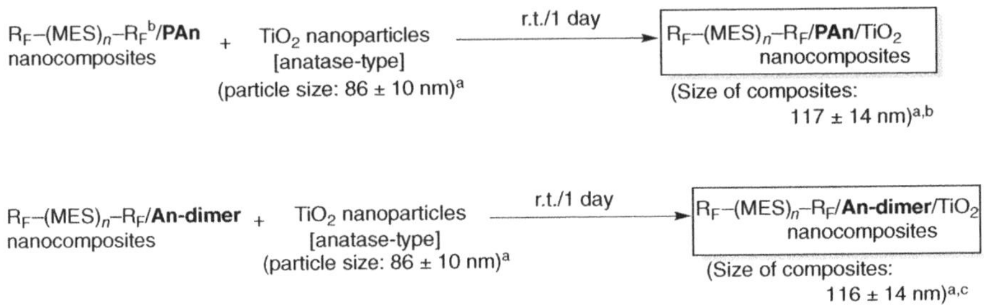

[a] Determined by dynamic light scattering(DLS) measurements
[b] Average particle size of parent of R_F–$(MES)_n$–R_F/**PAn** nanocomposites: 33 ± 5 nm
[c] Average particle size of parent of R_F–$(MES)_n$–R_F/**An-dimer** nanocomposites: 11 ± 2 nm

SCHEME 4.12 Preparation of R_F—$(MES)_n$—R_F/PAn/TiO_2 nanocomposites and R_F—$(MES)_n$—R_F/An-dimer/TiO_2 nanocomposites.

oligomer [R_F—$(AMPS)_n$—R_F]/PAn nanocomposites were prepared by the polymerization of aniline initiated by ammonium persulfate in the presence of the corresponding oligomers, respectively (see Scheme 4.11) [78]. These fluorinated oligomers were also applied to the preparation of the corresponding fluorinated oligomers/phenyl-capped aniline dimer (An-dimer: *N,N′*-diphenyl-1,4-diphenyldiamine) nanocomposites by the interaction of the fluorinated oligomers with An-dimer [78].

R_F—$(MES)_n$—R_F/polyaniline [PAn] nanocomposites and R_F—$(MES)_n$—R_F/An-dimer nanocomposites were effective for the preparation of colloidal stable fluorinated oligomer/PAn/TiO_2 and /An-dimer/TiO_2 nanocomposites (see Scheme 4.12) [79]. These two types of fluorinated TiO_2 nanocomposites can exhibit quite different photochromic behaviors: R_F—$(MES)_n$—R_F/PAn/TiO_2 nanocomposites can exhibit a reversible wavelength change for polaron absorptions around 760–820 nm by alternation of UV irradiation and storage in the dark; in contrast, R_F—$(MES)_n$—R_F/An-dimer/TiO_2 nanocomposites can exhibit a reversible color change from blue to colorless (a reversible absorbance change) by the similar treatment as shown in Figure 4.11 [79].

4.5 PREPARATION AND PROPERTIES OF FLUOROALKYL END-CAPPED OLIGOMERS/CALCIUM CARBONATE NANOCOMPOSITES

Calcium carbonate is a cheap, commercially available inorganic particle and has been widely used as an inorganic filler in the manufacture of paint, paper, rubber, and plastics and in the production of toothpaste and binding agents [80,81]. Usually, the commercially available micrometer-size calcium carbonates have been used to merely reduce the cost of expensive resins; however, the improvement in the various properties such as mechanical property of micrometer-size calcium carbonate-filled

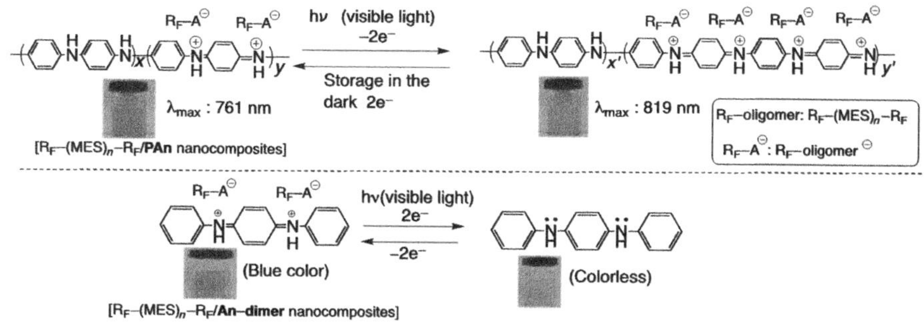

FIGURE 4.11 Controlling photochromic behaviors between R_F—$(MES)_n$—R_F/PAn nanocomposites and R_F—$(MES)_n$—R_F/An-dimer nanocomposites induced by UV-light-responsive titanium oxide nanoparticles were observed; R_F—$(MES)_n$—R_F/PAn nanocomposites can exhibit a reversible wavelength change by alternation of UV irradiation and storage in the dark; in contrast, R_F—$(MES)_n$—R_F/An-dimer nanocomposites can exhibit a reversible color change from blue to colorless (reversible absorbance change) by a similar treatment.

composites is believed to be poor due to the decrease of the polymer—filler interaction [82]. Thus, the preparation of nanocomposite particles by the encapsulation of calcium carbonate particles into organic polymers is an efficient method to prevent the agglomeration and improve a variety of properties such as mechanical and electrical properties, heat resistance, and radiation resistance of the parent polymers [83–92]. From this point of view, we tried to react calcium chloride with sodium carbonate in the presence of fluoroalkyl end-capped 2-methacryloyloxyethane sulfonic acid oligomers in aqueous solutions to afford the corresponding fluorinated oligomer/calcium carbonate composites as shown in Scheme 4.13 [93]. Cross-linked fluoroalkyl end-capped acrylic acid co-oligomer containing poly(oxyethylene) units was also applied to the preparation of new cross-linked fluorinated calcium carbonate nanocomposites under similar conditions (see Scheme 4.13) [93]. The obtained fluorinated calcium carbonate nanocomposites were found to have an extremely higher dispersibility in aqueous and organic media including fluorinated solvents [93].

In particular, it was verified that these fluorinated calcium carbonate nanocomposites are applicable to the dispersion above PMMA film surface. Interestingly, field-emission scanning electron microscopy (FE-SEM) images of the cross-section of the modified PMMA films showed that calcium carbonate particles dispersed into these PMMA films could be arranged regularly above the modified film surface (see Figures 4.12 and 4.13) [93].

We have studied the thermal stability of fluoroalkyl end-capped acrylic acid oligomer [R_F—$(ACA)_n$—R_F]/, R_F—$(MES)_n$—R_F/, fluoroalkyl end-capped *N,N*-dimethylacrylamide oligomer [R_F—$(DMAA)_n$—R_F]/, and fluoroalkyl end-capped acryloylmorpholine oligomer [R_F—$(ACMO)_n$—R_F]/$CaCO_3$ nanocomposites, in which each theoretical oligomer content in the composites is 19% by using thermal gravimetric analyses (TGA) measurements [94]. R_F—$(ACA)_n$—R_F/$CaCO_3$ nanocomposites, in which the theoretical oligomer contents in the composites

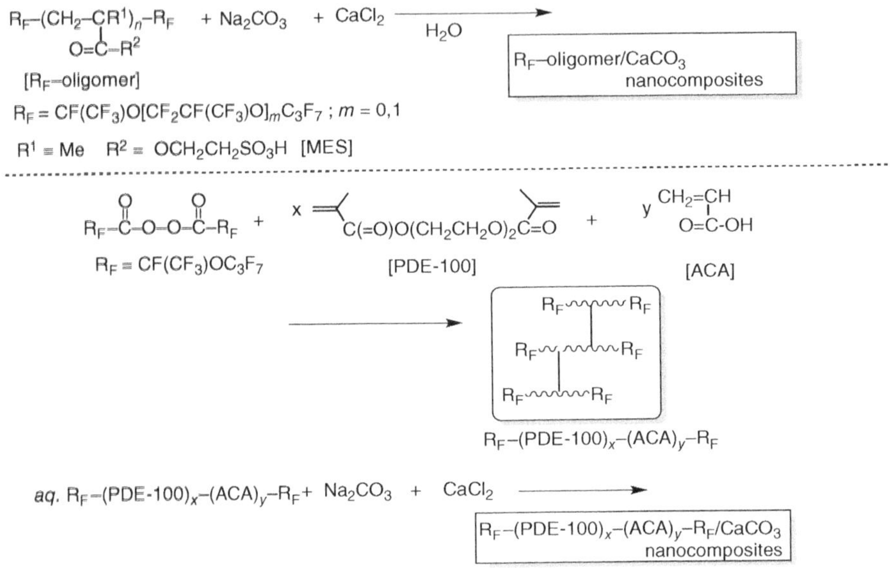

SCHEME 4.13 Preparation of fluoroalkyl end-capped oligomers/calcium carbonate nanocomposites.

are 36–53%, were found to exhibit a clear weight loss corresponding to the contents of oligomer in the composites after calcination at 800°C [94]. In contrast, R_F—$(MES)_n$—R_F/, R_F—$(DMAA)_n$—R_F/, and R_F—$(ACMO)_n$—$R_F/CaCO_3$ nanocomposites afforded a slight weight loss behavior after calcination under similar conditions [94]. It was clarified that each fluorinated oligomer/$CaCO_3$ nanocomposite possessing no weight loss behavior after calcination were applied to the surface modification of PMMA to exhibit a good oleophobic characteristic

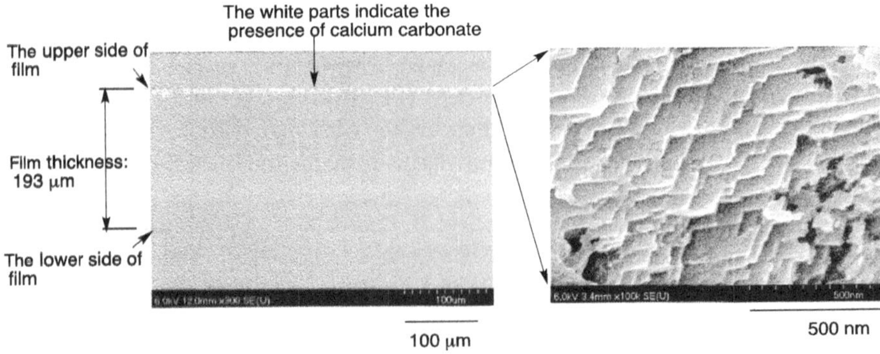

FIGURE 4.12 FE-SEM images of cross-section of the modified PMMA films treated with R_F—$(PDE\text{-}100)_x$-$(ACA)_y$—$R_F/CaCO_3$ nanocomposites.

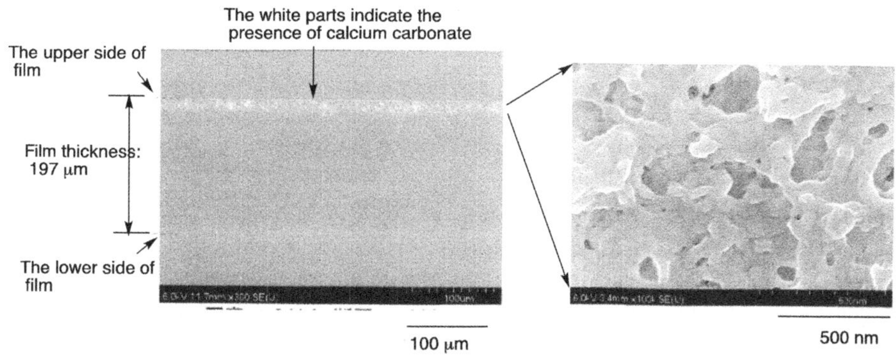

FIGURE 4.13 FE-SEM images of cross-section of the modified PMMA films treated with R_F—$(MES)_n$—$R_F/CaCO_3$ nanocomposites.

imparted by fluorine in the composites, as well as that before calcination [94]. Aromatic compounds possessing acidic hydroxyl protons such as bisphenol AF can exhibit a nonflammable characteristic even after calcination at 800°C in calcium fluoride nanocomposite matrices through not only the hydrogen bonding interaction between fluorine in calcium fluoride and acidic hydroxyl protons in Ar—OH but also the electrostatic interaction between the electronegative oxygen in Ar—OH and electropositive calcium atoms in calcium fluoride [95]. Therefore, it is suggested that the present fluorinated oligomer/calcium carbonate nanocomposites should exhibit a nonflammable characteristic even after calcination at 800°C, owing to the effect interaction (F—Ca) between fluorines in the oligomers and calcium moieties in the calcium carbonate nanocomposites matrices. Because, we can easily observe the bond-strengthening effect in Ca—F bond (132 kcal or 552 kJ/mol) [96], as well as the Si—F bond (129 kcal or 540 kJ/mol) [48, 97] in ammonium hexafluorosilicate. In this way, our present fluorinated oligomers/calcium carbonate nanocomposites have high potential for the development of novel nonflammable materials into a wide variety of fields.

4.6 PREPARATION AND PROPERTIES OF FLUOROALKYL END-CAPPED OLIGOMERIC NANOCOMPOSITES-ENCAPSULATED ORGANIC GUEST MOLECULES

Fluoroalkyl end-capped oligomeric aggregates can be applicable to the encapsulation of metal fine particles such as gold nanoparticles [98–101], magnetic nanoparticles [102], palladium nanoparticles [103], zinc oxide nanoparticles [104], copper nanoparticles [105, 106], and silver nanoparticles [107] to afford the colloidal stable fluorinated oligomeric nanocomposites-encapsulated these metal nanoparticles. In these nanocomposites, especially, encapsulated gold nanoparticles into these fluorinated oligomeric nanocomposite cores can afford the extremely red-shifted plasmon

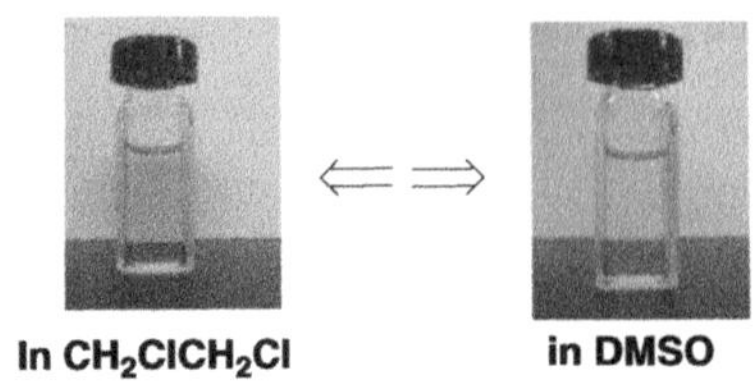

FIGURE 4.14 Photograph of R_F—(DOBAA)$_x$—(ACMO)$_y$—R_F/fluorescein nanocomposites in 1,2-dichloroethane and dimethyl sulfoxide (DMSO):R_F—(DOBAA)$_x$—(ACMO)$_y$—R_F/fluorescein nanocomposites can exhibit a coloring—decoloring behavior among protic solvents including protic-like solvents such as 1,2-dichloroethane and aprotic solvents such as DMSO, *N,N*-dimethylformamide, and tetrahydrofuran.

absorption band from 520 to around 960 nm [100]. Therefore, it is in particular interest to develop fluoroalkyl end-capped oligomers to the preparation of fluorinated oligomeric nanocomposites-encapsulated not only metal nanoparticles but also organic guest molecules such as organic dyes, from the developmental viewpoint of new fluorinated functional materials. In fact, amphiphilic fluoroalkyl end-capped *N*-(1,1-dimethyl-3-oxobutyl)acrylamide—acryloylmorpholine co-oligomers [R_F—(DOBAA)$_x$—(ACMO)$_y$—R_F] can interact with fluorescein as a guest molecule to afford the corresponding fluorinated cooligomer/fluorescein nanocomposites, and these fluorinated fluorescein nanocomposites can exhibit a coloring—decoloring behavior owing to a solvatochromic response as shown in Figure 4.14 [108].

We have very recently succeeded in the encapsulation of acetone in fluoroalkyl end-capped 2-acrylamido-2-methylpropanesulfonic acid oligomeric gel network cores to give the reddish-brown fluorinated 2-acrylamido-2-methylpropane sulfonic acid oligomer [R_F—(AMPS)$_n$—R_F]/acetone composite (see Scheme 4.14) [109]. It has been demonstrated that the color can arise from acetone encapsulated in the R_F—(AMPS)$_n$—R_F oligomeric gel network cores forming acetone polyaldol condensation products [109]. In addition, the retro-polyaldol condensation of acetone was also found to decolor the solution after 1 day at room temperature [109]. Especially, we have a good repeatability for the coloring—decoloring behavior between the polyaldol and retro-aldol condensations of acetone in the fluorinated oligomeric gel network cores as shown in Figure 4.15 [109].

Aromatic ketones such as 4′-methoxyacetophenone (MAP), acetophenone (AP), 4-acetylbiphenyl (ABP), and 2-acetyl-6-methoxynapthalene (AMN) interacted

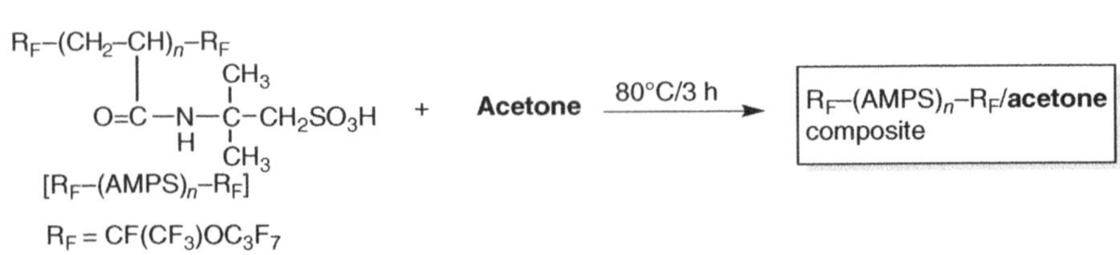

SCHEME 4.14 Preparation of R_F—(AMPS)$_n$—R_F/acetone composite.

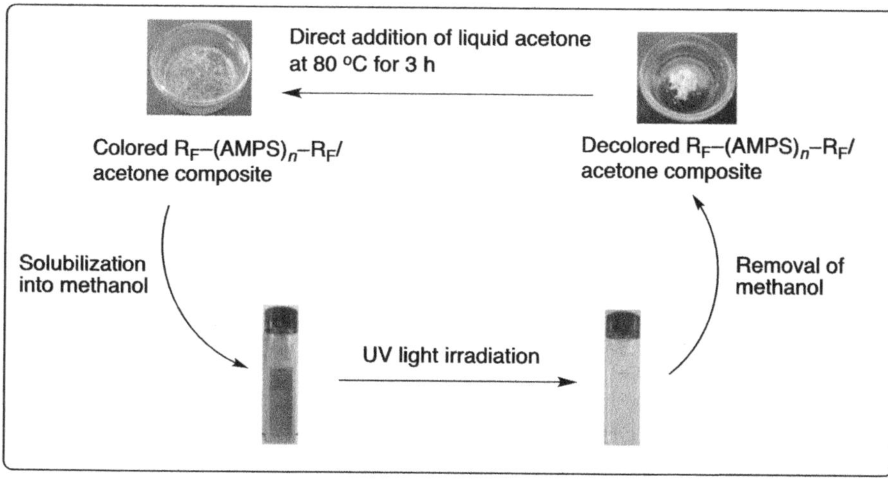

FIGURE 4.15 Coloring—decoloring behavior of R_F—$(AMPS)_n$—R_F/acetone composite possessing a good repeatability.

with fluoroalkyl end-capped 2-acrylamido-2-methylpropanesulfonic acid oligomer [R_F—$(AMPS)_n$—R_F] at 80°C for 3 hours to give the corresponding fluorinated oligomer/aromatic ketones composites (see Figure 4.16) [110]. In these composites, the R_F—$(AMPS)_n$—R_F/MAP and /AP composites were found to give the homoaldol condensation products of MAP and AP, respectively. In contrast, the corresponding nonfluorinated AMPS oligomer/MAP and sulfuric acid/MAP composites could not give the homoaldol product at all under similar conditions [110]. This suggests that the R_F—$(AMPS)_n$—R_F oligomer could provide the suitable fluorinated oligomeric gel network cores to interact with MAP or AP as a guest molecule, and the homoaldol

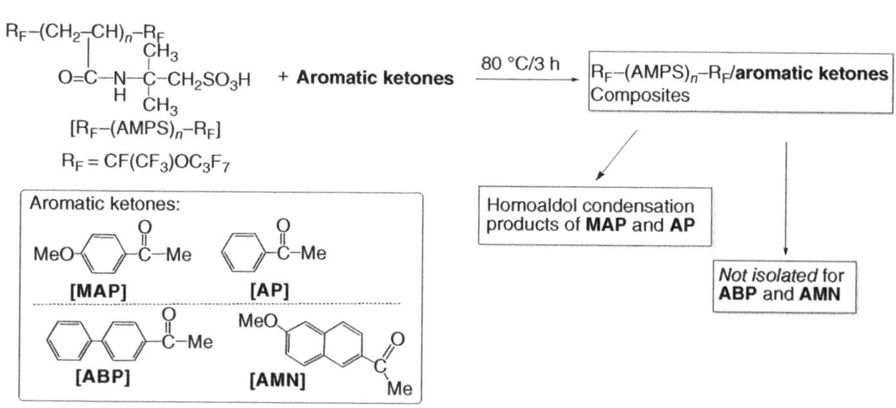

FIGURE 4.16 Homoaldol condensation of aromatic ketones in fluoroalkyl end-capped 2-acrylamido-2-methylpropane sulfonic acid oligomeric cores.

condensation of encapsulated MAP and AP should proceed smoothly in the fluorinated oligomeric gel network cores [110]. The R_F—$(AMPS)_n$—R_F/ABP and /AMN composites could not give the homoaldol products at all under similar conditions, indicating that the more bulky aromatic ketones than MAP or AP are not likely to be encapsulated as guest molecules into the fluorinated AMPS oligomeric gel network cores [110].

REFERENCES

1. Dolbier, W. R. *J. Fluor. Chem.* **2005**, *126*, 157.
2. Scheirs, J., Ed. *Modern Fluoropolymers*, Wiley: Chichester, 1997.
3. Ameduri, B. *Macromolecules* **2011**, *44*, 2394.
4. Johns, K.; Stead, G. *J. Fluor. Chem.* **2000**, *104*, 5.
5. Yano, S.; Okubo, N.; Takahashi, K. *Macromol. Symp.* **1996**, *108*, 279.
6. Fabbri, P.; Messori, M.; Montecchi, M.; Nannarone, S.; Pasquali, L.; Pilati, F.; Tonelli, C.; Toselli, M. *Polymer* **2006**, *47*, 1055.
7. Ameduri, B. *Chem. Rev.* **2009**, *109*, 6632.
8. Chen, Y.-C.; Tsai, C.-C.; Lee, Y.-D. *J. Polym. Sci. Part A: Polym. Chem.* **2004**, *42*, 1789.
9. Imae, T. *Curr. Opin. Colloid Interface Sci.* **2003**, *8*, 307.
10. Cho, J.-W.; Sul, K.-I. *Fibers Polym.* **2001**, *2*, 135.
11. Ameduri, B.; Boutevin, B. *J. Fluor. Chem.* **2000**, *104*, 53.
12. Ameduri, B.; Boutevin, B. *Well-Architectured Fluoropolymers: Synthesis, Properties and Applications*; Elsevier: Amsterdam, 2004; pp 231–348.
13. Berret, J.-F.; Calvet, D.; Collet, A.; Viguier, M. *Curr. Opin. Colloid Interface Sci.* **2003**, *8*, 296.
14. Imae, T.; Tabuchi, H.; Funayama, K.; Sato, A.; Nakamura, T.; Amaya, N. *Colloids Surf. A: Physicochem. Eng. Asp.* **2000**, *167*, 73.
15. Lebreton, P.; Ameduri, B.; Boutevin, B.; Corpart, J. M. *Macromol. Chem. Phys.* **2002**, *203*, 522.
16. Monteiro, M. J.; Adamy, M. M.; Leeuwen, B. J.; Van Herk, A. M.; Destarac, M. *Macromolecules* **2005**, *38*, 1538.
17. Feiring, A. E.; Wonchoba, E. R.; Davidson, F.; Percec, V.; Barboiu, B. *J. Polym. Sci. Part A: Polym. Chem.* **2000**, *38*, 3313.
18. Destarac, M.; Matyjaszewski, K.; Silverman, E.; Ameduri, B.; Boutevin, B. *Macromolecules* **2000**, *33*, 4613.
19. Shi, Z.; Holdcroft, S. *Macromolecules* **2005**, *38*, 4193.
20. Jankova, K.; Hvilsted, S. *J. Fluor. Chem.* **2005**, *126*, 241.
21. Laruelle, G.; Nicol, E.; Ameduri, B.; Tassin, J.-F.; Ajellal, N. *J. Polym. Chem. Sci. Part A: Polym. Chem.* **2011**, *49*, 3960.
22. Patil, Y.; Ameduri, B. *Polym. Chem.* **2013**, *4*, 2783.
23. Boschet, F.; Kostov, G.; Ameduri, B.; Jackson, A.; Boutevin, B. *Polym. Chem.* **2012**, *3*, 217.

24. Sawada, H. *Chem. Rev.* **1996**, *96*, 1779.
25. Sawada, H.; Kawase, T. *Kobunshi Ronbunshu* **2001**, *58*, 147.
26. Sawada, H.; Kawase, T. *Kobunshi Ronbunshu* **2001**, *58*, 255.
27. Sawada, H. *J. Fluor. Chem.* **2000**, *105*, 219.
28. Sawada, H. *J. Fluor. Chem.* **2003**, *121*, 111
29. Sawada, H. *Prog. Polym. Sci.* **2007**, *32*, 509.
30. Sawada, H. *Polym. Chem.* **2012**, *3*, 46
31. Sawada, H.; Itoh, N.; Kawase, T.; Mitani, M.; Nakajima, H.; Nishida, M.; Moriya, Y. *Langmuir* **1994**, *10*, 994.
32. Nakagawa, J.; Kamogawa, K.; Sakai, H.; Kawase, T.; Sawada, H.; Manosroi, J.; Manosroi, A.; Abe, M. *Langmuir* **1998**, *14*, 2055.
33. Nakagawa, J.; Kamogawa, K.; Momozawa, N.; Sakai, H.; Kawase, T.; Sawada, H.; Sano, Y.; Abe, M. *Langmuir* **1998**, *14*, 2061.
34. Sawada, H.; Ikeno, K.; Kawase, T. *Macromolecules* **2002**, *35*, 4306.
35. Sawada, H.; Narumi, T.; Kajiwara, A.; Ueno, K.; Hamazaki, K. *Colloid Polym. Sci.* **2006**, *284*, 551.
36. Zhang, S.-W.; Zhou, S.-X.; Weng, Y.-M.; Wu, L.-M. *Langmuir* **2005**, *21*, 2124.
37. Sawada, H.; Gong, Y.-F.; Minoshima, Y.; Matsumoto, T.; Nakayama, M.; Kosugi, M.; Migita, T. *J. Chem. Soc. Chem. Commun.* **1992**, 537.
38. Sawada, H.; Minoshima, Y.; Nakajima, H. *J. Fluor. Chem.* **1992**, *65*, 169.
39. Yoshida, M.; Moriya, K.; Sawada, H.; Kobayashi, M. *Chem. Lett.* **1985**, 755.
40. Sawada, H. *Polymeric Materials Encyclopedia—Synthesis, Properties and Applications*; Salamone, J. C., Ed.; CRC Press, Inc.: Boca Raton, FL, 1996; Vol. 4, p 2507.
41. Sawada, H.; Nakayama, M.; Kikuchi, O.; Yokoyama, Y. *J. Fluor. Chem.* **1990**, *50*, 393.
42. Iizuka, M.; Fukushima, S.; Goto, Y.; Okazaki, M.; Sawada, H.; Yoshida, M. *Colloid Polym. Sci.* **2013**, *291*, 595.
43. Sawada, H.; Kakehi, H.; Tashima, T.; Nishiyama, Y.; Miura, M.; Isu, N. *J. Appl. Polym. Sci.* **2009**, *112*, 3482.
44. Sawada, H.; Tashima, T.; Kodama, S. *Polym. Adv. Technol.* **2008**, *19*, 739.
45. Sawada, H.; Narumi, T.; Kodama, S.; Kamijo, M.; Ebara, R.; Sugiya, M.; Iwasaki, Y. *Colloid Polym. Sci.* **2007**, *285*, 977.
46. Sawada, H.; Tashima, T.; Kakehi, H.; Nishiyama, Y.; Kikuchi, M.; Miura, M.; Sato, Y.; Isu, N. *Polym. J.* **2010**, *42*, 167.
47. Sawada, H.; Tashima, T.; Nishiyama, Y.; Kikuchi, M.; Kostov, G.; Goto, Y.; Ameduri, B. *Macromolecules* **2011**, *44*, 1114.
48. Pauling, L. *The Chemical Bond*; Cornell University Press: New York, 1967.
49. Sawada, H.; Kikuchi, M.; Nishida, M. *J. Polym. Sci. Part A Polym. Chem.* **2011**, *49*, 1070.
50. Sawada, H.; Matsuki, Y.; Goto, Y.; Kodama, S.; Sugiya, M.; Nishiyama, Y. *Bull. Chem. Soc. Jpn.* **2010**, *83*, 75.
51. Goto, Y.; Takashima, H.; Takishita, K.; Sawada, H. *J. Colloid Interface Sci.* **2011**, *362*, 375.
52. Mills, A.; Lee, S.-K. *J. Photochem. Photobiol. A Chem.* **2002**, *152*, 33.

53. Mills, A.; Hill, G.; Bhopal, S.; Parkin, I. P.; O'Neill, S. A. *J. Photochem. Photobiol. A Chem.* **2003**, *160*, 185.
54. Arellano, M.; Mickel-Haciski, I.; Feke, D. L.; Manas-Zloczower, I. *J. Coat. Technol.* **1996**, *68*, 83.
55. Beecroft, L. B.; Ober, C. K. *Chem. Mater.* **1997**, *9*, 1302.
56. Nasu, A.; Otsubo, Y. *J. Colloid Interface. Sci.* **2007**, *310*, 617.
57. Egerton, T. A.; Everall, N. J.; Mattinson, J. A.; Kessell, L. M.; Tooley, I. R. *J. Photochem. Photobiol. A Chem.* **2008**, *193*, 10.
58. Chuang, F.-Y.; Yang, S.-M. *Synth. Met.* **2005**, *152*, 361.
59. Chandra, A.; Turng, L.-S.; Gong, S.; Hall, D. C.; Caufield, D. F.; Yang, H. *Polym. Compos.* **2007**, *28*, 241.
60. Ozerin, A. N.; Zelenetskii, A. N.; Akopova, T. A.; Pavlova-Verevkina, O. B.; Ozerina, L. A.; Surin, N. M.; Kechek'yan, A. S. *Polym. Sci. Ser. A.* **2006**, *48*, 638.
61. Kalappa, P.; Lee, J.-H. *Polym. Int.* **2007**, *56*, 371.
62. Chang, C.-M.; Chang, C.-L.; Chang, C.-C. *Macromol. Mater. Eng.* **2006**, *291*, 1521.
63. Pena, J.; Vallet-Regi, M.; San Roman, J. *J. Biomed. Mater. Res.* **1997**, *35*, 129.
64. Nakane, K.; Kurita, T.; Ogihara, T.; Ogata, N. *Compos. B.* **2004**, *35*, 219.
65. Perrin, F. X.; Nguyen, V. N.; Vernet, J. L. *Macromol. Chem. Phys.* **2005**, *206*, 1439.
66. Zheng, M.-P.; Jin, Y.-P.; Jin, G.-L.; Gu, M.-Y. *J. Mater. Sci. Lett.* **2000**, *19*, 433.
67. Perrin, F. X.; Nguyen, V. N.; Vernet, J. L. *Polym. Int.* **2002**, *51*, 1013.
68. Sawada, H.; Sawada, E.; Kakehi, H.; Kariya, T.; Mugisawa, M.; Chounan, Y.; Miura, M.; Isu, N. *Polym. Compos.* **2009**, *30*, 1848.
69. Sawada, E.; Kakehi, H.; Chounan, Y.; Miura, M.; Sato, Y.; Isu, N.; Sawada, H. *Compos. B.* **2010**, *41*, 498.
70. Hanaor, D. A. H.; Sorrell, C. C. *J. Mater. Sci.* **2011**, *46*, 855.
71. Peng, X.; Chen, A. *Adv. Funct. Mater.* **2006**, *16*, 1355.
72. Guo, S.; Yoshioka, H.; Kakehi, H.; Kato, Y.; Miura, M.; Isu, N.; Ameduri, B.; Sawada, H. *J. Colloid Interface Sci.* **2012**, *387*, 141.
73. Su, S.-J.; Kuramoto, N. *Synth. Met.* **2000**, *114*, 147.
74. Xia, H.; Wang, Q. *J. Appl. Polym. Sci.* **2003**, *87*, 1811.
75. Butterworth, M. D.; Corradi, R.; Johal, J.; Laschelles, S. F.; Maeda, S.; Armes, S. P. *J. Colloid Interface Sci.* **1995**, *174*, 510.
76. Gill, M.; Mykytiuk, J.; Armes, S. P. *Langmuir* **1993**, *8*, 2178.
77. Terrill, N. J.; Crowley, T.; Gill, M.; Armes, S. P. *Langmuir* **1993**, *9*, 2093.
78. Sawada, H.; Tsuzuki-ishi, T.; Kijima, T.; Iizuka, M.; Yoshida, M. *Colloid Polym. Sci.* **2011**, *289*, 1103.
79. Sawada, H.; Tsuzuki-ishi, T.; Kijima, T.; Kawakami, J.; Iizuka, M.; Yoshida, M. *J. Colloid Interface Sci.* **2011**, *359*, 461.
80. Rothon, R. N., Ed. *Particulate-Filled Polymer Composites*; Wiley: New York, 1995.
81. Reeder, R. J., Ed. Carbohydrates: Mineralogy and Chemistry. In: *Reviews in Mineralogy*; Mineralogy Society of America: Washington, DC, 1983; Vol. 11.
82. Chan, C.-M.; Wu, J.; Li, J.-X.; Cheung, Y.-K. *Polymer* **2002**, *43*, 2981.
83. Yang, Y.; Kong, X. Z.; Kan, C. Y.; Sun, C. G. *Polym. Adv. Technol.* **1999**, *10*, 54.

84. Bechthold, N.; Tiarks, F.; Willert, M.; Landfester, K.; Antonietti, M. *Macromol. Symp.* **2000**, *151*, 549.

85. Yu, J.; Yu, J.; Guo, Z.-X.; Gao, Y.-F. *Macromol. Rapid Commun.* **2001**, *22*, 261.

86. Wu, D.; Wang, X.; Song, Y.; Jin, R. *J. Appl. Polym. Sci.* **2004**, *92*, 2714.

87. Jiang, L.; Dan, Y. *Colloid Polym. Sci.* **2004**, *282*, 1374.

88. Xie, X.-L.; Liu, Q.-X.; Li, R. K.-Y.; Zhou, X.-P.; Zhang, Q.-X.; Yu, Z.-Z.; Mai, Y.-W. *Polymer* **2004**, *45*, 6665.

89. Rong, M. Z.; Zhang, M. Q.; Zheng, Y. X.; Zeng, H. M.; Walter, R.; Friedrich, K. *Polymer* **2001**, *42*, 167.

90. Rong, M. Z.; Zhang, M. Q.; Zheng, Y. X.; Zeng, H. M.; Walter, R.; Friedrich, K. *Polymer* **2001**, *42*, 3301.

91. Hasegawa, N.; Okamoto, H.; Kato, M.; Usuki, A. *J. Appl. Polym. Sci.* **2000**, *78*, 1918.

92. Levita, G.; Marchetti, A.; Lazzeri, A. *Polym. Eng. Sci.* **1989**, *19*, 39.

93. Sawada, H.; Shikauchi, Y.; Kakehi, H.; Katoh, Y.; Miura, M. *Colloid Polym. Sci.* **2007**, *285*, 499.

94. Saito, T.; Nishida, M.; Fukaya, H.; Kakehi, H.; Kato, Y.; Miura, M.; Isu, N.; Sawada, H. *Colloid Polym. Sci.* **2013**, *291*, 945.

95. Saito, T.; Kakehi, H.; Kato, Y.; Miura, M.; Isu, N.; Sawada, H. *Polym. Adv. Technol.* doi:10.1002/pat.3111

96. Blue, G.-D.; Green, J. W.; Bautista, R. G.; Margrave, J. L. *J. Phys. Chem.* **1963**, *67*, 877.

97. Walsh, R. *Acc. Chem. Res.* **1981**, *14*, 246.

98. Sawada, H.; Sasaki, A.; Sasazawa, K.; Kawase, T.; Ueno, K.; Hamazaki, K. *Colloid Polym. Sci.* **2005**, *283*, 583.

99. Sawada, H.; Sasaki, A.; Sasazawa, K. *Colloid Surf. A: Physicochem. Eng. Aspects* **2009**, *337*, 57.

100. Mugisawa, M.; Sawada, H. *Langmuir* **2008**, *24*, 9215.

101. Sawada, H.; Takahashi, K. *J. Colloid Interface Sci.* **2010**, *351*, 166.

102. Sawada, H.; Kijima, T.; Mugisawa, M. *Polym. J.* **2010**, *42*, 494.

103. Kijima, T.; Javakhishvili, I.; Jankova, K.; Hvilsted, S.; Kodama, S.; Sugiya, M.; Sawada, H. *Colloid Polym. Sci.* **2010**, *290*, 589.

104. Sasazawa, K.; Hirayama, Y.; Sawada, H. *Polym. Int.* **2009**, *58*, 177.

105. Sawada, H.; Furukuwa, R.; Sasazawa, K.; Toriba, K.; Ueno, K.; Hamazaki, K. *J. Appl. Polym. Sci.* **2006**, *100*, 1328.

106. Sawada, H.; Furukuwa, R.; Sasazawa, K.; Mugisawa, M.; Ohnishi, K. *Eur. Polym. J.* **2007**, *43*, 3258.

107. Sawada, H.; Sasaki, A.; Sasazawa, K.; Toriba, K.; Kakehi, H.; Miura, M.; Isu, N. *Polym. Adv. Technol.* **2008**, *19*, 419.

108. Sawada, H.; Izumi, S.; Sasazawa, K.; Yoshida, M. *J. Colloid Interface Sci.* **2012**, *377*, 76.

109. Kijima, T.; Nishida, M.; Fukaya, H.; Yoshida, M.; Sawada, H. *J. Polym. Sci. Part A Polym. Chem.* **2013**, *51*, 2555.

110. Kijima, T.; Nishida, M.; Fukaya, H.; Yoshida, M.; Sawada, H. *Polym. Adv. Technol.*, DOI: 10,1002/pat. 3229.

5

THERMAL DEGRADATION AND PYROLYSIS OF POLYTETRAFLUOROETHYLENE

GERARD PUTS AND PHILIP CROUSE
University of Pretoria
BRUNO AMEDURI
Ingénierie et Architectures Macromoléculaires Institut Charles Gerhardt

5.1 INTRODUCTION

Fluoropolymers possess excellent thermal stability and are nearly completely inert to chemical attack. Polytetrafluoroethylene (PTFE), discovered in 1938 by Roy Plunkett at I.E. du Pont de Nemours and Company, is the world's most used fluoropolymer. PTFE accounts for 59% of the fluoropolymers market and its production amounted to some 79,000 metric tons in 2004 [1]. Applications exploiting the extraordinary properties of the material are wide-ranging and include gaskets, thread-seal tape, pipe liners, bearings, gears, slide plates, filter bags, medical prosthetics, and architectural material [2, 3].

Due to the difficulties in producing the monomeric materials, and large market demand, fluoropolymers are high added-value products. The price for unprocessed PTFE amounted to $16,000 per ton in 2012, compared to $1500 per ton for high density polyethylene.

Unlike most other polymers, high-molecular-weight PTFE cannot be melt-processed [4] and PTFE products are prepared predominantly by mechanical working

Handbook of Fluoropolymer Science and Technology, First Edition. Edited by Dennis W. Smith Jr., Scott T. Iacono, and Suresh S. Iyer.

of a pressed and sintered preform, during which a considerable amount of waste material is produced. The mass of waste generated can account for as much as 50% of the final mass on products of complex geometries [5].

As a consequence, PTFE waste cannot be reprocessed via the usual methods employed for plastics recycling and presents both an economic loss and a serious ecological problem. The literature indicates that PTFE may be chemically recycled by pyrolysis to produce mixtures of various high value chemicals [6].

The objective of this chapter is to present a survey of the open literature to examine the various methods of PTFE pyrolysis so as to build a coherent image of the art, and to present and evaluate the known catalytic reactions of the pyrolysis products, with emphasis on those of TFE. This is essential to direct any research concerning the *selective beneficiation* of PTFE waste to high value chemicals. In areas where the information on the reactions of perfluorocarbons are lacking or nonexistent, reference is made to the reactions of fluorochlorocarbons or simple hydrocarbons for "by analog" information.

The literature on PTFE pyrolysis presented here is divided into three sections, namely, (i) the literature that deals directly with pyrolysis; (ii) material regarding the *autogenous* reactions of the pyrolysis products under pyrolysis conditions; (iii) and material that deals with the catalyzed reactions of the pyrolysis products. It should be noted that the reactions of the main pyrolysis products have traditionally been discussed alongside pyrolysis, but it is more appropriate to discuss them separately. The reactions of the various pyrolysis products with reactive gases are discussed along with the catalyzed reactions as there is considerable overlap between the two.

Information on the toxicity and physical hazards of the pyrolysis products dealt with in this chapter may be found in section 5.3.5.

5.2 PTFE PYROLYSIS

A brief overview of the various methods employed to pyrolyze PTFE is presented in section 5.2.1, while the actual mechanism by which the polymer decomposes is discussed in section 5.2.2.

5.2.1 Overview of Methods

Four methods of pyrolysis are reported in the literature, namely, (i) reduced pressure or vacuum pyrolysis, (ii) pressure pyrolysis, (iii) inert gas-assisted pyrolysis, (iv) and reactive gas-assisted pyrolysis.

5.2.1.1 Homogeneous Pyrolysis Vacuum pyrolysis and pressure pyrolysis are essentially the same operation, but these methods have traditionally been dealt separately due to the design of the laboratory setup. For industrial applications, this point is moot and these two methods may be grouped together as homogeneous pyrolysis, which may occur either as batch or continuous processes.

For batch processes, the reactor (either a quartz tube or a steel pressure vessel) is charged with a fixed mass of PTFE, flushed with an inert gas (usually argon or nitrogen) and then either evacuated or pressurized to the desired operating conditions. The pressurizing gas may be inert or may consist of pyrolysis gases. The reactor is then heated to the desired temperature by means of an electric furnace [5–11]. Only one instance of the flash pyrolysis (addition of PTFE after the reactor has reacted operating temperature) has been reported [12]. The gaseous pyrolysis products are collected in a liquid nitrogen cold trap and separated by distillation.

For continuous pyrolysis, only metal reactors (steel or nickel alloys) are used and PTFE is fed into the reactor with a screw feeder. The gaseous products are either collected using a scaled up cold trap, or fed directly into other processes [5, 13]. Continuous systems suffer from a drawback in that it is very difficult to maintain a pressure seal at the PTFE inlet, resulting in the inclusion of air in the reactor atmosphere, which induces PTFE combustion rather than pyrolysis and produces a mixture of mostly undesired products, such as CF_4, CO_2, and CF_2O.

Table 5.1 summarizes available data concerning the products obtained through homogenous pyrolysis of PTFE and their relative yields in relation to pyrolysis conditions. The data in the table were collated from several publications [7,9,10,13,14]. The data are as reported, and not normalized to 100%.

TABLE 5.1 Distribution of Products Obtained from the Pyrolysis of PTFE as Function of Temperature and Pressure

Temperature (°C)	TFE (%)	HFP (%)	OFCB (%)	PFIB (%)	Perfluorates (%)	Reference
101,325 Pa Absolute pressure						
600	15.9	25.7	54.3	4.1	–	[9]
83,600 Pa Absolute pressure						
600	32	26	42	–	–	[7]
700	52	24	18	5	–	[7]
30,000 Pa Absolute pressure						
600	30.79	30.22	6.89	–	32.1	[13]
700	12.17	28.32	3.76	15.47	40.28	[13]
800	1.17	40.6	–	6.93	51.3	[13]
5500 Pa Absolute pressure						
600	85.7	14.3	<4	–	–	[9]
700	82.1	17.9	<4	–	–	[9]
670 Pa absolute pressure						
500	96.8	2.9	–	–	–	[10]
600	97	3	–	–	–	[9]

On a side note, James and Rowsell [15] reported the synthesis of perfluorothiane species similar to those reported by Krespan (discussed in section 5.4.2.3) via the vacuum pyrolysis of tetrafluoroethylene–thiocarbonyl fluoride and tetrafluoroethylene–tetrafluorothiiran copolymers at temperatures between 500°C and 700°C.

It is clear that there is a severe lack of correlation between the various datasets presented in Table 5.1. Although much work has been done, a thorough study reporting the percentage yields at enough pressures and temperatures to allow for statistically meaningful analysis has not been conducted. The two main factors which make interpretation of the data in Table 5.1 difficult are the facts that the gaseous pyrolysis products self-react and that that there is a great likelihood that the materials of construction have a catalytic effect. Additional and more exact detail regarding these materials, along detail regarding reactor and system geometries and operating conditions are required for a definitive analysis.

A few authors have attempted to extract reaction kinetics from their experimental work. Kinetic expressions reported in the literature include the kinetics of both PTFE pyrolysis and various gas-phase reactions. A summary of these kinetics is presented in Table 5.2.

5.2.1.2 Inert Gas-Assisted Pyrolysis Inert gas-assisted pyrolysis concerns the continuous pyrolysis of PTFE using an inert gas as heat transfer and product transport medium. The reactor setup and products obtained for this method are similar to those reported for homogenous continuous pyrolysis. This method is preferred over homogeneous continuous pyrolysis as it allows for better control of the product residence time in the reactor and facilitates the exclusion of air from the pyrolysis system.

TABLE 5.2 Arrhenius Rate Constants and Reaction Orders for the Pyrolysis of PTFE and the Relevant Gas-Phase Reactions of the Pyrolysis Products

Reaction description	Reaction order	Frequency factor (s)	Activation energy (kJ/mol)	Reference
Dimerization of TFE to OFCB	Second order	10.3×10^{7}	106.3	[16]
Breakdown of OFCB to TFE	First order	8.9×10^{15}	310	[16]
HFP formation from TFE	Second order	9×10^{8}	126.7	[17]
PTFE depolymerization	First order	4.7×10^{18}	336	[10]
PTFE depolymerization	First order	3×10^{19}	347.3	[18]
PTFE depolymerization	First order	4.122×10^{21}	349.1	[19]
PTFE combustion	First order	2.038×10^{18}	299.6	[19]

It was reported by Simon and Kaminsky [6] that while any inert gas can theoretically be used, nitrogen is usually preferred as it does not condense in a cold trap and gives TFE yields upward of 75%. The same authors reported the use of steam as inert gas. They indicated that steam should be used for any inert gas-assisted pyrolysis as it gives TFE yields comparable to the nitrogen-assisted method, but allows for the easy separation of the product gases from the carrier medium.

While the work of Simon and Kaminsky [20] has resulted in a patent [20] and an indication of the intent to construct a pilot plant, no commercial recycling of PTFE waste has been conducted using either inert gas-assisted or homogeneous pyrolysis. Work similar to that of Simon and Kaminsky was reported for laboratory scale production of TFE by Arkles and Bonnett [21] and by Filatov et al. [22].

5.2.1.3 Reactive Gas-Assisted Pyrolysis Numerous research articles report the reactions of TFE with other gaseous species to produce high value products [23–29]. Since TFE is the major product in PTFE pyrolysis, it stands to reason that the use of a gas which is reactive toward TFE, such as SO_2, in place of an inert carrier gas will allow for the continuous production of high value chemicals from PTFE waste.

As yet, no mention has been made in the public-domain literature of this potential method, despite the fact that the procedure is very similar to that proposed for the recycling of PTFE waste by inert gas-assisted pyrolysis.

5.2.2 Mechanism of Pyrolysis

The literature contains considerable debate as to how exactly the polymer chain breaks down to gaseous compounds and why no solid residue remains after pyrolysis of pure PTFE.

Several workers have postulated that, since TFE is the main pyrolysis product, the PTFE chain snaps a C—C bond somewhere along its length to produce shorter, radically terminated chains which then cleave off TFE molecules, continuously shortening the chain in the process until the entire chain has "unzipped," as illustrated in Scheme 5.1 [6, 9, 10, 30].

This postulate, however, does not account for the formation of HFP, OFCB, or PFIB and provides no proper explanation for why only TFE should be liberated.

Confirmation of chain scission and, at least, liberation of a radical species arises from degradation studies achieved using filled PTFE. By electron spin resonance spectroscopy, Fock [31] found that the carbon black used in carbon-filled PTFE contained large numbers of unpaired electrons and that pyrolysis of carbon-filled PTFE showed vastly different decomposition rates to neat and bronze-filled PTFE. He concluded that the difference in degradation behavior is due to radical intermediates being stabilized by the unpaired carbon black electrons.

As illustrated by Figure 5.1, the fluorine atoms in a PTFE chain are tightly packed in a 13/6 helical arrangement around the carbon backbone, making the chain far more rigid than its polyethylene hydrocarbon analog.

The highly rigid, close packed structure and the great strength of the C—F bond (507 kJ/mol) imply that no fluorine abstraction or internal chain rearrangements of

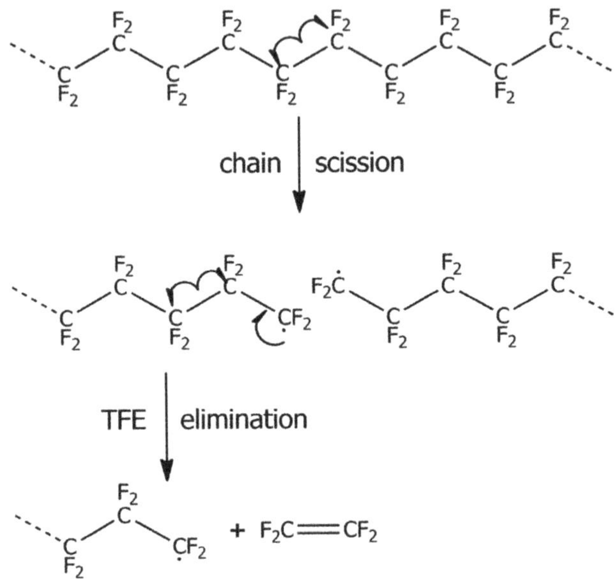

SCHEME 5.1 Postulated mechanism for releasing TFE during pyrolysis of PTFE [6, 9, 10, 30].

fluorine atoms can occur and products such as HFP and OCFB cannot be directly split from the chain. This is also the reason why no polymer cross-linking (which results in solid residue during polyethylene pyrolysis) occurs in PTFE pyrolysis [32].

The nominal bond strength of a C—C bond in a PTFE chain is 359 kJ/mol [6]. The calculations performed by Errede [33] indicate that the strength of the C—C bond falls from 359 kJ/mol to 184 kJ/mol when going from the centre of the fragment to the bond next to the radical carrying carbon, as illustrated in Figure 5.2.

As the weakest bonds will break first, it becomes evident that a [:CF_2] unit rather than a TFE unit will be cleaved from the radical fragment. These radicals are well studied and they have rather long half-lives compared to their hydrocarbon analogs due to the stabilization of the radical pair by the fluorine atoms. At temperatures below 700°C and at pressures of 1 mmHg, the difluorocarbene molecule has a half-life of about 1 second [34]. This half-life is sufficiently long for reactions other than

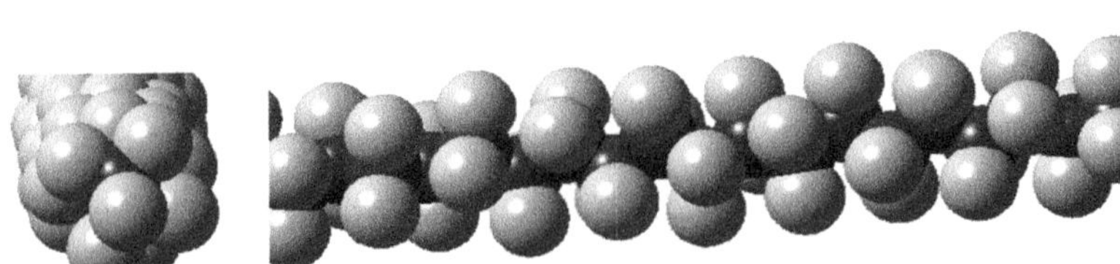

FIGURE 5.1 Front- and side-on view of the atomic arrangement in a PTFE chain.

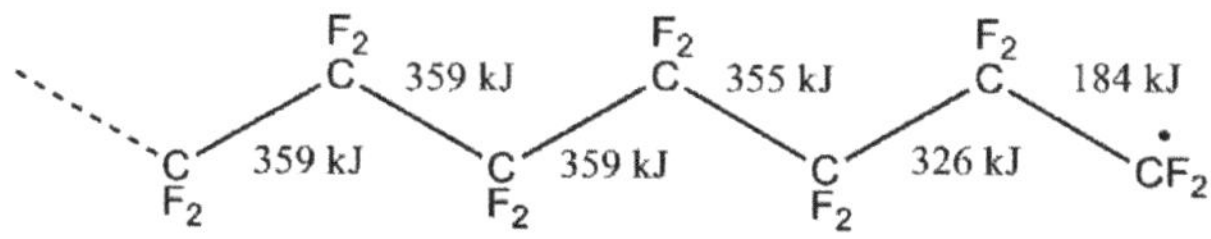

FIGURE 5.2 Bond strengths in a PTFE radical fragment [33].

recombination to occur, which accounts for the variety of pyrolysis products, as will be discussed in section 5.3.

The cleavage of difluorocarbene units is supported by the experimental observation that CF_2O is the major product of the combustion of PTFE in air [19, 35–37].

We postulate that, during pyrolysis, the PTFE chain undergoes random scission to produce radically terminated fragments which cleave off difluorocarbenes. This mechanism is summarized in Scheme 5.2.

The thermodynamics of the depolymerization process has not been studied in detail owing to the lack of thermodynamic data for the polymeric chain. Van der Walt [14] performed molecular modeling to predict the thermodynamic properties of the polymer chain and the radical intermediates. He calculated the heats of reaction for $C_{10}F_{21}$ releasing [$:CF_2$] or C_2F_4 as being 190.41 kJ/mol and 218.08 kJ/mol, respectively. These calculations only gave results for a system at 25°C, but provide a strong indication that the release of difluorocarbene is, from a thermodynamic point of view, more preferable than the direct release of TFE.

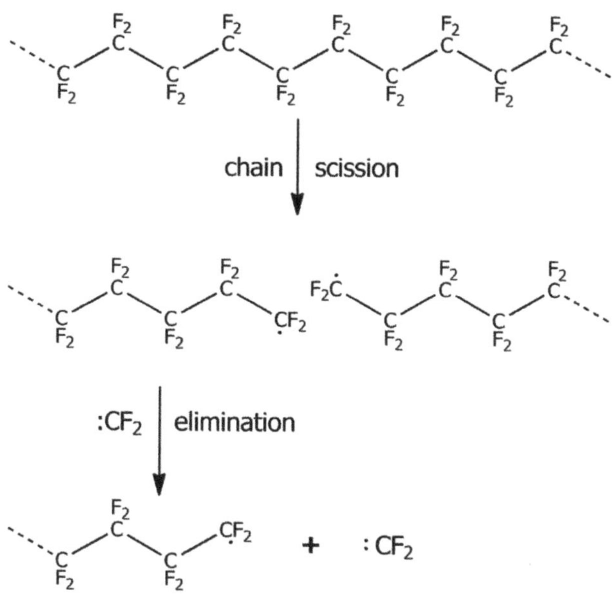

SCHEME 5.2 Proposed mechanism for the decomposition of PTFE during pyrolysis.

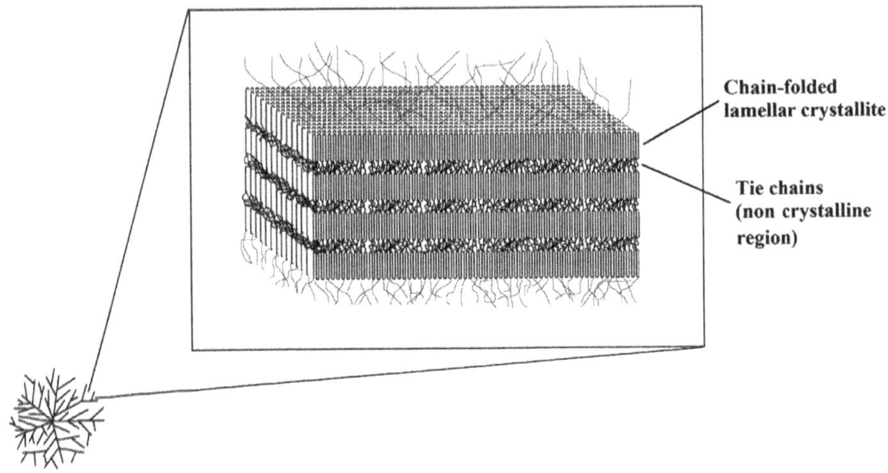

FIGURE 5.3 PTFE chain arrangements in crystalline PTFE.

As pointed out by Penski and Goldfarb [38], the effect of caging, that is, the solvation and retainment of gases inside the bulk polymer, cannot be ignored during pyrolysis. No studies on this topic other than the one by these authors have been published owing to the great complexity of the issue. However, it stands to reason that this effect only becomes important at high pressures. Under extreme vacuums, the diffusion of the gaseous species is simply too rapid.

Most solid PTFE exists in highly crystalline form. The crystalline phase consists of regular foldings of polymer chain that form parallel layers known as lamella, shown schematically in Figure 5.3 [3]. Bends in the polymer chain are accompanied by a disturbance of the helical arrangement, resulting in sites of higher steric strain and nonminimum electronic energy states.

This disturbance of the helix has been noted even in the molten state and the polymer melt has been shown by X-ray diffraction to be closely packed and entangled. Observations related to the crystallinity of processed PTFE confirm the presence of interlamellar entanglement [3].

Given this complex geometry in the molten state, the assumption that all parts of the chain are equally likely to break is suspect, but there has been no study done to determine whether there is a preference for scission to occur at specific points in the chain, such as bending sites in the lamellae. However, it stands to reason that sites of steric strain should be the place where scission primarily occurs.

5.3 AUTOGENOUS REACTIONS OF PYROLYSIS PRODUCTS

The difluorocarbene radicals produced by PTFE pyrolysis undergo a number of autogenous gas-phase reactions. These range from simple recombination to multistep

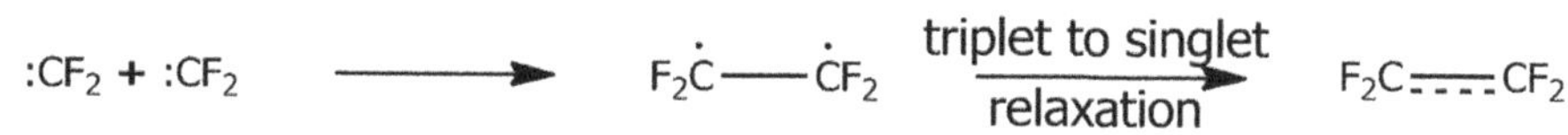

SCHEME 5.3 Mechanism of TFE formation from difluorocarbene.

rearrangements. Much literature exists on the topic and many conflicting mechanism have been proposed [1, 9, 30, 36, 39, 40].

A species by species account of the mechanism of formation of all products, taking into account the proposed mechanism of PTFE pyrolysis, is laid out in the following sections.

5.3.1 Mechanism of TFE Formation

As indicated by Table 5.1, the main pyrolysis product under vacuum conditions is TFE. It is formed by the recombination of two [$:CF_2$] units. Buravtsev's observation of a TFE biradicaloid during the decomposition of TFE and the recombination of difluorocarbene indicates that TFE is formed via the mechanism displayed in Scheme 5.3 [41, 42].

It should be noted that the biradicaloid is not the same as a 1,2-biradical. There is still sufficient sharing of the electrons between the carbons that the electrons cannot be said to be wholly localized to specific carbons. Then this bond takes on a character intermediate between that of a 1,2-biradical and a π-bond and is considerably more labile that the double bonds found in simple unsaturated hydrocarbons.

5.3.2 Mechanism of OFCB Formation

Atkinson and coworkers [16] discovered an equilibrium between OFCB and TFE. They postulated that TFE forms a dimer with itself which exists long enough for the double bond to open up and allow for cyclization to occur between the TFE units, as detailed in Scheme 5.4.

The dimer formation is extremely unlikely as it would require a side-on collision between TFE molecules. Being electron rich, the likelihood this occurring in such a way that a stable dimer could form before coulombic repulsion pushes them apart is very small and would result in low yields of OFCB.

Numerous studies by different authors shows that OFCB comprises a large percentage of the product when pyrolysis occurs at 500–700°C [6, 9, 10, 12]. They also

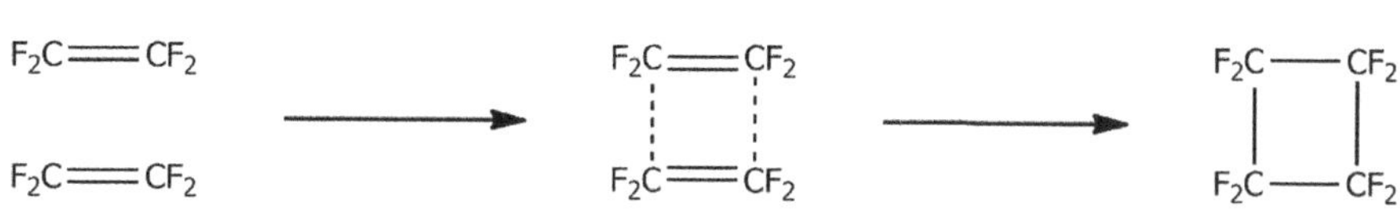

SCHEME 5.4 Atkinson's postulate for the mechanism of formation of OFCB from TFE [16].

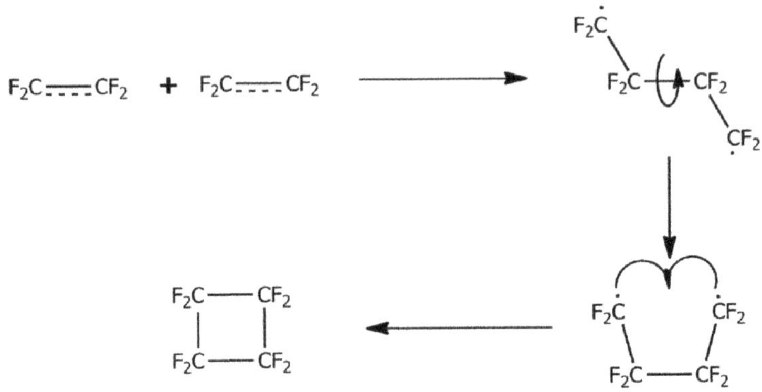

SCHEME 5.5 Buravtsev's mechanism for the formation of octafluorocyclobutane from TFE [44].

noted a decrease of OFCB yield with increase in reaction pressure, which, according to Atkinson's postulate would favor increasing OFCB yield. This indicates that the dimer postulate is incorrect.

An ab initio investigation by Wang and Borden [43] into the nature of the π-bond in TFE revealed that the 1,2-biradicaloid of TFE does not keep the theoretical planar configuration of the ground state. The carbon atoms only barely conform to sp^2 hybridization and have significant sp^3 character and take on a distorted tetrahedral form.

These facts indicate that OFCB is produced by collision of TFE biradicaloids to form 1,4-perfluorobutane biradicals which undergo cyclization via rotation of the bond between the second and third carbons followed by bonding of the terminal radicals, as detailed in Scheme 5.5.

As yet, no study has been conducted to determine the energy barrier for this rotation, nor how it correlates to the experimental activation energies reported for Atkinson's postulate. However, it is interesting that the activation energy for the reverse reaction in Atkinson's postulate is 310 kJ [16], which is only 40 kJ less than the energy required to rupture a C—C bond in a long $(CF_2)_n$ chain. It is well known from organic chemistry that cyclobutanes undergo great steric strain and will readily break a C—C bond to revert to a linear conformation. It stands to reason that the experimental activation energy for the reverse reaction is the energy required to rupture a C—C bond in OFCB.

Lacher and coworkers [45] studied the dimerization of TFE and trifluorochloroethylene (TFCE). The results indicated a difference of 3.8 kJ/mol in activation energy between the dimerization of TFE with itself and the dimerization of TFE with TFCE and a difference of 84 J between TFE–TFCE and TFCE–TFCE. It is expected that there will be an increase of the rotational energy barrier with increase in substituent size. As a result, the findings of Lacher and coworkers strongly support the existence of the mechanism detailed in Scheme 5.5.

5.3.3 Mechanism of Hexafluoropropylene Formation

Two major postulates for HFP formation have been reported in the literature [1,6,17, 40,44]. These postulates are summarized in Equations 5.1 and 5.2.

$$\text{OFCB} \rightarrow \text{HFP} + :CF_2 \tag{5.1}$$

$$\text{TFE} + :CF_2 \rightarrow \text{HFP} \tag{5.2}$$

No mechanisms have been reported for these postulates by the authors. The reaction in Equation 5.1 would, however, require fluorine rearrangement in the 1,4-biradical that intermediates OFCB formation, and the reaction in Equation 5.2 would require the simultaneous addition of carbene and rearrangement of fluorine in the 1,3-biradical which would form from the reaction of TFE and $:CF_2$.

No spectroscopic proof of the 1,4- and 1,3-radical rearrangements has been found at the relevant temperatures, despite a rather thorough investigation by Kolbanovskii et al. [46]. However, Kolbanovskii et al. did discover that a 1,2 fluorine shift occurs in TFE at ~750°C to produce $:CF{-}CF_3$.

It is known from organic chemistry that HFP is not sterically strained. It is thus more stable than OFCB, and should theoretically form first. This is, however, not the case [9, 12, 47–49]. Given that HFP yields become significant only above 750°C, a modification of the postulate in Equation 5.2 to include Kolbanovskii's observations is required to explain HFP formation. This mechanism is detailed in Scheme 5.6.

Gelblum et al. [50] claimed the intermediates presented in Scheme 5.9 in a patent concerning the optimized production of HFP, further substantiating the presented reaction scheme.

The double bond in HFP possesses a biradicaloid nature similar to that of TFE, but with a somewhat more pronounced electron division in favor of a localized electron on the centre carbon.

In a review of spectroscopic work on this topic, Buravtsev did not detect $:CF{-}CF_3$ during the decomposition of HFP and subsequently rejected the formation mechanism

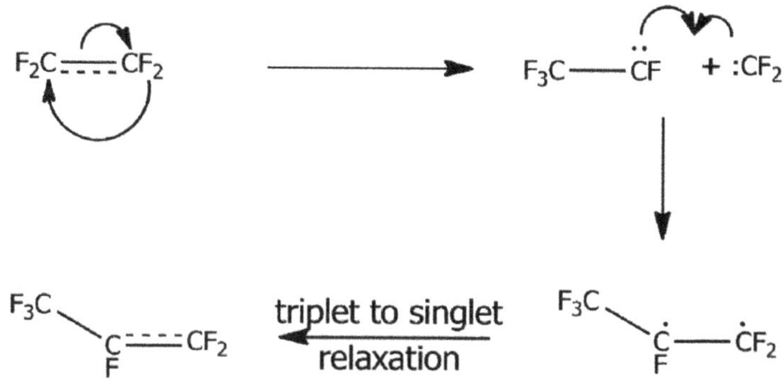

SCHEME 5.6 Proposed mechanism of HFP formation during the pyrolysis of TFE [46].

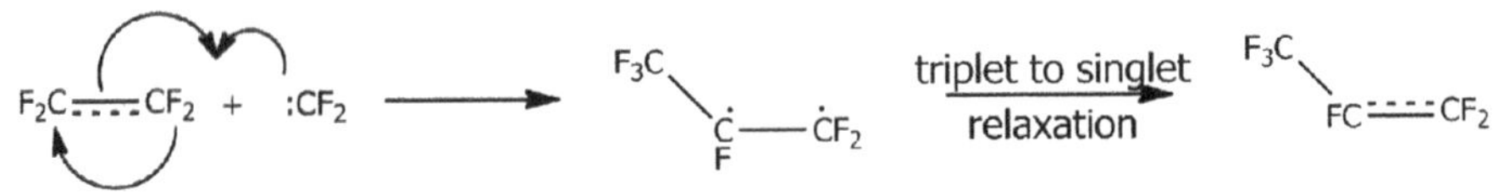

SCHEME 5.7 Hypothesis of HFP formation according to Buravtsev [44].

presented in Scheme 5.5 in favor of a mechanism that admits direct [$:CF_2$] addition to TFE biradicaloids and requires internal fluorine and electron rearrangement within the newly formed 1,3-biradicals to give 1,2-biradicaloids, which then relax to give HFP, as presented in Scheme 5.7 [39, 44]. He indicated that this mechanism was energetically more favorable than that in Scheme 5.6. However, he did not report the method by which his calculations were performed, making them potentially suspect.

His work fails to take into account that OFCB, which is less stable than HFP, is preferentially formed over HFP at lower temperatures and higher pressures during pyrolysis of PTFE, in an environment where high concentrations of both TFE and $:CF_2$ are present, which would favor HFP formation according to his postulate. Until this discrepancy is explained, the mechanism, as given in Scheme 5.5, will be accepted as valid.

5.3.4 Mechanism of Formation of Perfluorobutenes

No mechanism has as yet been proposed to account for PFIB formation. Given that its production only becomes significant at 900°C and 1 bar pressure, one can argue that 1,2-fluorine shift could occur within HFP to produce a high temperature radical intermediate stable enough to react with [$:CF_2$] and produce PFIB, as detailed by the postulate indicated in Scheme 5.8.

Two other butene isomers are formed besides PFIB, namely perfluoro-1-butene and perfluoro-2-butene. However, as with PFIB, a mechanism of formation from TFE or other pyrolysis products is lacking.

However, German et al. [51] did observe a 1,2 shift in perfluoro-2-butene that resulted in the isomerization of perfluoro-2-butene to perfluoro-1-butene, detailed in Scheme 5.9.

This isomerization proves the existence of an internal carbene radical and supports the mechanism in Scheme 5.8. This mechanism does not support Buravtsev's postulate for the formation of HFP as this mechanism is in two steps while Buravtsev's postulate requires simultaneous addition and fluorine rearrangement.

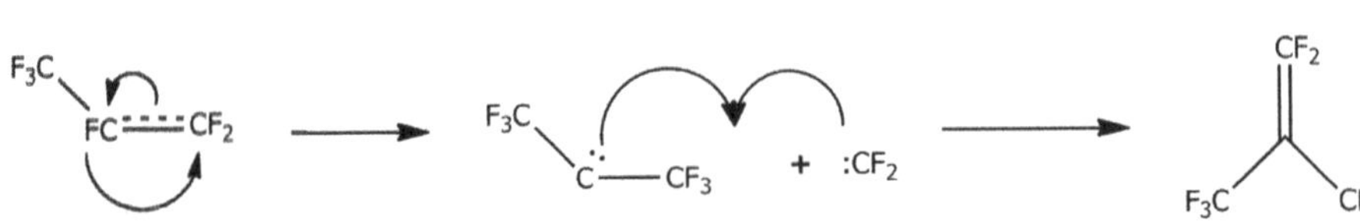

SCHEME 5.8 Proposed mechanism for PFIB formation during the pyrolysis of PTFE.

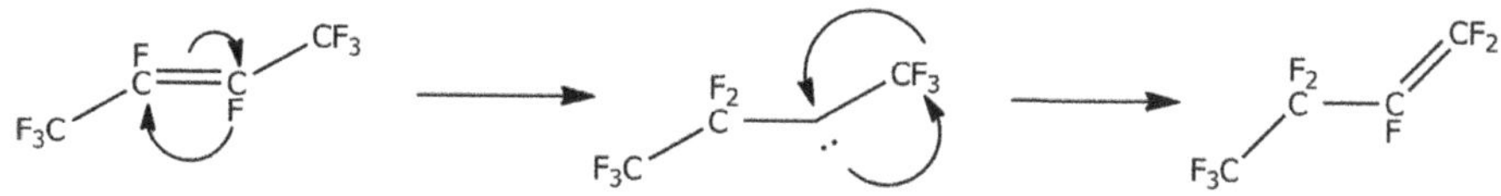

SCHEME 5.9 Mechanism of the isomerization of perfluoro-2-butene [51].

Additionally, they also indicated a possible mechanism for the formation of perfluoro-2-butene from TFE, presented in Scheme 5.10.

However, it must be made clear that this mechanism is a postulate only, and that no spectroscopic observation has yet been reported for this.

5.3.5 Toxicity and Safe Handling of Pyrolysis Products

While PTFE itself is nontoxic and physiologically inert [52] and poses a danger to human health only when inhaled as a microfine powder, its pyrolysis and combustion products can be lethal and precautions have to be taken when handling these products. The saturated perfluorocarbons generated during pyrolysis are similarly nontoxic and physiologically inert, their danger being only the potential to act as asphyxiants.

The unsaturated perfluorocarbons TFE, HFP, along with OCFB, on the other hand, are carcinogenic, producing tumors in the liver and kidneys. The chronic exposure threshold for TFE is 2 ppm [53]. No exposure limit has been set for HFP or OFCB, but it is assumed to be metabolized by the same mechanism as that of TFE and should therefore be treated similarly to TFE. No LC_{50} data appear have been reported for TFE, HFP, or OFCB.

Several incidents of polymer fume fever due to exposure to perfluoroisobutene have been reported, with four fatalities thus far [52, 54, 55]. Symptoms of PFIB exposure include nausea, coughing, and severe chest pain.

PFIB acts on the lung tissue and causes pulmonary edema. Mild PFIB exposure (concentrations higher than 0.014 ppm but lower than 0.042 ppm) may be treated by treating the subjects for edema. Immediate and complete immobilization of the exposed subject is recommended as exercise of any kind aggravates the damage to the lung tissue. Serious PFIB exposure (concentrations higher than 0.042 ppm) may be treated in a similar way to mild exposures, but with oral administration of *N*-acetylcysteine [56]. The LC_{50} of PFIB is taken as 0.25 ppm and the chronic exposure limit is set at 0.001 ppm.

Of all the gaseous products produced during pyrolysis of PTFE, TFE is the most unstable and extreme caution is required. Safe handling of the pyrolysis products

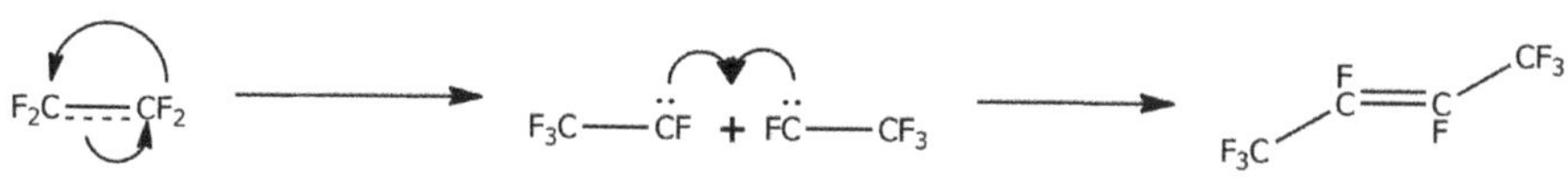

SCHEME 5.10 Postulated mechanism for the formation of perfluoro-2-butene from TFE [51].

should therefore focus on the safe handling and storage of TFE. TFE decomposes to CF_4 and carbon with the reaction occurring in the deflagration regime. Detail regarding minimum decomposition pressures can be found in a recent study by Lisochkin and Poznyak [57].

5.4 CATALYTIC REACTIONS OF PYROLYSIS PRODUCTS

TFE is the major product generated during PTFE pyrolysis and all attempts at catalytic beneficiation of pyrolysis off gas should focus on reformation or reaction of TFE. HFP is also considered as it is the second most abundant unsaturated compound produced.

An overview of the various catalytic reactions of TFE and HFP reported in literature is presented in the subsections below. A discussion on the mechanisms of catalysis is presented in its own section after the overview.

5.4.1 Catalytic Reformation of TFE

Only five investigations into catalytic reformation of TFE have been reported [7, 22, 48, 50, 58].

Filatov [22] pyrolyzed PTFE intimately mixed with CoF_3 and found that the yields of gaseous products were greatly reduced, but that the yield of perfluoroparaffins (6–26 carbons in length) was greatly improved.

Gelblum et al. [50] pyrolyzed various CHF gasses in a reactor line with the noble metals (gold, platinum, and palladium as well as the nickel alloy Inconel®) and found that there is a definite shift in equilibrium toward the production of HFP when using gold versus autogenous pyrolysis. Inconel acts as catalyst and decomposes the pyrolysis gases to carbon, CF_4, and C_2F_6.

Miller [48] attempted to catalyze TFE to HFP and OFCB by providing a high surface area for TFE to react using nuclear grade graphite as the catalyst. No catalytic action was observed other than some carbon deposits on the walls of the nickel tube.

Dmowski's [58] work required the use of a solvent, but showed that KF and CsF were capable of making dimeric and trimeric oligomers of TFE and HFP at 20°C.

Choi and Park [7] reported that there is a large increase in the yield of TFE (76–90%) at 700°C when using copper instead of Inconel as reactor tube material. A recent Mandarin language paper, which could not be obtained, also reported greatly increased yields of TFE when R-22 (CHClF) was pyrolyzed over a 60/40% Ni–Cu catalyst, supporting the findings of Choi and Park.

An analogous examination of the catalytic reactions of ethylene does not reveal any interesting information. Walker [59] found no catalytic reformation of ethylene to anything other than methane, hydrogen, and carbon.

Some of the catalysts tested for ethylene are SiO_2, $Na_2B_4O_7.10H_2O$, NaOH, $Ca(OH)_2$, Fe_2O_3, ZnO, PbO, Na, Ni, Co, Fe, and Cr_2O_3. Walker indicated two features: (i) that the decomposition reactions are mediated by a $:CH_2$ radical; (ii) but that this radical is too short-lived and too reactive to undergo any catalytic interaction.

In a somewhat unrelated patent, Karube et al. [60] showed that CCl_4 readily adds to TFE at 60°C in the presence of $AlCl_3$, indicating that TFE is at least reactive to other halocarbons.

Moore and Massey [61] detailed a reaction in which 1,1,2,2-tetrafluoroethane was isomerized to 1,1,1,2-tetrafluoroethane over chromia catalyst at 400°C, indicating that fluorine shifts may be catalyzed. The same authors also indicated that AlF_3 and NaF may also be used as catalyst. Similar work for TFE has not been reported, though it stands to reason that such a shift would be beneficial to the formation of HFP (See Scheme 5.7).

It is clear that little work has been done on the reactions of TFE over catalysts.

5.4.2 Reactions of TFE With Species Containing Hetero-Atoms

5.4.2.1 Reactions With Metals or Metallic Complexes There is considerably more literature regarding addition reactions of TFE with various other species. Most of these reactions occur in the gas phase.

Ohashi et al. [62] reported the addition of TFE to aryl zinc using palladium as the catalyst. However, this reaction did not proceed without the presence of LiI in the reaction mixture. They theorized that the reaction is catalyzed by a F^- which forms only in the presence of LiI.

Gas phase isomerization of tetrafluoroethane-β-sultone was effected by Barabanov et al. [23] with NH_3, NF_3, and trimethylamine as the catalysts [23]. Similar results were obtained by England and coworkers.

Takagi et al. [28] synthesized perfluoroamines from TFE and HFP using NF_3 as a nitrogen source and KF and NaF as catalysts. Tittle [29] added AsF_3 to TFE by using SbF_5 as catalyst. Furthermore, Fokin [25] synthesized perfluorothiols by addition of H_2S to TFE and HFP in the presence of alkaline catalysts.

5.4.2.2 Telomerization Reactions With Iodated Transfer Agents Syntheses involving TFE and I_2 were reported by Bedford and Baum [63], but at pressures of 20–27 bar and temperatures in the region of 200°C, also using a batch reactor. In these works, iodine was used both as a reagent and as a catalyst.

Kotov and coworkers [64] synthesized $C_2F_4I_2$ and $C_4F_8I_2$ directly from PTFE waste by copyrolysis with solid I_2 at atmospheric pressure and 285°C using a batch reactor. The yields were strongly dependent on the ratio of TFE to initial I_2.

Dindi and coworkers [65] patented a process of telomerizing TFE with $C_2F_4I_2$ as chain-transfer agent at high pressure and temperature in a batch reactor to produce $I–(C_2F_4)_n–I$, where $n \leq 6$. Their method requires TFE pressures of 2.6–2.9 MPa and temperatures in the range of 220°C. Their patent indicated that OFCB is formed during the reaction and must be periodically vented from the reactor as it inhibits product formation.

These iodine–TFE compounds are known as telechelics (or α, ω-difunctional derivatives) and find use as chain-transfer agents for the telomerization of TFE, HFP, or VDF, well reported in a book [66] and a review article [67]. No reports of HFP analogs to the iodine–TFE compounds could be found.

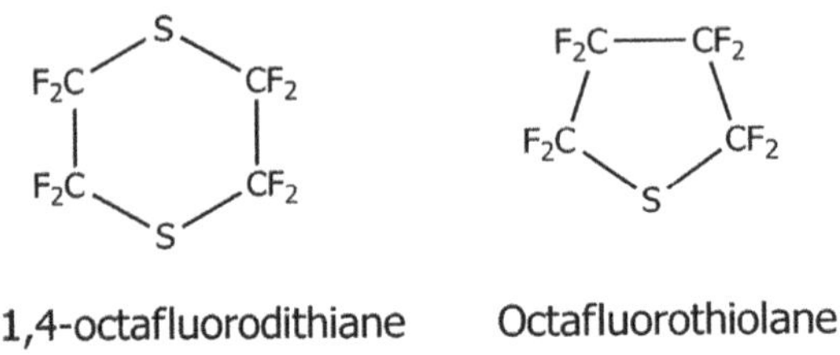

FIGURE 5.4 Cyclic perfluorothianes synthesized by Krespan and Langkammerer [26].

5.4.2.3 Reactions With Species Based on Sulfur, Phosphorous, and Selenium

Lastly, Krespan and Langkammerer [26] synthesized cyclic five and six member perfluorothianes using TFE and elemental sulfur in a batch reactor at temperatures ranging from 200°C to 300°C and under autogenous pressure. A maximum yield of 61% was obtained for 1,4-octafluorodithiane and a maximum yield of 15% was obtained for octafluorothiolane. The isolated structures are presented in Figure 5.4.

Both these compounds are volatile liquids, are stable at 300°C, do not readily react with oxygen, are resistant to attack by acids, bases, and free radicals and are reported to have high dielectric strength.

Krespan and Wheland [68] claimed in a patent that the above-mentioned compounds are highly desirable as solvents for the polymerization of perfluoromonomers. Interestingly, Krespan and Langkammerer [26] also found that 2-trifluoromethyl-pentafluoro-1,3-dithiolane, the structure of which is presented in Figure 5.5, was formed as the only by-product during the reactions where sulfur was in excess, constituting some 3% or less of the product mixture.

Krespan and Langkammerer also produced analogous compounds of selenium and phosphorus using iodine as catalyst in a batch process. The selenides are produced at temperatures in the range of 250°C as Krespan found that the selenides decompose at higher temperatures. Similarly, the phosphorous containing rings are also decomposed at temperatures above 250°C. The isolated products are presented in Figure 5.6.

Krespan noted that in all the phosphorus synthesis experiments, iodine satisfies the third valence of phosphorous. No third phosphorus–carbon bonds are formed and the six member phosphorus containing rings take on a puckered, hexane-like conformer. Krespan also indicated that the six member phosphorus containing rings are unreactive to water, but that the octafluorophospholanes is readily hydrolyzed to its phosphorous acid.

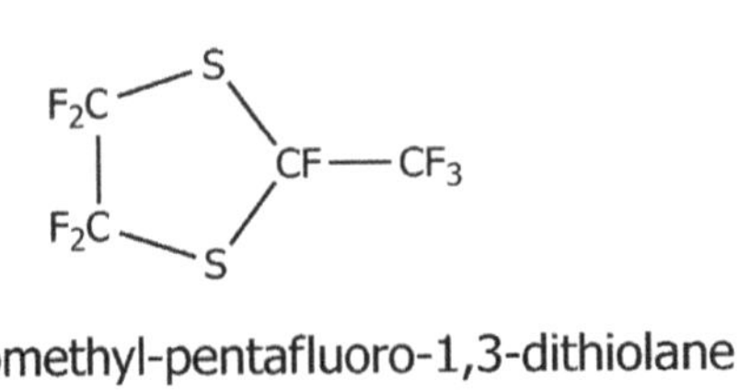

FIGURE 5.5 Molecular structure of 2-trifluoromethyl-pentafluoro-1,3-dithiolane obtained as by-product from the reaction of sulfur with TFE [26].

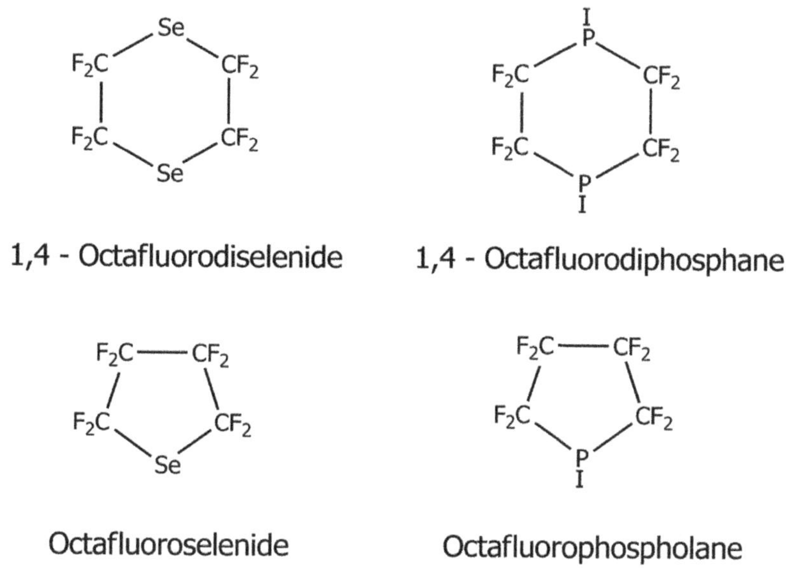

FIGURE 5.6 Cyclic perfluoroselenides and perfluorophosphanes synthesized by Krespan and Langkammerer [26].

Brown [69] reported the synthesis of the HFP analog of 1,4-octafluorodithiane by reacting HFP with sulfur at 300°C and 1 atmosphere pressure. The product had the empirical formula $C_6F_{12}S_2$. He identified this as a cyclic compound, but could not distinguish between the two chiral possibilities. The two possible structures are depicted Figure 5.7.

To the best of our knowledge no other work reporting similar syntheses using HFP has been reported.

5.4.3 Mechanism of Catalysis

The reports discussed above indicate that reactions of TFE and HFP were catalyzed by species that either have readily accessible electrons, as in the case of lone pair on

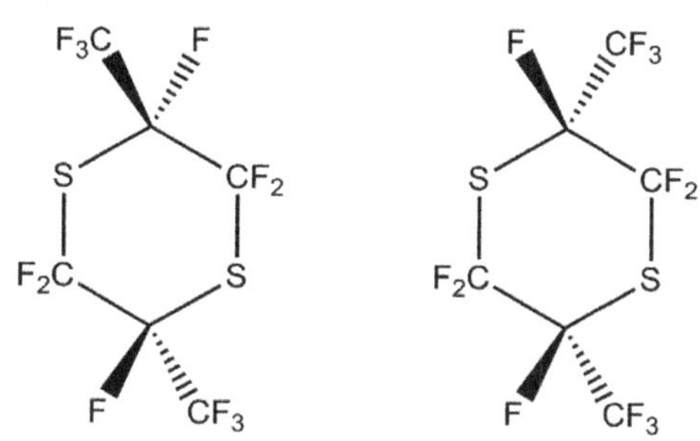

FIGURE 5.7 Structural isomers of bis(trifluoromethyl)–perfluorodithiane [69].

trimethylamine and NH_3, or are so polar that the metal atoms are essentially naked, as in the case of KF and NaF (solid bases).

It is known that the lone pair electrons of amines are capable of capturing a carbon radical long enough to convert it to a nonradical species, as is the case with hindered amine light stabilizers (HALS) [70]. It stands to reason that species with available lone pairs (Lewis bases) stabilize the TFE biradicaloid, allowing the end opposite to the catalyst to exist as a full radical and subsequently react. Barabanov's [23] observation of polymer formation is evidence for this terminal radical. Other perfluorinated species capable of forming biradicaloids would undergo similar radical stabilization.

The patent by Karube et al. [60] indicates that a variety of alkaline and alkaline earth metal fluorides and NiF_2 may be used as "promoters" in fluorocarbon rearrangements when doped onto chromia or alumina, but no explanation of their action is given in the literature.

A review by Pearson [71] on the hard and soft theory of acids and bases indicated that triethylamine, NH_3, and F^- are hard bases while Na^+ and K^+ are hard acids but that hardness decreases when going from Na^+ to K^+ to Cs^+. The implication is that the softer the metal is as a Lewis acid, the stronger the F^- will be as a Lewis base. It is obvious that, in light of the better catalysis achieved with CeF than with KF, the reaction proceeds via F^- catalyst and not via the metal. This is reinforced by the observations made by Ohashi et al. [62].

In the case of chromia, there is considerable electron richness on the chromium atoms, but the much more electron negative oxygen draws these electrons to itself and becomes almost completely ionic, in a sense, approaches O^{2-}. The oxygen anions are readily accessible at the surface of the crystal. Again, we conclude that the reactions are intermediated not by the hard acid Cr^{3+}, but by the strong basic O^{2-}.

The basic catalysis hypothesis is further reinforced by the work of Filatov et al. [22]. CoF_3 is a fluorinating agent and it is expected that it would give off fluorine to a fluorocarbon radical to produce small perfluorinated species. This is not the case as long chain fluorocarbons are formed during pyrolysis of PTFE, indicating CoF_3 catalyzes oligomerization.

No explanation could be found for the catalytic action of gold in the patent, but gold is classified as a very soft Lewis acid and it is assumed that the borderline cases such as gold interact weakly with carbon radicals.

As with gold, copper and nickel are classified as being soft to borderline Lewis acids and are assumed to interact weakly with fluorocarbon radicals. There is a slight difference in the electronegativity between copper and nickel and it stands to reason that an alloy of the two contain some basic metal sites due to this difference.

The addition of iodine and sulfur to TFE also occurs via a radical mechanism, but in their cases, the reagents act as the catalysts. As indicated by Bedford and Baum [63], addition of iodine to TFE occurs in three steps, detailed in Scheme 5.11.

Higher order α,ω-diiodoperfluoroalkanes are formed by radical addition of TFE to $I{-}CF_2{-}CF_2^{\cdot}$.

It is interesting to note that Kotov et al. reported that 1,4-diiodoperfluorobutane is formed in larger amounts than the 1,2-diiodoperfluoroethane (44% vs. 38%) but that

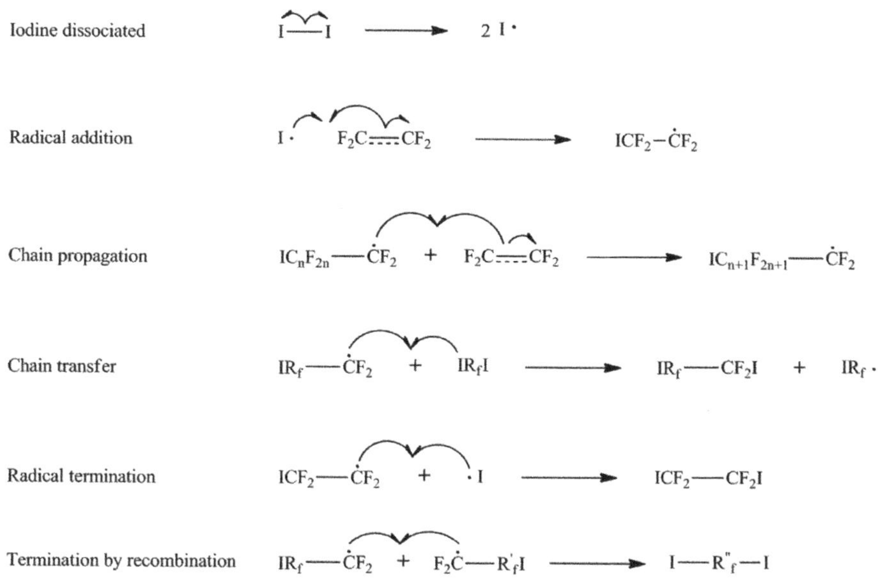

SCHEME 5.11 Postulated mechanism for iodine addition to TFE [63].

1,6-diiodoperfluorohexane and 1,8-diiodoperfluorooctane is formed in significantly smaller amounts (8% and 2%, respectively). This along with the observation by Dindi et al. [65] that OFCB forms in high quantities during the production of diiodoperfluoroalkanes is strongly supportive of the mechanism proposed in section 5.2.2 for OFCB formation.

It is also interesting to note that the yield of 1,6-diiodoperfluorohexane increased from 8% to 24% when the TFE to I_2 ratio was increased from 1 to 2, but that the yield of 1,8-diiodoperfluorooctane only increased to 8% from 2%. These results would indicate that the radically terminated carbon is not readily available to other TFE molecules and that some telomerization reactions occurred as detailed in scheme 5.11. This implies that the molecular rotation involved in the proposed OFCB formation mechanism happens quite rapidly.

Iodine transfer (co)polymerization of fluoroalkenes has been reviewed [66,67,72,73]. These processes have also involved TFE and have led to thermoplastic elastomers based on sequential copolymerization of fluoroalkenes that first yield a soft segment, followed a TFE (or VDF) monomer to enable the insertion of a "hard" PTFE or PVDF block as indicated in scheme 5.12:

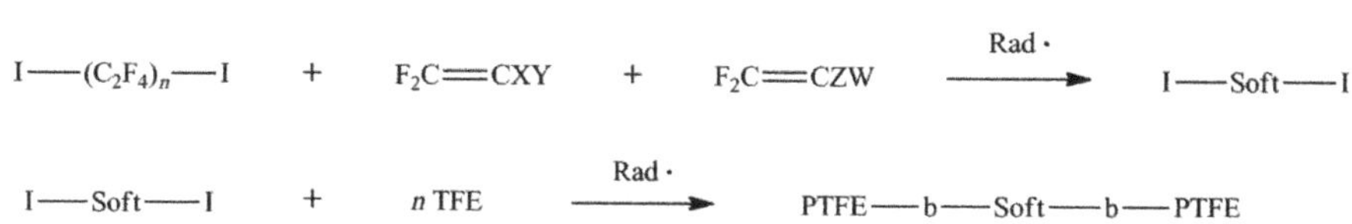

SCHEME 5.12 Thermoplastic elastomers based on Hard-b-Soft-b-Hard triblock by sequential iodine transfer copolymerization.

The hard sequence enables further physical cross-linking. The applications of such triblock thermoplastic elastomers are for seal, O-rings, shafts, for aeronautics (shuttles), and for the automotive industry (oil-resistant tubing).

Thermodynamic calculations for equilibrium composition of the gas phase during the thermal dissociation of iodine, detailed in Figure 5.8, indicate that there are very few iodine radicals present at 200°C when the gas is under pressure. This observation explains why Kotov et al. [64] reported yields in excess of 40% while other literature report low yields.

We may conclude that the iodine addition to TFE is catalyzed by the iodo-radicals generated by the thermal homolysis of iodine gas. This catalysis is analogous to the radical-initiated catalysis of polymerization.

Krespan and Langkammerer [26] proposed possible radical intermediate mechanisms for the addition of sulfur to TFE, detailed in Scheme 5.13. The central argument here is that sulfur forms a polyatomic radical which attaches itself to TFE while splitting off all the sulfur atoms save the atom attached to the CF_2. This new radical then undergoes further reaction.

It is known that gaseous sulfur is composed of many sulfur allotropes, but that the main component is disulfur and to a lesser extent, trisulfur [75]. At 1000 K and 1 mmHg pressure, S_2 makes up 99% of the gas phase.

This diatomic sulfur is known to exist as a triplet when in the ground state and readily reacts. Accordingly, the reaction as presented in Scheme 5.13 must be modified to admit that S_2 instead of a polyatomic sulfur chain is added to a TFE molecule.

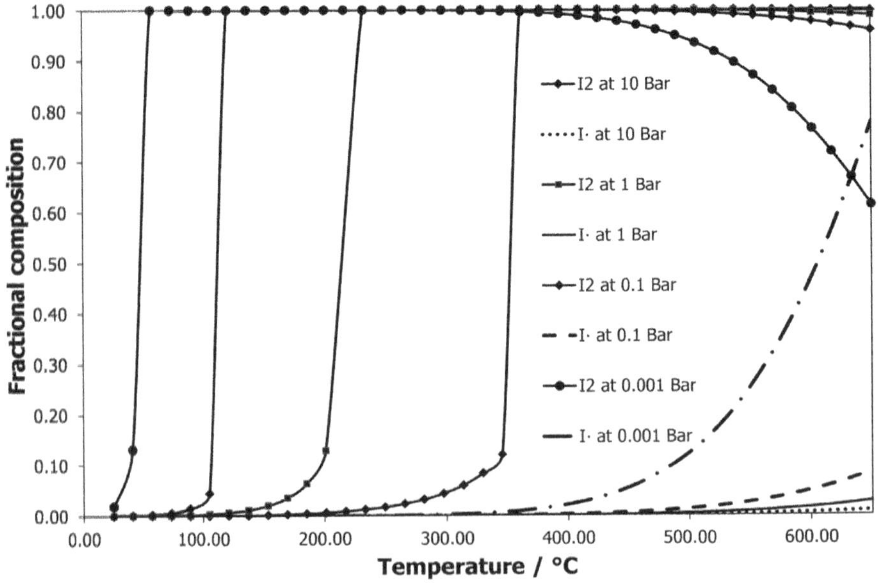

FIGURE 5.8 Temperature- and pressure-dependent equilibrium composition of iodine (calculated using [74]).

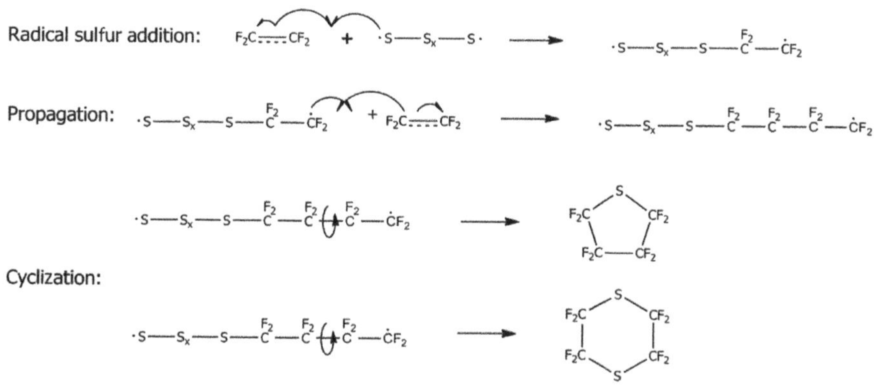

SCHEME 5.13 Krespan's mechanism for the addition of sulfur to TFE [68].

As indicated by Meyer [75], the S—S bond is much weaker than that of diatomic oxygen or the C—C bond and while no references could be found on the topic it stands to reason that when S_2 is attached to TFE, the S—S bond will break to afford $\cdot CF_2—CF_2—S\cdot$ and [:S] which undergoes further reaction to produce the thianes.

As was the case with iodine, we may conclude that the reaction of TFE and HFP with sulfur is catalyzed by sulfur radicals generated by thermal homolysis of S_8.

In summary, we may say that the reactions of TFE and other unsaturated perfluorocarbons always occur via radical intermediates and are catalyzed by strong bases or other species with unpaired or lone pair electrons.

NOMENCLATURE

HFP	Hexafluoropropylene
OFCB	Octafluorocyclobutane
PFIB	Perfluoroisobutylene
R-22	Chlorodifluoromethane
TFE	Tetrafluoroethylene
VDF	Vinylidene difluoride

REFERENCES

1. Garçia, A. N.; Viciano, N.; Font, R. *J. Anal. Appl. Pyrolysis* **2007**, *80*, 85.
2. Scheirs, J. *Modern Fluoropolymers*; Wiley: New York, 1997.
3. Drobny, J. G. *Technology of Fluoropolymers*; CRC Press: Boca Raton, FL, 2001.
4. Schildknecht, C. E. *Vinyl and Related Polymers*; Wiley: New York, 1952.
5. Meissner, E.; Wróblewska, A.; Milchert, E. *Polym. Degrad. Stab.* **2004**, *83*, 163.
6. Simon, C. M.; Kaminsky, W. *Polym. Degrad. Stab.* **1998**, *62*, 1.

7. Choi, S. K.; Park, J. D. *J. Korean Chem. Soc.* **1976**, *20*, 141.
8. Hunadi, R. J.; Baum, K. *Synthesis* **1982**, *1982*, 454.
9. Lewis, E. E.; Naylor, M. A. *J. Am. Chem. Soc.* **1947**, *69*, 1968.
10. Madorsky, J. S.; Hart, V. E.; Straus, S.; Sedlak, V. A. *J. Res. Natl. Bur. Stand.* **1953**, *51*, 327.
11. Marhevka, J. S.; Johnson, G. D.; Hagen, D. F.; Danielson R. D. *Anal. Chem.* **1982**, *54*, 2607.
12. Bhadury, P. S.; Singh, S.; Sharma, M.; Palit, M. *J. Anal. Appl. Pyrolysis* **2007**, *78*, 288.
13. Van Der Walt, I. J.; Grunenburg, A. T.; Nel, J. T.; Maluleke, G. G.; Bruinsma, O. S. L. *J. Appl. Polym. Sci.* **2008**, *109*, 264.
14. Van Der Walt, I. J. Recovery of Valuable Products from Polytetrafluoroethylene (PTFE) Waste. Ph.D. Dissertation, North West University, Potchefstroom, South Africa, **2007**.
15. James, R.; Rowsell, D. G. *J. Chem. Soc. Chem. Commun.* **1969**, 1274.
16. Atkinson, B.; Trenwith, A. B. *J. Chem. Soc.* **1953**, 2082.
17. Atkinson, B.; Atkinson, V. A. *J. Chem. Soc.* **1957**, 2086.
18. Siegle, J. C.; Muus, L. T.; Lin, T.-P.; Larsen, H. A. *J. Polym. Sci. A.* **1964**, *2*, 391.
19. Conesa, J. A.; Font, R. *Polym. Eng. Sci.* **2001**, *41*, 2137.
20. Hintzer, K.; Schöttle, T.; Staudt, H. J.; Weber, H. (Hoechst Aktiengesellschaft). Process for the Preparation of Fluorinated Monomers. European Patent 0,647,607 A1, 1995.
21. Arkles, B. C.; Bonnett, R. N. (Liquid Nitrogen Processing Corporation) Method for the Depolymerization of Polytetrafluoroethylene. US Patent 3,832,411, 1975.
22. Filatov, V. Y.; Murin, A. V.; Kazienkov, S. A.; Khitrin, S. V.; Fuks, S. L. *Russ. J. Appl. Chem.* **2011**, *84*, 147.
23. Barabanov, V. G.; Bispen, T. A.; Ilyin, A. N.; Kornilov, V. V.; Krivoshein, A. E.; Moldavsky, I. M.; Fenichev, I. M.; Vasilév, A. S.; Zubritskaya, N. G. *Fluor. Notes* **2011**, *2*, 1.
24. England, D. C.; Dietrich, M. A.; Lindsey, R. V. *J. Am. Chem. Soc.* **1960**, *82*, 6181.
25. Fokin, A. V.; Skladnev, A. A.; Studnev, Y. N.; Knunyants, L. L. *B. Acad. Sci. USSR Chem. Sci.* **1967**, *16*, 324.
26. Krespan, C. G.; Langkammerer, C. M. *J. Org. Chem.* **1962**, *27*, 3584.
27. Michaelsen, J. D.; Wall, L. A. *J. Res. Natl. Bur. Stand.* **1957**, *58*, 327.
28. Takagi, T.; Tamura, M.; Shibakami, M.; Quan, H.-D.; Sekiya, A. *J. Fluor. Chem.* **2000**, *101*, 15.
29. Tittle, B. *J. Fluor. Chem.* **1973**, *2*, 449.
30. Lonfei, J.; Jingling, W.; Shuman, X. *J. Anal. Appl. Pyrolysis* **1986**, *10*, 99.
31. Fock, J. *J. Polym. Sci. Part B Polym. Let.* **1968**, *6*, 127.
32. Florin, R. E.; Wall, L. A.; Brown, D. W.; Hymo, L. A.; Michaelsen, J. D. *J. Res. Natl. Bur. Stand.* **1954**, *53*, 121.
33. Errede, L. A. *J. Org. Chem.* **1962**, *27*, 3425.
34. Laird, R. K.; Andrews, E. B.; Barrow, R. F. *Trans. Faraday Soc.* **1950**, *46*, 803.
35. Baker, B. B., Jr.; Kasprzak, D. J. *Polym. Degrad. Stab.* **1993**, *42*, 181.
36. Odochian, L.; Moldoveanu, C.; Mocanu, A. M.; Carja, G. *Thermochim. Acta* **2011**, *526*, 205.

37. Williams, S. J.; Clarke, F. B. *Fire Mater.* **1982**, *6*, 161.
38. Penski, E. C.; Goldfarb, I. J. *J. Polym. Sci. Part B Polym. Let.* **1964**, *2*, 55.
39. Buravtsev, N. N.; Kolbanovskii, Y. A. *Russ. J. Appl. Chem.* **2002**, *75*, 598.
40. Morisaki, S. *Thermochim. Acta* **1978**, *25*, 171.
41. Buravtsev, N. N.; Kolbanovskii, Y. A.; Ovsyannikov, A. A. *Mendeleev Commun.* **1994**, *4*, 190.
42. Buravtsev, N. N.; Kolbanovskii, Y. A.; Ovsyannikov, A. A. *Mendeleev Commun.* **1994**, *4*, 48.
43. Wang, S. Y.; Borden, W. T. *J. Am. Chem. Soc.* **1989**, *111*, 7282.
44. Buravtsev, N. N.; Kolbanovsky, Y. A. *J. Fluor. Chem.* **1999**, *96*, 35.
45. Lacher, J. R.; Tompkin, G. W.; Park, J. D. *J. Am. Chem. Soc.* **1952**, *74*, 1693.
46. Kolbanovskii, Y. A.; Mamikonyan, E. R.; Matveyeva, L. N.; Shchipachev, V. S. *Sov. J. Chem. Phys.* **1990**, *7*, 885.
47. Moon, D. J.; Chung, M. J.; Kwon, Y. S.; Ahn, B. S. (Korea Institute of Science and Technology). Method of a Simultaneous Preparation of Hexafluoropropylene and Octafluorocyclobutane. US Patent 6,710,215 B2, 2004.
48. Miller W. T. *Preparation of Fluorocarbons by Polymerisation of Olefins* in Slesser, C., Schram, S. R., Eds., *Preparation, Properties and Technology of Fluorine and Organic Fluoro Compounds*, 1st ed.; McGraw-Hill: New York, 1951.
49. Harmon, J. (E.I. du Pont de Nemours & Company). Polyfluorinated Cycloparaffins and Process for Producing Them. US Patent 2,404,374, 1946.
50. Gelblum, G. P.; Herron, N.; Noelke, J. C.; Rao, M. N. V. (E.I. du Pont de Nemours and Company). Synthesis of Perfluoroolefins. European Patent 1,301,453 B1, 2010.
51. German, L. S.; Grigor'ev, A. S.; Kolbanovskii, Y. A.; Chepik, S. D. *B. Acad. Sci. USSR Chem. Sci.* **1987**, *36*, 1994.
52. Lee, C. H.; Guo, Y. L.; Tsai, P. J.; Chang, H. Y.; Chen, C. R.; Chen, C. W.; Hsiue, T. R. *Eur. Resp. J.* **1997**, *10*, 1408.
53. Lunn, R. M. *Twelfth Report on Carcinogens: National Toxicology Program*; US Department of Health and Human Services: Washington, DC, 2011.
54. Scheel, L. D.; Lane, W. C.; Coleman, W. E. *Am. Ind. Hyg. Assoc. J.* **1968**, *29*, 41.
55. Garrett, R. *Acute Exposure Guideline Levels for Perfluoroisobutylene*; US Environmental Protection Agency: Washington, DC, 2010.
56. Lailey, A. F. *Hum. Exp. Toxicol.* **1997**, *16*, 212.
57. Lisochkin, Y. A.; Poznyak, V. I. *Combust. Explo. Shock* **2006**, *42*, 140.
58. Dmowski, W.; Flowers, W. T.; Haszeldine, R. N. *J. Fluor. Chem.* **1977**, *9*, 94.
59. Walker, H. W. *J. Phys. Chem.* **1926**, *31*, 961.
60. Karube, D.; Chaki, T.; Shiotani, Y.; Sugiyama, A. (Daikin Industries Limited). Process for Production of 2,3,3,3-tetrafluoropropene. International Patent WO2010/013795 A1, 2010.
61. Moore, G. J.; Massey, H. M. (Imperial Chemical Industries). Tetrafluoroethane Isomerization. US Patent 5,091,600, 1992.
62. Ohashi, M.; Kambara, T.; Hatanaka, T.; Saijo, H.; Doi, R.; Ogoshi, S. *J. Am. Chem. Soc.* **2011**, *133*, 3256.
63. Bedford, C. D.; Baum, K. *J. Org. Chem.* **1980**, *45*, 347.

64. Kotov, S. V.; Ivanov, G. D.; Kostov, G. K. *J. Fluor. Chem.* **1988**, *41*, 293.
65. Dindi, H.; Hagedorn, J. J.; Hung, M.-H. (Du Pont Dow Elastomers). Process for Manufacturing Diiodoperfluoroalkanes. US Patent 6,825,389 B2, 2004.
66. Ameduri, B.; Boutevin, B. *Well-Architectured Fluoropolymers: Synthesis, Properties, and Applications*; Elsevier: Amsterdam, 2003.
67. David, J.; Boyer, C.; Tonnar, J.; Ameduri, B.; Lacroix-Desmazes, P.; Boutevin, B. *Chem. Rev.* **2006**, *106*, 3936.
68. Krespan, C. G.; Wheland, R. C. (E.I. du Pont de Nemours). Perfluoroalkyl Sulfide Polymer Solvents for Fluoromonomer Polymerization. US Patent 5,286,822, 1994.
69. Brown, H. *J. Org. Chem.* **1957**, *22*, 715.
70. Carlsson, D. J.; Jensen, J. P. T.; Wiles, D. M. *Macromol. Chem. Phys.* **1984**, *8*, 79.
71. Pearson, R. G. *J. Chem. Educ.* **1968**, *45*, 581.
72. Ameduri, B. *Chem. Rev.* **2009**, *109*, 6632.
73. Ameduri, B. *Macromolecules* **2010**, *43*, 10163.
74. Roine, A. HSC Chemistry, Version 7.0; Outotec, Pori, 2007.
75. Meyer, B. *Chem. Rev.* **1976**, *76*, 367.

6

MOLECULAR SIMULATION OF FLUOROPOLYMERS

Armand Soldera, François Porzio, and Nasim Anousheh
Université de Sherbrooke

6.1 INTRODUCTION

Molecular simulation can be considered as an experiment performed in a computer through a code. It aims to reveal properties at the molecular level. It should be differentiated from molecular modeling which concerns the theoretical aspect of quantum chemistry [1]. Thanks to the progress in computers and performance of algorithms, molecular simulation has become a powerful tool to complement experiments and theories. Fluoropolymers are no exception to the rule, and leverage this progress. In the book edited by Hougham, Cassidy, Johns, and Davidson, "Fluoropolymers 2," in 1999, one section was dedicated specifically for "Modeling and Simulation" [2]. In the preface, the authors wrote: "Much progress has been made in understanding the sometimes confounding properties of fluoropolymers. Computer simulation is now contributing to this with new fluorine force fields and other parameters, bringing realistic prediction within reach of the practicing physical chemist." Molecular modeling of PTFE (Chapter 10), PVDF (Chapter 11), and the use of the graph theory to compute the dielectric constant (Chapter 12) were the subjects of the three chapters constituting the simulation section of the book. One could ask: What is the current situation of molecular simulation for fluoropolymers? This chapter is thus aimed at providing an overview of the applicability of this technique for studying fluoropolymers, revealing the relevance of this powerful tool. To achieve this goal, an outline of computational models that will be discussed along the text is firstly exposed

Handbook of Fluoropolymer Science and Technology, First Edition. Edited by Dennis W. Smith Jr., Scott T. Iacono, and Suresh S. Iyer.

(section 6.2). Quantum calculations and atomistic methods are briefly described. The following section deals with the study of polymerization of fluorine compounds (section 6.3), which can be explicitly carried out at the quantum level. Developments of force fields which form the core of atomistic simulation for fluoropolymers are discussed in the next section (section 6.4). The fifth section deals with calculations on a polymer of great importance among fluoropolymers, PVDF (section 6.5). The last part treats some general applications (section 6.6) benefiting from the use of computational tools: characterization, environment, fuel cells, and coatings.

6.2 OVERVIEW OF COMPUTATIONAL METHODS

The very essence of computational chemistry is to solve the Schrödinger equation to provide electronic and atomic properties of a physical system [3–6]. Solving analytically the Schrödinger equation can be actually done for the Hydrogen atom (and He^+, Li^{2+} ...), since it contains only one electron. The Heisenberg uncertainty principle specifies that position and momentum of an electron cannot be measured simultaneously. Accordingly, adding another electron to the studied system makes analytically unsolvable the electron–electron interaction, since their positions must be known. Computational methods are then entirely focused on the handling of approximations. Quantum computational methods lead to wave functions that are approximated solutions of the Schrödinger equation. Ultimately, such methods are no more available for complex systems, and the Schrödinger equation cannot be applied to solve every problem such as the glass transition phenomenon [7]. The studied system must thus be modeled differently. Atoms are then depicted as particles which interact with the other atoms through the use of a force field. This atomistic scale is governed by Newton's laws. As the number of atoms or the time covered by simulation increases, dissipative effects must be taken into account. This coarser scale used to treat microscopic behavior refers to the mesoscale domain and is governed by the Langevin equation [8]. This will not be discussed in this chapter. The focus is on scales where fluorine atoms are explicitly pictured, that is, the quantum computational and atomistic domain. These two approaches are briefly described in this section in order for the nontechnical reader to gain the essential knowledge to follow the text.

6.2.1 Quantum Calculation Methods

Quantum computational methods are mainly used in systems for which electronic properties are of interest, such as molecular orbitals, Nuclear Magnetic Resonance (NMR) spectra, and polarizability. Usually an optimization process is carried out [3]. It consists in finding the structure which exhibits the lowest energy. The prerequisite in manipulating approximations to solve the Schrödinger equation gives rise to different approaches. They can be roughly classified into three major types: *ab initio* (Hartree–Fock method and derivatives), density functional theory (DFT), and semiempirical methods, *ab initio* is a Latin locution which means "from the beginning." In this method, only fundamental constants are used and no experimental data are introduced to compensate for the use of approximations. The Hartree–Fock method is the most commonly employed. It is based on the mean-field approximation. Since Coulomb interactions, or electrostatic interactions, between two electrons cannot be computed, each electron is subjected to an electric field averaged over all the other electrons. To get the final wave function, a self-consistency procedure is applied [3]. Nevertheless, initial wave functions are necessary. Hydrogenoid wave functions are definitely chosen since they correspond to the exact solutions of the Schrödinger equation applied to the hydrogen atom. For practical reasons, these wave functions are replaced by linear combinations of Gaussian type orbitals (GTO), or Slater orbitals [3]. A notation for GTO has been introduced by Pople's group [9] to indicate the number of Gaussian functions used to describe the different hydrogenoid orbitals, forming a basis set [10]. The essential of the basis set consists of functions that fit the hydrogenoid atomic orbitals such as 6-31g. 6-31g actually means that six Gaussian functions depict the core electrons, and the valence electrons are represented by a combination of two functions (Double Zeta), which represent 3 and 1 Gaussian functions, respectively; g stays for Gaussian orbitals kind. Specific functions can then be added, such as polarization functions (d, p) to enhance directional effects, or diffuse functions (+) to take into account long-range effects due to the presence of electronic delocalization in the studied system. Post-Hartree–Fock methods, such as the Møller–Plesset perturbation theory (MP), are aimed at refining this approach but are more central processing unit (CPU) and time-consuming. To improve computer efficiency, the DFT method offers a very interesting alternative. Within this approach, the total energy is expressed in terms of the total electron density, instead of the wave function that depends on atomic position. This functional, that is, the function of a function, is computationally or experimentally settled. The hybrid B3LYP functional is the most commonly used [11, 12]. Hybrid means that it also contains a Hartree–Fock term. However, the choice of the functional must be carried out carefully [13, 14] since a universal function is not yet available. To further increase computer time efficiency, CPU time consumers such as multiplication of overlap matrices can be removed. These omitted data must then be balanced by experimental data such as the measured heat of formation. The mostly employed semiempirical codes are Austin Model 1 (AM1) and Parameterized Model 3 (PM3). In the rest of the text, the level of theory will be specified for each case study.

The choice of the appropriate level of approximation usually results from a compromise between accuracy and availability in CPU. In fact, to avoid excessive computational time, the limit consisting in the use of a complete basis set is seldom reached [15]. However, the overlap of incomplete basis sets between two neighboring atoms leads to the basis set superposition error (BSSE) [16], thus resulting in a lowering of the energy. The extent of this mathematical artifact actually depends on the level of approximation and the computed property. Validation of calculations is thus a very important issue. The use of an inappropriate basis set can actually lead to important discrepancies with experimental data. For instance, if the studied system contains an anion, diffuse functions must be added to the basis set in order to account for the occurrence of longer range interactions [17].

In conclusion, the main drawback in using quantum calculations is the prohibitive demand in CPU. This problem increases with the number of atoms. As a typical example, the computer time in the Hartree–Fock approach raises as N^4 (which is the total number of integrals that must be evaluated) [6], where N is the number of basis function used to described the studied system. For DFT, the computer time raises as N^3. DFT is generally employed for large systems. In practice, optimization can be carried out using a low level basis set. It can then be followed by calculation carried out at a higher level of theory, without the optimization step, leading to single-point energy. Alternatives also exist to address problems stemming from larger systems such as polymers, where long range and large time scale are relevant factors in computing properties. Thus, instead of considering electrons as wave functions, the atomistic approach depicts atoms as particles with electronic interactions imbedded in a force field.

6.2.2 Atomistic Methods

6.2.2.1 Force Field Consider an atom in a specific environment, for instance a fluorine atom linked to an sp^3 carbon atom. This environment can be defined by a particular surface of potential energy (SPE). Such an SPE does not change greatly if the same environment is found in another system or molecule. Following the previous example, the number of molecules possessing this link between the fluorine and the carbon atoms is unlimited. A force field is then aimed at representing as accurately as possible this SPE [18]. To achieve this goal, a force field consists of equations in which terms tend to grasp specific interactions, and parameters that stem from the fit of these equations to the different SPE. To accurately depict the SPE, experiments or quantum calculations are carried out. The equation is usually separated in two terms: bonded and non-bonded equations. The different variables which will be defined in this paragraph are schematically displayed in Figure 6.1.

The bonded term depends on the internal coordinates of the molecule, that is, bonds, valence angles, and dihedral angles. Accordingly, its simplest expression consists of the sum of all the terms of stretching, bending, and torsion found in a molecule. The usual expressions are briefly described.

✓ Stretching term:

$$E_{\text{stretching}}(d) = k_s(d - d_0)^2 \tag{6.1}$$

where d is the atomic distance at which the stretching potential energy is computed; k_s and d_0 are the spring constant and the value of atomic distance for which the stretching potential energy is zero, respectively.

✓ Bending term:

$$E_{\text{Bending}}(\theta) = k_\theta(\theta - \theta_0)^2 \tag{6.2}$$

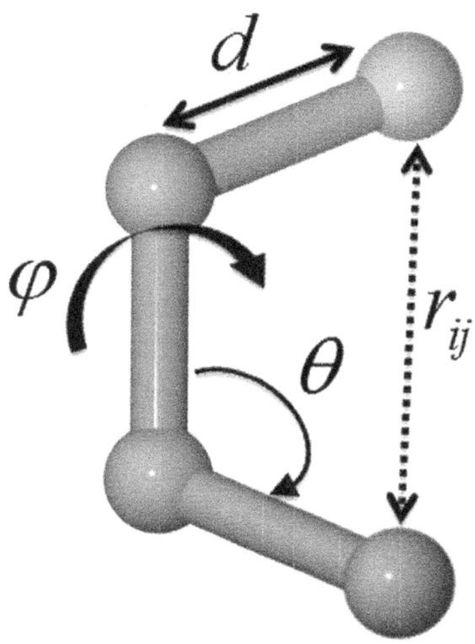

FIGURE 6.1 Representation of the variables discussed in Equations 6.1–6.5.

where θ is the valence angle at which the bending potential energy is computed; k_θ and θ_0 are the spring constant and the valence angle for which the bending potential energy is zero, respectively.

✓ Dihedral term:

$$E_{\text{dihedral}}(\phi) = \sum_{n=1}^{3} V_n \left[1 - \cos\left(n\phi - \phi_n^0\right)\right] \tag{6.3}$$

where ϕ is the torsion angle at which the dihedral potential energy is computed; V_n corresponds to the potential energy barrier, and ϕ_n^0 the phase. Other forms of the bonded term term exist, in particular the addition of cross terms, but they will not be developed in this chapter. For the interested readers, see, for example, works of Sun [19].

The non-bonded term expresses the interactions between atoms linked by three or more bonds, or from different molecules. It contains two terms.

✓ van der Waals interactions:

$$E_{\text{LJ}}^{ij}(r_{ij}) = \frac{A_{ij}}{r_{ij}^{12}} - \frac{B_{ij}}{r_{ij}^{6}} \tag{6.4}$$

where A_{ij} and B_{ij} are fitting parameters, and r_{ij} is the actual distance between atoms i and j. The equation displayed (Equation 6.4) corresponds to the Lennard–Jones potential: it is formed by a repulsive (in r_{ij}^{-12}) and a dispersive (in r_{ij}^{-6}) terms. While the latter term results from the London–Keesom–Debye interactions [20], the former term can have different forms. The r_{ij}^{-12} behavior can be substituted by an exponential decay, $\exp(-\rho r_{ij})$ to yield the Buckingham potential, or a less abrupt function, r_{ij}^{-9} [21].

✓ The electrostatic, or Coulomb potential energy:

$$E_{\text{Coulomb}}^{ij}(r_{ij}) = \frac{q_i q_j}{4\pi\varepsilon_o r_{ij}} \tag{6.5}$$

where q_i, q_j, and ε_o are the partial charges of atoms i and j, and the dielectric constant, respectively. The partial charges reproduce the first nonzero multipolar component.

Different equations are thus available to portray the SPE. The approach used to establish the SPE, from experiments or calculations, can lead to different fitting parameters. Moreover, due to the increasing computational efficiency, bonded cross terms can be added to better depict the SPE, leading to second generation force fields that cover a greater number of compounds. Thus they have a greater transferability, that is, the availability to transfer a force field for most systems. Accordingly, a great variety of force fields exists: PCFF, COMPASS, OPLS, MM2, MM4, UFF, AMBER, ESFF, CVFF, etc. [19, 22, 23]. Great efforts have been primarily fostered to formulate a reliable force field to accurately describe the SPE in fluoropolymers. Before discussing these strategies and their performance compared to more transferable force field in section 6.4, we are briefly exposing various ways of applying a force field. Monte Carlo method [24] and molecular mechanics consist of a direct use of a force field. With the latter approach, static properties such as energies [25], infrared spectra [26], and mechanical properties [27] can be computed. To reveal dynamics properties such as autocorrelation function [28, 29], and relaxation times [30], molecular dynamics (MD) is required.

6.2.2.2 Molecular Dynamics MD is a simulation technique that samples the conformational space dynamically [31]. Newton's equations of motion are applied to the studied system to yield a sequence of atomic positions as a function of time. This method is deterministic, unlike the Monte Carlo method where the conformational space is sampled stochastically [24]. To provide motion to atoms, the first equation of Newton where the force field appears is integrated. The state of the system is thus computed at a time step from the current state. The duration of this time step depends on the algorithm used to do the integration, and the SPE to be covered. Considering the widespread Verlet derivative algorithm and a full atomistic simulation, the time step is 1 fs, or 10^{-15} s [32]. This very low value is explained by the fact that the most rapid motion among atoms (without any external field), that is, the stretching between the carbon and hydrogen atoms which is in order of 10 fs, must be appropriately depicted. Relevant information can then be grasped from analyses of the time dependence of the studied properties. Integration of the Newton's equation is performed at constant energy (microcanonical ensemble). However, efficient algorithms exist to settle the system in the canonical ensemble (N, V, and the temperature T, are kept constant) or isobaric–isotemperature ensemble [31]. To run MD simulations and to exploit efficiently resulting data, knowledge in statistical physics is an essential asset [28].

6.3 POLYMERIZATION

Calculations used to reveal the polymerization are necessarily performed at the quantum level. It actually involves bond formation and rupture phenomena, which involve changes in the electronic environment.

6.3.1 Telomerization

Revealing the reaction pathway is an important issue in understanding a polymerization process. However, isolating the transition state remains an experimentally challenging task. Quantum chemistry combined with statistical physics concepts became an efficient tool to capture the underlying mechanism of any reactions. However, in the case of polymerization, the polymer chain length increases the level of difficulty. Studying telomerization offers a very interesting alternative. The radical telomerization of fluorinated olefins is experimentally well documented [33]. To complement those data, quantum calculations are of particular interest. Competition between the propagation and the transfer steps during the telomerization of vinylidene fluoride (VDF), with three chain transfer agents $Cl_3C{-}Z$ ($Z = H$, Cl, Br) was studied [34]. The most probable reaction pathway for each chain transfer was established at the DFT B3LYP/6-31(d,p) level of theory, using the intrinsic reaction coordinates (IRC) approach [3] (Figure 6.2a). This approach consists in starting calculations from the transition state, as well as steps toward reactants and products, through a minimum energy algorithm. However, far from the transition state, that is, for reactants and products, it was shown that energy values must be corrected for the BSSE. Relative internal energies of reactants, transition states, and products are thus reported versus the reaction coordinate in Figure 6.2. Internal energies of the reactants are set to zero and similar behavior as IRC results (Figure 6.2a) are observed. The transfer constant (TC), corresponding to the ratio between the transfer and the propagation rates was later computed. A very linear correlation between logarithm of experimental and simulated TC was revealed (Figure 6.2b). These results confirmed that the methylene side of the VDF is the active site in order for the telomerization to occur. The ability of a transfer agent to yield low molecular weight telomers can thus be predicted using this simulation technique.

6.3.2 Initiators

Influence of polymerization initiators have been studied using quantum chemistry methods. Several initiators have been tested in the copolymerization of tetrafluoroethylene (TFE) with functionalized norbornene derivatives. The semiempirical PM3 method was used [35]. Experimentally, it has been shown that the fluorinated initiator improves the alkaline solubility of the resulting fluoropolymer. The increased dissolution rate was thus argued to result from an increase in the acidity strength caused by the initiator. Accordingly, pKa have been computed for two compounds with different terminal groups, hydrocarbon and fluorocarbon. The simulated pKa was

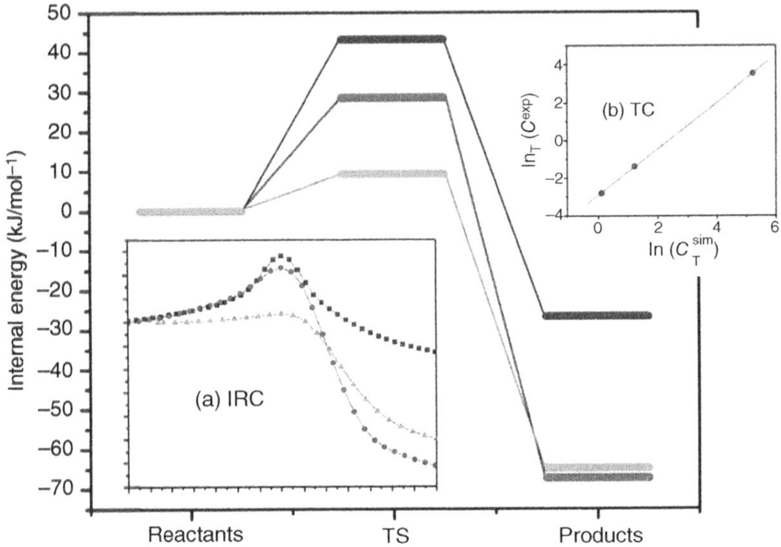

FIGURE 6.2 Relative internal energies corrected for BSSE, of reactants, transition states (TS), and products during the chain transfer between $CCl_3CH_2CC{\bullet}F_2$ radical and Cl_3C—H (■), Cl_3C—Cl (●), and Cl_3C—Br (▲) chain transfer agents. The oblique lines are used only as guidelines. Inset (a) is results from the IRC approach. Inset (b) displays the linear relationship between the logarithms of experimental and simulated TC.

6.0, and 5.0, respectively. The lower pKa for the fluorocarbon compound confirmed its role in increasing the acidity of the medium during polymerization.

Effects of intermediate and final products of diacyl peroxides decomposition in the initiation reaction of VDF polymerization were also studied using *ab initio* calculations. Diethyl peroxydicarbonate (DEPDC), trifluoroacetyl peroxide (TFAP), and diacetyl peroxide (AP) are polymerization initiators that decompose to yield intermediate radicals $C_2H_5OCOO{\bullet}$, $CF_3COO{\bullet}$, and $CH_3COO{\bullet}$, respectively. Hirunsit et al. were thus interested in ascertaining the role of these intermediate radicals in promoting the polymerization of VDF rather than the final radicals stemming from the complete decomposition of the polymerization initiators, $C_2H_5O{\bullet}$, $CF_3{\bullet}$, and $CH_3{\bullet}$, respectively [36]. Two approaches have been employed to disclose the reaction pathway between VDF and the six radicals. The first method consisted in the structural optimization of reactants, transition state, and products at the UHF/6-31g(d) level of theory, followed by a single-point energy calculation at the QCISD/6-311g(d,p). This latter level is very time-consuming. The second method involved the structural optimization at the B3LYP/6-31g(d) level of theory, followed by a single-point energy calculation at the B3LYP/6-311+g(3df, 2p) level of theory. Same regioselectivity, activation energies and reaction enthalpies have been computed independently of the method. Calculations showed that activation energies with intermediate radicals are smaller than with final radicals during the decomposition of the DEPDC, TFAP, and AP polymerization initiators. In conclusion, when the polymerization initiators DEPDC, TFAP, or AP are used, the addition reactions of intermediate radicals must not be neglected.

Regioselectivity during the addition step has been specifically examined. Addition of electrophilic $\bullet CF_3$ and nucleophilic $\bullet CH_3$ and $\bullet CH_2OH$ on four fluorinated propenes have been investigated using DFT B3LYP/6-31g(d,p) and the semiempirical PM3 level of theory. The propenes were $CF_3{-}CH{=}CH_2$, $CH_3{-}CH{=}CH_2$, $CF_3{-}CF{=}CF_2$, and $CH_3{-}CH{=}CF_2$ [37]. Agreement with experimental regioselectivity has been achieved by DFT calculations for the three former propenes. Both methods correctly characterized the $\bullet CF_3$ addition on $CH_3{-}CH{=}CF_2$. However, they yielded contradictory conclusions for the nucleophilic attacks. Unfortunately, for this last scenario, no experimental data can support any of these methods.

6.4 FORCE FIELDS FOR FLUOROPOLYMERS

The impressive development in molecular simulation stems from important improvement in both codes and computers. It made available accurate atomistic simulation of fluorine derivatives. Depiction of the SPE is most commonly achieved by quantum calculations. The major difficulty in describing interactions involving fluorine atoms actually lies in a correct description of the electrostatic effects [38]. Crystal unit cell dimensions [39, 40] and the thermal behavior of fluoropolymers [41–43] were thus originally difficult to reproduce. Dihedral potential energy (Equation 6.3) that plays a central role in the backbone dynamics was also incorrectly depicted. In this section, illustrative examples of force fields specifically dedicated to fluoropolymers and more transferable force fields are reviewed.

6.4.1 Force Fields for PTFE

6.4.1.1 Crystal Structure Holt et al. reported the development of two force fields for fluoropolymers with respectively a (6-9) and a classical (6-12) Lennard–Jones potential energy (Equation 6.4) [44]. Semiempirical AM1 calculations have been carried out on the perfluorohexadecane molecule. Force field parameters have been fitted in order to reproduce the helix deduced from crystallographic data analysis. Surprisingly, the best fit was obtained with no dihedral term (Equation 6.3). The van der Waals term (Equation 6.4) was then expected to compensate this shortcoming. Nevertheless, the lack of torsional barrier was undoubtedly unrealistic. *ab initio* calculations at the MP2/D95+(d) level of theory had been previously carried out on perfluoro-*n*-butane (PFB), perfluoro-*n*-pentane (PFP), and perfluoro-*n*-hexane (PFH) by Smith et al. [45]. They yielded more realistic values for torsional barriers and were thus used to scale the force field parameters. Despite the accuracy in the intramolecular geometry and intermolecular packing arrangements, the dihedral potential energy surface of fluoropolymers remained complex to establish.

Tokarev et al. computed the potential energy curves associated to the rotational barriers at a low level of theory. Quantum calculations were carried out at the semiempirical AM1 level of theory for geometry optimization and HF/STO-3G for single-point energies. They wanted to investigate the origin of the high free volume (and thus high permeability) of a 2,2-bis(trifluoromethyl)-4,5-difluoro-1,3-dioxole homopolymer (10 monomers) and its copolymers (8:1) with tetrafluoroethylene (Teflon AF® polymer) [46]. Careful construction and systematic investigation of the relationship between energy and the value of dihedral angles revealed a specific arrangement where two consecutive dioxole rings are successively coplanar and perpendicular. This arrangement leads to an increased stiffness in the lateral deformation of the polymer, thus enhancing the free volume.

6.4.1.2 Amorphous State Some force fields have been especially built up to deal with amorphous fluoropolymers [47]. The fitting parameters in the van der Waals term have been deduced from a fit of Equation 6.4 in order to reproduce the experimental density and heat of vaporization of four perfluoroalkanes (C_3F_8, C_4F_{10}, C_5F_{12}, and C_6F_{14}). Bonded fitting parameters have been established through the usual fitting procedure with *ab initio* calculations carried out at the HF/6-31g level of theory. To extract the dihedral and 1–5 non-bonded parameters, the central bond in C_4F_{10} was employed. The effective 1–5 non-bonded term was used to capture the steric hindrance that occurs between two fluorine atoms separated by four bonds. For these two parameters, the energy of the gauche conformation for which the torsion angle is 60° was inadequately characterized using the HF/6-31g level of theory. Data computed at the MP2/D95+(d) level of theory from Smith et al. [45] showed better results, and were thus employed for this particular torsion angle. MD simulations for 10 ns long were then carried out on PTFE polymers described by this force field. Density and radial distribution function (RDF) compared favorably with experimental data.

Coulomb potential energy involves long-range interactions since it is proportional to r^{-1} (Equation 6.5). In primary calculations, this term was thus eliminated to gain in computational efficiency [48]. Non-bonded interactions were then reduced to a Buckingham potential for repulsion, and the r^{-6}dispersion term. To get the fitting parameters of the repulsion term, calculations at the MP2/aug-cc-pvDz level of theory, followed by single-point energy calculations at the HF/aug-cc-pvDz level of theory, have been carried out for perfluoromethane dimers. The intermolecular energy was corrected for the BSSE using the counterpoise procedure. The dispersion parameters were scaled to match the experimental density of the C_7F_{16} liquid at 25°C. To get the bonded parameters, calculations at the B3LYP/aug-cc-pvDz, B3LYP/D95+(d), and MP2/aug-cc-pvDz levels of theory have been performed on conformers of C_4F_{10} and C_5F_{12}. The ensuing force field was first applied to compute the conformational energy behavior of C_6F_{14}. The resulting agreement with experimental data supported the transferability of this force field to longer perfluoroalkanes. MD simulation of C_4F_{10}, C_5F_{12}, C_6F_{14}, C_7F_{16}, C_9F_{20}, and $C_{20}F_{42}$ liquids were thus carried out under NPT (number of atoms, pressure, and temperature are kept constant) conditions. Relevant results are summarized hereafter. Experimental densities were correctly predicted. Static structure factor of C_9F_{20} showed excellent agreement with the experimental X-ray diffraction pattern. Conformer populations were consistent with populations extracted from Raman spectroscopy data. Moreover, the ortho conformer was proved to be energetically stable, as it was demonstrated by *ab initio* calculations from the same group [48]. This conformation stability was not observed using a more generalist force field, OPLS-AA [22]. Moreover, the torsional autocorrelation function computed with OPLS-AA revealed torsional dynamics slower than with the specific force field. Such differences can have great impacts in simulation of high molecular weight PTFE dynamics.

6.4.2 Force Fields for PVDF

6.4.2.1 Crystal Structure Karasawa and Goddard developed two force fields for PVDF. The MSXX force field derived from a previous Hessian biased force field. It has been built up to accurately reproduce experimental frequencies of the form I, or β, crystal [49]. It is based on the rigid ion model. Charges are centered on atoms characterizing the polar nature of the covalent bonding in PVDF. Calculations were performed at the Hartree–Fock 6-31g(d,p) level of theory to yield the force field parameters. The second force field, MSXXS, is actually the covalent shell model [50] which has been developed to account for electronic polarization calculation. The two force fields have been compared and nine stable structures of PVDF crystals have been revealed. They include the four structures experimentally observed: β, α, δ, and γ forms. The computed Young's modulus showed a higher value than the experimental one. The authors explained this discrepancy by the difficulty to experimentally get pure crystal from samples. Differences in the dielectric properties were then discussed, but could not be compared to experimental data. Carbeck et al. slightly modified these force fields, MSXX* and MSXXS*, to evaluate the long-range electrostatic and dispersive interactions [51].

6.4.2.2 Amorphous State The rotational isomeric state (RIS) model accurately describes conformations of polymer chains [52,53]. The strength of such an approach lies in the fact that complicated calculations involving multiple integrals are reduced to matrix multiplications. Partition function, averaged properties such as the radius of gyration, and the characteristic ratios, can be acquired. For this purpose, instead of considering the whole graph of the potential energy versus the torsion angle of one backbone bond, only rotameric states are taken into account. These states correspond to wells in the potential energy curve (Equation 6.3). In polyethylene, there are three rotameric states, trans (t), gauche (g), and gauche − (g−), that are located at 180°, 60°, and −60°, respectively. The actual probability of appearance of one rotameric state along the polymer chain depends on the rotameric state of the adjacent bond, resulting in the formation of matrices to take into account the most probable conformations. An accurate depiction of small-range interactions, the dihedral interactions, is therefore essential, as it has been outlined in section 6.4.1.2. The RIS matrices for PVDF have been computed at the SCF/D95(d,p) level of theory for the energy optimization of four PVDF oligomers, 1,1,1,3,3-pentafluorobutane (PFB), 1,1,1,3,3,5,5,5-octofluoropentane (OFP), 2,2,4,4-tetrafluoropentane (TFP), and 2,2,4,4,6,6-hexafluoroheptane (HFH) [54]. The MP2/D95(d,p) level of theory has been used for TFP [54]. Such calculations led to the building up of a force field [55]. MD has then been carried out. The simulated density and equation of state were found in good agreement with experimental data. The RIS matrices of PVDF, PTFE, and other polymers can be retrieved from Reference [56].

6.4.3 Specific Force Fields in Explosive Materials

Some force fields are designed specifically to address one problem for one kind of materials. Due to the importance of fluoropolymers as binders and plasticizers in high energy explosive (HEE), a distinctive force field has been built up [57]. Performances of HEE material actually depend on both explosive component, and binder. A force field was thus designed to describe interactions between an HEE, the octahydro-1,3,5,7-tetranitro-1,3,5,7-tetrazocine (HMX), and nine fluoropolymers, PVDF, PTFE, polychlorotrifluoroethene (PCTFE), polyhexafluoropropene (PHFP), four PVDF/PCTFE mixtures, and a PVDF/PHFP mixture (Viton-A®). The purpose of this study was to compute mechanical properties, which are an essential issue in explosive applications. The resulting force field consists of three sets of potential energy. One describes the SPE of HMX [58]. The second is the COMPASS force field to describe the SPE of fluoropolymers [19]. Lastly, interactions between HMX and fluoropolymers have been described using DFT GGA (Generalized Gradient Approximation) level of theory with the Projector Augmented Wave (PAW) pseudopotential [57]. The PAW method allows for greater CPU efficiency by using a pseudopotential with the DFT level of theory [59, 60]. These latter non-bonded terms correspond to a combination of different kinds of potential energy: Lennard–Jones, Buckingham, and Morse. MD simulations have then been carried out at 300 K on cells containing 70,000–130,000 atoms under periodic boundary conditions [32]. Mechanical properties have been computed: curves of energy versus strain and mechanical parameters for high speed loading (detonation). The plasticizing ability of fluoropolymers to HMX was also highlighted.

6.4.4 Current Status

Some existing general force fields have been built to take into account fluorine interactions. The OPLS [22], and COMPASS [19] force fields are presently used to treat fluorine molecules. The COMPASS force field was employed to compute the glass transition temperature (T_g) and elastic properties of PVDF. Simulated data have been found to match experimental data [61]. It was also used to simulate T_g of CTFE/VDF random copolymers and homopolymers [62].

To cover a greater phase space during an MD simulation, time step should be increased. For this purpose, rapid motions, especially C—H stretching, must be considered differently. In fact, the force field is modified in order to take into account those degrees of freedom which relax rapidly. Such a force field is referred to United Atoms (UA) force field. Murofushi et al. thus proposed a UA where CF_2 and CH_2 groups were substituted by a single atom, C_F^* and C_H^*, respectively [63]. The number of atoms and non-bonded interactions were thus reduced, and the time step was then of 3 fs. The studied fluoropolymers were PTFE, PVDF, and ETFE. The simulated glass transition temperature (T_g) was found in good agreement with the experimental T_g.

6.5 STUDIES ON PVDF

6.5.1 Electric Properties

Electric properties are an essential asset of fluoropolymers, in particular for PVDF. With the recent technological boom of touchpad, this polymer found additional applications in electronic haptic technology.

6.5.1.1 Ferroelectricity The very ferroelectric property of PVDF actually stems from a change in the crystal structure [64] and is thus, of relevant importance. As outlined in section 6.4.2, some force fields have been especially built up to address this issue. From DFT calculations and analysis of the cohesive energies, it has been shown that PVDF exhibits nine energetically stable crystalline forms [65]. Theoretically, they have been shown to be stable under specific pressure and temperature. However, only four of them have been experimentally observed. Among these crystal forms, the β-phase is particularly attractive from an electrical viewpoint. It exhibits a noncentrosymmetric unit cell [66] yielding a ferroelectric character in aligned macroscopic films. The existence of a spontaneous polarization under the Curie temperature is a characteristic of PVDF in the β-phase. By applying an electric field, switching between the two states of polarization is up to a few nanoseconds depending upon the strength of the applied field [39].

Atomistic simulation has also been carried out to calculate electric properties of PVDF. Two force fields, MSXX and MSXXS, were specifically built up to address this issue [49]. Dielectric constants and dielectric loss were thus computed for both the crystalline and amorphous phases of PVDF. In agreement with experimental data, the ensuing computational results showed that the amorphous phase exhibits a higher dielectric constant than the crystalline phase. This behavior arises from rapid changes in the torsion angles, leading to rapid modulation of the dipole moment perpendicular to the chain axis. It is stimulated by the presence of soliton-like defects that diffuse along the chain, as it was revealed by MD simulation [49].

The crystal polarization was especially investigated using the MSXX* and MSXXS* force fields [51]. It was shown that the repeat unit dipole increases by 0.9 Debye when it is computed from a single chain in vacuum to a chain packed in the crystal. By carrying out MD simulation, it was also observed that temperature leads to a decrease in the polarization. Piezoelectricity and the effect of trapped charges on the ferroelectric switching process have been investigated using a general force field MM+ from the HyperChem package [67, 68]. This force field is a modified version of the general MM2 (Molecular Mechanics 2) force field [18]. The simulated piezoelectric coefficient in the β-phase was found to be in good agreement with experimental data.

6.5.1.2 Electronic To have access to electronic properties, electronic structure and band symmetries, atomic orbitals approximation are no longer valuable [69]. Duan et al. used the full-potential linear-augmented plane-wave (FLAPW) method and the exchange-correlation potential treated in GGA, to reveal electronic properties of β-PVDF crystal [70]. The isotropic continuum dielectric theory was also shown to be unreliable in investigating the β-PVDF polarization [71]. Using the FLAPW method, band gap, dispersions, and band symmetries were in agreement with experimental data. Moreover, electronic structure of the copolymer P(VDF–TrFE) was also investigated, and revealed that the presence of –(CHF—CF_2)– moiety does not change the major features of the band structures.

6.5.2 Polymorphism in Crystalline PVDF

Melting of α- and β-PVDF forms has been simulated by Erdtman et al. [72] using the force field developed by Byutner and Smith [54]. The melting temperature, T_m, of β-PVDF was found to be higher than that of α-PVDF. It was observed that melting at the ends of the polymer chains precedes melting of the bulk crystal structure. Melting of α-PVDF initially occurs via transitions between the two gauche dihedral angles (g ↔ g′) followed by transitions between trans and gauche dihedral angles (t ↔ g/g′). Melting of β-PVDF begins with t → g/g′ transitions and through transitions from β-(tttt) to α-(tgtg′) quartets.

As outlined in section 6.5.1., the crystalline PVDF β-phase is needed for ferroelectric properties. However, it is not the predominant crystalline form. The nonpolar α-phase crystal structure is generally formed by cooling PVDF from the melt. Several experiments have therefore been deployed to get the β-phase. Imran-ul-haq et al. experimentally showed that end groups influence the form of the crystal phase, and the degree of crystallinity [73]. Molecular simulation then becomes an essential asset to reveal such an impact. Present studies are carried out in our group to address this goal. Using the *pcff* force field, we developed a protocol to deal with polyethylene nanocrystals [74]. Preliminary results about the stability of α- and β-phases according to iodine end group are shown in Figure 6.3. To reveal the melting temperature (T_m), nanocrystals of oligomer chains of PVDF were first built up. MD was then carried out. By reporting the van der Waals potential energy with respect to the temperature,

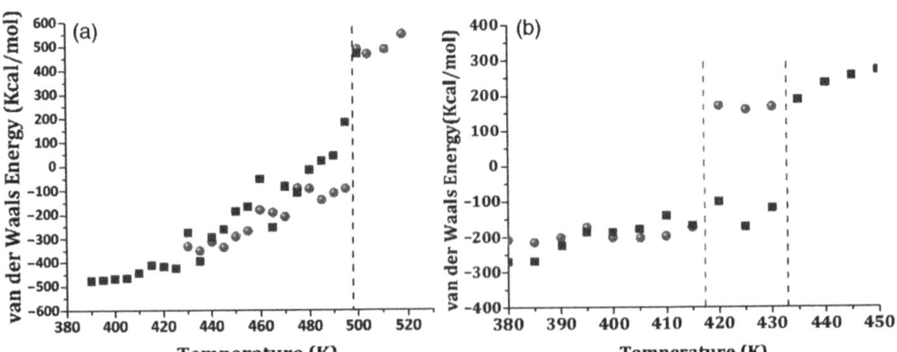

FIGURE 6.3 van der Waals potential energy with respect to the temperature for the α and β phases in (a) and (b), respectively, for pure nanocrystals (●) and 2.5% of iodine end groups (■).

the temperature at which a leap occurs corresponds to T_m (Figure 6.3). Behavior of pure PVDF nanocrystals (Figure 6.3a) and PVDF nanocrystals with 2.5% of iodine atoms at the end of the chain, that is, on the surface, can then be compared. The presence of iodine does not affect the value of T_m for the α phase, while an increase of 15 K is observed for the β phase. Further analyses are presently carried out to address this difference, and compare it to experimental data.

Experimentally, the simplest and most effective method to get the β phase is through mechanical drawing of the α phase. MD simulations were carried out to study the mechanism inducing a phase change from α- to β-PVDF using the COMPASS force field [62]. A strain was applied to the α-PVDF crystal along the molecular axis. It was found that a strain of 1.0475 is enough to enable such a change. Nevertheless, the ensuing β-PVDF phase exhibits molecules in antiparallel direction. The transformation of this antiparallel β-PVDF to the real β-PVDF structure can be induced by poling or by thermal annealing. It was also shown that, when no strains are applied to the α-PVDF crystal, a rather large electric field (at least 1010 V/m) is required to induce the crystal change. Direction of such an electric field is perpendicular to the molecular axis.

6.5.3 Presence of Defects

6.5.3.1 Crystals PVDF offers the particularity to possess along its backbone chain a relatively important percentage (5–10%) of regiodefects, that is, head-to-head or tail-to-tail connectivity, instead of the common head-to-tail propagation. Correia et al. used a protocol that combines the semiempirical approach under the complete neglect of differential overlap (CNDO) approximation with MD simulation where atomic forces are computed self-consistently. A decrease in the dipole moment was pointed out by increasing the concentration of defects, and no real changes in the molecular polarizability were noticed [75]. By lengthening the polymer chain, they also showed that the dipole moment reaches a constant after a certain chain length; it actually increases for β phase and decreases for the α form. No significant changes were observed in the molecular polarizability.

Any defect-mediated relaxation or transition entails two phenomena: the formation of a defect within the lattice and the propagation of this defect through the lattice. To confirm the specific dielectric behavior of the α-PVDF chain due to the inversion defects, Clark et al. computed the energy required to create a defect in a single chain. The calculated value was to the tune of 9 kcal/mol [76]. Such a value is clearly higher than the activation free energy computed by Miyamoto et al., that was less than 1 kcal/mol [77]. Despite the fact that both energies cannot be strictly compared, Clark et al. concluded that these defects cannot explain the particular dielectric behavior observed for α-PVDF.

6.5.3.2 Amorphous State Glass transition is a complex phenomenon that still remains unclear. It was actually quoted by *Science* magazine as "among the 100 most important questions in science" [78, 79]. Great experimental, theoretical, and simulation efforts have been devoted to give insight into this intricate problem [80].

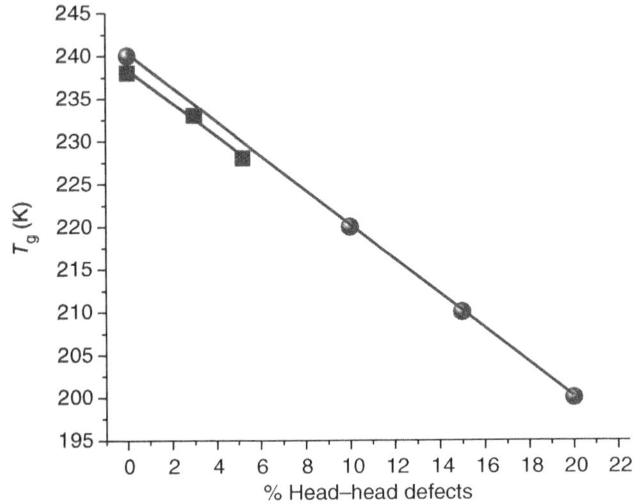

FIGURE 6.4 Experimental (■) and simulated (●) T_g with respect to the percentage of head–head defects.

Atomistic simulation gives an opportunity to study small molecular change that can affect the glass transition temperature (T_g). However, this property enters into the regime for which associated length scales could exceed the range of possibilities available with atomistic simulations [81]. Nevertheless, our group developed a protocol based on a reduced and guided representation of the configurational space, and mechanical equilibrium [82]. We thus obtained a linear relationship between experimental and simulated T_g [83].

Once the glass transition is accurately simulated, dynamical heterogeneities and relaxation times can be captured from MD trajectories. They are key parameters in polymer physics for better understanding glass-forming liquid systems. Moreover, they can be applied to reveal the presence of defects along the polymer chain. Our group recently studied the effect of regiospecific defects on the value of T_g [84]. Different percentages have been analyzed: 0, 10, 15, and 20%. In Figure 6.4, simulated and experimental T_g are shown in a same graph with respect to the percentage of regiodefects [85–87]. A nice correlation was observed between experimental and simulated data giving further insight on the role of inversion monomer defect on relaxation process.

A general procedure to address the influence of the regiodefects has been carried out by Carbeck using atomistic simulations. The force fields that have been used are MSXX* and MSXXS* [51]. Through the calculation of an adiabatic reaction pathway for a defect moving between two monomers, free energy was computed [88]. By revealing the reorientation of the dipole moment along the chain, it was shown that the divergent transition, where the dipole points away from the defect on either side of the defect point, moves along the chain by forming short, all-trans segment.

6.6 APPLICATIONS

6.6.1 Characterization Tool

*6.6.1.1 **NMR Investigations*** Interpretation of NMR spectra is an important issue in characterizing molecules. Calculations then become an efficient tool to assist experimentalists in such analysis. In the following examples, simulated NMR shielding constant have been computed using the gauge including atomic orbital (GIAO) scheme [89]. This approach is necessary to properly account for the magnetic field effects.

Experimental ^{19}F NMR spectrum of poly(trifluoroethylene) (PTrFE) reveals two axially symmetrical shielding anisotropy patterns, for the CF_2 and CFH moieties, respectively. ^{19}F NMR spectra have thus been computed using *ab initio* calculations, HF/3-21g and HF/6-31g(p) on two monomers with CF_3 end group, and with different configurations: syndiotactic and isotactic. The computed spectra have been compared to the experimental spectrum measured at 23°C, and thus close to T_g (31°C) [90]. Some discrepancies have been identified. The authors argued that these are due to the rapid averaging mode of motion occurring at ambient temperature. Thereafter, the length of the C—H bond distance was probed from experimental NMR data. The analysis highlighted the fact that the molecular motion at ambient temperature is an important issue that must definitely be addressed.

NMR spectra of copolymer have also been computed. NMR spectra from copolymerization of chlorotrifluoroethylene with ethyl vinyl ether (PCCE) have been experimentally characterized by Carnevale [91]. NMR spectra revealed an important complexity essentially due to the high sensitivity of ^{19}F shielding to stereochemical details. Carnevale computed NMR spectra of four different configurations to corroborate his signal assignment. Influence of neighboring monomers on central monomers and the effect of the inversion of the chiral centers have been inspected using DFT B3LYP/6-31+g(d) level of theory. Only the shielding from the central monomer was isolated. NMR shielding data were obtained through the GIAO and the continuum set of gauge transformation (CSGT) [92] schemes, with the same level of theory as the optimization. Both schemes gave similar results. The simulated data supported the experimental signals assignment and the four simulated configurations nicely mimicked the actual stereochemistry.

Quantum calculations have also been used to support the assignment of several different signals in the experimental spectrum of an amorphous cyclized perfluoropolymer, CYTOP® (from Asahi Glass Co.), which possesses a T_g at 108°C. Its NMR spectrum has been computed using the DFT B3LYP/6-31g(d) level of theory [93]. Four models, which represent four possible stereoisomers, have been specifically analyzed. ^{19}F and ^{13}C magnetic shielding constants were then computed. They compared favorably to the experimental chemical shifts measured at 35°C. This accuracy confirmed the original assignment of the NMR signals. Moreover, the effective experimental C—F distance deduced from experimental NMR data was found only slightly longer than the computed distance. In agreement with the fact that experimental probing was carried out far from T_g, there were no averages stemming from any molecular motion. This is opposite to the study case of Reinsberg et al. [90] in which T_g was much closer to the probing temperature, and thus more molecular motions were likely to occur.

Unlike other calculations based on clusters, Skachkov et al. developed a method to calculate NMR chemical shifts in periodic systems. They applied their technique to a series of organic compounds, using a specific functional and TZ2P level of theory [94]. Two fluoropolymers have been studied, PTFE and PVDF. Their structures originated from experimental X-ray data. Their chemical shifts have been referred to F_2, and $CFCl_3$ signals, respectively. From a calculation viewpoint, it was shown that calculations on three monomers inserted in a cell with periodic boundary conditions yielded comparable results to that of longer chains, that is, $F{-}(CF_2)_{28}{-}F$ and $H{-}(CH_2{-}CF_2)_9{-}CH_3$. However, important discrepancies with experimental chemical shifts have been observed. The authors argued that the level of theory was not sufficient for this particular kind of compounds, since calculations on other structures agreed well with experimental data. Nevertheless, from a computational perspective, the method is very promising.

6.6.1.2 ***Surface Treatment*** Lee et al. studied graphene transfer with the CYTOP® fluoropolymer as the supporting and doping layer [95]. The purpose was to develop a new method to simultaneously transfer and dope chemical vapor deposition-grown graphene onto a target substrate. Owing to this method, a strong graphene doping was obtained. To investigate the origin of this doping, *ab initio* calculations have been carried out. More specifically, the possibility that the doping comes from the contact between CYTOP® and graphene was studied. A CYTOP® chain of five monomers length was adsorbed on an 8×8 graphene sheet. The geometry of the CYTOP® chain was optimized at the semiempirical PM3 level of theory with a double-zeta polarized basis set (5d, 7f). The vdW–DFT and optB86b-vdW functionals were coupled with the PAW method, to calculate the band structure of the adsorbed chain on graphene. Conclusions stemming from calculations were that doping cannot result from the contact between the CYTOP® and the graphene.

6.6.2 Environment

Synergy between fluoropolymers and their environment can be identified according to two very opposite aspects: green chemistry where they are solutions to problem, and environmental chemistry where they are in some way the source of the problem.

6.6.2.1 ***Green Chemistry*** Green chemistry is a very active domain of research. Finding compounds labeled "Green Chemistry" is a rapidly growing field [96]. Quantum calculations can actually provide a strong role in assisting the design of such molecules. A meaningful example involves fluorinated molecules and their interactions with CO_2. To dissolve a compound in CO_2, it was originally stated that fluorination of molecules was the key point. However, this overly simplified assertion was called into questions [97]. The solubility actually depends on the quantity of fluorine as well as on the kind of molecule which has been fluorinated. The use of *ab* *initio* simulation to complete experimental data can thus lead to a better understanding of the solubility phenomenon. In fact, computational investigations in characterizing interactions between fluorine derivatives and CO_2 are a textbook case for paying special attention to the level of theory to consider.

Experimentally, it is known that ethane and hexafluoroethane are both miscible in CO_2. However, for high molecular weight (>500 kDa), some fluoropolymers are readily miscible in CO_2 while their hydrocarbon homologues are not. In the pioneering work of Cece et al., *ab initio* simulations at the HF/6-31g(d,p) level of theory have then been carried out in CO_2-polymer systems to study differences in solubility [98]. Two simple models have been examined: $(CO_2)_n/C_2H_6$ and $(CO_2)_n/C_2F_6$. Within each model, CO_2 molecules have been added one by one for up to four. After molecular optimization, the interaction energy was computed. It represents the difference between the energy of the whole system and the sum of energies of its molecules. Considering the hydrocarbon system, calculations showed that interactions between C_2H_6 and CO_2 molecules are small. Interactions between CO_2 molecules thus govern the behavior of the whole system giving rise to the poor solvation. Conversely, systems containing fluorocarbons and CO_2 molecules are strongly governed by solvent/solute interactions. CO_2 molecules organize themselves around the C_2F_6 molecule in a solvation shell. The high miscibility of fluoropolymers with CO_2 can thus be explained in terms of these favorable CO_2/fluorocarbon interactions. Nevertheless, some experimental results showed inconsistency with those conclusions [99], and optimizations at two different levels of theory HF/6-31g(d) and MP2/aug-cc-pVDZ, were carried out [100]. Several higher levels of theory and basis set were also used for single-point energy calculations. Diep et al. compared the results of raw, that is, without correction, single point energies, and BSSE corrected single-point energies. In addition to the previously studied systems, $(CO_2)_n/C_2H_6$ and $(CO_2)_n/C_2F_6$ [98], CO_2/CH_4 and CO_2/CF_4 systems have also been investigated. Value of the BSSE was found to be greater than the binding energies computed using HF/6-31g(d). The higher level of theory, MP2/aug-cc-pVDZ, with correction for the BSSE, revealed that the binding energies are usually slightly more attractive for the hydrocarbon model than for the fluorocarbon model. Accordingly, there are no specific CO_2/fluorocarbon interactions, comparatively to CO_2/hydrocarbons interactions. A simplified model for fluoropolymer did not capture requested information to unveil the enhanced solubility of fluorocarbons in CO_2.

Although some fluorocarbons exhibit high solubility in CO_2, this is not the case for PTFE which is insoluble. However, the random copolymer made up of PTFE and poly(vinyl acetate) (PVAc), a poorly CO_2-soluble polymer, becomes CO_2 soluble over a wide range of composition [101]. The CO_2 solubility of the copolymer reaches a peak for a TFE molar concentration of 19.3%. To better understand such a behavior, interaction energies have been computed using an *ab initio* approach. For computational reasons, the whole polymer chain could not be investigated since CPU time would be prohibitive. All calculations have been carried out using the second-order Møller–Plesset perturbation theory (MP2) method. Although the 6-31+g(d) basis set have been used for geometry optimizations, single-point binding energies have been computed with the larger aug-cc-pVDZ basis set. The binding

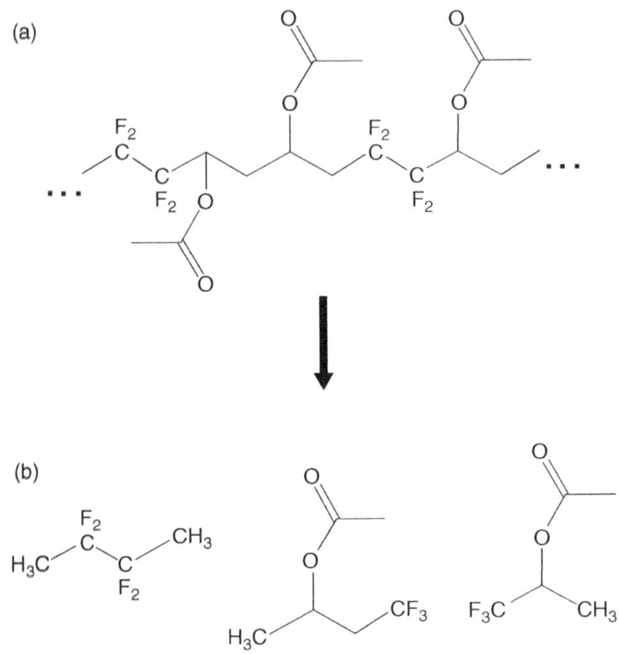

FIGURE 6.5 (a) Copolymer TFE-co-Vac. (b) Three fragments capturing the features of the copolymer (from left to right): tetrafluorobutane (FB), trifluoro-sec-butyl acetate (TFSBA), and trifluoroisopropyl acetate (TFIPA).

energies were corrected for BSSE. Since the approaches were very time-consuming, polymer fragments have been appropriately selected. The TFE-co-VAc copolymer (Figure 6.5a) chain was split into three small fragments (Figure 6.5b); tetrafluorobutane (FB), trifluoro-sec-butyl acetate (TFSBA), and trifluoroisopropyl acetate (TFIPA)). Each fragment exhibits a feature of the original copolymer. The scheme shown in Figure 6.5 is a very instructive example of an efficient use of *ab initio* calculations. In order to identify contributions of the fluorine atoms in the solubility behavior, the hydrocarbon homologues of each representative fragment were also simulated. Their interactions with CO_2 molecules have been analyzed. The enhanced CO_2 solubility was accounted for by structural features of the fluorine-containing dimers. More specifically, TFIPA and TFSBA dimers and their hydrocarbon homologues exhibited three similar binding configurations. Nevertheless, the fluorinated molecules displayed an additional stable binding configuration that was not observed for hydrocarbon molecules. Another key for the difference in the solubility between the two series of molecules came from the presence of more acidic hydrogens in fluorinated backbones, promoting stronger hydrogen bonding with the oxygen atom of CO_2. The CO_2 molecule acts as both a Lewis acid and a Lewis base when it interacts with the fluorinated copolymer backbone which also behaves as Lewis acid (H—O) and Lewis base (F—C) moieties.

6.6.2.2 Environmental Chemistry Long-chain perfluoroalkyl carboxylic acids (PFCAs, $C_xF_{2x+1}COOH$, where $x = 6–12$) have been observed in fish and mammals [102]. It is important to clarify the origin of this environmentally impacting compound. The presence of these fluoro-oligomers results from the degradation of perfluorinated aldehydes ($C_xF_{2x+1}CHO$) under atmospheric oxidation. To unveil the mechanism of degradation, the experimental IR spectra of the products of the Cl-initiated oxidation of $C_xF_{2x+1}CHO$ ($x = 1–4$) were studied by Andersen et al. The experiments have been conducted in the presence of NO_2, and under UV radiation. Concomitantly, the authors used the DFT B3LYP/6-311g(d,p) level of theory to compute the IR spectra of members of the $C_xF_{2x+1}C(O)OONO_2$ series ($x = 1–4$) [103]. Simulated frequencies were generally found in agreement with experimental ones. Accordingly, the nature of the experimental products could be evaluated giving insight into the actual mechanism underlying the atmospheric oxidation. The same way, an hydrate ($C_xF_{2x+1}CH(OH)_2$), obtained by adding a water molecule to a perfluorinated aldehyde reacted with OH radicals to yield a perfluorocarboxylic acid (PFCA). Further DFT studies have been performed on four $C_xF_{2x+1}CH(OH)_2$ ($x = 1–4$) hydrates [104]. Geometry optimizations have been done at the B3LYP/6-31g(d) level of theory, and IR spectra have been computed at the B3LYP/6-311+g(d,p) level of theory. Conformational isomers have been extensively explored. The relative abundance of each conformer was considered in order to merge the different computed IR spectra into one IR spectrum. This resulting IR spectrum accurately fitted the experimental IR spectra.

Simulation also offers a route to study the impact of the environment onto fluoropolymers. The atmosphere in the outer space, more precisely in the low earth orbit (LEO), is mostly constituted of neutral atomic oxygen (NAO). This highly reactive compound is probably at the origin of the degradation of materials used on the outside of spacecraft, in particular the degradation of fluorinated ethylene propylene (Teflon® FEP). These fluoropolymers are extensively used for spacecraft thermal control. Due to the numerous factors involved in such reaction, simulation offers the opportunity to isolate the different contributions. Gindulytė et al. employed HF/3-21g and MB3LYP/6-31+g(d,p) methods, to investigate damages on fluoropolymers, C_2F_6, C_3F_8, and C_5F_{12}, considering the collision of NAO with an incident energy compatible with LEO conditions [105]. Calculation revealed that the chain-breaking activation energy was equivalent for C_3F_8 and C_5F_{12}. The authors argued that same conclusions can be applied to larger fluorocarbons. They also found agreement between the activation energy of these bond-breaking reactions and the translational energy of the oxygen atom in LEO conditions. They concluded that degradation of fluoropolymers by NAO alone is likely to occur.

6.6.3 Fuel Cells

Proton exchange membranes (PEMs) are used in fuel cells to ensure transport of protons between the anode and the cathode. The perfluorinated polymer Nafion® (from DuPont) is extensively used as a PEM. This copolymer possesses a PTFE backbone and randomly distributed fluorinated ether side-chains each terminated with a sulfonic acid group. The hydrophilic component of the Nafion® side-chains controls the transport of protons through the membrane while its structural stability is governed by the PTFE backbone [106–110]. Characterizing its morphology remains an active area of research due to its high industrial potential [111]. Molecular simulation plays a significant role in such an investigation. However, an exhaustive review is out of the scope of this chapter. The purpose of this paragraph is rather to reveal the importance of molecular simulation in mimicking reality, leading to a better understanding of molecular properties through simple and efficient models. The works of Habenicht et al. are a very interesting example of this approach [112]. They studied the hydrophilic channel of the Nafion® membrane at low water content, using *ab initio* molecular dynamics (AIMD) with a Perdew–Burke–Ernzerhof (PBE) [113] exchange-correlation functional coupled with the PAW pseudopotential. These channels are in fact modelized as single-wall carbon nanotube (CNT) functionalized with CF_2SO_3H group. The use of CNTs is a very interesting approach. It offers the opportunity to mimic both the hydrophobic and hydrophilic domains of Nafion®, while investigating factors that influence the dissociation and the proton transport. These factors are the channel diameter, the relative position of the sulfonic group with respect to the main chain and the degree of hydration. This CNT model offers an alternative to circumvent experimental difficulties in separately probing these relevant factors. Different diameters for two series of functionalized CNTs (1.12, 1.32, and 1.41 nm) have been simulated. The difference between the two series lies in the inner wall. To reveal hydrophilic and hydrophobic characters, inner walls of the CNTs were either functionalized with fluorine atoms (more hydrophilic) or bare (more hydrophobic). Atoms of the CNTs walls were kept fixed while inside molecules, CF_2SO_3H and water molecules, were exposed to a MD at 300 K for 35 ps. Analysis of hydrogen bonding was then carried out. The ensuing conclusions were found in good agreement with analyses stemming from experimental IR spectra of Nafion [114]. Further investigations were then performed. The water–water coordination number was shown to be larger for systems with fluorinated CNT inner walls. Data from RDF between the sulfonate oxygen atoms and protons suggested that the fluorinated CNT inner wall stabilizes the dissociated proton. As soon as the number of water molecules reaches three per sulfonic acid group, RDF reveals that dissociation of all acidic protons is nearly completed. Using a different level of theory which takes into the long-range character of anion, that is, by adding diffuse function into the initial basis set leading to 6-31+g(d,p), the number of water molecules per sulfonic acid group attains five [17]. Fluorinated CNT inner walls have been shown to support stabilization of the hydronium ion (H_3O^+) while the Zundel ($H_5O_2{}^+$) ion configuration is promoted by nonfluorinated (more hydrophobic) CNT inner walls.

6.6.4 Coatings

Dielectric fluoropolymer film coatings are used on the surfaces of metal oxides materials in order to build micro- and nanoequipment. To analyze these contacts, quantum calculations were carried out to get the electronic state densities and bond energy between metal oxide and the fluoropolymer. Thus, processes that occur during the formation of fluoropolymer coatings on those oxides can be better understood.

Litinskii et al. investigated the contact between the fluoropolymer SKF_32 (fluorine-containing rubber) and three oxide surfaces: molybdenum oxide and solid solutions of molybdenum–niobium oxides and molybdenum–technetium oxide [115]. Contact appeared when a bond was created between a metal atom and a carbon that lost its fluorine atom. Molecular orbitals on clusters simulating the contact between the fluoropolymer and the oxide surfaces have been computed using the DFT B3LYP/3-21g. Nine contacts between the three available metal atoms (molybdenum, niobium, and technetium) and the three carbon atoms of the fluoropolymer have been studied. Electronic spectra revealed that the molybdenum atom leads to the most favorable contact with the SKF_32 fluoropolymer. A similar analysis has been carried out with ferric and nickel oxides [116].

REFERENCES

1. Ramachandran, K. I.; Deepa, G.; Namboori, K. *Computational Chemistry and Molecular Modeling*; Springer-Verlag: Berlin, Germany, 2008.
2. Hougham, G. G.; Cassidy, P. E.; Johns, K.; Davidson, T. *Fluoropolymers 2: Properties*; Kluwer Academic Publishers: New York, 1999.
3. Leach, A. R. *Molecular Modelling: Principles and Applications*, 2nd ed.; Prentice Hall: Harlow, UK, 2001.
4. Lewars, E. *Computational Chemistry: Introduction to the Theory and Applications of Molecular and Quantum Mechanics*; Kluwer Academic: Boston, MA, 2003.
5. Young, D. C. *Computational Chemistry: A Practical Guide for Applying Techniques to Real World Problems*; Wiley-Interscience: New York, 2001.
6. Cramer, C. J. *Essentials of Computational Chemistry: Theories and Models*, 2nd ed.; Wiley: Chichester, UK, 2004.
7. Anderson, P. W. *Science* **1972**, *177*, 393.
8. Berendsen, H. J. C. *Simulating the Physical World: Hierarchical Modeling from Quantum Mechanics to Fluid Dynamics*; Cambridge University Press: Cambridge, UK, 2007.
9. Ditchfield, R.; Hehre, W. J.; Pople, J. A. *J. Chem. Phys.* **1971**, *54*, 724.
10. Davidson, E. R.; Feller, D. *Chem. Rev.* **1986**, *86*, 681.
11. Becke, A. D. *J. Chem. Phys.* **1993**, *98*, 5648.
12. Lee, C.; Yang, W.; Parr, R. G. *Phys. Rev. B* **1988**, *37*, 785.
13. Zhao, Y.; Schultz, N. E.; Truhlar, D. G. *J. Chem. Theory Comput.* **2006**, *2*, 364.
14. Rondeau-Gagne, S.; Neabo, J. R.; Desroches, M.; Cantin, K.; Soldera, A.; Morin, J.-F. *J. Mater. Chem. C* **2013**, *1*, 2680.
15. Balabin, R. M. *J. Chem. Phys.* **2010**, *132*, 211103.
16. Kestner, N. R.; Combariza, J. E. Basis set superposition errors: Theory and practice. In *Reviews in Computational Chemistry*; Boyd, D. B., Lipkowitz, K. B., Eds.; Wiley-VCH, 1999; Vol. 13.
17. Laflamme, P.; Beaudoin, A.; Chapaton, T.; Spino, C.; Soldera, A. *J. Comput. Chem.* **2012**, *33*, 1190.
18. Burkert, U.; Allinger, N. L. Molecular Mechanics. In *ACS Monographs*; Caserio, M. C., Ed.; Library of Congress: Washington, DC, 1982; Vol. 177.
19. Sun, H. *J. Phys. Chem. B* **1998**, *102*, 7338.

20. Levine, I. N. *Physical Chemistry*, 6th ed.; McGraw-Hill: Boston, MA, 2009.
21. Warshel, A. *J. Chem. Phys.* **1970**, *53*, 582.
22. Jorgensen, W. L.; Maxwell, D. S.; Tirado-rives, J. *J. Am. Chem. Soc.* **1996**, *118*, 11225.
23. Cornell, W. D.; Cieplak, P.; Bayly, C. I.; Gould, I. R.; Merz, K. M.; Ferguson, D. M.; Spellmeyer, D. C.; Fox, T.; Caldwell, J. W.; Kollman, P. A. *J. Am. Chem. Soc.* **1995**, *117*, 5179.
24. Binder, K.; Heermann, D. W. *Monte Carlo Simulation in Statistical Physics: An Introduction*, 5th ed.; Springer-Verlag: Heidelberg, Germany, 2010.
25. Soldera, A. *Polymer* **2002**, *43*, 4269.
26. Soldera, A.; Monterrat, E. *Polymer* **2002**, *43*, 6027.
27. Metatla, N.; Soldera, A. *Macromol. Theor. Simul.* **2011**, *20*, 266.
28. McQuarrie, D. A. *Statistical Mechanics*; Harper & Row: New York, 1976.
29. Soldera, A.; Grohens, Y. *Macromolecules* **2002**, *35*, 722.
30. Palato, S.; Metatla, N.; Soldera, A. *Eur. Phys. J. E* **2011**, *34*, 90.
31. Haile, J. M. *Molecular Dynamics Simulation*; John Wiley & Sons: New York, 1992.
32. Allen, M. P.; Tildesley, D. J. *Computer Simulation of Liquids*; Clarendon Press: Oxford, UK, 1987.
33. Ameduri, B. *Well-Architectured Fluoropolymers: Synthesis, Properties and Applications*; Elsevier: Amsterdam, The Netherlands, 2004.
34. Laflamme, P.; Porzio, F.; Ameduri, B.; Soldera, A. *Polym. Chem.* **2012**, *3*, 652.
35. Toriumi, M. *Microelectron. Eng.* **2004**, *73–74*, 218.
36. Hirunsit, P.; Balbuena, P. B. *J. Phys. Chem. A.* **2008**, *112*, 4483.
37. Církva, V.; Böhm, S.; Paleta, O. *J. Fluorine Chem.* **2000**, *102*, 159.
38. Marutake, M. *Ferroelectrics* **1995**, *171*, 5.
39. Furukawa, T. *Phase Transit.* **1989**, *18*, 143.
40. Lovinger, A. J. *Dev. Cryst. Polym.* **1982**, *1*, 195.
41. Eby, R. K.; Clark, E. S.; Farmer, B. L.; Piermarini, G. J.; Block, S. *Polymer* **1990**, *31*, 2227.
42. Farmer, B. L.; Eby, R. K. *Polymer* **1981**, *22*, 1487.
43. Farmer, B. L.; Eby, R. K. *Polymer* **1985**, *26*, 1944.
44. Holt, D. B.; Farmer, B. L.; Macturk, K. S.; Eby, R. K. *Polymer* **1996**, *37*, 1847.
45. Smith, G. D.; Jaffe, R. L.; Yoon, D. Y. *Macromolecules* **1994**, *27*, 3166.
46. Tokarev, A. V.; Bondarenko, G. N.; Yampol'skii, Y. P. *Polym. Sci. Ser. A* **2007**, *49*, 909.
47. Okada, O.; Oka, K.; Kuwajima, S.; Tanabe, K. *Mol. Sim.* **1999**, *21*, 325.
48. Borodin, O.; Smith, G. D.; Bedrov, D. *J. Phys. Chem. B* **2002**, *106*, 9912.
49. Karasawa, N.; Goddard, W. A., III. *Macromolecules* **1992**, *25*, 7268.
50. Dick, R. G., Jr.; Overhauser, A. W. *Phys. Rev.* **1958**, *112*, 90.
51. Carbeck, J. D.; Lacks, D. J.; Rutledge, G. C. *J. Chem. Phys.* **1995**, *103*, 10347.
52. Mattice, W. L.; Suter, U. *Conformational Theory of Large Molecules: The Rotational Isomeric State Model in Macromolecular Systems*; Wiley: New York, 1994.
53. Flory, P. J. *Statistical Mechanics of Chain Molecules*; Interscience Pub.: Toronto, Canada, 1969.

54. Byutner, O. G.; Smith, G. D. *Macromolecules* **1999**, *32*, 8376.
55. Byutner, O. G.; Smith, G. D. *Macromolecules* **2000**, *33*, 4264.
56. Rehahn, M.; Mattice, W. L.; Suter, U. W. *Adv. Polym. Sci.* **1997**, *131/132*, 1.
57. Long, Y.; Liu, Y.-G.; Nie, F.-D.; Chen, J. *Coll. Polym. Sci.* **2012**, *290*, 1855.
58. Smith, G. D.; Bharadwaj, R. K. *J. Phys. Chem. B* **1999**, *103*, 3570.
59. Blöchl, P. E. *Phys. Rev. B* **1994**, *50*, 17953.
60. Laird, B. B.; Ross, R. B.; Ziegler, T. *Chemical Applications of Density-Functional Theory*; American Chemical Society: Washington, DC, **1996**.
61. Zeng, F.-L.; Sun, Y.; Zhou, Y.; Li, Q.-K. *Model. Simul. Mater. Sci.* **2011**, *19*, 025005.
62. Gee, R. H.; Fried, L. E. *J. Chem. Phys.* **2003**, *118*, 3827.
63. Gao, G. Large Scale Molecular Simulations With Application to Polymers and Nanoscale Materials. Ph.D. Thesis, California Institute of Technology, Pasadena, CA, USA, 1998.
64. Lovinger, A. J. *Science* **1983**, *220*, 1115.
65. Pei, Y.; Zeng, X. C. *J. Appl. Phys.* **2011**, *109*, 093514.
66. Hasegawa, R.; Takahashi, Y.; Chatani, Y.; Tadokoro, H. *Polym. J.* **1972**, *3*, 600.
67. Zhu, G.; Zeng, Z.; Zhang, L.; Yan, X. *Comput. Mater. Sci.* **2008**, *44*, 224.
68. Zhu, G.; Xu, J.; Yan, X.; Li, J.; Zeng, Z.; Shen, M.; Zhang, L. *Comput. Mater. Sci.* **2006**, *37*, 512.
69. Koch, W.; Holthausen, M. C. *A Chemist's Guide to Density Functional Theory*, 2nd ed.; Wiley-Interscience: New York, 2001.
70. Duan, C. G.; Mei, W. N.; Hardy, J. R.; Ducharme, S.; Choi, J.; Dowben, P. A. *Europhys. Lett.* **2003**, *81*.
71. Purvis, C. K.; Taylor, P. L. *Phys. Rev. B* **1982**, *26*, 4564.
72. Erdtman, E.; Satyanarayana, K. C.; Bolton, K. *Polymer* **2012**, *53*, 2919.
73. Imran-ul-haq, M.; Tiersch, B.; Beuermann, S. *Macromolecules* **2008**, *41*, 7453.
74. Metatla, N.; Palato, S.; Commarieu, B.; Claverie, J. P.; Soldera, A. *Soft Matter* **2012**, *8*, 347.
75. Correia, H. M. G.; Ramos, M. M. D. *Comput. Mater. Sci.* **2005**, *33*, 224.
76. Clark, J. D.; Taylor, P. L.; Hopfinger, A. J. *J. Appl. Phys.* **1981**, *52*, 5903.
77. Miyamoto, Y.; Miyaji, H.; Asai, K. *J. Polym. Sci., Polym. Phys. Ed.* **1980**, *18*, 597.
78. Kennedy, D.; Norman, C. *Science* **2005**, *309*, 75.
79. Ngai, K. L. *J. Non-Cryst. Solids* **2007**, *353*, 709.
80. Donth, E. *The Glass Transition*; Springer-Verlag: New York, 2001.
81. Buchholz, J.; Paul, W.; Varnik, F.; Binder, K. *J. Chem. Phys.* **2002**, *117*, 7364.
82. Soldera, A. *Mol. Simul.* **2012**, *38*, 762.
83. Soldera, A.; Metatla, N. *Phys. Rev. E* **2006**, *74*, 061803.
84. Soldera, A.; Porzio, F.; Anousheh, N. Presented at *Fluoropolymers*, Las Vegas (NA), **2012**, ACS Division of Polymer Chemistry, Oct. 14–17.
85. Kim, J.-W.; Cho, W.-J.; Ha, C.-S. *J. Polym. Sci. B: Polym. Phys.* **2002**, *40*, 19.
86. Gradys, A.; Sajkiewicz, P.; Adamovsky, S.; Minakov, A.; Schick, C. *Thermochim. Acta.* **2007**, *461*, 153.
87. Fujii, M.; Stannett, V.; Hopfenberg, H. B. *J. Macromol. Sci. Phys.* **1978**, *B15*, 421.

88. Carbeck, J. D. Temperature Dependence Structure and Properties of Crystalline Poly(vinylidene fluoride) from Molecular Simulation. Ph.D. Thesis, Massachusetts Institute of Technology (Harvard), Cambridge, MA, 1996.
89. Ditchfield, R. *Mol. Phys.* **1974**, *27*, 789.
90. Reinsberg, S. A.; Ando, S.; Harris, R. K. *Polymer* **2000**, *41*, 3729.
91. Carnevale, D. Nuclear Magnetic Resonance Spectroscopy and Computational Methods for the Characterization of Materials in Solution and the Solid State. Ph.D. Thesis, University of St Andrews, Scotland, 2010.
92. Keith, T. A.; Bader, R. F. W. *Chem. Phys. Lett.* **1993**, *210*, 223.
93. Ando, S.; Harris, R. K.; Hirschinger, J.; Reinsberg, S. A.; Scheler, U. *Macromolecules* **2001**, *34*, 66.
94. Skachkov, D.; Krykunov, M.; Kadantsev, E.; Ziegler, T. *J. Chem. Theory Comput.* **2010**, *6*, 1650.
95. Lee, W. H.; Suk, J. W.; Lee, J.; Hao, Y.; Park, J.; Yang, J. W.; Ha, H.-W.; Murali, S.; Chou, H.; Akinwande, D.; Kim, K. S.; Ruoff, R. S. *ACS Nano* **2012**, *6*, 1284.
96. Noyori, R. *Chem. Comm.* **2005**, 1807.
97. Tuminello, W. H.; Dee, G. T.; Mchugh, M. A. *Macromolecules* **1995**, *28*, 1506.
98. Cece, A.; Jureller, S. H.; Kerschner, J. L.; Moschner, K. F. *J. Phys. Chem.* **1996**, *100*, 7435.
99. Yee, G. G.; Fulton, J. L.; Smith, R. D. *J. Phys. Chem.* **1992**, *96*, 6172.
100. Diep, P.; Jordan, K. D.; Johnson, J. K.; Beckman, E. J. *J. Phys. Chem. A* **1998**, *102*, 2231.
101. Baradie, B.; Shoichet, M. S.; Shen, Z.; McHugh, M. A.; Hong, L.; Wang, Y.; Johnson, J. K.; Beckman, E. J.; Enick, R. M. *Macromolecules* **2004**, *37*, 7799.
102. Martin, J. W.; Smithwick, M. M.; Braune, B. M.; Hoekstra, P. F.; Muir, D. C. G.; Mabury, S. A. *Environ. Sci. Technol.* **2004**, *38*, 373.
103. Andersen, M. P. S.; Nielsen, O. J.; Hurley, M. D.; Ball, J. C.; Wallington, T. J.; Stevens, J. E.; Martin, J. W.; Ellis, D. A.; Mabury, S. A. *J. Phys. Chem. A* **2004**, *108*, 5189.
104. Kanno, N.; Tonokura, K.; Hurley, M. D.; Wallington, T. J. *J. Fluorine Chem.* **2008**, *129*, 1187.
105. Gindulytė, A.; Massa, L.; Banks, B. A.; Miller, S. K. R. *J. Phys. Chem. A* **2002**, *106*, 5463.
106. Laflamme, P.; Beaudoin, A.; Chapaton, T.; Spino, C.; Soldera, A. *J. Membr. Sci.* **2012**, 401–402, 56.
107. Hsu, W. Y.; Gierke, T. D. *J. Membr. Sci.* **1983**, *13*, 307.
108. Litt, M. H. Presented at, San Francisco, CA, USA, **1997**, Reevaluation of Nafion morphology, Proceedings of the 1997 ACS San Francisco Meeting, *38*, 80–81.
109. Haubold, H.-G.; Vad, T.; Jungbluth, H.; Hiller, P. *Electrochim. Acta* **2001**, *46*, 1559.
110. Rubatat, L.; Rollet, A. L.; Gebel, G.; Diat, O. *Macromolecules* **2002**, *35*, 4050.
111. Mauritz, K. A.; Moore, R. B. *Chem. Rev.* **2004**, *104*, 4535.
112. Habenicht, B. F.; Paddison, S. J.; Tuckerman, M. E. *J. Mater. Chem.* **2010**, *20*, 6342.
113. Perdew, J. P.; Burke, K.; Ernzerhof, M. *Phys. Rev. Lett.* **1996**, *77*, 3865.
114. Basnayake, R.; Peterson, G. R.; Casadonte, D. J.; Korzeniewski, C. *J. Phys. Chem. B* **2006**, *110*, 23938.
115. Litinskii, A. O.; Petrov, M. V.; Novikov, S. I. *J. Surf. Investig-X-Ra.* **2011**, *5*, 687.
116. Litinskii, A. O.; Novikov, S. I.; Petrov, M. V. *J. Surf. Investig-X-Ra.* **2010**, *4*, 12.

7

VAPOR DEPOSITION OF FLUOROPOLYMER SURFACES

Jose L. Yagüe and Karen K. Gleason
Massachusetts Institute of Technology

7.1 INTRODUCTION

Vapor deposition is a highly desirable method for modifying surfaces with fluoropolymers and has been extensively reviewed [1–6]. In chemical vapor deposition (CVD), one or more reactants are metered into a vacuum chamber. Inside the CVD reactor, monomer vapors undergo polymerization and thin film formation in a single step. In the case of vinyl polymerization, monomer vapors are introduced directly into the CVD chambers as reactants. Alternatively, less stable monomers, such as difluorocarbene (CF_2:), can be generated via *in situ* reaction (Table 7.1).

The attractive features of CVD for fluoropolymers include

- synthesis of true linear homopolymers, including polytetrafluoroethylene [(PTFE), Teflon®];
- no solvents required, favorable for depositing polymer coatings which do not dissolve in common solvents;
- no surfactants required, thus avoiding the need to use compounds which are persistent in the environment, such as perfluorooctanoic acid (PFOA), a common surfactant for PTFE.
- one-step formation of robust, insoluble, cross-linked networked layers from the copolymerization of fluorinated species, such as 1H,1H,2H,2H-perfluorodecyl

Handbook of Fluoropolymer Science and Technology, First Edition. Edited by Dennis W. Smith Jr., Scott T. Iacono, and Suresh S. Iyer.

TABLE 7.1 Examples of iCVD Fluoropolymer Reactants and Example Process Conditions

Fluoropolymer	Monomer	Initiator	T_f (°C)	T_s (°C)	P(mTorr)
Poly(tetrafluoroethylene) (PTFE)	Difluorocarbene (CF_2:) generated *in situ* from the pyrolysis of hexafluoropropylene oxide (HFPO)	Perfluorooctane sulfonyl fluoride (PFOS) or perfluorobutane sulfonyl fluoride (PFBS)	400	25	500
Poly(1H,1H,2H,2H-perfluorodecyl acrylate p(PFDA) and its copolymers	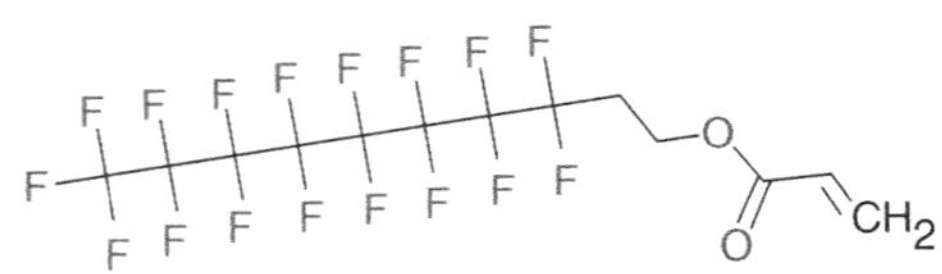	Tert butyl peroxide (TBPO)	200	35	125
Copolymers w/pentafluorophenyl methacrylate (PFM)	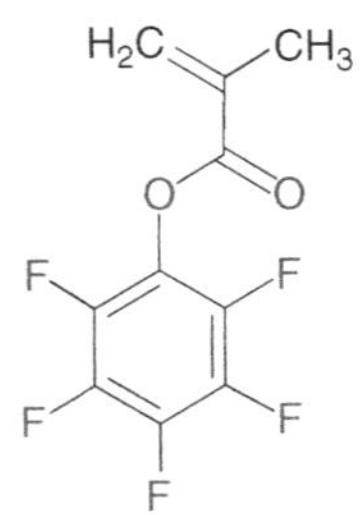	Tert butyl peroxide (TBPO)	280	30	200
Polyvinylidene fluoride (PVDF)	$CH_2{=}CF_2$	Difluorocarbene (CF_2:) generated from the pyrolysis of hexafluoropropylene oxide (HFPO)	500	−125	1000

TABLE 7.2 Different Types of CVD Polymerization Processes Based on the Excitation Source and the Need of Initiator

Excitation	Initiator feed	Nomenclature
Plasma	No	Plasma-enhanced CVD (PECVD)
Plasma (low power)	Yes	initiated PECVD (iPECVD)
Thermal (resistively heated wires/filaments)	No	Hot Wire CVD (HWCVD) or hot filament (HFCVD) or catalytic-CVD (cat-CVD)
Thermal (resistively heated wires/filaments)	Yes	Initiated CVD (iCVD)
UV	Yes, except if there is UV monomer fragmentation	Photo iCVD (piCVD)

acrylate (PFDA), with a cross-linking multivinyl monomer, such as ethylene glycol diacrylate (EGDA);

- coating objects which swell, dissolve or otherwise degrade during solution and melt processing;
- ability to form ultrathin contiguous films <20 nm thick;
- ability to form pinhole-free films >10 μm thick at high deposition rates;
- graded composition throughout the thickness of the film produced by varying the feed ratio of the reactants over the course of the deposition period;
- grafting to the substrate, by building films from the substrate up, resulting in robust adhesion;
- conformal coating over the entire surface of 3D objects and penetration to micro- and nano-scale features, resulting from the high diffusivity of vapor phase reactants at low pressure;
- automated and reproducible processing in reactors scaled up to dimensions of >1 m;
- real-time thickness monitoring and end-point control;
- the absence of residual solvent in CVD obviates the need for curing. Avoiding residual solvent is important for achieving biocompatibility and desirable dielectric properties.

Table 7.2 summarizes various subtypes of CVD polymerization processes. These methods differ in the means by which the CVD chemistry is driven (plasma, thermal, or UV). For hot wire chemical vapor deposition (HWCVD) and initiated chemical vapor deposition (iCVD), no plasma excitation or UV exposure is utilized during the polymerization, eliminating the possibility for forming defects in the films via these mechanisms. The CVD excitation produces reactive molecular fragments, including some free radical species. These fragments or other impurities in the CVD chamber can initiate polymerization. Alternatively, a free radical initiator can be metered into the CVD as a distinct reactant feed. In this case, the goal is to simultaneously achieve fragmentation of the intentionally introduced initiator species while avoiding decomposition of the monomeric species. For a fixed set of process conditions, the intentional use of an initiator often significantly increases the deposition rate. Alternatively, intentional initiation can allow reasonable CVD rates to be achieved at reduced excitation powers and temperatures, hence improving the odds that organic functional groups of the monomeric species will be retained in the deposited films. Thus, iCVD fluoropolymers will be the primary focus of this chapter.

Table 7.2 compares the several iCVD fluoropolymer chemistries, providing exemplary values of key processing parameters. The higher filament temperature, T_f, utilized for PTFE and Poly(vinylidene fluoride) PVDF, is required for the generation of difluorocarbene. The lower T_f values for PFDA and pentafluorophenyl methacrylate (PFM) preserve the structure of these vinyl monomers while producing the decomposition of the initiator. The substrate temperatures, T_s, are lower than T_f, so as to promote the absorption of the film forming species onto the substrate. For PVDF, a relatively high pressure and very cold substrate (approximately −125°C) is required for adsorption of the highly volatile monomer. The same reactor configuration (Figure 7.1) can be utilized for all the iCVD chemistries shown in Table 7.2.

7.2 POLYTETRAFLUOROETHYLENE (PTFE)

The first ever iCVD process utilized the initiator perfluorooctane sulfonyl fluoride (PFOS) for the deposition of PTFE thin films [7]. The addition of PFOS accelerated the deposition rates, which in some cases exceeded 1 μm/min. Thermal decomposition of the PFOS over the heated nichrome filament produces $CF_3(CF_2)_7\cdot$ -initiating radicals:

$$CF_3(CF_2)_7 \cdot SO_2F \rightarrow CF_3(CF_2)_7 \cdot + SO_2 + \cdot F \tag{7.1}$$

The iCVD synthesis of CF_3 end-capped PTFE chains was definitively confirmed by solid-state magic angle spinning ^{19}F nuclear magnetic resonance (NMR).

Ultrathin PTFE (<50 nm) films were deposited conformally by iCVD deposition over vertically aligned multiwalled carbon nanotube forests (mean diameter ~50 nm, mean height ~2 μm) (Figure 7.2a) [8]. In contrast, the conventional spray-and-bake PTFE coating is generally nonconformal and yields >10 μm thick films. The hexafluoropropylene oxide (HPFO) polymerization was initiated by PFOS over a stainless-steel filament. The iCVD PTFE-coated nanotube forest displayed advancing and receding contact angles with water of 170° and 160°, respectively. Essential spherical water drops were suspended on the surface (Figure 7.2b).

More recently, to avoid bioaccumulating C8 species, the C4 initiator, perfluorobutane sulfonyl fluoride ($CF_3(CF_2)_3SO_2F$, PFBSF) has been utilized [9]. With PFBSF

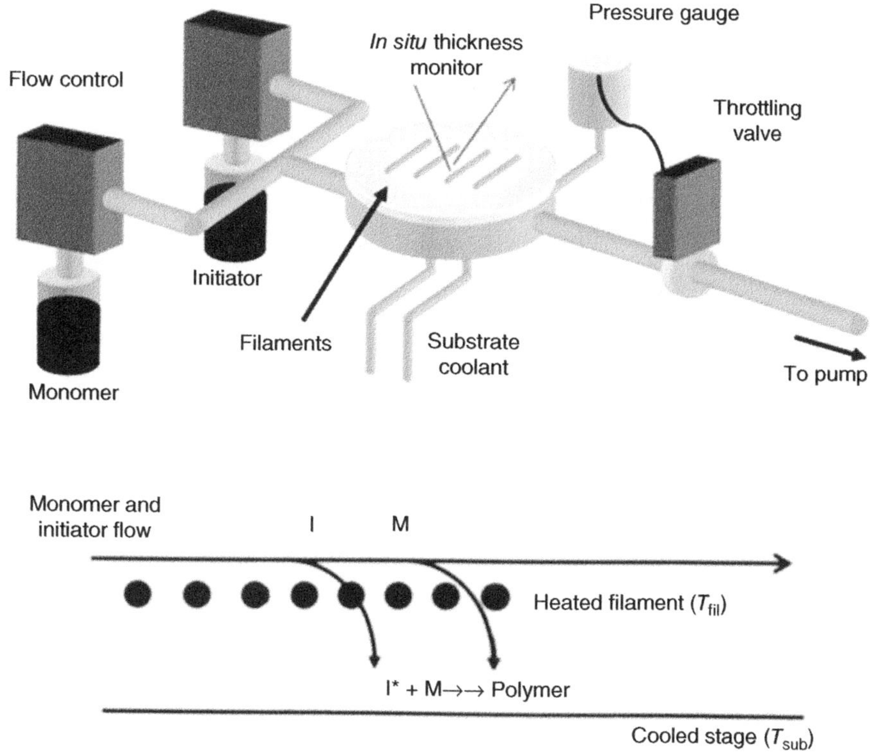

FIGURE 7.1 Schematic of a lab-scale 200 mm diameter iCVD reactor system. For a vinyl homopolymerization, a constant flow of monomer and initiator is metered into the "pancake"-style vacuum reaction chamber. An array of resistively heated wires, suspended a few centimeters above the substrate, heats the vapors. Laser interferometery provides real-time monitoring of the iCVD polymer thickness. The pressure of the chamber is controlled by a throttling value. Unreacted species and volatile reaction by-products are exhausted to a mechanical pump. For copolymerization, an additional monomer feed line would need to be added to the system (top image). Schematic cross-section of the iCVD reactor showing decomposition of the initiator by the heated filaments. Surface modification through polymerization of the monomer occurs on the actively cooled substrate (bottom image).

as the initiator, ultrathin iCVD PTFE conformal coatings of single-wall carbon nanotubes (SWCNT) show lamellar structures spaced on average 13 nm apart (Figure7.2c and 7.2d).

The invention of the iCVD PTFE process was preceded by extensive research on HWCVD PTFE. The HWCVD process utilizes the identical reactor system (Figure 7.1) but the only feed is HFPO since no initiator is used. Research on the HWCVD PTFE has been previously reviewed [10, 11].

For both the iCVD and HWCVD PTFE processes, the monomeric species is generated by reaction near the filament. The first-order thermal decomposition of the

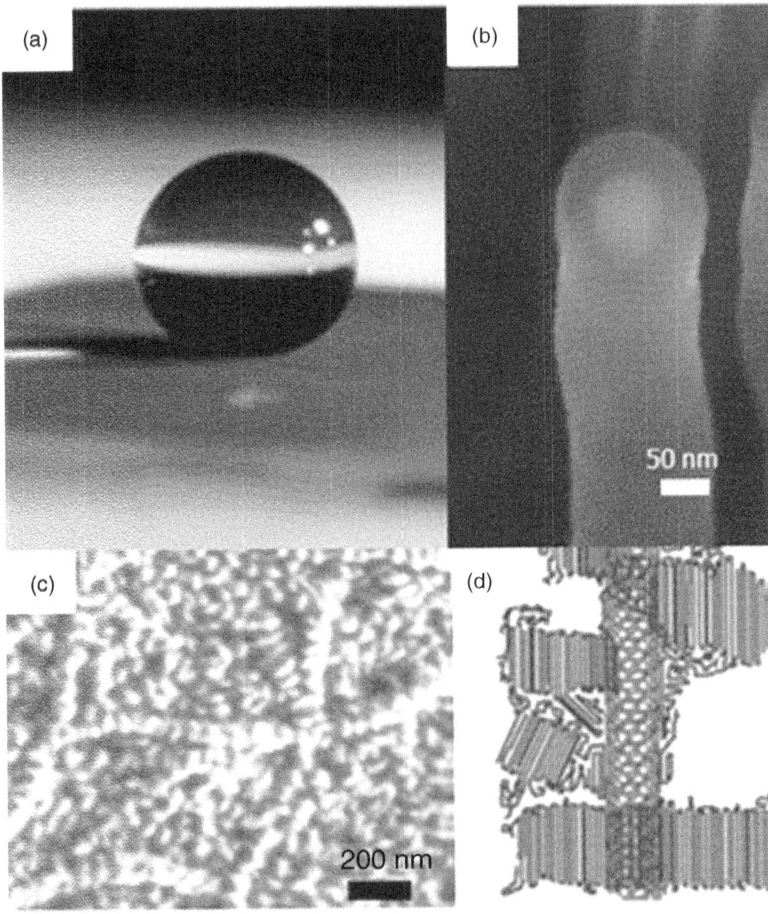

FIGURE 7.2 (a) Droplet of water with a near-spherical shape deposited on top of a carbon nanotubes forest PTFE coated by iCVD. (b) SEM image of a single carbon nanotube coated with a 40-nm thick PTFE film. (c) SEM image of SWCNT conformally coated with PTFE showing a "shish kebab" morphology. The SWCNT acts as a nucleation site for the growth of crystalline PTFE. (d) Scheme of the growth of the PTFE parallel to the carbon nanotube axis with lamellar structure.

reactant HFPO produces difluorocarbene (CF_2) with a measured rate coefficient of $10^{13.5}\exp(-18{,}270/T)\ s^{-1}$:

$$C_3F_6O \rightarrow CF_3(CO)F + CF_2 \tag{7.2}$$

UV absorption spectroscopy confirms the production of CF_2, dominantly in the more stable singlet state [12]. Difluorocarbene produced by HFPO decomposition over a heated filament was also observed by laser–induced fluorescence-based imaging of radicals [13]. In the absence of an intentionally introduced initiator species,

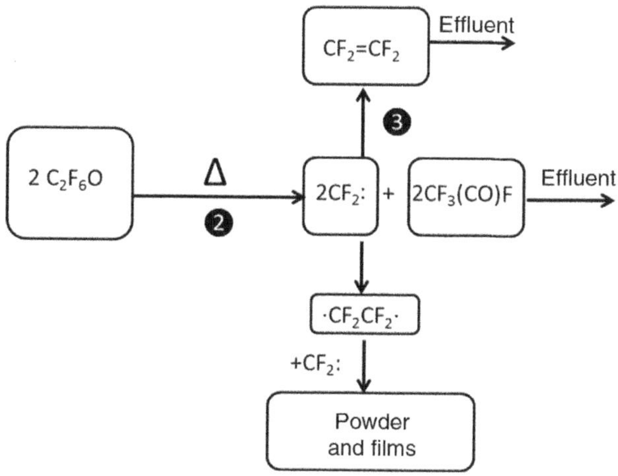

FIGURE 7.3 Overview of primary reactions in HWCVD PTFE growth in which no intentional initiator is utilized, with the labels 2 and 3 corresponding to eqs. 7.2 and 7.3, respectively.

density functional theory predicts that PTFE chain growth proceeds by the addition of singlet CF_2 to unterminated C_nF_{2n} chains [14].

Chain termination most likely occurs by reaction with the primary initiating radical or with another propagating chain. The calculated stability of any species containing a double bond precludes their participation in chain growth. Thus, the second-order recombination of difluorocarbene,

$$2CF_2 :\rightarrow CF_2 = CF_2, \tag{7.3}$$

having a rate constant of $4.15 \times 10^{-15}\ T^{-0.5} \exp(-11.5/T)\ \text{cm}^3\ \text{s}^{-1}$, to form tetrafluoroethylene (TFE), which does not react further. Indeed, TFE is one of only two species identified in gas phase Fourier transform infrared (FTIR) of the effluent of HWCVD from HFPO [15]. The second effluent species was trifluoroacetyl fluoride, the other primary product from thermally decomposing HFPO (Reaction 7.2). Decreasing the filament-to-substrate distance enhances the pathway to film formation over recombination to TFE (Figure 7.3) [16].

Analysis of trends in CF_2 concentration by UV absorption spectroscopy, effective sticking coefficients, deposition rate, and experimentally observed powder formation, all support the hypothesis that C_2F_{2n} oligomers formed in the gas phase contribute significantly to film deposition [12, 14, 16, 17].

In the first demonstration of HWCVD PTFE [18], the predominance of the $-CF_2-$ bonding environment in the films was shown by X-ray photoelectron spectroscopy (XPS). The FTIR spectra of HWCVD PTFE films are dominated by the CF_2 symmetric (1155 cm^{-1}) and asymmetric stretches (1215 cm^{-1}), which are also observed for conventionally synthesized bulk PTFE. Additionally, electron spin resonance (ESR) results found the dangling bond density $\sim 10^{18}$ spins/cm^3 for HWCVD PTFE, which is

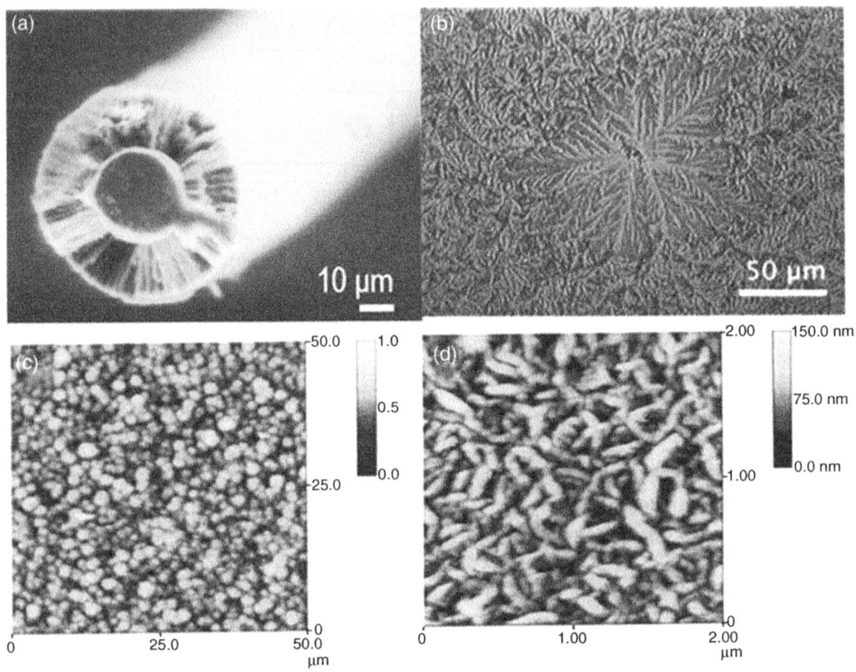

FIGURE 7.4 (a) Cross-section of a stainless-steel wire coated with a uniform 16-μm thick coating of PTFE. (b) Optical micrograph of crystalline spherulites of up to 1 mm in diameter produced by annealing of a PTFE film. AFM images of a PTFE film with spherical nodule morphology produced without pre-heating of the filaments (scan area 50 × 50 μm) (c) and with rod-like grain structure as a consequence of pre-heating the filaments (scan area 2 × 2 μm) (d).

low compared to plasma polymerized fluorocarbon films. Conformal coating, 16 μm thick, was achieved around the circumference of a 25-μm diameter stainless-steel wire (Figure 7.4a). Solid-state magic angle spinning ^{19}F NMR confirmed that the films were predominately $-(CF_2)_n-$ [19].

The presence of PTFE crystallites was observed by X-ray diffraction (XRD) [20]. Crystallization can be enhanced by annealing, producing spherulites up to 1 mm in diameter (Figure 7.4b) [21]. Corresponding changes in the intensity of the CF_2 wagging (641/629 cm^{-1}), deformation (555 cm^{-1}), and rocking (513/530 cm^{-1}) modes in the FTIR spectra provided additional evidence supporting changes in film crystallinity. Inefficient packing of anisotropic crystallites was observed to lead to micro- and nano-porous films [22]. Porosity up to 30 vol.% was quantified by variable angle spectroscopic ellipsometry (VASE). More rapidly deposited films had spherical nodules and lower porosity [15]. The difference in morphology is readily imaged by atomic force microscopy (AFM) (Figure7.4c and 7.4d).

Research continues on the thermal decomposition of HFPO to yield transparent, superhydrophobic PTFE films by CVD. Different filament materials have been

FIGURE 7.5 Commercial iCVD reactor from GVD Corporation designed to scale up the PTFE coating process with a door 1.2 m in width.

explored, including nickel [23–25]. An RF-sputtered PTFE film as a buffer layer was used to synthesize HWCVD PTFE having a contact angle of 150° [26]. Use of a heated cracking tube in place of filaments has also been demonstrated [27]. Ion Cyclotron Resonance Fourier Transform Mass-Spectrometry (ICR FT-MS) confirmed that the dominating bonding structure is $-(CF_2)_n-$ polymer chains, with a minor presence of oxygen and incorporated hydrocarbon fragments [28]. For high filament temperatures (>600°C), the HFPO yielded films exhibiting some cross-linking $-CF_2-$bonding configurations, while low substrate temperatures (−150°C) increased the degree of oxygen incorporation in the form of C=O/COO and OH [29, 30]. The HWCVD PTFE films can display a lower coefficient of friction (COF) than traditionally synthesized bulk PTFE [31].

Selective deposition of fluorinated films has emerged after the development of novel patterning processes. Patterning CVD PTFE films have been achieved through hard masks with 50 μm spaces [25]. E-beam lithography has resulted in line resolution as low as 1 μm. Furthermore, the integration of supercritical CO_2 as the developer has led to a new and friendly environmental process [32].

Pryce Lewis et al. have demonstrated the commercial viability of iCVD PTFE, obtaining deposition rates faster than 1 μm/min [33]. Figure 7.5 shows GVD Corporation's production batch coater, with a door 1.2 m wide.

7.3 POLY(VINYLIDENE FLUORIDE)

Rastogi and Desu [34] described the synthesis of PVDF by a low pressure chemical vapor polymerization (CVP) process. Due to the high vapor pressure of the VDF monomer (26,200 Torr at 25°C), the stage of the chamber was backcooled down to 150 K by passing nitrogen liquid to enhance the adhesion on the substrate. Once a layer of VDF has been physisorbed, a flow of HFPO is introduced into the reactor and the temperature of the filaments is raised to 500–600°C. Decomposition of the oxide generates CF_2 radicals, which, in turn, act as the initiator for the polymerization of the VDF. The polymerization can be carried out in continuous, as explained, or pulsed, alternating cycles of VDF condensation and formation of the radical initiator, to yield ferroelectric PVDF thin films.

7.4 POLY(1H,1H,2H,2H-PERFLUORODECYL ACRYLATE)

As a result of the their $-CF_3$ terminated side groups and comb-like structure, the typical range of surface energies for fluorinated acrylic polymers (5.6–7.8 mN/m) are even lower than that of PTFE (~20 mN/m) [35]. The p(PFDA) side chains with eight perfluorinated carbons, so-called "C8," provide highly desirable iCVD surface properties but there is concern over the bioaccumulation factor of the monomer [36], suggesting that future research would benefit from utilizing monomers with shorter perfluorocarbon side chains.

Chemical characterization of iCVD p(PFDA) synthesized using tert-butyl peroxide (TBPO) as the initiator, confirmed that the monomer's pendant fluorocarbon group is completely retained while the vinyl bond is completely consumed [37]. Deposition rate weight increased, up to 375 nm/min, with increasing P/P^{sat}, where P is the monomer's partial pressure and P^{sat} is its vapor pressure. The number average molecular weight, M_n, also increased up to 177,300 Da with increasing P/P^{sat}. The measured absorption isotherm of the monomer and model-derived rate constants of film deposition verified that p(PFDA) film growth kinetics are mechanistically similar to the free radical polymerization of other nonfluorinated iCVD polymers [38, 39].

Conformal coverage of p(PFDA) was achieved inside the cylindrical capillary pores in membranes. The pores were 3 μm in diameter and either 120 or 240 μm in length, corresponding to aspect ratios of 40:1 and 80:1, respectively [40]. Process conditions and time were used to vary the average thickness of the iCVD p(PFDA) film coating the pores from 10 to 150 nm. Electron microprobe analysis of the membrane cross-section confirmed the presence of coating down full length of the pore. Even when the edges of the membrane were fully taped down during deposition, iCVD p(PFDA) was detected on the backside of the membrane, confirming the ability of reactive species to diffuse down the entire length of the pore.

Similarly, p(PFDA) was used to functionalize the inner part of the 50-nm thick pores of a polycarbonate membrane to enhance selectivity during the separation procedure. First, a layer of iCVD polydivinylbenzene p(DVB) was conformally deposited to homogeneously narrow the pore size. After that, a very thin layer of p(PFDA) was deposited all the way down along the pores to result in a 5-nm pore diameter membrane with a hydrophobic character. Because of the small pore size, the molecular interaction with the inner walls of the pores is the main driving force in the separation mechanism. A maximum selectivity of 234 was achieved for a pair of molecules with similar size but different polarity [41].

Conformal coverage of iCVD p(PFDA) was also demonstrated on electrospun mats to create superhydrophobic fabrics [35]. Decreasing fiber diameter, from 2200 nm down to 600 nm, produced systematic increase in the static water contact angle, which agreed with predictions of theoretical model. Hierarchical morphology added in the form of beaded fibers, further increased the static water contact angle, up to a value of 175° and decreased the sliding angle to <2.5° for a 20-mg droplet. In one case, grade 8 oleophobicity was also exhibited by the iCVD p(PFDA)-coated electrospun mat. Adding nano-scale pores to the fibers is an alternative means of adding hierarchical structure. A 30-nm thin conformal iCVD coating preserved the nanopores and fiber structure, while successfully lowering surface energy [42].

Coating surfaces with complex geometries has supported the development of new interfaces and potential applications for these materials. Alternating regions of the hydrophobic p(PFDA) and the hydrophilic poly(4-vinylpyridine) (P4VP) polymer have been patterned on curved substrate using a photopolymerization approach [43]. Porous chromatography paper was patterned with p(PFDA) using metal salts. Since porous substrates do not allow using physical mask, metal salts were printed on the membranes inhibiting the polymerization process and thus, creating the pattern [44]. Ionic liquids (IL) were also modified using iCVD p(PFDA) [45, 46]. Due to the low vapor pressure of IL, the PFDA monomer is able to adsorb on its surface and encapsulate the IL.

Deposition conditions affect both the degree of crystallinity of iCVD p(PFDA) and the orientation of the crystallites (perpendicular or parallel to the surface) [47]. The XRD characterization shows a smectic B structure with a periodicity of 3.24 nm in the iCVD p(PFDA) films displaying crystallinity. This is the same crystal structure observed in conventionally synthesized p(PFDA). The perfluorinated side chains form alternating chains facing each other, forming a bilayer. The orientation of the side chains is orthogonal to the plane of the lamellar, which exhibits hexagonal packing with a lattice parameter of 0.64 nm.

At the highest filament temperature, 300°C, diffraction peaks are clearly seen when the X-ray beam is oriented perpendicular to the substrate (Figure 7.6b) but are absent when the X-ray beam is parallel to the substrate (Figure 7.6a), indicating vertical alignment of the perfluoro side chains. The converse is observed at the lowest filament temperature, 270°C, indicating a change to a horizontal alignment. Figure 7.6c schematically shows the two different crystalline textures. The difference in orientation is hypothesized to arise from a change in the dominant initiation mechanism. At low temperatures, the bulky tert butoxy radicals predominates, while at high temperatures, β-scission of the tert butoxy radicals leads to smaller methyl radicals becoming the dominant initiation species [48]. For films with parallel orientation, the highest observed advancing water contact angle is 130°. Films with perpendicular orientation have advancing water contact angles as high as 150°. The lowest degree of hysteresis in the water contact angle is observed in films with the highest degree

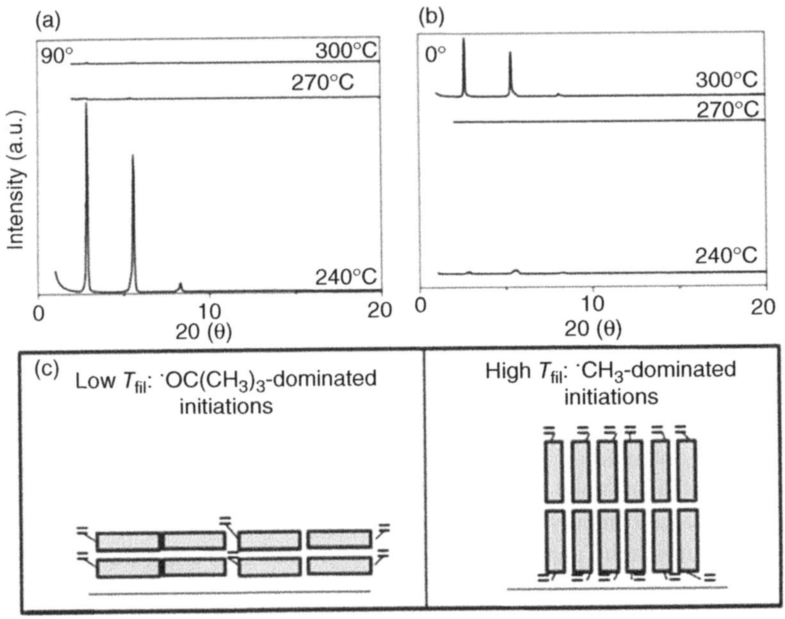

FIGURE 7.6 XRD profiles of iCVD p(PFDA) films synthesized using different filament temperatures (240, 270, and 300°C). At low temperatures, the perfluoro side chains arrange in a Smectic B structure parallel to the surface showing a diffraction pattern only for an irradiation angle of 90° with respect to the surface normal (a); while at high temperatures, the perfluoro side chains are aligned perpendicular to the surface and interacting only with the X-ray beam at 0° (b). (c) Scheme of the orientation of perfluoro side chains depending on the filament temperature.

of crystallinity. Grafting of iCVD p(PFDA) utilizing a vinyl silanol coupling agent pretreatment resulted in even higher advancing water contact angle of 160°, with a hysteresis of only 5°. The grafted film was also oleophobic, having an advancing contact angle with mineral oil of 120° [49].

Bifunctional surfaces displaying areas of hydrophobic iCVD p(PFDA) next to hydrophilic regions composed of iCVD hydrogel, polyhydroxyethylmethacrylate, p(HEMA) (Table 7.3), were patterned down to ~7 μm resolution using transmission electron microscope grids as lift-off masks, allowing control over microcondensation [63]. Patterning by colloidal lithography of grafted iCVD p(PFDA) resulted in hexagonally packed bowl structures 1 μm in diameter [64].

The substrate independent nature of iCVD enabled to encapsulate a paper-based photovoltaic (PV) array by deposition of a p(PFDA) layer to protect the device and extend its lifetime. Since no solvents are involved in the deposition method, the polymer electrodes were gently coated with the hydrophobic polymer. The coated array was folded and submerged into a water bath during the operation without affecting the performance of the device [65].

TABLE 7.3 Monomers Copolymerized with Fluorinated Monomers and Their Applications

Monomer	Chemical structure	Application
Ethylene glycol diacrylate (EGDA)	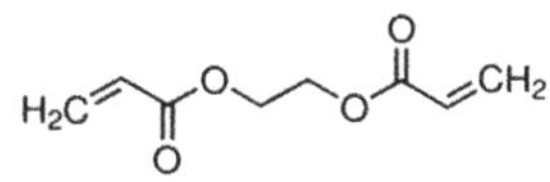	Barrier coating [50,51], microfluidics [52], mechanical properties [53]
2-Hydroxyethyl methacrylate (HEMA)	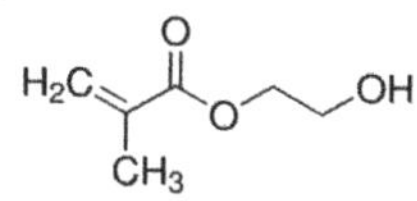	Antifouling [54,55], biomaterials [56]
Methacrylic acid (MMA)	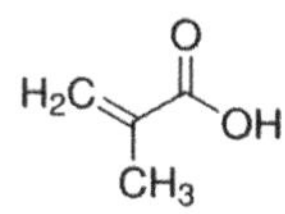	Fuel cells [57]
Divinylbenzene (DVB)	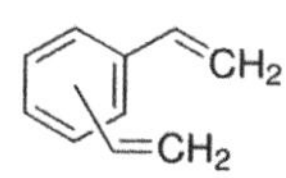	Photonics [58]
Glycidyl methacrylate (GMA)	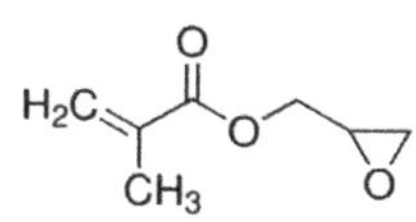	Mechanical properties [59], patterning [60]
Hexamethylcyclotrisiloxane (D_3)	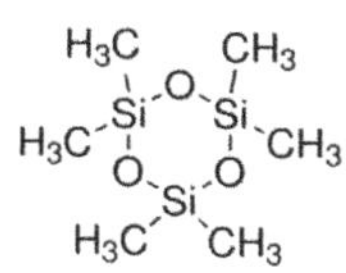	Mechanical properties [61]
(1,3,5,7-Tetravinyl-1,3,5,7-tetramethylcyclotetrasiloxane) (V4D4)	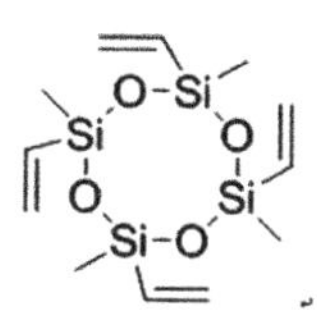	Chemical and mechanical robustness [62]

7.5 COPOLYMERIZATION OF FLUORINATED MONOMERS

Copolymerization of fluorinated monomers has resulted in the synthesis of new materials with very unique properties. Due to its solventless nature, CVD methods allow the use of monomers with a lack of a common solvent, such as the hydrophobic vinyl monomer PFDA, together with the hydrophilic monomer, 2-hydroxyethyl methacrylate (HEMA). Reacting two different monomers allows benefiting from the different properties of each one of them. As a consequence, the copolymer shows an enhancement in its physical and/or chemical properties to fulfill the needs in each different application. One way to study the composition of these copolymers is determining the reactivity ratios with the Fineman–Ross equation [66]. The reactivity ratio expresses the predilection of the radical species to react with the monomers and thus, the final composition of the copolymer. Precisely, previous research was focused on the compositional analysis of different fluorinated copolymers. The reactivity ratios of PFDA copolymerized with methacrylic acid (MMA), r_{PFDA} and r_{HEMA}, were found to be 0.35 and 0.85; in another study, 2,2,3,3,4,4,5,5,6,6,7,7-dodecafluoroheptyl acrylate (DFHA) was copolymerized with glycidyl methacrylate (GMA) obtaining reactivity ratios of 0.21 and 0.15. These values, between 0 and 1, indicate that the monomers are displayed randomly in the polymer structure.

7.5.1 Copolymers with 1H,1H,2H,2H-Perfluorodecyl Acrylate (PFDA)

EGDA is a cross-linker that has been employed repeatedly in iCVD polymerization to enhance the mechanical properties of the film [67, 68]. p(PFDA-*co*-EGDA) was deposited in channels of poly(dimethylsiloxane) (PDMS) microfluidic devices as a barrier coating to prevent diffusion and swelling [50] or to provide an adequate interface for the synthesis of nanoparticles in droplet-flow microfluidic reactors [52]. p(PFDA-*co*-EGDA) was used to robustly encapsulate IL marbles [51]. IL marbles were fabricated rolling off a droplet of IL in a Petri dish with PTFE microparticles. Next, the Petri dish was placed inside the vacuum chamber and the marbles were conformally coated over the entire surface.

Hydrophilic monomers were combined with PFDA to obtain a dual functionality in the polymer chain. Enhanced resistance to protein adsorption was achieved reacting HEMA and PFDA to yield a randomly distributed copolymer film p(HEMA-*co*-PFDA) [54]. This heterogeneity at molecular scale turns the surface energetically unfavorable inhibiting either hydrophilic or hydrophobic interactions. An analytical model and experimentation showed that the maximum protein rejection was observed for a 40% composition of PFDA in the copolymer. Reverse osmosis (RO) desalinization membranes were successfully modified with a 20-nm thick film of this copolymer composition to prevent fouling [55]. Fluorescence images indicate that microbial attachment is drastically reduced on the surface of the membrane. Furthermore, the salt rejection performance of the membrane was not affected by the iCVD coating. Another copolymer with dual functionality, p(PFDA-*co*-MMA), was synthesized to take advantage of the stability of PFDA and the high proton conductivity of MMA to design proton exchange membranes for fuel cells [57].

Fluoropolymers have attracted considerable attention for photonic applications as optical waveguides due to the ability to modulate its thermo-optical (TO) coefficient. PFDA copolymerized with divinyl benzene (DVB) resulted in a larger TO coefficient compared to the homopolymer improving the performance for athermal silicon circuits [58].

7.5.2 Copolymers with Pentafluorophenyl Methacrylate (PFM)

Due to its high reactivity toward amines, polypentafluorophenyl methacrylate p(PFM) is an excellent candidate for protein immobilization. However, its solubility in different solvents can hinder its potential qualities. O'Shaughnessy et al. [53] demonstrated that cross-linking the polymer with EGDA can prevent this undesired effect. Films of p(PFM-*co*-EGDA) were patterned through micro-contact printing with fluorescein 5-thiosemicarbazide. The fluorescent pattern was only immobilized when in contact with the p(PFM-*co*-EGDA) copolymer. PFM was also used to functionalize the biocompatible hydrogel p(HEMA) to show its characteristic reactivity [56]. PFM was copolymerized in the last 20 nm of the film. This graded composition allows maintaining the swelling properties of the hydrogel, while introducing the desire functionality for further reactivity.

7.5.3 Copolymers with 2,2,3,3,4,4,5,5,6,6,7,7-Dodecafluoroheptyl Acrylate (DFHA) and (Perfluoroalkyl)Ethyl Methacrylate (PFEMA)

Copolymerization of DFHA and (perfluoroalkyl)ethyl methacrylate (PFEMA) with GMA showed an enhancement of the mechanical properties of the films by carrying out a posterior vacuum annealing process to the as-deposited films. The thermal treatment entailed the ring opening of the epoxy groups and thus, the cross-linking among the polymer chains, which resulted in an improvement of the hardness and the elastic modulus. Furthermore, the cross-linking fixed a high concentration of fluorine groups on the outermost layer of the films showing a surface energy lower than PTFE [59]. In a parallel study, these two monomers, DHFA and PFEMA, were copolymerized with GMA and methacrylic acid (MAA) to develop negative and positive electron beam resists developable by supercritical CO_2 ($scCO_2$) obtaining patterns with features sizes of 300 nm [60].

7.5.4 Copolymers with Organosilicons

It should be noted that the possibility to copolymerize fluoromonomers with organosilicon monomers was explored first by means of HWCVD [61]. After heating the filaments to 620°C, the monomers HFPO and hexamethylcyclotrisiloxane (D_3) were decomposed into the active species difluorocarbene, dimethylsiloxane, and linear and cyclic units coming from the ring structure to form the copolymer. In addition, the cross-linking of the polymer was found to be due only to the siloxane units. Also, using an iCVD approach, PFDA was combined with (1,3,5,7-tetravinyl-1,3,5,7-tetramethylcyclotetrasiloxane) (V4D4) to form a stacked polymer covalently bound in the interface by flowing both monomers together for a short time during the transition from one polymer to the other [62]. Fabrics were coated with the stacked polymer showing superhydrophobicity and high resistance to chemical and mechanical attacks.

7.6 SUMMARY

CVD polymerization methods result in a versatile platform for the fabrication of high quality fluoropolymer films. In this chapter, we have given an insight into the mechanism of formation of these films, the ability to tailor their properties by tuning the experimental conditions and the chemistry associated with the multiple functionalities available. The absence of solvent and the conformal nature of CVD enable coating any material with complex geometry without damaging the substrate. The scale-up of these methods and the possibility to integrate them with vacuum processes currently used in the manufacturing industry permit to benefit from all the advantages of the fluorinated films in their commercialization. Continuous investigation in these materials has led to new applications in many different fields, such as optoelectronic, biomedicine or membrane separation systems, making a big impact in polymer thin-film technology.

ACKNOWLEDGMENTS

The authors gratefully acknowledge financial support from the MIT Institute for Soldier Nanotechnologies (ISN) under Contract DAAD-19-02D-0002 with the U.S. Army Research Office; the Masdar Institute; and the Center for Clean Water and Clean Energy at MIT and KFUPM, funded by the King Fahd University of Petroleum and Minerals in Dhahran, Saudi Arabia.

REFERENCES

1. Yagüe, J. L.; Coclite, A. M.; Petruczok, C.; Gleason, K. K. *Macromol. Chem. Phys.* **2013**, *214*, 302.
2. Alf, M. E.; Asatekin, A.; Barr, M. C.; Baxamusa, S. H.; Chelawat, H.; Ozaydin-Ince, G.; Petruczok, C. D.; Sreenivasan, R.; Tenhaeff, W. E.; Trujillo, N. J.; Vaddiraju, S.; Xu, J. J.; Gleason, K. K. *Adv. Mater.* **2010**, *22*, 1993.
3. Baxamusa, S. H.; Im, S. G.; Gleason, K. K. *Phys. Chem. Chem. Phys.* **2009**, *11*, 5227.
4. Im, S. G.; Gleason, K. K. *Aiche. J.* **2011**, *57*, 276.
5. Ozaydin-Ince, G.; Coclite, A. M.; Gleason, K. K. *Rep. Prog. Phys.* **2012**, 75.
6. Tenhaeff, W. E.; Gleason, K. K. *Adv. Funct. Mater.* **2008**, *18*, 979.
7. Lewis, H. G. P.; Caulfield, J. A.; Gleason, K. K. *Langmuir* **2001**, *17*, 7652.
8. Lau, K. K. S.; Gleason, K. K. *Surf. Coat. Tech.* **2007**, *201*, 9189.
9. Laird, E. D.; Bose, R. K.; Wang, W.; Lau, K. K. S.; Li, C. Y. *Macromol. Rapid. Comm.* **2013**, *34*, 251.
10. Lau, K. K. S.; Lewis, H. G. P.; Limb, S. J.; Kwan, M. C.; Gleason, K. K. *Thin Solid Films* **2001**, *395*, 288.
11. Lau, K. K. S.; Murthy, S. K.; Lewis, H. G. P.; Caulfield, J. A.; Gleason, K. K. *J. Fluorine. Chem.* **2003**, *122*, 93.

12. Cruden, B. A.; Gleason, K. K.; Sawin, H. H. *J. Phys. D Appl. Phys.* **2002**, *35*, 480.
13. Liu, D. P.; Martin, I. T.; Fisher, E. R. *Chem. Phys. Lett.* **2006**, *430*, 113.
14. Lau, K. K. S.; Gleason, K. K.; Trout, B. L. *J. Chem. Phys.* **2000**, *113*, 4103.
15. Lau, K. K. S.; Caulfield, J. A.; Gleason, K. K. *Chem. Mater.* **2000**, *12*, 3032.
16. Zhou, J.; Wolden, C. A. *Thin Solid Films* **2003**, *430*, 28.
17. Cruden, B. A.; Gleason, K. K.; Sawin, H. H. *J. Vac. Sci. Technol. B* **2002**, *20*, 690.
18. Limb, S. J.; Labelle, C. B.; Gleason, K. K.; Edell, D. J.; Gleason, E. F. *Appl. Phys. Lett.* **1996**, *68*, 2810.
19. Lau, K. K. S.; Gleason, K. K. *J. Fluorine. Chem.* **2000**, *104*, 119.
20. Limb, S.; Lau, K. S.; Edell, D.; Gleason, E.; Gleason, K. *Plasmas Polym.* **1999**, *4*, 21.
21. Lau, K. K. S.; Gleason, K. K. *J. Phys. Chem. B* **2001**, *105*, 2303.
22. Lau, K. K. S.; Caulfield, J. A.; Gleason, K. K. *J. Vac. Sci. Technol. A* **2000**, *18*, 2404.
23. Yasuoka, H.; Yoshida, M.; Sugita, K.; Ohdaira, K.; Murata, H.; Matsumura, H. *Thin Solid Films* **2008**, *516*, 687.
24. Cha, J. O.; Yeo, S. J.; Pode, R.; Ahn, J. S. *J. Nanosci. Nanotechno.* **2011**, *11*, 5829.
25. Takachi, M.; Yasuoka, H.; Ohdaira, K.; Shimoda, T.; Matsumura, H. *Thin Solid Films* **2009**, *517*, 3622.
26. Kim, H.-M.; Sohn, S.; Ahn, J. S. *Surf. Coat. Technol.* **2012**.
27. Wang, J.; Song, X.; Li, R.; Shen, J. P.; Yang, G. C.; Huang, H. *Appl. Surf. Sci.* **2012**, *258*, 9782.
28. Uvarov, A.; Uemura, K.; Alexandrov, S.; Murayama, H.; Soba, R. Proceedings of 10th International Conference on Modification of Materials with Particle Beams and Plasma Flows, 2010, p 500.
29. Rastogi, A. C.; Desu, S. B. *Polymer* **2005**, *46*, 3440.
30. Rastogi, A. C.; Desu, S. B. *Appl. Phys. A. Mater.* **2006**, *83*, 57.
31. Karnath, M. A.; Sheng, Q.; White, A. J.; Muftu, S. *Tribol. Lubr. Technol.* **2011**, *67*, 52.
32. Lewis, H. G. P.; Weibel, G. L.; Ober, C. K.; Gleason, K. K. *Chem. Vap. Deposition* **2001**, *7*, 195.
33. Lewis, H. G. P.; Bansal, N. P.; White, A. J.; Handy, E. S. *Thin Solid Films* **2009**, *517*, 3551.
34. Rastogi, A. C.; Desu, S. B. *Chem. Vap. Deposition* **2006**, *12*, 742.
35. Ma, M. L.; Mao, Y.; Gupta, M.; Gleason, K. K.; Rutledge, G. C. *Macromolecules* **2005**, *38*, 9742.
36. Kumar, V.; Pulpytel, J.; Arefi-Khonsari, F. *Plasma Process Polym.* **2010**, *7*, 939.
37. Gupta, M.; Gleason, K. K. *Langmuir* **2006**, *22*, 10047.
38. Lau, K. K. S.; Gleason, K. K. *Macromolecules* **2006**, *39*, 3688.
39. Lau, K. K. S.; Gleason, K. K. *Macromolecules* **2006**, *39*, 3695.
40. Gupta, M.; Kapur, V.; Pinkerton, N. M.; Gleason, K. K. *Chem. Mater.* **2008**, *20*, 1646.
41. Asatekin, A.; Gleason, K. K. *Nano Lett.* **2011**, *11*, 677.
42. Ma, M. L.; Gupta, M.; Li, Z.; Zhai, L.; Gleason, K. K.; Cohen, R. E.; Rubner, M. F.; Rutledge, G. C. *Adv. Mater.* **2007**, *19*, 255.
43. Petruczok, C. D.; Gleason, K. K. *Adv. Mater.* **2012**, *24*, 6445.
44. Kwong, P.; Flowers, C. A.; Gupta, M. *Langmuir* **2011**, *27*, 10634.

45. Haller, P. D.; Frank-Finney, R. J.; Gupta, M. *Macromolecules* **2011**, *44*, 2653.
46. Frank-Finney, R. J.; Haller, P. D.; Gupta, M. *Macromolecules* **2012**, *45*, 165.
47. Coclite, A. M.; Shi, Y. J.; Gleason, K. K. *Adv. Funct. Mater.* **2012**, *22*, 2167.
48. Ozaydin-Ince, G.; Gleason, K. K. *J. Vac. Sci. Technol. A* **2009**, *27*, 1135.
49. Coclite, A. M.; Shi, Y. J.; Gleason, K. K. *Adv. Mater.* **2012**, *24*, 4534.
50. Riche, C. T.; Marin, B. C.; Malmstadt, N.; Gupta, M. *Lab Chip* **2011**, *11*, 3049.
51. Bradley, L. C.; Gupta, M. *Langmuir* **2012**, *28*, 10276.
52. Lazarus, L. L.; Riche, C. T.; Marin, B. C.; Gupta, M.; Malmstadt, N.; Brutchey, R. L. *ACS Appl. Mater. Inter.* **2012**, *4*, 3077.
53. O'Shaughnessy, W. S.; Mari-Buye, N.; Borros, S.; Gleason, K. K. *Macromol. Rapid Comm.* **2007**, *28*, 1877.
54. Baxamusa, S. H.; Gleason, K. K. *Adv. Funct. Mater.* **2009**, *19*, 3489.
55. Ozaydin-Ince, G.; Matin, A.; Khan, Z.; Zaidi, S. M. J.; Gleason, K. K. *Thin Solid Films* **2013**, *539*, 181.
56. Montero, L.; Baxamusa, S. H.; Borros, S.; Gleason, K. K. *Chem. Mater.* **2009**, *21*, 399.
57. Coclite, A. M.; Lund, P.; Di Mundo, R.; Palumbo, F. *Polymer* **2013**, *54*, 24.
58. Raghunathan, V.; Yagüe, J. L.; Xu, J. J.; Michel, J.; Gleason, K. K.; Kimerling, L. C. *Opt. Express* **2012**, *20*, 20808.
59. Mao, Y.; Gleason, K. K. *Macromolecules* **2006**, *39*, 3895.
60. Mao, Y.; Felix, N. M.; Nguyen, P. T.; Ober, C. K.; Gleason, K. K. *Chem. Vap. Deposition* **2006**, *12*, 259.
61. Murthy, S. K.; Gleason, K. K. *Macromolecules* **2002**, *35*, 1967.
62. Yoo, Y.; You, J. B.; Choi, W.; Im, S. G. *Polym. Chem.-UK* **2013**, *4*, 1664.
63. Baxamusa, S. H.; Montero, L.; Borros, S.; Gleason, K. K. *Macromol. Rapid Comm.* **2010**, *31*, 735.
64. Trujillo, N. J.; Baxamusa, S. H.; Gleason, K. K. *Chem. Mater.* **2009**, *21*, 742.
65. Barr, M. C.; Rowehl, J. A.; Lunt, R. R.; Xu, J. J.; Wang, A. N.; Boyce, C. M.; Im, S. G.; Bulovic, V.; Gleason, K. K. *Adv. Mater.* **2011**, *23*, 3500.
66. Fineman, M.; Ross, S. D. *J. Polym. Sci.* **1950**, *5*, 259.
67. Chan, K.; Gleason, K. K. *Langmuir* **2005**, *21*, 8930.
68. Chan, K.; Kostun, L. E.; Tenhaeff, W. E.; Gleason, K. K. *Polymer* **2006**, *47*, 6941.

8

FUNCTIONALIZED AND FUNCTIONALIZABLE FLUOROPOLYMER MEMBRANES

Tao Cai, Koon-Gee Neoh, and En-Tang Kang
National University of Singapore

8.1 INTRODUCTION

Fluoropolymers, such as poly(vinyl fluoride) (PVF), poly(vinylidene fluoride) (PVDF), poly(chlorotrifluoroethylene) (PCTFE), poly(tetrafluoroethylene) (PTFE), poly(vinylidene fluoride-*co*-chlorotrifluoroethylene) (P(VDF–CTFE)), poly(vinylidene fluoride-*co*-hexafluoropropylene) P(VDF-*co*-HFP), poly(ethylene-*alt*-tetrafluoroethylene) (PETFE), and poly(ethylene-*alt*-chlorotrifluoroethylene) (PECTFE) have been widely studied [1, 2]. Fluoropolymers composed of flexible backbones and fluoroalkyl side chains have been of great interest due to their unique properties, including excellent chemical resistance, good mechanical properties, good thermal stability, and ferroelectric, piezoelectric, and dielectric properties, which originated from the low polarity, strong electronegativity and small van der Waals radius (1.32 Å) of the fluorine atoms, as well as the strong C–F bond (485 kJ/mol) [3–5].

Fluoropolymers can be readily processed into membranes which have found applications in ultrafiltration, microfiltration, wastewater treatment, protein adsorption and separation, proton conduction, stimuli–responsive and controlled deliveries, and biotechnology [6–10]. However, the practical applications of fluoropolymer membranes are limited to some extent by their hydrophobic and inert surface properties.

Handbook of Fluoropolymer Science and Technology, First Edition. Edited by Dennis W. Smith Jr., Scott T. Iacono, and Suresh S. Iyer.

Thus, the ability to manipulate and control the surface properties of fluoropolymer membranes is of crucial importance to extend and enhance their functionalities.

To date, remarkable progress has been made in the fabrication of fluoropolymer membranes with high performance for improving their practical applications [11–14]. The traditional approaches to the modification and functionalization of fluoropolymer membranes have included blending [14], ozonization [15], plasma treatment [16–18], γ-ray irradiation and electron beam (EB) exposure [19,20]. These approaches, though very useful, offer a somewhat limited opportunity for molecular engineering and design of the surface of fluoropolymer membranes. Fabrication of high performance and multifunctional fluoropolymer membranes remains a challenge. The present article aims to provide a comprehensive review of the recent progress in preparation of functionalized and functionalizable fluoropolymer membranes. The article also provides an insight into the selection of suitable controlled/living radical polymerization (CRP) techniques and click reactions for the synthesis of fluoropolymer membranes with tailored pore and surface functionalities and morphology. The highly robust and versatile CRP techniques and click chemistry are particularly suited for the preparation of functional membranes with stimuli–responsive, antifouling, antimicrobial, and ion exchange properties, as well as membranes with functionalizable surfaces, including pore surfaces.

8.2 PREPARATION OF FLUOROPOLYMER MEMBRANES

Microporous membranes are very similar in structure and function to conventional filters. It has a rigid, highly voided structure with randomly distributed, interconnected pores, on the order of 0.01–10 μm [21]. Separation of solutes by microporous membranes is mainly a function of solute size and membrane pore size distribution. Fluoropolymers are commonly used in the production of microporous membranes [22]. Several methods are employed in the fabrication of fluoropolymer membranes, including phase inversion, sintering, track etching, and electrospinning [23]. To date, most of the commercial membranes, including fluoropolymer membranes, are produced via phase inversion mainly because of the simplicity and flexibility to scale up production, resulting in a low cost of production.

8.2.1 Phase Inversion

Phase inversion is known to be an effective way to create porous structures in membranes, where a competitive mutual diffusion between solvent and nonsolvent occurs to yield the porous structure. Phase inversion can be described as a demixing process whereby the initially homogeneous polymer solution is transformed in a controlled manner from a liquid to a solid state [24]. Apart from immersion in a nonsolvent bath, or immersion precipitation (IP), a variety of related techniques, such as precipitation by solvent evaporation, precipitation by absorption of water from the vapor phase, and precipitation by air cooling, corresponding to thermally induced phase separation (TIPS), vapor-induced phase separation (VIPS), and air-casting phase separation

(ACPS), have also been used to fabricate microporous membrane [24–29]. However, among these techniques, TIPS and IP are the two most commonly employed methods in the fabrication of polymeric membranes including fluoropolymer membranes. One of the main difficulties that limit the successful casting of phase inversion membranes is the lack of a predictable and systematic method in the selection of solvent and nonsolvent systems [30].

Of all the techniques, IP is among the first to be commercially explored and is one of the most popular membrane formation methods because it allows for the preparation of membranes of different morphologies [24–26]. IP is a process where a polymer solution is cast on a suitable support or extruded through a die, then immersed in a nonsolvent liquid bath. The polymeric membrane solidifies through the exchange of solvent in polymer solution with the nonsolvent from the coagulation bath. Most of the commercially available fluoropolymer membranes, with the exception of fully fluorinated polymer membranes, are produced via the IP method because of the ease of dissolution of the polymers in common organic solvents. The phase-separation behavior of a fluoropolymer is often complicated by its semi-crystalline nature [31]. IP is more complicated than TIPS, because at least three components are involved, and because complex diffusion and convection processes play an important role. The research and development of fluoropolymer membranes prepared via the IP process have focused on the effect of preparation conditions on membrane morphology and performance, as well as the relationship between membranes and structure of polymers. The key factors that influence phase inversion membrane formation by IP include the solvent, polymer type and concentration, nonsolvent system type and composition, additives to the polymer solution, and membrane casting conditions (evaporation time, temperature, pH, among others).

Apart from IP, TIPS is another useful technique for the preparation of polymeric membranes by the phase inversion process [32]. In the preparation of membrane via TIPS, a homogeneous solution is prepared by dissolving the polymer in a high boiling point, low molecular weight diluent at high temperature. The solution is cast into desired shapes and then cooled to induce phase separation and solidification of the polymer. The diluent is subsequently removed to produce the microporous membrane structure. The major difference that distinguishes between IP and TIPS is that the latter requires removal of thermal energy to convert the solution into a two-phase mixture, while the former involves an exchange between solvent and nonsolvent for membrane formation.

ACPS and VIPS are two efficient methods to bring about highly porous fluoropolymer membranes with pore size in the micron and submicron ranges. ACPS uses air, and VIPS uses vapor (usually water vapor) as the nonsolvent phase to bring the composition of a cast membrane into the phase separation to generate the porous structures in the membrane. When a low volatility solvent is employed to prepare the cast solution, the evaporation of solvent is slower than the intake of nonsolvent, resulting in a condition of low polymer concentration near the membrane surface. The low polymer concentration near the membrane surface explains why ACPS and VIPS are efficient for preparing membranes with highly porous surface. Another feature of ACPS and VIPS is that the nonsolvent is taken into the cast membrane with a much slower rate, because the process introduces a gas phase mass transfer resistance to the membrane formation process. The slower nonsolvent intake rate in the ACPS and VIPS processes favors the polymer crystallization process over the liquid–liquid demixing process, resulting in highly porous structures.

As an alternative approach, supercritical CO_2 has recently been used as a nonsolvent to induce phase separation of polymer solution for the purpose of obtaining dry membranes [33–35]. Supercritical CO_2 is nontoxic, nonflammable and of low cost. Supercritical CO_2 can accelerate the drying process of membrane surfaces without collapsing the pore structures due to the absence of liquid–vapor interfaces. Thus, no additional post-treatment is needed for obtaining the dry membranes. This method can also be considered as environmentally friendly because the solvent that initially exists in the polymer solution can be easily recovered in the supercritical CO_2 phase after the exchange of solvent and CO_2. Although membranes prepared via this method showed porous and macrovoid-free structure, the obtained membrane pore sizes are relatively large, between 1 and 10 μm.

8.2.2 Track Etching

In the track etching method, polymer films or foils with discrete pores are formed through a combination of high energy particle (heavy metal ions) bombardment and chemical etching in acid or alkaline bath [36]. The massive particle bombardment results in the formation of latent tracks through the polymer films or foils. Controlled chemical etching of this modified zone leads to the formation of pores with a defined pore size. By choosing appropriate conditions for irradiation and etching, it is possible to prepare a wide range of polymer track-etched membranes, including fluoropolymer membranes. The membrane porosity is mainly determined by the duration of radiation, while the pore size is determined by the etching time and temperature.

8.2.3 Sintering

As a widely used method in the production of inorganic membranes, sintering has been introduced in the fabrication of symmetric PTFE and polypropylene (PP) membranes [26]. Because of their low solubility in organic solvents and high melting temperatures, PTFE and PP membranes cannot be formed by other conventional methods. The sintering method involves the compression of powder-containing particles of a given size and sintering at elevated temperatures.

8.2.4 Electrospinning

Electrospinning is a remarkably simple method for producing nanofibers of a wide range of polymers, including fluoropolymers [37]. In a typical electrospinning process, nanofibers are generated by the electrostatic force and sol–gel reaction. The nanofiber can be deposited on a substrate to form a nonwoven membrane. Electrospun nonwoven membranes have several unique properties, such as adjustable pore sizes ranging from nanometers to micrometers, high porosity, interconnected open pore structures, high permeability, large surface area per unit volume, and tailored membrane thickness. The necessary components of an electrospinning apparatus include a high voltage power supply, a capillary tube with a needle or pipette, and a collector which is usually composed of a conducting material.

8.3 FUNCTIONALIZATION OF FLUOROPOLYMER MEMBRANES

8.3.1 Modification by Blending

Polymer blends can improve the processing characteristics as a result of integrating the properties of constituent polymers. The interest in polymer blends lies in (i) obtaining properties different from those of parent homopolymers and copolymers, (ii) improving physical properties, (iii) ease of preparation, and (iv) a wide range of possible blends [4, 26]. Indeed, varying the blend composition can produce a great variety of products endowed with a wide range of properties. However, desirable blend properties depend highly on the degree of molecular interaction between the components. Many polymer pairs tend to form a multiphase mixture because of thermodynamic incompatibility, resulting in poor interfacial adhesion and unstable phase morphology.

8.3.1.1 Membranes from Polymer Blends Poly(ethylene glycol) (PEG) and poly(*N*-vinylpyrrolidone) (PVP) are two main hydrophilic homopolymer additives commonly used for blending with fluoropolymers [27, 35–38]. The main role of PEG or PVP in the modification and preparation of the fluoropolymer membranes by nonsolvent-induced phase inversion is more like a pore-forming agent, because these components are water soluble and can be extracted during membrane preparation and operation processes. Moreover, the addition of these components allows the adjustment of thermodynamics and kinetics of the casting solution to control the morphology, average pore size, and pore size distribution of the resulting membranes. The hydrophilic properties, as well as fouling resistance, are reduced to some extent as a result of removal of the hydrophilic polymer components, which are incompatible with the hydrophobic fluoropolymers.

To realize the long-term hydrophilicity of the blend polymer membranes, the key point is to improve the stability of the hydrophilic component on the membrane surface and matrix. To this end, amphiphilic polymers have been adopted to replace linear hydrophilic PEG and PVP in membrane preparation. Theoretically, improved interactions among the hydrophobic segments in amphiphilic polymers and the fluoropolymer can improve the compatibility of the blend, and thus enhance the stability of amphiphilic polymer in the membrane matrix. Three kinds of amphiphilic polymers, viz., triblock copolymer of poly(ethylene oxide)-*b*-poly(propylene oxide)-*b*-poly(ethylene oxide) (EPE), comb-like copolymer of polysiloxane with poly(ethylene oxide) and poly(propylene oxide) side chains (ACPS), and the hyperbranched star copolymer of polyester-*g*-methoxyl poly(ethylene glycol) (HPE-*g*-MPEG), have been respectively blended with PVDF to fabricate microporous membranes via phase inversion [39–42]. Under the same fabrication conditions, the introduction of each amphiphilic copolymer increased the porosity of the resulting blend membranes. The enrichment of amphiphilic copolymer on the membrane and pore surfaces gave rise to improved hydrophilicity of the blend membrane over that of the pure PVDF membrane. The blend membranes with the three amphiphilic copolymers also exhibited better stability and higher protein resistance than those of the PVDF/PEG blend membrane. PEO-*b*-PMMA amphiphilic copolymers were also used as modifiers for blending with PVDF [43]. PMMA is well known for its miscibility with PVDF and can provide long-term stability to the amphiphilic copolymer in the membrane matrix. The surface-enriched PEO chains endow the PVDF membrane with desired hydrophilicity and fouling resistance. The PVDF/PEO-*b*-PMMA blend membranes showed improved porous structure, electrolyte uptake, and ionic migration. The highest ionic conductivity of PVDF/PEO-*b*-PMMA membrane at room temperature reached 2.79×10^3 S/cm at 30 wt% of PEO-*b*-PMMA in the polymer matrix. In this state, the polymer electrolyte possessed the lowest activation energy for ion transportation. The blend membrane exhibited a good electrochemical stability up to 5.0 V (vs. Li/Li^+).

Chitosan has been identified as a hydrophilic, nontoxic, biocompatible, and antibacterial polymer. Modification of hydrophobic PVDF membrane with chitosan was performed in order to increase the hydrophilicity and to reduce protein fouling [44]. Cellulose acetate (CA) can also be used as a hydrophilic additive to enhance the performance of PVDF microfiltration membranes cast by the phase inversion process [45]. The highly hydrophilic nature of CA on the surface enhances the wettability and the antifouling ability of the fabricated PVDF membranes. The excellent antifouling property of PVDF microporous membranes with CA additives can be reused for several runs with high flux recovery.

Polyaniline (PANi) has been studied extensively for its electroactive characteristics and potential applications in electrical devices, such as polymer electrodes and sensors [46]. Semi-conductive membranes from PVDF/PANi blends in *N*-methyl-2-pyrrolidone (NMP) solutions were prepared by phase inversion in an aqueous solution of poly(styrenesulfonic acid) (PSSA) [47]. Entrapment of a "stoichiometric" amount of PSSA dopant molecules into the blend membrane occurred during phase inversion process and gave rise to a semi-conductivity membrane. At a PANi content of above 15 wt%, the entrapped PSSA chains were present in stoichiometric amount and dispersed evenly throughout the blend membrane. The membranes prepared by this method had an asymmetry structure with a dense skin layer and a porous inner layer. The surface resistance of the blend membrane decreased with the increase in PANi weight fraction. A surface resistance of about 10^5 Ω/cm was obtained for the PSSA-doped PVDF/PANi (65/35, w/w) membrane.

8.3.1.2 Composite Membranes Owing to their nanoscale sizes and ultra large surface areas, inorganic nanoparticles often possess unexpected electronic, magnetic, optical, thermal, and mechanical properties. Inorganic nanoparticles, including titanium dioxide (TiO_2), silicon dioxide (SiO_2), alumina (Al_2O_3), zinc oxide

(ZnO), zirconium dioxide (ZrO_2), and magnesium hydroxide ($Mg(OH)_2$), have been introduced to improve the fluoropolymer membrane performance [48–60]. Blending with inorganic nanoparticles is also expected to provide the fluoropolymer membranes with desired chemical, mechanical and thermal stability, adaptability to the harsh environments, and antifouling properties. Uniform dispersion of nanoparticles in the polymer matrix is of crucial importance to the fabrication of high performance organic–inorganic hybrid membranes.

Damodar et al. prepared the modified PVDF membranes by adding different amounts of TiO_2 nanoparticles to the casting solution, and investigated their photocatalytic, antifouling and antibacterial properties [48]. Addition of TiO_2 nanoparticles significantly affected the pore size and hydrophilicity of the membranes and thus improved the flux and permeability of the modified PVDF/TiO_2 composite membranes. TiO_2-doped PVDF membranes also exhibit better antifouling and bactericidal capabilities, as well as self-cleaning property arising from photocatalytic degradation of the foulants under UV exposure, in comparison with the pristine PVDF membrane. Similar results were obtained for the addition of nanosized ZnO into PVDF membranes [49]. The water permeability of the modified membrane almost doubled with the addition of 6.7 wt% ZnO nanoparticles, which was determined to be the optimum dosage for PVDF composite membrane. In addition, the mechanical strength of the modified membranes was found to be reinforced.

However, the beneficial effects of added nanoparticles are usually limited by aggregation and poor adhesion to the polymer matrix. High concentrations of nanoparticles often lead to poor mechanical stability and membrane formation. The entrapped nanoparticles have a possibility to plug the internal pores. Studies have revealed that TiO_2 nanoparticles deposited on PVDF membrane exhibited greater antifouling effect in comparison with that of TiO_2 nanoparticles-entrapped membrane [50]. Hydrogen-bonding interaction between TiO_2 nanoparticles and carboxylic acid groups of poly(styrene-*alt*-maleic anhydride) (SMA) allows the TiO_2 nanoparticles to tightly assemble onto the membrane surface. The resulting hybrid PVDF membranes exhibit enhanced hydrophilicity, improved permeability, and good fouling resistance in contrast to the PVDF/SMA blend membrane. Similar results were obtained in the case of deposition of nanosized Al_2O_3 on the PVDF membrane surface [51]. An optimum amount of nanosized Al_2O_3 (2 wt%) was used to acquire better mechanical and thermal properties, as well as hydrophilicity, permeability, and fouling resistance. Another strategy to improve the surface properties of fluoropolymer membranes lies in the addition of prefunctionalized nanoparticles to fluoropolymers during the phase inversion process. Surface segregation of PEG-functionalized SiO_2 nanoparticles during the phase inversion process caused the spontaneous migration of the nanoparticles toward polymer–water interface, leading to the surface enrichment of PEG-functionalized SiO_2 nanoparticles [52]. The flux recovery ratio after physical cleaning increased by 27% for the membrane with the addition of 0.5 wt% PEG-functionalized SiO_2 nanoparticles, in comparison with the pristine PVDF membrane. The stability of chemically modified SiO_2 nanoparticles in the membrane matrix, during membrane filtration and cleaning processes, prolongs the long-term usage of the composite membrane.

8.3.2 Modification by Graft Copolymerization

The practical applications of fluoropolymer membranes especially in the areas of purification and separation related to potable water production, wastewater treatment and bioprocessing, have been limited to some extent by their hydrophobic and inert surface properties. Among the different modification techniques, graft copolymerization of hydrophilic monomers, or inimers for further surface reactions, from fluoropolymers has been useful and effective in improving the physicochemical properties of the parent fluoropolymer with minimum alteration of their desirable bulk properties. Apart from fully fluorinated polymers, most of the partially fluorinated polymers can dissolve in polar organic solvents, such as *N,N*-dimethylformamide (DMF), *N,N*-dimethylacetamide (DMAc), NMP, and dimethyl sulfoxide (DMSO), but are insoluble in water, alcohols, and hydrocarbons.

Due to the absence of functionalities on the fluoropolymer backbone or chain ends, introduction of appropriate initiators or coupling agents to the fluoropolymers is the first step toward copolymerization via wet chemistry. Activation of fluoropolymers can be accomplished via several methods, including ozonization, γ-ray irradiation, EB exposure, and controlled/living radical polymerization (CRP).

8.3.2.1 Functionalized Fluoropolymers via Conventional Radical Graft Polymerization for Membrane Fabrication Ozone pretreatment is one of the most straightforward methods to activate the fluoropolymers. As a strong oxidizing agent, ozone can react with inorganic and organic compounds directly or indirectly via the formation of hydroxyl radicals [61, 62]. Ozone pretreatment of fluoropolymers, which can be used to introduce certain polar groups, such as carbonyl, ketone or peroxide groups, is an activation step for subsequent free-radical graft copolymerization. The ozone treatment is only effective for the partially fluorinated polymers. Generally, the amount of peroxides in fluoropolymers introduced by ozone treatment can be regulated by the treatment temperature, ozone concentration, and treatment time. Take PVDF for example; the peroxides content of the ozone-treated PVDF can be determined by assaying with 2,2-diphenyl-1-picrylhydrazyl (DPPH). The results are showed in Table 8.1 [62]. Under the particular treatment conditions, the desired time for ozone pretreatment is about 15 minutes and the half-life for decomposition of peroxides on the ozone-pretreated PVDF was estimated to be about 45 minutes at 60°C. Thus, the thermally induced graft copolymerizaiton time of 6 hours should be sufficient for the complete consumption of peroxides on the PVDF chains [61, 62].

Molecular modification of ozone-pretreated PVDF via thermally induced graft copolymerization with *N*-isopropylacrylamide (NIPAM) was carried out in NMP solution to produce the PVDF-*g*-PNIPAM copolymers [63]. The synthetic route is shown in Scheme 8.1. Microporous membranes were fabricated from the PVDF-*g*-PNIPAM graft copolymers by phase inversion in aqueous media. The surface composition and morphology of the resulting PVDF-*g*-PNIPAM membranes can be adjusted by the casting temperature, while the flux through the membrane exhibits temperature-responsive behavior. In addition to pure water flux, the copolymer membranes also exhibited reversible temperature-dependent permeability to calcein and

TABLE 8.1 Peroxides Content, Intrinsic Viscosity, and Water Contact Angle of Pristine and Ozone-Treated PVDF

Sample	Peroxides content[a] (mol/g of polymer)	$[\eta]^b$	Water contact angle (°) of the film cast from acetone solution	[O]/[C] ratio[c]
Pristine PVDF	0	1.63	133	0
5-min ozone-treated PVDF	5×10^{-5}	1.32	128	0.003
15-min ozone-treated PVDF	10×10^{-5}	1.17	120	0.007
30-min ozone-treated PVDF	13×10^{-5}	1.04	112	0.01

[a] Determined from reaction with DPPH.
[b] Intrinsic viscosity at 25°C in NMP.
[c] Determined from the corrected O1s and C1s X-ray photoelectron spectroscopy (XPS) core-level spectral area ratio of the respective sample (obtained at the photoelectron take-off angle, α, of 75°).

fluorescein isothiocyanate (FITC)-dextran in phosphate buffer solutions (PBS, pH 7.4), with the most drastic change in permeability being observed in the temperature range between 27°C and 32°C [64]. Extension of this synthetic strategy to the graft copolymerization of other functional monomers, including 4-vinylpyridine (4VP), 2-vinylpyridine (2VP), acrylic acid (AAc), *N,N′*-dimethyl-(methylmethacryloyl ethyl) ammonium propanesulfonate (DMAPS), and vinylbenzyl viologen (VBV), from

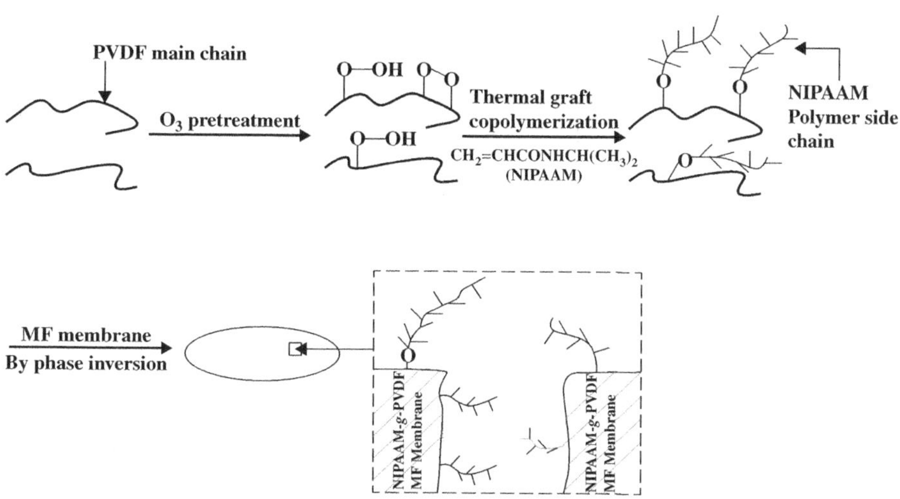

SCHEME 8.1 Schematic illustration of the processes of thermally induced graft copolymerization of NIPAM from the ozone-preactivated PVDF backbone and the preparation of the PVDF-*g*-PNIPAM microporous membrane by phase inversion. PVDF, poly(vinylidene fluoride); NIPAM, *N*-isopropylacrylamide. Reprinted with permission from Reference 63. Copyright 2002 American Chemical Society.

ozone-preactivated PVDF chains has also been carried out [65–74]. Microporous membranes were fabricated from the corresponding graft copolymers by phase inversion in aqueous media. All of the resulting membranes exhibit interesting stimuli-responsive fluxes in response to environmental stimuli, such as temperature, pH, ionic strength, and chemical cues. The AAc graft-copolymerized PVDF (PVDF-*g*-PAAc) membranes were further blended with PNIPAM homopolymer to obtain the temperature and pH-sensitive microporous membranes, or functionalized via covalent immobilization of a biologically active species, such as an enzyme [75, 76].

High energy irradiation-induced graft copolymerization is a commonly used technique for the modification of polymeric materials. High energy irradiations such as γ-ray and electron beam can be generated from commercially available source, and the grafting degree can be easily controlled by regulating the irradiation and reaction parameters. The free radicals generated by high energy irradiation are able to initiate graft polymerization of many monomers in various states, such as solution, bulk, emulsion, and even in the solid state, and under a wide range of temperature [19, 20]. Moreover, high energy irradiation-induced graft copolymerization can simplify the whole treatment process, leave behind no unfavorable residues, and reduce the cost of production. PVDF-*g*-PNVP graft copolymers were synthesized from simultaneous γ-ray irradiation-induced graft copolymerization of *N*-vinyl-2-pyrrolidone (NVP) from PVDF chains in a homogeneous system. The graft yield increased as the monomer concentration was increased under constant irradiation conditions [77]. Microporous membranes were fabricated from the PVDF-*g*-PNVP graft copolymers by phase inversion in an aqueous medium. The introduction of PNVP components to PVDF chains substantially improved the hydrophilicity of the PVDF membranes. The PVDF-*g*-PNVP membranes exhibit good antifouling properties with negligible cytotoxicity effects, making the membranes potentially useful for biomaterials and biomedical applications. Xu et al. prepared the PVDF-*g*-PNIPAM copolymers using graft copolymerization of NIPAM from γ-ray preactivated PVDF in water. The weight percentage of grafted PNIPAM chains in the graft copolymers increased almost linearly with the increase in monomer concentration in the reaction medium [78]. Due to surface segregation of PNIPAM chains during membrane formation, the PNIPAM concentration on the PVDF/PVDF-*g*-PNIPAM blend membrane surfaces was higher than that in the PVDF-*g*-PNIPAM copolymer. As the amount of PVDF-*g*-PNIPAM copolymer in the blend membrane was increased, the water flux increased drastically due to the pore-forming ability and enhanced membrane hydrophilicity of the amphiphilic additive. PVDF powders were also grafted with *N,N*-dimethylacrylamide (DMAA), acrylamide (AAm), AAc, and methacrylic acid (MAA) via the γ-ray irradiation-induced graft polymerization method [79–81]. The resulting PVDF-*g*-PDMAA, PVDF-*g*-PAAm, PVDF-*g*-PAAc, and PVDF-*g*-PMAA copolymers were processed into microporous membranes. The resulting copolymer membranes exhibited better antifouling properties due to enhanced hydrophilicity of the membrane surface.

8.3.2.2 Functionalized Fluoropolymers via Controlled/Living Radical Graft Copolymerization for Membrane Fabrication In general, bulk copolymerization via the conventional free-radical process usually leads to uncontrolled size distribution of the block/graft chains and complex macromolecular architectures. With the recent progress in polymerization methods, it has become possible to design and obtain well-defined fluoropolymers by CRP techniques, including nitroxide-mediated radical polymerization (NMRP), reversible addition–fragmentation chain transfer (RAFT) polymerization, atom transfer radical polymerization (ATRP), and iodine transfer polymerization (ITP) [4, 27, 82–88]. CRP techniques involve additives capable of temporarily deactivating the radical chains and drastically reducing the irreversible termination reactions via recombination and disproportionation. All along the process, a large majority of dormant chains are in equilibrium with a very small number of active ones. CRP techniques have provided a versatile route to the synthesis of well-controlled polymers with predetermined molecular weight (MW), narrow molecular weight distribution (MWD), well-defined architectures, and useful end-functionalities.

One of the distinct advantages of NMRP and RAFT polymerization lies in their relative simplicity and versatility, since conventional free-radical polymerization can be readily converted into NMRP or a RAFT-mediated process by the addition of appropriate nitroxide or chain transfer agent (CTA), while retaining other reaction parameters [82–85]. The mechanisms of NMRP and RAFT polymerization are illustrated in Scheme 8.2. NMRP or RAFT polymerization is governed by reversible transfer reactions between the growing radicals (active species) and nitroxides or CTAs (dormant species). The equilibrium is eventually reached between the propagating radical species and the dormant nitroxide-capped or CTA-containing species to give rise to well-defined polymer chains with narrow MWD [83–85].

Stable nitroxide-free radicals, such as 2,2,6,6-tetramethylpiperidinyl-1-oxy (TEMPO), are used as thermally labile capping agents, to permit reversible trapping of growing radicals to form dormant species (alkoxyamines) with a thermally labile C–OH bond. The NMRP strategy involved modification of PVDF by EB irradiation, and the free radicals formed were immediately quenched with TEMPO [89]. Graft copolymerization of styrene (St) from the nitroxide-capped macroinitiators was carried out via NMRP. After sulfonation of the grafted polystyrene (PSt) chains to PSSA, the obtained PVDF-*g*-PSSA copolymers were processed into membranes and were compared to those obtained from the conventional pre-irradiation grafting method. The former membranes have been used in a fuel cell for 930 hours at 70°C without any drop in current density while the latter membranes have failed within 150–200 hours under similar conditions. Though NMRP was restricted to a limited number of monomers in its early versions, notable advances have been accomplished with the appearances of novel nitroxides. Polymerization of various monomers including St, acrylates, acrylamides, and dienes and their derivatives, can be achieved with the aid of advanced NMRP techniques [83].

RAFT polymerization has also been applied to the modification of fluoropolymers, PVDF in particular [90–92]. The peroxides generated on the ozone-pretreated PVDF facilitated the thermally initiated graft copolymerization of poly(ethylene glycol methacrylate) (PEGMA) in the RAFT-mediated process. RAFT polymerization involves a reversible addition–fragmentation cycle, in which transfer of a dithioester

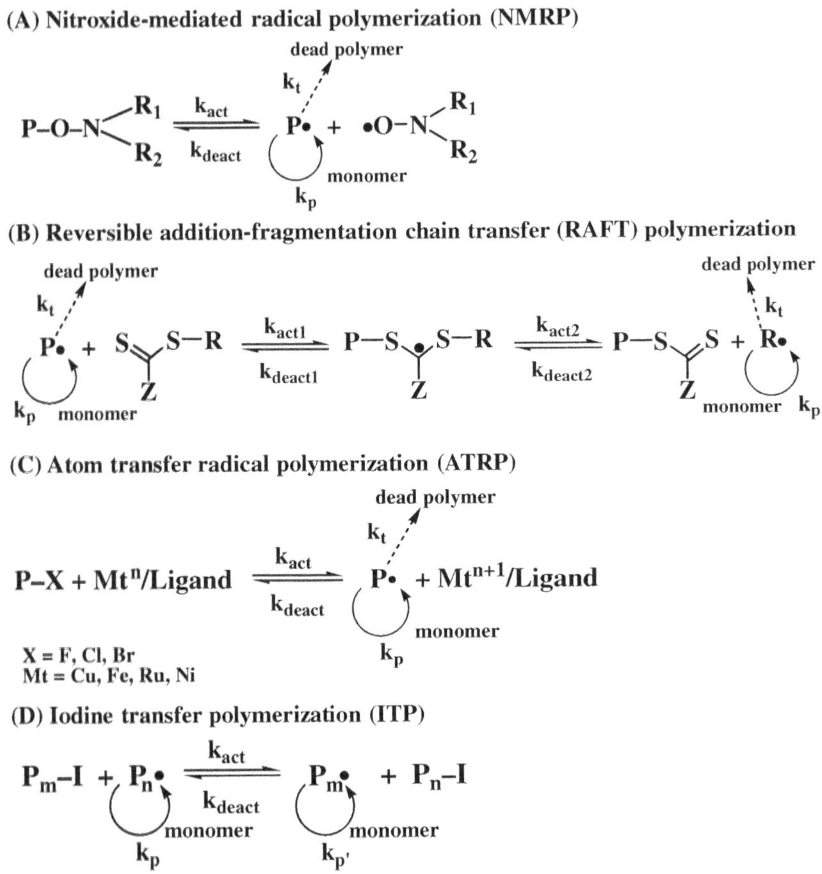

SCHEME 8.2 General schemes of NMRP, RAFT polymerization, ATRP, and ITP processes.

moiety between the active and dormant species maintains the controlled character of the polymerization process. Amphiphilic graft copolymers were prepared via living radical graft polymerization of PEGMA from PVDF in the RAFT-mediated process using 1-phenylethyl dithiobenzoate (PDB) as the CTA [91]. Microporous membranes fabricated from the PVDF-*g*-PPEGMA comb copolymers by phase inversion exhibited a uniform pore size distribution, good hydrophilicity, and substantial resistance to γ-globulin fouling. PVDF-*g*-PAAc copolymers have also been synthesized via a similar route and then cast into pH-sensitive microporous membranes with surface-tethered PAAc chains, which can be further functionalized via surface-initiated RAFT polymerization with NIPAM to obtain the PVDF-*g*-PAAc-*b*-PNIPAM microporous membranes, which can exhibit both pH- and temperature-dependent permeability to aqueous media (Scheme 8.3) [92].

ATRP is one of the methods in reversible-deactivation radical polymerization that produces well-defined vinyl polymers [86–88]. The mechanism of ATRP is shown

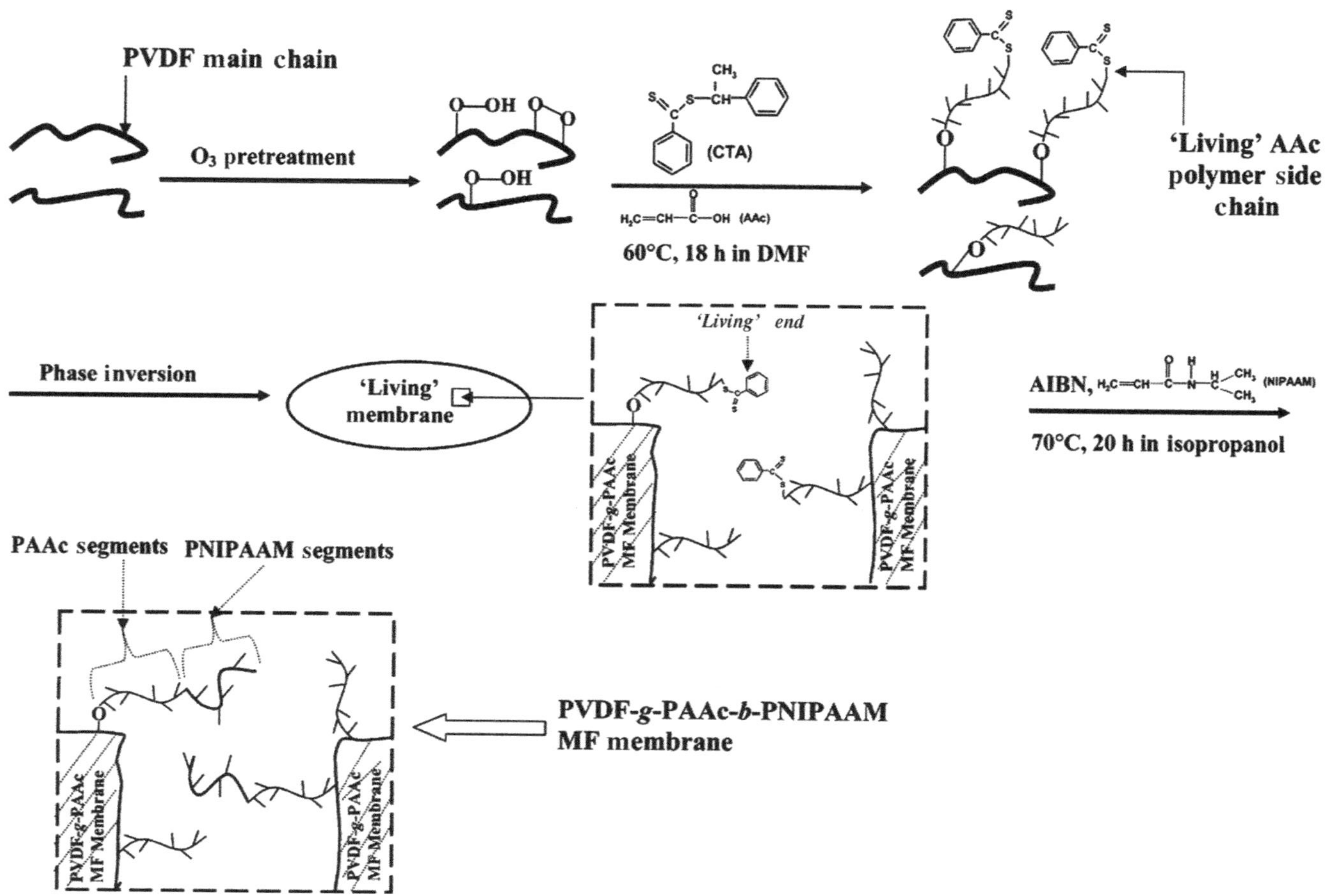

SCHEME 8.3 Schematic illustration of the synthesis of PVDF-*g*-PAAc copolymer by RAFT-mediated graft polymerization, the preparation of the PVDF-*g*-PAAc membrane with living surfaces, and the preparation of the pH- and temperature-sensitive PVDF-*g*-PAAc-*b*-PNIPAM microporous membrane via surface-initiated block copolymerization. PVDF, poly(vinylidene fluoride); AAc, acrylic acid; NIPAM, *N*-isopropylacrylamide. Reprinted with permission from Reference 92. Copyright 2004 American Chemical Society.

in Scheme 8.2. The catalyst complex establishes a dynamic equilibrium between the propagating radicals (active species) and dormant chain ends. When the concentration of propagating radicals is sufficiently low in comparison with the dormant species, the proportion of terminated chains from molecular recombination can often be neglected. Thus, by keeping the concentration of active species or propagating radicals sufficiently low, termination can be suppressed throughout the polymerization process to produce mostly well-defined polymer chains [86–88].

ATRP requires the introduction of suitable alkyl halides for initiation of polymerization of vinyl monomers from the fluoropolymers [93–95]. Graft copolymerization of ATRP inimers bearing alkyl halide groups, such as vinyl benzyl chloride (VBC), 2-(2-bromoisobutyryloxy)ethyl acrylate (BIEA), and 2-(2-bromoisobutyryloxy)ethyl methacrylate (BIEM), from fluoropolymers is a commonly used method to immobilize the ATRP initiators. The obtained copolymers allowed the initiation of ATRP of functional monomers from the copolymer and the copolymer membrane surfaces. Sundholm's group performed the synthesis of PVDF-*g*-PVBC graft copolymers by irradiation of PVDF with EB, followed by graft polymerization of VBC [93]. The resulting PVDF-*g*-PVBC graft copolymer acted as suitable macroinitiators for the ATRP of St, using CuBr/bipyridine as the catalytic system, to produce the well-defined PVDF-*g*-P[VBC-*g*-PSt] graft copolymers. The degree of PSt grafting increased linearly with time, regardless of the polymerization temperature, to reach 400% after 3 hours at 120°C. The driving force of this effort was the high degree of grafting obtainable that could not have been possibly achieved from conventional uncontrolled irradiation-induced grafting methods because of the presence of termination reactions. The PSt graft chains can be further sulfonated, leading to PVDF-*g*-P[VBC-*g*-PSSA] graft copolymers for use as proton exchange membranes in fuel cell applications. In another work, PVDF-*g*-PBIEA graft copolymer was synthesized via graft copolymerization of BIEA from ozone-pretreated PVDF in NMP solution [94]. An arborescent copolymer was prepared via ATRP of sodium 4-styrenesulfonate (NaSS) initiated from the PVDF-*g*-PBIEA copolymer side chains (PVDF-*g*-PBIEA-*g*-PNaSS). Microporous membranes were fabricated from the copolymer solution by phase inversion. In comparison with the PVDF-*g*-PBIEA-*g*-PNaSS copolymer membrane cast in water by phase inversion, the copolymer membrane cast in 1 M aqueous NaCl solution had enriched PNaSS side chains on the membrane and pore surface and larger pore size [95]. In addition, the PVDF-*g*-PBIEA-*g*-PPEGMA membrane was prepared via surface-initiated ATRP of PEGMA from the membrane and pore surfaces. Protein adsorption studies indicated that the PVDF-*g*-PBIEA-*g*-PPEGMA membrane exhibited substantially improved antifouling properties [95].

Fluoropolymers, such as PVDF, PCTFE, and their copolymers, comprise monomer units with pendent secondary fluorine/chlorine atoms, which are potential ATRP initiators. In principle, graft copolymerization initiated from these secondary halides should be possible to allow the preparation of functional graft chains, even though the reaction suffers from low initiation efficiency and requires high reaction temperature. Russell et al. reported the graft copolymerization of St and *tert*-butyl acrylate (*t*BA) directly from the secondary chlorine atoms in P(VDF-*co*-CTFE) via ATRP and illustrated the ability of fluorine atoms to activate the chlorine atoms toward atom transfer [96]. By hydrolyzing the P*t*BA side chains in the graft copolymers, amphiphilic P(VDF-*co*-CTFE) copolymers with grafted PAAc side chains were prepared. These amphiphilic fluoropolymers have potential applications for the preparation of filtration membranes with engineered surface properties. In addition, a series of P(VDF-*co*-CTFE)-*g*-PSSA graft copolymers, consisting of P(VDF-*co*-CTFE) backbone and PSSA side chains, have been synthesized via the similar ATRP process [97]. The obtained graft copolymers were converted into polymer electrolyte membranes. The ion exchange capacity (IEC), water uptake, and proton conductivity of the resulting membranes increased continuously with the increase in PSSA content. A great motivation for current research into functionalized fluoropolymers is the prospect for producing alternatives to the costly fuel cell membrane, Nafion. P(VDF-*co*-HFP)-*g*-PSSA and P(VDF-*co*-CTFE)-*g*-PSSA graft copolymers of different molecular and equivalent weights have been synthesized and the influence of structures on the electrochemical properties of resulting membranes has been investigated [98,99]. Difference in morphologies between these two copolymers provides an understanding of the swelling and conductivity behaviors, and proton mobility through the membranes.

Mayes' group demonstrated the direct preparation of amphiphilic graft copolymers by ATRP of poly(oxyethylene methacrylate) (POEM) monomers initiated from the secondary fluoride sites of PVDF backbone. Surface segregation of PVDF-*g*-POEM additives in PVDF was examined as a route to wettable and foul-resistant surfaces of PVDF microporous membranes [100–102]. ATRP of St, PEGMA, NaSS, 2-(*N*,*N*-dimethylamino)ethyl methacrylate (DMAEMA), butyl methacrylate (BMA), and 2VP, initiated directly from the vinylidene fluoride (VDF) units of PVDF main chains, has also been reported [103–109].

Activator generated by electron transfer (AGET) ATRP has been successfully applied to graft polymerize PEGMA and DMAEMA from different polymers and substrates in the presence of limited amounts of air. AGET–ATRP utilizes an appropriate reducing agent, such as stannous octoate ($Sn(Oct)_2$) or L-ascorbic acid, to convert the higher oxidation state deactivator (i.e., Cu(II)) to lower oxidation state transition metal activator (Cu(I)), instead of relying on radicals from initiator decomposition [110–113]. For AGET–ATRP, the amount of metal catalyst can be reduced by 10^3 times from that of standard ATRP. This technique eliminates the requirement of deoxygenation of reaction mixtures and storage of metal catalysts, and greatly simplifies the ATRP process. PVDF-*g*-PDMAEMA copolymers were synthesized via VDF-initiated normal ATRP of DMAEMA, using CuBr as the catalyst, *N*,*N*,*N*′,*N*′,*N*″-pentamethyl diethylene triamine (PMDETA) as the ligand. It can be converted into the AGET–ATRP process by simply replacing CuBr with $CuBr_2$ and adding L-ascorbic acid as the reducing agent (Scheme 8.4) [114]. The PVDF-*g*-PDMAEMA copolymers obtained by these two ATRP methods have a highly hydrophobic PVDF main chain and hydrophilic PDMAEMA side chains, and can be cast into multistimuli–responsive microporous membranes with surface-enriched "living" PDMAEMA graft chains. In comparison with the membrane cast from PVDF-*g*-PDMAEMA copolymers with similar [–DMAEMA–]/[$–CH_2–CF_2–$], but synthesized by thermally induced free-radical graft copolymerization of DMAEMA from ozone-pretreated PVDF chains,

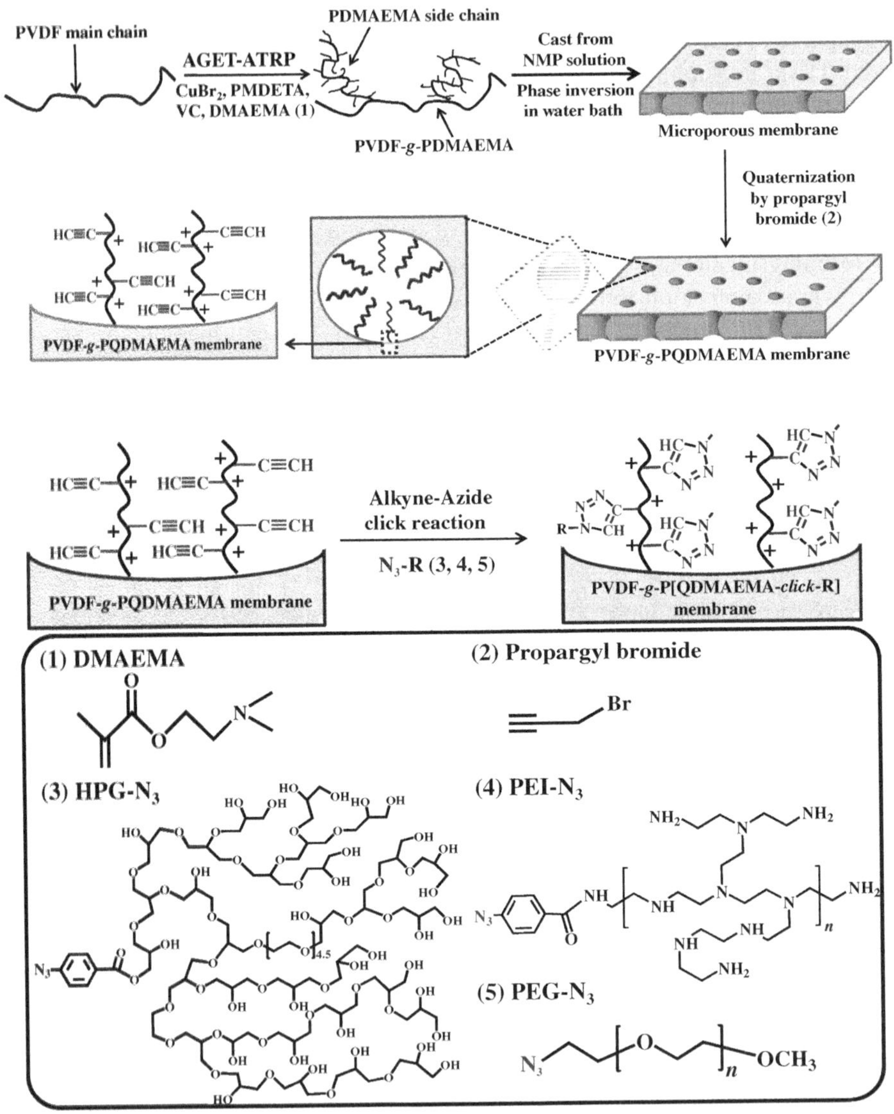

SCHEME 8.4 Schematic illustration of the processes involved in the graft copolymerization of DMAEMA from PVDF main chains via AGET–ATRP, preparation of PVDF-*g*-PDMAEMA membrane by phase inversion, and quaternization of PVDF-*g*-PDMAEMA membrane by propargyl bromide to produce PVDF-*g*-PQDMAEMA membrane with pendant alkyne groups, covalent immobilization of hyperbranched polymer HPG-N_3 or PEI-N_3 onto the PVDF-*g*-PQDMAEMA membrane surface via surface alkyne-azide click reaction. PVDF, poly(vinylidene fluoride); DMAEMA, 2-(*N,N*-dimethylamino)ethyl methacrylate; VC = L-ascorbic acid; PMDETA = *N,N,N′,N′,N′′*-pentamethyl diethylene triamine; HPG-N_3 = *azido*-hyperbranched polyglycerols; PEI-N_3 = *azido*-polyethylenimine. Reprinted with permission from Reference 114. Copyright 2013 American Chemical Society.

the membranes cast from AGET–ATRP copolymers have lower values of standard deviation and relative standard deviation in average pore size. PVDF-*g*-PDMAEMA with more regular PDMAEMA graft chain can provide better control over the morphology and pore size distribution of the PVDF membranes during their casting by phase inversion. The PDMAEMA polymer brushes on the PVDF-*g*-PDMAEMA membrane and pore surfaces can be further quaternized through surface reaction with 3-bromopropionic acid and 1,3-propane sulfone into polyzwitterions [115]. Changes in membrane morphology, when zwitterions are formed, are responsible for the decrease in water flux and increase in water retention.

In view of the inherent good biocompatibility and low toxicity of iron salts, AGET–ATRP of PEGMA directly from the secondary fluorine atoms on the PVDF backbone has been carried out in the presence of a limited amount of air, using $FeCl_3 \cdot 6H_2O$ as the catalyst, triphenylphosphine (PPh_3) as the ligand and L-ascorbic acid as the reducing agent [116]. The PVDF-*g*-PPEGMA copolymers can be cast into hydrophilic microporous membranes by phase inversion in an aqueous medium. The PVDF-*g*-PPEGMA membranes exhibited substantial resistance to bovine serum albumin (BSA) fouling, in comparison with the pristine hydrophobic PVDF membrane.

In comparison with NMRP, ATRP, and RAFT polymerization, ITP is well known as a simple and robust method, which does not use metal catalyst, for the synthesis of well-defined polymers from the less reactive monomers, such as fluoromonomers and phosphoric monomers [117–121]. In the case of ITP of fluoroalkenes, the terminal active bond is the C–I bond originating from the initial iodine-containing CTA and monomer, as shown in Scheme 8.2. ITP is based on the degenerative chain transfer process between dormant chains and propagating radicals, in which the iodine compounds are used as a control agent. Various fluorinated monomers, including tetrafluoroethylene (TFE) and VDF, have been successfully polymerized by ITP. The MW of the resulting fluoropolymers can reach up to 1×10^7 g/mol and yet MWD remains narrow (1.2–1.3). The dormant C–I groups at the fluoropolymer chain ends can be reactivated for another round of ITP of other functional monomers. ITP is suitable for the synthesis of fluoropolymers of various architectures, especially for block copolymers. However, the high cost of purifying the gaseous monomers, unusual process of polymerization, and small scale of production have largely limited the production of fluoropolymers, as well as fluoropolymer membranes, by ITP [117].

8.3.3 Surface Modification and Functionalization of Fluoropolymer Membranes

Surface modification of the polymeric membranes via molecular design is one of the most versatile means to improve the surface properties without affecting bulk properties. Surface modification of fluoropolymer membranes, especially for fully fluorinated polymer membranes, such as PTFE membranes, has been of particular interest, due to their physical and chemical inertness. Surface modification of fluoropolymer membranes can be classified into two categories: surface coating and surface grafting.

8.3.3.1 Surface Coating Surface coating is one of the simplest ways to improve the surface properties of fluoropolymer membranes through coating or deposition of a thin functional layer, via physical adsorption and interaction (including van der Waals forces, electrostatic interaction, or hydrogen bonding), onto the membrane surface. However, passive adsorption could induce competitive adsorption of other materials in the system. In addition, the surface coated layer is susceptible to erosion during the operation and cleaning process due to the relatively weak physical forces between the fluoropolymer membrane and the coated layer.

Recently, the biomimetic approach to form multifunctional polymer coatings through simple dip-coating of substrates in an aqueous solution of dopamine has been widely studied [122–124]. Mimicking the adhesive proteins of mussel, dopamine, which contains catechol and amine functional groups, is oxidized in an alkaline environment and it spontaneously undergoes self-polymerization to form an adherent polydopamine (PDA) coating onto a wide range of substrates, including metal, metal oxide, ceramic and polymer surfaces, with the accompanied oxidation of catechol groups to the quinone form. Versatile and facile approaches to fabricate fluoropolymer membranes with stable structure have been carried out by employing dopamine as the starting material for transition layer construction. The thickness of such a biomimetic coating can be precisely controlled by changing the coating conditions such as the dipping time, solution pH and dopamine concentration. Furthermore, the oxidized quinone form of catechol groups in coated PDA layer can undergo reactions with various functional groups, including thiol, amine, and quinone itself, via Michael addition or Schiff base reaction to form covalently grafted functional layer in secondary modification [122–124].

PDA-coated microporous membranes, including PTFE and PVDF membranes, had been prepared, and secondary modification by heparin grafting has also been carried out. The PDA coating enhances the hydrophilicity of the resultant microporous membranes [125, 126]. Another strategy involves the immobilization of ATRP initiators on the membrane and pore surfaces through the coupling reaction of hydroxyl groups on the PDA coating with 2-bromoisobutyryl bromide (BIBB). DMAEMA and 2-hydroxylethyl methacrylate (HEMA) brushes were grafted from the membrane surface via consecutive surface-initiated ATRP [127]. The antibacterial effects against *Staphylococcus aureus* (*S. aureus*) of the quaternized PVDF-*g*-PQDMAEMA-*b*-PHEMA membranes were examined and the antibacterial efficacy can reach up to 98%.

8.3.3.2 Surface Functionalizing via "Grafting From" and "Grafting To" Approaches Surface grafting is another promising method for modifying the membrane through covalent bonding of the grafted chains to the surface [128]. Covalent attachment offers long-term chemical stability of the grafted chains on the membrane surface, in comparison with that of the physical coating and surface adsorption processes.

Surface modification and functionalization of fluoropolymer membranes can be achieved via two main approaches, viz., (a) the "grafting from" and (b) the "grafting to" approaches (Scheme 8.5). The "grafting from" method incorporates active sites

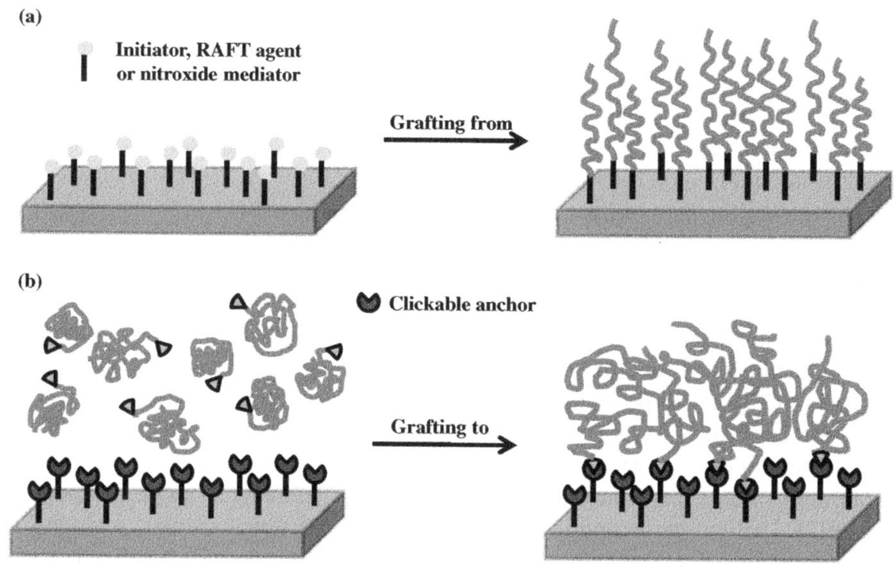

SCHEME 8.5 Functionalization of polymeric membranes via (a) surface-initiated controlled/living radical polymerization (CRP) ("grafting from" approach) and (b) surface click reactions ("grafting to" approach). Reprinted with permission from Reference 129. Copyright 2009 American Chemical Society.

along the backbone from which polymerization can be initiated. Polymerization from these initiating sites results in the formation of polymeric graft chains to afford the graft copolymer. The "grafting from" approach allows the control of grafting degree, chain lengths, and structure through the variation in initiating methods, monomer concentration, reaction temperature, solvent, additive, and other reaction conditions. In order to achieve the ultimate aim of modifying the fluoropolymer membrane with improved functions, synergy between the excellent intrinsic properties of the fluoropolymer membrane (e.g., mechanical strength, chemical and thermal stability) and the novel functionalities of the grafted layer (e.g., hydrophilicity and fouling resistance) must be carefully tailored. Pre-activation of the fluoropolymer membrane surfaces can be achieved by ozone and plasma treatments, high energy irradiation, and CRP.

Due to their physical and chemical inertness, fluoropolymers are resistant to UV irradiation in the wavelength range of 100–400 nm. It is difficult to directly activate the fluoropolymer membrane surface via UV initiation due to its inert property. A photo-initiator, such as benzophenon, is usually needed to initiate the photo graft polymerization process from the fluoropolymer membrane surfaces [130]. Benzophenon was decomposed into radicals. The free radicals are transferred to the fluoropolymer membrane and hydrogen abstraction leads to the generation of initiating radicals. Benzophenon can be coated onto the membrane surface by dipping the membrane in benzophenon solution in an adsorption process [130, 131]. This method could minimize the homopolymerization of monomer and enhance the grafting efficiency. Both monomer concentration and polymerization time can influence the grafting degree. PVDF microporous membranes have been modified by graft polymerization of DMAEMA and 2-acrylamido-2-methyl-1-propanesulfonic acid (AMPS) using benzophenon as the photo initiator to demonstrate the antibacterial action against *Escherichia coli* (*E. coli*) [130].

Surface modification of fluoropolymer membranes by plasma treatment is industrially attractive, as the technique is easy to implement, reliable, and cost effective. Plasma treatment affects the polymer surface to an extent of several hundred to several thousand angstroms. The bulk properties of fluoropolymer membranes, therefore, remain unchanged. Apart from being a surface-sensitive modification technique, plasma treatment does not give rise to toxic waste problems as in the case of chemical treatment [132]. The type of functionalization imparted can be varied by selection of plasma gas (Ar, N_2, O_2, H_2O, CO_2, NH_3, among others) and operation parameters (pressure, power, time, and gas flow rate). Inert gases are most commonly used to introduce radical sites on the polymer surface for subsequent surface modification [132]. Defluorination and incorporation of new functional species, such as the oxygen- and nitrogen-containing species, are generally observed as a result of plasma treatment. When the Ar-plasma-preactivated membranes were exposed to air, these radicals react with oxygen in air to form peroxides [133–135]. PVDF microporous membranes with surface-immobilized PEG were prepared by the Ar plasma-induced grafting of PEG [133]. PEG was precoated on the membrane and pore surfaces, by dipping the membrane in a PEG/$CHCl_3$ solution, prior to the Ar plasma exposure. The flux decreased with increasing surface concentration of the grafted PEG polymer, while the pore size remained almost unchanged. Protein adsorption experiments revealed that the PVDF-*g*-PEG membrane with a PEG graft concentration ([C–O]/[–CF_2–] molar ratio) above 3.2 exhibited good antifouling property. It was suggested that the grafting sites were produced by the cleavage of weaker C–H bond (413 kJ/mol) by Ar plasma. Another strategy involved the preparation of PVDF-*g*-PAAc membranes by plasma-induced grafting of PAAc. Heparin (an antithrombin material) was then immobilized onto the PVDF-*g*-PAAc membranes either by the dry or by the immersion precipitation method [134]. Blood compatibility tests indicated that the immobilized heparin could effectively inhibit platelet adhesion on membrane surfaces. Poly(glycidyl methacrylate) (PGMA) was also grafted onto the PVDF membrane by means of plasma-induced graft polymerization of GMA [135]. Then, L-lysine or 1,6-hexanediamine was reacted with the epoxy groups in the grafted PGMA chains to create an extra cellular matrix (ECM) environment suited for cell culture.

In comparison with the initiating ability of UV irradiation and plasma treatment, high energy irradiation (e.g., electron beam and γ-ray) is another efficient way to activate fluoropolymer membrane and initiate the graft copolymerization of functional monomer from the membrane and pore surfaces [136–138]. Ochoa et al. have studied the PVDF membrane surfaces modified by γ-ray-induced graft copolymerization of GMA [136]. In order to obtain charged membranes, the oxirane rings of PGMA graft chains on the PVDF-*g*-PGMA membrane was efficiently sulfonated by reaction with sodium sulfite. The resulting charged membranes exhibited very low irreversible fouling, high oil retention, and low chemical oxygen demand (COD) to the permeate stream.

Hydrophilic modification of the PTFE membranes is performed for treatment of potable water and waste water. Xu et al. reported the graft copolymerization of AAc and NaSS binary monomers from the high energy EB pre-irradiated PTFE microporous membrane [137]. The AAc/NaSS binary monomers exhibited synergistic effect and improved the graft polymerization reaction with the pre-irradiated PTFE membrane. The water flux decreased with the increase in NaSS concentration. The grafted polymer brushes on the PTFE membrane decreased the water contact angle and produced a stable hydrophilic surface.

Among the different CRP techniques that are available, surface-initiated ATRP and RAFT have been most extensively used to grow polymer brushes from the membrane and pore surfaces [129, 139–143]. Both methods can tailor and manipulate the structure and length of the grafted chains on the membrane and pore surfaces, which in return, are able to tune the pore size and pore size distribution to avoid pore blocking. Furthermore, ATRP and RAFT can produce more uniform and smoother surfaces, which have a more prominent effect on protein adsorption and fouling during the microfiltration or ultrafiltration process, than conventional radical polymerization.

In order to introduce polymer brushes, the membrane surfaces need to be modified with an appropriate ATRP initiator, which can be introduced either from the corresponding fluoropolymers during the phase inversion process, or directly to the fluoropolymer membrane surfaces. Surface initiators can be immobilized on the PVDF membrane by surface hydroxylation and esterification of the hydroxyl groups covalently linked to the surface with 2-bromoisobutyrate bromide (BIBB) [141]. Thereafter, polymer brushes of MMA and PEGMA were grafted from the PVDF membrane surfaces by surface-initiated ATRP. Because of the higher reactivity of alkyl bromide in the initiation of ATRP, bromide can be introduced onto the PVDF membrane via ozone treatment and subsequent bromination reaction. The PVDF membrane grafted with PDMAPS and PPEGMA brushes exhibited improved fouling resistance to BSA filtration [142]. Moreover, it has also been reported that network-like PPEGMA on the PVDF membrane surface adsorbed less BSA than brush-like PPEGMA because of the higher hydration capacity [142].

Besides ATRP, RAFT polymerization can also be employed to graft hydrophilic polymer brushes from the PVDF membrane surface. The PVDF membrane was firstly functionalized with hydroxyl groups, which can react with 4,4′-azobis(4-cyanopentanoic acid) (ACPA) to produce the activated PVDF membrane to allow the subsequent graft copolymerization of MMA or PEGMA in the presence of a CTA via RAFT polymerization [143]. The modified membrane can be further graft copolymerized with DMAEMA to produce the diblock copolymer brushes on the PVDF membrane surface.

On the other hand, the "grafting to" approach involves the attachment of a prefabricated polymer with suitable functionality via covalent bonding to reactive sites on the membrane surface (Scheme 8.5). In comparison with direct graft copolymerization of the corresponding monomers from the membrane surfaces, which sometimes requires harsh reaction conditions or involves undesirable side reactions, the "grafting to" approach allows the covalent attachment of well-defined polymer chains onto the membrane surfaces in a controlled manner and under mild conditions. Due to steric congestion that can occur when trying to couple polymer chains onto a substrate, the grafting density of polymer brushes introduced by the "grafting to" approach via conventional coupling reactions is typically low [144]. One strategy used to increase the grafting density is by employing highly efficient coupling reactions that will maximize the extent of reaction between the polymer graft chains and membrane surface [145–148]. The first successful example is the Cu(I)-catalyzed 1,3-dipolar cycloaddition reaction of an azide and an alkyne, which has been applied in nearly all areas of modern chemistry from biology to materials science due to its high coupling efficiency, high selectivity, mild reaction conditions, functional group tolerance, and no extraneous byproducts. In addition, the ease of synthesis of alkyne and azide functionalities on polymer chains and substrates allows the direct and controlled functionalization of polymeric membranes via one-step surface-initiated "click" ("grafting to") reactions [149–154].

Several other highly efficient coupling reactions have been reported to be of "click" in nature since they fulfilled all or some of the click chemistry criteria. Several types of metal-free click reactions have been developed. The more prominent ones are thiol-ene, thiol-yne, thiol-*para*-fluoro, hetero Diels–Alder (HDA) coupling reactions, and pyridyl disulfide exchange reactions (Scheme 8.6) [155–162]. Each of them requires specific initiators, monomers, and/or polymerization conditions. Therefore, the synthetic route should be carefully designed to prepare the desired functional polymers or solid substrates.

The combination of click reactions with CRP techniques has enabled the access to a multitude of complex and sophisticated polymeric architectures. The great success of this approach owes to the availability of different CRP techniques which allows the introduction and transformation of specific end groups. The use of functional initiators in CRP processes lead to end-functionalized polymers in one step. Thiol-terminated polymers can be easily obtained by RAFT polymerization of a wide range of monomers and subsequent cleavage of the CTA. Alternatively, disulfide containing bifunctional ATRP initiators can be used for the preparation of thiol functionalized polymers. Polymers with an azide end group can be easily prepared by converting the bromide end group obtained from the ATRP process, into an azide [163–166].

In comparison with St derivatives, MMA derivatives bearing clickable functionalities, such as propargyl methacrylate (PMA), 2-azidoethyl methacrylate (AEMA), allyl methacrylate (AMA), and furfuryl methacrylate (FMA), are much more favorable for modification of fluoropolymers, due to their hydrophilicity and low reaction temperature. Thermally induced graft copolymerization of these inimers from ozone-preactivated PVDF chains was carried out in NMP solution [167, 168]. The inimers incorporated into the PVDF chains, which have also been brought to the membrane and pore surfaces during the phase inversion process in an aqueous medium, serve as a versatile platform for further functionalization of the membrane surface via one-step surface-initiated click reaction. For example, the AMA graft-copolymerized PVDF (PVDF-*g*-PAMA) membranes were further functionalized via thiol-ene click reaction of 3-mercaptopropionic acid (MPA) to obtain the pH-sensitive PVDF-*g*-P[AMA-*click*-MPA] membranes. Due to pH-induced conformational transition of the P[AMA-*click*-MPA] side chains, the permeability of aqueous solutions through the PVDF-*g*-P[AMA-*click*-MPA] membranes displayed pronounced pH dependence. Similarly, the PMA graft-copolymerized PVDF (PVDF-*g*-PPMA) membranes were further functionalized via thiol-yne click reaction of 3-mercapto-1-propanesulfonic acid sodium salt (MPS) to obtain the PVDF-*g*-P[PMA-*click*-MPS] membranes, or via alkyne-azide click reaction of mono(6-azido-6-desoxy)-β-cyclodextrin (*azido*-β-CD) to obtain the PVDF-*g*-P[PMA-*click*-β-CD] membranes (Scheme 8.7) [168]. Due to the antipolyelectrolyte effect, the permeability of aqueous solutions through the PVDF-*g*-P[PMA-*click*-MPS] membranes exhibited a dependence on the ionic strength, and the flow rate was observed to decrease as the ionic strength of permeate was increased. Host–guest recognition between β-CD and adamantine groups (AD) end capped PEO has been used to immobilize PEO polymer onto the PVDF-*g*-P[PMA-*click*-β-CD] membrane surfaces [168]. The hydrophilicity of PEO polymers,

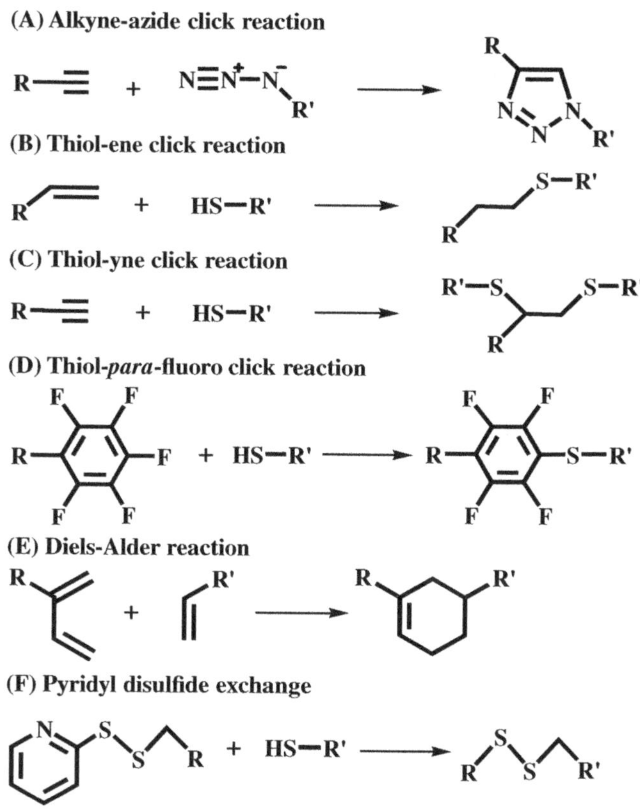

SCHEME 8.6 Schematic representation of click reactions that have been used in combination with CRP techniques.

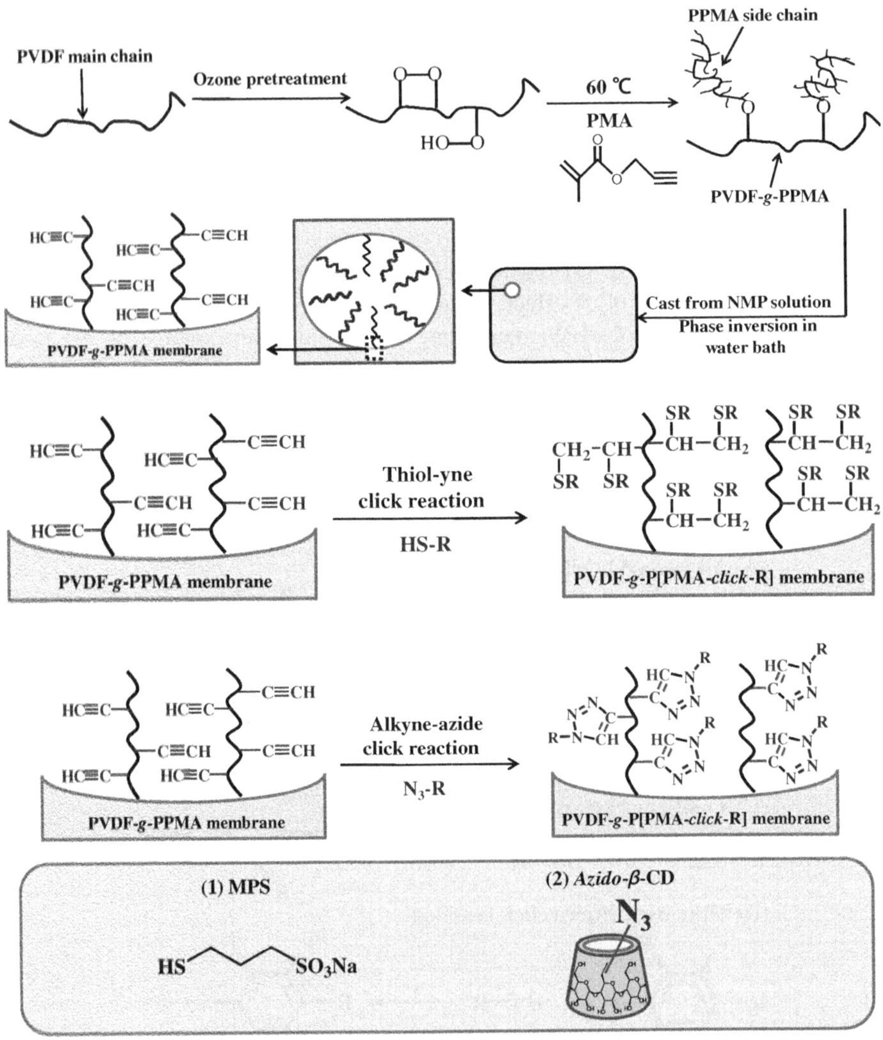

SCHEME 8.7 Schematic illustration of the process of ozone pretreatment, graft copolymerization PVDF with PMA, preparation of PVDF-*g*-PPMA membrane with "clickable" surface by phase inversion, preparation of functional PVDF-*g*-P[PMA-*click*-MPS] membranes via surface thiol-yne click reaction of thiols on the PVDF-*g*-PPMA membrane, and preparation of PVDF-*g*-P[PMA-*click*-β-CD] membranes via surface alkyne-azide click reaction on the PVDF-*g*-PPMA membranes. PVDF, poly(vinylidene fluoride); PMA, propargyl methacrylate; MPS, 3-mercapto-1-propanesulfonic acid sodium salt; *azido*-β-CD, mono(6-azido-6-desoxy)-β-cyclodextrin. Reprinted with permission from Reference 168. Copyright 2011 American Chemical Society.

in turn, imparts the resulting PVDF-*g*-P[PMA-*click*-β-CD-*guest*-PEO] membrane with enhanced fouling resistance to protein permeates.

Postmodification reactions are used to convert latent functional groups into clickable units, whereby the polymer chains can be modified either at pendant or at terminal positions. For this purpose, commercially available monomers, such as glycidyl methacrylate (GMA), DMAEMA, HEMA, and VBC can be used as latent clickable monomers. Thermally induced graft copolymerization of these monomers, from ozone-preactivated PVDF chains, was carried out in NMP solution [113, 169]. Microporous membranes were fabricated from the corresponding graft copolymers by phase inversion in an aqueous medium. The oxirane rings in the PGMA graft chains of the PVDF-*g*-PGMA copolymer or the corresponding membranes can react readily with sodium azide in ring opening reaction to produce the PVDF-*g*-P[GMA-(N_3)(OH)] copolymer [164]. Alkyne-functionalized PNIPAM, prepared a priori by ATRP, was used for the click reaction with azide-containing PVDF-*g*-P[GMA-(N_3)(OH)] copolymer as well as the corresponding membrane surfaces. The surface composition and morphology of the PVDF-*g*-P[GMA-*click*-PNIPAM] membranes can be adjusted by the temperature of casting medium, while the flux through both types of membranes exhibits thermoresponsive behavior. The PVDF-*g*-PDMAEMA membranes can be quaternized by propargyl bromide to immobilize the propargyl groups onto the membrane and pore surfaces. The resulting PVDF-*g*-PQDMAEMA membrane bearing pendant propargyl moieties could be further functionalized via surface-initiated alkyne-azide click reaction of *azido*-terminated hyperbranched polyglycerols (HPG-N_3) to form the PVDF-*g*-P[QDMAEMA-*click*-HPG] membranes (Scheme 8.4). The PVDF-*g*-P[QDMAEMA-*click*-HPG] membranes exhibit better resistance to protein adsorption and fouling than that of PVDF-*g*-P[QDMAEMA-*click*-PEG] membrane, and that of starting PVDF-*g*-PDMAEMA membrane [170]. It has also been predicted theoretically that, at an equivalent area per molecule, a branched polymer will be more efficient in protein rejection than its linear counterpart [171]. Alternatively, alkyne-azide click reaction of *azido*-terminated polyethylenimine (PEI-N_3) on the PVDF-*g*-PQDMAEMA membranes, followed by quaternization with 1-bromohexane, produce the PVDF-*g*-P[QDMAEMA-*click*-QPEI] membranes which is effective in reducing bacterial growth and proliferation under continuous-flow conditions.

Through the intelligent choice of monomers with appropriate functional groups, new molecular functionalities can be incorporated onto the activated fluoropolymer membrane surfaces via surface-initiated CRP or click reactions. PVDF-*g*-[PBIEM-*co*-PPMA] copolymer with grafted copolymer side chains of ATRP inimer (BIEM) and "clickable" inimer (PMA) was synthesized via graft copolymerization of BIEM and PMA from the ozone-preactivated PVDF backbones (Scheme 8.8) [172]. Microporous membranes were fabricated from the PVDF-*g*-[PBIEM-*co*-PPMA] copolymers by phase inversion in an aqueous solution. The PBIEM and PPMA graft chains on the membrane and pore surfaces provide functionalities for the respective surface-initiated ATRP of the viologen-containing monomer (BVbpy) and alkyne-azide click reactions with *azido*-terminated PNIPAM. The resulting PVDF-*g*-[P(BIEM-*g*-PBVbpy)-*co*-P[PMA-*click*-PNIPAM] membrane exhibited both

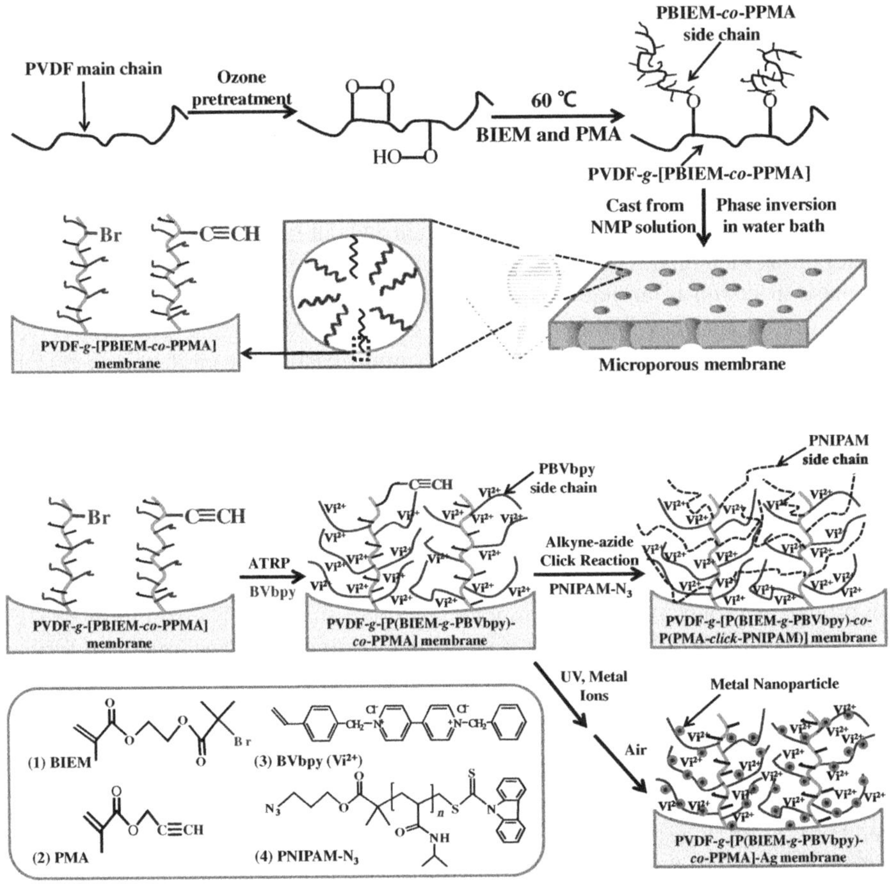

SCHEME 8.8 Schematic illustration of the process of ozone pretreatment and graft copolymerization of PVDF with BIEM and PMA, preparation of PVDF-*g*-[PBIEM-*co*-PPMA] membrane with dual functionalities on the surface by phase inversion. Preparation of the redox-sensitive PVDF-*g*-[P(BIEM-*g*-PBVbpy)-*co*-PPMA] membrane via surface-initiated ATRP of BVbpy from the PVDF-*g*-[PBIEM-*co*-PPMA] membrane, preparation of the redox- and temperature-sensitive PVDF-*g*-[P(BIEM-*g*-PBVbpy)-*co*-P(PMA-*click*-PNIPAM)] membrane via surface alkyne-azide click reaction of PNIPAM-N_3 on the PVDF-*g*-[P(BIEM-*g*-PBVbpy)-*co*-PPMA] membrane, and deposition of metal nanoparticles (Ag) on the PVDF-*g*-[P(BIEM-*g*-PBVbpy)-*co*-PPMA] membrane. PVDF, poly(vinylidene fluoride); BIEM, 2-(2-bromoisobutyryloxy)ethyl methacrylate); PMA, propargyl methacrylate; BVbpy, *N*-benzyl-*N*′-(4-vinylbenzyl)-4,4′-bipyridium dichloride; PNIPAM-N_3, *azido*-poly(*N*-isopropylacrylamide). Reprinted with permission from Reference 172. Copyright 2011 The Royal Society of Chemistry.

redox- and temperature-dependent permeability to aqueous solutions. Alternatively, chemical interconversion of the viologen-containing PBVbpy brushes can be coupled to the metal reduction process for immobilizing metal nanoparticles, such as Ag, Au, or Pt nanoparticles, on the PVDF-*g*-[P(BIEM-*g*-PBVbpy)-*co*-P[PMA-*click*-PNIPAM] membrane. A significant reduction in microbial adhesion and biofilm formation was observed on the PVDF-*g*-[P(BIEM-*g*-PBVbpy)-*co*-PPMA]-Ag membrane, arising from the combined action of the quaternary ammonium ions of the PBVbpy units and the Ag nanoparticles. The PVDF microporous membranes with both functionalities for the "grafting from" and "grafting to" the surface thus provide a versatile platform for multifunctionalization of the fluoropolymer membrane surface toward application in water treatment, aqueous separation and controlled release.

8.4 SUMMARY

The traditional approaches to the modification of fluoropolymer membranes have included blending, ozonization, plasma treatment, γ-ray irradiation, and electron beam exposure. These approaches, though very useful, offer a somewhat limited opportunity for molecular engineering and design of the fluoropolymer membranes. This review summarizes the recent attempts to incorporate the feature of surface molecular design and engineering into the traditional approaches. Most of these attempts are based on the intelligent choice of functional monomers for graft copolymerization on the fluoropolymers as well as fluoropolymer membranes, which have been pre-activated or pre-modified by the traditional approaches. The advent of various CRP techniques and click chemistry has made it possible to produce fluoropolymer membranes with an unprecedented level of control over composition, structure, and properties. Due to the controlled character of CRP techniques and the quantitative yield of click reactions, CRP techniques, in combination with click chemistry, will continue to play an important role in the design and synthesis of precursor polymers and copolymers with precisely controlled architecture and tailored properties to impart novel functionalities to the fluoropolymer membranes.

REFERENCES

1. Souzy, R.; Ameduri, B.; Boutevin, B. *Prog. Polym. Sci.* **2004**, *29*, 75.
2. Junginckel, B. J.; Salamone, J. C. *The Polymeric Materials Encyclopedia*; CRC Press: Boca Raton, FL, 1996; Vol. 7, pp 7114–7123.
3. Seiler, D. A. PVDF in the Chemical Process Industry. In *Modern Fluoropolymers;* Scheirs, J., Ed.; John Wiley & Sons, Inc.: New York, 1997; Chapter 25, pp 487–506.
4. Ameduri, B. *Macromolecules* **2010**, *43*, 10163.
5. Ameduri, B. *Chem. Rev.* **2009**, *109*, 6632.
6. Ameduri, B. *Prog. Polym. Sci.* **2001**, *26*, 105.
7. Souzy, R.; Ameduri, B. *Prog. Polym. Sci.* **2005**, *30*, 644.

8. Kang, E. T.; Zhang, Y. *Adv. Mater.* **2000**, *12*, 1481.
9. Hirao, A.; Sugiyama, K.; Yokoyama, H. *Prog. Polym. Sci.* **2007**, *32*, 1393.
10. Lee, H.; Pietrasik, J.; Sheiko, S. S.; Matyjaszewski, K. *Prog. Polym. Sci.* **2010**, *35*, 24.
11. Wandera, D.; Wickramasinghe, S. R.; Hussom, S. M. *J. Membr. Sci.* **2010**, *357*, 6.
12. Stamatialis, D. F.; Papenburg, B. J.; Girones, M.; Saiful, S.; Bettahalli, S. N. M.; Schmitmeier, S.; Wessling, M. *J. Membr. Sci.* **2008**, *308*, 1.
13. Mishra, A. K.; Bose, S.; Kuila, T.; Kim, N. H.; Lee, J. H. *Prog. Polym. Sci.* **2012**, *37*, 842.
14. Mucha, M. *Prog. Polym. Sci.* **2003**, *28*, 837.
15. Mansour, O. Y.; Nagaty, A. *Prog. Polym. Sci.* **1985**, *11*, 91.
16. Schnabel, W.; Sotobayashi, H. *Prog. Polym. Sci.* **1983**, *9*, 297.
17. Forsythe, J. S.; Hill, D. J. T. *Prog. Polym. Sci.* **2000**, *25*, 101.
18. Nasef, M. M.; Guven, O. *Prog. Polym. Sci.* **2012**, *37*, 1597.
19. Denes, F. S.; Manolache, S. *Prog. Polym. Sci.* **2004**, *29*, 815.
20. Dargaville, T. R.; George, G. A.; Hill, D. J. T.; Whittaker, A. K. *Prog. Polym. Sci.* **2003**, *28*, 1355.
21. Gryta, M.; Barancewicz, M. *J. Membr. Sci.* **2010**, *358*, 158.
22. Xu, Z.; Li, F.; Wu, F.; Tan, S.; Zhang, Z. *J. Membr. Sci.* **2005**, *255*, 125.
23. Rahimpour, A.; Madaeni, S. S.; Amirinejad, M.; Mansourpanah, Y.; Zereshki, S. *J. Membr. Sci.* **2009**, *330*, 189.
24. Guillen, G. R.; Pan, Y. J.; Li, M. H.; Hoek, E. M. V. *Ind. Eng. Chem. Res.* **2011**, *50*, 3798.
25. Van de Witte, P.; Dijkstra, P. J.; van den Berg, J. W. A.; Feijen, J. *J. Membr. Sci.* **1996**, *117*, 1.
26. Liu, F.; Hashim, N. A.; Liu, Y.; Moghareh Abed, M. R.; Li, K. *J. Membr. Sci.* **2011**, *375*, 1.
27. Uragami, T.; Naito, Y.; Sugihara, M. *Polym. Bull.* **1981**, *4*, 617.
28. Munari, S.; Bottino, A.; Capannelli, G. *J. Membr. Sci.* **1983**, *16*, 181.
29. Li, X.; Wang, Y.; Lu, X.; Xiao, C. *J. Membr. Sci.* **2008**, *320*, 477.
30. Klein, E.; Smith, J. K. *Product R&D* **1972**, *11*, 207.
31. Wienk, I. M.; Boom, R. M.; Beerlage, M. A. M.; Bulte, A. M. W.; Smolders, C. A.; Strathmann, H. *J. Membr. Sci.* **1996**, *113*, 361.
32. Li, C.-L.; Wang, D.-M.; Deratani, A.; Quemener, D.; Bouyer, D.; Lai, J.-Y. *J. Membr. Sci.* **2010**, *361*, 154.
33. Cao, J.-H.; Zhu, B.-K.; Ji, G.-L.; Xu, Y.-Y. *J. Membr. Sci.* **2006**, *266*, 102.
34. Huang, S.; Wu, G.; Chen, S. *J. Membr. Sci.* **2007**, *293*, 100.
35. Reverchon, E.; Cardea, S. *Ind. Eng. Chem. Res.* **2006**, *45*, 8939.
36. Komaki, Y. *Nucl. Tracks* **1979**, *3*, 33.
37. Agarwal, S.; Greiner, A.; Wendorff, J. H. *Prog. Polym. Sci.* **2013**, *38*, 963.
38. Simone, S.; Figoli, A.; Criscuoli, A.; Carnevale, M. C.; Rosselli, A.; Drioli, E. *J. Membr. Sci.* **2010**, *364*, 219.
39. Rajabzadeh, S.; Liang, C.; Ohmukai, Y.; Maruyama, T.; Matsuyama, H. *J. Membr. Sci.* **2012**, *423*, 189.

40. Cui, Z.-Y.; Xu, Y.-Y.; Zhu, L.-P.; Wang, J.-Y.; Xi, Z.-Y.; Zhu, B.-K. *J. Membr. Sci.* **2008**, *325*, 957.
41. Yuan, Z.; Dan-Li, X. *Desalination* **2008**, *223*, 438.
42. Zhao, Y.-H.; Qian, Y.-L.; Zhu, B.-K.; Xu, Y.-Y. *J. Membr. Sci.* **2008**, *310*, 567.
43. Xiao, Q.; Wang, X.; Li, W.; Li, Z.; Zhang, T.; Zhang, H. *J. Membr. Sci.* **2009**, *334*, 117.
44. Boributh, S.; Chanachai, A.; Jiraratananon, R. *J. Membr. Sci.* **2009**, *342*, 97.
45. Mu, C.; Su, Y.; Sun, M.; Chen, W.; Jiang, Z. *J. Membr. Sci.* **2010**, *350*, 293.
46. Wang, P.; Tan, K. L.; Kang, E. T.; Neoh, K. G. *Appl. Surf. Sci.* **2002**, *193*, 36.
47. Kang, E. T.; Neoh, K. G.; Tan, K. L. *Prog. Polym. Sci.* **1998**, *23*, 277.
48. Oh, S. J.; Kim, N.; Lee, Y. T. *J. Membr. Sci.* **2009**, *345*, 13.
49. Liang, S.; Xiao, K.; Mo, Y.; Huang, X. *J. Membr. Sci.* **2012**, *394*, 184.
50. Li, J.-H.; Xu, Y.-Y.; Zhu, L.-P.; Wang, J.-H.; Du, C.-H. *J. Membr. Sci.* **2009**, *326*, 659.
51. Yan, L.; Li, Y. S.; Xiang, C. B.; Xianda, S. *J. Membr. Sci.* **2006**, *276*, 162.
52. Wu, H.; Mansouri, J.; Chen, V. *J. Membr. Sci.* **2013**, *433*, 135.
53. Yu, L.-Y.; Xu, Z.-L.; Shen, H.-M.; Yang, H. *J. Membr. Sci.* **2009**, *337*, 257.
54. Zuo, X.; Yu, S.; Xu, X.; Xu, J.; Bao, R.; Yan, X. *J. Membr. Sci.* **2009**, *340*, 206.
55. Madaeni, S. S.; Zinadini, S.; Vatanpour, V. *J. Membr. Sci.* **2011**, *380*, 155.
56. Dong, C.; He, G.; Li, H.; Zhao, R.; Han, Y.; Deng, Y. *J. Membr. Sci.* **2012**, *387*, 40.
57. Song, H.; Shao, J.; He, Y.; Liu, B.; Zhong, X. *J. Membr. Sci.* **2012**, *405*, 48.
58. Razmjou, A.; Arifin, E.; Dong, G.; Mansouri, J.; Chen, V. *J. Membr. Sci.* **2012**, *415*, 850.
59. Bottino, A.; Capannelli, G. *Desalination* **2002**, *146*, 35.
60. Liu, X.; Peng, Y.; Ji, S. *Desalination* **2008**, *221*, 376.
61. Liu, Y.; Lee, J. Y.; Kang, E. T.; Wang, P.; Tan, K. L. *React Funct. Polym.* **2001**, *47*, 201.
62. Wang, P.; Tan, K. L.; Kang, E. T.; Neoh, K. G. *J. Mater. Chem.* **2001**, *11*, 783.
63. Ying, L.; Kang, E. T.; Neoh, K. G. *Langmuir* **2002**, *18*, 6416.
64. Ying, L.; Kang, E. T.; Neoh, K. G.; Kato, K.; Iwata, H. *J. Membr. Sci.* **2004**, *243*, 253.
65. Zhai, G. Q.; Toh, S. C.; Tan, W. L.; Kang, E. T.; Neoh, K. G. *Langmuir* **2003**, *19*, 7030.
66. Ying, L.; Wang, P.; Kang, E. T.; Neoh, K. G. *Macromolecules* **2002**, *35*, 673.
67. Ying, L.; Zhai, G. Q.; Winata, A. Y.; Kang, E. T.; Neoh, K. G. *J. Colloid Interf. Sci.* **2003**, *265*, 396.
68. Zhai, G. Q.; Ying, L.; Kang, E. T.; Neoh, K. G. *J. Mater. Chem.* **2002**, *12*, 3508.
69. Zhai, G. Q.; Ying, L.; Kang, E. T.; Neoh, K. G. *Macromolecules* **2002**, *35*, 9653.
70. Zhai, G. Q.; Kang, E. T.; Neoh, K. G. *J. Membr. Sci.* **2003**, *217*, 243.
71. Zhai, G. Q.; Ying, L.; Kang, E. T.; Neoh, K. G. *Surf. Interface Anal.* **2004**, *36*, 1048.
72. Chiang, Y.-C.; Chang, Y.; Higuchi, A.; Chen, W.-Y.; Ruaan, R.-C. *J. Membr. Sci.* **2009**, *339*, 151.
73. Kaur, S.; Ma, Z.; Gopal, R.; Singh, G.; Ramakrishna, S.; Matsuura, T. *Langmuir* **2007**, *23*, 13085.
74. Liu, X.; Neoh, K. G.; Kang, E. T. *Macromolecules* **2003**, *36*, 8631.
75. Ying, L.; Kang, E. T.; Neoh, K. G. *J. Membr. Sci.* **2003**, *224*, 93.
76. Ying, L.; Kang, E. T.; Neoh, K. G. *J. Membr. Sci.* **2002**, *208*, 361.

77. Qin, Q.; Hou, Z.; Lu, X.; Bian, X.; Chen, L.; Shen, L.; Wang, S. *J. Membr. Sci.* **2013**, *427*, 303.
78. Yu, J.-Z.; Zhu, L.-P.; Zhu, B.-K.; Xu, Y.-Y. *J. Membr. Sci.* **2011**, *366*, 176.
79. Yang, X.; Zhang, B.; Liu, Z.; Deng, B.; Yu, M.; Li, L.; Jiang, H.; Li, J. *J. Mater. Chem.* **2011**, *21*, 11908.
80. Deng, B.; Yu, M.; Yang, X.; Zhang, B.; Li, L.; Xie, L.; Li, J.; Lu, X. *J. Membr. Sci.* **2010**, *350*, 252.
81. Yang, X.; Deng, B.; Liu, Z.; Shi, L.; Bian, X.; Yu, M.; Li, L.; Li, J.; Lu, X. *J. Membr. Sci.* **2010**, *362*, 298.
82. Barner-Kowollik, C., Ed. *Handbook of Radical Polymerization*; Wiley-VCH: Weinheim, Germany, 2008.
83. Sciannamea, V.; Jerome, R.; Detrembleur, C. *Chem. Rev.* **2008**, *108*, 1104.
84. Boyer, C.; Bulmus, V.; Davis, T. P.; Ladmiral, V.; Liu, J. Q.; Perrier, S. B. *Chem. Rev.* **2009**, *109*, 5402.
85. Destarac, M. *Polym. Rev.* **2011**, *51*, 163–187.
86. Braunecker, W. A.; Matyjaszewski, K. *Prog. Polym. Sci.* **2007**, *32*, 93.
87. Siegwart, D. I.; Oh, J. K.; Matyjaszewski, K. *Prog. Polym. Sci.* **2012**, *37*, 18.
88. Debuigne, A.; Hurtgen, M.; Detrembleur, C.; Jerome, C.; Barner-Kowollik, C.; Junkers, T. *Prog. Polym. Sci.* **2012**, *37*, 1004.
89. Holmberg, S.; Holmlund, P.; Nicolas, R.; Wilen, C.-E.; Kallio, T.; Sundholm, G.; Sundholm, F. *Macromolecules* **2004**, *37*, 9909.
90. Peng, Q.; Lu, S.; Chen, D.; Wu, X.; Fan, P.; Zhong, R.; Xu, Y. *Macromol. Biosci.* **2007**, *7*, 1149.
91. Chen, Y. W.; Lei, Y.; Yu, W. H.; Kang, E. T.; Neoh, K. G. *Macromolecules* **2003**, *36*, 9451.
92. Ying, L.; Yu, W. H.; Kang, E. T.; Neoh, K. G. *Langmuir* **2004**, *20*, 6032.
93. Holmberg, S.; Holmlund, P.; Wilen, C. E.; Kallio, T.; Sundholm, G.; Sundholm, F. *J. Polym. Sci. A: Polym. Sci.* **2002**, *40*, 591.
94. Zhai, G. Q.; Kang, E. T.; Neoh, K. G. *Macromolecules* **2004**, *37*, 7240.
95. Zhai, G. Q.; Shi, Z. L.; Kang, E. T.; Neoh, K. G. *Macromol. Biosci.* **2005**, *5*, 974.
96. Zhang, M. F.; Russell, T. P. *Macromolecules* **2006**, *39*, 3531.
97. Kim, Y. W.; Lee, D. K.; Lee, K. J.; Kim, J. H. *Eur. Polym. J.* **2008**, *44*, 932.
98. Tsang, E. M. W.; Zhang, Z. B.; Yang, A. C. C.; Shi, Z. Q.; Peckham, T. J.; Narimani, R.; Frisken, B. J.; Holdcroft, S. *Macromolecules* **2009**, *42*, 9467.
99. Tsang, E. M. W.; Zhang, Z. B.; Shi, Z. Q.; Soboleva, T.; Holdcroft, S. *J. Am. Chem. Soc.* **2007**, *129*, 15106.
100. Akthakul, A.; Salinaro, R. F.; Mayes, A. M. *Macromolecules* **2004**, *37*, 7663.
101. Hester, J. F.; Banerjee, P.; Won, Y. Y.; Akthakul, A.; Acar, M. H.; Mayes, A. M. *Macromolecules* **2002**, *35*, 7652.
102. Akthakul, A.; Hochbaum, A. I.; Stellacci, F.; Mayes, A. M. *Adv. Mat.* **2005**, *17*, 532.
103. Hashim, N. A.; Liu, F.; Li, K. *J. Membr. Sci.* **2009**, *345*, 134.
104. Su, Y.-H.; Liu, Y.-L.; Wang, D.-M.; Lai, J.-Y.; Sun, Y.-M.; Chyou, S.-D. *J. Membr. Sci.* **2009**, *345*, 13.

105. Sauguet, L.; Boyer, C.; Ameduri, B.; Boutevin, B. *Macromolecules* **2006**, *39*, 9087.
106. Zhang, Z. C.; Chalkova, E.; Fedkin, M.; Wang, C. M.; Lvov, S. N.; Komarneni, S.; Chung, T. C. M. *Macromolecules* **2008**, *41*, 9130.
107. Samanta, S.; Chatterjee, D. P.; Layek, R. K.; Nandi, A. K. *Macromol. Chem. Phys.* **2011**, *212*, 134.
108. Samata, S.; Chatterjee, D. P.; Manna, S.; Mandal, A.; Garai, A.; Nandi, A. K. *Macromolecules* **2009**, *42*, 3112.
109. Xue, J.; Chen, L.; Wang, H. L.; Zhang, Z. B.; Zhu, X. L.; Kang, E. T.; Neoh, K. G. *Langmuir* **2008**, *24*, 14151.
110. Jakubowski, W.; Min, K.; Matyjaszewski, K. *Macromolecules* **2006**, *39*, 39.
111. Zhu, W. P.; Zhong, M. J.; Li, W. W.; Dong, H. C.; Matyjaszewski, K. *Macromolecules* **2011**, *44*, 1920.
112. Oh, J. K.; Min, K.; Matyjaszewski, K. *Macromolecules* **2006**, *39*, 3161.
113. Luo, R.; Sen, A. *Macromolecules* **2008**, *41*, 4514.
114. Cai, T.; Yang, W. J.; Neoh, K. G.; Kang, E. T. *Ind. Eng. Chem. Res.* **2012**, *51*, 15962.
115. Yi, Z.; Zhu, L.-P.; Xu, Y.-Y.; Gong, X.-N.; Zhu, B.-K. *J. Membr. Sci.* **2011**, *385*, 57.
116. Zhao, T.; Zhang, L. F.; Zhang, Z. B.; Zhou, N. C.; Cheng, Z. P.; Zhu, X. L. *J. Polym. Sci. A: Polym. Chem.* **2011**, *49*, 2315.
117. Boyer, C.; Ameduri, B.; Hung, M. H. *Macromolecules* **2010**, *43*, 3652.
118. Boyer, C.; Ameduri, B.; Boutevin, B.; Dolbier, W. R.; Winter, R.; Gard, G. *Macromolecules* **2008**, *41*, 1254.
119. Boyer, C.; Valade, D.; Sauguet, L.; Ameduri, B.; Boutevin, B. *Macromolecules* **2005**, *38*, 10353.
120. Valade, D.; Boyer, C.; Ameduri, B.; Boutevin, B. *Macromolecules* **2006**, *39*, 8639.
121. Laruelle, G.; Nicol, E.; Ameduri, B.; Tassin, J.-F.; Ajellal, N. *J. Polym. Sci. A: Polym. Chem.* **2011**, *49*, 3960.
122. Faure, E.; Falentin-Daudre, C.; Jerome, C.; Lyskawa, J.; Fournier, D.; Woisel, P.; Detrembleur, C. *Prog. Polym. Sci.* **2013**, *38*, 236.
123. Fan, X.; Lin, L.; Dalsin, J. L.; Messersmith, P. B. *J. Am. Chem. Soc.* **2006**, *127*, 15843.
124. Lee, H.; Dellatore, S. M.; Miller, W. M.; Messersmith, P. B. *Science* **2007**, *318*, 426.
125. Xi, Z.-Y.; Xu, Y.-Y.; Zhu, L.-P.; Wang, Y.; Zhu, B.-K. *J. Membr. Sci.* **2009**, *327*, 244.
126. McCloskey, B. D.; Park, H. B.; Ju, H.; Rowe, B. W.; Miller, D. J.; Freeman, B. D. *J. Membr. Sci.* **2012**, *413*, 82.
127. Sui, Y.; Gao, X.; Wang, Z.; Gao, C. *J. Membr. Sci.* **2012**, *394*, 107.
128. Kato, K.; Uchida, E.; Kang, E. T.; Uyama, Y.; Ikada, Y. *Prog. Polym. Sci.* **2003**, *28*, 209.
129. Barbey, R.; Lavanant, L.; Paripovic, D.; Schuwer, N.; Sugnaux, C.; Tugulu, S.; Klok, H. A. *Chem. Rev.* **2009**, *109*, 5437.
130. Rahimpour, A.; Madaeni, S. S.; Zereshki, S.; Mansourpanah, Y. *Appl. Surf. Sci.* **2009**, *255*, 7455.
131. Denes, F. S.; Manolache, S. *Prog. Polym. Sci.* **2004**, *29*, 815.
132. Bhattacharya, A.; Misra, B. N. *Prog. Polym. Sci.* **2004**, *29*, 767.
133. Wang, P.; Tan, K. L.; Kang, E. T.; Neoh, K. G. *J. Membr. Sci.* **2002**, *195*, 103.

134. Lin, D.-J.; Lin, D.-T.; Young, T.-H.; Huang, F.-M.; Chen, C.-C.; Cheng, L.-P. *J. Membr. Sci.* **2004**, *245*, 137.
135. Young, T.-H.; H.-H. Chang, D.-J. Lin, L.-P. Cheng, *J. Membr. Sci.* **2010**, *350*, 32.
136. Masuelli, M. A.; Grasselli, M.; Marchese, J.; Ochoa, N. A. *J. Membr. Sci.* **2012**, *389*, 91.
137. Xi, Z.-Y.; Xu, Y.-Y.; Zhu, L.-P.; Zhu, B.-K. *J. Membr. Sci.* **2009**, *339*, 33.
138. Chang, Y.; Chang, W. J.; Shih, Y. J.; Wei, T. C.; Hsiue, G. H. *ACS Appl. Mater. Interfaces* **2011**, *3*, 1228.
139. Chen, Y. W.; Liu, D. M.; Deng, Q. L.; He, X. H.; Wang, X. F. *J. Polym. Sci. A: Polym. Chem.* **2006**, *44*, 3434.
140. Singh, N.; Husson, S. M.; Zdyrko, B.; Luzinov, I. *J. Membr. Sci.* **2005**, *262*, 81.
141. Liu, D.; Chen, Y.; Zhang, N.; He, X. *J. Appl. Polym. Sci.* **2006**, *101*, 3704.
142. Chang, Y.; Ko, C. Y.; Shih, Y. J.; Quemener, D.; Deratani, A.; Wei, T. C.; Wang, D. M.; Lai, J. Y. *J. Membr. Sci.* **2009**, *345*, 160.
143. Chen, Y.; Sun, W.; Deng, Q.; Chen, L. *J. Polym. Sci. A: Polym. Chem.* **2006**, *44*, 3071.
144. Roy, D.; Semsarilar, M.; Guthrie, J. T.; Perrier, S. *Chem. Soc. Rev.* **2009**, *38*, 2046.
145. Nebhani, L.; Schmiedl, D.; Barner, L.; Barner-Kowollik, C. *Adv. Funct. Mater.* **2010**, *20*, 2010.
146. Nebhani, L. Sinnwell,S.; Inglis, A. J.; Stenzel, M. H.; Barner-Kowollik, C.; Barner, L. *Macromol. Rapid Commun.* **2008**, *29*, 1431.
147. Carlmark, A.; Malmstrom, E. *Biomacromolecules* **2003**, *4*, 1740.
148. Hansson, S.; Trouillet, V.; Tischer, T.; Goldmann, A. S.; Carlmark, A.; Barner-Kowollik, C.; Malmstrom, E. *Biomacromolecules* **2013**, *14*, 64.
149. Huisgen, R. *Angew. Chem. Int. Ed.* **1963**, *2*, 633.
150. Huisgen, R. *Angew. Chem. Int. Ed.* **1968**, *7*, 321.
151. Kolb, H. C.; Finn, M. G.; Sharpless, K. B. *Angew. Chem. Int. Ed.* **2001**, *40*, 2004.
152. Lewis, W. G.; Green, L. G.; Grynszpan, F.; Radic, Z.; Carlier, P. R.; Taylor, P.; Finn, M. G.; Sharpless, K. B. *Angew. Chem. Int. Ed.* **2002**, *41*, 1053.
153. Meldal, M.; Tornoe, C. W. *Chem. Rev.* **2008**, *108*, 2952.
154. Hawker, C. J.; Wooley, K. L. *Science* **2005**, *309*, 1200.
155. Hoyle, C. E.; Bowman, C. N. *Angew. Chem. Int. Ed.* **2010**, *49*, 1540.
156. Hoyle, C. E.; Lowe, A. B.; Bowman, C. N. *Chem. Soc. Rev.* **2010**, *39*, 1355.
157. Tasdelen, M. A. *Polym. Chem.* **2011**, *2*, 2133.
158. Mansfeld, U.; Pietsch, C.; Hoogenboom, R.; Becer, C. R.; Schubert, U. S. *Polym. Chem.* **2010**, *1*, 1560.
159. Iha, R. K.; Wooley, K. L.; Nystrom, A. M.; Burke, D. J.; Kade, M. J.; Hawker, C. J. *Chem. Rev.* **2009**, *109*, 5620.
160. Golas, P. L.; Matyjaszewski, K. *Chem. Soc. Rev.* **2010**, *39*, 1338.
161. Sheiko, S. S.; Sumerlin, B. S.; Matyjaszewski, K. *Prog. Polym. Sci.* **2008**, *33*, 759.
162. Gregory, A.; Stenzel, M. H. *Prog. Polym. Sci.* **2012**, *37*, 38.
163. Zhao, H. Y.; Kang, X. L.; Liu, L. *Macromolecules* **2005**, *38*, 10619.
164. Tsarevsky, N. V.; Bencherif, S. A.; Matyjaszewski, K. *Macromolecules* **2007**, *40*, 4439.
165. Xia, Y.; Yin, X. C.; Burke, N. A. D.; Stover, H. D. H. *Macromolecules* **2005**, *38*, 5937.

166. Galvin, C. J.; Genzer, J. *Prog. Polym. Sci.* **2012**, *37*, 871.
167. Cai, T.; Wang, R.; Kang, E. T.; Neoh, K. G. *Polym. Chem.* **2011**, *2*, 1849.
168. Cai, T.; Kang, E. T.; Neoh, K. G. *Macromolecules* **2011**, *44*, 4258.
169. Cai, T.; Kang, E. T.; Neoh, K. G.; Teo, S. L. M. *Langmuir* **2011**, *27*, 2936.
170. Voit, B. I.; Lederer, A. *Chem. Rev.* **2009**, *109*, 5924.
171. Xu, Y.; Takai, M.; Ishihara. *Biomacromolecules* **2009**, *10*, 267.
172. Cai, T.; Wang, R.; Yang, W. J.; Lu, S. J.; Neoh, K. G.; Kang, E. T. *Soft Matter* **2011**, *7*, 11133.

9

POLY[METHYL(3,3,3-TRIFLUOROPROPYL)SILOXANE]

Michael J. Owen
Michigan Molecular Institute

9.1 INTRODUCTION

The element fluorine, usually in the form of a fluorocarbon group, is generally incorporated into polymers to confer hydrophobicity, enhanced thermal, chemical, and oxidative stability, reduced adhesion, increased solubility, improved biocompatibility, and increased permeability [1]. With the exception of chemical stability at extremes of pH, silicones such as polydimethylsiloxane (PDMS) already adequately possess such characteristics so it is not immediately obvious what benefits a fluorosilicone polymer might exhibit. Fluorosilicone is a rather imprecise term that could embrace a wide variety of polymer structures as is evident from a recent review by Pasquet and coworkers [2]. However, in practice, it generally means poly[methyl(3,3,3-trifluoropropyl)siloxane] (PMTFPS; see Figure 9.1), accounting for a predominant 96% share of the fluorosilicone market [3]. Sometimes the polymer is referred to as PTFPMS although IUPAC rules indicate that two different substituents should be listed in alphabetical order. Even more rarely, the IUPAC structure-based name of poly(oxymethyl-3,3,3-trifluoropropylsilylene) is encountered.

The principal intent of this chapter is to compare and contrast the properties of this most common fluorosilicone with those of PDMS, the predominant polymer of

Handbook of Fluoropolymer Science and Technology, First Edition. Edited by Dennis W. Smith Jr., Scott T. Iacono, and Suresh S. Iyer.

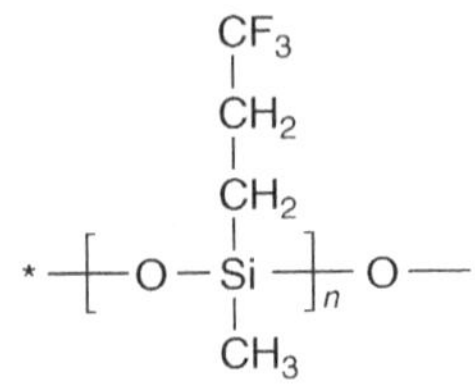

FIGURE 9.1 The structure of PMTFPS. ∗End-groups not shown. They are usually $(CH_3)_3Si$– trimethylsilyl groups.

the silicone industry. This is the purpose of most of the tables but, where appropriate, other polymers including fluoropolymers such as polytetrafluoroethylene (PTFE) are included. The examples are not meant to be exhaustive but are chosen to illustrate significant similarities and differences between the properties of the two polymers, much of which is explicable simply by the bulkier trifluoropropyl group detracting from the high flexibility of the siloxane backbone. Substituting one of the methyl groups on silicon in PDMS hardly changes the wettability by water of the polymer, nor is its thermal or chemical stability improved. The primary value of fluorosilicones such as PMTFPS is in their resistance to solvents, particularly at high temperatures. For example, a PMTFPS fluorosilicone elastomer with a breaking elongation of 550% after being aged in JP-4 jet fuel for 28 days at 260°C still had 250% elongation [4]. These elastomers were first introduced as a commercial product in 1957 [5], over half a century ago. They are mostly used in the auto and aerospace industries as seals, gaskets, and valves because of their good oil and fuel resistance combined with good thermal stability and low temperature flexibility resulting in one of the broadest useful temperature range of any fluorocarbon-based elastomer.

9.2 MANUFACTURE

PMTFPS is one example of the large, general class of organofunctional silicones where the side chain, methyl in the case of PDMS, is modified to obtain special properties or reactivity not obtainable with PDMS. Many organofunctional siloxanes are copolymers with PDMS and even for the homopolymers it is usual to modify only one of the two methyl groups on silicon to retain desirable silicone-like characteristics such as polymer-chain flexibility and solubility in common solvents and also to use more available intermediates and ensure that an adequate degree of polymerization can be achieved. Boutevin and coworkers [6] have surveyed the field of side-group-modified polysiloxanes in general and Pasquet et al. [2] have focused on what they term pendant fluorosilicones in particular.

There are three general routes to these materials [6]: modification of polysiloxanes, particularly polyhydrogenmethylsiloxane (PHMS), $[CH_3(H)SiO]_n$, polycondensation of bis(silafunctional) monomers, and ring-opening polymerization of

TABLE 9.1 Properties of Typical Commercial Siloxane Elastomers [9]

Property	PDMS	PMTFPS
Hardness (shore A)	30–80	20–80
Tensile strength (MPa)	1.55–9.0	5.5–11.7
Elongation (%)	430–725	100–600
Compression set (%) (22 h/177°C)	~10	10–40
Tear strength (die B) (kN/m)	4.9–37.7	10.5–46.6
Bashore resilience (%)	30–65	10–40
Service temperature (°C)	−55 to +260	−68 to +232

functional cyclosiloxanes. This last approach is the most common route for the preparation of PMTFPS. The cyclic trimer 1,3,5-trimethyl-1,3,5-tris(3′,3′,3′-trifluoropropyl)-cyclotrisiloxane (usually designated F_3) is polymerized by acid-catalyzed (e.g., trifluorosulfonic acid) or base-catalyzed, (e.g., *n*-butyl lithium or KOH) ring-opening polymerization. Pasquet et al. [2] noted that trimethylsilylmethyllithium in the presence of cryptands or a mixture of H_2O/phosphazene bases are now preferred industrially, since they polymerize the F_3 trimer so rapidly that the final conversion and the molecular weights are controlled by the kinetics. Kinetic control, whereby the reaction is quenched when maximum molecular weight is achieved before equilibrium, is critical because when fully equilibrated at 383 K with KOH catalyst, 82% cyclics and 18% polymer are formed. When diluted with cyclohexanone solvent, the mixture approaches 100% cyclics [7].

The F_3 cyclic species is obtained through hydrolysis of the corresponding silane, methyl(3,3,3-trifluoropropyl)dichlorosilane. This silane is the hydrosilylation adduct of 3,3,3-trifluoropropene and methyldichlorosilane, CH_3HSiCl_2, a minor product of the Rochow–Muller direct process for manufacturing methylchlorosilanes. The hydrosilylation reaction is catalyzed with transition metal complexes of platinum or rhodium such as Speier's catalyst and hexachloroplatinic acid in isopropanol. The reaction conditions are very similar to those employed with unfluorinated reagents, complicated only by the fact that 3,3,3-trifluoropropene is a gas.

Most commercial PMTFPS polymers are used in the manufacture of peroxide-curable high consistency fluorosilicone elastomer compounds [8]. These elastomers have high molecular weight in 1–2 million range with 0.2–1.0 mol% methylvinylsiloxane to improve the effectiveness of the peroxide free-radical vulcanization. As with all silicone elastomers, they contain a reinforcing filler which is usually a high surface area fumed silica. Also present are a small amount of low molecular weight fluorosilicone diol processing fluid and any pigments, thermal stability enhancers, and extending fillers required by the specific application. The performance of the final elastomer is obviously dependent on these formulation and compounding variables. However, in practice, the properties of typical commercial PMTFPS elastomers are often deliberately tailored to be similar to those of PDMS elastomers. Table 9.1 shows such a comparison of typical elastomer property ranges.

9.3 PROPERTIES

The effect of replacing one methyl group with a trifluoropropyl group in PDMS is quite considerable. Compared to PDMS, PMTFPS exhibits

- an approximately 10-fold increase in viscosity at the same degree of polymerization;
- improved resistance to nonpolar liquids;
- increased specific gravity;
- lower refractive index;
- improved solubility in polar fluids such as esters and ketones;
- improved lubrication behavior;
- increase in glass transition temperature;
- elimination of polymer crystallization temperature;
- slightly reduced high temperature stability;
- increased solubility parameter;
- reduced gas diffusion; and
- surprising surface tension properties.

Some of these changes, such as increased glass transition temperature and reduced gas diffusion are the expected consequence of the bulkier trifluoropropyl group detracting from the marked flexibility of the siloxane backbone. In what follows, the emphasis is on the surface properties of PMTFPS, as these are not so predictable.

9.3.1 Liquid Surface Tension

As noted above, PMTFPS exhibits somewhat surprising surface properties. According to our measurements, its liquid surface tension is higher than that of PDMS, its critical surface tension using *n*-alkane contact angle test liquids is similar to the PDMS value, while its solid surface tension judged by the Owens and Wendt [10] approach with water and diiodomethane liquids has a value considerably less than that of PDMS.

There can be little doubt about the validity of the liquid surface tension (γ_{LV}) values. This property can be readily measured directly, provided adequate time is allowed for higher viscosity polymers to come to surface equilibrium. For a homologous series of liquid polymers, the surface tension usually varies with molecular weight owing to end-group and other effects, so it is wise to compare polymers of similar size or seek a means of eliminating the molecular weight variable. LeGrand and Gaines [11] provide such an opportunity with their equation:

$$\gamma_{LV} = \gamma_{\infty} - K/M_n^{2/3}, \quad (9.1)$$

where γ_{∞} is the surface tension at infinite molecular weight, K is a constant, and M_n is the number average molecular weight. Figure 9.2 shows our data [12] for PMTFPS

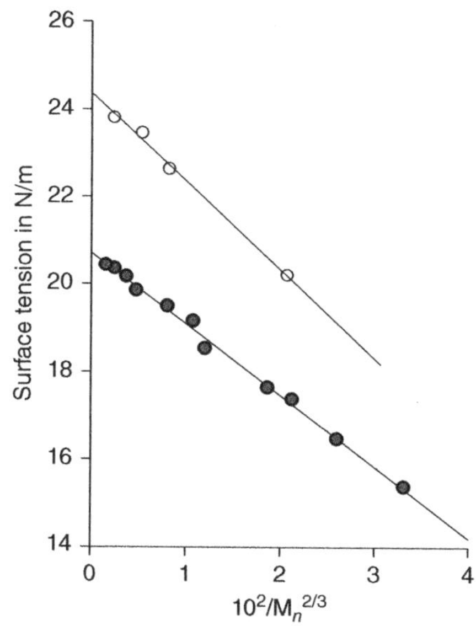

FIGURE 9.2 Liquid surface tension as a function of molecular weight for PDMS (solid circles) and PMTFPS (open circles). Reproduced with permission from Reference 12. Copyright Wiley-VCH Verlag GmbH & Co. KGaA.

and PDMS. All values were obtained by the pendant drop method. The values in the low viscosity range agree well with data obtained by the Wilhelmy plate method. The γ_∞ value for PMTFPS, 24.4 mN/m, is significantly higher than the 20.7 mN/m value for PDMS.

These polymers are commercially available (Dow Corning FS 1265 and Dow Corning Toray Silicone SH 200 fluids) so one remaining objection might be that manufacturing contaminants influence the values obtained. An anomalously low value rather than a high value would be anticipated thermodynamically but it is possible that incompatibility drives higher surface energy impurities to the surface. That this is unlikely is demonstrated by the studies of Nakamura and coworkers [13]. They also started with commercial samples (Gelest Inc.) but took pains to exhaustively solvent extract the polymers, 2 weeks in acetone for PDMS and 3 weeks in hexane for PMTFPS. Purity was essential for their intended use in retinal detachment surgeries but, as can be seen in Table 9.2, this treatment made marginal difference to refractive index (RI), density (ρ), or surface tension.

The unexpectedly high surface tension of liquid PMTFPS is evident in Figure 9.3 which shows surface tensions of fluorosilicones $\{[F(CF_2)_X(CH_2)_2]CH_3SiO\}_n$ with increasing pendant perfluorinated side-chain lengths from the review by Pasquet et al. [2]. The diamonds and triangles show a variety of homopolymer data that are in good agreement. The stars are data for fluorosilicone/PDMS copolymers.

Considering Zisman's order of the effect of substituent groups [14] developed from solid surface tension measurements, when we go from PDMS to PMTFPS, we

TABLE 9.2 Effect of Extraction on PDMS and PMTFPS Properties

Polymer	Treatment	RI (25°C)	ρ g/cm³ (21°)	γ_{LV} (21°)
1000 cSt PDMS	Unextracted	1.404	0.966	20.9
1000 cSt PDMS	Extracted	1.404	0.967	21.1
12,500 cSt PDMS	Unextracted	1.404	0.970	21.2
12,500 cSt PDMS	Extracted	1.404	0.971	21.3
1000 cSt PMTFPS	Unextracted	1.381	1.267	24.1
1000 cSt PMTFPS	Extracted	1.381	1.282	24.1
10,000 cSt PMTFPS	Unextracted	1.382	1.293	24.0
10,000 cSt PMTFPS	Extracted	1.382	1.293	24.2

RI, refractive index.

remove one methyl and replace it with one trifluoromethyl of lower surface energy and two methylene groups of higher surface energy. The net effect in this liquid state is a higher surface tension for PMTFPS compared to PDMS. An uncompensated dipole has also been introduced between the fluorocarbon and hydrocarbon entities which might also increase overall surface energy but it is rapidly screened with increasing X as the length of the perfluorinated portion has almost no influence on the liquid surface tension above that of PMTFPS. We endorse the comment of Pasquet et al. that it would be of particular interest to measure the surface tension of the fluorosilicone with $X = 2$, a study which to their and our knowledge has not yet been reported. It would also be intriguing to study bis(trifluoropropyl)siloxane in this context to eliminate any effect of the methyl group. Holbrook [15] has described the preparation of cyclic bis(trifluoropropyl)siloxanes. A homopolymer of significant

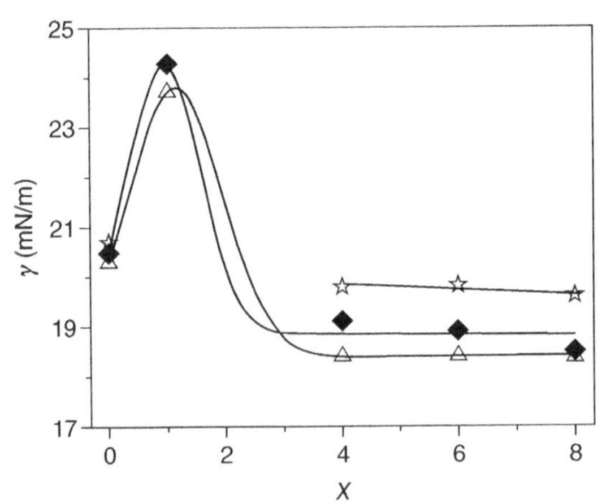

FIGURE 9.3 Liquid surface tensions of fluorosilicones as a function of fluorocarbon side-chain length (solid diamond M_n ~7000; open triangle M_n 20–30,000; open star copolymer of PMTFPS and PDMS). Reprinted with permission from Reference 2. Copyright 2012 Springer Science+Business Media B.V.

TABLE 9.3 XPS (Atomic%) Composition of PMTFPS Surfaces

Sample type	F%	O%	C%	Si%
Unfilled, cleaved	38.1	11.1	40.7	10.1
Unfilled, cleaved	33.9	13.5	43.2	9.5
Unfilled, molded against PTFE	38.3	11.3	39.8	10.5
Filled, cleaved	35.7	13.7	39.0	11.6
Filled, molded against PTFE	37.3	11.9	39.8	11.1
PMTFPS theory	33.3	11.1	44.4	11.1
PTFE theory	66.7	0	33.3	0

molecular weight is difficult to achieve but it is claimed that these cyclics can be copolymerized with other siloxanes including methyltrifluoropropylsiloxanes. No surface tension data were provided in this patent.

9.3.2 Solid Surface Tension

Solid surface energies are usually derived from contact angle studies. Given the various choices of test liquids and theoretical approaches available, interpretation and comparison between results of different researchers are not straightforward.

Filled silicone elastomers are generally in the roughness range of R_a (arithmetic mean) 2–8 nm, so we can be confident that contact angles on such substrates are not complicated by micro-roughness considerations. The unfilled, cleaved surfaces are even smoother. Under sideways tension, PMTFPS gums and elastomers can be smoothly cleaved with a little more pressure than the weight of a sharp, clean knife. In addition, the X-ray photoelectron spectroscopy (XPS) atomic composition data for PMTFPS surfaces shown in Table 9.3 give encouragement that these fluorosilicone surfaces are not contaminated. In particular, no other unexpected elements are detected and no PTFE, used as a release sheet, seems to have transferred to the fluorosilicone during molding. The data in Table 9.3 are taken from Owen [16] except for the second unfilled, cleaved sample which comes from Perz and coworkers [17].

Table 9.4 shows solid surface energy data for the first PMTFPS sample listed in Table 9.3. Also included for comparison are PDMS from She et al. [18] and PTFE data taken from the compilation by Owen [19]. The She measurements are for a very thin film of PDMS, bonded to a smooth silicon wafer, that had very low contact angle hysteresis, only 3° difference between the advancing and receding water contact

TABLE 9.4 Solid Surface Energy Data

Polymer	θ_{water} (°)	γ_C (mN/m)	γ_S^d (mN/m)	γ_S (mN/m)
PDMS	108	22.7	21.3	22.8
PMTFPS	104	21.4	18.4	13.6
PTFE	108	18.3	19.3	19.1

angle values. Note also that the three critical surface tension of wetting values in Table 9.4 were determined using only *n*-alkanes as the contact angle test liquids.

In this table, γ_S^d is calculated from the Girifalco–Good–Fowkes–Young equation [20] using *n*-hexadecane as the sole contact angle test liquid:

$$\gamma_S^d = \gamma_{LV}(1 + \cos\theta)^2/4 \quad (9.2)$$

The γ_S values are obtained by the Owens and Wendt approach [10] (Equation 9.3) using water and diiodomethane, where γ_S is the sum of the dispersion force component of surface tension, γ^d, and the polar component, γ^p, and the subscripts LV and SV refer to the liquid/vapor and solid/vapor interfaces, respectively.

$$\gamma_{LV}(1 + \cos\theta) = 2(\gamma_{LV}^d \gamma_{SV}^d)^{0.5} + 2\left(\gamma_{LV}^p \gamma_{SV}^p\right)^{0.5} \quad (9.3)$$

Three aspects of Table 9.4 are notable. First, taking the water contact angle as the simplest index of hydrophobicity, all three polymers are quite similar in this respect. Second, unlike the liquid surface tension situation, PMTFPS does not have a higher solid surface tension than PDMS; it has a slightly lower critical surface tension of wetting, a dispersion force component that is lower still and a markedly lower total solid surface tension using the Owens and Wendt method. Third, since this last quantity is the sum of the dispersion force component and the polar component, it ought to be larger, not smaller, than the dispersion force component alone. However, one should not expect different equations and different contact angle test liquids to give convergent results. Figure 9.4 taken from Pasquet and coworkers [2] shows that the same situation is encountered with fluorosilicones having longer fluorocarbon

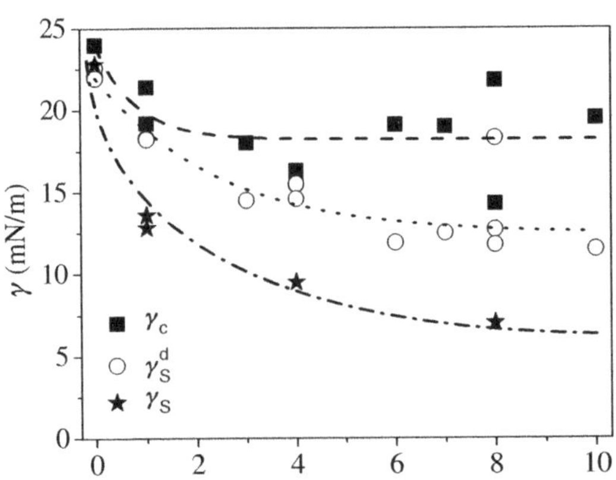

FIGURE 9.4 Solid surface tensions of fluorosilicones as a function of fluorocarbon side-chain length. Reprinted with permission from Reference 2. Copyright 2012 Springer Science+Business Media B. V.

TABLE 9.5 Solid Surface Properties of Fluorosilanes

Silane	θ_{water} (°)	γ_C (mN/m)
Methyltrimethoxysilane	88	22.5
Trifluoropropyltrimethoxysilane	84	33.5
Nonafluorohexyltriethoxysilane	103	23.0
Tridecafluorooctyltriethoxysilane	110	14
Heptadecafluorodecyltrimethoxysilane	115	12
Heneicosafluorododecyltrichlorosilane	118	6

side chains. Chaudhury [21] has shown that a $CF_3(CF_2)_7(CH_2)_2SiCl_3$-treated plasma-oxidized PDMS surface has a surface energy of 16 mN/m measured by the Johnson, Kendall, and Roberts (JKR) [22] contact mechanics approach. This method does not suffer the ambiguities of contact angle approaches. Inspection of Figure 9.4 in this context suggests that *n*-alkanes are better choices for determining fluorosilicone surface tension than are water and diiodomethane.

Hydrolyzed cross-linked trialkoxysilane films are quite different from fluoropolymers such as PMTFPS. No methyl groups are involved but it is interesting to note that a higher than expected deviation in surface tension by the trifluoropropylfunctional silane is once again evident, this time in the critical surface tension of wetting. Table 9.5 is taken from the compilation of smooth surface studies from various sources by Arkles and coworkers [23]. Note that the γ_C value for trifluoropropyltrimethoxysilane is from Lee [24] and the θ_{water} value for methyltrimethoxysilane is from Burck [25]. Lee commented that "Contrary to Zisman's prediction, the trifluoropropyl group increases the wettability instead. Presumably, CF_3 in this case behaves as an electronegative substituent as evidenced by the relatively high γ^p_{SV}," this value being given by Lee as 18.7 mN/m. His value for γ^d_{SV} was 16.0 mN/m giving a total γ_S of 34.7 mN/m. Note that Lee used water and a series of other hydroxyl-containing liquids rather than water and diiodomethane as used for the data presented in Table 9.4.

9.3.3 Polymer/Water Interface

Given the polar interaction between the siloxane backbone and water, it is expected that the interfacial tensions of PDMS and PMTFPS with water would be lower than the 50–55 mN/m range typically measured for *n*-hydrocarbon/water interfacial tension and this turns out to be the case. Table 9.6 shows some such interfacial tension measurements.

The first four entries in Table 9.6 come from Nakamura and coworkers [26] using the du Nuoy ring technique; the final entry is from Princen et al. [27] measured from the shape of a rotating drop. The purification process of Nakanamura and coworkers consisted of washings with ethanol and water, heating in vacuum and filtration through activated charcoal to remove possible impurities. Judging by the quite close agreement with Princen's data, there are very little surface-active impurities in the commercial product. Note also that the PMTFPS interfacial tension with water is lower than that

TABLE 9.6 Silicone/Water Interfacial Tension

Polymer	Type	Viscosity (cSt)	Interfacial tension (mN/m)
PDMS	Medical grade	1000	45.5
PDMS	Medical grade	12,500	47.3
PMTFPS	Purified	1000	38.8
PMTFPS	Purified	10,000	39.9
PMTFPS	Commercial	1000	40.3

of PDMS despite the expectation that alignment of Si–O dipoles to the water surface would be easier for PDMS than it would be for PMTFPS. Perhaps there is some C–F dipole interaction with water that compensates for any restriction caused by the relatively bulky trifluoropropyl group.

Another aspect of polymer/water interaction worthy of comment is their Langmuir trough behavior. For the same thermodynamic reason that water forms a finite contact angle on PDMS and PMTFPS, so too will these liquid polymers spread out on a water surface enabling surface pressure/area (Π/A) isotherms to be obtained. PDMS and PMTFPS behave quite similarly in this regard. Their isotherms have a very low surface pressure region at a low area. The surface pressure then increases at a liftoff point through a region of increasing surface pressure to a turnover point where the slope again levels off into a plateau-like region. The liftoff point for PDMS is at ~0.2 nm^2/monomer unit and at ~0.27 nm^2/monomer unit for PMTFPS. At this liftoff point, it is generally assumed that all Si and O atoms are in contact with the water substrate and the chains are tightly close-packed. These area per monomer unit values are in good agreement with this postulate. The turnover point to the plateau region is at $\Pi =$ 8.4 mN/m for PDMS and a little less at 7.5 mN/m for PMTFPS. One difference is that the PMTFPS plateau is featureless, whereas there is a further small step in the PDMS plateau. The configurational changes the polymers must undergo during compression and expansion on the Langmuir trough have been the subject of considerable debate since the original studies in 1947. Esker and Yu [28] have recently reviewed this topic. At the turnover point, it is proposed that either helices or multilayers form. The latter is more likely considering the bulkiness of the trifluoropropyl groups.

9.3.4 X-Ray Photoelectron Spectra

XPS is the surface analysis technique most widely used to provide chemical information about the top few nanometers of silicone surfaces. High resolution spectra of PMTFPS performed by Leadley and coworkers [29] are shown in Figure 9.5. The Si 2p and O 1s spectra are very similar to those of PDMS which is to be expected as they share the same polymer backbone. The F 1s binding energy of 688 eV is significantly lower than the 689.7 eV value for fluorine in PTFE. The C 1s spectrum is the most interesting feature of these PMTFPS spectra. Note the high binding energy component at 292.4 eV from the terminal CF_3 group which has a similar binding energy to the CF_2·C 1s of PTFE, and the peak at 286 eV which comes from the CH_2

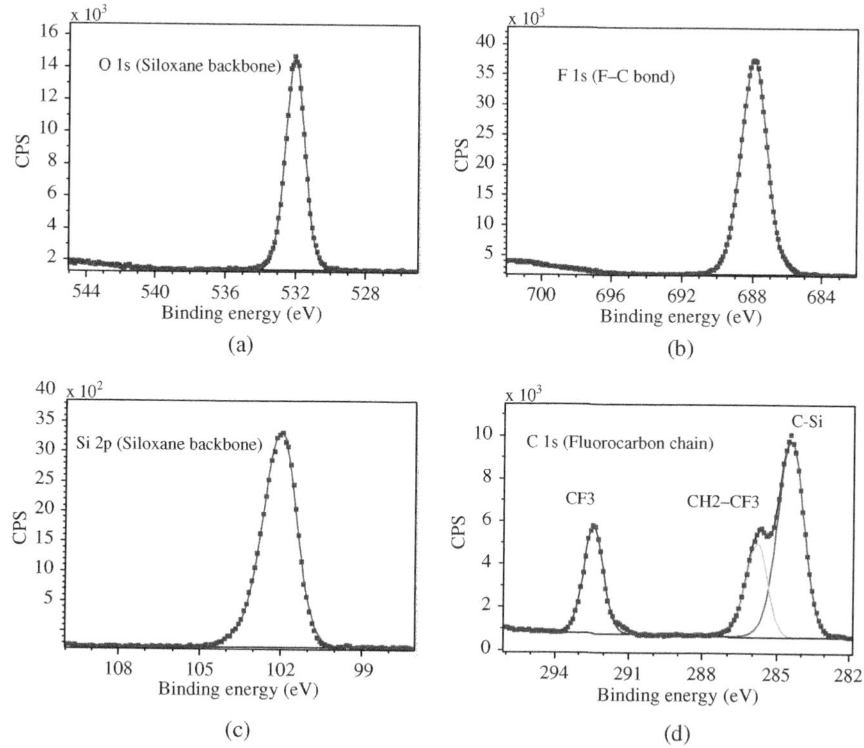

FIGURE 9.5 High resolution XPS spectra of PMTFPS. Reprinted with permission from Reference 29. Copyright 2012 Springer Science+Business Media B. V.

group adjacent to the CF_3 group whose fluorine atoms are the cause of this shift to a higher binding energy than the CH_2 group adjacent to Si. This latter CH_2 group signal is the same as the methyl group signal at 284.4 eV. This C–Si component is expectedly about twice the intensity of the terminal CF_3 group and its adjacent CH_2 group.

Perz and coworkers [17] earlier published Si 2p and C 1s spectra of untreated and various plasma-treated samples of PMTFPS gum and elastomer. PMTFPS is affected by RF plasma treatment in much the same way as PDMS. Significant improvements in wettability are possible but the treatments are temporary and hydrophobic recovery is mostly complete within 24 hours. A related topic to preformed fluoropolymers is the introduction of fluorocarbon gases into the plasma to create fluorinated surface layers. In the specific example of the use of a CF_4 plasma to treat PDMS surfaces [30], a fluorine content as high as 47% can be achieved. Unlike, for example, an oxygen plasma that produces a brittle surface, the CF_4 treatment is not brittle. However, contrary to expectations, the CF_4 treatment resulted in a decreased water contact angle, from 119° to 101° at 100 W, 300 s. This was attributed to possible side effects such as surface oxidation, chemical heterogeneity, and surface reorganization.

Plasma and corona treatment can have interesting effects. We [31] studied PDMS elastomers with 5% added PMTFPS fluid ($M_n = 950$). No fluorine could be detected by XPS on the surface before discharge, consistent with the higher liquid surface tension of PMTFPS. After discharge, which raises the surface tension of PDMS above that of PMTFPS, the fluorine concentration on the surface increased markedly to 21%. The rate of hydrophobic recovery of the PMTFPS-containing PDMS elastomers after discharge ceased was much faster than that of similarly treated PDMS elastomers containing comparable quantity and molecular weight of free PDMS fluid. One possible explanation is that the PMTFPS is fragmenting under the discharge conditions but this was not proven.

9.4 OTHER PROPERTIES

9.4.1 Swelling Resistance

Fluorosilicone elastomers have good resistance to swelling by jet fuels, oils, and hydrocarbon solvents and fuels as shown in Table 9.7 taken from data in [8] and [32].

It was to meet the need for hydrocarbon, oil-resistant seals, gaskets, O-rings, and so on in the automotive and aerospace industries that PMTFPS was first developed. PDMS is an oleophilic polymer, quite unsuited to this application. All the *n*-alkane values used in the determination of solid surface tension of PDMS earlier in this chapter are instantaneous values or taken as soon after drop deposition as possible. All *n*-alkane contact angle values on PDMS decline inexorably to zero as the drop swells the polymer and eventually disappears. The CF_3– cluster of PMTFPS was found to combine excellent solvent resistance with adequate thermal stability for the minimum degree of fluorination, critical to furnishing a cost-effective product. The

TABLE 9.7 PMTFPS Elastomer Swelling by Various Fluids

Fluid	Immersion conditions (days/°C)	Volume change (%)
ASTM No. 1 oil	3/150	0
Crude oil 7 API	14/135	5
Hydraulic fluid Oronite 8515	3/190	9
ASTM ref. fuel B	3/65	20
JP-4 fuel	3/25	10
JP-5 fuel	7/25	4
Benzene	7/25	25
Carbon tetrachloride	7/25	20
70:30 Isooctane/toluene	3/70	18
Methanol	14/25	4
Ethanol	7/25	5
Hydrochloric acid (10%)	7/25	0
Nitric acid (70%)	7/25	5
Sodium hydroxide (50%)	7/25	0

ethylene bridge between the CF_3– and the silicon atom was found to be essential to maintain the required thermal stability and longer hydrocarbon bridges would only dilute the fluorocarbon character.

Although PMTFPS performs well with low volume swells in alcohol/fuel blends, higher swells are observed in ketones and esters [33], which we attribute to permanent dipole/dipole interaction between the C–F and C=O bonds, and is reflected in the higher solubility parameter of PMTFPS (17.9 $[MPa]^{0.5}$) compared to PDMS [15.1 $(MPa)^{0.5}$]. Gomez-Anton and coworkers [30] further argue that the polar trifluoropropyl side chain, being articulated, effectively shields the chain from interactions with the solvent unless the solvent molecules are sufficiently polar.

9.4.2 Glass Transition Temperature

PMTFPS has a glass transition temperature (Tg) of –70°C (203 K). Unlike PDMS, it does not exhibit a low temperature crystallization due to the inability of the polymer chains to pack into a crystalline lattice. PMTFPS is ordinarily an atactic polymer. However, stereoregular versions have been prepared by Kuo et al. [34] and these do show a melting temperature by differential scanning calorimetry (DSC) of 269 K for the polymer made from the *trans* isomer (100%) and 321 K from the *cis* isomer (96%). They also reported glass transition temperatures of 204 and 207 K, respectively, slightly higher than that of the atactic polymer. Madkour and Mark [35] have carried out Monte Carlo simulations of these stereoregular PMTFPS polymers to further understand the nature of the stereoregular sequence distributions, degree of crystallinity, melting points, interfacial-free energies, and moduli. Note that the interfacial-free energy referred to here is the excess free energy per mole associated with a unit at the end of a crystallite and not the surface energies reviewed earlier in this chapter. Because of the low Tg and absence of a low temperature crystallization, conventional atactic PMTFPS elastomers remain flexible at lower temperatures than many other fluoropolymer elastomers. The brittleness temperature (ASTM D 746B) has been measured as –59°C and the temperature of retraction at which an elongated, frozen specimen retracts by 10% (ASTM D 1329) as –50°C [36].

9.4.3 Viscosity

Table 9.8 shows the molecular weight of various PDMS and PMTFPS fluids taken from a Gelest Inc. publication [37]. Both types of fluid are trimethylsiloxy terminated.

There is an approximately fivefold reduction in molecular weight of the PMTFPS fluids at comparable viscosities to the PDMS fluids and since the molecular weight of each MTFPS unit (156) is a little over twice that of a DMS unit (74), this equates to approximately a 10-fold increase in viscosity at the same degree of polymerization. The viscosity temperature coefficient, measured as 1 – (viscosity at 99°C/viscosity at 38°C) is 0.51 for PDMS and 0.84 for PMTFPS. The very small change in viscosity with temperature is a notable feature of PDMS and it is not surprising that it is greater for PMTFPS with its bulkier side-group inhibiting chain flexibility. It should also be noted that the activation energy for viscous flow of these polymers is low, 14.2 kJ/mol

TABLE 9.8 Viscosity/Molecular Weight Data for PDMS and PMTFPS

PDMS viscosity (cSt)	PDMS molecular weight	PMTFPS viscosity (cSt)	PMTFPS molecular weight
100	5970	80–120	900–1000
350	13,650	300–350	2400
500	17,250	400–500	3000
1000	28,000	1000	4600
10,000	62,700	10,000	14,000

for PDMS and 33 kJ/mol for PMTFPS. These values imply that only small frictional forces are associated with the translational flow of the molecules relative to each other [38]. As with the viscosity temperature coefficient, it is expected that the PDMS value would be lower.

9.4.4 Permeability

Gas permeability is a property where PMTFPS elastomers differ considerably from PDMS elastomers. For example, PMTFPS is considerably less permeable to oxygen than PDMS as can be seen in Table 9.9 [39]. This difference is not unexpected considering that PDMS is one of the most permeable elastomers known. Presumably the bulkier trifluoropropyl groups hinder the passage of gases more than methyl groups. However, PMTFPS has a higher permeability than most other elastomers, see, for example, the value for natural rubber included in the table, and is much more permeable than some other fluoropolymers such as PTFE. For this reason, it is not ordinarily considered for applications requiring low gas permeability. Nevertheless, it has some special military applications, particularly at low temperatures and where oil resistance is required. In such applications, fillers such as mica have been found to considerably reduce the gas permeability of PMTFPS elastomers [40].

9.4.5 Electrical Properties

Table 9.10 shows some key polymer electrical property data taken from the compilation of Ku and Liepens [41] with the exception of the PMTFPS data reported in [8].

TABLE 9.9 Oxygen Permeability of Various Polymers

Polymer	Permeability coefficient ($\times 10^9$) cm^3 cm/(s cm^2 cm Hg)
PDMS	60
PMTFPS	11
Natural rubber	2.4
PTFE	0.0004

TABLE 9.10 Electrical Properties of Various Polymers

Polymer	E_B (60 Hz) (V/mil)	ε (100 Hz)	tanδ (100 Hz)	ρ_V (ohm·cm)
PDMS	552	2.9	0.00025	5.3×10^{14}
PMTFPS	350	7.0	0.020	1.0×10^{14}
Viton	351	8.6	0.040	4.1×10^{11}
Natural rubber	665	2.4	0.0024	1.1×10^{15}

In this table, Viton is a copolymer of vinylidene fluoride and hexafluoropropylene, and natural rubber is cross-linked *cis*-plyisoprene. E_B is the dielectric strength, ε is the dielectric constant, tanδ is the tangent of dielectric loss angle or dissipation factor, and ρ_V the resistivity. The two fluoropolymers are quite similar with lower dielectric strength, higher dielectric constant, higher dissipation factor, and lower volume resistivity than PDMS. Technical interest in the electrical properties of PMTFPS is a consequence of its relatively low dielectric loss and high electrical resistivity coupled with nonelectrical characteristics such as low levels of water absorption, resistance to oils and solvents and other harsh environments, plus retention of properties over a broad temperature range.

Baird and Sengupta [42] have examined the dielectric relaxation behavior of PMTFPS. They observed two relaxation processes, an α-relaxation between −80°C and −30°C and a β-relaxation in the glassy region, −190°C to −130°C. These processes were interpreted in terms of side-group and backbone motions with the α-relaxation attributed almost entirely to reorientational motions of the trifluoropropyl side-group.

9.4.6 Lubrication

Although useful as a lubricant for plastic materials, PDMS fluids are a poor lubricant for metal on metal, particularly at high loadings. Gelest Inc. data [37] give scar diameters in the range of 2–3 mm in steel-on-steel four-ball wear tests for PDMS fluids and 0.8 mm for PMTFPS under the same conditions—1 hour at 75°C at a 40 kg load. The low intermolecular forces and backbone flexibility of PDMS are not conducive to the formation of a good boundary layer. The considerable effectiveness of PMTFPS as a metal lubricant is presumably the result of the bulkier trifluoropropyl side-group providing a thicker, harder-to-displace boundary film. It is also possible that at the highest use temperatures, decomposition and surface reaction–enhancing lubrication may occur in a similar manner to the mode of action of halogen-containing extreme pressure additives.

9.5 CONCLUDING REMARKS

It is a straightforward concept that the relative bulkiness of the trifluoropropyl group can impede siloxane backbone flexibility and influence the physical properties of the PMTFPS polymer when we substitute one of the methyl groups in the Me_2SiO repeat units with a trifluoropropyl one. Not so obvious is the effect such a substitution has on intermolecular interactions between the polymer chains. When we declare that a trifluoromethyl group has a lower surface energy than a methyl group, we essentially mean that the molecular forces between the former groups are lower than those between the latter ones. However, the CF_3– group is quite polar with the C–F bond having a high dipole moment (1.41 D) on account of the high electronegativity of fluorine. This causes some literature reports to consider PMTFPS to be a polar polymer despite it being simultaneously a hydrophobic polymer ($\theta_{water} > 90°$).

It must follow that the inert gas electron environment around the fluorines more than compensates for any polarity conferred by the C–F dipoles. Kirsch [43] notes that although fluorine itself is the most reactive of all elements, some fluoroorganic compounds have chemical inertness like that of the noble gases. He points out that the apparently contradictory observation that perfluorocarbons are among the most nonpolar compounds in existence despite the strong C–F dipole can be explained by the fact that all local dipole moments within the same molecule cancel each other, leading in total to a nonpolar compound. In some semifluorinated compounds, some local dipole moments are not compensated, leading to some unexpected properties such as is seen with the surface properties of PMTFPS. Although not so for PMTFPS, note that the noncompensated dipole effect is most marked when bonds to F and H arise from the same carbon atom. In this case, the C–H bonds are polarized to the extent that they can act as hydrogen-bond donors with the fluorine as the acceptor.

REFERENCES

1. Schiers, J. *Modern Fluoropolymers*; John Wiley and Sons, Inc.: New York, 1997; p xv.
2. Pasquet, C.; Longuet, C.; Hamdani-Devarennes, S.; Ameduri, B.; Ganachaud, F. Comparison of surface and bulk properties of pendant and hybrid fluorosilicones. In *Silicone Surface Science*; Owen, M. J.; Dvornic, P. R., Eds.; Springer, 2012; Chapter 5.
3. Ameduri, B.; Boutevin, B. *J. Fluor. Chem.* **2005**, *126*, 221–229.
4. Rhein, R. A. *Thermally Stable Elastomers: A Review*, NWC Technical Publication 6372, China Lake CA, 1983.
5. Pierce, O. R. Fluorinated alkyl organosilicon compounds. U.S. patent 2,894,969. 1959 (Filed March 1957).
6. Boutevin, B.; Guida-Pietrasanta, F.; Ratsimihety, A. Side Group Modified Polysiloxanes. In *Silicon-Containing Polymers*; Jones, R. G., Ando, W., Chojnowski, J. Eds.; Kluwer: Dordrecht, The Netherlands, 2000; Chapter 3.
7. Wright, P.; Semlyen, J. A. *Polymer* **1970**, *11*, 462–471.
8. Maxson, M. T.; Norris, A. W.; Owen, M. J. Fluorosilicones. In *Modern Fluoropolymers*; Schiers, J., Ed.; John Wiley and Sons, Inc.: New York, 1997; Chapter 20.
9. Owen, M. J. Elastomers: siloxane. In *Encyclopedia of Materials: Science and Technology*, Elsevier Science Ltd, 2001; p 2480.
10. Owens, D. K.; Wendt, R. C. *J. Appl. Polym. Sci.* **1969**, *13*, 1741–1747.
11. LeGrand, D. G.; Gaines, G. L. *J. Colloid Interface Sci.* **1969**, *31*, 162–167.

12. Kobayashi, H.; Owen, M. J. *Makromol. Chem.* **1993**, *194*, 1785–1792.
13. Nakamura, K.; Refojo, M. F.; Crabtree, D. V.; Leong, F-L. *Invest. Ophthalmol. Vis. Sci.* **1990**, *31*, 2059–2069.
14. Zisman, W. A. Relation of the equilibrium contact angle to liquid and solid constitution. In *Contact Angle, Wettability and Adhesion*; Fowkes, F. M., Ed.; Adv. Chem. Ser. Vol. 43; American Chemical Society: Washington, DC, 1964; p 1.
15. Holbrook, G. W. Cyclic bistrifluoropropyl siloxanes. U.S. patent 3,070,617. 1957.
16. Owen, M. J. *J. Appl. Polym. Sci.* **1988**, *35*, 895–901.
17. Perz, S. V.; McMillan, C. S.; Owen, M. J. Wettability of fluorosilicone surfaces. In *Fluorinated Surfaces, Coatings, and Films*; Castner, D. G., Grainger, D. W., Eds., ACS Symp. Ser. 787; American Chemical Society: Washington DC, 2001; Chapter 9.
18. She, H.; Chaudhury, M. K.; Owen, M. J. Surface properties of thin film polydimethylsiloxane. In *Silicones and Silicone-Modified Materials*; Clarson, S. J., Fitzgerald, J. J., Owen, M. J., Smith, S. D., Eds.; ACS Symp. Ser. 729; American Chemical Society: Washington, DC, 2000; Chapter 21.
19. Owen, M. J. Surface and interfacial properties. In *Physical Properties of Polymers Handbook*; Mark, J. E., Ed.; American Institute of Physics: Woodbury, New York, 1996; Chapter 48.
20. Good, R. J.; Girifalco, L. A. *J. Phys. Chem.* **1960**, *64*, 561–565.
21. Chaudhury, M. K. *J. Adhes. Sci. Technol.* **1993**, *7*, 669–675.
22. Johnson, K. L.; Kendall, K.; Roberts, A. D. *Proc. R. Soc. Lond. A.* **1971**, *324*, 301–313.
23. Arkles, B.; Pan, Y.; Kim, Y. M. The role of polarity in the structure of silanes employed in surface modification. In *Silanes and Other Coupling Agents*; Mittal, K. L., Ed.; VSP/Brill: Leiden, 2009; Vol. *5*, p 55.
24. Lee, L-H. Wettability of functional polysiloxanes. In *Adhesion Science and Technology*; Lee, L-H., Ed.; Plenum: New York, 1975; Vol. *9B*, p 647.
25. Burck, P. W. The Influence of Wettability on Scaling the Capillary Pressure-Saturation Relationship. M.Sc. Thesis, New Mexico Institute of Mining and Technology, 1993.
26. Nakamura, K.; Refojo, M. F.; Crabtree, D. V. *Invest. Ophthalmol. Vis. Sci.* **1990**, *31*, 647–656.
27. Princen, H. M.; Zia, I. Y. Z.; Mason, S. G. Measurement of interfacial tension from the shape of a rotating drop. *J. Colloid Interface Sci.* **1967**, *23*, 99–107.
28. Esker, A. R.; Yu, H. Langmuir monolayers of siloxanes and silsesquioxanes. In *Silicone Surface Science*; Owen, M. J., Dvornic, P. R., Eds.; Springer, 2012; Chapter 7.
29. Leadley, S.; O'Hare, L-A.; McMillan, C. Surface analysis of silicones. In *Silicone Surface Science;* Owen, M. J., Dvornic, P. R., Eds.; Springer, 2012; Chapter 12.
30. Cordeiro, A. L.; Nitschke, M.; Janke, A.; Helbig, R.; Souza, F. D.; Donnelly, G. T.; Willemsen, P. R.; Werner, C. *Exp Polym. Lett.* **2009**, *3*(2), 70–83.
31. Kim, J.; Chaudhury, M. K.; Owen, M. J.; Orbeck, T. *J. Colloid Interface Sci.* **2001**, *244*, 200–207.
32. Montermoso, J. C. *Rubber Chem. Technol.* **1961**, *34*, 1521–1552.
33. Gomez-Anton, M. R.; Masegosa, R. M.; Horta, A. *Polymer* **1987**, *28*, 2116–2121.
34. Kuo, C-M.; Saam, J. C.; Taylor, R. B. *Polym. Int.* **1994**, *33*, 187–195.
35. Madkour, T. M.; Mark, J. E. *Macromolecules* **1995**, *28*, 6865–6870.

36. Norris, A. M.; Fiedler, M. D.; Knapp, T. L.; Virant, M. S. *Automot. Polym. Des.* **1990**, *19*, 12.
37. Gelest Inc. publication. *Silicone Fluids: Stable Inert Media*, **2012**.
38. Dvornic, P. R. In *Silicon-Containing Polymers*; Jones, R. G., Ando, W., Chojnowski, J., Eds.; Kluwer: Dordrecht, The Netherlands, 2000; Chapter 7.
39. Robb, W. L. *Ann. NY. Acad. Sci.* **1968**, *146*, 119–137.
40. Williams, J. A. *Oil-Resistant Elastomers with Low Permeability to Nitrogen.* US Army Weapons Command, Technical Report SWERR-TR-72-58, August 1972.
41. Ku, C. C.; Liepins, R. *Electrical Properties of Polymers: Chemical Principles*; Hanser: Munich, Germany, 1987; p 326.
42. Baird, M. E.; Sengupta, C. R. *Polymer* **1974**, *15*, 608–609.
43. Kirsch, P. *Modern Fluoroorganic Chemistry: Synthesis, Reactivity, Applications*, Wiley/VCH, 2004; p 8.

10

FUNCTIONAL FLUOROUS COPOLYOXETANE POLYMER SURFACE MODIFIERS

KENNETH J. WYNNE, PINAR KURT, UMIT MAKAL, TOMOKO FUJIWARA, KENNARD BRUNSON, AND ASIMA CHAKRAVORTY

Virginia Commonwealth University

LYNN WOOD AND DENNIS E. OHMAN

Virginia Commonwealth University
McGuire Veterans Affairs Medical Center

10.1 INTRODUCTION

The findings described in this chapter focus on fluorous copolyoxetane soft block surface modifiers. These copolyoxetane polyurethane modifiers bring about new surface properties due to surface concentration and/or phase separation of the soft block. An important theme concerns the effectiveness of a fluorous "chaperone" in conferring a surface function to a commodity or "base" polyurethane at 2 wt% or less. Thus, from one perspective, this chapter describes literally and figuratively the interface between specialty fluoropolymers and commodity polyurethanes. A significant applications niche would result if only a small fraction of the $55 billion polyurethane market (2016) [1] employed specialty fluorous modifiers.

Handbook of Fluoropolymer Science and Technology, First Edition. Edited by Dennis W. Smith Jr., Scott T. Iacono, and Suresh S. Iyer.

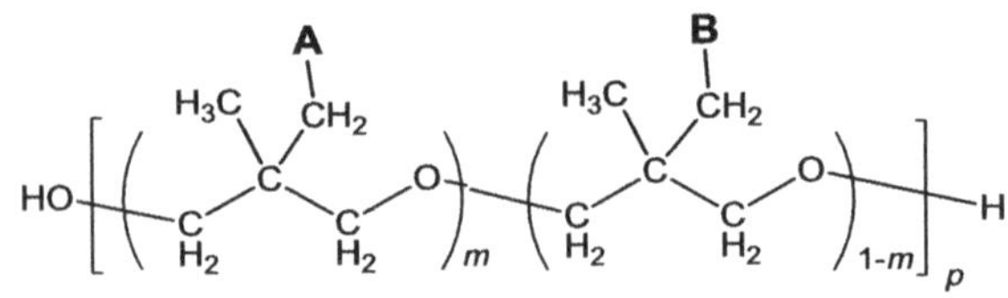

STRUCTURE 10.1 P[AB] [10].

Virtually all commodity polyurethanes employ homopolyether or polyester soft blocks [2,3]. Practical methods for surface modification have received relatively little attention, although the surface is the "frontier" for interactions such as adhesion to an over-layer, resistance to solvents, and biocompatibility of medical devices [4,5].

Our approach to surface modification is the incorporation of P[AB] copolyoxetanes into polyurethanes (Structure 10.1). To describe these polymers, A and B (and sometimes C) are used to designate soft block repeat units containing A and B side chains. Conventional polyurethane designations are employed, for example, HMDI/BD(40)/P[AB-x:y-5.0], where HMDI is H_{12}MDI, (4,4′-(methylene-*bis*-cyclohexyl isocyanate), BD is butanediol, P denotes ring-opened oxetane structures for repeat units A and B, and 5.0 is the soft block molecular weight (M_n, kDa). Figure 10.1 contains monomer structures and acronyms. Results herein often describe utilization of small quantities of polyurethanes containing a P[AB] soft block as surface modifiers for conventional polyurethanes. To simplify, notation-modified polyurethane compositions are designated by 2 wt% U–P[AB] or similar notation when the composition is clear in context.

Functional surface model. Interesting and often unexpected surface properties result from pairing a fluorous side chain (A) with a second side chain (B). A model for surface composition of P[AB] copolyoxetane polyurethanes is shown in Figure 10.2.

A combination of features distinguishes the P[AB] soft block approach for concentrating a desired functional group at a polymer surface.

a. Surface concentration of soft blocks in polyurethanes [6–9].
b. Multiple P[AB] side chains that act as pseudo-chain ends [10].
c. Differing solubility parameters that enhance phase separation for the P[AB] soft block vis-à-vis other coating constituents.
d. A low P[AB] soft block T_g that may provide kinetic enhancement for conjugation with a target receptor.
e. Optimized functional effectiveness by varying mole fractions in the P[AB] copolyoxetane.
f. A and B synergistic interactions that produce a new surface property.
g. Fluorous A acting as a "chaperone" to surface-concentrate B groups which, in the absence of A, would not be surface concentrated [11].

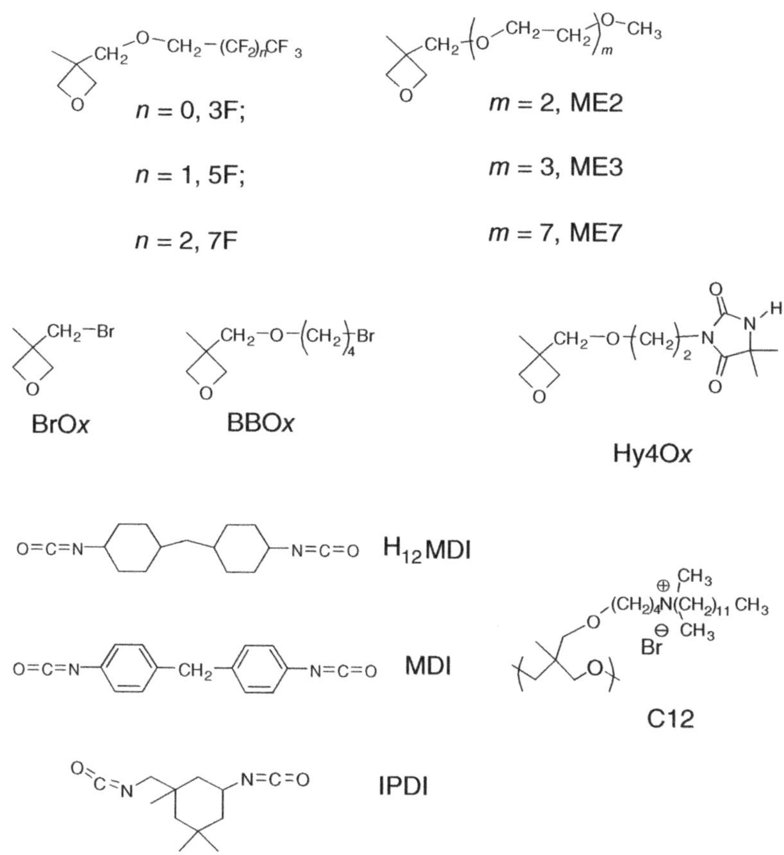

FIGURE 10.1 Structures and designations for monomers, diisocyanates, and repeat units.

h. A "green" method for surface concentrating functionality B, as a multiplicity of short ($<C_{4f}$), environmentally acceptable perfluorinated (R_f) side chains are employed. This approach avoids the use of mixed fluorous tails [12] that contain eight-carbon (R_{8f}) moieties, which are known to be perfluorooctanoic acid (PFOA) precursors [13–15].

i. Compositional economy by using U–P[AB] as a minor surface modifier constituent ($\leq$2 wt%) in a blend such that the polymer modifier defines the surface properties and the matrix polymer defines bulk mechanical properties and adhesion to substrate.

For U–P[AB], features **a–c** provide a combination of thermodynamic driving forces that enhance soft block surface concentration. Of considerable interest is **f**, a feature that has been observed in the majority of new U–P[AB]. That is, some new and unexpected surface characteristics such as morphology or wetting behavior have

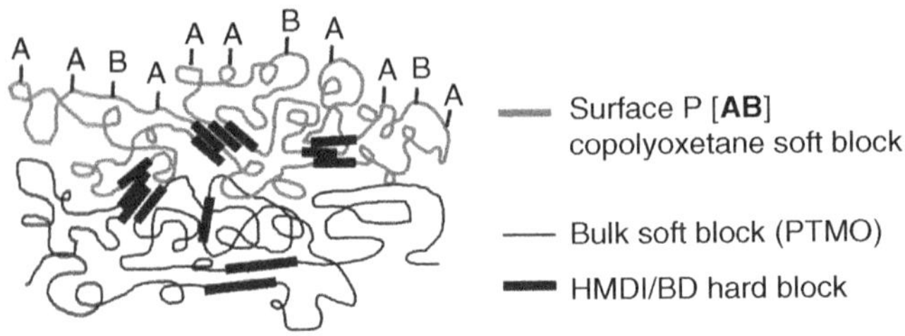

FIGURE 10.2 Depiction of soft block nanoscale surface concentration in a P[AB] soft block polyurethane [16, 17].

been found that provide a rich basis for control or tuning polymer surface behavior. Feature **g** has been observed for 2 wt% U–P[AB] wherein the presence of the fluorous A plays an important "chaperone" role in surface concentration of functional B [11]. As will be seen in the sections 10.2–10.5, function in 2 wt% U–P[AB] coatings run the gamut from surface properties that are indistinguishable from "pure" (neat) U–P[AB] to the opposite, where the surface function of the modifier is not observed at all.

Five topics are covered in sections 10.2–10.6 that exemplify the different features described in **a–i** above.

Section 10.2: Different nanoscale phase separation and wetting behavior for U–P [A-*co*-B] having random versus block soft blocks: **f**; **i** [16, 17].

Section 10.3: U–P[AB] where a copolymer effect enhances hydrophobicity: **f**; **i** [18].

Section 10.4 (and 10.2): Contraphilic wetting, that is, surfaces that are reversibly hydrophilic when dry and hydrophobic when wet: **f**; **i** [17, 19, 20].

Section 10.6: Introducing surface charge via U–P[AB] modifiers that results in contact antimicrobial effectiveness: **f**; **i** [22].

Oxetane monomers and telechelics. 3-Bromomethyl-3-methyloxetane, BrOx, served as the starting material for the fluorous and other oxetane monomers described in sections 10.2–10.5 (Figure 10.1). The synthesis of BrOx is conveniently carried out on a large scale from 1,1,1-tris(hydroxymethyl) ethane as shown in Equations 10.1 and 10.2 [23–25]. Fluorous and other oxetane monomers were prepared from BrOx by variations of the Williamson synthesis (Equation 10.3) in a manner similar to that described by Malik [26, 27].

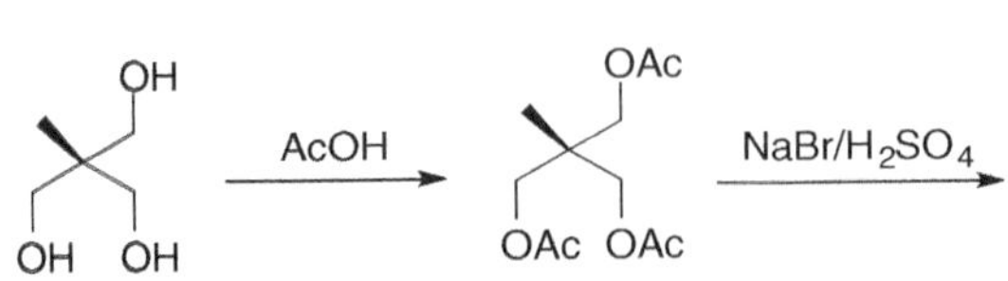

(10.1)

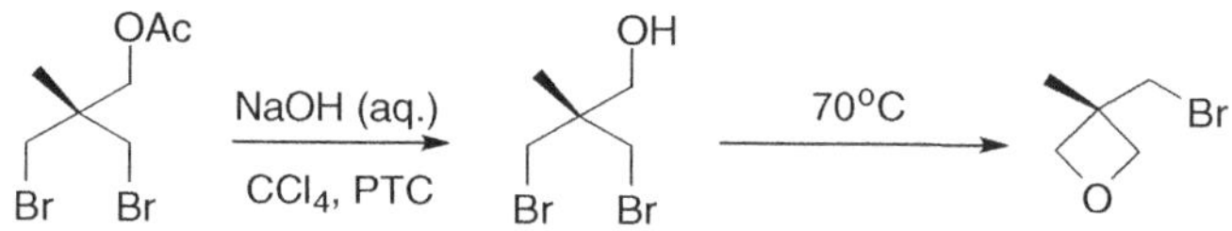

(10.2)

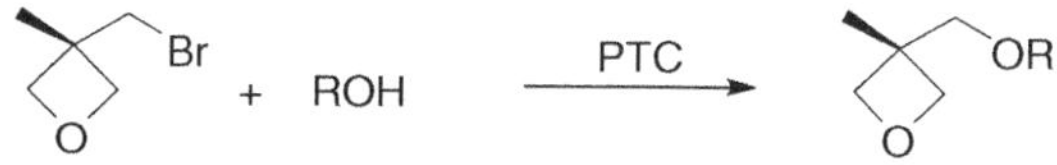

(10.3)

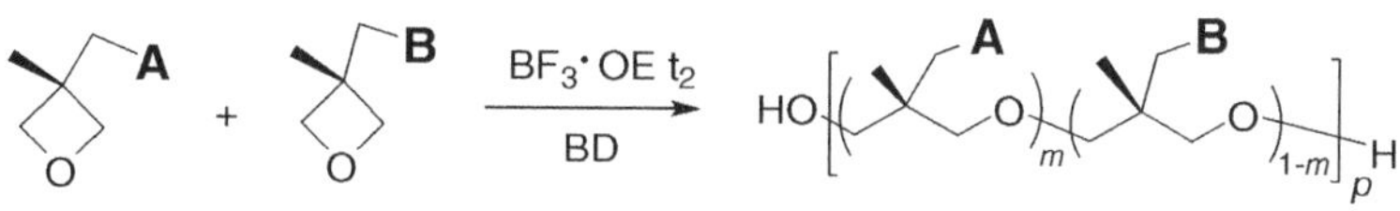

(10.4)

Several papers describe the synthesis of copolyoxetanes and the evaluation of relative reactivity of differently substituted aliphatic oxetanes [28–31]. Adapting previous work [26, 32], $BF_3(Et_2O)$ catalyst and 1,4-butanediol cocatalyst were used for cationic ring–opening polymerization of A and B oxetanes (Equation 10.4).

Kubisa summarized the propagation mechanisms of cyclic ethers with hydroxyl groups [33, 34]. An active chain-end mechanism is typical for cationic polymerizations. Alternatively, polymerization may occur by an activated monomer mechanism. That is, in the presence of alcohol or diol, monomers compete with alcohol groups as transfer agents. An important factor influencing the mechanism is the instantaneous [M]/[OH], which depends on the rate of monomer addition and/or temperature. Well-controlled living-like polymerization of cyclic ethers using BF_3—OEt_2/1,4-butanediol system has been reported to occur only through the activated monomer mechanism [31, 33].

There are uncertainties about mechanistic details for oxetane cationic ring–opening polymerization. In our experience, molecular weight distributions are broad and molecular weights are not particularly sensitive to monomer/cocatalyst ratio. Nevertheless, the degrees of polymerization achieved are at least as high as or higher than those required for utilization as soft blocks in segmented copolymers.

P[AB] copolyoxetanes are viscous oils with low glass transition temperatures. Atactic structures preclude crystallization, which stands in contrast to commonly employed polyether glycols, such as polytetramethylene oxide, PTMO-2000 (T_m = 28–40°C). Lack of crystallization is helpful for maintaining P[AB] soft block polyurethane surface properties over wide temperature ranges.

U–P[AB]. Polyurethanes were synthesized by a conventional two-step, "soft block first" method (Scheme 10.1) [35, 36]. Isophorone diisocyanate (IPDI), methylene diphenyl diisocyanate (MDI), or HMDI were used to generate hard blocks (Figure 10.1). U–P[AB] molecular weights are generally high enough for film formation and conventional purification procedures. Mechanical properties are normally a major concern for polyurethanes [37]. Because U–P[AB] are employed for surface modification at 2 wt% or less, mechanical properties are not a concern as these are determined by the majority polymer. Coatings are simply generated by dipping glass

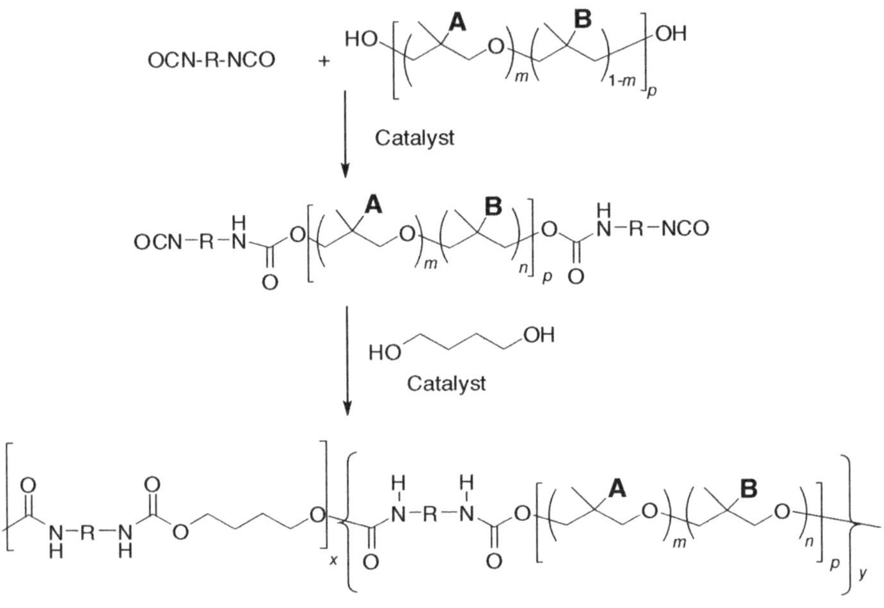

SCHEME 10.1 "Soft block first" polyurethane synthesis.

slides or coverslips into solutions (~15 wt% solids) of the majority polymer and U–P[AB]. Solvent is removed at 50–60°C. Coatings are optically transparent.

Surface science. Surface characterization studies included those for base polyurethane segmented block copolymers and base polymer modified by ≤2 wt% U–P[AB]. Surface analytical techniques included tapping mode atomic force microscopy (TM-AFM), X-ray photoelectron spectroscopy (XPS), wetting behavior by dynamic contact angle (DCA) analysis and sessile drop measurements, and attenuated total reflectance infrared spectroscopy (ATR-IR). Contact antimicrobial behavior of U–P[AB] containing hydantoin and alkylammonium B side chains was determined by spray-on and sandwich tests previously described in detail elsewhere [11, 22, 38].

10.2 DIFFERENT NANOSCALE PHASE SEPARATION AND WETTING BEHAVIOR FOR RANDOM VERSUS BLOCK P[A-*co*-B] SOFT BLOCKS

Copolyoxetanes. Structure 10.2 shows amphiphilic copolyoxetanes having fluorous A (hydrophilic/oleophobic) and poly(ethylene glycol) (PEG-like) hydrophilic B side chains. These copolyoxetanes are remarkable in that random or block architectures may be obtained depending on the order of addition of monomers [17, 39]. The discussion below focuses on P[A-*co*-B] **2** with 3F and ME3 pendant groups. Table 10.1 provides compositions, M_n, and modulated differential scanning calorimetry

(MDSC) data. Molecular weights (M_n = 3–7 kDa) were obtained through end-group analysis [25] and were not very sensitive to monomer/cocatalyst ratio.

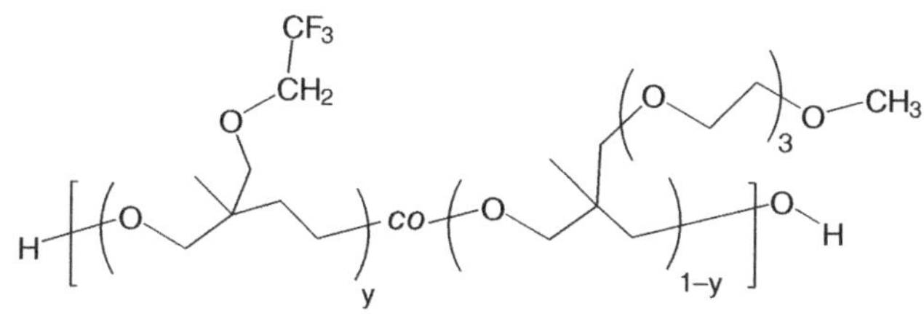

STRUCTURE 10.2 P[3F-*co*-ME3][22].

When the 3F monomer feed was added first in the presence of BF_3–OEt_2 and BD and the second monomer ME3 was then added, P[3F] and P[ME3] were obtained as two-phase viscous liquids. However, when ME3 monomer was added first followed by 3F, the product was a one-phase viscous liquid. After the ME3 reaction, M_n was 2.6 kDa by end-group analysis with trifluoroacetic anhydride. Then, after slow addition of 3F monomer, M_n = P[3F-*b*-ME3-3:2], was obtained (Table 10.1). The systematic increase in M_n together with the observation of a one-phase, viscous liquid indicated the formation of a block copolyoxetane. A random P[A-*r*B] copolyoxetane was obtained by simultaneous addition of monomers.

DSC was employed to study thermal transitions (Table 10.1). Due to immiscibility, a physical mixture of 3F and ME3 homopolymers has two T_gs near those of the homopolyoxetane diols P[3F] and P[ME3], −47°C and −74°C, respectively. T_g for P[3F-*r*-ME3-1:1] is −62°C, which is between the two of homopolymers as predicted

TABLE 10.1 3F/ME3 Polyoxetanes: Synthesis and Thermal Properties

Polyoxetane telechelics	Molar feed ratio	Observed molar ratio[a]	Molecular weight[b]	T_g (MDSC)[c]
	3F/ME3	3F/ME3	M_n kDa	(°C)
PTMO	NA	NA	2.2	−78.1
P[3F]	1/0	1/0	7.1	−47.4
P[ME3]	0/1	0/1	3.0	−74.2
P[3F] + P[ME3] (3:2)[d]	–	–	–	−69.8/−57.9
P[3F-*r*-ME3-1:1]	1/1	0.96/1	3.1	−61.8
P[3F-*b*-ME3-3:2]	2/1	1.45/1	4.2	−59.7

NA, not applicable.
[a]From ^{1}H-NMR spectra.
[b]M_n from end-group analysis by ^{1}H-NMR integration [39].
[c]Modulation amplitude ±0.5°C, modulation period 30 s, and heating rate 3°C/min. T_gs designated at midpoint of inflection.
[d]Physical mixture of P[3F] and P[ME3].

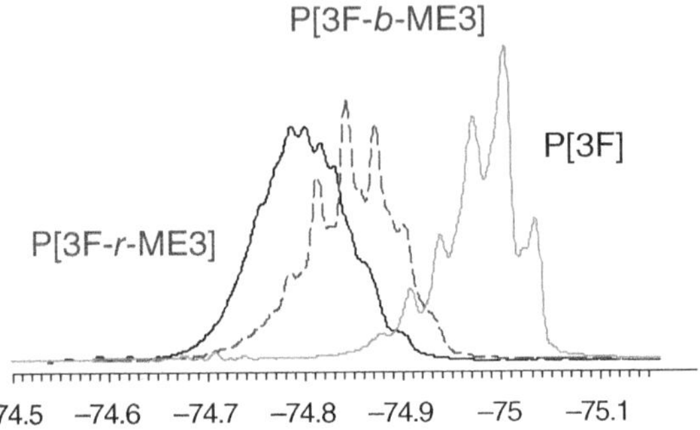

FIGURE 10.3 ^{19}F-NMR spectra for (a) P[3F], (b) P[3F-*b*-ME3-3:2], and (c) P[3F-*r*-ME3-1:1] in $CDCl_3$.

by the Fox equation [16]. P[3F-*b*-ME3-3:2] (−60°C) also has only one T_g. From these results, the block copolymer is not a mixture of homopolymers but has a T_g similar to the random copolymer.

^{19}F-NMR spectra were obtained to investigate random and block structures (Figure 10.3) [17]. The 3F CF_3− peaks in both block and random copolymers shift to low field relative to 3F homopolymer due to a solvent effect. Homo- and block-telechelics show a series of well-resolved peaks with $J_{1H-19F} = 8$ Hz. In contrast, the random copolymer peak is broad with little resolvable structure supporting the notion that the random-copolyoxetane is comprised of random sequences with many sequence distributions. In contrast, the block copolyoxetane contains $(3F)_n$ sequences paralleling those in the homotelechelic. Hence, P[3F] and P[3F-*b*-ME3] have similar ^{19}F-NMR spectra. Copolymers prepared from 3F and ME2 monomers were also block or random depending on order of addition [16].

P[AB] Polyurethanes. MDI was used as the isocyanate hard block precursor for this series (Figure 10.1). For these polyurethanes, designations are MDI/BD (wt%)/P[3F-*sequence*-ME3-mole ratio-(M_n)], with M_n in kDa. MDSC on polyurethanes with either P[3F-*r*-ME3] or P[3F-*b*-ME3] soft blocks have T_gs in the vicinity of −40°C. Based on subsequent experience with other polyurethanes with PEG-like side chains (section 10.6), the T_g increase of ~20°C is ascribed to a moderate degree of phase mixing of ME3 with the MDI/BD hard block.

Figure 10.4 shows TM-AFM phase images for (a) MDI/BD(27)/P[3F-*r*-ME3-1:1-(3.1)] and (b) MDI/BD(32)/P[3F-*b*-ME3-3:2-(4.2)]. The phase image of the polyurethane containing the random soft block is featureless (Figure 10.4a). This is consistent with a surface morphology where the random soft block predominates. With increased tapping force ($A/A_0 = 0.5$–0.6), a phase-separated structure emerges (data not shown), reflecting the presence of near-surface hard block domains.

In contrast, TM-AFM of the polyurethane containing the block soft segment (Figure 10.4b) shows clear nanoscale (~20 nm) phase separation that is attributed to two block domains, namely, 3F and ME3. The conventional interpretation of modulus-sensitive phase images at light tapping assigns the lighter colored domains to those

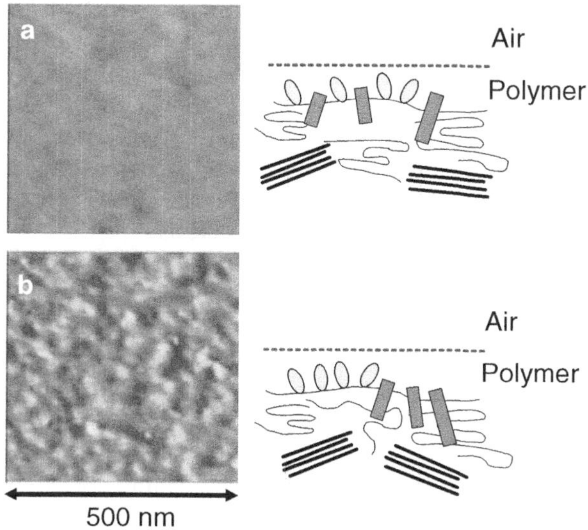

FIGURE 10.4 TM-AFM phase images for (a) MDI/BD(27)/P[3F-*r*-ME3-1:1-(3.1)] and (b) MDI/BD(32)/P[3F-*b*-ME3-3:2-(4.2)]. Settings: (a) $r_{sp} = A/A_0 = 0.83$, phase angle = 20°; (b) $r_{sp} = A/A_0 = 0.92$, phase angle = 20°. R_q (nm): (a) 0.3; (b) 0.9.

of higher modulus [40, 41]. Accordingly, the lighter false-colored nanodomains are assigned to those containing 3F, as the 3F homopolymer has a $T_g > 30°C$ higher than ME3. This nanoscale phase separation for MDI/BD(32)/P[3F-*b*-ME3-3:2(4.2)] reflects the immiscibility of 3F and ME3 blocks and the short block copolyoxetane sequences. At ambient temperature, there is no thermal barrier to phase separation as the respective blocks are 75°C (3F) and 100°C (ME3) above T_g.

The clear resolution of nanoscale phase separation by AFM stands in contrast to the single T_g observed in DSC. Precedent exists for resolving this dichotomy based on consideration of TM-AFM for nanolayers generated by forced assembly [42, 43]. Multiple layers of two immiscible amorphous polymers produced by coextrusion show separate T_gs when the layer spacing is above ≈20 nm. Below this spacing, separate layers can be observed by AFM but only one T_g is observed by DSC. A broadened heat capacity inflection is seen for the P[A-*b*-B] soft block polyurethane in a manner analogous to that for sub-20 nm glassy nanolayers.

The interesting difference in nanoscale surface phase separation for U–P[AB] is reflected in contrasting wetting behavior. For evaluation of surface wetting properties, DCA analysis by the Wilhelmy plate method was used [44, 45]. MDI/BD(36)/PTMO(2.0) was examined as a reference: θ_{adv} of 93° and θ_{rec} of 49°. From previous work [4, 46], θ_{adv}, θ_{rec}, and contact angle hysteresis [$\theta_\Delta = \theta_{adv} - \theta_{rec} = 44°$] are fairly typical values for PTMO polyurethanes. Consistent with the soft block domain occupying the surface, the moderate θ_Δ (44°) is attributed principally to rapid surface reorganization of the PTMO soft block.

Homotelechelic composition and contact angles were: MDI/BD(29)/P[3F], θ_{adv}, 110°, θ_{rec}, 70°; MDI/BD(37)/P[ME3], θ_{adv}, 93°, θ_{rec}, 32°. Analysis of MDI/BD(27)/P[3F-*r*-ME3-1:1(3.1)] gave $\theta_{adv} = 104°$, $\theta_{rec} = 39°$, and $\theta_{\Delta} = 65°$. This U–P[A-*r*-B] is hydrophobic in air due to fluorous groups with θ_{adv} similar to the P[3F]-polyurethane, but is hydrophilic after immersion in water (θ_{rec}, 39°) with a receding contact angle closer to U–P[ME3] (32°) than to U–P[3F] (70°). Extensive surface reorganization occurring in water favoring hydrophilic PEG-like side chains at the water–polymer interface accounts for the large contact angle hysteresis.

For MDI/BD(32)/P[3F-*b*-ME3-3:2(4.2)], θ_{adv} (106°) was also close to θ_{adv} for U–P[3F]. However, θ_{rec} (56°) was 17° higher than that U–P[A-*r*-B] (θ_{rec}, 39°) resulting in a smaller contact angle hysteresis ($\theta_{\Delta} = 50°$) for the *block* compared to 65° for the *random*. This result indicates that U–P[A-*b*-B] is hydrophobic in air and only moderately hydrophilic in water with wetting behavior more like the U–P[3F]. Over the limited timescale investigated (four DCA cycles take about 10 minutes), the self-assembly responsible for fluorinated surface nanodomains apparently inhibits access of a significant fraction of near-surface, more hydrophilic polyether side chains to water. These results demonstrated for the first time the effect of soft block sequence distribution on polyurethane surface morphology and wetting behavior.

2 wt% P[AB] polyurethanes. For modification of base polyurethane MDI/BD/PTMO(2.2) with 2 wt% U–P[3F-*co*-ME3], the focus is "feature **i**," that is, the determination of the extent to which the surface properties of 2 wt% modified compositions replicate the neat P[A-*co*-B] soft block polyurethane.

Table 10.2 shows XPS results for 2 wt% MDI/BD(27)/P[3F-*r*-ME3-1:1(3.1)] and 2 wt% MDI/BD(32)/P[3F-*b*-ME3-3:2(4.2)]. The high atom% F attests to the surface concentration of the P[AB] soft block. The atom% F is higher than that calculated for the soft block alone, while the atom% O is consistently low. The atom% F for 2 wt% P[A-*b*-B] is much higher than random analog. This suggests a model wherein the outermost surface layer in air is dominated by the fluorous side chain.

TM-AFM imaging was carried out to determine whether the nanoscale morphologies of U–P[A-*r*-B] and U–P[A-*b*-B] were comparable to 2 wt% modified compositions. The phase images for the 2 wt% U–P[A-*r*-B] and U–P[A-*b*-B] (Figures 10.5c and 10.5d, respectively, bear a striking resemblance to the corresponding neat

TABLE 10.2 C, O, and F XPS Data (55 Take Off Angle) for 2 wt% MDI/BD(27)/P[3F-*co*-ME3] Polyurethanes

	Soft block atom%					
	Calculated			Observed		
Compositions	C	O	F	C	O	F
2 wt% MDI/BD(27)/P[3F-*r*-ME3-1:1(3.1)]	66	24	10	62.4	21.8	13.8
2 wt% MDI/BD(32)/P[3F-*b*-ME3-3:2(4.2)]	64	23	13	59.4	18.8	21.1

MDI, methylene diphenyl diisocyanate; BD, butanediol; XPS, x-ray photoelectron spectroscopy.

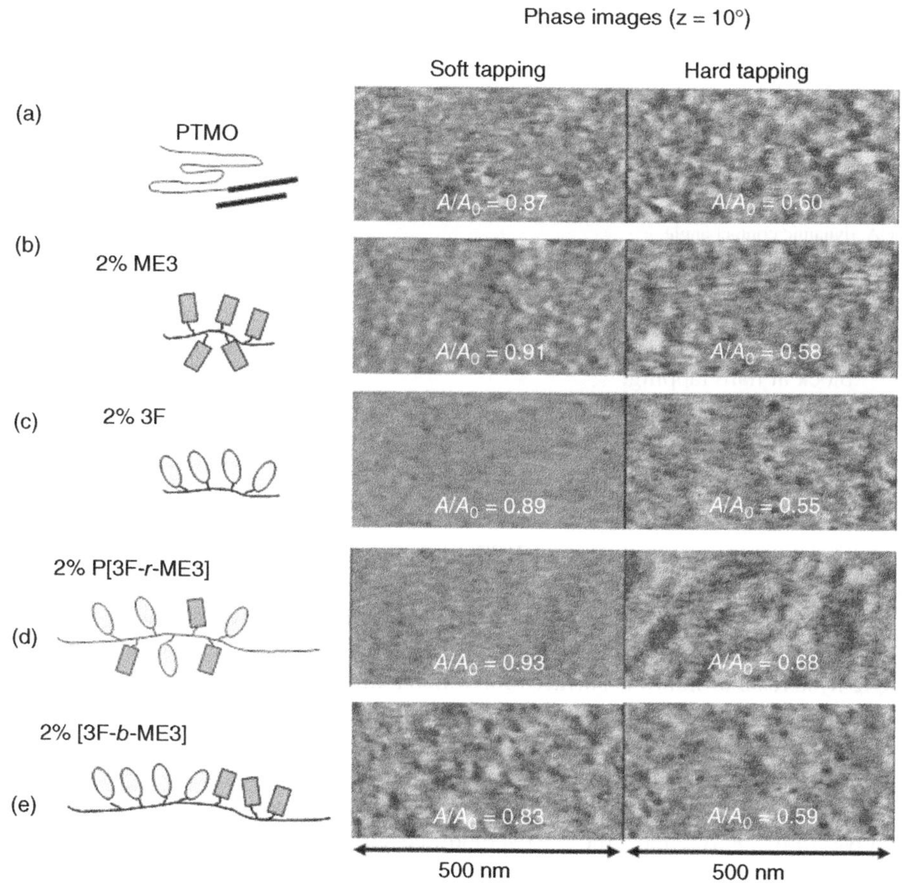

FIGURE 10.5 TM-AFM phase images for (a) MDI/BD(36)/PTMO(2.0), (b) 2 wt% MDI/BD/P[ME3], (c) 2 wt% MDI/BD/P[3F], (d) 2 wt% MDI/BD(27)/P[3F-*r*-ME3-1:1(3.1)], and (e) 2 wt% MDI/BD(32)/P[3F-*b*-ME3-3:2(4.2)].

U–P[A-*b*-B]. For reference, the phase images for related U–P[A], U–P[B], and 2 wt% U–P[AB] modified MDI/BD(36)/PTMO(2.0) are also shown.

- Phase images for 2 wt% P[A-*r*-B] and P[A-*b*-B] (Figures 10.5d and e, respectively) reflect modifier soft block surface concentration and account for similarities in nanomorphology with images for the corresponding U–P[AB] (Figure 10.4).
- 2 wt% U–P[3F] lacks discernable phase separation at soft tapping, but near-surface hard block at hard tapping [18].
- The dependence of near-surface nanomorphology on the setpoint ratio for 2 wt% MDI/BD(27)/P[3F-*r*-ME3-1:1(3.1)] (Figure 10.5d) is similar to that of 2 wt%

TABLE 10.3 Dynamic Contact Angle Data for Neat and 2 wt% U–P[3F-*co*-ME3]

Composition	θ_{adv1}	θ_{rec1}	θ_{adv2}	θ_{rec2}	θ_{adv3}	θ_{rec3}
MDI/BD/P[3F-*r*-ME3-1:1][a]	104	39	105	44	101	50
2 wt% MDI/BD/P[3F-*r*-ME3-1:1]	99	46	98	45	98	45
MDI/BD/P[3F-*b*-ME3-3:2][a]	106	56	101	57	99	57
2 wt% MDI/BD/P[3F-*b*-ME3-3:2]	101	53	100	52	100	52

DCA, dynamic contact angle.
[a]Water contamination after DCA analysis; estimated standard deviation ±2–3°.

U–P[3F] (Figure 10.5c), namely, featureless at soft tapping, near-surface hard block at hard tapping.

- Both MDI/BD(36)/PTMO(2.2) and 2 wt% MDI/BD/P[ME3] (Figure 10.5a and b, respectively) are phase; evidence of near-surface hard block is apparent at soft and hard tapping.

Surface composition in vacuum (XPS) and morphology in air (TM-AFM) clearly demonstrate surface concentration of the P[AB] modifier soft block. Wetting was evaluated to compare advancing and receding contact angles with those of the neat U–P[AB] (Table 10.3). All of these materials show the characteristic high contact angle hysteresis of fluorous surfaces for which a model has been proposed [47]. An additional contribution to the low θ_{rec} is the presence of hydrophilic PEG-like side chains. Overall, neat U–P[A-*r*-B] is somewhat more hydrophilic than the U–P[A-*b*-B] analog. This is consistent with XPS data that reflects a higher surface concentration of fluorous moieties for the P[A-*b*-B] soft block polyurethane.

Over three DCA cycles, the 2 wt% modified surfaces have θ_{adv} wetting behavior that tracks those of the neat modifiers. However, the 2 wt% MDI/BD/P[3F-*r*-ME3] has a very stable receding contact angle (45°) compared to the rapidly increasing contact angle for the neat U–P[A-*r*-B] (39–50° over three cycles). The neat and 2 wt% P[A-*b*-B] polyurethanes have similar and somewhat less hydrophilic (θ_{rec}) characteristics.

From AFM, XPS, and DCA analyses, it is clear that surface nanomorphology affects macroscopic wetting behavior. These methods provide a coherent set of data indicating that the nanoscale phase separation for the neat U–P[A-*b*-B] is found in the 2 wt% composition. The more hydrophobic wetting behavior follows from XPS analysis and nanoscale morphology via AFM imaging. Similarly, neat U–P[A-*r*-B] has unique XPS, nanoscale morphology, and wetting behavior consistent with the P[A-*r*-B] soft block. These results give insight into a new approach to control subtle aspects of surface composition that may be important in areas such as adhesion or biocompatibility.

Increasing the length of the PEG side chain: U-P[3F-co-ME7]. ME7 has an average of seven ethylene oxide repeats in the PEG-like side chain [17]. Surface modification with U[3F-*co*-ME7-U] resulted in dry surface morphology and composition again driven by 3F surface concentration. However, wetting behavior contrasted with ME3 analogs. 2 wt% U–[3F-*r*-ME7] was amphiphilic, with high contact angle hysteresis but also modestly *contraphilic*. By contraphilic is meant that the dry coating is more hydrophilic (lower $\theta_{adv} = \theta_{adv1}$) than the wet coating [19, 20]; that is, $\theta_{adv\text{-}1} < \theta_{adv2}$.

The magnitude of the contraphilic effect is defined by $\Delta c = -(\theta_{adv\text{-}1} - \theta_{adv\text{-}2})$. Interestingly, both U–P[3F-*b*-ME7] and 2 wt% U[3F-*b*-ME7] exhibited contraphilic wetting with Δc for 2 wt% U–P[3F-*b*-ME7] being higher (8°). Contraphilic wetting and the measurement of dynamic contact angles is described in more detail in section 10.4 for systems that have a Δ_C as high as 37°.

Like the 3F analogs, θ_{rec} is imposed by the U–P[3F-*co*-ME*n*] modifier independent of swelling, which is controlled by the bulk polyurethane. Independent control is important for applications where retention of bulk mechanical properties and adhesion to substrate are desired while separately tuning surface interactions with the environment.

10.3 U–P[AB] WHEREIN A COPOLYMER EFFECT ENHANCES HYDROPHOBICITY

Polyurethanes. U–P[AB] having copolyoxetane soft blocks with structure 10.3 constitute model surface modifiers [36]. The preparation of P[*n*F-*co*-BrO*x*] by cationic ring–opening polymerization is described by Equation 10.4 [36]. The discussion in this section focuses on P[3F-*co*-BrO*x*-1:1]. The 3F:BrO*x* ratio is easily determined from the ^{1}H-NMR spectrum by integration of methyl peaks at 0.92 ppm (**C**$\mathbf{H_3}$, 3F) and 1.05 ppm (**C**$\mathbf{H_3}$, BrO*x*). Copolyoxetanes **3** have low T_gs (−33°C to −39°C) and molecular weights in a range typical for soft blocks (M_n = 3–5 kDa).

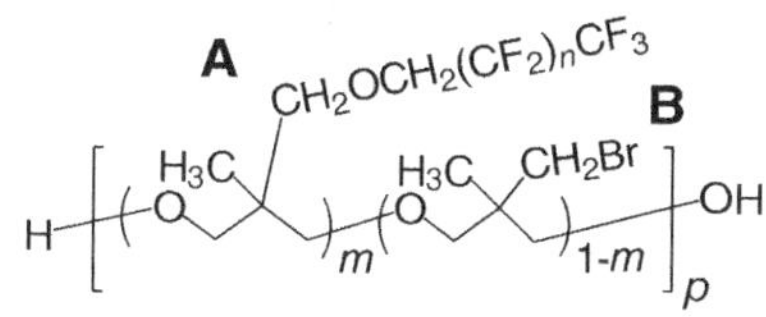

STRUCTURE 10.3 [36]

U–P[3F-*co*-BrO*x*] were synthesized using IPDI and BD for the hard block by the soft block first method described above (Scheme 10.1) [36]. Characterization of polyurethane composition and bulk properties utilized ^{1}H-NMR, temperature MDSC, and gel permeation chromatography (GPC). The molecular weights were modest (30–60 kDa, Table 10.4) but all formed smooth, optically transparent coatings and free-standing films. Subsequently, these polyurethanes were used in a "reaction on polymer" approach to generate the hydantoin substituted polymers whose contraphilic wetting behavior is discussed in section 10.4.

Because bulk phase separation determines the degree of surface concentration of soft block [9], thermal analysis was performed with MDSC as well as conventional

TABLE 10.4 Molecular Weights and Glass Transitions Temperatures of PTMO, P[BrOx], P[3F], and P[3F-*co*-BrOx] Polyurethanes

Designation	M_n (kDa)	M_w (kDa)	PDI	T_g^a (ss)	T_g^b (hs)	Phase separation[c]
IPDI–BD(50)/PTMO(2.0)	23.3	52.5	2.26	−46	38	0.62
IPDI–BD(40)/P[BrO*x*](2.8)	19.4	42.9	2.21	−10	56	0.81
IPDI–BD(40)/P[3F](3.4)	17.5	37.4	2.14	−37	46	0.84
IPDI–BD(40)/P[3F-*co*-BrO*x*-2:1-(4.7)]	18.9	46.0	2.43	−29	73	0.89
IPDI–BD(40)/P[3F-*co*-BrO*x*-1:1-(4.7)]	17.9	36.8	2.05	−29	62	0.90
IPDI–BD(40)/P[3F-*co*-BrO*x*-1:2-(3.4)]	16.5	33.9	2.06	−24	56	0.89
IPDI–BD(40)/P[3F-*co*-BrO*x*-2:1-(2.5)]	18.9	40.1	2.12	−27	57	0.88
IPDI–BD(40)/P[3F-*co*-BrO*x*-1:1-(4.1)]	29.6	61.2	2.07	−25	64	0.89
IPDI–BD(40)/P[3F-*co*-BrO*x*-1:2-(3.4)]	16.6	33.8	2.04	−29	64	0.89
IPDI–BD	17.2	31.8	1.85	NA	85	NA

NA, not applicable.

[a]Soft segment (ss) glass transition temperature.

[b]Hard segment (hs) transition temperature designated T_g, alternatively described as an order–disorder temperature for the hard block.

[c]The maximum error for T_g measurement is ±2.7°, which leads to an estimated deviation of ±0.03 in estimating pure soft phase content using the Fox equation.

DSC. Table 10.4 summarizes MDSC data for PTMO, P[BrO*x*], P[3F], and P[3F-*co*-BrO*x*] polyurethanes. U–P[BrO*x*], U–P[3F], and U–P[3F-*co*-BrO*x*] soft block T_gs are shifted to higher temperatures (5–14°) compared to the telechelics [36] while the hard segment T_gs are lower (12–37°) than the T_g of pure hard block (IPDI–BD). These changes are attributed to phase mixing of the urethane soft and hard segments [48,49].

Knowing the T_gs of pure hard and soft segments and their respective T_gs and weight fractions, phase mixing was estimated by applying the Fox equation [37, 50]. The calculated values in Table 10.4 are expressed as weight fraction of soft block in the soft-segment phase. The "pure" soft block phase thus accounts for about 90% of the soft segment for U–P[BrO*x*], U–P[3F], and U–P[3F-*co*-BrO*x*]. According to this analysis, the U–P[3F-*co*-BrO*x*] are more phase separated than the P[BrO*x*]-polyurethane and marginally better phase separated than the P[3F]-polyurethane.

Surface science. TM-AFM phase images for IPDI/BD(40)/P[3F-*co*-BrO*x*-1:1-(4.7)] are shown in Figure 10.6 (topmost). The image on the left is for soft tapping (r_{sp} = 0.93) while that on the right is for hard tapping (r_{sp} = 0.70). The featureless phase image at soft tapping supports the notion that the surface is dominated by the low T_g soft block. Harder tapping reveals the presence of near-surface hard block as was observed by Runt for a well phase-separated MDI/ED/PTMO urethane–urea (ED = ethylene diamine) [9]. Phase images (Figure 10.6, E-1, E-2) for the base polyurethane IPDI/BD(50)/PTMO(2.0) show similar phase images at soft and hard tapping, which is consistent with the lower degree of phase separation indicated by the DSC analysis discussed above. The intermediate images in Figure 10.6 are discussed in the "≤2 wt%" section below.

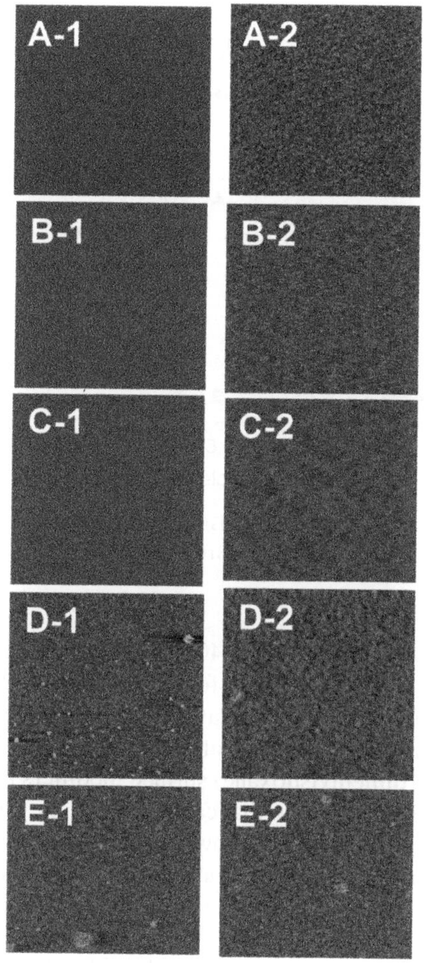

FIGURE 10.6 TM-AFM phase images (z = 20°), 2 × 2 μm (left, soft tapping, r_{sp} > 0.90; right, hard tapping, $r_{sp} \leq 0.71$): A, U–P[AB]; B, 2.0 wt%; C, 1 wt%; D, 0.5 wt%, E, IPDI-base. "2 wt%" designates 2 wt% IPDI–BD(40)/P[3F-*co*-BrOx-1:1-4.7] and 98% IPDI–BD(50)/PTMO(2.0).

From DSC and TM-AFM, IPDI/BD(40)/P[3F-*co*-BrOx-1:1-(4.7)] is well phase-separated compared to IPDI/BD(50)/PTMO(2.0). The average degree of polymerization for P[3F-*co*-BrOx-1:1-(4.7)] and PTMO(2.0) is similar (D_p = 33 and 28, respectively). Hence, the high degree of bulk phase separation for IPDI/BD(40)/P[3F-*co*-BrOx-1:1-(4.7)] is due to limited miscibility of the fluorous soft block with the hard block (feature **b**) and multiple low surface-free energy side chains that act as pseudo-chain ends (feature **a**) [10].

XPS was used to analyze the outer few nanometers of the P[AB]-polyurethanes [47]. Only data for IPDI/BD(40)/P[3F-*co*-BrOx-1:1-(4.7)] are summarized here. The

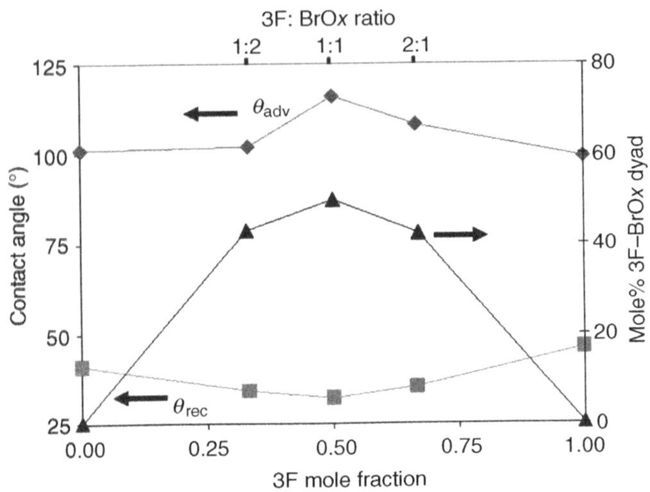

FIGURE 10.7 Contact angles for P[3F-*co*-BrO*x*] polyurethanes correlated with 3F–BrO*x* dyad content. ♦, θ_{adv}; ■, θ_{rec}; ▲, mole% 3F–BrO*x* dyad.

observed atom% F (17.7%) for IPDI/BD(40)/P[3F-*co*-BrO*x*-1:1-(4.7)] may be compared with the bulk value (9.1%) and that for the P[3F-*co*-BrO*x*] soft block (16.8%). The good agreement of atom% F with that calculated for the soft block alone provides evidence for surface concentration of the soft block, as has been observed for semifluorinated or poly(dimethylsiloxane) surface modifiers [23, 51–57].

The atom% Br (2.9%) is closer to that calculated for the bulk (2.5%) than for the soft block alone (4.7%). The relatively low value for atom% Br suggests —CH_2Br groups are not concentrated in the outermost surface. Because photoelectron emission is an exponential function of depth, subsurface partitioning would result in a considerable decrease in atom% Br. Thus, fluorous moieties act as chaperones for concentrating —CH_2Br groups near the surface, but CF_3CH_2O- groups preferentially occupy the outermost surface compared to —CH_2Br.

DCA analysis was used to interrogate wetting behavior [18]. Interestingly, P[3F-*co*-BrO*x*]-polyurethanes have higher θ_{adv} and lower θ_{rec} than the homotelechelic polyurethanes. Figure 10.7 shows that the unusual wetting behavior for P[(3F)(BrO*x*)] polyurethanes is correlated with AB dyad content. For IPDI/BD(40)/ P[(3F)(BrO*x*)-1:1], θ_{adv} (116°) is higher and θ_{rec} (32°) is lower ($\Delta\theta$, 84°) than any other homo- or copolyoxetane polyurethane. These values are constant over three DCA cycles wherein force–distance curves superpose [47]. A reversible H-bonding mechanism was proposed to explain the results [47].

2 wt% U–P[AB]. TM-AFM, XPS, and DCA analysis were utilized to estimate the surface modifier concentration vis-à-vis neat U–P[3F-*co*-BrO*x*]. XPS data were acquired for several weight percents IPDI–BD(40)-P[3F-*co*-BrO*x*-1:1-(4.7)] modifiers, but the results for 0.5 wt% are representative and better test the limits of surface modification.

The observed atom% F (18.2%) for 0.5 wt% IPDI–BD(40)-P[3F-*co*-BrO*x*-1:1-(4.7)] is somewhat higher than that calculated for the P[AB] soft block alone (16.8%) attesting to the surface concentration of the P[AB] soft block. The XPS results may be compared with the phase images shown in Figure 10.6 where successively lower weight percentage compositions are interrogated. TM-AFM for 0.5 wt% U–P[AB] (Figure 10.6D-1) at light tapping shows the presence of near-surface hard domain and a morphology almost identical to the base polymer (Figure 10.6, E-1). Thus, TM-AFM does not provide morphological evidence of U–P[AB] modifier surface concentration. However, at the 5-nm depth resolution of XPS, the surface of 0.5 wt% IPDI–BD(40)–P[3F-*co*-BrO*x*-1:1-(4.7)] is indistinguishable from the base polymer alone. These results are a reminder that TM-AFM is an imaging method relying on mechanical–physical interactions of the AFM tip with the outermost surface. XPS provides direct physical chemical analysis with depth sensitivity at length scales comparable to the root mean square end to end distance of soft blocks (2–3 nm). Thus atomic composition determined by XPS is strongly influenced by even a "single layer" of soft block domain.

As for neat IPDI–BD(40)–P[3F-*co*-BrO*x*-1:1-(4.7)], the observed atom% Br for 0.5 wt% IPDI–BD(40)–P[3F-*co*-BrO*x*-1:1-(4.7)] is about half (2.2%) that calculated for the soft block alone (4.7%). While the surface concentration of Br shows the chaperone effect of fluorous side chains, near-surface partitioning of the $-CH_2Br$ groups is proposed to explain the low atom% Br compared to the soft-block-alone value.

In contrast to polymer surface modifiers containing the 3F side chain, bromine is barely detected for 0.5 wt% IPDI/BD(40)/P[(BrO*x*)-2.8] (0.1 atom% Br). These data provide a negative control that emphasizes the importance of fluorous side chains in bringing the $-CH_2Br$ groups to the surface, which is illustrated in Figure 10.8b.

Dynamic contact angles as a function of weight percent IPDI/BD(40)/P[3F-*co*-BrO*x*-1:1-(4.7)] are shown in Figure 10.8a. The threshold for maximum hydrophobic behavior is between 0.5 and 1 wt% U–P[AB] modifier. θ_{adv} for 0.5 wt% IPDI/BD(40)/P[3F-*co*-BrO*x*-1:1-(4.7)] (104°) was much greater than θ_{adv}

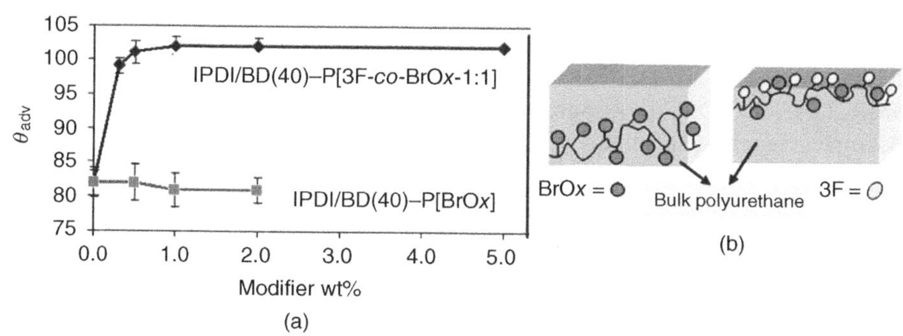

FIGURE 10.8 Advancing contact angles (fdc-1) for IPDI/BD(50)/PTMO(2.0) modified with IPDI/BD(40)–P[3F-*co*-BrO*x*]-1:1-(4.7)] (◆) and IPDI–BD(40)–P[BrO*x*-2.8] (■).

for IPDI/BD(50)/PTMO(2.0) (85°) but 12° less than neat IPDI/BD(40)–P[3F-*co*-BrO*x*-1:1-(4.7)] (116°). The P[3F-*co*-BrO*x*-1:1-(4.7) soft block is clearly surface-concentrated as illustrated in Figure 10.8b, but the unusually high water contact angle for the neat U–P[AB] is not retained in the modified coating.

IPDI/BD(40)/P[(BrO*x*)-2.8], which has a high θ_{adv} (102°), is not surface concentrated as a modifier. Wetting behavior for 0.5 wt% IPDI/BD(40)/P[(BrO*x*)-2.8] is similar to the bulk polyurethane (Figure 10.8a) and only trace atom% Br is detected by XPS.

Calculations showed that the solubility parameter for P[BrO*x*] (8.3) is close to that for PTMO (8.6) [20, 58, 59]. Similar solubility parameters for the modifier and PTMO soft blocks and a short side chain that does not mimic an end-group may explain the absence of P[BrO*x*] surface concentration.

In summary, there are two entropically driven processes for surface concentration of the P[3F-*co*-BrO*x*] soft block. 3F repeat units are surface concentrated by virtue of low surface energy, immiscibility, and side chain length that mimics an end-group. BrO*x* repeat units are miscible with the PTMO domain of the bulk phase and lack an end-group-like architecture. The latter provides a contravening effect that works against surface concentration in the chosen bulk polyurethane.

10.4 SURFACES THAT ARE REVERSIBLY HYDROPHILIC WHEN DRY AND HYDROPHOBIC WHEN WET, OR "CONTRAPHILIC"

Polyurethanes. Worley and Sun showed that the chloramide (—CON—Cl) function provides rapid oxidative kill of a wide range of pathogenic bacteria [60–64]. Chloramide and related functionality is renewed periodically by treatment of a parent amide (—CON—H) with bleach, which is convenient for applications such as water disinfection and treatment of fabrics used in nosocomial environments [63, 65]. Interest in these results led us to introduce the hydantoin moiety into copolyoxetane soft blocks [11]. The strategy for this work was to use a P[AB] fluorous "chaperone" to surface-concentrate hydantoin (Figure 10.2).

A reaction-on-polymer approach was used to introduce 5,5-dimethylhydantoin starting with fluorous/BrO*x* polyurethanes described in section 10.3. The substituted repeat unit is designated HyOx, while the fluorous segment was either 3F or 5F (Figure 10.1). IPDI-based polyurethanes were characterized by ^{1}H-NMR, GPC, and DSC [20]. Substitution of BrO*x* by 5,5-dimethylhydanotoin was incomplete resulting in lengthy polyurethane designations; here, IPDI/BD(40)/P[*co*-5FOx/HyOx/BrO*x*)-2.0:0.7:0.3-(5.0)] is designated U–P[5F-*co*-HyOx-66:23] with the remaining 11 mol% BrO*x* not shown. This U–P[ABC] exhibited surprising and unprecedented wetting behavior.

The contraphilic effect for 2 wt% U–P[3F-*b*-ME7] (8°) was described in section 10.2. The common structural feature of PEG and fluoroalkyl groups for a system investigated by Kornfeld led to a similar manifestation of contraphilic wetting, namely increased contact angles with increased humidity [66]. This effect by which wetted surfaces have higher contact angles than dry ones was first discovered with novel 3F and 5F copolyoxetanes with HyOx segments that were prepared to study oxidative antimicrobial characteristics [19]. All U–P[3F-*co*-HyOx] displayed the contraphilic effect to varying extents.

DCA measurements are described briefly before the description of contraphilic wetting for U–P[3F-*co*-HyOx]. DCAs are "operator independent" and provide a good deal of related information that is unobtainable by the sessile drop method. The DCA method is based on the work of Wilhelmy in the nineteenth century [67]. Reports concerning implementation of this method to polymer films and coatings are sparse but include work from the laboratories of Andrade [68], Brittain [69], and DeSimone [70]. A description of the method is given in a text by Adamson and Gast [71].

In a typical determination, a coated slide is attached to the instrument electrobalance via a clip and a stage with the water reservoir is automatically raised and lowered to allow water to impinge upon the slide. By analyzing the resulting force versus distance curves (fdcs) for immersion and emersion cycles, θ_{adv} and θ_{rec} are obtained using Equation 10.5, where F = force, m = mass of the coated slide (electrobalance), g = acceleration of gravity, P = coated slide perimeter, γ_L = surface tension of the liquid, θ = contact angle, and F_b = buoyancy force acting on the coated slide.

$$F = mg + P\gamma_L \cos\theta + F_b \tag{10.5}$$

Extrapolating the fdc to the point of immersion and initially taring the mass of the coated slide leads to Equation 10.6, from which θ_{adv} and θ_{rec} are easily obtained.

$$\cos\theta = mg/P\gamma_L \tag{10.6}$$

Previously, DCA facilitated measurements of intrinsic contact angles for model PDMS networks. By "intrinsic" is meant contact angles free from commonly encountered water contamination that changes water surface tension and results in inaccurate determinations [45]. Several advantages for the Wilhelmy plate method have been noted including interrogation of a large surface area compared to the sessile drop method [69]. Another advantageous feature is facile testing of water surface tension after sample analysis [45].

Returning to fluorous/HyOx copolyoxetanes, the wetting behavior for U–P[5F-*co*-HyOx (66:23)] is discussed as this composition has the highest $\Delta_C = -(\theta_{adv1} - \theta_{adv2}) = 37°$. Fdcs, drop profiles, and contact angles are shown in Figure 10.9. The initial force–distance curve for U–P[5F-*co*-HyOx (66:23)], designated Adv-fdc-1, shows an apparent mass *gain* on immersion characteristic of hydrophilic surfaces ($\theta_{adv1} = 69°$). Water "wicks" up onto the coated surface driven by a gain in enthalpy through formation of hydrogen bonds.

Remarkably, an apparent mass *loss* is found for the second immersion cycle (Adv-fdc-2) reflective of a hydrophobic surface ($\theta_{adv2} = 106°$). Now, water–hydrogen bonding interactions are stronger than water–surface interactions and water resists wetting the surface. Contact angles measured by drop profile analysis are close to those observed for DCA.

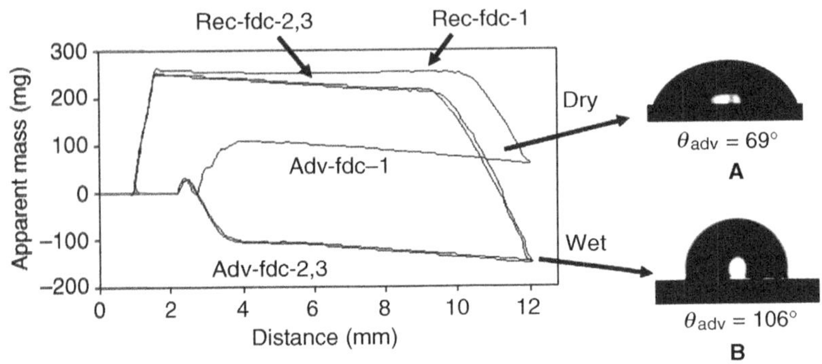

FIGURE 10.9 DCA force–distance curves and goniometer drop profiles for U–P[5F-*co*-HyOx (66:23)]: (a) sessile drop on dry surface; (b) wet surface.

The short-term kinetic stability of switched areas was explored. A water drop was placed on a U–P[5F-*co*-HyOx-66:23] coating resulting in spreading drop profile **A**, Figure 10.9. A water drop was placed again on the same spot giving high contact angle profile **B**, Figure 10.9. The volume of this drop was increased so that the drop perimeter crossed the boundary set by the first drop. On crossing, the drop suddenly "jumped" from profile **B** (>100°) to **A** (~70°) (Figure 10.9). This switching from hydrophilic to hydrophobic character means the water-wetted area (hydrophobic) is stable and does not spread, at least over the course of a few minutes.

Switching kinetics was also evidenced by the horizontal first receding force–distance curve (Figure 10.9, Rec-fdc-1). On initial emersion, a relatively high mass was recorded as the upper area of the coating was in water for a short time (~20 seconds). In contrast, at the point where the coated coverslip is removed from water, the recorded mass was similar to that for Rec-fdc-2. The virtually horizontal Rec-fdc-1 is due to the contravening effects of a hydrophilicity gradient and decreased buoyancy (Equation 10.5). That is, the surface is not completely converted to hydrophobicity on initial emersion, but the contribution of buoyancy is maximized. Near the point of removal, the contribution of buoyancy to apparent mass is minimal, but the bottom of the coated slide has been in water for a relatively long time (2 minutes) so that hydrophobicity is maximized.

Mechanism. For the dry surface, enthalpically driven hydrogen bonding of 5,5-dimethylhydantoin amide groups to acidic —$\mathbf{CH_2}$— of semifluorinated groups is proposed to disrupt normal fluorous surface concentration (Figure 10.10a). The triad sequence shown is favored for U–P[5F-*co*-HyOx-66:23], which has the highest Δ_C for compositions investigated. Polar groups such as —N—H (shown) result in a hydrophilic surface.

Upon exposure to water (Figure 10.10b), amide groups switch from hydrogen bonding to —$\boldsymbol{CH_2}CF_2CF_3$ to stronger hydrogen bonding with water. As a result, the —$CH_2CF_2CF_3$ is "released" and the surface becomes hydrophobic. After drying, the initial hydrophilic wetting character is restored. Cycling from the dry hydrophilic

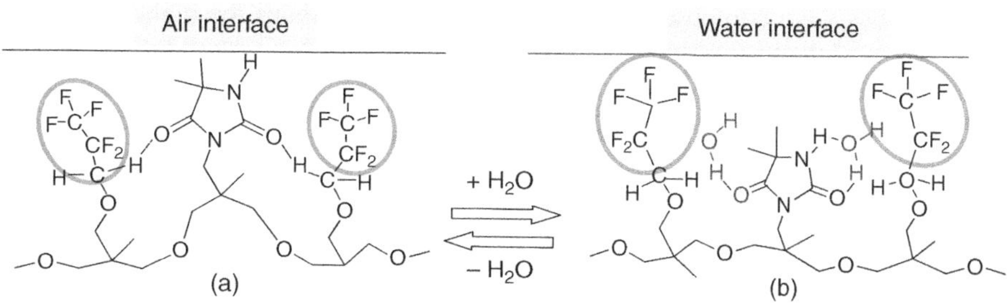

FIGURE 10.10 Proposed mechanism for contraphilic wetting: (a) hydrophilic dry surface; (b) hydrophobic wet surface.

state to the wet hydrophobic state may be repeated many times. If the coating is left at ambient humidity, θ_{adv} has a value intermediate between the dry and wet states.

According to the proposed switching mechanism, if the hydantoin hydrogen bonding is eliminated, the contraphilic effect should disappear. The conversion of near-surface amide (—CON—H) to chloramide (—CO—NCl) has received considerable study, as chloramide has a strong oxidative antimicrobial character [72, 73]. Accordingly, a U–P[5F-*co*-HyOx-66:23] coating was immersed in dilute bleach for 1 hour, after which $\theta_{adv1} = 101.3 \pm 1.7°$, $\theta_{rec1} = 41.6 \pm 0.6°$, regardless of the number of cycles. The conversion of amide to chloramide did indeed eliminate the contraphilic effect. Immersing the coating in thiosulfate for 1 hour reduced chloramide back to amide. Contraphilic wetting behavior was restored and was indistinguishable from the pristine coating.

The contraphilic effect stands in contrast to the wetting behavior of other polar polymers that instead become more hydrophilic on immersion in water [74, 75]. Upon exposure to water, the surface of these other polymers rearranges thereby opening paths to more polar hydrophilic moieties so that hydrophilicity often increases with time [76, 77]. For these polymers, the migration of hydrophilic polar groups to the polymer surface decreases θ_{adv}.

The serendipitous discovery of the contraphilic effect for U–P[5F-*co*-HyOx-66:23] and other fluorous/HyOx polyurethanes demonstrated that "water makes it hydrophobic." Elegant work that provides compositional control of core-shell nanoparticles has shown that it is possible to "nanoengineer" purposefully the contraphilic effect [78, 79]. As described in section 10.2, "contraphilic surface modifiers" were subsequently found, with 2 wt% U[3F-*b*-ME7] exhibiting modest contraphilic wetting ($\Delta_C = 8°$) [17]. An investigation of U–P[5F-*co*-HyOx] as surface modifiers gave unexpected results described below.

U–P[5F-co-HyOx] copolyoxetanes as surface modifiers. Validation of the model shown in Figure 10.2 was explored with 2 wt% U–P[5F-*co*-HyOx-66:23] and other fluorous-hydantoin soft block polyurethanes as surface modifiers for IPDI–BD–PTMO or HMDI–BD–PTMO base polyurethanes. Contraphilic wetting was not found. Rather, the resulting 2 wt% U–P[5F-*co*-HyOx-66:23] showed hydrophobic wetting behavior typical of polyurethanes with fluorous soft blocks. The base polyurethanes were composed of 50 wt% hard block. Perhaps the high hard block composition of the base polymer results in preferential hydrogen bonding of hydantoin to near-surface hard block or soft block and disrupts the inter–intra molecular interactions that result in contraphilic behavior.

Returning to the original goal of research described in this section, 2 wt% U–P[5F-*co*-HyOx polyurethanes were examined for oxidative biocidal effectiveness, which was demonstrated for HyOx and related systems by Worley [60–64, 80–83]. Interestingly, when 2 wt% U–P[5F-*co*-HyOx-66:23] coatings were treated with bleach for 1 hour to effect conversion of amide (—CON—H) to chloramide (—CON—Cl), efficient contact antimicrobial behavior via oxidative kill was demonstrated [11]. Thus, as modifiers, 2 wt% U–P[5F-*co*-HyOx-66:23] failed to provide contraphilic behavior, but based on oxidative antimicrobial effectiveness was a success. These and related findings [21, 38] on oxidative kill for polyurethanes modified with fluorous/HyOx systems will not be reviewed in this article. Instead the use of fluorous "chaperones" to effect surface concentration of quaternary (quat) polycations along with antimicrobial effectiveness via contact kill is described in section 10.5.

10.5 INTRODUCING SURFACE CHARGE VIA P[AB] SOFT BLOCK POLYURETHANES WITH ALKYLAMMONIUM B SIDE CHAINS

Infection is a serious medical complication associated with health care environments. The 5–10% incidence of infections for hospital patients is well documented [84,85] Sources of bacteria include healthcare equipment such as cell phones, pagers, computer keyboards, doorknobs, and electrocardiograph leads [86]. Modes of transmission of antibiotic-resistant pathogens occur in obvious ways such as touches, coughs, and sneezes. Biocidal polymers offer promise in helping curb the spread of infections by providing coatings for biomedical devices or molded articles. The growing problem of antibiotic-resistant bacteria was the motivation for our initial work in hydantoin-based polymer surface modifiers [11, 21, 38] and led to further work with alkylammonium modifiers described below.

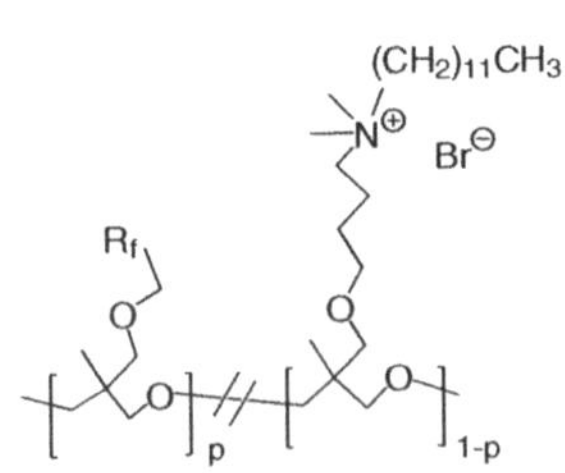

STRUCTURE 10.4 [108]

As noted in section 10.4, the model shown in Figure 10.2 was validated by surface concentration of hydantoin functionality and subsequent demonstration of oxidative bacterial kill after bleach conversion of amide to chloramide [11, 21, 38]. However, due to cytotoxicity, oxidative antimicrobials are unlikely candidates for biomaterials applications such as fighting infections and/or foreign body reactions associated with catheters and hemodialysis tubing [87–91]. With these applications in mind, we turned to polycations that mimic antimicrobial peptides and are therefore promising candidates for biocidal action [92, 93]. In solution, the mode of action for polycations is thought to be bacterial cell disruption that limits possibilities for buildup of resistance [94, 95]. Accessible charge density at the bacteria–nanosurface interface is an important parameter for biocidal effectiveness, suggesting that a membrane disruption mechanism is also operative for thin films [96]. Considering the potential advantages (a–i, introduction), the potential impact of effective antimicrobial surface modifiers led to an investigation of the applicability of the P[AB] copolyoxetane model (Figure 10.2) for surface concentration of quaternary charge in polyurethane coatings.

Contact antimicrobial coatings via 2 wt% U–P[3F-co-C12] modification. Results from an evaluation of modifier antimicrobial effectiveness are discussed first [22]. Next, physical surface characterization data are presented in light of biocidal activity.

The "first-generation" fluorous modifier was based on an H_{12}MDI–BD(50) hard block and Structure 10.4, with $p = 0.89$. This modifier is designated U–P[3F–C12(11)], where the weight percent C12 is indicated (M_n = 5.1 kDa). Coatings were generated by dipping or spreading tetrahydrofuran (THF) or THF–isopropanol (THF–IPA) solutions of base polyurethane and U–[3F–C12(11)]. H_{12}MDI–BD(50)–PTMO-1000 was selected as the bulk polyurethane (designations, Figure 10.1). The compositional rationale and other details may be found elsewhere [22].

Two weight percent U–[3F–C12(11)] coatings were tested against aerosol challenges (10^6 CFU/mL) of Gram-negative (*Pseudomonas aeruginosa* and *Escherichia coli*) and Gram-positive (*Staphylococcus aureus*) bacteria. After 30 minutes residence, bacteria were stripped, serially diluted, and plated for colony counts. Two weight percent U–[3F–C12(11)] coatings effected 100% kill (3.6–4.4 log reduction) of Gram-negative or Gram-positive bacteria; that is, there were no live bacteria in the culture plates. A zone of inhibition test was carried out by placing a coated glass disk on a "lawn" of bacteria on agar. Only the bacteria directly under the disk were killed. This demonstrated *contact kill* rather than biocide release that causes a "halo" of dead bacteria around the disk. The model for concentration of quaternary charge function (Figure 10.2) was validated. This encouraging result was attributed to good hard block/soft block phase separation, a cation/corepeat group ratio mimicking natural biocidal proteins [92], and a semifluorinated "chaperone" or low surface energy moiety that aids in alkylammonium surface-concentration.

Biocidal testing on the same coatings as a function of time was carried out to assess stability of contact kill. Figure 10.11 shows aerosol test results for 2 wt% U–P[3F–C12(11)] coatings within a few days and after 2 weeks storage in air at ambient temperature. To our surprise, the modified coating had lost biocidal effectiveness after 2 weeks. Several sets of tests showed that contact biocidal kill for 2 wt% U–P[3F–C12(11)] coatings ranged from 0% to 50%.

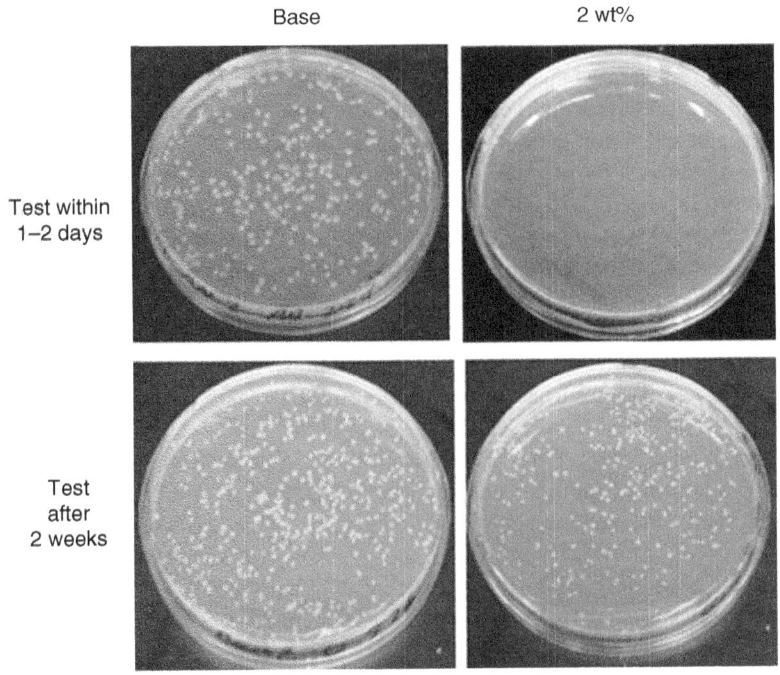

FIGURE 10.11 Agar plates after a spray challenge of *Pseudomonas aeruginosa*: Left, base polyurethane control; Right, 2 wt% U–P[3F–C12(11)].

TM-AFM imaging. To gain insight into loss of contact biocidal effectiveness, coating topology and morphology were investigated by TM-AFM imaging. Figure 10.12 shows the 3D height image for a 2 wt% U–P[3F–C12(11)] coating after 2 days, 2 weeks, and 8 weeks. One to two days after the coating was generated, the image in Figure 10.12a was almost featureless [22]. However, an extensive topological change was found after 2 weeks followed by an even more radical change at 8 weeks (Figures 10.12b and c). Complex rugosity developed with the formation of features up to ~10 μm wide and 100–200 nm high. XPS analysis and related studies suggest that the source of this complex topology is near-surface phase separation of the U–P[3F–C12(11)] modifier.

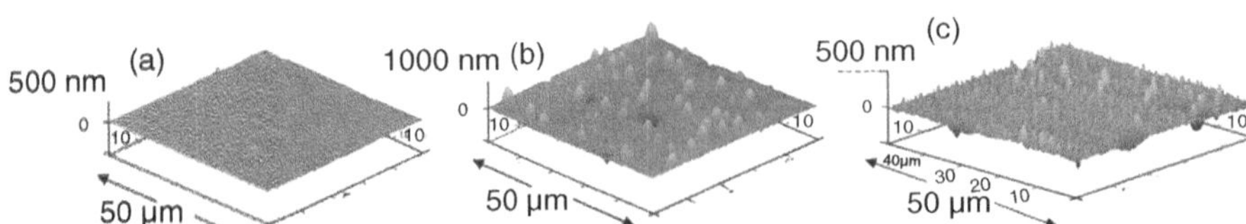

FIGURE 10.12 TM-AFM 3D height images for a 2 wt% U–P[3F–C12(11)] coating after: (a) 2 days; (b) 2 weeks; and (c) 8 weeks at ambient temperature.

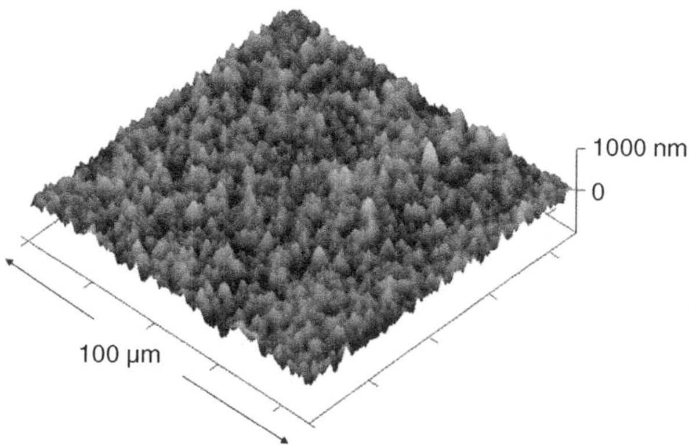

FIGURE 10.13 TM-AFM 3D image for 2 wt% U–P[3F-*r*-ME3]: $r_{sp} = 0.8$; $R_q = 47$ nm.

The novel wetting behavior and surface phase separation for 2 wt% U–P[3F-*co*-ME3] blends was discussed in section 10.1. A complex surface topology was also observed for these modified systems [97]. The immiscibility of 3F again drives the formation of dense peaked 10–20 μm wide features with heights ~50 nm (Figure 10.13).

Topological complexity is often observed for fluorinated polymers as a result of thermodynamically driven processes such as crystallization [98] and minimization of surface energy [99, 100]. A topological model for surface phase separation for U–P[3F–ME2]- and U–P[3F–C12]-modified base polyurethane is suggested in Figure 10.14. The surface concentration driven by 3F immiscibility results in nano-clusters that aggregate to form micron-scale features. Curvature of these features into peaked structures is proposed to arise from synergistic driving forces comprised of (a) 3F immiscibility, low surface energy, and low P[AB] T_g facilitating reorganization

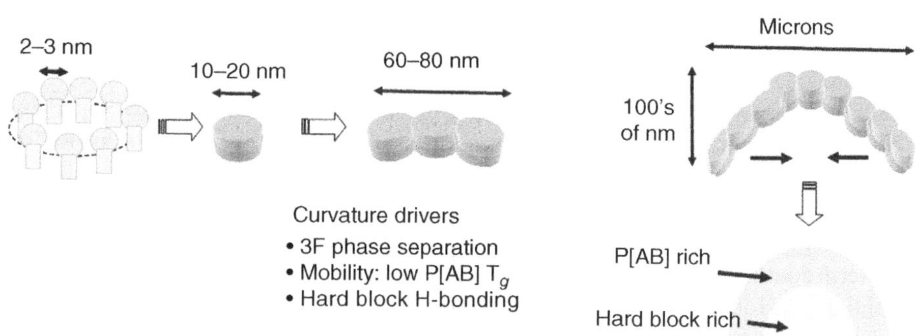

FIGURE 10.14 Model for surface phase separation of U–P[3F-*co*-ME2] and U–P[3F-*co*-C12]-modified base polyurethane.

together with (b) enthalpically favored self-organization of HMDI–BD hard block domains via H-bonding.

Based on biotesting and AFM imaging, the solution coating process for U–P[3F–C12(11)] results in a metastable state. Quaternary charge is accessible and contact antimicrobial effectiveness is high but with time (in air), phase separation occurs (Figures 10.12 and 10.14). Phase-separated domains must sequester quaternary charge leading to ineffective contact antimicrobial kill.

Streaming potentials. Solution test methods provide quantitative estimates of antimicrobial effectiveness including minimum inhibitory concentrations (MICs), minimum biocidal concentrations (MBCs), and kill kinetics [101]. However, methods for estimating biocidal effectiveness of surface-bound polycation coatings vary widely and often do not permit unambiguous comparisons [102]. The spray and zone of inhibition tests described above are tedious and carry an element of personal risk, as Gram negative (*P. aeruginosa* and *E. coli*) and Gram positive (*S. aureus*) are opportunistic pathogenic bacteria. The most often cited model for bacterial kill is membrane disruption by the polycation. This model is consistent with a compositional threshold for biocidal effectiveness of block copolymer nanofilms containing poly(2-(dimethylamino)ethyl methacrylate) and poly(3-(trimethoxysilyl)propyl methacrylate) segments [96]. The number of bacteria killed increased with increasing quat density. In view of the importance of accessible charge density and unanticipated surface instability of accessible quaternary charge, a rapid quantitative measurement was sought for U–P[AB]-modified coatings.

Measurement of quaternary charge density has often employed fluorescein dye binding and release of bound dye by an ion exchange surfactant such as dodecyl trimethyl ammonium chloride [103–107]. This method failed for polyurethanes modified with quat U–P[AB] surface modifiers because of nonspecific dye adsorption and slow desorption [108].

Streaming potential measurements have long been known for quantifying surface charge, especially for colloidal particles [109]. Assessing surface accessible charge for thin films formed by alternating polyelectrolyte deposition was elegantly demonstrated by Adamczyk et al. [110]. Alternating positive and negative potentials attested to the charge of the last polyelectrolyte deposited. Streaming potentials have also been employed for studying antimicrobial function of polyelectrolyte multilayers comprising poly(allyl amine) hydrochloride and poly(sodium 4-styrene sulfonate [111].

For 2 wt% U–P[3F–C12(11)] coatings, the microfluidic method developed by Alvarez was adapted. Streaming potentials were obtained by coating the inside of 100 μm capillaries with 2 wt% U–P[3F–C12(11)] coatings using a 1 mM KBr solution [108]. Initial streaming potentials were above 100 mV for a 1 wt% U–3F–C12(11) coating (Figure 10.15). The high near-surface accessible charge agrees with excellent near-term contact antimicrobial effectiveness. However, a rapid decrease in streaming potentials was found after successive pumping cycles (Figure 10.15) [108]. Streaming potentials scale to charge density so that the results for the 2 wt% U–3F–C12(0.11) coating indicate a precipitous decrease (~50%) in accessible near-surface quaternary charge in <2 minutes.

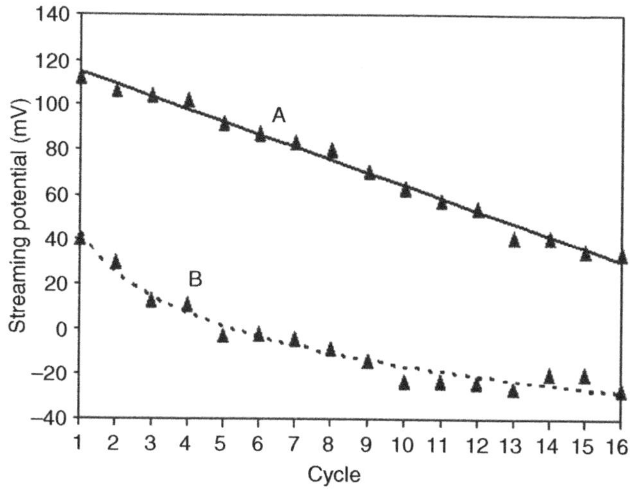

FIGURE 10.15 Streaming potentials for a 1 wt% U–P[3F–C12(11)] coating as a function of capillary pumping cycle (≈20 s per cycle): A, ambient temperature cure/solvent removal; B, annealed at 60°C.

Compared to the relatively slow (2 weeks) deactivation of quaternary charge density in air (biotesting, AFM), streaming potential measurements show loss of surface accessible positive charge for 2 wt% U–3F–C12(0.11) takes only a few minutes. Apparently, water plasticization of the near-surface U–P[AB] results in an acceleration of phase separation that sequesters quat function.

Contact antimicrobial coatings via 2 wt% U–P[4F-co-C12] modification. The unexpected temporal instability of 2 wt% U–P[3F–C12(11) has led to the development of alternative surface modifiers that will achieve long-term contact antimicrobial effectiveness. One of these is based on the replacement of $-CF_3$ by $-CF_2CF_2H$ (Figure 10.4, $R_f = CF_2CF_2H$). H-bonding was proposed by Ellison to explain decreased surface tension upon replacement of terminal $-CF_3$ by $-CF_2CF_2H$ [112]. Thus replacement of 3F with amphiphilic 4F seemed attractive based on the notion that CF_2**H** hydrogen bonding might provide an enthalpic contribution to enhance the free energy of mixing, while "taming" the tendency of CF_3-based soft blocks to phase separate, but not eliminating the "chaperone" ability of a fluorinated moiety.

The copolyoxetane for the polyurethane modifier U-P[4F-C12(34)] was prepared by a method paralleling that described previously for the 3F analog (Figure 10.4, $R_f = -CF_2CF_2H$, p = 0.64) [22]. The weight fraction of C12 was increased to 0.34 compared to 0.11 for the first generation modifier based on preliminary experiments that showed enhanced modified surface stability. By ^{1}H-NMR end-group analysis, M_n for P[4F–C12(34)] diol was 6 kDa. Copolyoxetane diol and polyurethane solution characterization (^{1}H-NMR, GPC) will be provided in a separate publication.

Unlike the analogous 3F polyurethane modifiers, U–P[4F–C12(34)] was insoluble in THF. DMAC (dimethyl acetamide) was used as an alternative solvent for deposition of 2 wt% U–P[4F–C12(34)] coatings having 98 wt% HMDI-BD(50)-PTMO(1000). An overnight thermal treatment in vacuum (120°C) removed solvent and gave optically transparent coatings that were typically 160 μm thick.

Based on the hypothesis that stable, accessible near-surface charge would be a predictor of long-term antimicrobial effectiveness, zeta potential measurements were a priority. As noted above, a capillary method was developed for this purpose [108]. However, coating the inside of 100 μm capillaries is tedious and subsequent characterization of the coating surface virtually impossible.

In a new approach, zeta potential measurements were carried out on drip-coated microscope slides using a SurPASS electrokinetic analyzer and clamping cell (Anton PAAR). Each run consisted of a forward and backward pumping cycle taking ~15 minutes. Three runs (~45 minutes) provided a much more demanding test of stability in dilute polyelectrolyte (10^{-3} M KBr) compared to the capillary method used previously (~2 minutes) [108]. Zeta potentials for polymer coatings are known to be affected by pH [113]; initial measurements reported here were carried out at pH 7.

Compared to HMDI–BD(50)–PTMO(1000) (−37 ± 4 mV), zeta potentials for 2 wt% U–P[4F–C12(34)] were positive. Table 10.5 lists zeta potentials relative to base polyurethane at 1 week and 4 weeks. Zeta potentials for 2 wt% U–[4F–C12(34)] were diminished 5–10% after 1 month storage in air at ambient temperature. Each set of zeta potential tests amounted to ~2 hours immersion in 1 mM NaBr. Interestingly, the reduction observed over the course of the three runs for the first test at <1 week (42.1 → 38.8 mV) was restored to some degree for the first run after 4 weeks in air (39.8 mV). However, the final value after 4 weeks "aging" in air (35.8 mV) was 10% less than the corresponding value after 1 week.

The stability of zeta potentials for 2 wt% U–P[4F–C12(34)] supports the hypothesis that $-CF_2H$ hydrogen bonding provides enthalpic stabilization of the modified surface. However, a detailed study is warranted to confirm these initial results.

Spray challenges of *P. aeruginosa*, *E. coli*, and *S. aureus* were carried out as described previously [22]. The residence time for each bacterial challenge (10^6 cfu/mL) was 1 hour. Table 10.6 shows percent kill for the pathogen challenges after 1-hour residence time. Within experimental error, complete kill was observed within 1 week and after 4 weeks storage at ambient temperature. These results are important in validating the model shown in Figure 10.2 for surface concentration of the U–P[4F–C12(34)] modifier. Importantly, the biocidal effectiveness after

TABLE 10.5 Zeta Potentials for 2 wt% U–P[4F–C12(34)] Relative to Base Polyurethane

	Zeta potential (mV)	
Run	1 week	4 week
1	42.1	39.8
2	41.0	37.6
3	38.8	35.8

TABLE 10.6 Percent Kill of Spray Challenges for U–4F–C12-34

Challenge	<1 wk	>4 wk
Pseudomonas aeruginosa	100	95
Escherichia coli	100	100
Staphylococcus aureus	99	90

Residence Time for Each Challenge (10^6 cfu/mL) was 1 h.

storage for at least a month under ambient conditions of temperature and humidity also provide support for the use of zeta potentials as a predictor for polycation contact kill.

Utilization of 4F as a novel "chaperone" has only recently been established. A complete study will be published separately. In summary, 3F and 4F "chaperones" result in contrasting morphological and functional stability.

10.6 SUMMARY

The U–P[AB] copolyoxetane soft block nanostructure shown in Figure 10.1, where A is a fluorous side chain, has resulted in new surface phenomena and surface concentration of B function. The surface modifier model depicted in Figure 10.2 has been validated for concentrating function and biocidal quaternary charge at ≤2 wt% modifier. Unexpected interactions between A and B soft block side chains have resulted in novel "contraphilic wetting" and new surface morphologies. XPS, TM-AFM, and dynamic contact angle analysis provide surface analytical information that facilitates understanding of the model depicted in Figure 10.2.

The problem of temporal and thermal instability of the U–P[3F–C12-(11)] modifier was solved by utilizing a $-CF_2CF_2H$ (4F) fluorous side chain and increasing the C12 mole fraction. Biocidal effectiveness was demonstrated after storage for at least a month under ambient conditions of temperature and humidity. This stability was correlated with zeta potential measurements of accessible near surface charge. These results suggest zeta potential measurements hold promise as a physical method for assessing effectiveness of polycation contact kill and polycation stability in contact with aqueous systems.

Polyurethanes with P[AB] copolyoxetane soft blocks are effective and potentially processing-compatible surface modifiers for conventional polyurethanes. This surface modification approach is "green" in the sense that only a minimal amount of surface modifier (2 wt% or less) is leveraged so as to effect a radical change in surface function. In addition, the 4F chaperone is "green" in avoiding the use of perfluorinated eight carbon moieties that are likely PFOA precursors.

Looking ahead, simpler surface modification is being sought that will leverage the power of P[AB] surface modification but will avoid the synthesis of P[AB] polyurethane modifiers. A broader perspective is being employed that includes *nanosurface* modification (≈1 nm), which is essential for tailoring properties such as contact antimicrobial efficiency, protein adhesion, and wetting behavior, but also innovative ways to tailor *mesosurface* characteristics (≈1000 nm) that affect adhesion of rigid objects [114].

ACKNOWLEDGMENTS

K.J.W. thanks the National Science Foundation (DMR-grants DMR-0207560, DMR-0802452, and DMR-1206259) and the VCU School of Engineering Foundation for support of this research. This work was also supported by Public Health Service Grant AI-19146 from the National Institute of Allergy and Infectious Disease (D.E.O.), and in part by Veterans Administration Medical Research Grant I01BX000477 (D.E.O.).

REFERENCES

1. MarketsandMarkets. http://www.marketsandmarkets.com, accessed 27 February 2014.
2. Oertel, G., Ed. *Polyurethane Handbook*; Hanser Publishers, Munich, Germany, 1985.
3. Szycher, M. *Handbook of Polyurethanes*; CRC Press, Boca Raton, FL, 1999.
4. Lamba, N. M. K.; Woodhouse, K. A.; Cooper, S. L. *Polyurethanes in Biomedical Applications*; CRC Press: Boca Raton, FL, 1998.
5. Ratner, B. D.; Hoffman, A. S.; Schoen, F. J.; Lemons, J. E. *Biomaterials Science*; Elsevier Academic Press: Boston, 2004.
6. Ratner, B. D.; Cooper, S. L.; Castner, D. G.; Grasel, T. G. *J. Biomed. Mater. Res.* **1990**, *24*, 605.
7. Tingey, K. G.; Andrade, J. D. *Langmuir* **1991**, *7*, 2471.
8. Garrett, J. T.; Runt, J.; Lin, J. S. *Macromolecules* **2000**, *33*, 6353.
9. Garrett, J. T.; Siedlecki, C. A.; Runt, J. *Macromolecules* **2001**, *34*, 7066.
10. Jalbert, C.; Koberstein, J. T.; Hariharan, A.; Kumar, S. K. *Macromolecules* **1997**, *30*, 4481.
11. Makal, U.; Wood, L.; Ohman, D. E.; Wynne, K. J. *Biomaterials* **2006**, *27*, 1316.
12. Ernsting, M. J.; Bonin, G. C.; Yang, M.; Labow, R. S.; Santerre, J. P. *Biomaterials* **2005**, *26*, 6536.
13. Ellis, D. A.; Mabury, S. A.; Martin, J. W.; Muir, D. C. G. *Nature* **2001**, *412*, 321.
14. Giesy, J. P.; Kannan, K. *Environ. Sci. Technol.* **2001**, *35*, 1339.
15. Martin, J. W.; Whittle, D. M.; Muir, D. C. G.; Mabury, S. A. *Environ. Sci. Technol.* **2004**, *38*, 5379.
16. Fujiwara, T.; Wynne, K. J. *Macromolecules* **2004**, *37*, 8491.
17. Zhang, W.; Fujiwara, T.; Taskent, H.; Zheng, Y.; Brunson, K.; Gamble, L.; Wynne, K. J. *Macromol. Chem. Phys.* **2012**, *213*, 1415.
18. Makal, U.; Fujiwara, T.; Cooke, R. S.; Wynne, K. J. *Langmuir* **2005**, *21*, 10749.
19. Makal, U.; Wynne, K. J. *Langmuir* **2005**, *21*, 3742.
20. Makal, U.; Uslu, N.; Wynne, K. J. *Langmuir* **2007**, *23*, 209.
21. Grunzinger, S. J.; Kurt, P.; Brunson, K. M.; Wood, L.; Ohman, D. E.; Wynne, K. J. *Polymer* **2007**, *48*, 4653.

22. Kurt, P.; Wood, L.; Ohman, D. E.; Wynne, K. J. *Langmuir* **2007**, *23*, 4719.
23. Malik, A. A.; Carlson, R. P.; US Patent 5668251, **1997**.
24. Ho, T.; Wynne, K. J.; Malik, A. A. *Abstr. Pap. Am. Chem. Soc.* **1995**, *209*, 319.
25. Malik, A. A.; Manser, G. E.; Archibald, T. G.; US Patent 5,650,483, **1995**.
26. Malik, A. A.; Archibald, T. G.; U.S. Patent 6037483, **2000**.
27. Malik, A. A.; Archibald, T. G.; Carlson, R. P.; Wynne, K. J.; Kresge, E. N.; US Patent 6479623, **2002**.
28. Saegusa, T.; Fujii, H.; Kobayashi, S.; Ando, H.; Kawase, R. *Macromolecules* **1973**, *6*, 26.
29. Bucquoye, M.; Goethals, E. J. *Eur. Polym. J.* **1978**, *14*, 323.
30. Kim, K. N.; Kwon, Y. W.; Jang, C. W.; Kim, M. J.; Lim, T. K.; Jin, J. I. *Liq. Cryst.* **2007**, *34*, 535.
31. Jutier, J. J.; de Gunzbourg, A.; Prud'homme, R. E. *J. Polym. Sci. Pol. Chem.* **1999**, *37*, 1027.
32. Fujiwara, T.; Makal, U.; Wynne, K. J. *Macromolecules* **2003**, *36*, 9383.
33. Kubisa, P. *J. Polym. Sci. Pol. Chem.* **2003**, *41*, 457.
34. Bednarek, M.; Kubisa, P.; Penczek, S. *Macromolecules* **2001**, *34*, 5112.
35. Grasel, T. G.; Cooper, S. L. *Biomaterials* **1986**, *7*, 315.
36. Makal, U.; Uilk, J.; Kurt, P.; Cooke, R. S.; Wynne, K. J. *Polymer* **2005**, *46*, 2522.
37. Velankar, S.; Cooper, S. L. *Macromolecules* **1998**, *31*, 9181.
38. Grunzinger, S. J.; Wynne, K. J. *Polymer* **2006**, *47*, 4230.
39. Fujiwara, T.; Makal, U.; Uilk, J.; Wynne, K. J. *Abstr. Pap. Am. Chem. Soc.* **2003**, *225*, U534.
40. Magonov, S. N.; Elings, V.; Whangbo, M. H. *Surf. Sci.* **1997**, *375*, L385.
41. Uilk, J.; Johnston, E. E.; Bullock, S.; Wynne, K. J. *Macromol. Chem. Phys.* **2002**, *203*, 1506.
42. Liu, R. Y. F.; Jin, Y.; Hiltner, A.; Baer, E. *Macromol. Rapid Commun.* **2003**, *24*, 943.
43. Liu, R. Y. F.; Bernal-Lara, T. E.; Hiltner, A.; Baer, E. *Macromolecules* **2005**, *38*, 4819.
44. Tingey, K. G.; Andrade, J. D. *Langmuir* **1991**, *7*, 2471.
45. Uilk, J. M.; Mera, A. E.; Fox, R. B.; Wynne, K. J. *Macromolecules* **2003**, *36*, 3689.
46. Herbert, C. B.; Hernandez, A. M.; Hubbell, J. A. *Biotechnol. Bioeng.* **1996**, *52*, 81.
47. Wynne, K. J.; Makal, U.; Kurt, P.; Gamble, L. *Langmuir* **2007**, *23*, 10573.
48. Koberstein, J. T.; Galambos, A. F. *Macromolecules* **1992**, *25*, 5618.
49. Li, Y. J.; Gao, T.; Liu, J.; Linliu, K.; Desper, C. R.; Chu, B. *Macromolecules* **1992**, *25*, 7365.
50. Koberstein, J. T.; Galambos, A. F.; Leung, L. M. *Macromolecules* **1992**, *25*, 6195.
51. Ho, T.; Wynne, K. J. *Polym. Adv. Technol.* **1994**, *6*, 25.
52. Tezuka, Y.; Fukushima, A.; Matsui, S.; Imai, K. *J. Colloid Interface Sci.* **1986**, *114*, 16.
53. Thomas, R. R.; Anton, D. R.; Graham, W. F.; Darmon, M. J.; Sauer, B. B.; Stika, K. M.; Swartzfager, D. G. *Macromolecules* **1997**, *30*, 2883.
54. Wang, L. F.; Ji, Q.; Glass, T. E.; Ward, T. C.; McGrath, J. E.; Muggli, M.; Burns, G.; Sorathia, U. *Polymer* **2000**, *41*, 5083.

55. Thomas, R. R.; Anton, D. R.; Graham, W. F.; Darmon, M. J.; Stika, K. M. *Macromolecules* **1998**, *31*, 4595.
56. Kim, Y. S.; Lee, J. S.; Ji, Q.; McGrath, J. E. *Polymer* **2002**, *43*, 7161.
57. Thanawala, S. K.; Chaudhury, M. K. *Langmuir* **2000**, *16*, 1256.
58. Rudin, A. *The Elements of Polymer Science and Engineering*; Academic Press: Orlando, FL, 1982.
59. Chanda, M. *Advanced Polymer Chemistry*; Marcel Dekker: New York, 2000.
60. Worley, S. D.; Sun, G. *Trends Polym. Sci.* **1996**, *4*, 364.
61. Eknoian, M. W.; Worley, S. D. *J. Bioact. Compat. Polym.* **1998**, *13*, 303.
62. Eknoian, M. W.; Worley, S. D.; Bickert, J.; Williams, J. F. *Polymer* **1999**, *40*, 1367.
63. Cerkez, I.; Worley, S. D.; Broughton, R. M.; Huang, T. S. *Prog. Org. Coat.* **2013**, *76*, 1082.
64. Kenawy, E. R.; Worley, S. D.; Broughton, R. *Biomacromolecules* **2007**, *8*, 1359.
65. Sun, G.; Worley, S. D. *J. Chem. Educ.* **2005**, *82*, 60.
66. Mackel, M. J.; Sanchez, S.; Kornfield, J. A. *Langmuir* **2007**, *23*, 3.
67. Wilhelmy, L. *Ann. Phys. Chem. Leipzig* **1863**, *119*, 177.
68. Hogt, A. H.; Gregonis, D. E.; Andrade, J. D.; Kim, S. W.; Dankert, J.; Feijen, J. *J. Colloid Interface Sci.* **1985**, *106*, 289.
69. Lander, L. M.; Siewerski, L. M.; Brittain, W. J.; Vogler, E. A. *Langmuir* **1993**, *9*, 2237.
70. Guo, J.; Resnick, P.; Efimenko, K.; Genzer, J.; DeSimone, J. M. *Ind. Eng. Chem. Res.* **2008**, *47*, 502.
71. Adamson, A. W.; Gast, A. P. *Physical Chemistry of Surfaces*, 6th ed.; John Wiley & Sons: New York, 1997.
72. Lin, J.; Winkelmann, C.; Worley, S. D.; Kim, J. H.; Wei, C. I.; Cho, U. C.; Broughton, R. M.; Santiago, J. I.; Williams, J. F. *J. Appl. Polym. Sci.* **2002**, *85*, 177.
73. Sun, Y.; Sun, G. *J. Appl. Polym. Sci.* **2002**, *84*, 1592.
74. Tretinnikov, O. N.; Ikada, Y. *Langmuir* **1994**, *10*, 1606.
75. Holmes-Farley, S. R.; Reamey, R. H.; McCarthy, T. J.; Deutch, T. J.; Whitesides, G. H. *Langmuir* **1985**, *1*, 724.
76. Senshu, K.; Yamashita, S.; Mori, H.; Ito, M.; Hirao, A.; Nakahama, S. *Langmuir* **1999**, *15*, 1754.
77. Lemieux, M.; Usov, D.; Minko, S.; Stamm, M.; Shulha, H.; Tsukruk, V. V. *Macromolecules* **2003**, *36*, 7244.
78. Christodoulakis, K. E.; Vamvakaki, M. *Langmuir* **2010**, *26*, 639.
79. Horecha, M.; Senkovskyy, V.; Schneider, K.; Kiriy, A.; Stamm, M. *Colloid Polym. Sci.* **2011**, *289*, 603.
80. Kocer, H. B.; Worley, S. D.; Broughton, R. M.; Huang, T. S. *React. Funct. Polym.* **2011**, *71*, 561.
81. Cerkez, I.; Kocer, H. B.; Worley, S. D.; Broughton, R. M.; Huang, T. S. *React. Funct. Polym.* **2011**, *72*, 673.
82. Cerkez, I.; Kocer, H. B.; Worley, S. D.; Broughton, R. M.; Huang, T. S. *J. Appl. Polym. Sci.* **2012**, *124*, 4230.
83. Cerkez, I.; Kocer, H. B.; Worley, S. D.; Broughton, R. M.; Huang, T. S. *Cellulose* **2012**, *19*, 959.

84. Friedman, N. D.; Korman, T. M.; Fairley, C. K.; Franklin, J. C.; Spelman, D. W. *J. Infect.* **2002**, *45*, 47.

85. Hacek, D. M.; Suriano, T.; Noskin, G. A.; Krusynski, J.; Reisberg, B.; Peterson, L. R. *Am. J. Clin. Pathol.* **1999**, *111*, 647.

86. Borer, A.; Gilad, J.; Smolyakov, R.; Eskira, S.; Peled, N.; Porat, N.; Hyam, E.; Trefler, R.; Riesenberg, K.; Schlaeffer, F. *Emerging Infect. Dis.* **2005**, *11*, 1160.

87. Von Eiff, C.; Kohnen, W.; Becker, K.; Jansen, B. *Int. J. Artif. Organs* **2005**, *28*, 1146.

88. Johnson James, R.; Kuskowski Michael, A.; Wilt Timothy, J. *Ann. Intern. Med.* **2006**, *144*, 116.

89. Drekonja, D. M.; Kuskowski, M. A.; Wilt, T. J.; Johnson, J. R. *Expert Rev. Med. Devices* **2008**, *5*, 495.

90. Trinh, T. T.; Chan, P. A.; Edwards, O.; Hollenbeck, B.; Huang, B.; Burdick, N.; Jefferson, J. A.; Mermel, L. A. *Infect. Control Hosp. Epidemiol.* **2011**, *32*, 579.

91. Woodward, S. *Br. J. Nurs.* **2012**, *21*, 655.

92. Tew, G. N.; Liu, D. H.; Chen, B.; Doerksen, R. J.; Kaplan, J.; Carroll, P. J.; Klein, M. L.; DeGrado, W. F. *Proc. Natl. Acad. Sci. U. S. A.* **2002**, *99*, 5110.

93. Kuroda, K.; DeGrado, W. F. *J. Am. Chem. Soc.* **2005**, *127*, 4128.

94. Tew, G. N.; Clements, D.; Tang, H.; Arnt, L.; Scott, R. W. *Biochim. Biophys. Acta Biomembr.* **2006**, *1758*, 1387.

95. Sovadinova, I.; Palermo, E. F.; Urban, M.; Mpiga, P.; Caputo, G. A.; Kuroda, K. *Polymers* **2011**, *3*, 1512.

96. Huang, J. Y.; Koepsel, R. R.; Murata, H.; Wu, W.; Lee, S. B.; Kowalewski, T.; Russell, A. J.; Matyjaszewski, K. *Langmuir* **2008**, *24*, 6785.

97. Zhang, W.; Henke, D.; Presnall, D.; Chakrabarty, S.; Wang, C.; Wynne, K. J. *Macromol. Chem. Phys.* **2012**, *213*, 1225.

98. Zheng, Y.; Zhang, W.; Gupta, M.; Kankanala, S.; Marks, C.; Carpenter, E.; Carroll, K.; Wynne, K. J. *J. Polym. Sci. Part B: Polym. Phys.* **2010**, *48*, 1022.

99. Ming, W.; Melis, F.; van de Grampel, R. D.; van Ravenstein, L.; Tian, M.; van der Linde, R. *Prog. Org. Coat.* **2003**, *48*, 316.

100. van Ravenstein, L.; Ming, W.; van de Grampel, R. D.; van der Linde, R.; de With, G.; Loontjens, T.; Thune, P. C.; Niemantsverdriet, J. W. *Macromolecules* **2004**, *37*, 408.

101. Chakrabarty, S.; King, A.; Kurt, P.; Zhang, W.; Ohman, D. E.; Wood, L. F.; Lovelace, C.; Rao, R.; Wynne, K. J. *Biomacromolecules* **2011**, *12*, 757.

102. Kugel, A.; Stafslien, S.; Chisholm, B. J. *Prog. Org. Coat.* **2011**, *72*, 222.

103. Goddard, J. M.; Hotchkiss, J. H. *Prog. Polym. Sci.* **2007**, *32*, 698.

104. Kugler, R.; Bouloussa, O.; Rondelez, F. *Microbiology-SGM* **2005**, *151*, 1341.

105. Ledbetter, J. W., Jr.; Bowen, J. R. *Anal. Chem.* **1969**, *41*, 1345.

106. Murata, H.; Koepsel, R. R.; Matyjaszewski, K.; Russell, A. J. *Biomaterials* **2007**, *28*, 4870.

107. Tiller, J. C.; Liao, C. J.; Lewis, K.; Klibanov, A. M. *Proc. Natl. Acad. Sci. U. S. A.* **2001**, *98*, 5981.

108. Gupta, M. L.; Brunson, K.; Chakravorty, A.; Kurt, P.; Alvarez, J. C.; Luna-Vera, F.; Wynne, K. J. *Langmuir* **2010**, *26*, 9032.

109. Stanley, J. S. *J. Phys. Chem.* **1954**, *58*, 533.

110. Adamczyk, Z.; Zembala, M.; Warszynski, P.; Jachimska, B. *Langmuir* **2004**, *20*, 10517.
111. Lichter, J. A.; Rubner, M. F. *Langmuir* **2009**, *25*, 7686.
112. Ellison, A. H.; Fox, H. W.; Zisman, W. A. *J. Phys. Chem.* **1953**, *57*, 622.
113. Kirby, B. J.; Ernest F. Hasselbrink, J. *Electrophoresis* **2004**, *25*, 203.
114. Chakrabarty, S.; Wang, C.; Zhang, W.; Wynne, K. J. *Macromolecules* **2013**, *46*, 2984.

11

SELF-ORGANIZING SEMIFLUORINATED METHACRYLATE COPOLYMERS

DORIS POSPIECH AND DIETER JEHNICHEN
Leibniz-Institut für Polymerforschung Dresden e.V.

11.1 INTRODUCTION AND OVERVIEW

Polymers with semifluorinated (*sf*) segments have gained much attention over the last 30 years due to their unique properties particularly originating from phase separation between alkyl and perfluoroalkyl segments.

The phase separation between thermodynamically immiscible, but covalently linked structural segments has often been used in polymer chemistry to create materials with property features arising from demixing of immiscible segments. Block copolymers (BCPs) with two or more blocks arranged in different architecture types form the most prominent and most exploited class of polymers exhibiting phase separation (see, e.g., References 1–3). Their phase separation [4–6] generates periodic, regular nanostructured morphologies in bulk, but also in coatings or thin films, thus forming the base for numerous applications in nanotechnology (see, e.g., References 7–12). Typical examples of the simplest BCP structure of a diblock copolymer A-*b*-B are as follows: poly(styrene-*b*-methyl methacrylate), (PS-*b*-PMMA); poly(styrene-*b*-isoprene), (PS-*b*-PI); poly(styrene-*b*-butadiene), (PS-*b*-PB); poly(styrene-*b*-2-vinylpyridine), (PS-*b*-2VP); and poly(pentyl methacrylate-*b*-methyl methacrylate), (PPMA-*b*-PMMA). Their blocks are highly immiscible, phase separate strongly, and thus form very regular nanostructures.

Handbook of Fluoropolymer Science and Technology, First Edition. Edited by Dennis W. Smith Jr., Scott T. Iacono, and Suresh S. Iyer.

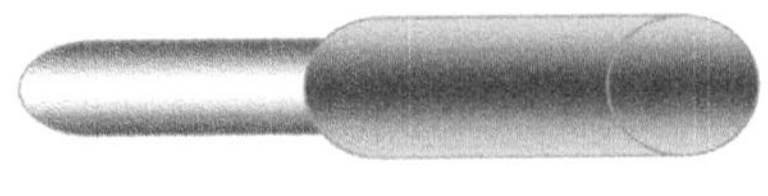

$R_1 —(CH_2)_n — (CF_2)_m — R_2$

FIGURE 11.1 General chemical structure of semifluorinated compounds.

On the other hand, perfluorinated polymers like poly(tetrafluoroethylene), (PTFE), as one of the most important examples, represent a polymer class which is in particular characterized by its special surface and adhesion properties. Perfluorinated segments in polymers often result in strongly nonpolar surfaces with reduced wettability of water or organic molecules, reduced adhesion and yielding, excellent gliding, and antisticking properties, and, in particular, chemical inertness [13, 14].

Linking these perfluoroalkyl segments with alkylene segments by C—C bonds results in semifluorinated compounds. Here, we use this term exclusively for the structural type characterized by $-(CH_2)_n-(CF_2)_m-$, in contrast to many papers referring semifluorinated" polymers to partially fluorosubstituted units. The simplest example of semifluorinated compounds are semifluorinated alkanes. Their general structure is depicted in Figure 11.1. The semifluorinated central part is flanked by two terminal groups which can be chemically equal or different. Often, the structure is abbreviated by H*n*F*m*, *n* reflecting the number of CH_2 groups (H), and *m* reflecting the number of CF_2 groups (F).

The general structure of *sf* compounds reflects a diblock copolymer and can thus be treated as such. The Flory–Huggins interaction parameter χ indicates the "strength" of demixing of the blocks and can be determined, for example, by incremental methods [15, 16]. Using Fedors' concept and increments given by Van Krevelen [17], a χ-parameter of 0.59 between $-(CH_2)-$ and $-(CF_2)-$ can be calculated which is rather high compared to values for other polymer pairs, for instance, 0.065 for PMMA-*b*-PPMA, 0.054 for PS-*b*-PMMA, and 0.064 for PS-*b*-PI. The strong demixing tendency of $-(CH_2)_n-$ and $-(CF_2)_m-$ can already be derived from the value of χ.

Phase separation can be understood by means of the phase diagram for diblock copolymers (plot of χN over Φ_A) [18], as illustrated in Figure 11.2 with the phase diagram from Reference 19. The position of the spinodal dividing phase-separated and nonphase-separated region depends on the flexibility of the blocks linked with each other. The "standard" spinodal resulting from mean field calculations [20] represents the coupling of two flexible, amorphous blocks in the molten state. The minimum occurs at equal volume fraction of each block. This also applies for *sf* diblocks. Taking the value of $\chi N = 10.5$ at $\Phi_A = 0.5$ (molar fraction of block A) and the known χ into account, the critical number of repeating units in both blocks (N_{crit}) necessary for phase separation can be derived. With $\chi = 0.56$ for *sf* compounds and $\Phi_A = 0.5$, $N_{crit} = 18$ results. However, the demixing of *sf* compounds was observed experimentally [21] at lower *N*. The effect of crystallization of alkyl and perfluoroalkyl segments can be discussed to enhance phase separation. Crystallization processes as additional

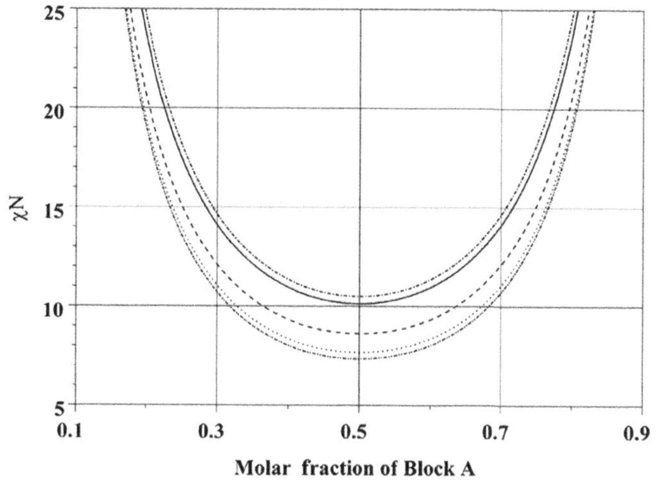

FIGURE 11.2 General phase diagram of diblock copolymers and shift of the spinodals depending on the stiffness of the blocks: symmetrical spinodal diagrams of diblock copolymers with increasing stiffness (flex: $l = b$, semi: $l = Nb/10$, stiff: $1 = Nb$ of both blocks): flex–flex (solid), semi–semi (dashed), stiff–stiff (dotted) including the limits Gauss–Gauss (dash–dot) and rod–rod (dash–dot–dot), with l, segment length; b, monomer length, N, number of segments. Reproduced from Reference 19 with permission from John Wiley & Sons, Inc.

driving force result in the shift of spinodals to lower χN (illustrated in Figure 11.2 by the spinodals of diblocks rod–rod) [19]. In conclusion, the phase separation of *sf* compounds occurs at lower N. If the composition is asymmetric, the required N raises.

Inserting *sf* diblocks into polymer architectures will mutually raise the tendency of phase separation even further due to geometric restrictions. This coupling offers the opportunity to combine basic properties of polymers (e.g., visco-elastic properties, good processability and film formation, excellent mechanical properties and durability) with the special features of fluorocompounds, as discussed above. *Sf* segments may be incorporated into polymer architectures in different manner, as schematically illustrated in Figure 11.3.

A variety of *sf* polymers has already been reported. In the following, only the most important architectures will be mentioned. Alkylene chains which are flanked by perfluoroalkyl segments on both sides give triblock copolymers (Figure 11.3a). Linking *sf* segments in the main chain (Figure 11.3b) results in *sf* main chain polymers [22–24]. However, most semifluorinated polymers contain the *sf* segments attached as substituent in the side chain (Figure 11.3c). In this case, each repeating unit possesses the *sf* substituent. Alternatively, *sf* side chains are distributed along the backbone which is typically the case in random copolymers (Figure 11.3d). Besides these polymers, *sf* substituents can also be incorporated into one block of a di- or triblock copolymer, as depicted schematically in Figure 11.3e for a diblock copolymer. Also other architectures, for instance, with dendritic *sf* substituents attached to a

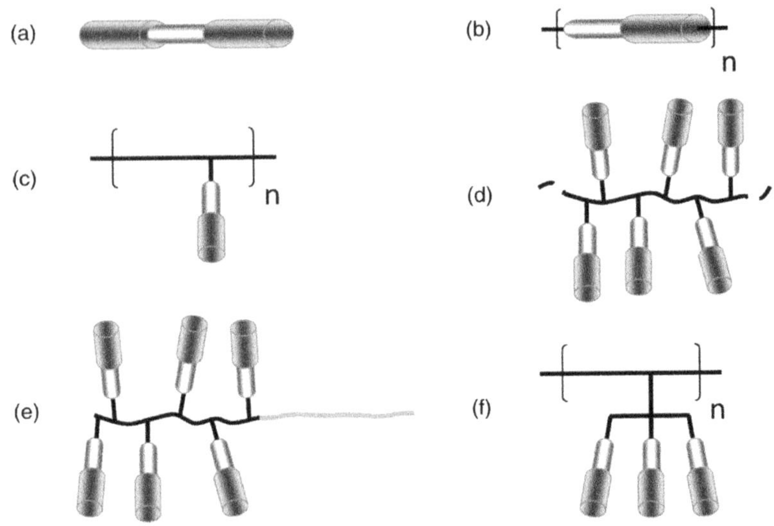

FIGURE 11.3 Schematic architectures of polymers with semifluorinated segments: (a) *sf* triblock copolymers; (b) *sf* main chain polymers; (c) *sf* side chain polymers; (d) polymers with randomly distributed *sf* side chains; (e) di- and triblock copolymers with one *sf* block; (f) copolymers with dendritic *sf* segments.

polymer backbone to achieve a dense dendritic shell, have been reported (Figure 11.3f) [25–28].

Semifluorinated segments with varying number of alkyl and perfluoroalkyl groups have been used as substituents on many different polymeric backbones over the years, among them polyesters [29, 30], cross-linkable acrylics [31], polyurethanes [32], polyglutamates [33], poly(vinyl ether)s [34], polystyrene [35], as well as polyacrylates [36], polymethacrylates [37], and *sf* poly(styrene-*alt*-maleimide) copolymers [38]. The main difference between these polymers is the flexibility and dynamics of the polymer backbone.

All efforts to prepare structures with *sf* substituents followed the concept that *sf* side chains attached to polymer backbones give the *sf* segments certain mobility and support surface segregation of the fluorinated parts, which was discussed as a necessary requirement for the generation of coatings with extremely low surface free energy. They are characterized by outer surfaces consisting of densely packed CF_3 groups, as schematically shown in Figure 11.4a. This structural concept has a clear advantage over the attachment of only CF_3 groups directly to a polymer backbone (Figure 11.4b). In this case, the CF_3 groups are hidden in the polymer bulk material and the geometric restrictions do not allow extensive surface segregation.

In this contribution, we will only consider polymers with longer *sf* segments (alkyl and perfluoroalkyl units directly linked via C—C— bond, not by functional groups like ester, amide, and urethane as, for instance, in References 39 and 40). Only *sf* segments able to self-organize will be discussed. We will comment only briefly on polymers with short *sf* groups such as trifluoroethyl groups. Reviews about this polymer type

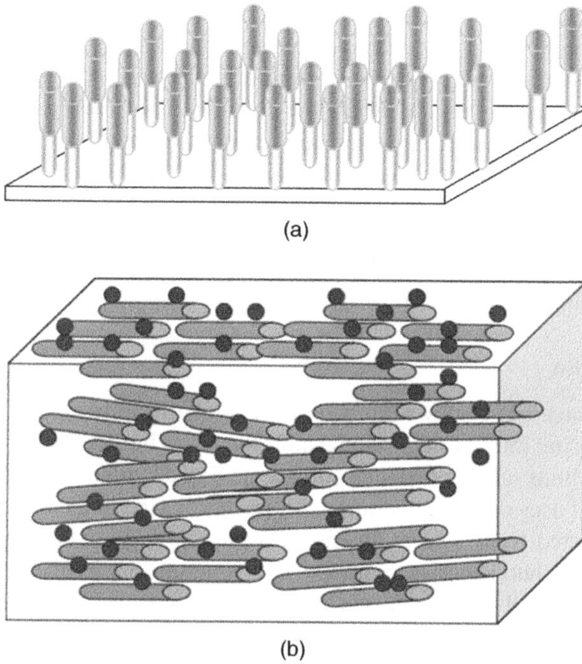

FIGURE 11.4 Schematic illustration of (a) surfaces with ordered CF_3 groups as top layer and (b) CF_3 groups directly attached to polymer chains.

can be found elsewhere [41–43]. Furthermore, we will also not discuss polymers with perfluoropolyether side chains [44–47] and partially hydrogenated perfluoropolyether side chains [48], as well as cross-linked polymer networks containing incorporated or dangling *sf* segments (e.g., Reference 49). The reader is also here referred to the literature.

After summarizing synthesis methods for *sf* polymers (Section 1.2), we will discuss their phase separation behavior in bulk and thin films (Section 1.3), derive the consequences of phase separation on the macroscopic wetting behavior (Section 1.4), and propose applications for these interesting polymer class (Section 1.5). The main focus will always be on *sf* methacrylate copolymers. Polymethacrylates are an important base for coating resins and different types of binders. PMMA itself is characterized by an amorphous structure yielding high transparency and brilliance, hardness, scratch resistance, and UV/ozone stability. Combining these beneficial properties with those of *sf* compounds will offer newly designed property profiles including hydrophobicity, oleophobicity, and antibiofouling/antibacterial properties. The main aim of the work discussed here was to provide materials for low surface free energy coatings, to understand their structure–property relationships, and to explore future applications. To achieve this, the whole chain of knowledge from synthesis of the polymers, chemical characterization, analysis of solid state structure, preparation and characterization of thin films (coatings) to prove whether or not self-organization also takes place in the restricted confinement of the thin film should be covered. These basic examinations were followed by characterization of the macroscopic properties of the materials (preferentially macroscopic wetting behavior) and continued by exploring applications, for example, ultrahydrophobic coatings. The investigations should also allow answering the question for which applications self-organization is a necessary requirement, and for which not.

11.2 SYNTHESIS METHODS FOR SEMIFLUORINATED COPOLYMERS

11.2.1 Overview

Sf compounds have been known for long time. One of the basic reactions that was exploited by Feiring [50] was the radical addition of perfluoroalkyl iodides to olefins catalyzed by sodium salts of benzene sulfinic acid under mild conditions. The most common method uses radical generators, for instance, azo compounds. The probable mechanism adapted from Ren et al. [51] is shown in Figure 11.5. The reaction is initiated by the isobutyronitrile radical (a), abstracting the iodine from R_fI which yields an R_f radical (b), that adds to the double bond (c). The radical formed is able to be transferred to R_fI to proceed the chain reaction (d). In a second step, the formed iodine compound has to be reduced by a strong reducing agent, for example, by tri-*n*-butyl tin hydride finally yielding the *sf* compound [52]. The usage of this reaction to synthesize *sf* alkanes was also demonstrated by Russell et al. ([21] and references herein). The reaction with α,ω-dialkenes yields triblock structures [53].

Polymerizable sf monomers. Höpken et al. [54] employed the basic reaction to ω-functionalized olefins (shown in Figure 11.5 by "R") and generated functionalized *sf* compounds opening a wide range of possibilities for the synthesis of *sf* polymers.

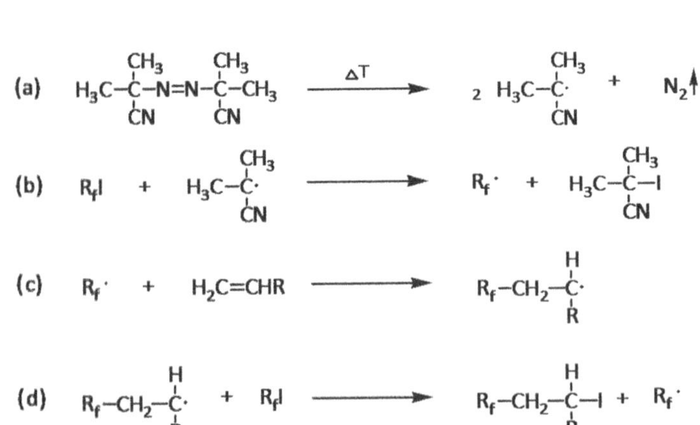

FIGURE 11.5 Synthesis of semifluorinated alkanes by radical addition of perfluoroalkyl iodides to olefins, derived from Reference 51: (a) initiation; (b) iodine abstraction from R_f; (c) addition of R_f radical to the double bond; (d) chain propagation.

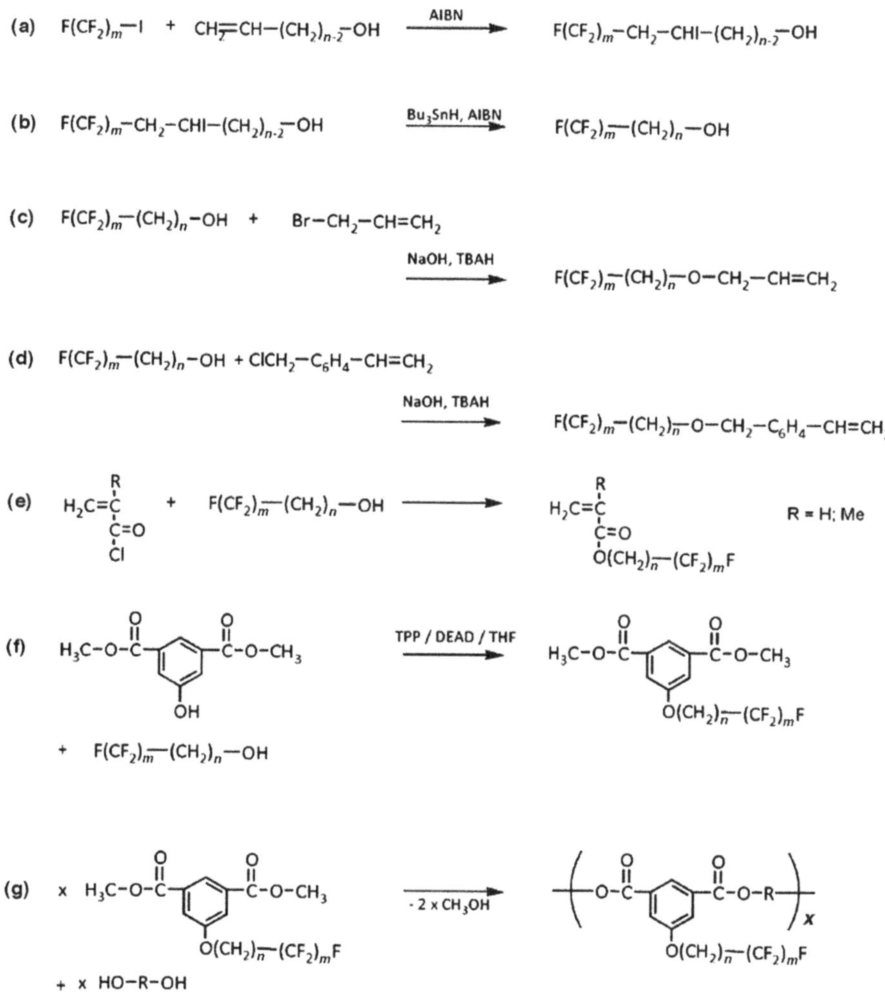

FIGURE 11.6 Synthesis of functionalized semifluorinated compounds: (a) radical addition of R_fI to the olefin; (b) reduction of the iodide. Conversion into monomers: (c) allyl ether; (d) *sf* styrene; (e) *sf* (meth)acrylate; (f) 5-*sf* dimethyl isophthalate as polycondensation monomer; (g) polycondensation of (f) with diols.

The synthesis strategy for *sf* alcohols is depicted in Figure11.6a and 11.6b. *Sf* alcohols were converted into monomers or attached to polymer chains by several subsequent reactions. Etherification with allyl bromide resulted in *sf* allyl ethers [54] (Figure 11.6c). Phase transfer-catalyzed etherification of *sf* alcohols and *p*-(chloromethyl) styrene yielded *sf* styrene [35] (Figure 11.6d). By reaction with (meth)acryloyl chloride, *sf* (meth)acrylates were obtained, a method that was widely employed [36, 55, 56] (Figure 11.6e). Besides, the reaction of isocyanatoethyl (meth)acrylate with *sf* alcohols to give *sf* (meth)acrylates was often used [57,58]. Starting from 5-hydroxy-dimethylisophthalate, the *sf* side chain was attached via Mitsunobu etherification with *sf* alcohols [59, 60] which generated a monomer (Figure 11.6f) useful for polycondensation (Figure 11.6g) [29, 61]. Furthermore, a variety of other synthetic pathways using *sf* alcohols has been reported. Conversion of *sf* alcohols into bromo compounds gave educts that were afterward used for ionenes [62].

Attachment of sf side chains to polymers by polymer-analogous reactions. In the literature, numerous experiments to attach *sf* side chains to polymers have been reported. A few examples are outlined in the following. *Sf* allyl ethers as shown in Figure 11.6c were attached to poly(methyl hydrosiloxane) by hydrosilylation to give *sf* polysiloxanes (Figure 11.7a) [63]. Iyengar et al. [64] converted the double bonds in the PI block of PS-*b*-PI diblock copolymers by hydroboration/oxidation into hydroxyl groups. These OH groups were afterward reacted with heptafluorobutyryl chloride to give diblock copolymers with one fluorinated block. Wang et al. [65] used these OH groups for the reaction with *sf* carboxylic acid chlorides to achieve PS-*b*-PI with one *sf* block (Figure 11.7b). To increase the surface density of *sf* side chains and the surface stability, PS-*b*-PI diblock copolymers with *sf* monodendron side chains attached at the PI block were reported [28].

11.2.2 Semifluorinated Methacrylate Copolymers

General survey. The most common and often applied synthesis method for *sf* poly(meth)acrylate polymers and copolymers is radical (co)polymerization, as depicted in Figure 11.8. In general, radical polymerization is easy to perform, cheap, and reproducible.

In many cases, azobis(isobutyronitrile) (AIBN) is employed as radical initiator. The polymerization conditions, in particular solvent, depend mainly on both, solubility of the starting *sf* monomers and choice of comonomer. To give just a few examples, copolymers of dodecafluoroheptyl methacrylate with methacrylic acid could be synthesized in dioxane due to the solubilizing effect of methacrylic acid [66], copolymers of sfMA-H2F8 and sfMA-H2F4 with styrene could be prepared in toluene [35], and copolymerizations of *sf* methacrylates with butyl acrylate, hydroxybutyl acrylate, and styrene were performed using *tert*-butyl peroxyacetate as initiator in methyl amyl ketone [31].

However, monomers with long self-organizing *sf* side chains interesting for this contribution do not dissolve easily in common solvents. The longer the *sf* unit, the inferior is the solubility of the monomer. Thus, other solvents had to be figured out. Höpken et al. [37] used bulk polymerization for the synthesis of poly(1H,1H,2H,2H-perfluorodecyl methacrylate), eventually adding a very small amount of solvent to avoid diffusion problems during polymerization. Later on, bis(trifluoromethyl) benzene (BFMB) was proposed by the same working group as solvent for radical copolymerization of sfMAs [67] and their copolymers with various comonomers like itaconic acid, maleic anhydride, hydroxyethyl methacrylate (HEMA) [68]. The relative molar masses obtained depended strongly on the conditions used for size exclusion chromatography (SEC) and calibration standards, and may thus not be compared to

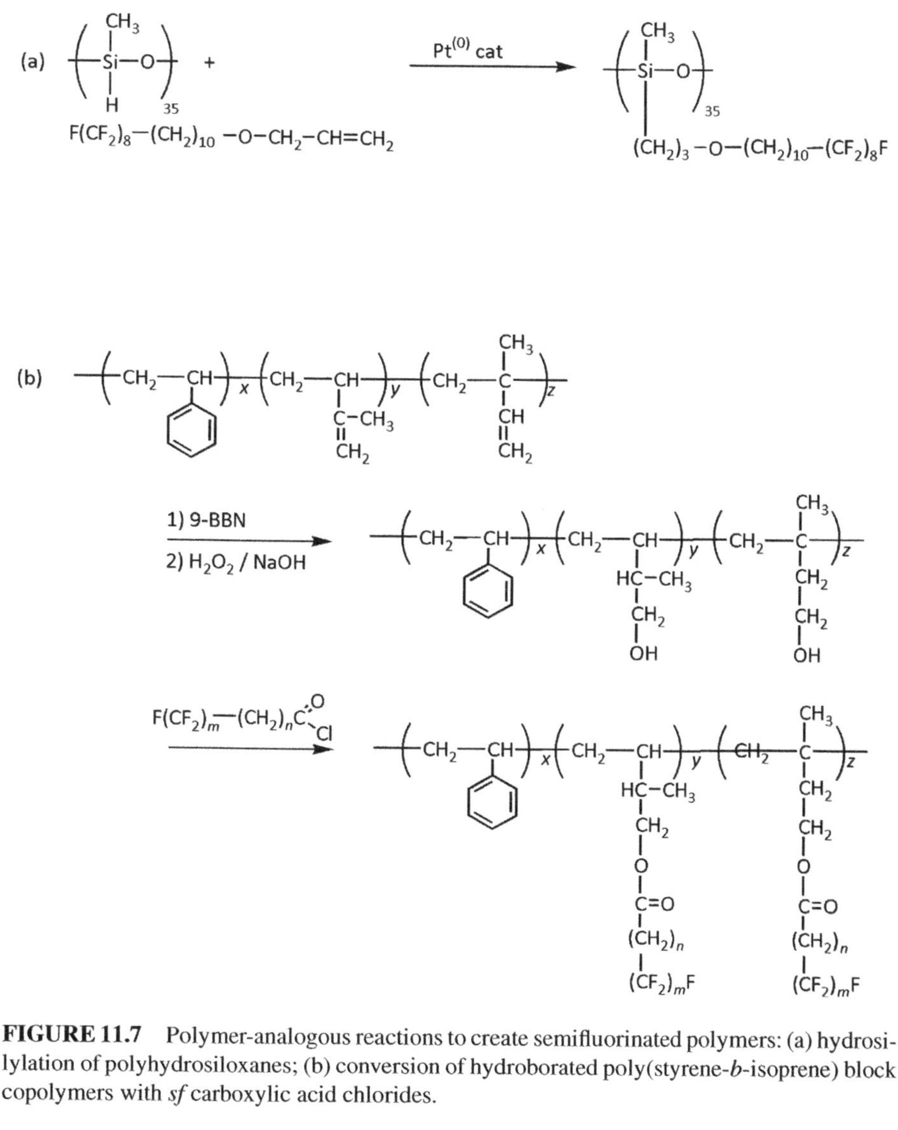

FIGURE 11.7 Polymer-analogous reactions to create semifluorinated polymers: (a) hydrosilylation of polyhydrosiloxanes; (b) conversion of hydroborated poly(styrene-*b*-isoprene) block copolymers with *sf* carboxylic acid chlorides.

each other. The molar mass distributions M_w/M_n were in the range of 2, as expected for radical copolymerization.

Another very useful solvent for polymerization of (semi)fluorinated monomers in homogeneous solution is supercritical carbon dioxide ($scCO_2$) due to its high ability to dissolve fluorinated compounds. CO_2 reaches the critical point at T_c = 31°C and p_c = 73.75 bar. Thus, high pressure cells are needed and special equipment, such as tubular reactors described by Beuermann et al. [69], has to be used. This type of reactor was also applied for the (controlled) radical polymerization of *sf*

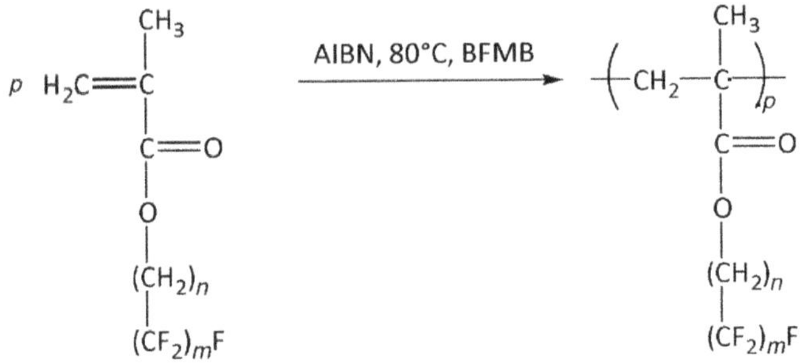

FIGURE 11.8 Synthesis of *sf* methacrylate copolymers by radical polymerization.

methacrylate BCPs [70]. Due to the required equipment, polymerizations in $scCO_2$ are rather seldom for synthesis, but widely employed for processing of thin films, coatings, or in lithography [71–73]. The very first examples of polymerizations were given by De Simone et al. [74–76]. Hwang et al. [77] reported the synthesis of random copolymers of oligo(ethylene glycol) methacrylate and 1H,1H,2H,2H-perfluorooctyl methacrylate (sfMA-H2F6) in $scCO_2$ at 345 bar and 65°C by radical polymerization with AIBN.

A smart solution was proposed by Storsberg and Ritter [78] who generated a cyclodextrin–monomer complex of randomly methylated β-cyclodextrin and the *sf* monomer (sfMA-H1F2, sfMA-H2F6, and sfMA-H2F8) in aqueous solution. Then, the monomer was polymerized within this complex in water with water-soluble initiators (either a water-soluble azo initiator or the redox system $K_2S_2O_8/NaHSO_3$). The molar mass distributions obtained were broader than in standard radical polymerization and increased with increasing length of the *sf* side chain, thus reflecting that the reaction in such systems is more complicated, in particular in copolymerization.

To take higher control on molar mass and molar mass distribution, techniques of *controlled polymerizations* should be employed even if copolymers with random monomer distribution are the desired product. One of the first controlled (and in this case also living) polymerizations of *sf* (meth)acrylates was group transfer polymerization (GTP) with 1-methoxy-2-methyl-1-trimethylsilyloxy-1-propene as initiator and transfer agent, as reported by Krupers and Möller in 1997 [55, 79] (see Figure 11.9a). Successful anionic polymerization of fluoromethacrylates with MMA, trimethylsilyl-protected HEMA, and tert-butyl MA was also demonstrated [80, 81]. At this point, it has to be noted that controlled techniques are particularly favored for synthesis of sfMA-containing BCPs. BCP with (semi)fluorinated blocks will not be discussed in this contribution, but the techniques of controlled radical polymerization (CRP) can basically also be applied to synthesis of random copolymers. Therefore, only a short glance on the possibilities for CRP of *sf* monomers will be given in the following taking the fact into account that a couple of comprehensive reviews are already available in the field (see, for instance [41]). CRP is characterized by significant reduction of the radical concentration compared to standard radical polymerization, which is normally achieved by reversible deactivation of the growing macroradicals (the so-called persistent radical effect) [82, 83], thus significantly suppressing termination

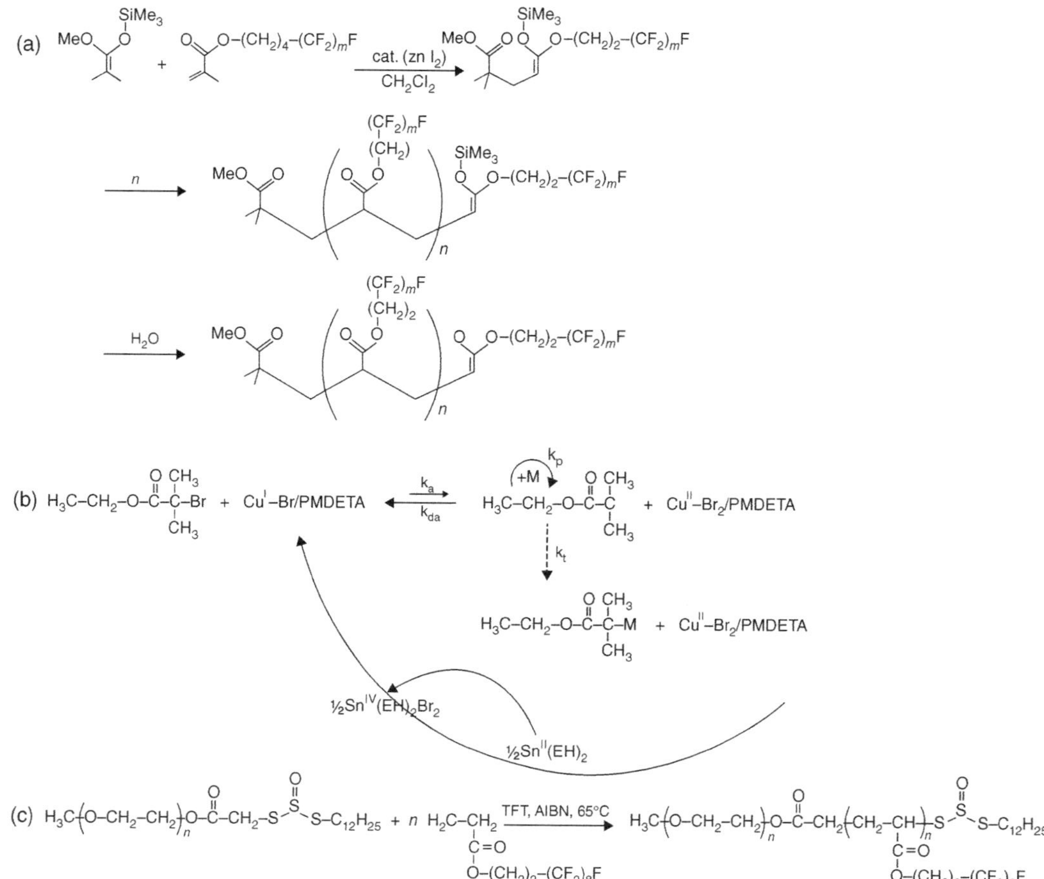

FIGURE 11.9 Synthesis of *sf* methacrylate (co)polymers by living and controlled polymerizations: (a) group transfer polymerization according to Reference 55; (b) ARGET ATRP; (c) polymerization of fluoroacrylates by RAFT according to References 87 and 98.

reactions. Since the general principles are well established, the reader is referred to standard literature on CRP. The procedure chosen to create persistent radicals originates the name of the method: nitroxy-mediated CRP (NMRP) uses stable nitroxide radicals [84,85]; atom transfer CRP (ATRP) employs the electron transfer between an alkyl halide and a transition metal complex (preferably Cu(I)) to activate/deactivate the radicals [86]; reversible addition/fragmentation transfer polymerization (RAFT) uses chain transfer agents. In addition, many other special, but important methods are known, for example, using outer electron transfer and single electron transfer [83].

The three main methods of CRP, NMRP, ATRP, and RAFT, have been numerously applied to (semi)fluorinated (meth)acrylates. Excellent reviews are given by Ameduri and Boutevin [41], Hansen et al. [87], and Ameduri [88]. Hansen et al. summarized fluorinated and *sf* methacrylate monomers with different structure of the *sf* side chain that have been polymerized via ATRP in trifluorotoluene as solvent at 110°C with methyl-2-bromopropionate as initiator, CuBr as catalyst, and 2,2′-bipyridine as ligand for the copper salt. These polymerizations yielded rather high polydispersities M_w/M_n from 1.3 to 1.7, which reflects that the control over the reaction is not very high. The critical point again is to find a suitable solvent for the fluorinated monomer. The structure of the initiator may be adjusted to improve solubility and polymerization result, as shown by Perrier et al. [89] by using a fluorinated ATRP initiator and toluene as solvent. The controlled nature of the polymerization was proven by linearity of the plot of M_n versus conversion. In the synthesis of BCP, solubility problems can be circumvented by choosing the monomer yielding the block with better solubility (e.g., poly(ethylene gylcol) (PEO)) as first block and macroinitiator. This procedure normally enhances the solubility of the final BCP with fluorinated block. In this manner, a high number of polar/nonpolar (fluorinated) BCP was accessed. Hussain et al. [90] synthesized PEO-*b*-PsfMA-H2F8 diblock copolymers and PsfMA-H2F8-*b*-PEO-*b*-PsfMA-H2F8 triblock copolymers [91] by ATRP in butyl acetate as solvent, either starting from PEO macroinitiator or from a bifunctional PEO macroinitiator. The reported polydispersity was rather high (1.2–1.5 at relative molar masses from 10,000 to 30,000 g/mol). The use of $scCO_2$ as solvent in ATRP was demonstrated and required additionally fluorinated ligands with good solubility in $scCO_2$ [70, 92]. The polydispersity was comparable to Hussain's result [91]. Decreasing the amount of copper catalyst by adding an excess of a reducing agent, for example, tin(II)-2-ethyl hexanoate, results in ARGET ATP (activators regenerated by electron transfer) [70]. The principle of ARGET ATRP derived from Jakubowski et al. [93] is shown in Figure 11.9b. Synthesis of PsfA-H2F8 by ARGET ATRP in anisole gave low molar masses, while synthesis in $scCO_2$ enhanced both, molar mass, but also polydispersity (1.7–1.9). The best result (1.2) was obtained with a fluorinated bipyridyl ligand.

NMRP was only seldom applied for synthesis of fluorinated poly(methacrylate)s due to the competing disproportionation reaction where abstraction of a proton from the methyl group leads to an alkene at the polymer chain end, while corresponding *sf* acrylates (e.g., 1H,1H,2H,2H-perfluorodecyl acrylate) gave good results in *N*-methyl pyrrolidone as solvent and *N*-*tert*-butyl-1-diethylphosphono-2,2-dimethylpropyl nitroxide as radical scavenger [94]. Iodotransfer polymerization is another prominent method to increase control in radical polymerization [88]. It works excellent in polymerization of fluoroalkenes, but is not useful to polymerize *sf* methacrylates.

In contrast, RAFT has already been used manifold to prepare (semi)fluorinated polymers and BCP with fluorinated blocks [87]. Almost all monomers for which free radical polymerization works can also be polymerized by RAFT, assumed the right choice of solvent, transfer agent, and initiator. Again, the method was mainly applied to BCP synthesis, for instance, by Laschewsky and coworkers [95–97], as depicted in Figure 11.9c.

Own work with self-organizing sf methacrylates. Self-organization of *sf* polymers is necessary to achieve stable surfaces with low surface free energy. Thus, it is extremely important to find out at which conditions with respect to chemical structure of the polymers as well as with respect to preparation conditions self-organization can be achieved. To lighten this topic for *sf* methacrylate copolymers, three approaches were pursued:

1. varying the length of the *sf* side chain and comparison of the sfMA homopolymers;
2. use of strongly self-organizing H10F10 side chains which were well established in former work [29, 61, 99], using the monomer 1H,1H,2H,2H copolymers with methyl methacrylate (MMA) with stepwise varied concentration of sfMA-H10F10 to figure out the self-organization "threshold" of H10F10; and
3. systematic shortening of H10F10 side chains, investigated by using sfMA-H2F8, synthesis of a series with stepwise varied sfMA-H2F8 content.

The resulting copolymers of all three topics were compared with respect to their self-organizing tendency in bulk and thin films and with respect to their macroscopic properties. The monomer sfMA-H10F10 was prepared according to Höpken et al. [54] and Krupers and Möller [55] as shown in Figures 11.6a, 11.6b, and 11.6e, while sfMA-H2F8, sfMA-H2F4, and sfMA-H1F1 (trifluoroethyl methacrylate) monomers are commercially available.

All homo- and copolymers were synthesized by radical polymerization at 80°C for 4 hours using AIBN (2–4 mol% with respect to the total monomer amount) as radical initiator. The solvent used depended on the monomer; sfMA-H1F1 and sfMA-H2F4 were polymerized in chloroform, while sfMA-H10F10 and sfMA-H2F8 copolymerizations with MMA required BFMB as solvent. The polymerizations were carried out under stirring in a two-necked Schlenk flask adapted to a Schlenk line after three freeze-vacuum thaw cycles in nitrogen atmosphere. After polymerization, the reaction mixture was diluted with BFMB/$CHCl_3$ and precipitated into the 10-fold excess of methanol, filtered off, washed, and dried in a vacuum oven at 50°C.

Inline-NMR measurements of copolymerizations of sfMA-H2F8 with MMA in BFMB were carried out according to a procedure described by Alvares et al. [100]. The measurements in the NMR tube showed that complete conversion was reached after 150 minutes [101], as illustrated in Figure 11.10 for one of the examined monomer ratios. ^{1}H NMR spectroscopy allows simultaneous determination of

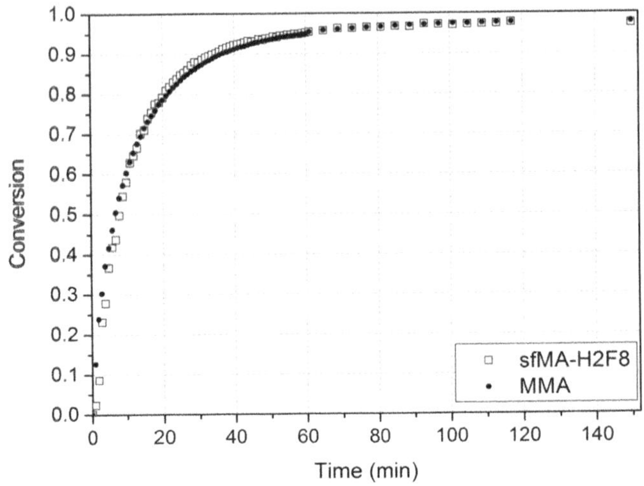

FIGURE 11.10 Conversion–time curve of the copolymerization of sfMA-H2F8 (18 mol%) and MMA (82 mol%) in BFMB (60 wt%) at 80°C as obtained from inline-^{1}H NMR monitoring experiments.

monomer concentrations (i.e., conversion) and additionally the copolymer composition (for details, see [101]). Figure 11.10 shows the conversion–time curves of MMA and sfMA-H2F8 in the copolymerization of 18 mol% sfMA-H2F8 and 82 mol% MMA. The conditions are comparable to the synthesis conditions for the copolymers. Copolymerization at 80°C proceeds fast and thus, determination of *r*-parameters by linearization of the copolymerization equation at low conversion (usually <10%) was not possible. Therefore, *r*-parameters for the solution polymerization in BFMB were determined by the extended Kelen–Tüdös method which allows to use data pairs (monomer concentration vs. copolymer composition) up to higher conversions of about 50% [102, 103]. Here, we used conversions up to 40% for the calculations. The *r*-parameters were obtained graphically from the plot of the Kelen–Tüdös parameters η versus ξ [103].

The copolymerization parameters *r* for MMA/sfMA-H2F8 (obtained by inline-NMR monitoring), and for MMA/sfMA-H1F1 in dimethyl acetamide (DMAc) resulting from the copolymerization diagram at low conversions [104] are summarized in Table 11.1.

TABLE 11.1 Copolymerization Parameters *r* for the Radical Polymerization of sfMA-H2F8 with MMA and sfMA-H1F1 with MMA

Sample	r_{MMA}	r_{sfMA}	$r_{MMA} \cdot r_{sfMA}$
MMA/sfMA-H2F8, BFMB	1.27	0.66	0.83
MMA/sfMA-H1F1, DMAc	1.15	0.92	1.06

The product of r_{MMA} and r_{sfMA} is for both monomer pairs in the range of 1, indicating that random copolymers with statistically distributed repeating units were achieved under the conditions used. The slightly higher value of r_{MMA} shows that addition of MMA to the MMA chain-ends as well as to the sfMA chain-ends is slightly preferred over the addition of sfMA-H2F8 to both. A good correlation between monomer ratios in the feed and in the copolymer was observed and an obvious solvent influence could not be noticed comparing the *r*-parameters for MMA/sfMA-H2F8 in BFMB and MMA/sfMA-H1F1 in DMAc. A comparable result for the copolymerization of 2,2,2,-trifluoroethyl methacrylate (MATRIFE) and 2,2,2-trichloroethyl α-fluoroacrylate (FATRICE) in acrylonitrile with $r_{\text{MATRIFE}} = 1.52$ and $r_{\text{FATRICE}} = 0.61$ was obtained by Cracowski et al. [105]. The result for the copolymerization in BFMB shown in Table 11.1 correlates nicely to *r*-parameters for MMA/octadecyl methacrylate (ODMA) obtained under almost comparable conditions (in benzene, at 80°C, with 3 wt% AIBN). Alvares et al. [100] found $r_{\text{MMA}} = 1.37$ and $r_{\text{ODMA}} = 1.17$, reflecting a comparable situation as for MMA/sfMA-H2F8. Thus, it is reasonable to assume random copolymerization also for the MMA/sfMA-H10F10.

Determination of molar masses and polydispersity of *sf* polymers is often complicated because of the limited solubility particularly of highly fluorinated polymers. To avoid different conditions for the polymers and to ensure a certain comparison, SEC with a mixture of pentafluorophenol (PFP)/chloroform (1/3 vol/vol) as eluent (even if some of the samples were soluble in chloroform or tetrahydrofurane, THF), refractive index (RI) detection, and PMMA standards for calibration and determination of relative molar masses was employed. For some samples, molar masses could not be detected due to isorefractive behavior (no signal due to same RI of sample and eluent). This was observed in the series for P(MMA-*co*-sfMA-H10F10) at 20–25 mol% sfMA-H10F10 and for P(MMA-*co*-sfMA-H2F8) at 20–30 mol% sfMA-H2F8. The chemical characterization of selected copolymers used for the present study is given in Table 11.2.

The relative molar masses of the PMMA control relative to PMMA calibration standards in Table 11.2 measured in PFP/$CHCl_3$ is slightly lower than the one obtained in THF. The same trend is found for P(MMA-*co*-sfMA-H2F8) and P(MMA-*co*-sfMA-H10F10) copolymers with low *sf* content up to 25 mol%. The data given in Table 11.2 also refer to a molar mass drop with increasing sfMA content. Simultaneously, the solution viscosity drops. From the course of polymerization and the kinetic studies, there is no obvious reason for that. Keeping the restricted solubility of highly fluorinated polymers into account, it may be hypothesized that the polymer molecules are not dissolved properly and the hydrodynamic volume therefore does not reflect the "real" molar mass. In such case, estimation of M_n by means of the ^{1}H NMR spectrum may be employed, supposed that the protons of the isobutyronitrile start group can be separated from the protons of the polymer chain. This was indeed possible for P(MMA-*co*-sfMA-H2F8), as illustrated in Figure 11.11, but it was not possible for P(MMA-*co*-sfMA-H10F10).

As it can be seen from Table 11.3, the molar masses obtained from NMR do not reflect a dependence on the sfMA content. All M_n are in the range or above 10,000 g/mol. Thus, they are high enough to ensure generation of polymer properties.

TABLE 11.2 Molar Masses, Molar Mass Distributions and Chemical Compositions of Selected Samples Used in the Present Study

Sample denotation	Molar content sfMA (mol%)	Solution viscosity[a] η_{inh} (dL/g)	$M_{n,SEC}$[b] (g/mol)	$M_{w,SEC}$[b] (g/mol)	M_w/M_n
PMMA	0	0.50	29,800	51,900	1.74
PsfMA-H1F1	100	0.15	31,600	45,500	1.44
PsfMA-H2F4	100	0.44	Isorefractive	–	–
PsfMA-H2F8	100	0.05	1800	2500	1.39
P(MMA-*co*-sfMA-H2F8)-20/80	80	0.08	2800	4500	1.61
P(MMA-*co*-sfMA-H2F8)-29/71	71	0.08	3600	5000	1.39
P(MMA-*co*-sfMA-H2F8)-40/60	60	0.09	5400	10,700	1.98
P(MMA-*co*-sfMA-H2F8)-50/50	50	0.08	6200	15,100	2.44
P(MMA-*co*-sfMA-H2F8)-59/41	41	0.09	7000	18,100	2.59
P(MMA-*co*-sfMA-H2F8)-75/25	25	0.09	6400	15,700	2.45
P(MMA-*co*-sfMA-H2F8)-85/15	15	0.17	14,000	28,700	2.05
P(MMA-*co*-sfMA-H2F8)-90/10	10	0.17	12,400	23,300	1.88
PsfMA-H10F10	100	0.14	not sol.	not sol.	not sol.
P(MMA-*co*-sfMA-H10F10)-17/83	83	0.09	1800	3700	2.09
P(MMA-*co*-sfMA-H10F10)-51/49	49	0.12	4300	19,800	4.98
P(MMA-*co*-sfMA-H10F10)-62/38	38	0.11	4600	22,000	4.84
P(MMA-*co*-sfMA-H10F10)-69/31	31	0.16	32,300	63,800	1.98
P(MMA-*co*-sfMA-H10F10)-76/24	24	0.19	26,400	50,100	1.90
P(MMA-*co*-sfMA-H10F10)-89/11	11	0.18	16,900	31,400	1.86

[a] $\eta_{inh} = (\ln\eta_{rel})/c$, $c = 0.5$ g/dL; measured in trifluoroacetic acid/$CHCl_3$ (1/1 vol/vol), AVS 470 Viscometer (Schott); Ubbelohde capillary no. 0.4 mm; 25°C.
[b] SEC: Knauer modul GPC; Eluent: PFP/$CHCl_3$ (1/2 vol/vol), 0.3 mL/min; Plgel 5 MiniMix-D separation column (Agilent Technologies), RI detection (Knauer): PMMA calibration standard.

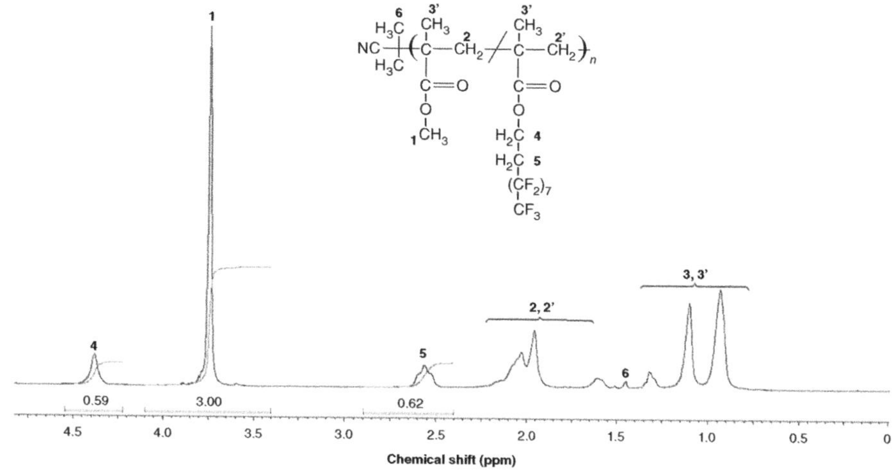

FIGURE 11.11 ^{1}H NMR spectrum of P(MMA-*co*-sfMA-H2F8) (79/81 mol/mol) in TFA-d/$CDCl_3$ (1/1 vol/vol).

Significant deviation between NMR and SEC values caused by the obvious reduction of hydrodynamic volume in PFP/$CHCl_3$ (1/2 vol/vol) due to restricted solubility starts above the isorefractive compositions. The molar mass distributions of all polymers are in the range expected for radical polymerization (1.4–2.6).

A few copolymerization experiments with MMA and sfMA-H2F8 were performed under ATRP conditions according to Figure 11.9b with a procedure given by Haddleton et al. [106], but adapted to BFMB as a solvent in order to prove the

TABLE 11.3 Comparison of Relative Number-Averaged Molar Masses of P(MMA-*co*-sfMA-H2F8) Copolymers Obtained by SEC in PFP/$CHCl_3$ with PMMA Standards and ^{1}H NMR Spectroscopy

Molar content sfMA-H2F8 in copolymer (mol%)	$M_{n,SEC}$^{a} (g/mol)	$M_{n,NMR}$ (g/mol)
5	15,900	11,200
10	12,400	10,200
15	14,000	10,600
21	Isorefractive	10,600
25	6400	7900
41	7000	8700
50	6200	9400
60	5400	13,500
71	3600	14,400
80	2800	16,400
100	1800	–

^{a}SEC: Knauer modul GPC; Eluent: PFP/$CHCl_3$ (1/2 vol/vol), 0.3 mL/min; Plgel 5 MiniMix-D separation column (Agilent Technologies), RI detection (Knauer): PMMA calibration standard.

applicability. Ethyl-2-bromoisobutyrate (EBiB) was applied as initiator, CuBr as catalyst, and *N,N,N′,N′′,N′′*-pentamethyldiethylene triamine (PMDETA) as ligand. Selected results are given in Table 11.4. The composition of the copolymer matches that of the feed. The polydispersity is lower than in free radical polymerization, while the control over the molar mass in these experiments has to be improved. Large excess of the ligand PMDETA with respect to CuBr in the molar ratio 2:1 (used by Haddleton et al. [106] for alkylpyridylmethanimine ligands) seems to lower, in our case, the molar mass below the calculated one. Using only slight excess results in molar masses not too far from the calculated values. The polydispersities achieved are in the range of results with other *sf* (meth)acrylates reported in literature [107].

To summarize, it can be stated that both free radical copolymerization and ATRP can be applied successfully to synthesize random copolymers of MMA with sfMA. First copolymerization experiments with ARGET-ATRP in BFMB as well as in $scCO_2$ according to Schreiber et al. [70] showed the applicability of the method; further optimization is necessary.

11.3 SELF-ORGANIZATION BEHAVIOR OF SEMIFLUORINATED POLYMERS IN BULK AND IN THIN FILMS

11.3.1 Methods to Observe Self-Organization

Self-organization phenomena take place in many natural and man-made systems of at least two different components of all combinations of aggregate states. Self-organization is a process which can be initiated by local interactions between the components. In achieving an equilibrium structure, the separation into distinct phases (domains of involved components) is characterized by an area-minimization of the interfaces between the domains (ruled by volume fraction). Starting from disordered state, this mostly spontaneous process leads to local or global ordering principles in one, two, or three space dimensions if physical, chemical, and biological systems are in consideration. Here, the interest is directed on the self-organization behavior in polymer systems under the special aspect of the presence of *sf* segments.

The main reason for polymeric phase separation at intermolecular as well as intramolecular scale is the incompatibility of chemically different segments in the polymer chain. Chemically identical and compatible segments merge in one phase. The molecular scenario is an interplay between internal physical and chemical driving forces, which can be influenced by external fields (e.g., electrostatic fields) additionally. Mediated by their internal geometry (i.e., chain conformation and degree of rigidity) and interaction forces (i.e., electrical charges), the polymer system tries to enter a new state in which a part or all segments are arranged in phases with ordered distance relations. Typically for polymers, such disorder–order transitions are based on segment movements and are hindered consequently by the polymer itself at temperatures below the glass transition T_g. Initial conditions like higher temperatures (annealing) or interaction with an appropriate solvent (vapor annealing) play often an important role to trigger the self-organizing process.

TABLE 11.4 Results of Copolymerizations of MMA with sfMA-H2F8 Under ATRP Conditions in BFMB at 90°C

MMA/sfMA-H2F8										
Calc. (NMR) (mol/mol)	Found (NMR) (mol/mol)	$[M_0]/[I_0]$	$M_{n,calc}$ (g/mol)	[CuBr] (g/mol)	[PMDETA] (mol)	[EBiB] (mol)	$[M_0]$ (mol)	$M_{n,SEC}$[a] (g/mol)	$M_{w,SEC}$[a] (g/mol)	M_w/M_n
70/30	68/37	83.9	19,300	1.0	2.0	1	83.9	4150	4850	1.20
70/30	70/30	71.5	16,400	0.7	1.0	1	71.5	25,550	32,650	1.28
30/70	30/70	78.2	29,300	0.9	2.4	1	78.2	9650	12,300	1.43

[a]SEC: Knauer modul GPC; Eluent: PFP/$CHCl_3$ (1/2 vol/vol), 0.3 mL/min; Plgel 5 MiniMix-D separation column (Agilent Technologies), RI detection (Knauer): PMMA calibration standard.

An overview about the most important analyzing techniques to observe and describe the self-organizing process in solid state (bulk) in its duration or as a result is presented in Table 11.5. Table 11.6 summarizes surface-relevant methods. For more detailed information about surface and interface characterization methods for polymers, see Reference 108. Particularly useful is the combination/hyphenation of methods.

Self-organization in the bulk can be observed either directly by microscopic methods [109, 110], or indirectly by thermal transitions with differential scanning calorimetry (DSC), by X-ray scattering methods [111] or by the dynamic behavior (dielectric spectroscopy) [112]. These methods are extended for examination of thin films by scanning electron microscopy (SEM) and atomic force microscopy (AFM) as direct methods [108], and by X-ray methods XPS [113, 114] and NEXAFS (spectroscopy) [115, 116], and XR and GISAXS [117, 118] as scattering methods. Recent developments also allow detecting thermal transitions of thin films on a chip calorimeter (nanocalorimetry) [119]. Particularly for fluorinated surfaces, different kinds of wetting methods are employed [120, 121].

11.3.2 Overview about Self-Organized Semifluorinated Alkanes and Polymers

The first description of self-organization in semifluorinated alkane solids dates back to 1986 when Russell et al. [21] reported structural models for H*n*F12-iodide outlining different packing possibilities in dependence on the number of CH_2 groups (H*n*). One of the proposed models is illustrated in Figure 11.12 showing alternating layers of alkane and perfluoroalkane parts. This layered structure was proven to have typical liquid crystalline (LC) temperature behavior. That feature was rediscovered some years ago by Broniatowski et al. [e.g., References 122 and 123].

After these reports the semifluorinated "hype" started and new structures with more LC phases and phase transitions appeared in the literature, not only covering the bulk-state structure, but also the special arrangements of the structures at different surfaces. *Sf* alkanes have recently been demonstrated to form discrete self-assembled surface domains at the surface of water as well as on the surface of solids. These arrangements are by a "chain-up" orientation with respect to the surface [124]. The author mentioned the outstanding self-aggregation tendency. This was also underlined by de Vigeurie et al. [125] who observed regular surface structures on water. Terminated with one thiol group, self-organization of *sf* molecules can be used to generate self-assembled monolayers (SAMs) on gold [126]. Phosphonate-terminated *sf* alkanes form SAMs on metals [127].

From the early beginning, it was clear that a certain number of n and m in the $-(CH_2)_n-(CF_2)_m-$ units is necessary to generate self-organization. Derived from the phase diagram (Figure 11.2), phase separation is enhanced at symmetrical ratios, and deviation from symmetry required higher numbers [21, 128, 129].

Connection of *sf* units to different polymer backbones does only slightly change the basic feature of phase separation, which will be discussed in the next section. Comparing the variety of *sf* side chain polymers, for example, all-aromatic polyesters

TABLE 11.5 Methods to Observe Self-Organization in Solid State

Technique	Interaction process	Main information	Limitations	Comments
DSC	Heat transfer	Enthalpies and temperatures of phase transitions	Not cross-linked material	
XRD (SAXS and WAXS)	X-ray scattering and diffraction	Bulk structures on morphology and molecular scale	Coherent length, phase contrast	Also in dependence on T
Polarization microscopy	Polarized light	Areas of spherulites	Dedicated polymer structures only	T-dependence
TEM	Electron beam absorption/ scattering	Bulk structures on morphology scale	Phase contrast, cryo-microtoming	Additionally contrasting, in vacuum
Dielectric relaxation spectroscopy	Dielectric polarization	Specific relaxation behavior of solid states in different ordering levels	Polarizability	Different measuring modes

TABLE 11.6 Surface Methods to Observe Self-Organization in Thin Films

Technique	Interaction process	Main information	Limitations	Comments
Nano-calorimetry	Ultrafast heat transfer	Enthalpies and temperatures of phase transitions	Films <1 μm	
XR	X-ray scattering and diffraction	Film structures (layers), surface roughness	Coherent length, phase contrast, flat films	Also in dependence on T
GISAXS	X-ray scattering and diffraction	Lateral structures, surface roughness	Coherent length, phase contrast, flat films	Also in dependence on T
SEM	Emission of secondary electrons	Topography	Cutting artifacts	In vacuum
AFM (SFM)	Atomic forces by cantilever	Topography, toughness	Flat surfaces	Different measuring modes
XPS	Ionization of atoms + emission of characteristic X-rays	Surface enrichment of segments (atoms)	Less than ~3 nm surface depth	In vacuum
NEXAFS	Characteristic absorption of X-rays	Binding energies in nonoccupied states	Sensitivity of deeper areas	Orientational information by using polarized X-rays
Contact angle	Interaction to a liquid probe, three-phase angle	Surface tension	Type of liquid, roughness, chemical heterogeneity	Static and dynamic experiments

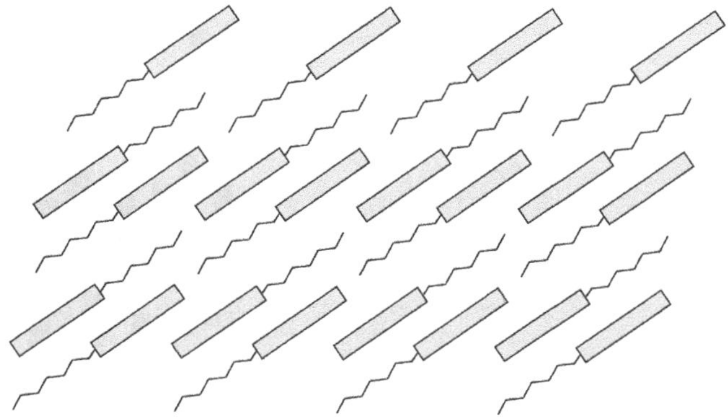

FIGURE 11.12 Structural model for semifluorinated alkanes following Russell et al. [21], showing the interdigitated structure of the *sf* alkane molecules.

[29, 99, 130], PS-*b*-PI diblock copolymers [65], poly(fumarate)s or poly(acrylate)s [131], lamellar structures are found in nearly all cases if the number of *m* is not too small. The backbone and its flexibility seem to have only marginal influence. In simple terms, the strong phase separation into areas with fluorinated segments and areas with the nonfluorinated rest seems to be the first principle in structure formation. Whereas in *sf* alkanes besides lamellar structures also hexagonally packed cylinder structures are possible [128], the attachment of the *sf* segments as side chain onto the polymer backbone prevents structure modifications with other symmetries.

11.3.3 Self-Organization in Polymers with Long Semifluorinated Side Chains

Xiang et al. [28] reported that *sf* side chains with eight and more CF_2 groups generally lead to ordered states characterized by a smectic B-type structure. This result is valid for a special side chain arrangement with three-armed monodendrons. Own investigations of aromatic polyesters with different *sf* side chain lengths (H*n*F*m*: 6-6, 6-8, 6-10, 10-8, 10-10) [99] showed that a lamellar structure was found starting with $m = 8$. However, the prediction of a defined smectic structure as done by Xiang et al. [28] is not reliable. Depending on the thermal history of sample preparation, position of the glass transition and the other transitions, different smectic phases can occur—(a) smectic types without positional order of the mesogens (i.e., *sf* side chain) inside the layer: S_A (orthogonal) and S_C (tilted); (b) smectic types with lateral positional order of mesogens, for example, S_B (orthogonal) and S_F or S_I (tilted, with hexagonal arrangement). For characterization of the special smectic type, a combined temperature-dependent X-ray scattering experiment in the small-angle region (sensitive for layer dimensions) and in the wide-angle region (sensitive for the mesogen arrangement inside the layer) is recommended. Almost complete information can be achieved by investigation of highly oriented samples and use of area detection systems.

The requirements for the onset of phase separation in *sf* segments lead us to the conclusion to use very long H10F10 units as substituents for different polymer

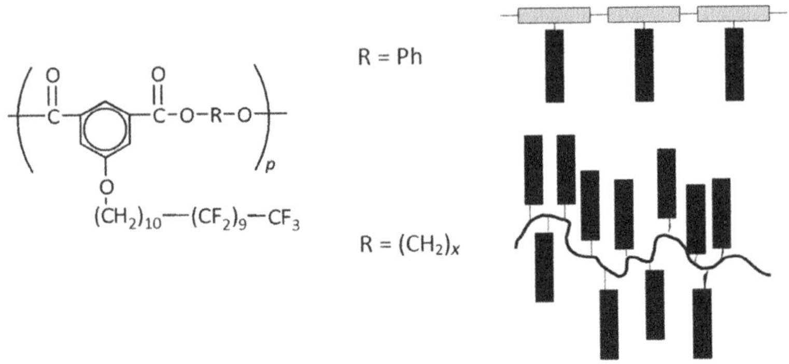

FIGURE 11.13 Chemical structure of semifluorinated H10F10 polyesters.

backbones to elucidate the influence of backbone flexibility onto the self-organization behavior. The self-organization of the H10F10 basic segments could be proven, in agreement to literature reports [21, 55, 65].

The investigations started with attachment of H10F10 to an aromatic polyester main chain (sample IPA-H10F10-HQ outlined in Figure 11.13 with $R = \mathrm{Ph}$) [29] and were continued by aromatic–aliphatic polyesters with varying number of methylene groups in the polyester backbone (outlined in Figure 11.13 with $R = (CH_2)_x$).

The aromatic–aromatic polyester IPA-H10F10-HQ is highly crystalline. The well-ordered structure was formed in solution polymerization which enhances the mobility of polymer segments. The structure simulated by molecular modeling could be adapted by Rietveld refinement to the experimental wide-angle X-ray (WAXS) curve. The resulting model is shown in Figure 11.14 [132]. It is characterized by the arrangement of smectic layers containing side-by-side arranged *sf* H10F10 side chains and

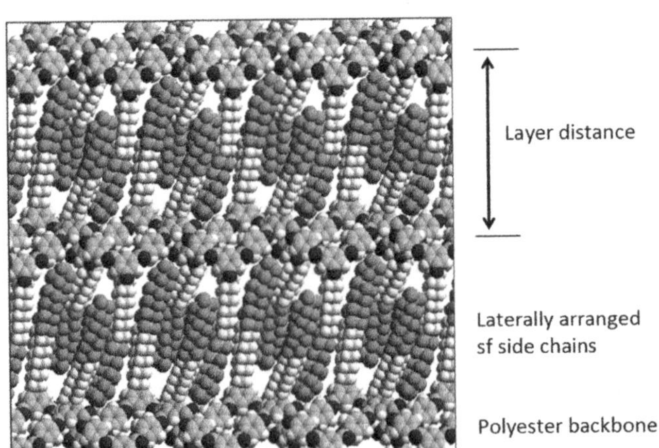

FIGURE 11.14 Structural model of the semifluorinated H10F10 polyester. Reprinted from Reference 132 with permission from John Wiley & Sons, Inc.

one polyester backbone. The typical *d*-spacing is 3.76 nm. From the atomic lengths of the H10F10 chain and the *d*-spacing, it can be derived that the *sf* side chains are tilted. The perfluorinated parts of the side chains of neighboring backbones are interdigitated.

In principle, this basic model of the aromatic *sf* polyester can be applied to most of the polymers with H10F10 side chains that we prepared. Only slight variations of the *d*-spacing at room temperature as well as variations of the typical temperature behavior were observed. It has to be noted that most structural investigations of *sf* polymers were performed on samples with comparable thermal history, that is, the data acquirement was executed after a first heating/cooling procedure. That means, the generation of ordered structures started from molten state in which the mobility of the polymer segments is lower compared to a solution polymerization. This is the reason for higher *d*-spacings of melt-polymerized polyesters ($d = 4.65$ nm).

Further examinations showed the temperature dependence of the structure by employing combined/hyphenated temperature-dependent measurements. By combination of T-WAXS, T-SAXS [133], temperature-dependent ^{19}F solid state NMR, rheology (complex viscosity as function of temperature/frequency) [134] as well as dielectric spectroscopy [135] proved the probability of the model and the change of the structure with temperature.

The T-dependency of the layer distance below melting is extremely small (weakly positive expansion coefficient). The typical PTFE-like structure of the interdigitated perfluorinated parts disappears upon melting at about 70–80°C. This results in *sf* side chains with higher dynamics, allowing crystalline to smectic B/smectic A phase transitions. The side chains are still confined between the rigid polyester backbones. Above glass transition, T_g, of the polymer, the dynamics increases and the layer distance decreases. The prominent negative linear expansion coefficient of the layer distance in the range between T_m and T_i is characteristic of all polymers with long *sf* side chains (compare the two examples in Figure 11.15). At the temperature T_m, the *d*-spacing function has a bend. At further heating, the ordered structure disappears (isotropization transition). Melt rheology of the *sf* polyester in this temperature range indicated that complete isotropization does not take place [134]. The obviously remaining near-range order leads to complete reverse behavior upon cooling without supercooling effects, a behavior characteristic for liquid crystals. The complete heating/cooling cycle is demonstrated in the T-SAXS contour plots (SAXS curves over temperature) in Figure 11.15a. The behavior is completely reversible over more than 10 heating/cooling cycles, thus indicating the reliability of the structural behavior, which is of the prerequisite for a later application. The *d*-value 4.65 nm of the layer fits well to that of PS-*b*-PI diblock with *sf* PI blocks for the morphology within the *sf* block ($d = 4.63$ nm) [62]. In the case of *sf* PSU with 85 mol% H10F10 [136, 137] T-dependent SAXS measurements showed a slightly enlarged layer distance of 5.2 nm which can be understood in terms of an enlarged required volume of the polymer backbone.

It has to be taken into account that the basic structure model of H10F10 side chain polyesters obtained via solution polymerization (see Figure 11.14) reflects thermodynamic equilibrium. The reduced mobility of the polymer segments in melt

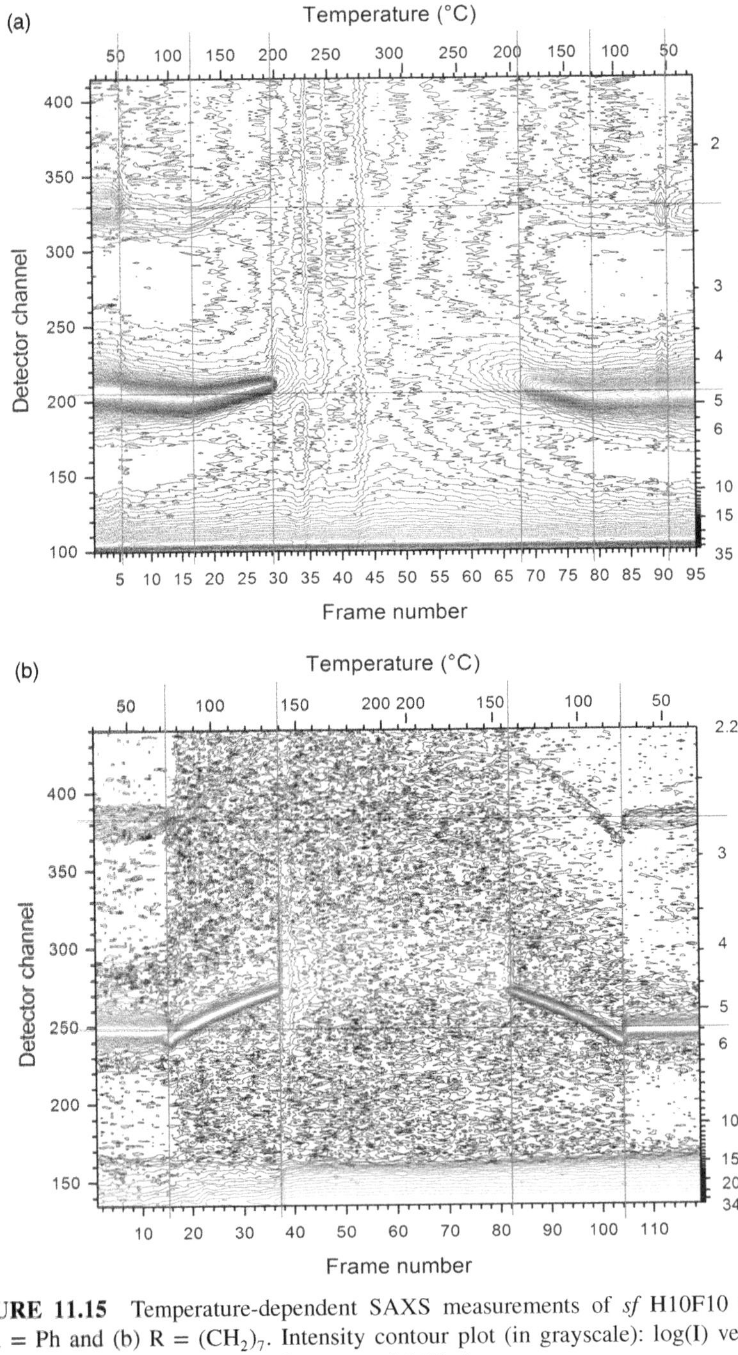

FIGURE 11.15 Temperature-dependent SAXS measurements of *sf* H10F10 polyesters: (a) R = Ph and (b) R = $(CH_2)_7$. Intensity contour plot (in grayscale): log(I) versus T and versus d, first run with heating/cooling rates of 3 K/min.

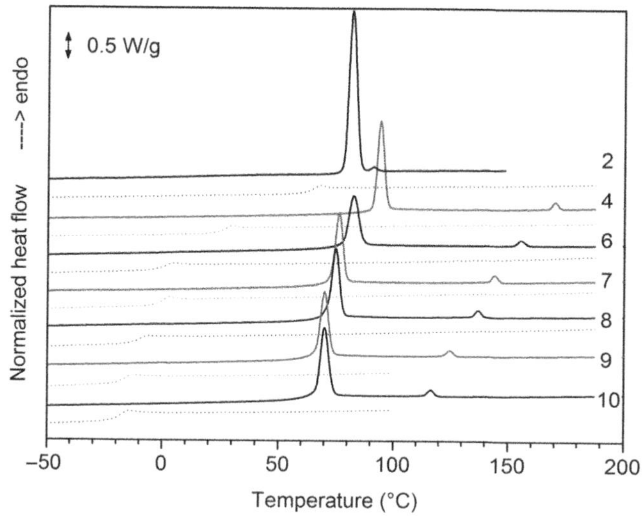

FIGURE 11.16 DSC curves (second heating) of aromatic–aliphatic H10F10 polyesters with R = $(CH_2)_x$ (x = 2, 4, 6, 7, 8, 9, 10). Solid line, (IPA–H10F10—O—R—O)$_n$; dashed line, (IPA—O—R—O)$_n$.

cannot prevent the strong self-organizing of long *sf* side chains. However, it has to be noted that ruled by the cooling conditions, the final solid state of these polymers is far from thermodynamic equilibrium.

Figure 11.15b shows the T-SAXS contour plot of one of the aliphatic–aromatic polyesters of the series according to Figure 11.13 with R = $(CH_2)_7$, having the same *sf* side chains. These polyesters have a higher flexible backbone than the aromatic polyesters, which is indicated by the reduction of the T_g of the backbone in the range of the *sf* side chain melting. This can only be derived from comparison of DSC curves with those of the nonsubstituted polyesters. Figure 11.16 shows the DSC curves of the second heating for the series of aromatic–aliphatic polyesters with and without H10F10 side chains. The nonfluorinated polyesters have indeed T_g's in the range of the *sf* side chain melting.

The negative linear expansion coefficient for *sf* aliphatic–aromatic polyesters is higher (a typical relation is visible in Figures 11.15a and 11.15b). As a characteristic, at the temperature T_m, the *d*-value function has a jump to larger *d*-values of the (now) smectic layer. This *sf* polymer class shows such T-dependent layer behavior already for very short aliphatic spacers (x = 2) in the main chain. In this case, the *d*-spacing jump is considerably smaller [133, 138].

11.3.4 Semifluorinated Methacrylate Copolymers

In order to find out the influence of *sf* side chain length on the methacrylate polymers under study, the number of *n* and *m* in the side chain was varied and the resulting

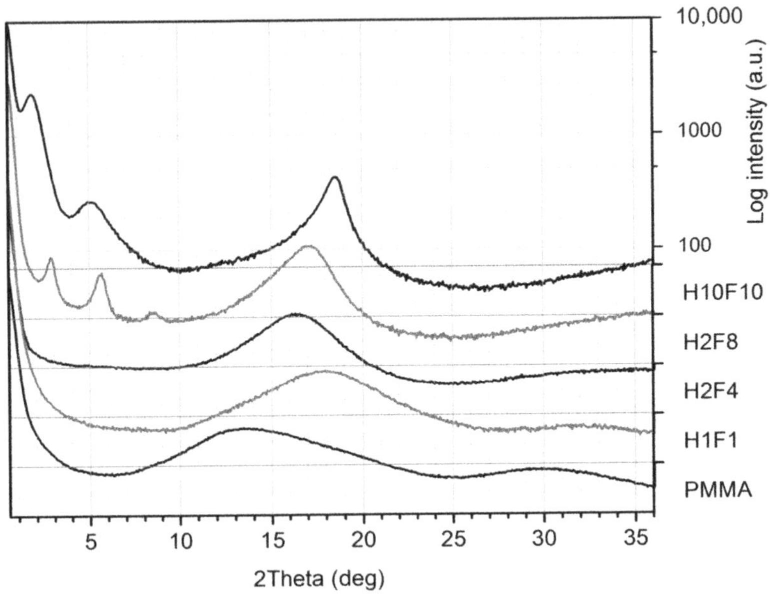

FIGURE 11.17 WAXS curves of as-synthesized *sf* methacrylate homopolymers with different sfMA.

sf homopolymers were examined by WAXS and DSC. Taking previous results on the variation of *sf* side chains on aromatic polyesters [29] and the WAXS curves shown in Figure 11.17 into account, it becomes clear that indeed long *sf* substituents are required to induce phase separation between polymer backbone and side chains. While aromatic polyesters required H6F8 for phase separation, the polymethacrylate with H2F8 reveals distinct phase separation. The same result can be derived from DSC measurements, too (not shown here).

Based on own knowledge of a series of *sf* polymers with other backbones, H10F10 and H2F8 side chains were selected for copolymers with MMA. Möller et al. [37] found a strong dependence of the transition temperatures and entropies of perfluoroalkyl methacrylates on the length of the fluorinated segment, and a rather weak variation caused by the nonfluorinated one. In contrast to these monomers, *sf* methacrylate homo- and copolymers show a considerably different behavior. By combination of DCS and different T-dependent X-ray scattering methods (WAXS, SAXS, XR, GISAXS), and a broad variation of the *sf* molar ratio, the next three figures containing our results will demonstrate the structure of these *sf* MMA-based polymers.

Figure 11.18 compares selected copolymers of both series. Samples with a low *sf* molar ratio exhibit only the appreciable movement of the glass transition temperature T_g to subsequent lower values with increasing ratio.

The first transition temperature (crystalline to smectic) is different for the polymers with different *sf* side chains. For H10F10 polymers, this transition is visible starting from about 25 mol% sfMA, whereas for H2F8 polymers at least 67–70 mol% sfMA

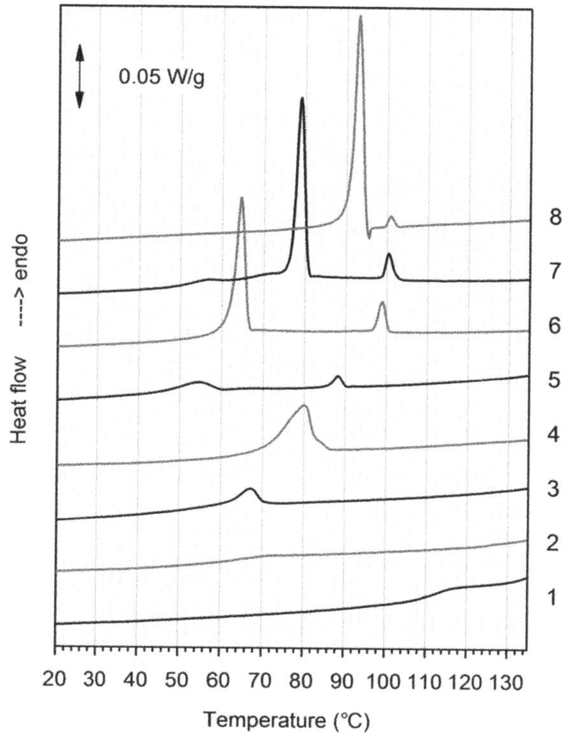

FIGURE 11.18 DSC curves (second heating) of selected samples of two series of P(MMA-*co*-sfMA) copolymers (P(MMA-*co*-sfMA-H10F10) and P(MMA-*co*-sfMA-H2F8)): (1) PMMA; (2) 25 mol% H2F8; (3) 70 mol% H2F8; (4) 100 mol% H2F8; (5) 27 mol% H10F10; (6) 38 mol% H10F10; (7) 66 mol% H10F10; (8) 100 mol% H10F10.

content is necessary to form a crystalline structure in the solid state. Melting (T_m) as well as crystallization (T_c) temperatures increase also with increasing *sf* content. The difference between the melting temperatures of H10F10 and H2F8 polymers with comparable sfMA content is only 13 K. In contrast, melting temperatures of perfluoroalkanes are more significantly influenced by the number of CF_2 groups and show differences of about 50 K with increasing number of CF_2 groups [37]. The same trend can be observed for the melting enthalpies. Additionally, in the case of H10F10 polymers, a transition into the isotropic (liquid) state can be proven. In case of H2F8 polymers, this transition is merged with the melting transition.

Figure 11.19 compares the T-SAXS contour plots for one example of each series. It was demonstrated, in accordance to earlier results with H10F10 polymers, that P(MMA-sfMA-H10F10) copolymers showed an extremely strong tendency to self-organization. This tendency of the *sf* side chains to form smectic layers as reported earlier for *sf* polyesters [133] is significantly higher in P(MMA-*co*-sfMA-H10F10) than in P(MMA-*co*-sfMA-H2F8). As suggested by SAXS, only 27 mol% sfMA-H10F10 are enough to generate the fully organized copolymer structure (not shown

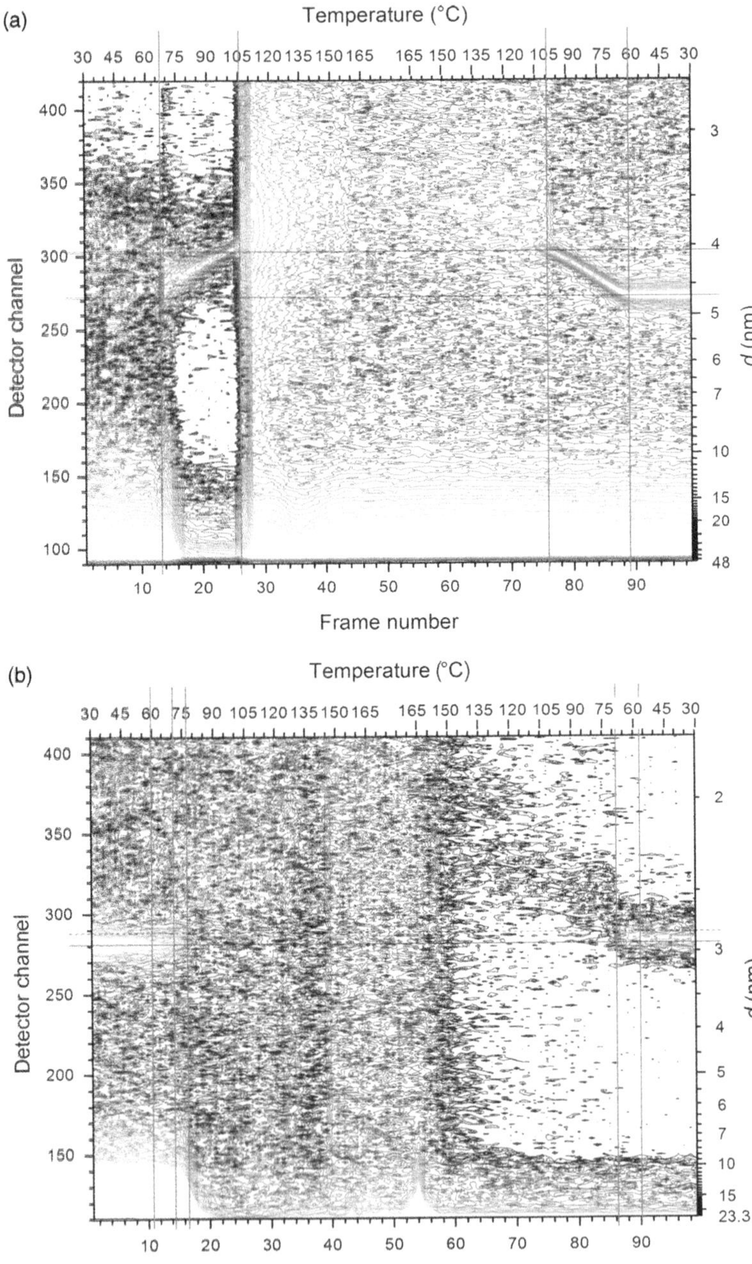

FIGURE 11.19 Temperature-dependent SAXS curves (as contour plot) of two series of P(MMA-*co*-sfMA) copolymers: (a) P(MMA-*co*-sfMA-H10F10) with 49 mol% *sf*; (b) P(MMA-*co*-sfMA-H2F8) with 77 mol% *sf*. Intensity contour plot (in grayscale): log(I) versus *T* and vs. *d*, first run with heating/cooling rates of 3 K/min.

here, but also indicated by the onset of side chain melting in the DSC curves as seen in Figure 11.18), while in the PMMA-sfMA-H2F8 series, ~70 mol% of *sf* monomer are required for structure formation. MMA-based *sf* polymers show scattering patterns reflecting the layered structure caused by the phase separation between *sf* side chains and the remaining "backbone" (alternating layers), as discussed for the *sf* polyester (see Figure 11.14). The layer distances are in the order of 4.7 nm, only weak increasing with the *sf* ratio. For $T > T_m$ a negative expansion behavior can be found again (for H2F8, very weak and in a small T range). At T_m, the d-value function has a bend, too, in contrast to *sf* aromatic–aliphatic polyesters [138] (except for 100 mol% H10F10—there is a jump of the d-value to a smaller value). The d-values of both series in comparison with the relevant monomer size (side chain length) calculated by molecular modeling suggest different arrangements of the side chains inside the crystalline or smectic layer. In the case of H2F8, the monomer diameter is ~1.5 nm. The layer distances measured (~3.0 nm) correlate roughly to the twofold value of the monomer diameter. The most supposable arrangement is two head-to-head side chains which are perpendicular to the backbone. On the other side, modeling of the H10F10 monomer provides a size of ~2.5 nm. But the layer distance is smaller as the twofold monomer size. At $T = T_i$ and for higher *sf* molar ratios, the layer distance is merely in the order of 4 nm. This suggests that the H10F10 side chains are not arranged head-to-head, but interdigitated as shown in Figure 11.12.

The phase separation in solid state (found by combination of "bulk" methods) is reflected in the same tendency in the surface structure found in thin films by X-ray reflectometry, which is seen as reason for the high water contact angles (CAs) at low contents of *sf* comonomer (compare Figure 11.23). In the copolymer series, glass transition and melting of side chains are in the same range, allowing a high mobility of the polymer chains right after melting of the side chain structure, indicating a high potential for the generation of ordered structures at the surface.

Thin films spin-coated from diluted BMBF solution onto precleaned Si wafer were prepared with different thickness and different *sf* molar ratio, as well. The films were stepwise annealed, fast cooled, and examined offline. Figure 11.20 shows the XR-curves obtained on copolymer samples with different sfMA concentration and film thickness. It gives an impression about the influences on the structure formation in thin films, that is, in confined space dimensions. Most important influences are the kind of substrate and surface, preparation, annealing and cooling conditions, as well as the film thickness.

In the temperature range of T_m, the generation of the layered structure shows a different behavior dependent on the side chain type and concentration. In the case of H10F10 side chains (Figure 11.20b for lower H10F10 molar ratio), the layer arrangement starts as in the bulk at $T_m < T < T_i$. The layer distance found for the second annealing step (80°C) is 4.1 nm which is slightly smaller than in the bulk. For higher annealing temperatures (up to T = 160°C), frozen-in states were found accompanied by degradation damages of the film causing higher roughness. In principle, this behavior is confirmed by films having higher H10F10 concentration, but much lower film thicknesses (~15 nm), as seen in Figure 11.20a. Certainly, the discussion of details is difficult due to the restricted film thickness. Only about three

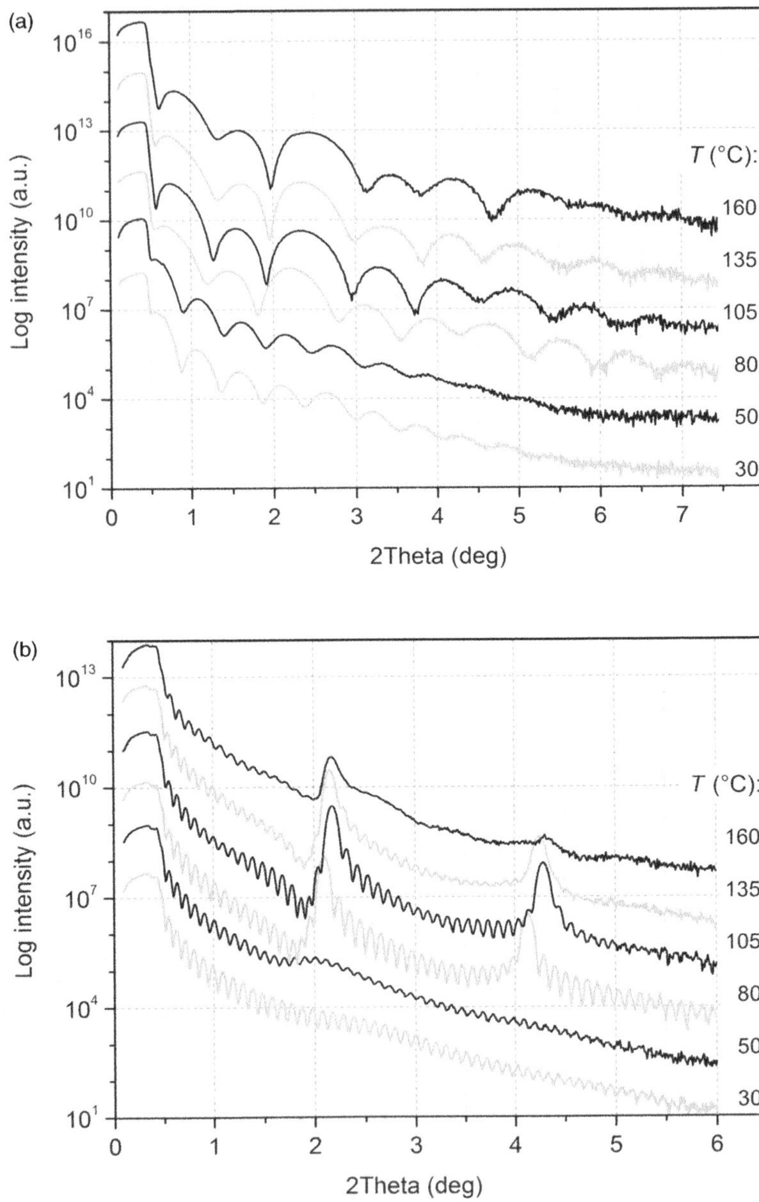

FIGURE 11.20 XR-measurements on films of P(MMA-*co*-sfMA-H10F10): (a) with 54 mol% *sf*, film thickness ~15 nm and (b) with 36 mol% *sf*, film thickness ~87 nm, annealed at different temperatures. Reflectivity curves after off-line annealing and cooling.

layers can be formed by phase separation. The lowest is in interaction to the substrate, the highest in interaction to air. This can be discussed as reason for the deviation of the *d*-spacing compared to the bulk.

As expected from bulk results, H2F8 copolymer films show a different behavior (not shown here). In the case of lower H2F8 molar ratio (e.g., 37 mol%), the typical layer structure is not generated during the annealing process. The curves are published in Reference 139. Only a very weak and broad scattering maximum occurs. At $T > T_g$ film parameters reflect thermal degradation, again. The layer arrangement in films with higher molar ratio (e.g., 72 mol%) starts at $T \sim 50°C$. The layer generation is significantly stronger compared to the bulk and appears at lower temperatures.

For both types of *sf* side chains, it was found that the layer distance in thin films is significantly decreased. The effect is stronger in the case of H10F10 side chains. The reasons can be (a) significant influence of the confined volume and/or (b) the bulk values represent a (isotropic) structure generation from the melt, whereas the film was prepared from solution by spin coating. Spin coating causes orientation. Both preparations are far from equilibrium even after annealing procedures. To get more information about oriented structures in thin films of P(MMA-*co*-sfMA-H10F10), T-dependent GISAXS experiments were performed [139]. In Table 11.7 *d*-values of the layer distances of H10F10 copolymers with different sfMA content during the heating/cooling cycle are collected for selected temperatures of continuous measurements.

The as-prepared sample state shows weak lateral correlation peaks at 4.72 nm and a broad halo in the detector scan. During heating, the lateral peaks disappear and an increasing Bragg reflection is shaped in the detector scan. For $T = 84°C$, the Bragg value is 4.69 nm and the reflection shows a marked sickle-like azimuthal intensity distribution proving the unidirectional layer orientation (layers lying parallel to the substrate) and narrow deviation, as well. The small amount of formerly standing layers is newly oriented. From this, the existence of nearly perpendicular standing *sf* side chains can be concluded. This can be explained by an interplay of micro(nano)phase separation, the smectic ordering of the *sf* side chains, and the orientational wettability

TABLE 11.7 Structure Parameters of T-GISAXS Experiments—Layer Distances (nm) in Films of P(MMA-*co*-sfMA-H10F10): With 54 mol% *sf*, Film Thickness ~15 nm and With 36 mol% *sf*, Film Thickness ~87 nm

Temperature (at measurement) (°C)	54 mol% H10F10	36 mol% H10F10
30	4.70	4.72_{broad}
66	4.70	4.70
84	4.70	4.69
105 ... 190 ... 90	–	–
84	4.80	4.80
66	4.70	4.76
30	4.50	4.52

of the *sf* segments to the silicon substrate. At $T > 105°C$ ($T > T_i$), no scattering structures can be detected. In the cooling run, the generation of a highly oriented layered structure starts again at $T < 90°C$ with *d*-values decreasing from 4.80 to 4.52 nm. For $T < 66°C$, a marked orientation distribution around substrate, normal and weak lateral correlation peaks occur again.

In difference to the XR results (*d*-values measured at room temperature), the values obtained by T-GISAXS are online acquired data. The latter correspond much better than XR with those found by T-SAXS (from bulk behavior). Additionally, the effect of orientation of the *sf* layers parallel to the Si wafer surface is strongly increased if the sample temperature was raised to $T > T_m$.

The phase separation in P(MMA-*co*-sfMA-H2F8) both in bulk and thin films can be enhanced by changing the polymer architecture into diblock copolymers PMMA-*b*-PsfMA-H2F8 [140]. Details of this study under consideration of a further phase separation caused by the diblock of immiscible segments will be discussed separately.

11.4 CONSEQUENCES OF SELF-ORGANIZATION FOR THE MACROSCOPIC WETTING BEHAVIOR

Wetting and adhesion are two of the most important macroscopic properties of polymers considered for coatings. These properties are also important for blends (physical mixtures of polymers) either on their surface or at the interfaces. In this chapter, only wetting on materials by liquids on their outer surface is discussed.

In the last part, the ability of polymers (also methacrylate copolymers) with self-organizing *sf* side chains H10F10 and, limitedly, also H2F8, to form well-ordered surfaces and the structural change with temperature was demonstrated. Very early it was shown by Zisman [141] that chemical composition as well as order on the very top surface of a film has clear consequences on its wetting behavior. The surface free energy (in the work of Zisman surface tension γ_c) decreases in dependence of the chemistry in the following order of γ_c: $CH_3 > CH_2 > CF_2 > CF_3$, assumed that surface is occupied with the given groups, which could be shown in other early works [142]. Bernett and Zisman [143] outlined that fluorinated side chains have a higher impact to lower the surface tension than fluorinated segments in the main chain of a polymer. Working with SAMs with different chemistry, Whitesides and coworkers (e.g., Reference 144) proved the conclusions by Zisman et al. and moreover showed that self-organization in the top layer of a surface generated by well-ordered SAMs, which were prepared by Langmuir–Blodgett technique, additionally had an impact on the wetting behavior. The lowest even possible surface free energies were found on well-ordered polyelectrolyte–surfactant complexes with sodium[bis(perfluoro-hexylethyl)2-sulfosuccinate] [145] or perfluorodecanoic acid [146, 147], where the authors detected mesomorphous (LC) order. Similar observations were made on surfaces of SAMs from *sf* thiols on gold [127]. Wang et al. [62] were the first to show that the surface structure of BCPs with *sf* side chains underlies the same conditions as SAMs. Wang derived a correlation in *sf* polymers between the order of CF_3 groups at the top surface and the critical surface tension γ_c obtained

TABLE 11.8 Correlation Between Length of *sf* Side Chain in PS-PI Diblock Copolymers, Surface Structure, and Surface Tension as Obtained by Wang et al. [65]

Sf side chain H*n*F*m*	Surface order at room temperature	γ_c (mN/m)
$n > 4, m > 10$	Crystalline	6
$n > 4, m = 8$	Smectic B	8–9
$n > 4, m = 6$	Smectic A	10–11
$n > 4, m < 6$	Isotropic	12–16

by a Zisman plot. Wang's results are summarized in Table 11.8. The surface tensions decreased with increasing degree of order at the surface.

If the degree of order depends on the temperature which is true for the ordered phases discussed, then also the wetting behavior depends on the temperature. This was reported by Sheiko et al. [148] also for H2F8 and H6F10 homo- and copolymethacrylates, indicating the reduction of CAs after isotropization of the surface.

We found a comparable behavior on surfaces of aromatic polyesters bearing H10F10 side chains [130] with the only difference that they had different transition temperatures compared to Sheiko's copolymethacrylates. The ordered structure on this H10F10 polyester was proven by NEXAFS and X-ray reflectivity [61]. All polyesters with H10F10 side chains independent of the nature of the main chain have once more extremely low surface free energy, as depicted in Figure 11.21 for aromatic–aliphatic polyesters with varying number of CH_2 groups in the main chain. While the nonsubstituted isophthalate polyesters surface free energies range between 30 and 35 mN/m, H10F10-substituted polyesters with comparable main chain all have surface free energies of about 10 mN/m (calculated by Neumann's equation-of-state [149]). All these surfaces are self-organized. This is illustrated by the fact that the

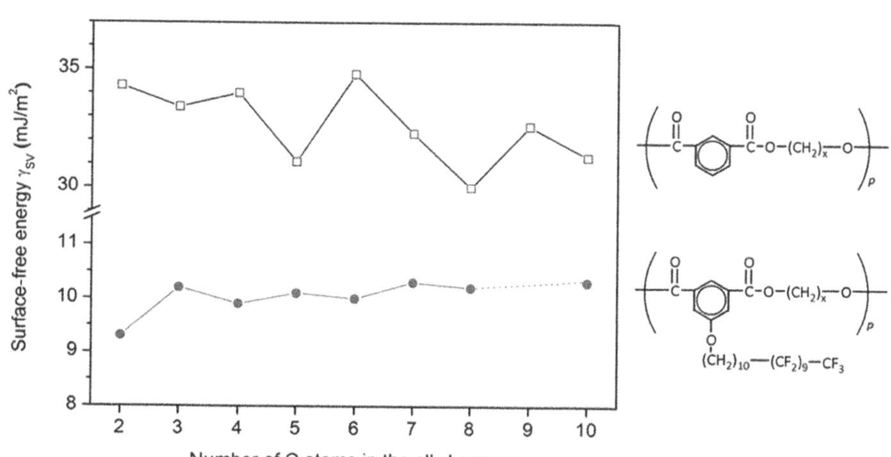

FIGURE 11.21 Surface free energies of aromatic–aliphatic polyesters with different spacer length in the main chain and H10F10 side chains.

space requirement for the repeat units of the polymers as measured by Langmuir–Blodgett technique by means of pressure–area isotherms on water (polymers spread from trifluoroacetic acid/chloroform (1/4 vol/vol) mixtures as reported in Reference 150) does only less depend on the number of alkyl groups in the main chain. The required space per basic unit is 37.5 Å^2 for $x = 2$, 41.0 Å^2 for Ph and goes only slightly up to 51 Å^2 for $x = 10$. From those results, it becomes obvious that the main influence on self-aggregation comes from the *sf* side chains.

The absolutely necessary base for drawing the above discussed conclusions are accurately measured by advancing CAs with different measuring liquids which are correlated to the surface free energy by Young's equation [149]. Further requirements are very smooth surfaces with roughness in the nanometer range, chemical homogeneity of the surface, and correct measurements of the CA. CA measurements were subsequently improved over time, for instance, using the ADSA (Axisymmetric drop-shape analysis) technique which takes the shape of the three-phase contact line and the whole drop into account [149, 151]. Additionally, it has to be noted that besides correctness of the experimental measurement the mode of calculation, that is, the theoretical concept to calculate the surface free energy from the CA, can yield different results [145]. In our work [120], exclusively Neumann's equation-of-state was applied to calculate surface free energies. Note that the surface free energies γ_s obtained by Neumann's equation-of-state were found to be 1–3 mN/m higher than dispersive surface free energies γ_s^d obtained by Girifalco–Good–Fowkes–Young equation, as shown by Thünemann et al. [147]. Thus, before comparing values given in literature, one should always make sure that comparable experimental conditions and calculations exist, which in most cases are not given. Also the preparation conditions as well as the thermal history have to be taken into account and can influence the result dramatically.

Now, the surface properties of *sf* polymethacrylates with respect to differences and similarities will be discussed. Figure 11.22 shows the surface image of a blend from PMMA with PsfMA-H2F8 homopolymer. Due to the high interaction parameter χ between both, the blend is thermodynamically immiscible and macrophase separation on the micrometer scale can be recognized in the Raman imaging picture. The parts consisting of PMMA are brighter than the parts of PsfMA-H2F8. CAs on such demixed surfaces are not stable, do not give clear results, and depend strongly on preparation and annealing conditions.

Macroscopic phase separation and nonstable surfaces are main reasons to synthesize and use random copolymers instead of employing blends.

The advancing CAs of both water and hexadecane on surfaces of *sf*-polymethacrylate copolymers with different side chains and molar content of the *sf*-methacrylate show clear influences of both (Figures 11.23a and 11.23b). Note that all thin films on silicon wafer with very low roughness as ensured by ellipsometry and X-ray reflectivity were prepared under comparable conditions by spin-coating from BFMB or chloroform, followed by drying in a vacuum oven at 140°C for several hours.

Copolymers with nonself-organizing side chains H1F1 and H2F4 show a continuous increase of water and hexadecane advancing CAs with semifluorinated molar

FIGURE 11.22 Surface of a homopolymer blend PMMA/PsfMA-H2F8 (70/30), mixed by solution casting from BFMB into methanol, dried, and melt-pressed. Raman imaging (area: 175 µm × 175 µm, measuring point distance 2 µm) of the film surface. White: PMMA represented by the CH_2 rocking vibration at 820 cm^{-1}; black: PsfMA-H2F8 represented by the CF_2 symmetric stretching vibration at 740 cm^{-1}.

content. In the H1F1 series, only the homopolymer reaches certain hydrophobicity (CA of about 90°, very comparable to that of polypropylene [152]). The H2F4 homopolymer is found in the range of PTFE (CA of about 110°C), the CA of copolymers are lower. H2F8 copolymers reach earlier than expected from their self-organizational behavior very high CA of about 120°. Self-organization in the bulk and at the surface can be proven with about 70–80 mol% of *sf* comonomer, while constant water CA of 120° are observed at about 50 mol% sfMA-H2F8. This may be caused by the annealing conditions of the film; annealing at 100°C allows first the surface enrichment of fluorinated segments (well known from literature, e.g., Reference 153), and afterward self-organization of the side chains upon cooling to room temperature. This behavior is detected both in bulk (in T-SAXS) as well as in thin films (in GISAXS). The same is true for H10F10 copolymethacrylates. The onset of self-organization in bulk as well as at the surface was found here at 27 mol% sfMA-H10F10. The water CAs indeed reveal the highest value achievable on a smooth surface (124°) exactly for the sample with 27 mol% sfMA-H10F10. Increasing concentration of H10F10 does not alter them anymore. The same observation can be made for the hexadecane advancing CA. Only self-organized samples (the same as discussed above) show CA of about 80°, a value which is very high and allows to claim significant oleophobicity of the surfaces, too.

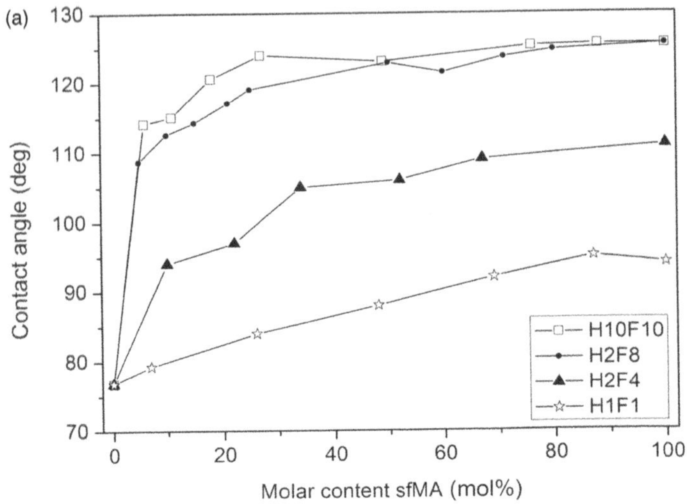

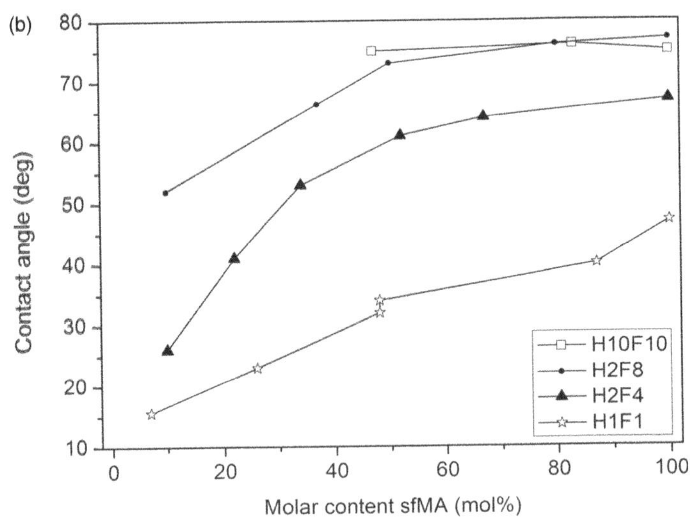

FIGURE 11.23 Dynamic contact angles—comparison of *sf* P(MMA-*co*-sfMA-H*n*F*m*) copolymers with different *sf* side chains. Advancing contact angles with (a) water and (b) hexadecane.

In summary, it has to be stated that self-organization is the required assumption for polymer surfaces on which extremely high water and hexadecane CAs are found (i.e., extremely low surface free energies). Thus, they provide the basis for coatings with enhanced antisticky, antigraffiti, and antiadhesion properties. Self-organization can prevent surface reconstruction which changes hydrophobic properties dramatically after long exposure time to the respective liquids [154]. *Sf* copolymers without self-organizing side chains show only moderate hydrophobic and olephobic properties.

11.5 POSSIBLE APPLICATIONS OF SEMIFLUORINATED METHACRYLATES

Semifluorinated compounds X-H*n*F*m*-Y which are not bound to polymer backbones are perfect nonionic amphiphilic surfactants [155]. These substances form micelles and arrange at the interface between polar and nonpolar liquids. They are currently applied in widespread fields, for instance, in pharmaceutical compounds for topical administration, as carrier for the delivery of active ingredients (anti-inflammatory agents, retoids, immunosuppressants) into deep layers of the skin [156], or in formulations to increase the oxygen transport capacity of the blood [157]. *Sf* amphiphiles act here as liquid vehicles with significantly higher density than physiologic liquids, opening applications also in ophthalmology for eye treatment [158], for example, to reduce high intraocular pressure [159], which works well due to the extremely low surface tensions [141]. *Sf* compounds have also been reported to be useful excipients for inhalative drug delivery when brought into aerosol form [160]. Impressively, they find applications as carriers for drug targeting in acute respiratory failure causing pulmonary recruitment in acute lung injury, for example, in intensive care, due to their high lipophilicity [161]. Self-organized *sf* alkanes can also be incorporated into phospholipids and form vesicles [162]. These applications indicate that *sf* compounds show a certain biocompatibility, or at least bioinertness.

Biomaterial applications. The biocompatibility of *sf* compounds should give rise to comparable behavior when linked to polymers. Thus, use of *sf* poly(meth)acrylates with short, not self-organizing *sf* side chains, for biomedical applications is reported, too: copolymers of 2-methacryloyloxyethyl phosphorylcholine (MPC) and sfMA-H1F1 were prepared by ATRP to combine the biocompatible properties of MPC with the hydrophobicity of sfMA-H1F1 in biomaterials with reduced adsorption of protein (bovine plasma fibrinogen and bovine serum γ-globulin) [163]. In contrast, our results for protein adsorption (human serum albumin and fibrinogen) on surfaces composed of *sf* polyester materials did not prove this concept and showed high protein adsorption on hydrophobic surfaces [164], as reported also before [165] and explained by enhanced hydrophobic–hydrophobic interaction between material and protein. Fluoromethacrylate copolymers obviously do not show toxicity because they find use in consumer-close products in cosmetics industry to provide water-repellant and makeup-holding properties to cosmetic powders [166], in copolymers for powder foundation [167] to cause reduced wetting of the treated skin by water and salina, to yield low water uptake to dental composites [168, 169], and even for restorative dentistry and treatment of tooth surfaces to inhibit leakage of liquids into capillary spaces [170]. Copolymers of MMA with mixtures of different sfMA monomers, mainly 1H,1H,2H,2H-perfluoroheptyl-MA and 1H,1H,2H,2H-perfluorononyl-MA, were processed into contact lenses with RI of 1.4 and high CO_2 permeability [171]. Copolymerization of hexafluoropropyl-MA with PEG-MA enhanced the oxygen permeability of hard contact lenses [172]. Also soft hydrogel contact lenses often contain fluoromethacrylate comonomers, demonstrated in Reference 173 for copolymers of hexafluoroisopropyl methacrylate (0.6 wt%), 2-hydroxyethyl methacrylate (79.4 wt%), *N*-vinylpyrrolidone (10 wt%), and methacrylic acid (10 wt%).

All these applications would not be possible if *sf* compounds possessed toxicity or bioincompatibility. On the other hand, other perfluorinated compounds, such as the most prominent examples perfluorooctane sulfonate and perfluorooctanoic acid (PFOA), are under heavy discussion with regard to biotoxicity, and bioaccumulation in living organisms and environment [174]. Bioaccumulation of perfluorinated substances with eight and more carbons is reasoned by the fact that the degradation times are very long. For PFOA, a half-life in the human body of 3.8 years is reported [175]. The advantage of semifluorinated compounds over perfluorinated ones can be seen in that CH_2 groups adjacent to CF_2 allow degradation of perfluoroalkyl segments by biodegradation with certain microorganisms step by step [176]. These results suggest that even H*n*F8 and H*n*F10 may be degraded stepwise under suitable conditions. However, shortening of the perfluorinated part while maintaining self-organization of *sf* side chains is a goal consistent with more sustainability for this type of polymers.

Antibiofouling applications. The term "biofouling" covers a wide range of processes and is usually used for generation and colonization of living and/or artificial structures on surfaces. The environment of these surfaces defines the term used [177]: marine biofouling, freshwater fouling, fouling of biomaterials by body fluids, weathering, and membrane fouling, to name just a few. One of the first steps in fouling processes is almost always the adsorption of proteins, pointing out the high importance of those. Briand [177] pointed out that no biofouling assay would ever be able to reflect all the complicated processes going on. The results obtained will always depend on kind and concentration of foulers, conditions, and even from the experience of the experimenter. Thus, it is extremely unlikely to achieve results which can truly be compared, only rough trends may be derived.

Among the variety of antibiofouling materials, for example, silver, zinc, antimony, and copper compounds, quaternary ammonium compounds [178], organic biocides [177, 179], and enzymes [180], fluoropolymers play a particular role. Older literature claims a remarkable antibiofouling effect of fluorinated and *sf* polymers which was mainly explained by the prevention of adhesion of foulers due to the low surface free energy. Linder, for instance, demonstrated reduced marine biofouling on surfaces coated with polysiloxanes, polymethacrylates, and polyacrylates with perfluorinated side chains [181, 182]. Yoshino et al. [183, 184] reported that surface modifiers for teeth consisting of *sf* silanes could, in a first step, inhibit plaque formation and, in a second step, were able to support plaque detachment from glass and resin composite surfaces *in vitro*. Modern developments [185] try to combine several functions in one material: (a) biopassive layers to avoid settlement of proteins, bacteria, micro- and macrofoulers; (b) biocides: if foulers succeed to settle, they will be killed by biocides, and finally, (c) fouling release: the dead material will be released by soft and hydrophobic parts with low E-modulus. In the latter, *sf* polymers (or segments in polymers) play a major role. Marabotti et al. [186] demonstrated the effect of poly(dimethyl siloxane)-modified poly(methyl methacrylate-*co*-1H,1H,2H,2H-perfluorodecyl acrylate) on the reduction of the settlement of *Ulva* sporelings and adult barnacles. Park et al. [187] employed copolymers with randomly distributed hydrophilic (PEG) and hydrophobic (H10F10) side chains to suppress attachment of *Ulva* sporelings and diatoms. However, it was shown to work best on surfaces with high content of PEG.

In our work, we followed the concept of combination of different properties in one material for antibiofouling coatings for glass and metal in freshwater applications. Random copolymers with monomers giving different functions to the coating were prepared [188]—MMA as first comonomer to give mechanical stability, sfMA-H2F8 as second comonomer to add hydrophobic properties in combination with low E-modulus because it is rather soft to introduce fouling-release properties, 2-acetoacetoxyethylmethacrylate (AAMA) as third comonomer with the ability to form complexes with copper salts to act as biocide, and 4-benzoylphenylmethacrylate as fourth, UV-cross-linkable comonomer to enhance the stability of coatings further. Coatings stable after UV irradiation in freshwater for longer time were obtained. Before exposure to different foulers, the cross-linked coatings were treated with solutions of copper acetate in THF to generate copper complexes between the acetoacetate units and copper ions, followed by thoroughly rinsing with water. Thereby, the concentration of toxic copper ions (the biocide) can be kept very low. The copper complex should prevent uncontrolled release into the environment, in contrast to pure physical mixtures in copper paints. Table 11.9 illustrates the results obtained after exposure of several coatings to *Staphylococcus aureus* in freshwater.

Coverage of glass with standard PMMA already lowers the number of grown *S. aureus* cells by 40%. On the copolymer with 80 mol% sfMA-H2F8 (Copo1) without complexing units only 15% of the original *S. aureus* concentration is detected, thus proving that the highly hydrophobic surface (advancing water CA: 126°) indeed yields reduced bacteria adhesion and growth. Additional incorporation of 10 mol% AAMA in Copo3 leads to further drop of the number of adhered cells to 2.2% of the original value. Here, an additional influence of the *sf* methacrylate cannot be proven directly. It seems that the presence of the copper complex which could be detected by XPS is more important. However, a general positive influence of the sfMA comonomer seems to exist.

Ultrahydrophobic surfaces. Ultrahydrophobicity, synonymously called "superhydrophobicity," "Lotus® effect," or "self-cleaning effect" was first discovered and observed on the lotus plant *(Nelumbo nucifera)* by Barthlott et al. [189], but is not restricted to this. "Ultrahydrophobicity" means that drops of water do not wet the surface. If this surface is only slightly tilted, the drops roll away, taking with adhered particles (dust) or whatever sticks to the surface (in this sense, it is not a real "self-cleaning" because external liquids are required). Since the very first discovery and even before a clear scientific understanding of ultrahydrophobicity was provided, many different strategies were created to transfer the effect from biology to technical products. Some among them are use of nano- and microstructured surfaces with hydrophobic top-coatings [190], combination of roughness created by *in-situ* sol–gel procedures with hydrophobic components [191], creation of roughness by core–shell particles with hydrophobic shell [192], creation of roughness by latex particles (resulting in the well-known "Lotusan®" paint [193]), and others. Hydrophobic components are in most of the cases different fluoropolymers, making them once more very important for everything which is related to nonwetting, nonsticking, and so on. At almost the same time, the conditions for ultrahydrophobicity expressed by physical parameters were found out. While Barthlott et al. [189] already showed in his first reports that a hierarchic surface roughness, that is, a macro-roughness covered

TABLE 11.9 Antibacterial Effect of Methacrylate Copolymers as Probed with *Staphylococcus aureus*

Sample	MMA (mol%)	AAMA (mol%)	sfMA-H2F8 (mol%)	BPMA (mol%)	Advancing water contact angle (deg)	*S. aureus* (10^4 cells/cm^2)	Comment
Glass control	–	–	–	–	20	440	Purified
PMMA control	100	0	0	0	78	258	Nontreated with Cu^{2+}
Copo1	10	0	80	10	126	66	Without Cu, only *sf*
Copo2	80	10	0	10	84	2.0	With Cu^{2+}, no *sf*
Copo3	10	10	80	0	126	5.6	With Cu^{2+} and *sf*

FIGURE 11.24 Ultrahydrophobicity: a drop of water on an ultrahydrophobic P(MMA-*co*-sfMA-H10F10) surface (O. Fischer, IPF Dresden).

by a micro- to nano-roughness is necessary to achieve the Lotus effect. Further work indicated [194] that the optimal surface is a fractal surface [195], which can mathematically be described by the parameter "fractal dimension" with a required value of at least 2. The optimal coverage of this surface is a hydrophobic, mostly fluorinated polymer. Structuring of PTFE surfaces by plasma also yields ultrahydrophobicity [196].

Sf polymers (in this case *sf* polyester with H10F10 side chains) were also used as top-coatings on aluminum surfaces preroughened by anodic oxidation [197]. These surfaces revealed extremely high water CAs ($\sim$150°) without CA hysteresis (as another basic requirement for ultrahydrophobicity), together with low roll-off angles. Nakajima et al. [198] showed that CA hysteresis and sliding angle are proportional assumed that the deformation of the water droplet at sliding is constant. Similar effects were achieved by using the *sf* methacrylate copolymers reported here. Figure 11.24 illustrates a drop of water on an ultrahydrophobic PMMA-*co*-sfMA-H10F10 surface on top of anodic oxidized aluminum doped with 1 wt% magnesium. Although the durability of the coatings is still not high enough at the time present, the photograph outlines the potential of the copolymers. The durability may be increased in the same manner as for the antibiofouling surfaces by incorporation of cross-linkable stabilizing groups. This concept was followed by Belov [68] using maleic anhydride and itaconic acid cross-linkable units. The stability may also depend on the stability of the substrate (roughened aluminum seems to be mechanically not too stable).

It has to be noted that all technical products which claim to show the "Lotus effect" suffer from certain drawbacks. One of the most important is the fact that these surfaces do not show the same behavior against oily liquids which destroys the effect significantly. Moreover, incrustation by algae cannot be suppressed, again reflecting the importance of this topic. Thus merging these two scientific topics and enhancing

TABLE 11.10 Refractive Indices of *sf* Methacrylate Homopolymers Measured on Thin Films at λ = 632.8 nm by Ellipsometry

Polymer	Film thickness (nm)	Refractive index	Reference
PMMA	165	1.495	This work
PsfMA-H1F1	155	1.418	This work
PsfMA-H2F4	203	1.384	This work
PsfMA-H2F8	104	1.391	This work
PsfMA-H10F10	202	1.413	This work
PTFE	(bulk)	1.35 ... 1.38	[a]
Highly fluorinated CHF layers after plasma polymerization	50	1.38	[b]

[a]From Reference 203.
[b]From Reference 204.

the durability of the coating by cross-linking represents an interesting topic for the future.

Refractive properties. As in the case of contact lenses [199], the special refractive properties of fluoromethacrylate copolymers allow application as plastic optical fibers [200] and optical waveguides. The application potential of fluoromethacrylate containing copolymers for graded-index plastic optical fibers was recently demonstrated in Reference 201 using copolymers of trifluoroethyl-MA and 2,5-bis(trifluoromethyl)styrene. Copolymerization and variation of comonomer concentrations allow to tune the RI widely (see e.g., Reference 202). Most of the described fluoropolymers are highly transparent.

RI of thin films of the sfMA copolymers under study with film thicknesses between 100 and 200 nm on silicon wafer were measured at λ = 632.8 nm with ellipsometry as described in Reference 204 to find out which range can be covered. The results are summarized in Table 11.10.

With the *sf* homo- and copolymers listed in Table 11.10, RI values between that of PMMA (1.495) and that reported for plasma-deposited CHF layers (1.380) [204] can be covered depending on the composition selected. The latter is close to the value reported for PTFE with perfluoropolyether loops [205].

Fluoromethacrylate copolymers as lubricants. As all the known fluorine-containing polymers, also fluorinated methacrylate copolymers show the typical lubricating behavior which is originated by surfaces densely occupied by fluorine atoms [13]. Test specimens prepared of butyl acrylate-*co*-3,3,4,4,5,5,6,6,7,7,8,8,9,9,10,10-heptadecafluorodecyl methacrylate-*co*-MMA copolymers demonstrated high lubricity durability [206]. 1H,1H-pentadecafluorooctyl methacrylate monomer was added to a lubricating oil (to 1–2 wt%) or to gasoline (0.1–0.5 wt%). The monomers are presumed to polymerize in the engine, thus forming lubricant coatings which finally reduced fuel consumption by 1.9–7.3% [207].

Sf methacrylate copolymers as compatibilizers for polymer blends with PMMA. Both, the calculation of interaction parameters and experimental results showed

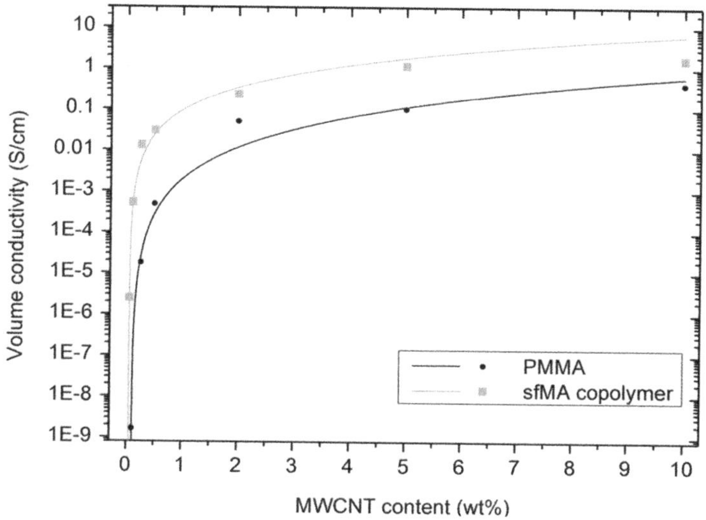

FIGURE 11.25 Electrical volume conductivity of (a) PMMA and (b) P(MMA-*co*-sfMA-H2F8) (30/70 mol/mol) with different contents of MWCNT Nanocyl NC3150. Melt-pressed samples with thickness of 0.5 mm.

the macroscopic demixing of sfMA homopolymers and PMMA. Thus it results in strongly demixed surfaces as already illustrated in Figure 11.22. The surface properties of such surfaces is characterized by high heterogeneity, that means such mixtures cannot be used successfully as coatings. It could be shown that random copolymers of MMA and sfMA have enhanced compatibility with PMMA, but no thermodynamic miscibility. Thus, they may be considered as compatibilizers to improve adhesion of coatings containing semifluorinated constituents on PMMA. Details will be published separately.

Sf methacrylate copolymers as matrix for multiwalled carbon nanotubes. Under suitable preparation conditions (from common solvent), the electrical percolation threshold of nanocomposites of P(MMA-*co*-sfMA-H2F8) (30/70 mol/mol) with multiwalled carbon nanotubes (MWCNT) (type Nanocyl™ NC3150) is found to be below 0.1 wt%, that means extremely low compared to other polymer matrices. Details will be published elsewhere. The electrical volume conductivities of PMMA and a P(MMA-*co*-sfMA-H2F8) copolymer with MWCNT are plotted in Figure 11.25 versus the concentration of MWCNT. At similar contents of MWCNT, the percolation threshold of P(MMA-*co*-sfMA-H2F8)/MWCNT is lower and the electrical conductivity is higher than that of PMMA/MWCNT nanocomposites. We discuss this in terms of better compatibility of the fluorinated polymer with MWCNT, which was also indicated in former work [208].

The main advantage of the *sf* copolymer over PMMA is their limited solubility. At higher concentrations of *sf* monomer, they are not soluble anymore in solvents like THF, acetone, chloroform, and others. Thus, they can be utilized as MWCNT

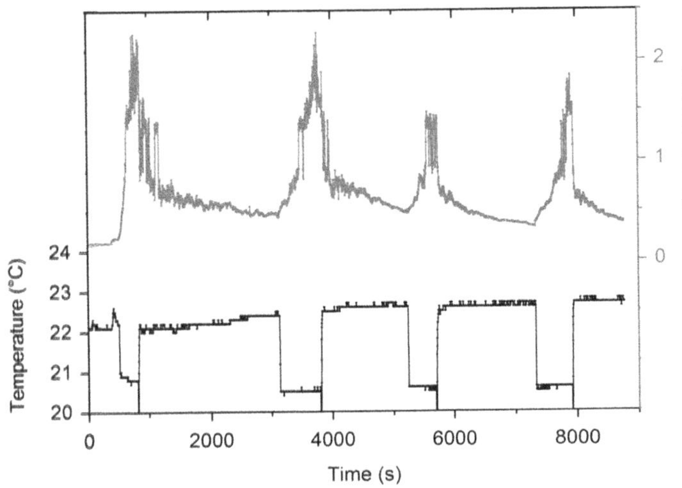

FIGURE 11.26 Liquid sensing with PsfMA-H2F8 homopolymer with 5 wt% MWCNT: electrical conductivity in repeated cycles of immersing into chloroform followed by receding.

nanocomposite for liquid [209] and gas [210] sensing applications of these solvents, without damaging the polymer matrix. Figure 11.26 shows the conductivity of a PsfMA-H2F8 homopolymer nanocomposite with 5 wt% MWCNT in repeated cycles of immersing into chloroform followed by receding. The curves reflect the high potential of this class of materials.

Self-healing properties. Due to the unique surface ordering behavior and reorientation behavior, *sf*-containing diblock copolymers have been proposed after simulations to provide self-healing behavior [211]. Although the term "self-healing" implies self-detection of damages and self-repair after detection [212], at least a certain potential for thermally self-repairing [213] coatings by using the self-organization as well as very low melt viscosity of *sf* copolymethacrylates is foreseen. This potential should be explored more intensively in the future.

11.6 SUMMARY AND OUTLOOK

Polymers with long *sf* side chains are characterized by a distinct twofold phase separation between polymer backbones and *sf* side chains as well as between alkyl- and perfluoroalkyl part of the semifluorinated alkane side groups. Long, symmetric H10F10 substituents provide the highest degree of phase separation, while shortening to H2F8 and its asymmetry results in considerably weaker structural order, both in bulk and at surfaces. This weaker order is not immediately reflected in the measured advancing CAs which are still high. However, it is argued that less ordered surfaces are subject to surface reconstruction after certain exposure time.

Regarding the widespread applications of low molar mass *sf* compounds and for *sf* polymers, it can be emphasized that polymers with self-organizing groups will open an even wider range of future prospects. The suitability of the methacrylate copolymers with self-organized *sf* side chains discussed in this contribution for ultrahydrophobic, self-cleaning surfaces, coatings with tailored RI, and as matrix for nanocomposites with carbon nanotubes could be demonstrated.

In summary, the use of the property combinations as well as synergies of polymer backbone (here polymethacrylate) and substituent in semifluorinated copolymers allows to design a very special type of materials with unique property profile, thus proceeding on the way of a science for tailor-made innovative materials [214].

ACKNOWLEDGMENTS

Financial support of projects about semifluorinated materials by Leibniz Institut für Polymerforschung Dresden e.V., DAAD (German Exchange Service, Vigoni and Ikyda Programs), SMWK (Sächsisches Staatsministerium für Wissenschaft und Kunst), and DESY Hamburg, Hasylab, is gratefully acknowledged.

The authors are grateful to Prof. C. K. Ober, Cornell University, Ithaca, NY, for inspiration to start this work during the 1-year visit of D.P. at Cornell University, Mat. Sci. & Eng., which raised subsequently a lot of scientific pleasure to all people involved. Fantastic scientific collaborations enhanced our knowledge about semi-fluorinated polymers over the years. We would like to thank here in particular our partners at DESY Hamburg, Hasylab, Dr S. Funari and Dr J. Perlich for the possibility to perform measurements at beamlines A2 and BW4 and for scientific support, Prof C. Papadakis, Universität München, Germany, Prof. P. Pissis and his group, National Technical University of Athens, Greece, Prof. C. Schick and his group, University of Rostock, Germany, Prof. F. Pilati/Prof. M. Messori, and their group, University of Modena and Reggio Emilia, Italy, Prof. R. Bongiovanni, Prof. M. Sangermano, Polytechnic Turin, Italy.

Furthermore, we would like to express many thanks to our colleagues at IPF for long-lasting and fruitful collaboration—Dr. H. Komber (NMR spectroscopy), Dr. P. Friedel (simulations), L. Häußler and K. Arnhold (thermal analysis), Dr. K. Grundke (wetting measurements), Dr. F. Simon (surface analysis by XPS), Dr. K.-J. Eichhorn (ellipsometry), and Dr. P. Pötschke and Dr. B. Krause (MWCNT nanocomposites, liquid sensing).

This work would not have been realized without the valuable experimental work of Mrs. M. Dittrich, Mrs. Dipl.-Ing. K. Eckstein and Mr. A. Korwitz – many thanks.

We also like to thank all the students that contributed with their experimental work and graduation thesis. Many thanks to Oliver Fischer (Diploma Thesis, TU Dresden, 2009), Andreas Berndt (BEng Thesis, University of Applied Sciences Lausitz, Senftenberg, 2010), Larissa de Morais Schmittgens (MSc Thesis, TU Dresden, 2011), Felix Müller, BSc Thesis (TU Dresden, 2011), Romy Selinger (Internship Thesis, HTW Dresden, 2010), Tobias Bünker (Diploma Thesis, TU Dresden, 2012), Sebastian Stein (BSc Thesis, TU Dresden, 2012).

The group of Prof S. Beuermann (especially Dr M. Drache, Mrs Abisheva, Dr R. Vukicevic), Technical University of Clausthal, Germany) is gratefully acknowledged for performing experiments with supercritical CO_2 (at University of Potsdam) and for radical copolymerization of trifluoroethyl-methacrylate and 1H,1H,2H,2H-perfluorohexyl methacrylate copolymers with MMA. Thanks to Dr M. Opitz and R. Kruspe (IDUS Biologisch Analytisches Umweltlabor GmbH, Ottendorf-Okrilla, Germany) for excellent collaboration and biofouling tests.

REFERENCES

1. Hadjichristidis, N. *Block Copolymers: Synthetic Strategies, Physical Properties, and Applications*; John Wiley & Sons Inc.: Chichester, UK, 2002.
2. Hadjichristidis, N.; Pitsikalis, M.; Pispas, S.; Iatrou, H. *Chem. Rev.* **2001**, *101*, 3747–3792.
3. Matyjaszewski, K. *Macromolecular Engineering: Precise Synthesis, Materials Properties, Applications—Volume 1. Synthetic Techniques*; Wiley-VCH: Weinheim, Germany, 2007.
4. Bates, F. S.; Fredrickson, G. H. *Phys. Today* **1999**, *52*, 32–38.
5. Rosedale, J. H.; Bates, F. S.; Almdal, K.; Mortensen, K.; Wignall, G. D. *Macromolecules* **1995**, *28*, 1429–1443.
6. Park, C.; Yoon, J.; Thomas, E. L. *Polymer* **2003**, *44*, 6725–6760.
7. Leibler, L. *Macromolecules* **1980**, *13*, 1602–1617.
8. Cheng, J. Y.; Ross, C. A.; Smith, H. I.; Thomas, E. L. *Adv. Mater.* **2006**, *18*(19), 2505–2521.
9. Nandan, B.; Kuila, B. K.; Stamm, M. *Eur. Polym. J.* **2011**, *47*, 584–599.
10. Kim, H. C.; Park, S. M.; Hinsberg, W. D. *Chem. Rev.* **2010**, *110*, 146–177.
11. Albert, J. N. L.; Epps, T. H. *Mater. Today* **2010**, *13*, 24–33.
12. Yoon, J.; Lee, W.; Thomas, E. L. *MRS Bull.* **2005**, *30*, 721–726.
13. Thomas, R. R. Material Properties of Fluoropolymers and Perfluoroalkyl-Based Polymers. In *Fluoropolymers, 2. Properties*; Hougham, G., Cassidy, P., Johns, K., Davidson, T., Eds.; Kluwer Academic/Plenum Publishers: New York, 1999; pp 47–67.
14. Scheirs, J. Structure/Property Considerations for Fluoropolymers and Fluoroelastomers to Avoid In-service Failure. In *Modern Fluoropolymers, High Performance Polymers for Diverse Applications*; Scheirs, J., Ed.; John Wiley & Sons: Chichester, UK, 1997; pp 1–69.
15. Hildebrand, J. H.; Scott, R. L. *The Solubility of Nonelectrolytes*, 3rd ed.; Van Nostrand-Reinhold: Princeton, NJ, 1950.
16. Pospiech, D.; Gottwald, A.; Jehnichen, D.; Friedel, P.; John, A.; Harnisch, C.; Voigt, D.; Khimich, G.; Bilibin, A. Yu. *Colloid Polym. Sci.* **2002**, *280*, 1027–1037.
17. Van Krevelen, D. W. *Properties of Polymers, Their Correlation with Chemical Structure; Their Numeric Estimation and Prediction from Additive Group Contributions*. Elsevier: Amsterdam, The Netherlands, 1990; pp 189–224.
18. Benoit, H.; Hadziioannou, G. *Macromolecules* **1988**, *21*, 1449–1464.
19. Friedel, P.; John, A.; Pospiech, D.; Jehnichen, D.; Netz, R. R. *Macromol. Theory Simul.* **2002**, *11*(7), 785–793.

20. Benoit, H.; Hadziioannou, G. *Macromolecules* **1988**, *21*, 1449–1464.

21. Russell, T. P.; Rabolt, J. F.; Twieg, R. J.; Siemens, R. L.; Farmer; B. L. *Macromolecules* **1986**, *19*, 1135–1143.

22. Napoli, M. *J. Fluor. Chem.* **1996**, *79*, 59–69.

23. Grest, G. S.; Pierce, F.; Perahia, D. Molecular Dynamics Simulations of Responsive Semi-fluorinated Interfaces. *Polym. Prepr. Am. Chem. Soc. Div. Polym. Mater.: Sci. & Eng.* 2010, PMSE-323 (240th ACS Natl. Meeting, Boston, MA, USA, August 22–26, 2010).

24. Wilson, L. M. *Macromolecules* **1995**, *28*, 325–330.

25. Sivaniah, E.; Genzer, J.; Hexemer, A.; Kramer, E.; Xiang, M.; Xuefa, L.; Ober, C. K.; Magonov, S. *Macromolecules* **2008**, *41*(24), 9940–9945.

26. Ober, C. K.; Xiang, M.; Char, K.; Genzer, J.; Sivaniah, E.; Kramer, E. J.; Fisher, D. *Polym Prepr. Am. Chem. Soc. Div. Polym. Mat.: Sci. & Eng.* **1999**, *80*, 416–417.

27. Percec, V.; Bera, T. K. *Tetrahedron* **2002**, *58*(20), 4031–4040.

28. Xiang, M.; Li, X.; Ober, C. K.; Char, K.; Genzer, J.; Sivaniah, E.; Kramer, E. J.; Fischer, D. A. *Macromolecules* **2000**, *33*, 6106–6119.

29. Pospiech, D.; Jehnichen, D.; Häußler, L.; Voigt, D.; Grundke, K.; Ober, C. K.; Körner, H.; Wang, J. *Polym. Prepr. Am. Chem. Soc. Div. Polym. Chem.* **1998**, *39*(2), 882–883.

30. Wilson, L. M.; Griffin, A. C. *Macromolecules* **1994**, *27*, 1928–1931.

31. Thomas, R. R.; Anton, D. R.; Graham, W. F.; Darmon, M. J.; Sauer, B. B.; Stika, K. M.; Swartzfager, D. G. *Macromolecules* **1997**, *30*, 2883–2890.

32. Ho, T.; Wynne, K. J. *Macromolecules* **1992**, *25*, 3521–3527.

33. Dessipri, E.; Tirrell, D. A.; Atkins, E. D. T. *Macromolecules* **1996**, *29*, 3545–3551.

34. Percec, V.; Lee, M. *J. Macromol. Sci. Pure Appl. Chem.* **1992**, *A29*(9), 723–740.

35. Höpken, J.; Möller, M. *Macromolecules* **1992**, *25*, 1461–1467.

36. Castelvetro, V.; Aglietto, M.; Montagnini di Mirabello, L.; Toniolo, L.; Peruzzi, R.; Chiantore, O. *Surf. Coat. Int.* **1998**, *11*, 551–556.

37. Höpken, J.; Faulstich, S.; Möller, M. *Mol. Cryst. Liq. Cryst.* **1992**, *210*, 59–73.

38. Appelhans, D.; Wang, Z.-G.; Zschoche, S.; Zhuang, R.-C.; Häußler, L.; Friedel, P.; Simon, F.; Jehnichen, D.; Grundke, K.; Eichhorn, K.-J.; Komber, H.; Voit, B. *Macromolecules* **2005**, *38*, 1655–1664.

39. Schneider, H. A.; Steinhauser, N.; Mülhaupt, R. *Polym. Bull.* **1994**, *32*, 339–346.

40. Chapman, T. M.; Gribbin Marra, K. *Macromolecules* **1995**, *28*, 2081–2085.

41. Ameduri, B.; Boutevin, B. *Well-Architectured Fluoropolymers: Synthesis, Properties and Applications*; Elsevier: Amsterdam, The Netherlands, 2004; pp 353.

42. Raihane, M.; Ameduri, B. *J. Fluor. Chem.* **2006**, *127*, 391–399.

43. Guyot, B.; Ameduri, B.; Boutevin, B.; Melas, M.; Viguier, M.; Collet, A. *Macromol. Chem. Phys.* **1998**, *199*, 1879–1885.

44. Bongiovanni, R.; Malucelli, G.; Lombardi, V.; Priola, A.; Siracusa, V.; Tonelli, C.; Di Meo, A. *Polymer* **2001**, *42*, 2299–2305.

45. Bongiovanni, R.; Lombardi, V.; Priola, A.; Tonelli, C.; Di Meo, A. *Surf. Coat. Int.B: Coat. Transact.* **2003**, *86*(B1), 53–57.

46. Bongiovanni, R.; Malucelli, G.; Sangermano, M.; Priola, A. *Prog. Org. Coat.* **1999**, *36*, 70–78.

47. Casazza, E.; Mariani, A.; Ricco, L.; Russo, S. *Polymer* **2002**, *43*, 1207–1214.
48. Hung, M.-H.; Farnham, W. B.; Feiring, A. E.; Rozen, S. *J. Am. Chem. Soc.* **1993**, *115*, 8954–8959.
49. Sangermano, M.; Bongiovanni, R.; Priola, A.; Pospiech, D. *J. Polym. Sci.A: Polym. Chem.* **2005**, *43*, 4144–4150.
50. Feiring, A. E. *J. Org. Chem.* **1985**, *50*, 3269–3274.
51. Ren, Y.; Lodge, T. P.; Hillmyer, M. A. *Macromolecules* **2001**, *34*, 4780–4787.
52. Wilson, L. M.; Griffin, L. M. *Macromolecules* **1993**, *26*, 6312–6314.
53. Twieg, R. J.; Rabolt, J. F. *Macromolecules* **1988**, *21*, 1806–1811.
54. Höpken, J.; Möller, M.; Boileau, S. *New Polym. Mater.* **1991**, 2(4), 339–356.
55. Krupers, M. J.; Möller, M. *J. Fluor. Chem.* **1997**, *82*, 119–124.
56. Ameduri, B.; Bongiovanni, R.; Lombardi, V.; Pollicino, A.; Priola, A.; Recca, A. *J. Polym. Sci. A: Polym. Chem.* **2001**, *39*(24), 4227–4235.
57. Bongiovanni, R.; Beamson, G.; Mamo, A.; Priola, A.; Recca, A.; Tonelli, C. *Polymer* **2000**, *41*, 409–414.
58. Bongiovanni, R.; Malucelli, G. A.; Tonelli, C.; Simeone, G.; Priola, A. *Macromol. Chem. Phys.* **1998**, *199*, 1099–1105.
59. Mitsunobu, O. The Use of Diethyl Azodicarboxylate and Triphenylphosphine in Synthesis and Transformation of Natural Products. In *A Review of the Mitsunobu Reaction*; Georg Thieme Verlag: Stuttgart, Germany, 1981; pp 1–28.
60. Nemoto, N.; Miyata, F.; Nagase, Y.; Abe, J.; Hasegawa, M.; Shirai, Y. *Macromolecules* **1996**, *29*, 2365–2371.
61. Pospiech, D.; Häußler, L.; Eckstein, K.; Voigt, D.; Jehnichen, D.; Gottwald, A.; Kollig, W.; Janke, A.; Grundke, K.; Werner, C.; Kricheldorf, H. R. *Macromol. Symp.* **2001**, *163*, 113–126.
62. Wang, J.; Ober, C. K. *Macromolecules* **1997**, *30*, 7560–7567.
63. Gabor, A. H.; Lehner, E. A.; Long, T. E.; Mao, G.; Rauch, E. C.; Schell, B. A.; Ober, C. K. *Polym. Prepr. Am. Chem. Soc. Div. Polym. Chem.* **1993**, *3*, 284–285.
64. Iyengar, D. R.; Perutz, S. M.; Dai, C. A.; Ober, C. K.; Kramer, E. J. *Macromolecules* **1996**, *29*, 1229–1234.
65. Wang, J.; Mao, G.; Ober, C. K.; Kramer, E. J. *Macromolecules* **1997**, *30*, 1906–1914.
66. Jin, L.; Xu, Q. *Jingxi Huagong Bianjibu* **2006**, *23*(7), 683–686.
67. Krupers, M.; Slangen, P.-J.; Möller, M. *Macromolecules* **1998**, *31*, 2552–2558.
68. Belov, N. V. Reactive Fluoropolymers, Synthesis and Application. Ph.D. Thesis, DWI at RWTH Aachen. http://www.dwi.rwth-aachen.de/; 2011.
69. Beuermann, S.; Buback, M.; Gadermann, M.; Günzler, F.; Jürgens, M. *Macromol. Symp.* **2004**, *206*(1), 229–240.
70. Schreiber, U.; Hosemann, B.; Beuermann, S. *Macromol. Chem. Phys.* **2011**, *212*, 168–179.
71. Weibel, G.; Lewis, H. G. P.; Gleason, K.; Ober, C. K. *Polym. Prepr. Am. Chem. Soc. Div. Polym. Chem.* **2000**, *41*(2), 1838–1839.
72. Rastogi, A.; Paik, M. Y.; Ober, C. K. *ACS Appl. Mater. Interfaces* **2009**, *1*(9), 2013–2020.
73. Sundararajan, N.; Yang, S.; Ogino, K.; Valiyaveetil, S.; Wang, J.; Zhou, X.; Ober, C. K. *Chem. Mater.* **2000**, *12*(1), 41–48.

74. DeSimone, J. M.; Guan, Z.; Elsbernd, C. S. *Science* **1992**, *257*, 945–947.

75. Guan, Z.; Combes, J. R.; Menceloglu, Y. Z.; DeSimone, J. M. *Macromolecules* **1993**, *26*, 2663–2669.

76. Hsiao, Y. L.; Maury, E. E.; DeSimone, J. M. *Macromolecules* **1995**, *28*, 8159–8166.

77. Hwang, H. S.; Kim, H. J.; Jeong, Y. T.; Gal, Y.-S.; Lim, K. T. *Macromolecules* **2004**, *37*, 9821–9825.

78. Storsberg, J.; Ritter, H. *Macromol. Chem. Phys.* **2002**, *203*(5/6), 812–818.

79. Krupers, M.; Möller, M. *Macromol. Chem. Phys.* **1997**, *198*, 2163–2179.

80. Ishizone, T.; Sugiyama, K.; Sakano, Y.; Mori, H.; Hirao, A.; Nakahama, S. *Polym. J.* **1999**, *31*(11–12), 983–988.

81. Hirao, A.; Sugiyama, K.; Yokoyama, H. *Prog. Polym. Sci.* **2007**, *32*(12), 1393–1438.

82. Patten, T. E.; Matyjaszewski, K. *Adv. Mater.* **1998**, *10*(2), 901–915.

83. Rosen, B. M.; Percec, V. *Chem. Rev.* **2009**, *109*, 5069–5119.

84. Malmström, E. E.; Hawker, C. J. *Macromol. Chem. Phys.* **1998**, *199*(6), 923–935.

85. Hawker, C. J.; Hedrick, J. L. *Macromolecules* **1995**, *28*, 2993–2995.

86. Patten, T. E.; Matyjaszewski, K. *Acc. Chem. Res.* **1999**, *32*, 895–903.

87. Hansen, N. M. L.; Jankova, K.; Hvilsted, S. *Eur. Polym. J.* **2007**, *43*, 255–293.

88. Ameduri, B. *Macromolecules* **2010**, *43*(24), 10163–10184.

89. Perrier, S.; Jackson, S. G.; Haddleton, D. M.; Ameduri, B.; Boutevin, B. *Tetrahedron* **2002**, *58*, 4053–4059.

90. Hussain, H.; Busse, K.; Budde, H.; Kressler, J. *Macromol. Chem. Phys.* **2003**, *204*(7), 936–946.

91. Hussain, H.; Budde, H.; Höring, S.; Busse, K.; Kressler, J. *Macromol. Chem. Phys.* **2002**, *203*(14), 2103–2112.

92. Xia, J. H.; Johnson, T.; Gaynor, S. G.; Matyjaszewski, K.; DeSimone, J. *Macromolecules* **1999**, *32*(15), 4802–4805.

93. Jakubowski, W.; Min, K.; Matyjaszewski, K. *Macromolecules* **2006**, *39*, 39–45.

94. Lacroix-Desmazes, P.; Andre, P.; DeSimone, J. M.; Ruzette, A. V.; Boutevin, B. *J. Polym. Sci. A: Polym. Chem.* **2005**, *42*(14), 3537–3552.

95. Skrabania, K.; von Berlepsch, H.; Böttcher, C.; Laschewsky, A. *Macromolecules* **2010**, *43*, 271–281.

96. Marsat, J.-N.; Heydenreich, M.; Kleinpeter, E.; von Berlepsch, H.; Böttcher, C.; Laschewsky, A. *Macromolecules* **2011**, *44*, 2092–2105.

97. Skrabania, K.; Laschewsky, A.; von Berlepsch, H.; Böttcher, C. *Langmuir* **2009**, *25*(13), 7594–7601.

98. Ma, Z.; Lacroix-Desmazes, P. *J. Polym. Sci. A: Polym. Chem.* **2004**, *42*(10), 2405–2415.

99. Pospiech, D.; Jehnichen, D.; Gottwald, A.; Häußler, L.; Scheler, U.; Friedel, P.; Kollig, W.; Ober, C. K.; Li, X.; Hexemer, A.; Kramer, E. J.; Fischer, D. A. *Polym. Prepr. Am. Chem. Soc. Div. Polym. Mater.: Sci. & Eng.* **2001**, *84*, 314–315.

100. Alvares, D. R. S.; Menezes, S. M. C.; Lucas, E. F. *Polym. Int.* **2004**, *53*, 1639–1643.

101. De Morais Schmittgens, L.; Pospiech, D.; Komber, H.; Voit, B. *Manuscript in prep.*

102. Kelen, T.; Tüdos, F. *J. Macromol. Sci. Chem.* **1975**, *A9*(1), 1–27.

103. Tüdös, F.; Kelen, T.; Foldesberezsnich, T.; Turcsanyi, B. *J. Macromol. Sci. Chem.* **1976**, *A10*(8), 1513–1540.

104. Abisheva, Z.; Drache, M.; Beuermann, S. Personal communication.

105. Cracowski, J. M.; Montembault, V.; Ameduri, B. *J. Polym. Sci. A: Polym. Chem.* **2010**, *48*, 2154–2161.

106. Haddleton, D. M.; Crossman, M. C.; Dana, B. H.; Duncalf, D. J.; Heming, A. M.; Kukulj, D.; Shooter, A. J. *Macromolecules* **1999**, *32*, 2110–2119.

107. Li, K.; Wu, P.; Han, Z. *Polymer* **2002**, *43*, 4079–4086.

108. Stamm, M., Ed. *Polymer Surfaces and Interfaces—Characterization, Modification and Application*; Springer-Verlag: Berlin, Germany, 2008.

109. Galli, G.; Andruzzi, L.; Chiellini, E.; Li, X.; Ober, C. K.; Hexemer, A.; Kramer, E. J. *Surf. Coat. Int. B: Coat. Trans.* **2004**, *87*(2), 77–82.

110. Seidel, A., Ed. *Characterization and Analysis of Polymers*; John Wiley & Sons, 2008.

111. Nozue, Y.; Shinohara, Y.; Amemiya, Y. *Polym. J.* **2007**, *39*(12), 1221–1237.

112. Kremer, F.; Schönhals, A., Eds. *Broadband Dielectric Spectroscopy*. Springer-Verlag: Berlin, Germany, 2003.

113. Beamson, G.; Alexander, M. R. *Surf. Interface Anal.* **2004**, *36*, 323–333.

114. Grundke, K.; Pospiech, D.; Kollig, W.; Simon, F.; Janke, A. *Colloid Polym. Sci.* **2001**, *279*, 727–735.

115. Hähner, G.; Kinzler, M.; Wöll, Ch.; Grunze, M.; Scheller, M. K.; Cederbaum, L. S. *Phys. Rev. Lett.* **1991**, *67*(7), 851–854.

116. Li, X.; Andruzzi, L.; Chiellini, E.; Galli, G.; Ober, C. K.; Hexemer, A.; Kramer, E. J.; Fischer, D. A. *Macromolecules* **2002**, *35*, 8078–8087.

117. Busch, P.; Rauscher, M.; Smilgies, D.-M.; Posselt, D.; Papadakis, C. M. *J. Appl. Cryst.* **2006**, *39*, 433–442.

118. Busch, P.; Posselt, D.; Smilgies, D. M.; Rauscher, M.; Papadakis, C. M. *Macromolecules* **2007**, *40*, 630–640.

119. Minakov, A. A.; Schick C. *Rev. Sci. Instr.* **2007**, *78*(7), 073902–073910.

120. Grundke, K. Wetting, Spreading and Penetration. In *Handbook of Applied Surface and Colloid Chemistry*; Holmberg, K., Ed.; John Wiley & Sons: Chichester, UK, 2001; pp 119–142.

121. Tavana, H.; Yang, G.; Yip, C. M.; Appelhans, D.; Zschoche, S.; Grundke, K.; Hair, M. L.; Neumann, A. W. *Langmuir* **2006**, *22*, 628–636.

122. Broniatowski, M.; Minones, J.; Dynarowicz-Latka, P. *J. Coll. Interface Sci.* **2004**, *279*(2), 552–558.

123. Broniatowski, M.; Dynarowicz-Latka, P.; Witko, W. *J. Fluor. Chem.* **2005**, *126*(1), 79–86.

124. Krafft, M. P. *Acc. Chem. Res.* **2012**, *45*(4), 514–524.

125. De Viguerie, L.; Keller, R.; Jonas, U.; Berger, R.; Clark, C. G.; Klein, C. O.; Geue, T.; Müllen, K.; Butt, H. J.; Vlassopoulos, D. *Langmuir* **2011**, *27*(14), 8776–8786.

126. Chinwangso, P.; Lee, T. R. Self-assembled Nanoscale Coatings Derived from Unsymmetrical Partially Fluorinated Spiroalkanedithiols. *Polym. Prepr. Am. Chem. Soc. Div. Polym. Chem.* 2010, COLL-165 (239th ACS Natl. Meeting, San Francisco, CA, USA, March 21–25, 2010).

127. Fukushima, H.; Seki, S.; Nishikawa, T.; Takiguchi, H.; Tamada, K.; Abe, K.; Colorado, R.; Graupe, M.; Shmakova, O. E.; Lee, T. R. *J. Phys. Chem.* **2000**, *B104*, 7417–7423.

128. Höpken, J.; Möller, M. *Macromolecules* **1992**, *25*, 2482–2489.

129. Sheiko, S.; Turetskii, A.; Höpken, J.; Möller, M. Molecular Organization of Polystyrene and Polymethylmethacrylate With Fluorocarbone Side Chains. In *Macromol. Engineering—Recent Advances*; Mishra, M. K. et al., Eds.; Plenum Press: New York, 1995; pp 219–227.

130. Jehnichen, D.; Pospiech, D.; Janke, A.; Friedel, P.; Häußler, L.; Gottwald, A.; Kummer, S.; Kollig, W.; Grundke, K. *Mater. Sci. Forum* **2001**, *378–381*(2), 378–382.

131. Volkov, V. V.; Platé, N. A.; Takahara, A.; Kajiyama, T.; Amaya, N.; Murata, Y. *Polymer* **1992**, *33*(6), 1316–1320.

132. Friedel, P.; Pospiech, D.; Jehnichen, D.; Bergmann, J.; Ober, C. K. *J. Polym. Sci. B: Polym. Phys.* **2000**, *38*(12), 1617–1625.

133. Jehnichen, D.; Pospiech, D.; Häußler, L.; Friedel, P.; Gottwald, A.; Kummer, S. *Mater. Sci. Forum* **2004**, *443–444*(2), 223–226.

134. Gottwald, A.; Pospiech, D.; Jehnichen, D.; Häußler, L.; Friedel, P.; Pionteck, J.; Stamm, M.; Floudas, G. *Macromol. Chem. Phys.* **2002**, *203*(5–6), 854–861.

135. Tsuwi, J.; Hartmann, L.; Kremer, F.; Pospiech, D.; Jehnichen, D.; Häußler, L. *Polymer* **2006**, *47*(20), 7189–7197.

136. Pospiech, D.; Komber, H.; Voigt, D.; Jehnichen, D.; Häußler, L.; Gottwald, A.; Kollig, W.; Eckstein, K.; Baier, A.; Grundke, K. *Macromol. Symp.* **2003**, *199*, 173–186.

137. Tsuwi, J.; Pospiech, D.; Jehnichen, D.; Häußler, L.; Kremer, F. *J. Appl. Polym. Sci.* **2007**, *105*(1), 201–207.

138. Jehnichen, D.; Pospiech, D.; Häußler, L.; Friedel, P.; Funari, S. S.; Tsuwi, J.; Kremer, F. *Z. Kristallogr. Suppl.* **2007**, *26*, 605–610.

139. Jehnichen, D.; Friedel, P.; Selinger, R.; Korwitz, A.; Wengenmayr, M.; Berndt, A.; Pospiech, D. *Powder Diffr.* **2013**, *28*(S2), 144–160.

140. Jehnichen, D.; Pospiech, D.; Friedel, P.; Funari, S. S. *Z. Kristallogr. Proc.* **2011**, *1*(2), 487–492.

141. Zisman, W. A. Relation of the Equilibrium Contact Angle to Liquid and Solid Constitution. In *Contact Angle, Wettability, and Adhesion*; Fowkes, F. M., Ed.; Advances in Chemistry; American Chemical Society: Washington, DC, 1964; pp 1–51.

142. Grajeck, E. J.; Petersen, W. H. *Text. Res. J.* **1962**, *32*, 320–331.

143. Bernett, M. K.; Zisman, W. A. *J. Phys. Chem.* **1962**, *66*, 1207–1208.

144. Bain, C. D.; Whitesides, G. M. *Angew. Chem.* **1989**, *101*(4), 522–528.

145. Thünemann, A. F.; Lochhaas, K. H. *Langmuir* **1999**, *15*, 4867–4874.

146. Thünemann, A. F.; Lochhaas, K. H. *Langmuir* **1998**, *14*, 4898–4903.

147. Thünemann, A. F.; Lochhaas, K. H. *Langmuir* **1999**, *15*, 6724–6727.

148. Sheiko, S.; Lermann, E.; Möller, M. *Langmuir* **1996**, *12*, 4015–4024.

149. Tavana, H.; Neumann, A. W. *Adv. Colloid Interf. Sci.* **2007**, *132*, 1–32.

150. Pospiech, D.; Jehnichen, D.; Komber, H.; Voigt, D.; Häußler, L.; Friedel, P.; Grundke, K.; Busch, P.; Große, H.; Kremer, F.; Tsuwi, J.; Menzel, H.; Helmecke, O. *e-Polymers* **2005**, *E_002*, 1–2.

151. Tavana, H.; Jehnichen, D.; Grundke, K.; Hair, M. L.; Neumann, A. W. *Adv. Colloid Interf. Sci.* **2007**, *134–135*, 236–248.
152. Schulze, U.; Pospiech, D.; Komber, H.; Häußler, L.; Voigt, D.; Eschner, M. *Eur. Polym. J.* **2008**, *44*, 694–703.
153. Bottini, F. A.; Di Pasquale, G.; Pollicino, A.; Pilati, F.; Toselli, M.; Tonelli, C. *Macromolecules* **1998**, *31*, 7814–7819.
154. Ober, C. K.; Wang, J.; Kramer, E. J. Semifluorinated Side Chain-Containing Polymers. US Patent 5,907,017, 1999.
155. Kissa, E. Fluorinated Surfactants: Synthesis, Properties, Applications. In *Surfactant Science Series*; Schick, M. J., Fowkes, F. M., Eds.; Marcel Dekker, Inc.: New York, 1994; Vol. 50, p 362.
156. Guenther, B.; Theisinger, B.; Theisinger, S.; Scherer, D. Topical Pharmaceutical Composition Based on Semifluorinated Alkanes. PCT Int. Appl. WO 2012160179 A2 20121129; 2012.
157. Seiffge, D. J.; Lapina, N. E.; Tsagogiorgas, C.; Theisinger, B.; Henning, R. H.; Schilling, L. *Exp. Neurol.* **2012**, *237*(1), 18–25.
158. Wilson, C. G. Liquid Pharmaceutical Composition Comprising Semi-fluorinated Alkane Carriers for the Treatment of a Posterior Eye Disease. Europ. Pat. Appl. EP 2462921 A1 20120613, 2012
159. Theisinger, B.; Theisinger, S.; Guenther, B. Pharmaceuticals Containing Semifluorinated Alkane Carriers for Treatment of Increased Intraocular Pressure. PCT Int. Appl. WO 2011113855 A2 20110922, 2011.
160. Tsagogiornas, C.; Jung, T.; Krebs, J.; Theisinger, B.; Beck, G.; Yard, B. A.; Quintel, M. *Int. J. Pharm.* **2012**, *422*(1–2), 194–201.
161. Dembinski, R.; Bensberg, R.; Marx, G.; Rossaint, R.; Quintel, M.; Kuhlen, R. *Exp. Lung Res.* **2010**, *36*(8), 499–507.
162. Sabin, J.; Ruso, J. M.; Gonzalez-Perez, A.; Prieto, G.; Sarmiento, F. *Colloids Surf.* **2006**, *47*(1), 64–70.
163. Inoue, Y.; Watanabe, J.; Takai, M.; Ishihara, K. *J. Biomat. Sci. Polym. Ed.* **2004**, *15*(9), 1153–1166.
164. Bunk, J.; Pospiech, D.; Eichhorn, K.-J.; Müller, M.; Werner, C.; Bellmann, C.; Simon, F.; Pleul, D.; Grundke, K. *Colloids Surf. A: Physicochem. Eng. Aspects* **2010**, *362*, 47–57.
165. Haynes, C. A.; Norde, W. *Colloids Surf.* **1994**, *B2*, 517–566.
166. Nagashima, K.; Morita, M.; Kubo, M. (Daikin Industries, Ltd., Japan). Cosmetic Powders Treated With Reactive Fluoro(meth)acrylate Copolymer, and Their Manufacture. Jpn. Kokai Tokkyo Koho, JP 2000290137, A 20001017, 2000.
167. Morita, M.; Kubo, M. (Daikin Industries Ltd., Japan). Copolymers for Cosmetics. PCT Int. Appl., WO 9855078, A1 19981210, 1998.
168. Washimi, A.; Yoshida, T.; Kimura, K. (Toa Gosei Chemical Industry Co., Ltd., Japan). Dental Cements With Fluorine-Containing (Meth)acrylates. Jpn. Kokai Tokkyo Koho, JP 03255011, A 19911113, 1991.
169. Craig, R. G.; Douglas, W. H. *Macromol. Symp.* **1982**, *28*, 386.
170. Craig, R. G.; Douglas, W. H. (University of Michigan, USA). Hydrophobic Composite Restorative Materials and Their Use in Tooth Treatment. US Patent, US 4292029, A 19810929, 1981.

171. Cleaver, C. S. (du Pont de Nemours, E. I., and Co., USA). Contact Lens Material. Britain. Patent, GB 1374670, 19741120, 1974.

172. Sawamoto, T.; Ishida, K.; Tarumi, J.; Izumitani, T. (Hoya Corp., Japan). Methacrylate Polymer Hard Contact Lenses With Improved Oxygen Permeability. Jpn. Kokai Tokkyo Koho, JP 63159820, A 19880702, 1988.

173. Salamone, J. C. Fluorine Containing Soft Contact Lens Hydrogels. PCT Int. Appl., WO 9211300, A1 19920709, 1992.

174. Lindstrom, A. B.; Strynar, M. J.; Libelo, E. L.; Field, J. A. *Environ. Sci. Technol.* **2011**, *45*, 7951–7953.

175. Lindstrom, A. B.; Strynar, M. J.; Libelo, E. L.; Field, J. A. *Environ. Sci. Technol.* **2011**, *45*, 7954–7961.

176. Wang, N.; Szostek, B.; Buck, R. C.; Folsom, P. W.; Sulecki, L. M.; Capka, V.; Berti, W. R.; Gannon, J. T. *Environ. Sci. Technol.* **2005**, *39*(19), 7516–7528.

177. Briand, F. *Biofouling* **2009**, *25*(4), 297–311.

178. Charnley, M.; Textor, M.; Acikgoz, C. *React. Funct. Polym.* **2011**, *71*, 329–334.

179. Fay, F.; Carteau, D.; Linossier, I.; Delbury, M.; Vallée-Réhel, K. *Colloids Surf. B: Biointerfaces* **2013**, *102*, 569–577.

180. Cordeiro, A.; Werner, C. *J. Adh. Sci. Technol.* **2011**, *25*, 2317–2344.

181. Lindner, E. Low Surface Free Energy Fouling Resistant Coatings. In *Recent Dev. Biofouling Control*; Thompson, M. F., Ed.; Balkema: Rotterdam, NL, 1994; pp 305–311.

182. Lindner, E. *Biofouling* **1992**, *6*, 193–205.

183. Yoshino, N.; Yamauchi, T.; Kondo, Y.; Kawase, T.; Teranaka, T. *React. Funct. Polym.* **1998**, *37*, 271–282.

184. Yoshino, N.; Teranaka, T. *J. Biomater. Sci. Polym. Ed.* **1997**, *8*(8), 623–653.

185. Krishnan, S.; Weinman, C. J.; Ober, C. K. *J. Mat. Chem.* **2008**, *18*, 3405–3413.

186. Marabotti, I.; Morelli, A.; Orsini, L. M.; Martinelli, E.; Galli, G.; Chiellini, E.; Lien, E. M.; Pettitt, M. E.; Callow, M. E.; Callow, J. A.; Conlan, S. L.; Mutton, R. J.; Clare, A. S.; Kocijan, A.; Donik, C.; Jenko, M. *Biofouling* **2009**, *25*(6), 481–493.

187. Park, D.; Weinman, C. J.; Finlay, J. A.; Fletcher, B. R.; Paik, M. Y.; Sundaram, H. S.; Dimitriou, M. D.; Sohn, K. E.; Callow, M. E.; Callow, J. A.; Handlin, D. L.; Willis, C. L.; Fischer, D. A.; Kramer, E. J.; Ober, C. K. *Langmuir* **2010**, *26*(12), 9772–9781.

188. Starke, S.; Bünker, T.; Pospiech, D.; Frenzel, R.; Simon, F.; Opitz, M.; Kruspe, R.; Voit, B. *Macromol. Chem. Phys.* **2013**, *214*(2), F69.

189. Barthlott, W.; Neinhuis, C. *Planta* **1997**, *202*(1), 1–8.

190. Thieme, M.; Frenzel, R.; Schmidt, S.; Simon, F.; Hennig, A.; Worch, H.; Lunkwitz, K.; Scharnweber, D. *Adv. Engin. Mat.* **2001**, *3*(9), 691–695.

191. Taurino, R.; Fabbri, E.; Messori, M.; Pilati, F.; Pospiech, D.; Synytska, A. *J. Coll. Interf. Sci.* **2007**, *325*, 149–156.

192. Synytska, A.; Ionov, L.; Grundke, K.; Stamm, M. *Langmuir* **2009**, *25*(5), 3132–3136.

193. www.lotusan.de (Ispo Co., now Sto AG, Germany), 2014.

194. Hennig, A.; Grundke, K.; Frenzel, R.; Stamm, M. *Tenside Surfact. Det.* **2002**, *39*, 243–246.

195. Synytska, A.; Stamm, M.; Grundke, K. *Polym. Mat. Sci. Eng.* **2006**, *51*, 194–195.

196. Grundke, K.; Nitschke, M.; Minko, S.; Stamm, M.; Froeck, C.; Simon, F.; Uhlmann, S.; Pöschel, K.; Motornov, M. Merging Two Concepts: Ultrahydrophobic Polymer Surfaces and Switchable Wettability. In *Contact Angle, Wettability and Adhesion;* Mittal, K. L., Ed.; VSP: Utrecht, The Netherlands, 2003; Vol. 3, 267–291.

197. Pospiech, D.; Jehnichen, D.; Gottwald, A.; Häußler, L.; Kollig, W.; Grundke, K.; Janke, A.; Schmidt, S.; Werner, C. *Surf. Coat. Int.B: Coat. Trans.* **2003**, *86*(B1), 43–52.

198. Nakajima, A.; Saiki, C.; Hashimoto, K.; Watanabe, T. *J. Mater. Sci. Lett.* **2001**, *20*, 1975–1977.

199. Fukuda, K.; Kuwabara, T. (Asahi Chemical Ind, Japan). Manufacture of Multifocal Contact Lens. Jpn. Kokai Tokkyo Koho, JP 05-313107, A 19931126 Heisei, 1993.

200. Yamamoto, T.; Shimada, K.; Onishi, H. (Mitsubishi Rayon K. K., Japan). Plastic Optical Fibers. Jpn. Kokai Tokkyo Koho, JP 04051205, A 199 20219 Heisei, 1992.

201. Teng, H.; Okamoto, Y. Copolymer Systems for Graded Index Plastic Optical Fibre Materials. *Polym. Prepr., Am. Chem. Soc. Div. Polym. Mater. Sci. Eng.* 2010, PMSE-280 (240th ACS Natl. Meeting, Boston, MA, USA, August 22–26, 2010).

202. Teng, H.; Wan, Z.; Koike, Y.; Okamoto, Y. *Polym. Adv. Technol.* **2013**, *24*(5), 520–523.

203. *Polymer Data Handbook*, James F. Mark, Ed.; Oxford University Press, Oxford, UK, 1999; pp 842–847.

204. Werner, C.; Eichhorn, K.-J.; Grundke, K.; Simon, F.; Grählert, W.; Jacobasch, H.-J. *Colloids Surf. A: Physicochem. Engin. Aspects* **1999**, *156*, 3–17.

205. Okamoto, Y. Perfluoropolymers Based on Perfluorodioxolane Compounds—A Most Transparent Polymer System. "Fluoropolymer 2012"—*Workshop of Am. Chem. Soc. Div. Polym. Chem.*, Las Vegas, CA, USA, October 14–17, 2012.

206. Toyoka, Y.; Matsumoto, K.; Koizumi, H.; Kaneda, M.; Okano, K. (Mitsubishi Rayon Co., Ltd., Japan). Fluoroacrylic Polymer having Lubricating Effect and Thermoplastic Resin Composition. Eur. Pat. Appl., EP 449274, A1 19911002, 1991.

207. Rowe, C. N. (Mobil Oil Corp., USA). Reducing Fuel Consumption With a Fluorinated Compound. US Patent 4394134, A 19830719, 1983.

208. Vukicevic, R.; Vukovic, I.; Stoyanov, H.; Korwitz, A.; Pospiech, D.; Kofod, G.; Loos, K.; ten Brinke, G.; Beuermann, S. *Polym. Chem.* **2012**, *3*, 2261–2265.

209. Villmow, T.; Pegel, S.; Pötschke, P.; Heinrich, G. *Polymer* **2011**, *52*, 2276–2285.

210. Pandis, C.; Peoglos, V.; Kyritsis, A.; Pissis, P. *Proc. Engin.* **2011**, *25*, 243–246.

211. Pierce, F.; Shtrestra, U.; Perashia, D.; Clarson, S. J.; Grest, G. S. Interfacial Behavior of Fluorine Containing Di-block Copolymers. Simulations and Experiments. Abstracts of Papers, POLY-026; 238th ACS Natl. Meeting, Washington, DC, USA (August 16–20, 2009).

212. Wu, D. Y.; Meure, S.; Solomon, D. *Progr. Polym. Sci.* **2008**, *33*, 479–522.

213. Fischer, H. *Nat. Sci.* **2010**, *2*(8), 873–901.

214. Ober, C. K.; Cheng, S. Z. D.; Hammond, P. T.; Muthukumar, M.; Reichmanis, E.; Wooley, K. L.; Lodge, T. P. *Macromolecules* **2009**, *42*, 465–471.

12

SYNTHESIS OF FLUOROPOLYMERS USING BORANE-MEDIATED CONTROL RADICAL POLYMERIZATION FOR ENERGY STORAGE APPLICATIONS

T.C. MIKE CHUNG
The Pennsylvania State University

12.1 INTRODUCTION

Fluoropolymers are commonly used in high-end applications, such as aerospace, automotive, textile finishing, and microelectronics [1]. They exhibit a unique combination of properties, including thermal stability, chemical inertness (acid and oxidation resistance), low water and solvent absorptivities, self-extinguishing, excellent weatherability, very interesting surface properties, and becoming important electric and electronic properties. However, fluoropolymers also have some drawbacks, including limited processibility, poor adhesion to substrates, limited cross-linking chemistry, and inertness to chemical modification, which limit their applications when interactive and reactive properties are paramount. Functionalization of fluoropolymers, having specific functional groups, has been a constant research interest in the past decades. Most of the research approaches have been focusing on copolymerization of fluorinated monomers with functional comonomers to form functional fluoro-copolymers containing pendent functional groups [2–6]. Few reports discussed the preparation of fluoropolymers containing terminal functional groups [7–10]. Recently, Saint-Loup

Handbook of Fluoropolymer Science and Technology, First Edition. Edited by Dennis W. Smith Jr., Scott T. Iacono, and Suresh S. Iyer.

et al. [11] prepared telechelic VDF/HFP elastomers containing two opposing hydroxy terminal groups by using hydrogen peroxide as an initiator. However, many side reactions occur in this polymerization, and the final product contains not only hydroxy terminal groups but also carboxylic acid terminal groups, as well as some unsaturated terminal groups.

Living radical polymerization provides a very useful method to prepare a wide range of polymers with well-defined molecular structures, that is, narrow molecular weight distribution, control molecular weight, and desirable polymer chain ends. Early attempts to realize living radical polymerization involved the concept of reversible termination of the growing polymer chains by iniferters [12, 13], such as *N,N*-diethyldithiocarbamate derivatives, with some success. The first living radical polymerization was observed in reactions involving a stable nitroxyl radical, such as TEMPO [14, 15], which does not react with monomers but forms a reversible end-capped propagating chain end. Usually, the reactions have to be carried out at an elevated temperature (>100°C) to obtain a sufficient concentration of propagating radicals for monomer insertion. Subsequently, several research groups have replaced the stable nitroxyl radical with transition metal species or reversible chain transfer agents as the capping agents to mediate living free radical systems. These polymerization reactions follow the mechanisms of atom transfer radical polymerization (ATRP) [16, 17] or reversible addition-fragmentation chain transfer (RAFT) [18], respectively. Overall, these systems have a central theme-reversible termination via equilibrium between the active and dormant chain ends at an elevated temperature.

In our group, we have been studying a new free radical initiation system based on the oxidation adducts of organoborane and oxygen, which contains boroxyl radical—a mirror image of the stable nitroxyl radicals—as shown in Scheme 12.1. Our early interest in the borane/oxygen radical initiator stemmed from the desire to develop a new effective route in the functionalization of polyolefins [19–22] (i.e., PE, PP) and block/graft copolymers [23–26], which has been a longstanding scientific challenge area with great potential for industrial applications.

The unexpected good control in the incorporation of borane groups to polyolefin by metallocene catalysis and the subsequent radical chain extension by the incorporated borane groups prompted us to examine this free radical polymerization mechanism in greater details. Several relatively stable borane-based radical initiators were discovered, which exhibited living radical polymerization characteristics, with a linear relationship between polymer molecular weight and monomer conversion [27] and producing block copolymers by sequential monomer addition [28]. This stable radical

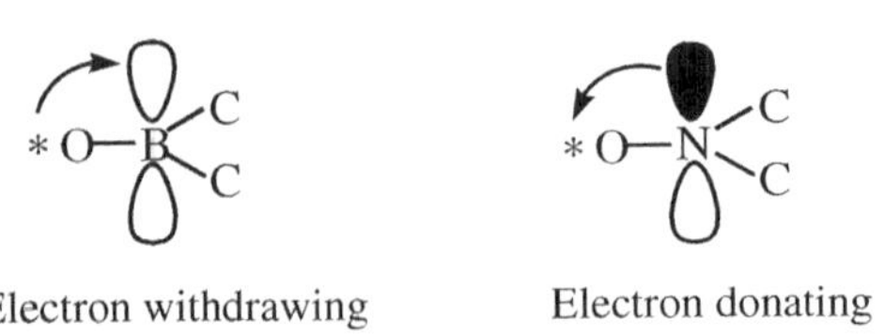

SCHEME 12.1 Boroxyl and nitroxyl radicals.

initiator system was recently extended to the polymerization of fluorinated monomers, which can effectively occur in bulk and solution conditions. Some interesting ferroelectric fluoro-terpolymers [29], showing large electromechanical response, have been prepared with high molecular weight and controlled polymer structure with narrow molecular weight and composition distributions. In this chapter, I will focus on the application of this stable borane initiator technology to prepare fluoropolymers having one or two terminal functional groups, which includes the published results in fluorinated acrylate [30, 31] and new observation in VDF-based polymers.

One of the interests in this borane/O_2 initiator is the preparation of fluoro-copolymers with well-controlled molecular structures that can be fabricated into the dielectric thin films in capacitors or proton exchange membranes (PEMs) for energy storage applications. Opposite to battery technology, which has high energy density and low power density, capacitors [32–34] usually exhibit high powder density but very low energy density that limits their applications. The capacitor energy density is directly governed by the dielectric material that separates the opposite static charges. The energy density can be estimated by the equation (J/cm^3) $= \frac{1}{2}\,\varepsilon\, E^2 = \frac{1}{2}\,\varepsilon\, (V/d)^2$, wherein ε is the dielectric constant of the dielectric and E is the applied electric field. The ideal dielectric material should exhibit high dielectric constant and high breakdown strength. On the other hand, the proton exchange membrane fuel cell (PEMFC)—a promising power resource without green house gas emission—has promoted considerable research interests in chemistry, physics, and theory [35–39]. The PEM is a key component in PEMFC, which serves as both an electrolyte and separator. Due to the demanding environment and functions, an ideal PEM material [35–37] requires a combination of chemical and physical properties; long-term chemical and electrochemical stability in the reducing environment at the cathode, and the harsh oxidative environment at the anode; good mechanical strength and dimension stability in tight PEM stacks; and high proton conductivity under various operation conditions (i.e., temperatures and relative humidity).

12.2 BORANE OXIDATION/POLYMERIZATION MECHANISM

Despite the known autoxidation reaction of trialkylborane (BR_3) by oxygen which results in quantitative formation of the three corresponding alcohols (R—OH) after hydrolysis [40, 41], the detailed oxidation mechanism is complicated. Following the initial formation of a B—O—O—C moiety, one suggested mechanism is asymmetric cleavage of the C—O bond to produce B—O—O* and C* radicals [42–44]. Some reports suggest an intramolecular rearrangement of the R_2B—O—O—R intermediate to form a RB(O—R)$_2$ molecule [45–47]. Others point to homolytic cleavage of the O—O bond to form B—O* and R—O* radicals [48,49]. The problem relates to the complexity of the oxidation products from three reactive B—C bonds and the formation of two possible moieties (B—O—R and B—O—O—R). The reaction intermediates are too unstable to permit isolation. In our study, we compared two oxidation adducts of $B(C_2H_5)_3$ and $B(OCH_3)(C_2H_5)_2$ and carried out the oxidation reactions in the presence of VDF monomers to form PVDF polymers. The idea is to capture the

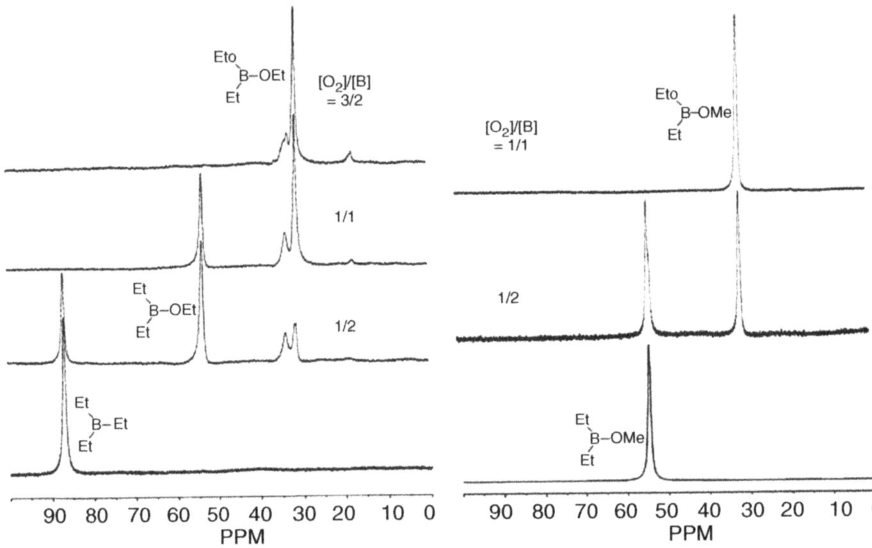

FIGURE 12.1 ^{11}B NMR spectra of the oxidation adducts of (left) $B(Et)_3$ and (right) $B(OMe)(Et)_2$ by varying amount of oxygen at ambient temperature.

unstable intermediate radicals that *in situ* initiate the polymerization and are incorporated in the beginning of polymer chains.

The oxidation reactions were conducted in a 100-mL flask equipped with a magnetic stirrer. After adding 2 mmol of borane, either $B(C_2H_5)_3$ or $B(OCH_3)(C_2H_5)_2$ and 20 mL ethyl ether under nitrogen, about 8 mmol of O_2 was then injected into the flask to start the oxidation reaction. The reaction was carried out at room temperature under vigorous agitation for 4 hours. The solution was directly subjected to ^{11}B NMR measurements [50]. Figure 12.1 compares two sets of the ^{11}B NMR spectra of the oxidation adducts of $B(Et)_3$ and $B(OMe)(Et)_2$ by varying amounts of oxygen at ambient temperature.

The oxidation adducts are quite complicated in $B(Et)_3$ yet are better controlled in $B(OMe)(Et)_2$. With an incremental amount of oxygen, two new chemical shifts are located at 56 and 32 ppm, corresponding to $B(OEt)(Et)_2$ and $B(OEt)_2Et$, and a minor peak at 36 ppm associated with B—O—B. Increasing $[O_2]/[B(Et)_3] > 1$, the $B(Et)_3$ completely disappears, and the main oxidation products become di-oxidation adducts with a very small amount of tri-oxidization one at around 19 ppm. On the other hand, the incremental oxidation of $B(OMe)(Et)_2$ with oxygen ($[O_2]/[B(OMe)(Et)_2] < 0.5$) is a clean reaction. Only a new chemical shift at 32 ppm for B(OMe)(OEt)Et is shown with the presence of the unreacted $B(OMe)(Et)_2$ at 56 ppm. After increasing oxygen concentration to a stoichiometric amount, the $B(OMe)(Et)_2$ is completely converted to B(OMe)(OEt)Et without any detectable tri-oxidization adducts. Almost no peroxyl species of B(O—O—Et)Et_2 (II) and B(OEt)(OOEt)Et (IV) were observed in both cases, although they should form before intermolecular reactions, as illustrated

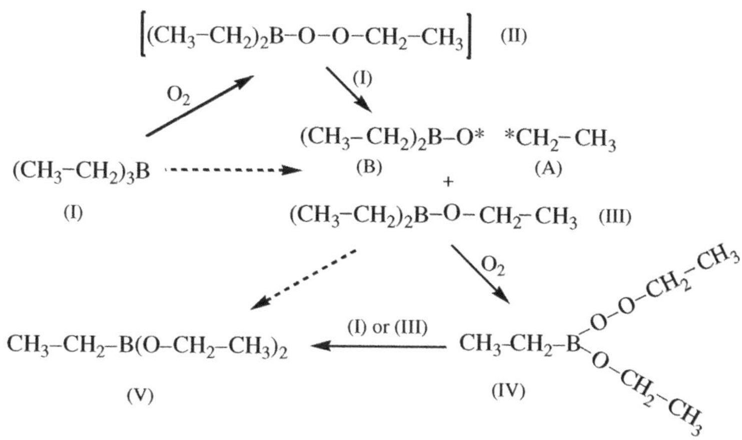

SCHEME 12.2 Oxidation mechanism of triethylborane.

in Scheme 12.2. In one-shot addition of excess oxygen ([oxygen]/[borane] = 4/1) to both $B(Et)_3$ and $B(OMe)(Et)_2$ they result in almost identical ^{11}B NMR spectra with a chemical shift at 34 ppm, corresponding to B(OR)(OOEt)(Et) (IV) intermediate species, which were captured before having a chance to engage in an intermolecular reaction with B—C bonds. However, the mono-oxidized peroxide $B(OOEt)Et_2$ (II) is too unstable to survive during this oxidation process.

The best way to observe the initial oxidation adduct of $B(OOEt)(Et)_2$ (II) and its subsequent reaction with $B(Et)_3$ (I) is to carry out this control oxidation in the presence of large amounts of VDF monomers ($[VDF]/[B(Et)_3] > 100$). The unstable intermediates containing free radicals during the intermolecular reaction should *in situ* initiate the radical polymerization to form poly(vinylidene fluoride) (PVDF) and deposit the adduct in the beginning of the PVDF chain. Table 12.1 summarizes the experimental results. The polymerization is effective at ambient temperature to obtain PVDF with an almost quantitative yield in 5 hours. Comparing a set (A-1–A-6), the molecular weight increases with the monomer conversion. However, plots of the polymer molecular weight vs. the monomer conversion, the line is above the theoretical line for living radical polymerization. Some uncontrolled radical polymerization may take place in the beginning of polymerization. The protecting borinate radical (B) (Scheme 12.1), may not have a sufficient quantity to mediate all propagating radicals. Usually, excess mediates are needed to have a successful control radical polymerization [51–53]. The polymer molecular weight reaches Mv > 63,000 g/mol in about 5 hours. The relatively slow propagating rate, compared to that of the regular free radical polymerization mechanism, removes the safety concern for heat transport and temperature control usually associated with bulk or solution polymerization of fluoromonomers. In addition, this process produces fluoropolymers with high purities, without any contaminants from surfactants or suspension agents in most of the commercial fluoropolymers, which are difficult to remove but detrimental in electric applications.

TABLE 12.1 Summary of VDF Polymerization Conditions and Results by $B(Et)_3/O_2$ Radical Initiator at Room Temperature

					End groups				Molecular weight	
Run	TEB/O_2 (mmol)	Time (h)	Yield (g)	Cov (%)	$CH_3CH_2{}^*$ CH_2CF_2 and CF_2CH_2	CH_3CH_2—O—	CF_2CH_2–OH	CF_2COX (X=F or OH)	Mn^a (NMR)	Mv^b
A-1	2.4/1.6	0.25	1.6	11.9	1.58 (95.2%)	0.08 (4.8%)	ND	ND	9,700	14,500
A-2	2.4/1.6	0.5	3.2	23.9	1.49 (96.1%)	0.06 (3.9%)	ND	ND	13,900	29,000
A-3	2.4/1.6	0.75	5.1	38.1	1.48 (95.5%)	0.07 (4.5%)	ND	ND	14,100	27,457
A-4	2.4/1.6	1	5.6	41.8	1.51 (96.8%)	0.05 (3.2%)	ND	ND	14,900	31,000
A-5	2.4/1.6	2.5	7.5	56.0	1.30 (97.0%)	0.04 (3.0%)	ND	ND	17,100	66,000
A-6	2.4/1.6	5	9.1	67.9	1.49 (98.0%)	0.03 (2.0%)	ND	ND	17,500	78,000
B-1	2.4/0.8	5	5.4	40.3	1.38 (96.5%)	0.05 (3.5%)	ND	ND	18,300	47,000
B-2	2.4/1.6	5	9.1	67.9	1.49 (98.0%)	0.03 (2.0%)	ND	ND	17,500	78,000
B-3	2.4/2.4	5	11.5	85.8	1.39 (92.7%)	0.11 (7.3%)	ND	ND	14,000	63,000
B-4	2.4/2.8	5	12.8	95.5	1.34 (91.2%)	0.13 (8.8%)	ND	ND	28,600	100,000
B-5	2.43.2	5	12.0	89.6	1.33 (85.8%)	0.15 (9.7%)	0.04 (2.6%)	0.03 (1.9%)	36,600	101,000
B-6	2.4/3.6	5	10.8	80.6	1.33 (82.6%)	0.17 (10.6%)	0.06 (3.7%)	0.05 (3.1%)	42,200	100,000
B-7	2.4/4.0	5	6.0	44.8	1.40 (64.2%)	0.20 (9.2%)	0.25 (11.5%)	0.33 (15.1%)	49,000	127,000
B-8	2.4/4.4	5	4.9	36.6	1.17 (40.2%)	0.34 (11.7%)	0.60 (20.6%)	0.80 (27.5%)	48,400	97,000
B-9	2.4/4.8	5	4.7	35.1	1.17 (41.5%)	0.37 (13.1%)	0.50 (17.7%)	0.78 (27.7%)	35,600	110,000
B-10	2.4/5.2	5	4.5	34.3	1.23 (62.4%)	0.27 (13.7%)	0.25 (12.7%)	0.22 (11.2%)	55,900	101,000
B-11	2.4/6.0	5	0	0	–	–	–	–	–	–

Source: Redrawn from Reference 50. Copyright 2006 American Chemical Society.

ND, Not detectable.

[a]Mn is based on the chain end group analysis, obtained from NMR spectrum.

[b]Mv is determined by intrinsic viscosity in dimethylacetamide at 25°C.

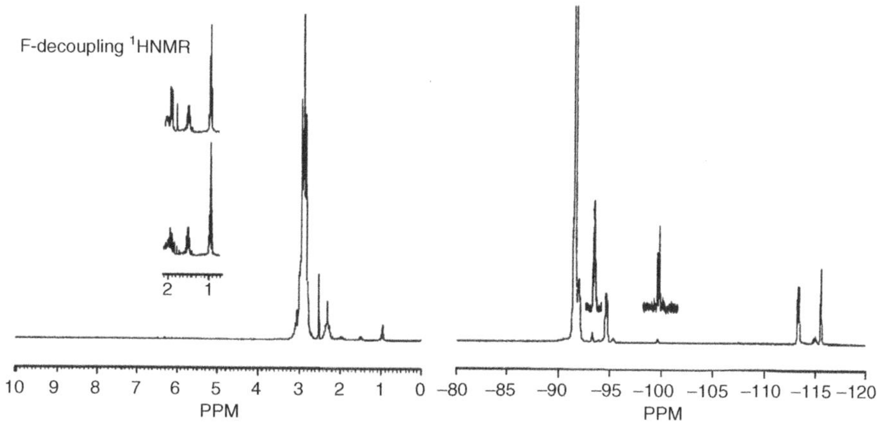

FIGURE 12.2 ^{1}H (left) and ^{19}F NMR (right) spectra of a PVDF polymer (run A-6) prepared by a $B(Et)_3/O_2$ initiator (mole ratio = 3/2).

Figure 12.2 shows the ^{1}H and ^{19}F NMR spectra of a typical PVDF polymer (run A-6). The distinctive chemical shifts of PVDF, separated from those of borane oxidation fragments, greatly help the end group analysis by the NMR technique. In addition to two major chemical shifts at 2.9 and 2.3 ppm, corresponding to (CF_2—$\mathbf{CH_2}$—CF_2—$\mathbf{CH_2}$) and (CF_2—$\mathbf{CH_2}$—$\mathbf{CH_2}$—CF_2), respectively, three minor chemical shifts at 1, 1.5, and 2.0 ppm are associated with three types of protons at the $\mathbf{CH_3}$—$\mathbf{CH_2}$—$\mathbf{CH_2}$—CF_2— end group, which is originated from the oxidation adduct (CH_3—CH_2*) of $B(Et)_3$. The same chain end assignments were observed in the ^{19}F NMR spectrum. In addition to several main chain peaks at near 91, 95, 113, and 116 ppm, there were several weak chain end peaks near 93 ppm ($\mathbf{CF_2}$—CH_2—CH_2—CH_3) and 100 ppm ($\mathbf{CF_2}$—CH_2—CH_3). In fact, all reaction runs with $[O_2]/[B(Et)_3] < 1$ (Table 12.1) show the same CH_3—CH_2—polymer chain end structure. There is no detectable CH_3—CH_2—O— group and borane fragment existing at the polymer chain end.

Scheme 12.2 illustrates the possible oxidation mechanism of $B(Et)_3$ (I) that can explain all our observations. In the early stage of the oxygen reaction ($[O_2]/[B(Et)_3]$ <1), the first oxidation adduct shall be $B(OOEt)(Et)_2$ (II) that is very unstable and immediately reacts with the unreacted $B(Et)_3$ (I) to form a $B(OEt)(Et)_2$ (III) and two intermediate radicals, including CH_3—CH_2* radical (A) and borinate radical (B). Without monomers presence, two radicals just combine to form another $B(OEt)(Et)_2$ (III). With VDF monomers, the CH_3—CH_2* radical (A) *in situ* initiates the radical polymerization of VDF at ambient temperature. However, borinate radical (B) may be too stable to react with the monomer due to electron back-donating to the empty p-orbital of boron. But it may serve as the end-capping agent to form a reversible bond with the growing PVDF chain end to prolong the propagating process. Evidently, the mono-oxidation product $B(OEt)(Et)_2$ (III) is still very reactive, which is subjective to further oxidation, either by oxygen to form B(OOEt)(OEt)Et (VI), or by another $B(OOEt)(Et)_2$ (II) to form $B(OEt)_2Et$ (V). After the di-oxidation reaction, the peroxyl group in B(OOEt)(OEt)Et (VI) is still unstable but is a little bit more stable than that in the mono-oxidized $B(OOEt)(Et)_2$ (II). The observed amount of B(OOEt)(OEt)Et (VI) is dependent on the availability of B—C bonds in $B(Et)_3$ (I) and $B(OEt)(Et)_2$ (III). With the further increase of oxygen concentration and reaction temperature, all three B—C bonds in $B(OMe)(Et)_2$ and $B(Et)_3$ are eventually oxidized to form tri-oxidation species.

The combined information from oxidation adducts and the polymer chain end structure presents a complete picture of oxidation and the initiation mechanism of borane/oxygen-mediated radical polymerization of fluoromonomers at ambient temperature. This borane/oxygen initiator offers a safe and effective route to prepare fluoropolymers with high molecular weights and high purities. It also provides the control of polymer chain end structure by initiator.

12.3 FLUOROPOLYMER DIELECTRIC FOR HIGH ENERGY DENSITY CAPACITORS

This borane/oxygen initiator was applied to prepare PVDF-based random copolymers for energy storage capacitors [54, 55]. The chemistry shows advantages of narrow molecular weight and composition distribution [56]. For this study, we have prepared a family of PVDF polymers (Scheme 12.3) that contains vinylidene difluoride (VDF), trifluoroethylene (TrFE), and chlorotrifluoroethylene (CTFE) units. The consecutive VDF units provide strong polarization; the randomly distributed TrFE units (co-crystallizable with VDF units) direct the VDF sequence to all-trans (polar) conformation; and the small amount of bulky Cl atoms in CTFE units serve as the kinks to reduce the crystalline size without significantly reducing the overall crystallinity. Some terpolymers exhibit high dielectric constant and relaxed ferroelectric behaviors [57–59]. It is logical to think that this type of terpolymer with a particular composition might show high energy density (due to high dielectric constant and breakdown voltage) and low energy loss (due to relax ferroelectric behavior).

Table 12.2 summarizes two comparative sets of VDF/TrFE/CTFE terpolymers with a systematic variation of TrFE and CTFE contents, respectively. To obtain high quality terpolymers (high purity and narrow molecular weight and composition distributions), we employed borane/oxygen-mediated control radical polymerization in

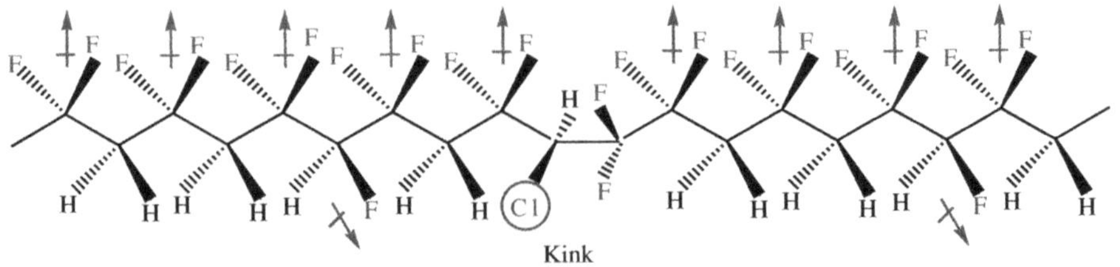

SCHEME 12.3 Chain conformation of VDF/TrFE/CTFE terpolymer.

TABLE 12.2 Summary of Two Comparative Sets of VDF/TrFE/CTFE Terpolymers Prepared by Triethylborane/Oxygen Radical Initiator

Run	VDF/TrFE/CTFE (%)	T_m (°C)	T_c (°C)	ΔH (J/g)
A-1	58.4/34.2/7.4	121.5	19.8	20.6
A-2	65.6/26.7/7.7	123.6	23.8	22.0
A-3	80.7/11.6/7.7	120.1	NA	27.9
A-4	92.0/0.0/8.0	145.0	NA	25.9
B-1	63.3/36.7/0.0	151.7	64.6	30.1
B-2	58.3/37.4/4.3	134.6	31.2	23.3
B-3	56.2/37.0/6.8	121.3	18.8	21.3
B-4	58.4/34.2/7.4	121.5	19.8	20.6
B-5	57.1/31.4/11.5	106.4	NA	17.0

NA, Not applicable.

a homogeneous acetonitrile solution at ambient temperature. The thermal transitions of terpolymers were examined by DSC [54,55]. The relatively sharp melting peak in all co- and terpolymers implies a relatively uniform molecular structure and morphology. The increase of CTFE units (by substituting TrFE units) systematically reduces both melting (T_m) and Curie (T_c) temperatures but maintaining high heat of fusion (ΔH) that starts its diametrical descent after the CTFE content reaches >8 mol%. The terpolymers with ~7.5 mol% CTFE units and various VDF/TrFE mole ratios maintain an overall high crystallinity. The VDF/TrFE (63.3/36.7 mol%) copolymer shows an expected Curie temperature at 65°C. With ~7.5 mol% of CTFE content, the terpolymer Curie temperature shifts to near ambient temperature and the intensity becomes very weak, indicating a relaxed ferroelectric behavior.

Figure 12.3 compares the dielectric constant between VDF/CTFE (92/8 mol%) copolymer and VDF/TrFE/CTFE (65.6/26.7/7.7 mol%) terpolymer. The dielectric

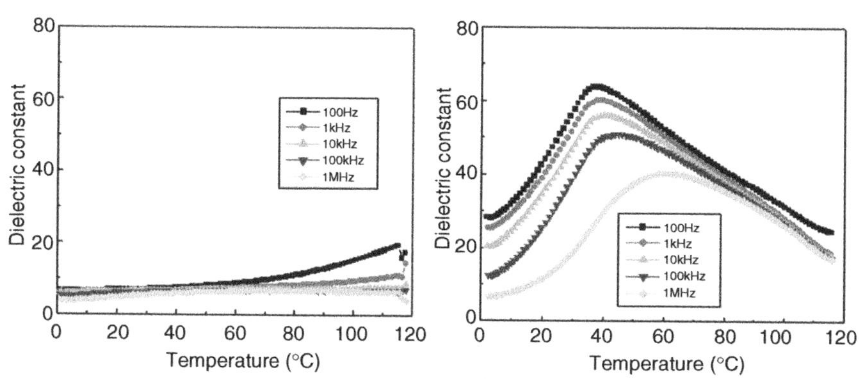

FIGURE 12.3 Dielectric constant of (left) VDF/CTFE (92/8 mol%) copolymer (run A-1) and (right) VDF/TrFE/CTFE (65.6/26.7/7.7 mol%) terpolymer (run A-3).

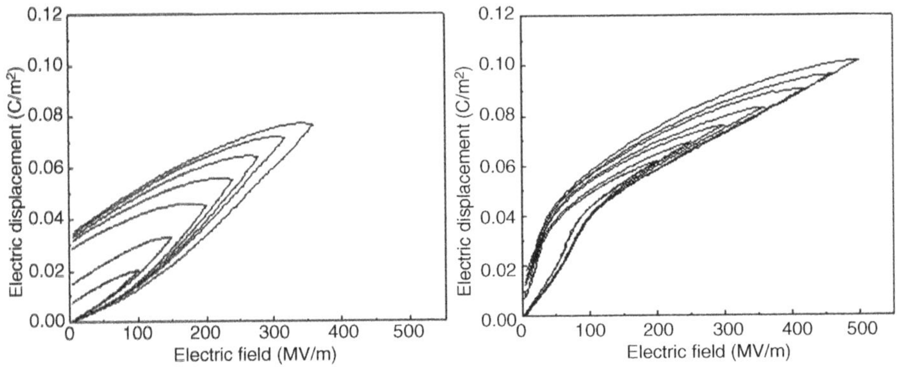

FIGURE 12.4 Unipolar D–E hysteresis curves of (left) VDF/CTFE (92/8 mol%) copolymer (run A-1) and (right) VDF/TrFE/CTFE (65.6/26.7/7.7 mol%) terpolymer (run A-3).

constant profile provides a sharp Curie transition temperature. The polymers containing less than 15 mol% TrFE units show no Curie temperature. It is necessary to have more than 20 mol% of TrFE units to observe the Curie temperature, which shifts rapidly toward the ambient temperature as the TrFE content increases toward 35%. In other words, the dielectric constant profile (polarization of CF_2 dipoles) can be effectively controlled by terpolymer composition and thermal energy. It is engrossing to understand the similar polarization–depolarization effect under the applied electric field.

Figure 12.4 compares unipolar D–E (charge displacement vs. unipolar electric field) hysteresis curves of the same set of VDF/CTFE and VDF/TrFE/CTFE polymers, with similar CTFE content (7.7–8.0 mol%). The charging–discharging cycles were first applied to 100 MV/m, then increasing 50 MV/m intervals until reaching the breakdown electric field. It is important to note that the thin film quality has a great effect on the breakdown electric field. To assure uniform and defect-free thin films for side-by-side comparison at equilibrium state, all polymers were solution cased into approximately 30–40-μm thick films. The films were then conditioned in a vacuum oven at melting temperature for a few hours before slowly cooling to room temperature in order to obtain transparent uniform films 10–15 μm thick. Without TrFE unit, the breakdown electric field for VDF/CTFE (92/8 mol%) copolymer (A-1) is typically 350 MV/m, and the copolymer shows a large hysteresis in polarization–depolarization curves with high energy loss and large remnant polarization at the end of each cycle (zero electric field). The initial increase of the TrFE content seems to have a very small effect to the D–E hysteresis curves. The significant changes were only observed while the TrFE content reached >20 mol%. The VDF/TrFE/CTFE (65.6/26.7/7.7) terpolymer (A-3), with the Curie temperature at around 35–40°C, shows the highest breakdown electric field at >500 MV/m and slim polarization–depolarization curves with low energy loss and almost no remnant polarization. The electric displacement reaches to about 0.1 C/m^2 at 500 MV/m. Compared with the theoretical electric displacement of 0.13 C/m^2 for the fully polarized (β-phase) PVDF

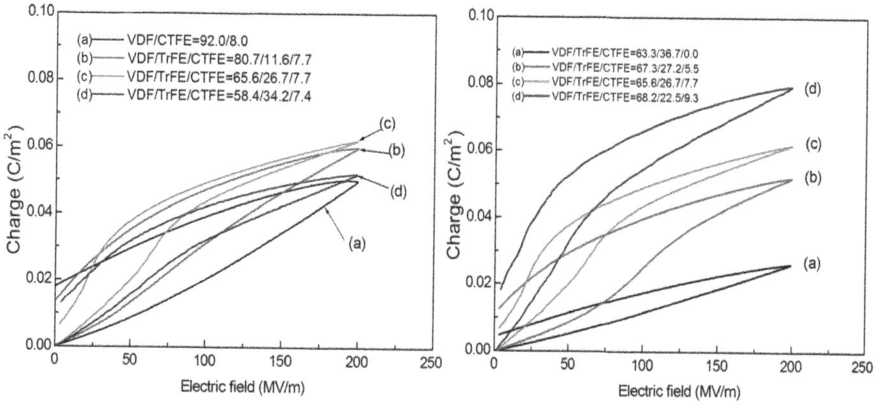

FIGURE 12.5 Comparison of unipolar D–E hysteresis curves of two sets of VDF/TrFE/CTFE terpolymers that contain (left) a fixed CTFE content (~7.5 mol%) and various VDF/TrFE mole ratios, including (a) 92/0 mol%, (b) 80.7/11.6 mol%, (c) 65.6/26.7 mol%, and (d) 58.4/34.2 mol%, and (right) a fixed VDF content (~65 mol%) and various TrFE/CTFE mole ratios, including (a) 36.7/0 mol%, (b) 27.2/5.5 mol%, (c) 26.7/7.7 mol%, and (d) 22.5/9.3 mol%.

homopolymer [60], this terpolymer is almost at its limit with almost all VDF and TrFE units polarized along the electric field. With the further increase of the TrFE content (>30%), the polarization–depolarization curves broaden again.

Figure 12.5 compares two sets of unipolar D–E loops of PVDF co- and terpolymers with a fixed CTFE or TrFE content, respectively. The results show that certain content of TrFE is necessary to keep the polymer confirmation in polar phase, which makes the crystal easier to be polarized and depolarized. Introduction of CTFE is reducing the crystal phase content as well as cutting the crystal in small pieces, which allows the polar crystals to be polarized under high electric field and leaves enough free space for them to rotate back to the original orientation.

Figure 12.6 compares releasing energy density and energy loss of the polarization–depolarization cycle (with the maximum electric field) for the same set of co- and terpolymers with CTFE content ~7.5 mol%. In most cases (except A-4, discussed later), the energy density closely follows the same master curve. The releasing energy density increases with the increase of the applied electric field, and the higher breakdown electric field offers the higher energy density. Both VDF and TrFE units in the polymer chain contribute to the electric displacement and energy density. However, the energy loss is dramatically different. For the VDF/CTFE (92/8 mol%) copolymer (A-1) and VDF/TrFE/CTFE (80.7/11.6/7.7) (A-2), with no and low TrFE content and without observing the Curie temperature, the energy loss is very high (even higher than the energy released in all cycles), which is unacceptable in the capacitor design. A dramatic improvement is shown in the VDF/TrFE/CTFE (65.6/26.7/7.7) terpolymer (A-3). It shows not only the highest breakdown electric field (>500 MV/m) and highest energy density (13 J/cm^3) but also lowest energy loss. On the other hand,

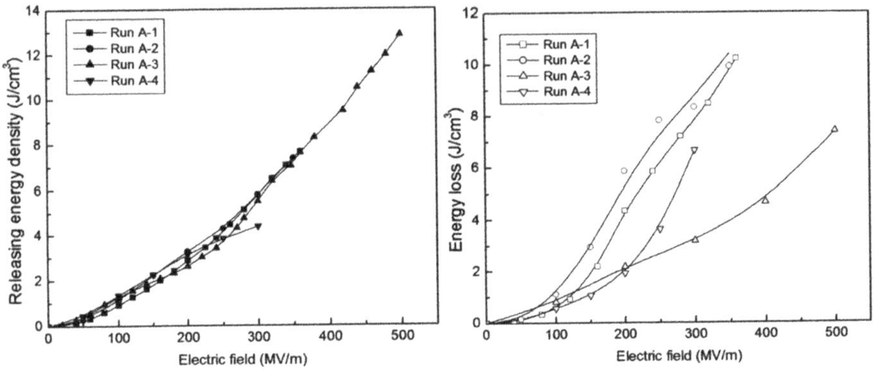

FIGURE 12.6 (Left) Releasing energy density and (right) energy loss of VDF/CTFE copolymer and three VDF/TrFE/CTFE terpolymers (runs A-2, A-3, A-4). The solid curves are only for guiding the eyes in order to see the trend.

the energy density curve of the VDF/TrFE/CTFE (58.4/34.2/7.4) terpolymer (A-4) shows a clear deviation from the master curve, with saturation of about 200 MV/m and maximum releasing energy density of only about 4 J/cm^3. The energy loss also shapely increases after 200 MV/m, indicating that all the energy charged is lost after 200 MV/m.

Overall, we have systematically studied PVDF-based co- and terpolymers to tune its polarization profile for capacitor applications. The desirable terpolymer (VDF/TrFE/CTFE = 65.6/26.7/7.7), exhibiting a high dielectric constant ($\varepsilon \sim 60$) at the Curie temperature (35°C), shows a high breakdown electric field >500 MV/m, high energy density (releasing >13 J/cm^3), relatively small energy loss, and almost no remnant polarization at zero electric field.

12.4 FUNCTIONAL FLUOROPOLYMER FOR PROTON EXCHANGE MEMBRANES

A series of poly(vinylidene fluoride)-*g*-sulfonated polystyrene (PVDF-*g*-SPS) well-defined graft copolymers were systematically synthesized and examined with the focus of understanding how the polymer microstructure (backbone molecular weight, graft density, graft length, sulfonic acid concentration, ion exchange capacity (IEC), etc.) affects their morphology, water uptake, and proton conductivity under various environmental conditions (temperature and relative humidity) [61]. As illustrated in Scheme 12.4, the chemistry involves two control radical polymerization mechanisms, including the borane-mediated radical polymerization to prepare PVDF random copolymers containing a few mole percent of CTFE units, then the subsequent ATRP graft-from reaction to incorporate several polystyrene side chains. The last step is the sulfonation reaction on the PS side chains to form PVDF-*g*-SPS graft copolymer.

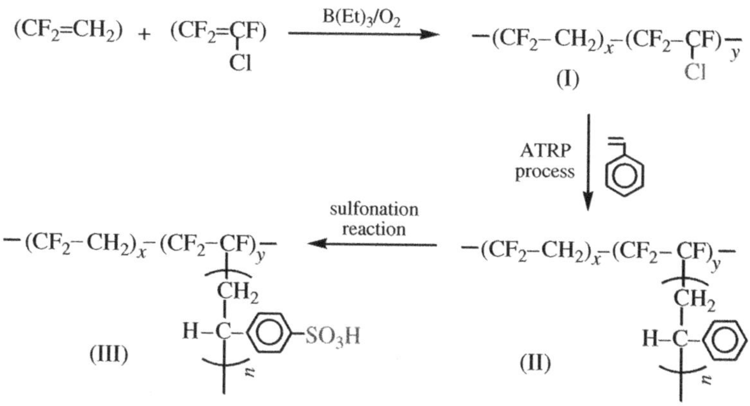

SCHEME 12.4 Synthesis of poly(vinylidene fluoride)-*g*-sulfonated polystyrene.

Table 12.3 summarizes the experimental results of graft copolymers with systematical variation of graft density, graft length, degree of sulfonation (DS), and IEC. We started with VDF/CTFE copolymers containing 1–6 mol% of CTFE units, which provided the graft points in the corresponding graft copolymers. During the ATRP graft reaction, various amount of styrene (St) were added to control the graft length in the PVDF-*g*-PS graft copolymer. The graft length (average styrene repeating units per PS graft) in the PVDF-*g*-PS copolymer was calculated from the styrene incorporation (^{1}H NMR) and graft density (40% of the CTFE units in the starting polymer; estimated by ^{19}F NMR). The same graft density and graft length are expected in the corresponding PVDF-*g*-SPS graft copolymer. On the other hand, the DS value is determined from the ^{1}H NMR spectrum of the PVDF-*g*-SPS copolymer, by estimating the relative peak intensity between aromatic protons in styrene (St) units and sulfonated styrene (SSt) units. It is interesting to note that DS of the PVDF-*g*-PS copolymer is quite dependent on its PS content and graft density. The copolymer with a lower PS content or higher grafting densities requires a longer sulfonation period or rigorous sulfonation condition. To achieve a high DS in these samples, multiple sulfonation reaction processes have been applied. The IEC is then calculated from the DS value by estimating the sulfonic acid concentration in the final PVDF-*g*-SPS graft copolymer. Overall, we have applied three well-controlled reactions: copolymerization (or hydrogenation), ATRP graft reaction, and sulfonation, in order to prepare a wide range of PVDF-SPS graft copolymers with various molecular weight, graft density, graft length, and sulfonic acid concentration. This offers us the opportunity to systematically examine the effects of the polymer microstructure on proton conductivity and water uptake under various conditions (temperature and relative humidity).

Since the graft copolymers contain a crystalline backbone and amorphous side chains, it is convenient to apply DSC and TEM techniques to understand their morphologies and thermal properties. Figure 12.7 compares DSC curves of three PVDF-*g*-PS graft copolymers (II) (Set A1 in Table 12.1), which have the same (0.4 mol%)

TABLE 12.3 A Summary of PVDF-g-PS and PVDF-g-SPS Graft Copolymers

	VDF-co-CTFE		PVDF-g-PS			PVDF-g-SPS		
Run	VDF/CTFE (mole ratio)	St (g)	VDF/St (mole ratio)	Graft density[a]	Graft length[b]	VDF/St/SSt[c] (mole ratio)	DS (%)	IEC (mmol/g)
A1-1	99.0/1.0	1	99.0/5.3	0.4	13.2	99.0/2.2/3.1	58.0	0.43
A1-2	99.0/1.0	3	99.0/17.7	0.4	44.3	99.0/4.2/13.5	76.1	1.44
A1-3	99.0/1.0	3	99.0/17.7	0.4	44.3	99.0/2.1/15.6	88.0	1.64
A1-4	99.0/1.0	6	99.0/31.5	0.4	78.8	99.0/11.3/20.2	64.2	1.78
A1-5	99.0/1.0	6	99.0/31.5	0.4	78.8	99.0/10.3/21.2	67.4	1.86
A1-6	99.0/1.0	6	99.0/31.5	0.4	78.8	99.0/0.0/31.5	100	2.57
A2-1	98.0/2.0	1.5	98.0/13.8	0.8	17.3	98.0/5.1/9.9	71.9	1.12
A2-2	98.0/2.0	1.5	98.0/13.8	0.8	17.3	98.0/2.5/12.5	83.0	1.38
A2-3	98.0/2.0	3	98.0/21.5	0.8	26.9	98.0/3.7/18.1	83.0	1.77
A2-4	98.0/2.0	3	98.0/21.5	0.8	26.9	98.0/2.2/19.3	89.6	1.88
A2-5	98.0/2.0	3	98.0/21.5	0.8	26.9	98.0/1.6/19.9	92.7	1.93
A2-6	98.0/2.0	5	98.0/54.2	0.8	67.8	98.0/12.5/41.7	76.9	2.70
A3-1	96.6/3.4	1.8	96.6/14.1	1.4	10.1	96.6/6.5/7.6	46.1	0.88
A3-2	96.6/3.4	1.8	96.6/14.1	1.4	10.1	96.6/2.8/11.3	80.2	1.26
A3-3	96.6/3.4	2.4	96.6/23.6	1.4	16.9	96.6/7.6/16.0	67.7	1.55
A3-4	96.6/3.4	2.4	96.6/23.6	1.4	16.9	96.6/4.9/18.7	79.2	1.78
A3-5	96.6/3.4	3	96.6/36.4	1.4	26.0	96.6/8.9/27.5	75.7	2.19
A3-6	96.6/3.4	3	96.6/36.4	1.4	26.0	96.6/6.8/29.6	81.2	2.32
A4-1	95.4/4.6	1.8	95.4/15.1	1.8	8.4	95.4/7.4/7.7	50.8	0.87
A4-2	95.4/4.6	1.8	95.4/15.1	1.8	8.4	95.4/5.0/10.1	67.1	1.12
A4-3	95.4/4.6	2.3	95.4/22.5	1.8	12.5	95.4/7.4/15.1	67.1	1.48
A4-4	95.4/4.6	2.3	95.4/22.5	1.8	12.5	95.4/1.4/21.1	93.6	1.98
A4-5	95.4/4.6	2.5	95.4/25.5	1.8	14.2	95.4/0.6/24.9	97.5	2.21

Source: Redrawn from Reference 61. Copyright 2008 American Chemical Society.

[a]Number of PS grafts per 100 VDF units in the backbone.

[b]Number of styrene units in each graft.

[c]St, styrene; SSt, sulfonated styrene.

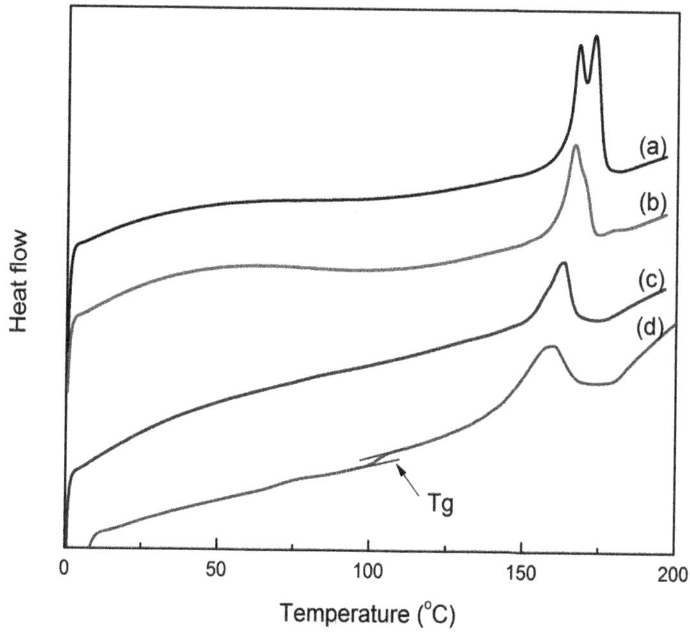

FIGURE 12.7 Comparison of DSC curves between (a) a starting P(VDF-*co*-CTFE) copolymer (I) and three corresponding PVDF-*g*-PS graft copolymers (II) with 0.4 mol% graft density and various PS graft length (b) 13.2, (c) 44.3, and (d) 78.8 average styrene units per graft, respectively.

graft density and various PS graft length (13.2, 44.3, and 78.8 average styrene units per graft, respectively). The starting copolymer (I) shows a melting temperature (T_m) at 162.5°C. The increase of PS content leads to only a less than 4°C decrease in T_m, and the decrease of heat of fusion (ΔH) is mainly due to the dilution effect. As the molecular weight of the PS graft increases, the T_g of the PS domain appears at 100°C. The coexistence of the T_m for the P(VDF-*co*-CTFE) backbone and the T_g for the PS grafts indicates a clear microphase separation morphology in this PVDF-*g*-PS (99/31.5 mole ratio) graft copolymer. Overall, with low graft density the PS graft length, residing in the amorphous phase, has only a small effect to the PVDF crystalline phase.

Figure 12.8 shows TEM micrographs of the cross-sectional slices of three representative PVDF-*g*-SPS graft copolymers (runs A1-6 and A3-6). In general, they are not as well ordered as the typical morphologies in diblock copolymers. Comparing the first two graft copolymers (III), they have the same low molecular weight PVDF backbone, similar SPS contents (VDF/SSt ~100/30 mole ratio), and different graft densities (0.4 and 1.4 mol%, respectively) and graft lengths (79 and 26 SSt units, respectively). The membrane morphology is clearly affected by graft density and graft length. As predicted by theory [62, 63], the morphology of the graft copolymer is controlled by a combination of volume fraction, graft density, and graft length. The

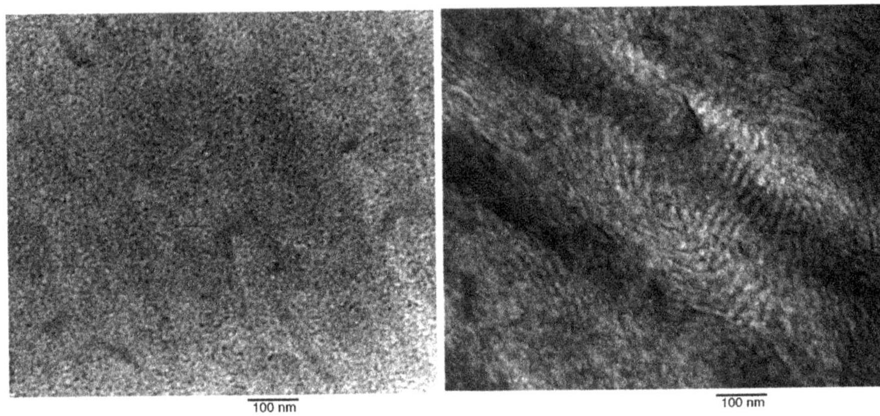

FIGURE 12.8 TEM micrographs of three PVDF-*g*-SPS graft copolymers including (left) run A3-6, (right) run A1-6.

A3-6 sample with higher graft density and lower graft length (Figure 12.8a) shows a cluster-network morphology with a cluster diameter ~5 nm, similar to that of Nafion. On the other hand, Figure 12.8b shows many long-range ionic channels (10–15 nm in width) with no specific spatial ordering in the A1-6 sample that possesses low graft density and high graft length. The morphology seems to contain largely lamella structures, and it may also include some cylinder structures. Overall, it is not as well ordered a morphology as those in the block copolymers.

Figure 12.9 shows the plots of proton conductivity (in-plane) vs. relative humidity (RH) under various environmental conditions (i.e., temperature and humidity) for the PEM based on the high molecular weight PVDF-*g*-SPS sample. Since the graft copolymer exhibits isotropy continuous SPS ionic channels, it is expected to show similar in-plane and through-plane conductivity [64]. For comparison, Nafion 112 (a random copolymer) was also measured under similar conditions. As expected, the conductivity exponentially increases with RH, and it requires more than 60% RH to show significant proton conductivity. In the RH >60% range, the PVDF-*g*-SPS membrane shows notably higher proton conductivity than Nafion 112 in all three measured temperatures (30°C, 80°C, and 120°C), but Nafion 112 performs better under low RH conditions, especially at 120°C. This conversion may be related to the water content. The short-range cluster-network morphology in Nafion may be better in retaining the absorbed water molecules in PEM under high temperature and low humidity conditions. However, under high humidity conditions, the water evaporation is no longer a major issue. The conductivity should directly relate to the IEC value, water content, morphology, and temperature. The graft copolymer—with the combination of high molecular weight, low graft density, and isotropic long-range hydrophilic–hydrophobic phase separation—allows the incorporation of high sulfonic acid content (IEC=2.75 mmol/g) without showing the negative consequence of excessive water swelling for high proton conductivity under high RH conditions.

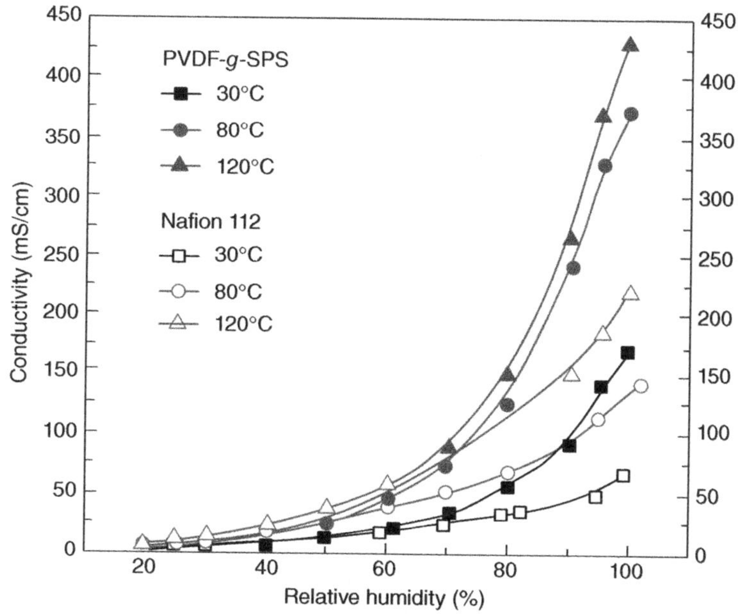

FIGURE 12.9 Comparison of proton conductivity vs. relative humility between PVDF-*g*-SPS graft copolymers and Nafion 112 at 30°C, 80°C, and 120°C (the lines are used solely for guiding purposes).

In details, the proton conductivity of the PVDF-*g*-SPS PEM increases quickly as testing temperature is elevated from 30°C to 80°C, showing only an incremental increase from 80°C to 120°C. As reported [65, 66], the increase in conductivity with temperature is attributed to the activation barriers for proton motion. Once the temperature is sufficiently high to overcome the activation barrier, further increasing the temperature has much less effect to the proton conductivity. It seems that 80°C is sufficient for the sample PVDF-*g*-SPS, but not enough for Nafion 112, which still leads to increased proton conductivity after 80°C. The temperature dependence can be approximated by an Arrhenius law, which yields the activation energy of 12.4 kJ/mol for Nafion 112 in the 30–120°C temperature range and 12.9 and 3.9 kJ/mol for PVDF-*g*-SPS in the temperature ranges of 30–80°C and 80–120°C, respectively. Similar trends are also observed in all comparative measurements under high humidity (>60%) conditions. The significant reduction of activation energy of PVDF-*g*-SPS at high temperature range is consistent with the better continuous network of the ionic conductive channels under high humility conditions.

In order to further understand the influence of the polymer microstructure on proton conductivity of the membrane, a series of high molecular weight PVDF-*g*-SPS graft copolymers with similar IEC ~2.20 mmol/g and different graft densities (0.8, 1.4, 1.7, and 2.4 mol%) are compared in Figure 12.10. The major difference between these polymers is their microstructures, and therefore different morphologies are

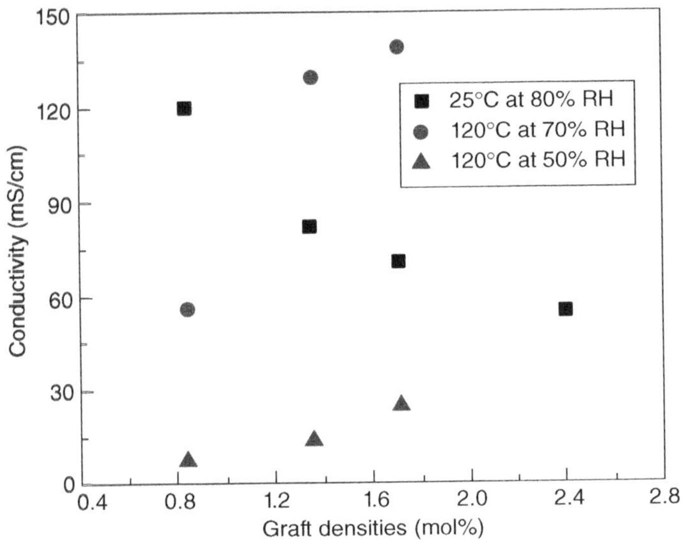

FIGURE 12.10 Comparison of proton conductivity vs. graft density of four PVDF-*g*-SPS graft copolymers having IEC ~2.20 mmol/g and different graft densities 0.8, 1.4, 1.7, and 2.4 mol%, respectively, under 25°C (80% RH) and 120°C (70% and 50% RH).

formed in the corresponding membranes. Proton conductivity at a low temperature (25°C) and high humidity (80%) decreases from 120 mS/cm to 55 mS/cm when grafting density is increased from 0.8 to 2.4 mol%. However, at a higher temperature (120°C) and similar humidity (70%), a completely inverse result was observed; in this instance, the polymer with a higher graft density exhibited higher proton conductivity. A similar trend was also observed at the same high temperature (120°C) and lower humidity (50%). Overall, in this PVDF-*g*-SPS graft copolymer set with the same sulfonic acid content, the lower graft density and longer graft length lead to a larger-scaled hydrophilic domain and lower activation energy. As a result, this polymer shows high conductivity at a low temperature and quickly reaches maximum conductivity at a relatively low temperature and even starts to exhibit decreased conductivity at a higher temperature [67]. On the other hand, the corresponding graft copolymer with a higher graft density and a shorter graft length forms cluster-network morphology with higher activation energy. The proton conductivity is relatively low at low temperature and increases consistently as the temperature is elevated until it reaches a maximum at a relatively high temperature. Generally speaking, the high molecular weight PVDF-*g*-SPS copolymer with a high IEC value and very low graft density (0.3 and 0.8 mol%) is suitable for PEM under high RH and a wide range of temperatures, while the corresponding higher graft density (1.4 and 1.7 mol%) ones show increased conductivity at higher temperatures. However, the graft copolymer with too high of a graft density (2.4 mol%) lost most of the PVDF crystallinity, and the PEM became too soft to maintain its integrity at 120°C.

12.5 FUNCTIONAL FLUOROPOLYMER FOR PROTON-CONDUCTIVE COMPOSITE MEMBRANES

We have also developed a new family of functionalized fluoropolymer containing vinylidene fluoride (VDF) and CTFE units, two terminal silane [$Si(OC_2H_5)_3$] cross-linkers for forming 3D network, and several functional silane groups for chemical or physical interactions between the polymer network and inorganic phase [68]. Scheme 12.5 illustrates the synthesis route. The chemistry for preparing the chain end functionalized PVDF polymers is centered on a specifically designed functional borane initiator containing a silane group. We have developed a simple chemical route to prepare the borane initiator by hydroboration of commercially available vinylsilane and boronhydride at elevated temperature. This chemistry is advantaged by its simplicity and mild reaction conditions. Different trifluorovinyl functional comonomers, containing $Si(CH_3)_2(OR)$ group have been prepared in a high-purity state. Suitable coupling agents that can effectively terminate and link several propagating fluoropolymer chains to form a linear, star, or long chain branch structure were studied. Catalyzed by the borane functional initiator, VDF, CTFE, and functional comonomers were copolymerized to the resultant polymers with terminal $Si(OC_2H_5)_3$ groups and certain content of side chain $Si(CH_3)_2(OR)$ groups. This low molecular weight polymer can be cross-linked under heating (during the compress molding process in forming composite membranes) to form stable 3D Teflon-like network with all the desirable Teflon properties. The cross-linking density and, therefore, the water uptake of final composite membrane could be controlled by varying the concentration of functional silicon groups in fluoropolymer.

VDF/CTFE copolymer has thermostability similar to that of VDF/HFP copolymer but possesses advantages of high copolymerization reactivity and wide variation in molar VDF/CTFE ratio, which could effectively reduce the cost of a membrane and give more options to control physical properties of the polymer. By varying CTFE

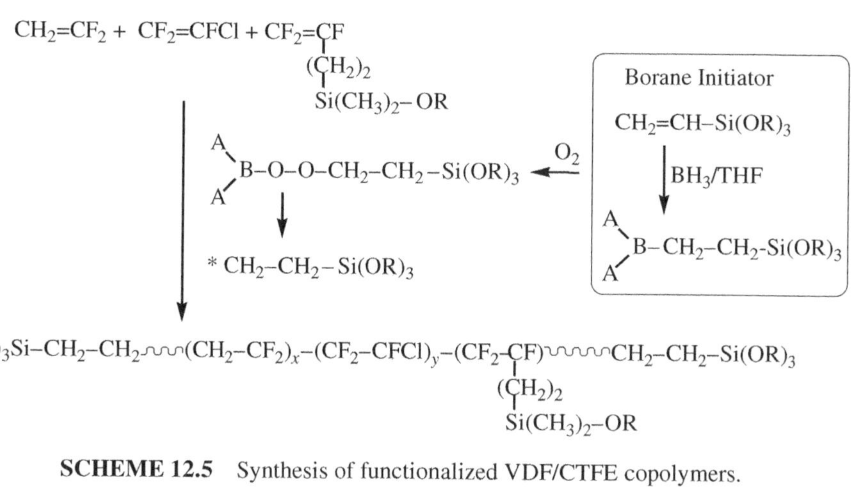

SCHEME 12.5 Synthesis of functionalized VDF/CTFE copolymers.

content, the copolymer crystallinity, and accordingly flexibility can be well controlled, which is very important for the introduction of inorganic particles. When 16 mol% of CTFE was introduced, the melting point of copolymer almost disappeared on the DSC curve, indicating a transition from semi-crystalline plastic to completely amorphous elastomer.

In order to understand the role of inorganic material in composite membranes, we conducted the synthesis of an amorphous zirconium phosphate as well as a layered alpha zirconium phosphate. The surface area of synthesized amorphous zirconium phosphate was 325 m^2/g, while $H_3OZr_2(PO_4)_3$ had a surface area of only 8.7 m^2/g. For membrane preparation, a certain amount of Nafion and cross-linkable P(VDF–CTFE) precursor with silane end groups were dissolved in acetone, then inorganic particles were introduced into the solution, and the suspension was vigorously stirred and cast on a Teflon-coated glass slide. After the complete solvent evaporation, the composite membrane was peeled off and compressed at 180°C. Membranes with different types of inorganic proton conductors were synthesized for conductivity measurements. Based on the SEM cross-sectional images of composite membranes, the distribution of inorganic particles inside the membranes was uniform. Figure 12.11 shows a cross-sectional SEM image of one of the composite membranes—20%(VDF–CTFE)/50% Nafion/20% $H_3OZr_2(PO_4)_3$. It demonstrates an uniform distribution of rhombohedral $H_3OZr_2(PO_4)_3$ particles about 500 nm in size through the whole cross section of the membrane.

The conductivity tests were carried out for a number of P(VDF–CTFE)/ Nafion/Inorganic membranes: 3D phase zirconium phosphate, 3D porous titanosilicate, amorphous zirconium phosphate (AZP), mesoporous alumina, and a silica-based commercial molecular sieve (MS). Figure 12.12 shows the conductivity of various PEM composite membranes under various humidity conditions. The inorganic phase

FIGURE 12.11 Cross-sectional SEM image of a 20% (VDF–CTFE)/50% Nafion/20% $H_3OZr_2(PO_4)_3$ composite membrane.

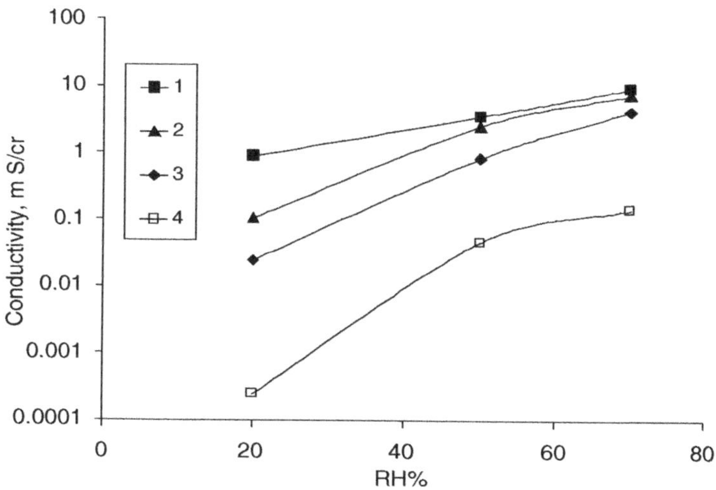

FIGURE 12.12 Conductivity of P(VDF–CTFE)/Nafion/inorganic membranes with different inorganic additives at 120°C and different RH. (1–30% P(VDF–CTFE)/30%Nafion/40% $H_3OZr_2(PO_4)_3$; 2–20% P(VDF–CTFE)/20%Nafion/60% AZP; 3–30% P(VDF–CTFE)/30%Nafion/40% $H_2(SiTi_2O_7)1.5H_2O$; 4–30% P(VDF–CTFE)/30%Nafion/40% MS.)

has a great effect on membrane conductivity. The membranes with the same (VDF–CTFE)/Nafion ratio and content of inorganic component had very different conductivity depending on the type of inorganic phase. Alumina-based membranes had three order of magnitude lower conductivity than the membranes with 3D phase zirconium phosphate, which turned out to be more beneficial for the membrane conductivity than other inorganic including amorphous zirconium phosphate.

12.6 CONCLUSION

This chapter discusses two closely relative areas, including the borane-mediated control radical polymerization to prepare some unique fluoropolymer structures and the resulting fluoropolymers for energy storage applications. The system of borane/oxygen control radical initiators is surprisingly effective for initiating polymerization of fluoromonomers, such as vinylidene fluoride (VDF), TrFE, CTFE, at ambient temperature. Mechanistic study indicates that the *in situ* formed alkyl radical (C*) is responsible for the initiation, and the borinate radical (B—O*) forms a reversible bond with the propagating radical to prolong the polymerization. The control polymerization is characterized by predictable molecular weight, narrow molecular weight distribution, formation of chain end functionalized polymers and diblock copolymers, and tolerance to many functional groups that usually cause chain transfer reactions in regular free radical polymerization. The other aspect of this chapter is to apply this control polymerization to prepare a new family of PVDF copolymers. By systematically tuning the polymer chain conformation and crystal structure, we have identified the most suitable fluoropolymer for thin film capacitor that exhibits high energy density and low energy loss. On the other hand, we have prepared a series of poly(vinylidene fluoride)-*g*-sulfonated polystyrene graft copolymers for PEMs. The systematical study reveals the effects of polymer microstructure (backbone molecular weight, graft density, graft length, sulfonic acid concentration, IEC, etc.) to morphology, water uptake, and proton conductivity under various environmental conditions (temperature and relative humidity). The PVDF-*g*-SPS graft copolymer with a combination of a high PVDF backbone molecular weight, low SPS graft density, and high graft length (SPS content >30 mol%) presents an interesting PEM material. This graft copolymer self-assembles into a microphase-separated morphology with randomly oriented long-range lamella/cylinder ionic channels (with small widths) that are imbedded in the hydrophobic semicrystalline PVDF matrix. This graft copolymer morphology offers a high IEC (2.75 mmol/g) and resistance to excessive water swelling, which yields notable higher proton conductivity than Nafion under 30–120°C and high humidity conditions. We have also prepared cross-linkable PVDF copolymer with terminal and side chain silane groups for polymer electrolyte composite membranes. The resulting P(VDF–CTFE)/Nafion/$H_3OZr_2(PO_4)_3$ composite membranes show high proton conductivity even under low (20%) humidity conditions.

ACKNOWLEDGMENT

The author would like to thank the Office of Naval Research for the financial support.

REFERENCES

1. Scheirs, J. *Modern Fluoropolymers*; John Wiley & Sons, Ltd: Australia, 1997.
2. Yamabe, M. Y. *Makromol. Chem., Macromol. Symp.* **1992**, *64*, 11.
3. Takakura, T. *CTFE-Vinyl Ethers Copolymers in Modern Fluoropolymers*; Scheirs, J., Ed.; John Wiley & Sons, Inc.: New York, 1997; Chapter 29, p 557.
4. Boutevin, B.; Ameduri, B. *Macromol. Symp.* **1994**, *82*, 1.
5. Kostov, G. K.; Matsuda, O.; Machi, S.; Tabata, Y. *J. Polym. Sci. Part A, Polym. Chem.* **1979**, *17*, 3991.
6. Kostov, G.; Ameduri, B.; Boutevin, B. *J. Fluorine Chem.* **2002**, *114*, 171.
7. Rice, D. E.; Sandberg, C. L. US Patent 3,461,155 (1969).
8. Maxwell Robinson, I.; Kochi, J. K. *Macromolecules* **1983**, *16*, 526.
9. Oka, M.; Tatemoto, M. *Contemp. Top. Polym. Sci.* **1984**, *4*, 763.
10. Tatemoto, M.; Nakagawa, T. US Patent 4,158,678 (1979).
11. Saint-Loup, R.; Manseri, A.; Ameduri, B.; Lenret, B.; Vignane, P. *Macromolecules* **2002**, *35*, 1542.
12. Otsu, T.; Yoshida, M. *Makromol. Chem., Rapid Commun.* **1982**, *3*, 127.

13. Otsu, T.; Kuriyama, A. J. *Macromol. Sci., Chem.* **1984**, *A21*, 921.
14. Georges, M. K.; Veregin, P. R. N.; Kazmaier, P. M.; Hamer, G. K. *Macromolecules* **1993**, *26*, 2987.
15. Hawker, G. J. *J. Am. Chem. Soc.* **1994**, *116*, 11185.
16. Wang, J. S.; Matyjaszewski, K. *J. Am. Chem. Soc.* 1995, *117*, 5614.
17. Kato, M.; Kamigaito, M.; Sawamoto, M.; Higashimura, T. *Macromolecules* **1995**, *28*, 1721.
18. Chiefari, J.; Chong, Y. K.; Ercole, F.; Krstina, J.; Jeffery, J.; Le, T.; Mayadunne, R.; Meijs, G. F.; Moad, C. L.; Moad, G.; Rizzardo, E.; Thang, S. H. *Macromolecules* **1998**, *31*, 5559.
19. Chung, T. C. US Patents 4,734,472 (1988) and 4,751,276 (1988).
20. Chung, T. C. *Macromolecules* **1988**, *21*, 865.
21. Chung, T. C.; Jiang, G. J. *Macromolecules* **1992**, *25*, 4816.
22. Chung, T. C. *Functionalization of Polyolefins*; Academic Press: London, 2002.
23. Chung, T. C.; Jiang, G. J.; Rhubright, D. *Macromolecules* **1993**, *26*, 3467.
24. Chung, T. C.; Rhubright, D.; Jiang, G. J. US Patent 5,286,800 (1994).
25. Chung, T. C.; Lu, H. L.; Janvikul, W. *Polymer* **1997**, *38*, 1495.
26. Xu, G.; Chung, T. C. *J. Am. Chem. Soc.* **1999**, *121*, 6763.
27. Chung, T. C.; Lu, H. L.; Janvikul, W. *J. Am. Chem. Soc.* 1996, *118*, 705.
28. Chung, T. C. *Prog. Polym. Sci.* **2002**, *27*, 39.
29. Chung, T. C.; Petchsuk, A. *Macromolecules* **2002**, *35*, 7678.
30. Hong, H.; Chung, T. C. *Macromolecules* **2004**, *37*, 6260.
31. Chung, T. C.; Hong, H.; Oka, M.; Kubo, K. US Patent 6,911,509 (2005).
32. Winter, M.; Brodd, R. J. *Chem Rev.* **2004**, *104*, 4245.
33. Sarjeant, W. J. *IEEE Trans. Electr. Insul.* **1990**, *25*, 861.
34. Sarjeant, W. J.; Zirnheld, J.; MacDougall, F. W. *IEEE Trans. Plasma Sci.* **1998**, *26*, 1368.
35. Rikukawa, M.; Sanui, K. *Prog. Polym. Sci.* **2000**, *25*, 1463.
36. Brandon, N. P.; Skinner, S.; Steele, B. C. H. *Annu. Rev. Mater. Res.* **2003**, *33*, 183.
37. Hickner, M. A.; Ghassemi, H.; Kim, Y. S.; Einsla, B. R.; McGrath, G. E. *Chem. Rev.* **2004**, *104*, 4587.
38. Li, Q.; He, R.; Jensen, J. O.; Bjerrum, N. J. *Chem. Mater.* **2003**, *15*, 4896.
39. Zhou, Z.; Dominey, R. N.; Rolland, J. P.; Maynor, B. W.; Pandya, A. A.; DeSimone, J. M. *J. Am. Chem. Soc.* **2006**, *128*, 12963.
40. Brown, H. C. *Organic Synthesis via Boranes*; Wiley-Interscience: New York, **1975**.
41. Mikhailov, B. M.; Bubnov, Y. N. *Organoboron Compounds in Organic Synthesis*; Harwood Academic Publishers: Switzerland, **1984**.
42. Davies, A. G.; Roberts, B. P. *J. Chem. Soc. B* **1969**, 311.
43. Davies, A. G.; Ingold, K. U.; Roberts, B. P.; Tudor, R. *J. Chem. Soc. B*, **1971**, *60*, 698.
44. Arimoto, F. S. *J. Polymer Sci.* **1966**, *4A*, 275.
45. Davies, A. G.; Hare, D. G.; White, F. M. *J. Chem. Soc.* **1960**, 1040.
46. Mirviss, S. B. *J. Am. Chem. Soc.* **1961**, *83*, 3051.
47. Davies, A. G.; Hare, D. G.; White, F. M. *J. Chem. Soc.* **1961**, 341.
48. Hansen, R. L.; Hamann, R. *J. Phys. Chem.* **1963**, *67*, 2868.

49. Welch, F. J. *J. Poly. Sci.* **1962**, *61*, 243.
50. Zhang, Z. C.; Chung, T. C. *Macromolecules* **2006**, *39*, 5187.
51. Hawker, C. J.; Bosman, A. W.; Harth, E. *Chem. Rev.* **2001**, *101*, 3661.
52. Kamigaito, M.; Ando, T.; Sawamoto, M. *Chem Rev.* **2001**, *101*, 3689.
53. Wang, J. S.; Matyjaszewski, K. *J. Am. Chem. Soc.* **1995**, *117*, 5614.
54. Zhang, Z. C.; Chung, T. C. *Macromolecules* **2007**, *40*, 783.
55. Zhang, Z. C.; Chung, T. C. *Macromolecules* **2007**, *40*, 9391.
56. Wang, Z. M.; Zhang, Z. C.; Chung, T. C. *Macromolecules* **2006**, *39*, 4268.
57. Chung, T. C.; Petchsuk, A. US Patent 6,355,749 (2002).
58. Chung T. C.; Petchsuk A. *Ferroelectrics Lett.* **2001**, *28*, 135.
59. Chung T. C.; Petchsuk A. *Macromolecules* **2002**, *35*, 7678.
60. Furukawa, T. *Adv. Colloid Interface Sci.* **1997**, *71*, 183.
61. Zhang, Z.; Chalkova, E.; Fedkin, M.; Wang, C.; Lvov, S. N.; Komarneni, S.; Chung, T. C. *Macromolecules* **2008**, *41*, 9130.
62. Milner, S. T. *Macromolecules* **1994**, *27*, 2333.
63. Xenidou, M.; Beyer, F. L.; Hadjichristidis, N.; Gido, S. P.; Tan, N. B *Macromolecules* **1998**, *31*, 7659.
64. Tsang, E. M. W.; Zhang, Z. B.; Shi, Z. Q.; Soboleva, T.; Holdcroft, S. *J. Am. Chem. Soc.* **2007**, *129*, 15106.
65. Damay, F.; Klein, L. C. *Solid State Ionics* **2003**, *162–163*, 261.
66. Alberti, G.; Casciola, M.; Massinelli, L.; Bauer, B *J. Membrane Sci.* **2001**, *185*, 73.
67. Xu, K.; Li, K.; Khanchaitit, P.; Wang, Q. *Chem. Mater.* **2007**, *19*, 5937.
68. Wang, C.; Chalkova, E.; Lute, C.; Fedkin, M.; Komarneni, S.; Chung, T. C.; Lvov, S. *Electrochem. Soc. Trans.* **2008**, *16*, 1451.

13

FLUOROPOLYMERS IN SUPERCRITICAL CARBON DIOXIDE: PHASE BEHAVIOR, SELF-ASSEMBLY, AND STABILIZATION OF WATER/CO_2 EMULSIONS

ETIENNE GIRARD, JEAN-DANIEL MARTY, AND MATHIAS DESTARAC
University Paul Sabatier of Toulouse

13.1 INTRODUCTION AND OVERVIEW

In the development of green chemistry solutions, the replacement of conventional organic solvents by "greener" ones has been identified as a key criterion to address the environmental challenges imposed by chemical synthesis, processing, and separations [1, 2]. In this perspective, supercritical carbon dioxide (sc-CO_2) has been considered a solvent of high potential to reduce environmental damage due to its low inflammability, cost, toxicity, reactivity, and critical temperature and pressure (31.1°C, 73 bar). However, its use in chemical processes has been limited until now, mainly because of its feeble solvating power which results in the low or nonsolubility of a large variety of polar or ionic reactants, catalysts, and macromolecules. Therefore, the seek for novel generations of hydrophilic/CO_2-philic (macro)molecular species for the stabilization of water/sc-CO_2 emulsions is highly topical and relevant.

Most commodity polymers possess poor solubilities in sc-CO_2 under mild conditions ($P < 350$ bar, $T < 100$°C) [3, 4]. The prominent role of polymer–polymer interactions in the solubility of polymers in sc-CO_2 mainly explains this limitation

Handbook of Fluoropolymer Science and Technology, First Edition. Edited by Dennis W. Smith Jr., Scott T. Iacono, and Suresh S. Iyer.

[3,5]. Hence, polymers with a low cohesive energy density such as fluoropolymers and polysiloxanes exhibit high solubilities, while highly cohesive polymers do not. Due to their lower self-interactions, fluoropolymers constitute the reference polymers in terms of CO_2-philicity although their cost may represent a limitation for some of the envisioned industrial applications. But challengingly enough, some novel nonbioaccumulable fluorooligomer and polymer surfactants might be considered replacement solutions for banned persistent and toxic long-chain fluorosurfactants [6]. This chapter aims at exhaustively presenting the physico-chemical properties in sc-CO_2 of the different classes of fluoropolymers that have been studied in literature. This includes their solubility and self-assembly in sc-CO_2, as well as their use as building blocks to design macromolecular surfactants for water/sc-CO_2 emulsions.

13.2 SOLUBILITY OF (CO)POLYMERS IN sc-CO_2

13.2.1 Interactions of CO_2 with Fluorine Atoms

The nature and strength of the interactions of CO_2 with different functional groups are central key parameters to understand the solvation of molecules in CO_2 and to design highly CO_2-philic polymer materials. While the interactions with carbonyl groups seem qualitatively well understood in the literature, a persistent question concerning CO_2 has been whether or not it interacts with fluorinated substituents which are known to induce significant solubility in CO_2.

Using high-pressure NMR spectroscopy, Dardin et al. studied solute–solvent interactions between fluorinated compounds and CO_2 [7]. While the absence of site-specific interactions between *n*-hexane and CO_2 through ^{1}H NMR experiments was noted, excess chemical shifts effects in ^{19}F NMR that were independent of magnetic susceptibility were evidenced in the case of *n*-perfluorohexane. This was attributed to site-specific Van der Waals interactions with CO_2 [8]. However, in a conflicting paper, Yonker et al. could not find any specific interactions between fluoromethane, trifluoromethane, and CO_2 through ^{1}H, ^{19}F experiments, and molecular dynamics simulations [9]. This work supported previous *ab initio* calculations performed by Diep et al. [10]. In their simulations performed at an MP2/aug-cc-pVDZ level of theory, the binding energies of methane/CO_2 and tetrafluoromethane/CO_2 were respectively computed to −0.88 and −0.79 kcal/mol, thus suggesting no specific interaction between the fluorinated compound and CO_2. To conceal these contradictory results, Raveendran et al. studied the effects of stepwise fluorination on methane [11]. Comparing the partial charges on the hydrogen atoms of methane and the fluorine atoms of tetrafluoromethane at an MP2/aug-cc-pVDZ level of theory, they computed charges of opposite signs for each molecule. Methane with positively charged hydrogen atoms was thus thought to interact with the negatively charged oxygen of CO_2 through acid/base Lewis interactions, while repulsive interactions would be expected with the negatively charged fluorine atoms of tetrafluoromethane. However, the dissociation energies were of same magnitude, that is, 0.63 kcal/mol for the CH_4/CO_2 dimer and 0.58 kcal/mol for the CF_4/CO_2 one. By means of comparison, these interactions are weaker than the interaction of a CO_2 dimer in which energy was estimated at 1.10 kcal/mol with a similar level of theory [12]. As for the partially fluorinated compounds, the dissociation energies of the complexes were evaluated at 1.87, 1.78, and 1.43 kcal/mol CH_3F, CH_2CF_2, and CHF_3, respectively. In CH_3F, the carbon–fluorine bond was more polarized resulting in enhanced LA–LB interactions and hydrogen atoms with increased positive charges. Upon increasing fluorination, the interactions between hydrogen atoms and the oxygen atom of CO_2 were increasingly stronger and their distances decreased. However, the dissociation energies decreased as a result of the increasing competition between fluorine atoms and the weakening of their Lewis basicity. This study thus revealed that fluorine atoms can actually interact with CO_2 in specific cases.

13.2.2 Phase Behavior of (CO)Polymers in CO_2

This section is dedicated to the description of the main family of CO_2-philic fluorinated polymers. Their solubility in sc-CO_2, the underlying physical reasons, as well as the existing strategies of improvement are presented.

13.2.2.1 From Polyethylene to Fluorinated Polyolefins

13.2.2.1.1 Solubility of Homopolymers Polyethylene (PE) is a semi-crystalline and nonpolar hydrocarbon polymer. Because of its apolar nature, this polymer is practically insoluble in sc-CO_2 in conditions as harsh as 270°C and 2750 bar [4]. The interchange energy is then dominated by quadrupolar–quadrupolar interactions of CO_2. To promote dipole–quadrupole interactions, the progressive introduction of fluorine atoms onto the polymer backbone has been studied through poly(vinyl fluoride) ($-(CH_2-CHF)_n-$), poly(vinylidene fluoride) ($-(CH_2-CF_2)_n-$), and polytetrafluoroethylene ($-(CF_2-CF_2)_n-$) (see Figure 13.1).

Due to a high melting temperature of 187°C and strong polymer–polymer interactions, a high molecular weight poly(vinyl fluoride) having an M_w of 125,000 g/mol was found insoluble at temperatures up to 300°C and pressures of 2750 bar [4, 13]. Poly(vinylidene fluoride) (PVDF) exhibited enhanced solubilities in sc-CO_2 as 5 wt% of a sample with an M_w of 181,000 g/mol was fully soluble at 1650 bar and 140°C [14]. Investigations on the effect of higher molecular weights showed little effects on the cloud point pressures (see Figure 13.2). However, the pressure cloud point

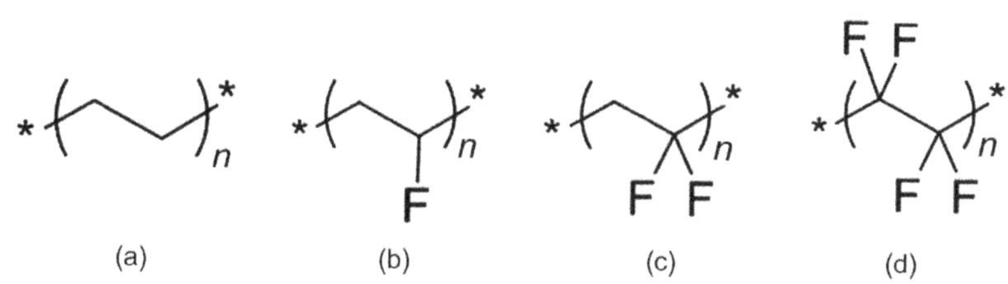

FIGURE 13.1 Molecular structures of (a) polyethylene, (b) poly(vinyl fluoride), (c) poly(vinylidene fluoride), and (d) polytetrafluoroethylene.

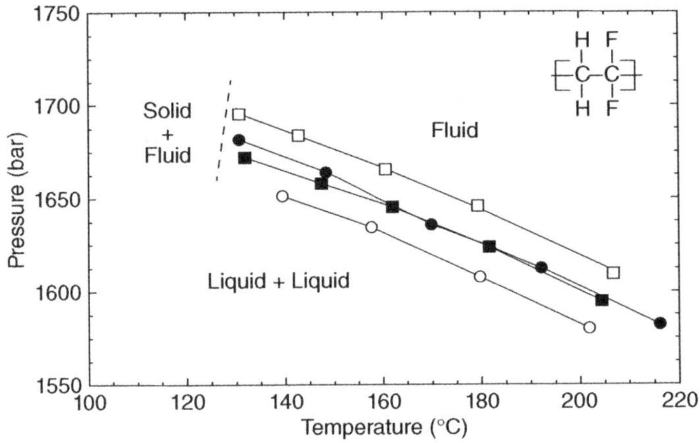

FIGURE 13.2 Pressure–temperature phase diagrams of binary mixtures of sc-CO_2 and PVDF with increasing M_w: (○) 181,000 g/mol, (■) 269,000 g/mol, (●) 275,000 g/mol, and (□) 329,000 g/mol. Reproduced from Reference 14.

was lowered by 200 bar upon decreasing molecular weight (i.e., for an M_n around 3000 g/mol) [15]. Under milder conditions of 150 bar and 65°C, only PVDF oligomers having a number of repeating units less than approximately 20 (i.e., an M_n close to 1300 g/mol) exhibited high solubility in sc-CO_2 [16]. The maximum soluble chain length was determined to be 35 (i.e., around 2200 g/mol). Finally, a high molecular weight poly(tetrafluoroethylene) (PTFE) was insoluble at 207 bar and 35°C, probably due to its high crystallinity [17].

These results demonstrate that an optimal polarity exists through partial fluorination of the polyethylene backbone. In PVDF, this creates a dipolar moment that promotes favorable dipolar–quadrupolar interactions with CO_2.

However, the solubility of homopoly(fluorinated olefins) is extremely limited as a result of strong polymer–polymer interactions which are experimentally evidenced by their high melting points, degrees of crystallinity, and high surface tensions (in the range of 30–33 mN/m) [17].

Finally, the poor solubility of fluorinated polyolefins can also be exemplified by the phase behavior of fluorinated polyisoprenes and polybutadienes in sc-CO_2 (see Figure 13.3) [18]. A pressure of 1000 bar and a temperature of 60°C were thus

FIGURE 13.3 Molecular structures of (a) fluorinated polyisoprene and (b) fluorinated polybutadiene.

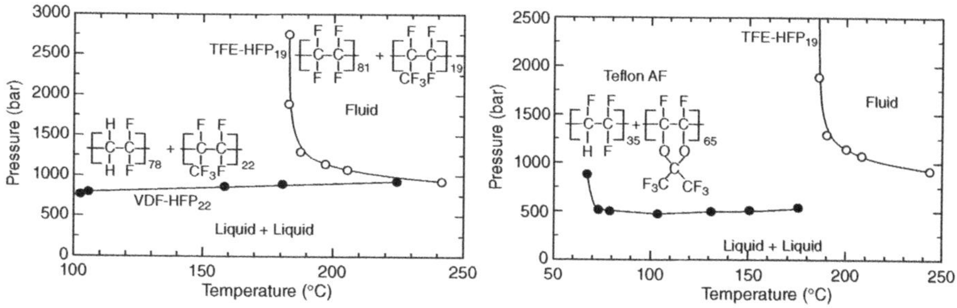

FIGURE 13.4 Pressure–temperature phase diagrams for the VDF-*co*-HFP, TFE-*co*-HFP, and Teflon AF copolymers. Reprinted from Reference 4.

required to obtain a single phase of CO_2 and approximately 2–4 wt% of a fluorinated poly(isoprene) with a molecular weight of 45,000 g/mol. Harsher conditions of pressure and temperature, respectively 1200 bar and 80°C, were even necessary to dissolve a sample of fluorinated polybutadiene having a molecular weight of 31,200 g/mol.

13.2.2.1.2 Solubility of Copolymers A usual strategy to enhance the solubility of PVDF- or PTFE-based polymers in sc-CO_2 is the copolymerization of VDF (or TFE) with a comonomer such as hexafluoropropylene (HFP) [4], dioxole monomers [4], or vinyl acetate (VAc) [19,20]. The microstructure of the resulting copolymer is affected in order to decrease or suppress the crystalline domains and incorporate strong Lewis base sites to promote interactions with CO_2.

Hexafluoropropylene is undoubtedly the most studied comonomer for this purpose. The incorporation of 22 mol% of HFP into a VDF–HFP fluoropolymer renders the materials amorphous. The copolymer with an M_w of 85,000 g/mol gets soluble in milder conditions of approximately 750 bar and 100°C (see Figure 13.4) [4, 21]. However, this effect is less pronounced with copolymers of TFE and HFP. An equivalent amount of HFP only decreases the melting temperature of the copolymer to 147°C, which makes the copolymer (M_w equal to 210,000 g/mol) soluble at temperatures and pressures higher than 185°C and 1000 bar. The different shape of the curves was attributed to the stronger dipole moment of VDF units and the favored quadrupole–dipole interactions at lower temperatures. Similar remarks can be made from the phase behavior of a Teflon AF fluoropolymer synthesized from TFE and a perfluorinated cyclooxyaliphatic substituted ethylenic monomer (see Figure 13.4) [4]. The presence of this bulky comonomer incorporated at 65 mol%—suppressed the crystalline domains of PTFE, which lowered the cloud points of this PTFE-based copolymer to approximately 500 bar at 100°C. In this case, the sharp increase at lower temperatures was thought to result from the dipole interactions of Teflon AF. Recent works also suggest perfluoromethyl vinyl ether as a comonomer to design CO_2-philic VDF-based copolymers with M_n comprised between 1500 and 6000 g/mol and extremely low values of T_g [22].

In addition to affecting the microstructure, the incorporation of vinyl acetate into PTFE or PVDF chains through free radical copolymerization is expected to promote Lewis interactions with CO_2. Baradie et al. investigated the influence of TFE mol% in TFE-*co*-VAc copolymers [20]. A copolymer having 63.3 mol% of TFE (M_w = 290,000 g/mol) was insoluble even at 2100 bar and 144°C. Reducing the molar fraction of TFE resulted in the solubility of the copolymers due to the likely partial or total disappearance of crystalline domains and the enhancement of polymer–CO_2 interactions. Thus, a VAc-*co*-TFE copolymer with 46.7 mol% of TFE and an M_w equal to 180,000 g/mol exhibited a cloud point at 740 bar and 75°C for a polymer content of 5 wt%. Then, different copolymers with 26.5, 19.3, and 11.6 mol% of TFE and M_w ranging between 140,000 and 160,000 g/mol were fully soluble at around 550 bar and 25°C for equivalent polymer contents in CO_2. The enhanced solubilities were thought to arise from higher interactions of the partially fluorinated polymer backbone with CO_2 and stronger hydrogen bonding as a consequence of the enhanced acidity of protons neighboring fluorine atoms. In another study, they built on the enhanced solubility of such fluorocarbon-VAc copolymers to copolymerize VAc with vinylidene fluoride, chlorotrifluoroethylene, or tetrafluoroethylene in a surfactant-free polymerization process under mild pressure conditions of 200–230 bar [19].

13.2.2.2 Perfluoropolyethers Similarly to fluorinated olefins, the introduction of fluorine atoms onto the poly(ethylene oxide) (PEO) backbone has been explored through the family of perfluoropolyethers (sometimes coined perfluoroalkylethers or perfluoropolyalkylethers). Functionalized poly(hexafluoropropylene oxides) are representative perfluoropolyethers that actually attracted early attention for their solubility in sc-CO_2 (see Figure 13.5). These polymers are produced by anionic ring-opening polymerization of HFP oxide and commercialized as industrial lubricants by Dupont™ under the name of Krytox®.

In their pioneering work on CO_2 thickeners, Hoefling et al. studied the solubility of different Krytox-class compounds [23]. At a constant temperature of 295 K (i.e., 22°C), a Krytox polymer having an M_w of 13,000 g/mol exhibited cloud points at around 75, 125, and 150 bar for respective polymer contents of 1, 5, and 10 wt%. Another Krytox polymer having a carboxylic acid chain end and an M_w of 2500 g/mol (also known as Krytox 157 FSL) was also evaluated. This polymer surfactant was fully soluble at 110 bar and 40°C, for polymer contents up to 20 wt%. The sodium carboxylate version of this polymer was also studied with varying molecular weights [24]. A notable increase in pressure with increasing M_w was observed since 3 wt% of polymers of 5000 and 7500 g/mol were roughly soluble at 210 and 300 bar, respectively (see Figure 13.6), while around 4 wt% of polymer with an M_w of 2500 g/mol was soluble at 150 bar and 40°C.

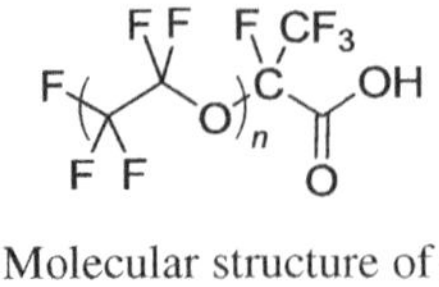

FIGURE 13.5 Molecular structure of a Krytox polymer.

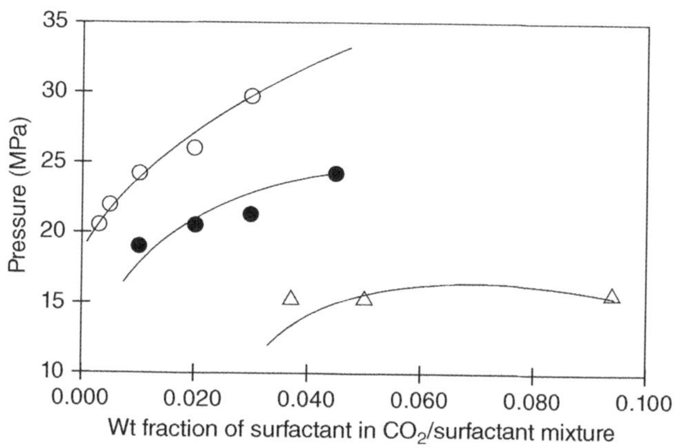

FIGURE 13.6 Pressure–composition phase diagram at 40°C for binary mixtures of CO_2 and sodium carboxylate–Krytox polymers with an M_w of (Δ) 2500 g/mol, (●) 5000 g/mol, and (O) 7500 g/mol. Reproduced from Reference 24.

The high solubility of Krytox polymers in sc-CO_2 was attributed to their weak polymer–polymer interactions, which is illustrated by their low surface tension in the range of 16–20 mN/m [25]. Interestingly, the second virial coefficient was also determined by neutron scattering experiments for a Krytox polymer with an M_w of 16,000 g/mol [26]. A value of $0 \pm 0.2 \times 10^{-4}$ mol.cm^3/g^2 was found, which suggested that the polymer chain adopted an unperturbed chain dimension in sc-CO_2.

As a consequence of their high solubility, perfluoropolyethers could be used to synthesize highly soluble amphiphilic copolymers. This will be described in the following sections of this chapter.

13.2.2.3 Fluorinated Poly(vinyl Esters) Poly(vinyl acetate) (PVAc) is a nonfluorous standard CO_2-philic polymer. Its solubility mainly arises from its strong Lewis-based interactions with CO_2 [12]. A strategy to increase the solubility of PVAc-based comes with the copolymerization of VAc with a fluorinated vinyl ester such as vinyl trifluoroacetate (see Figure 13.7a) [5]. As a result of the lowering of polymer–polymer interactions with increasing fluoromonomer contents, the solubility of such copolymers could be significantly enhanced. Thus, 0.2 wt% of a P(VAc$_{50}$-*stat*-VTFAc$_{50}$)

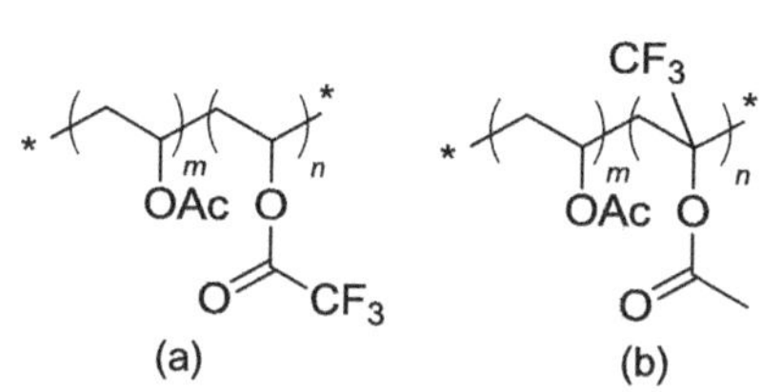

FIGURE 13.7 Molecular structures of (a) poly(vinyl acetate-*stat*-vinyl trifluoroacetate) and (b) poly(vinyl acetate-*stat*-(1-trifluoromethyl) vinyl acetate) copolymers.

copolymer having an M_n of 4100 g/mol was soluble at 177 bar and 40°C, that is, 100 bar lower than a PVAc equivalent sample. Beyond this specific composition, the reinforcement of polymer–polymer interactions through the existence of crystalline zones considerably hampered the solubility of the P(VAc-*stat*-VTFAc) copolymers.

Due to the high sensitivity of vinyl trifluoroacetate toward hydrolysis, (1-trifluoromethyl) vinyl acetate (CF_3VAc) has been suggested as an alternative fluorinated vinyl ester to design CO_2-philic PVAc-based copolymers (see Figure 13.7b) [27]. In this series, the minimal cloud point was observed at 222 bar, corresponding to a copolymer with an M_n of 4500 g/mol and a VAc:CF_3VAc composition of 67:33. This composition optimum marked the balance between polymer–polymer interactions and entropy of mixing. However, the position of the trifluoromethyl group on the polymer backbone affected the self-interactions of the resulting copolymers to a lesser extent than the P(VAc-*stat*-VTFAc) copolymers.

13.2.2.4 Fluorinated Poly(meth)acrylates Comparing fluorinated and hydrogenated poly(methacrylates), McHugh et al. studied the solubility of poly(1,1,2,2-tetrahydroperfluorohexyl methacrylate) (PFHMA, M_n = 200,000 g/mol), poly(1,1,2,2-tetrahydroperfluorooctyl methacrylate) (PFOMA, M_n = 292,000 g/mol), and poly(1,1,2,2-tetrahydroperfluorodecyl methacrylate) (PFDMA, M_n = 196,000 g/mol) in sc-CO_2 [28]. They observed a dramatic drop in the pressures and temperatures required to solubilize the fluorinated materials (see Figure 13.8). The reduction to a shorter fluorinated side chain—such as in poly(2,2,2-trifluoroethyl methacrylate)—gave equivalent cloud point pressures (M_n = 268,000 g/mol). For a polymer content of 3.9 wt%, the single phase transition was thus observed at 323 bar and 60°C [29].

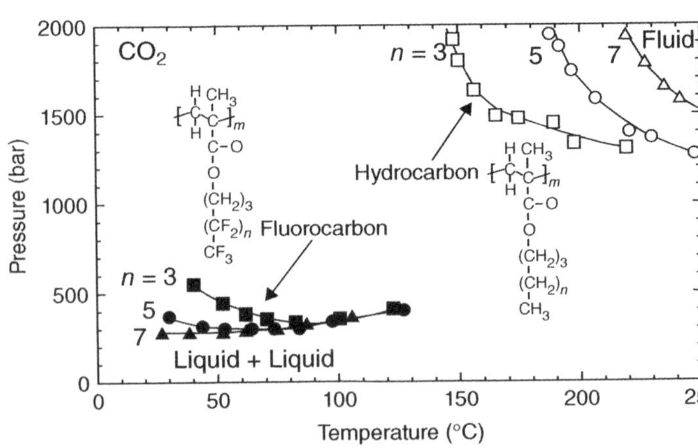

FIGURE 13.8 Pressure–temperature diagram comparing the solubility of fluorinated poly(methacrylates) and their nonfluorous analogous (polymer content of 4 wt%). Reproduced from Reference 28.

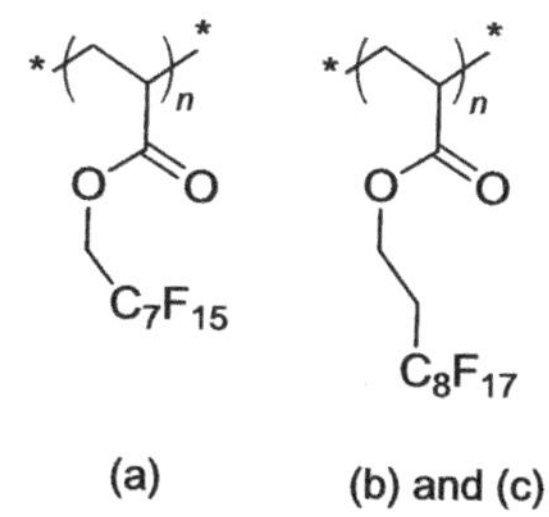

FIGURE 13.9 Molecular structures of (a) poly(1,1-dihydroperfluorooctyl acrylate) (PFOA), (b) poly(1,1,2,2-tetrahydroperfluorodecyl acrylate) (PFDA), and (c) poly(1,1,2,2-tetrahydroperfluorodecyl acrylate) with an average value of fluorinated carbon equal to 8 (PTAN).

In this family of fluorinated poly((meth)acrylates), investigations on sc-CO_2 solubility mostly focused on poly(1,1-dihydroperfluorooctyl acrylate) (PFOA), poly(1,1,2,2-tetrahydroperfluorodecyl acrylate) (PFDA), or poly(1,1,2,2-tetrahydroperfluorodecyl acrylate) with an average value of fluorinated carbon equal to 8 (PTAN) (see Figure 13.9).

Hsiao et al. reported a cloud point pressure at approximately 150–170 bar and 40°C for 4 wt% of a PFOA sample (M_w = 1,000,000 g/mol) [30]. An LCST-type behavior was also noted as a result of the positive dependency of pressure upon increasing temperatures. Through neutron scattering experiments, this high solubility of PFOA could be related to a positive second virial coefficient (A_2) in conditions as mild as 207 bar and 45°C in the case of sample with an M_w equal to 36,500 g/mol [31]. However, the theta pressure was not determined. In another set of experiments, the positive A_2 was found to be a decreasing function of molecular weight, as expected from the case of polymer solutions in incompressible solvents [32]. A_2 was also estimated to be proportional to M_w at the power of -0.4.

Although PFDA has a longer side chain, its solubility in sc-CO_2 is equivalent to amorphous PFOA [33]. Thus, 4 wt% of a PFDA (M_w = 254,000 g/mol) was fully soluble at pressures around 110 and 200 bar at 25°C and 50°C, respectively. Lacroix-Desmazes and coworkers even determined lower cloud point pressures of 85 and 160 bar under the same conditions of polymer content and temperature but for a lower M_w of 120,000 g/mol [34, 35]. The influence of decreasing molecular weight on solubility also appeared negligible [36, 37]. From static light-scattering experiments, the theta pressure of this polymer at 25°C was determined at 160 bar (i.e., a theta density around 0.89 g/cm^3), which was above the observed cloud point pressure at 85°C [35]. Above this pressure of 160 bar, the second virial coefficient was positive (see Figure 13.10). The difference between the theta pressure and the cloud point ones was attributed to unfavorable interactions with CO_2 originating from the entropic contribution of the smaller chains. The theta pressure is actually the extrapolation of the cloud point pressures at infinite molecular weight. Finally, the length dispersity of the fluorinated chains in PTAN did not affect its solubility compared to those of PFDA and PFOA (see Figure 13.9) [28]. Thus, 4.9 wt% of a

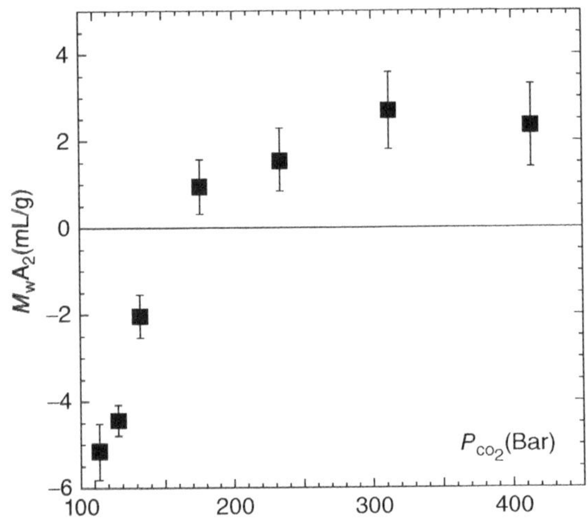

FIGURE 13.10 Weight average molecular weight times second virial coefficient as a function of CO_2 pressure at 25°C for a PFDA sample (≈ 4 wt% in CO_2). Reproduced from Reference 35.

PTAN (unspecified M_w) was fully soluble at 190 bar and 50°C although a melting point was detected at 78°C [38]. This may be due to the liquid crystalline character of PTAN, rather than a semi-crystalline one.

On the contrary, the introduction of a methyl group on the polymer backbone—thus changing acrylates into methacrylates—has a much greater impact on the solubility of fluorinated methacrylates compared to their acrylate analogues. Considering the case of poly(1,1,2,2-tetrahydroperfluorodecyl methacrylate) (PFDMA), the cloud point pressures were approximately higher of 100 bar at 40°C (see Figure 13.11) compared to PFDA. This was related to the higher steric hindrance and the repulsive interactions provided by the methyl group. However, this effect was less pronounced at higher temperatures. Similarly to PFOA and PFDA, the second virial coefficient could also be evaluated via static light-scattering measurements. The theta density at 25°C was evaluated at 0.88 ± 0.02 g/mL for a PFDMA sample having an M_w of 300,000 g/mol [39].

Similarly to perfluoropolyethers, the widely accepted explanation for the high solubility of fluorinated poly(acrylates) in sc-CO_2 is provided by their low cohesive energy densities and weak self-interactions. Johnston and coworkers reported surface tension values of 10–15 mN/m, which is actually lower than the values for perfluoropolyether and poly(vinyl acetate), the other aforementioned CO_2-philic references.

Thanks to their "true" CO_2-philicity, macromolecular engineering using fluorinated poly((meth)acrylates) as CO_2-philic building blocks has been largely investigated by CO_2 researchers as it will be developed later in the chapter [34,40–42].

13.2.2.5 Poly(perfluorodiitaconates) Following the high solubility of fluorinated poly((meth)acrylates), McHugh and coworkers investigated the CO_2-philicity of poly(perfluorodiitaconates) which differ from poly((meth)acrylates) by an additional

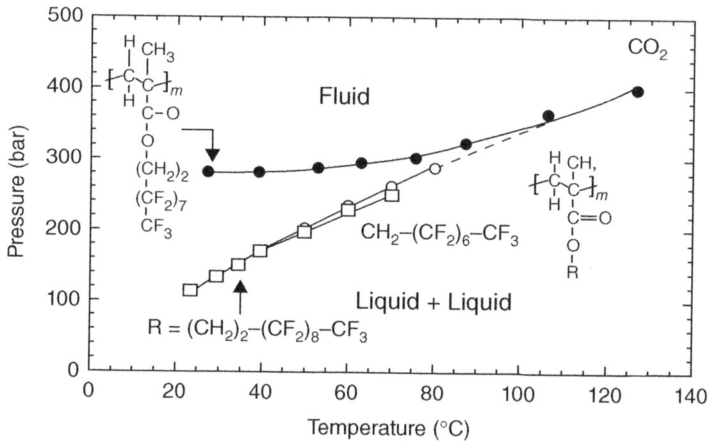

FIGURE 13.11 Pressure–temperature phase diagrams for (□) PTAN (○) PFOA, and (●) PFDMA samples with polymer contents ranging between 3 and 4 wt%. Reproduced from Reference 28.

fluorinated ester functionality replacing the methyl group (see Figure 13.12) [43]. Irrespective of the different side chain lengths and polymer content, these polymers (M_w near 150,000 g/mol) exhibited equivalent behaviors at low temperatures. Thus, their cloud point pressures were located near 150 bar at 40°C, which is somehow comparable to those of fluorinated poly(acrylates) but lower than fluorinated poly(methacrylates).

13.2.2.6 Miscellaneous Polymers With Grafted Perfluorinated Side Chains Following the high solubility of fluorinated poly((meth)acrylates), the grafting of long perfluorinated chains has been applied to a variety of CO_2-phobic polymers including poly(styrene), poly(para-phenylenes), polysiloxanes, and polyphosphazenes to enhance their solubility in sc-CO_2.

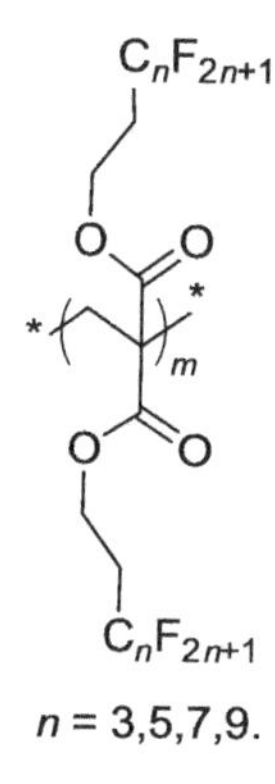

n = 3,5,7,9.

FIGURE 13.12 Molecular structure of poly(perfluorodiitaconates).

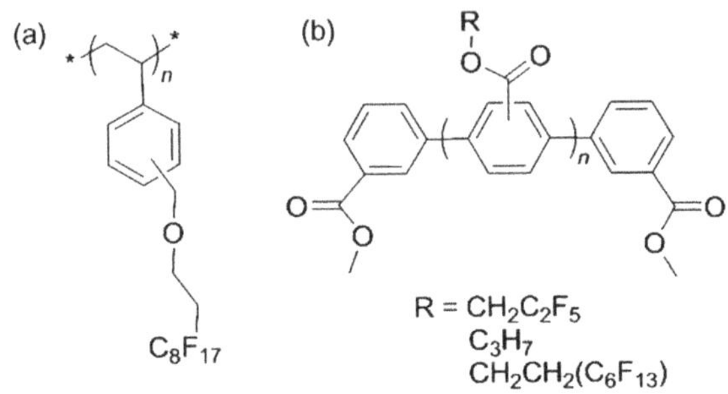

FIGURE 13.13 Molecular structure of (a) poly(perfluorooctyl-ethyleneoxymethylstyrene) (PFDS) and (b) poly(para-phenylenes).

13.2.2.6.1 Polystyrene With Fluorinated Grafts Due to strong polymer–polymer interactions and a high T_g, polystyrene is practically insoluble in CO_2 [4]. However, the etherification of chloromethylstyrene with 1,1,2,2-tetrahydroperfluorodecanol and the consecutive polymerization of the monomer provided high solubility in sc-CO_2 to the resulting poly(perfluorooctyl-ethyleneoxymethylstyrenes) (PFDS) (see Figure 13.13a) [34]. For a polymer content of 4 wt% at 40°C, a PFDS having an M_n of 43,600 g/mol thus exhibited a cloud point around 200 bar. This value was higher of approximately 60 bar compared to a PFDA having an M_n of 52,600 g/mol.

13.2.2.6.2 Poly(para-phenylene) With Fluorinated Grafts The same paradigm was applied to poly(para-phenylenes) (see Figure 13.13b) [44]. While samples with fluorous or nonfluorous propyl side chains (unknown M_n) were insoluble in CO_2 even at 190°C and 2400 bar, the extension to a 1,1,2,2-tetrahydroperfluorohexyl chain allowed the complete solubilization of 1.9 wt% of materials around 140 bar and 40°C.

13.2.2.6.3 Poly(siloxanes)-Based Copolymers With Fluorinated Side Groups A decrease in polymer–polymer interactions was also investigated synthesizing methylpropenoxyperfluoroalkyl siloxane homopolymers (see Figure 13.14) [28]. For a

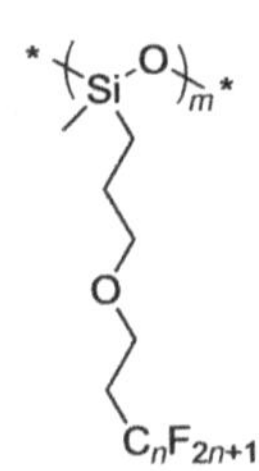

FIGURE 13.14 Molecular structure of methylpropenoxyperfluoroalkyl siloxane homopolymers (n = 5 or 7).

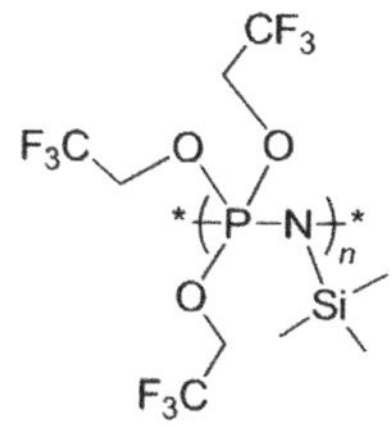

FIGURE 13.15 Molecular structure of poly[*P*-tris(trifluoroethoxy)-*N*-trimethylsilyl] phosphazene.

polymer content of 3.5 wt% and a temperature of 25°C, poly(siloxanes) functionalized with chains of either five or seven fluorinated carbons ($M_w = 14{,}600$ and 17,700 g/mol) exhibited lower cloud point pressures around 100 bar, while their hydrogenated equivalents could not be solubilized under mild conditions of pressure and temperature ($P < 500$ bar and $T < 100$°C). By means of comparison, propyl acetate-functionalized poly(siloxane) oligomers of 25 repeating units were soluble at around 150 bar and 22°C, for a polymer content of 3.5 wt% [45]. Lowering self-interactions with the help of fluorinated grafts thus revealed to be more efficient than increasing polymer–solvent interactions through the incorporation of carbonyl groups.

13.2.2.6.4 Fluorinated Poly(phosphazenes) Polyphosphazenes are inorganic polymers with a polymer backbone made of phosphorus and nitrogen atoms (see Figure 13.15). Like poly(siloxanes), they exhibit a high chain flexibility, and their side chains can be tuned to induce CO_2-philicity. The solubility of poly[*P*-tris(trifluoroethoxy)-*N*-trimethylsilyl]phosphazene (M_w = 96,000 g/mol) has been investigated although no data were published on a nonfluorous equivalent [3]. Interestingly, this polymer demonstrated a high solubility in sc-CO_2 at 35°C since a cloud point was observed around 185 bar for a polymer content of 4 wt%.

13.3 SELF-ASSEMBLY OF AMPHIPHILIC COPOLYMERS IN sc-CO_2

13.3.1 Block Copolymers

Amphiphilic block copolymers that are soluble in CO_2 comprise a CO_2-philic block and a CO_2-phobic one and are thus prone to form complex structures in solution such as micelle-like aggregates and polymersomes. As a result of their high solubility, fluorinated poly(acrylates) and polysiloxanes mainly constitute the CO_2-philic building blocks of choice to design CO_2-soluble amphiphilic copolymers.

In their seminal work, McClain et al. designed a series of poly(styrene)-*b*-poly(1,1-dihydroperfluorooctyl acrylate) (PS-*b*-PFOA) through a photopolymerization process using a tetraethylthiuram disulfide (see Figure 13.16a) [46]. The chain length of each block was varied with regard to their self-assembly behavior. At 65°C and 340 bar, these copolymers self-assembled into polydisperse spherical core-shell structures as

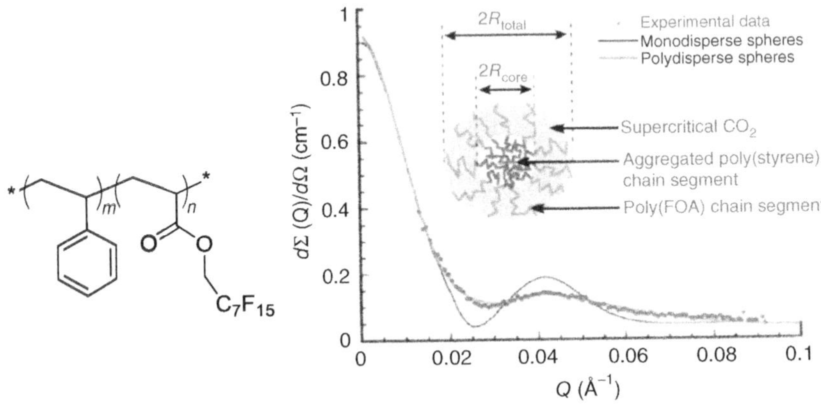

(a) (b)

FIGURE 13.16 (a) Molecular structure of poly(styrene)-*b*-poly(1,1-dihydroperfluorooctyl acrylate) (PS-*b*-PFOA) (b) SANS plot for a solution a PS-*b*-PFOA (with respective M_n of 3700 and 16700 g/mol, at 4 wt%) in sc-CO_2 at 65°C and 340 bar. Reproduced from McClain et al. [46].

evidenced by small-angle neutron scattering (SANS) (see Figure 13.16b). For a PS block with a constant M_n of 3700 g/mol, an increase in the PFOA chain length resulted in an increase of the radius of the shell and a decrease of the swelling of the core. Inversely, the radius of the core and the aggregation number (N_{agg}) were constant. Increasing the radius of the core could be achieved by an increase of the chain length of the PS block. As a consequence, the scaling laws for block copolymer aggregates were equivalent in CO_2 and usual organic solvents. Finally, the influence of the solvent strength on the self-assembly of PS-*b*-PFOA surfactants was also evaluated. At a lower temperature (i.e., a higher density of 0.934 g/mL), the solvation of both blocks was enhanced, which increased the radius of both core and shell but decreased N_{agg}. The aggregation of PS-*b*-PFOA amphiphilic block copolymers were also corroborated by small-angle X-ray scattering (SAXS) [47] measurements. Finally, the dynamic but slow exchange between unimers and copolymers aggregates was evidenced by DOSY-NMR experiments [48].

Using poly(tert-butyl methacrylate) as the CO_2-phobic block, Lo Celso et al. studied the self-assembly of poly(tert-butyl methacrylate)-*b*-poly(1,1-dihydroperfluorooctyl methacrylate) (PtBMA-*b*-PFOMA) in sc-CO_2 at 45°C (see Figure 13.17a) [49]. Through SAXS and SANS experiments, a transition from a random coil to a polydisperse sphere morphology was evidenced suggesting the existence of a critical micelle density (CMD) in the range of 0.85–1.05 g/mL. From different chain lengths of both blocks, the CO_2-phobic part appeared to be the major influencing factor concerning the value of the CMD. Note that the CMD was a decreasing function of temperature. This transition was characterized by a two-step process where a phase separation between the solvent and the polymer chain first occurred upon decreasing pressure. Then, the CO_2-philic moiety reoriented toward

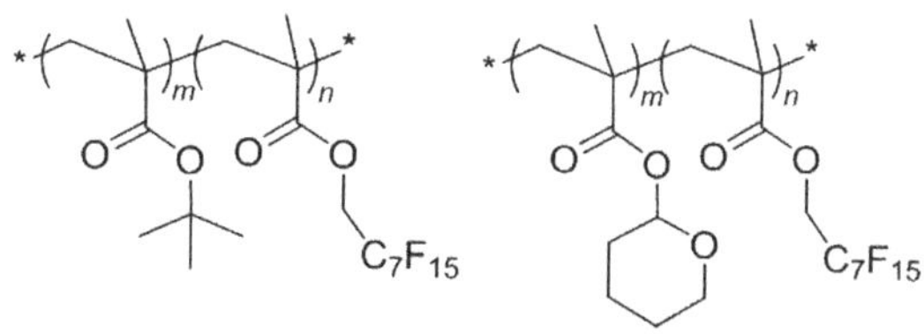

FIGURE 13.17 Molecular structures of (a) poly(tert-butyl methacrylate)-*b*-poly(1,1-dihydroperfluorooctyl acrylate) (PtBMA-*b*-PFOMA) and (b) poly(2-tetrahydropyranyl methacrylate)-*b*-poly(1,1-dihydroperfluorooctyl methacrylate) (PHPMA-*b*-PFOMA).

the solvent-rich domains, which led to the formation of aggregates. The kinetics of the second step was revealed to be slower and depended on the pressure jump.

The unimer-to-aggregates transition was spotted by SAXS experiments on a solution of poly(2-tetrahydropyranyl methacrylate)-*b*-poly(1,1-dihydroperfluorooctyl methacrylate) (PHPMA-*b*-PFOMA) in sc-CO_2 at 52°C (see Figure 13.17b) [50]. Three distinctive behaviors were identified. While the copolymer was insoluble for pressures lower than 121 bar, the radius of gyration (R_g) gradually increased with pressures increasing from 121 to 220 bar, suggesting the association of unimers into aggregates. The values of R_g were then constant around 3 nm for pressures between 220 and 335 bar, and decreased for higher pressures as the aggregates partially dissolved.

Perfluoropolyethers were also used as CO_2-philic building blocks to design amphiphilic block copolymers. Lodge and coworkers synthesized two examples of poly(lactide)-*block*-poly(hexafluoropropylene oxide) (PLA-*b*-PFPO, M_n around 10,000 g/mol) with varying CO_2-philic/CO_2-phobic balances (see Figure 13.18) [51]. The self-assembly of these copolymers in CO_2 at 25°C and 400 bar was studied by dynamic light scattering (DLS). A 1 wt% solution of a copolymer ($M_{n,PLA}$ = 4000 g/mol, $M_{n,PFPO}$ = 6000 g/mol) was found to self-assemble into spherical aggregates with a hydrodynamic radius (R_h) of 4 nm. Interestingly, the R_h of a copolymer with a lower CO_2-philic/CO_2-phobic balance ($M_{n,PLA}$ = 5000 g/mol, $M_{n,PFPO}$ = 4000 g/mol) was 135 nm although the maximum chain length was evaluated at 30 nm for this copolymer. Upon freezing of the solution and redispersion in a PFPO selective solvent, observations via transmission electron microscopy (TEM) revealed the formation of polymersomes (i.e., polymer vesicles). To date, this constitutes the unique example of such structures in CO_2.

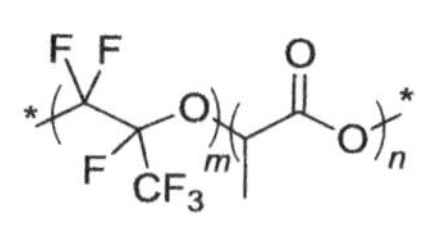

FIGURE 13.18 Molecular structure of poly(hexafluoropropylene oxide)-*block*-poly(lactide) copolymers.

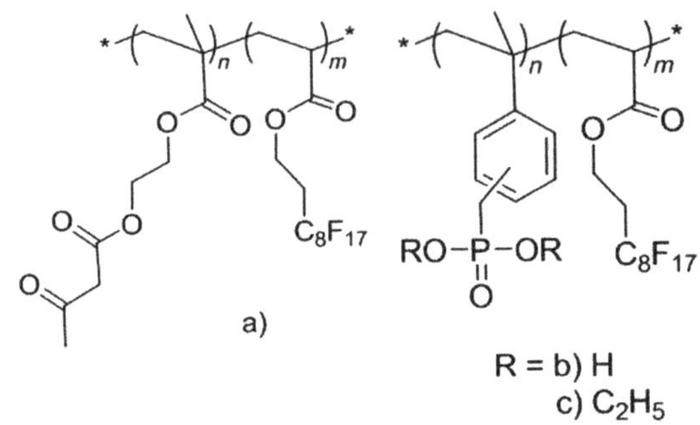

FIGURE 13.19 Molecular structure of gradient copolymers synthesized by Ribaut et al. [36].

13.3.2 Amphiphilic Gradient Copolymers

Contrary to statistical and block copolymers, gradient copolymers exhibit a composition gradient along their chain. The properties of gradient copolymers are expected to be intermediate between those of block and statistical copolymers. In particular, their kinetics of aggregation and exchange can be faster than in the case of block copolymers.

Building on the lower reactivity ratios of 1,1,2,2-tetrahydroperfluorodecyl acrylate (PFDA) compared to methacrylates and styrenics, Ribaut et al. synthesized amphiphilic gradient copolymers incorporating either acetoacetoxyethyl methacrylate, vinyl benzylphosphonic acid diethylester or vinyl benzylphosphonic acid by controlled radical polymerization (see Figure 13.19) [36]. The gradient structures could be confirmed by monitoring the individual conversions of each monomer. At similar conditions of composition, molecular weight, and polymer content, these gradient copolymers exhibited enhanced solubilities in CO_2 compared to equivalent block copolymers [52]. Interestingly, the nature of the CO_2-phobic monomer hardly had any effect on the solubility of gradient copolymers contrary to diblock copolymers where it dominated over compositional effects.

Their self-assembly behavior in sc-CO_2 was then revealed by SANS experiments [52]. As a general trend, both composition and molecular weight of the gradient copolymers had little influence of R_g and N_{agg}. Comparing the organization of equivalent block and gradient copolymers above their respective cloud points, R_g was found lower for the block sample whereas N_{agg} was similar. Parallel to the cloud point observations, this suggested the lower solvation of block copolymer aggregates as a result of a better microphase separation. Experimental parameters such as pressure, temperature, and density were also investigated (see Figure 13.20) [52]. In the case of acetoacetoxyethyl methacrylate-based gradient copolymers at 4 wt% in sc-CO_2, an increase of pressure from 140 to 210 bar resulted in N_{agg} and R_g, respectively, decreasing from 2.9 to 1 and from 3.3 to 2.5 nm. This corresponded to the transition from small aggregates to unimers resulting from the enhancement of solvent strength of CO_2. This evidenced a pressure-induced aggregation. Similarly,

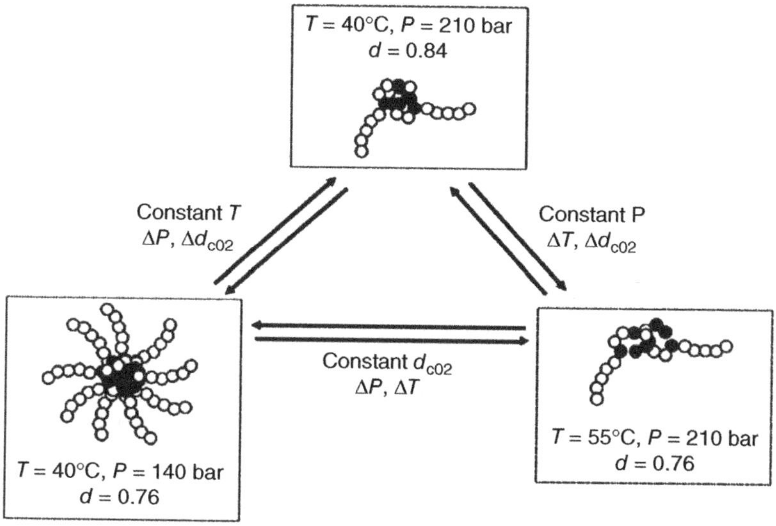

FIGURE 13.20 Experimental parameters investigated for the solubility of gradient amphiphilic copolymers in sc-CO_2. Reproduced from Reference 52.

a decrease in temperature from 55°C to 40°C at a constant pressure of 210 bar resulted in a better solvent quality, that is, a higher density and increased R_g. This was characteristic of a temperature-induced aggregation. However, the existence of hydrogen bonding between poly(acetoacetoxyethyl methacrylate) blocks limited the enhancement of aggregation compared to a purely density-induced phenomenon. The same remark on H-bonds applied when changing both temperature and pressure at a constant CO_2 density of 0.76 g/mL. The aggregates were actually more swollen due to temperature effects.

13.3.3 Double CO_2-Philic Block Copolymers

Similarly to double hydrophilic block polymers, double CO_2-philic copolymers are made of two chemically different blocks soluble in CO_2. At high pressures, they behave as fully soluble block copolymers. Their amphiphilic behavior can be induced by a change in pressure or temperature which renders CO_2 a selective solvent for one of the blocks. The resulting amphiphilic copolymers with a CO_2-philic outer shell and a CO_2-phobic core can be expected to self-assemble into aggregates or polymersomes. To date, the only example of double CO_2-philic copolymers concerns block copolymers comprising fluorinated poly(acrylate) and poly(vinyl acetate) blocks (see Figure 13.21).

In their work, Zhou and Chu used DLS to study the pressure-induced aggregation of a poly(1,1-dihydroperfluorooctyl acrylate)-*b*-poly(vinyl acetate) (PFOA-*b*-PVAc) block copolymer under an isothermal process at 65°C [53]. The respective molecular weights of the blocks were 43,100 and 10,300 g/mol. The block polymer exhibited

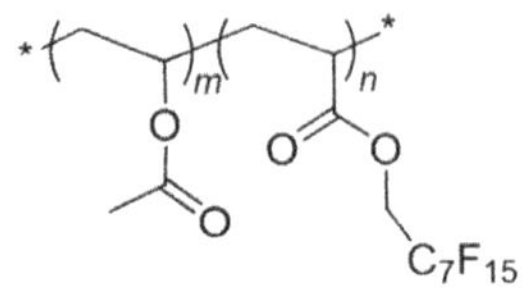

FIGURE 13.21 Molecular structure of a double CO_2-philic block copolymer comprising PFOA (or PTAN) and PVAc blocks.

distinctive aggregation states depending on pressure and concentration. For a concentration of 0.057 g/mL, a first transition was observed at 148 bar, which corresponded to the appearance of well-defined unimers with a hydrodynamic radius of 2.2 nm in solution. Then, a small fraction of polydisperse aggregates with an $\langle R_h \rangle$ around 500 nm coexisting with unimers were formed at 225 bar. With increasing pressures of 242, 310, 345, and 379 bar, the large aggregates disappeared at the benefit of a high proportion of narrowly distributed micelles with an $\langle R_h \rangle$ of 23 nm. Within this range of pressure, the micelle-to-unimer intensity ratio and the number aggregation, respectively, decreased from 30:1 to 11:1, and from 18 to 9. Inversely, the calculated critical micelle concentration increased from 0.021 to 0.033 g/mL. At higher pressures, aggregates formed again, and the micelles gradually disappeared at the benefit of both aggregates and unimers until reaching a pressure of 552 bar where all micelles disappeared. This value was taken as the critical micelle pressure (CMP). The complex behavior of this copolymer was attributed to the composition heterogeneity of the polymer sample, resulting from its dispersity.

Using the same polymer sample, Chu and coworkers also studied the influence of temperature on the aggregation states under an isobaric process at 225 bar and for a polymer concentration of 0.024 g/mL [50]. At 65°C, ill-defined aggregates with an $\langle R_h \rangle$ and unimers were concomitantly observed. With a decrease of solvent quality at 55°C, the disappearance of unimers was observed while aggregates and micelles having an $\langle R_h \rangle$ around 20 nm coexisted. Lowering the temperature to 35°C and 30°C resulted in the increase of the aggregate:micelle ratio and, finally, the only existence of large particles. The critical micelle temperature (CMT) was estimated at around 33°C for these particular conditions of concentration and pressure.

Parallel to CO_2-philic homopolymers and their upper critical solution density, the existence of both CMT and CMP transitions is actually indicative of a CMD. This concept was introduced by Buhler et al. [31], performing static and dynamic light-scattering studies on a poly(1,1-dihydroperfluorooctyl acrylate)-*b*-poly(vinyl acetate) (PTAN-*b*-PVAc) block copolymer (see Figure 13.22). The respective molecular weights of the blocks were 43,100 and 4400 g/mol. Through the examination of aggregation numbers (N_{agg}), this copolymer was found insoluble at densities lower than 0.814 g/cm^3 (N = 120) and fully soluble at densities higher than 0.95 g/cm^3 (N_{agg} = 1). Although N decreased with increasing density resulting from the better solvation of both blocks, spherical micelles with a constant $\langle R_h \rangle$ around 15 nm were observed in between. This appeared as a consequence of compensating effects of the excluded volume parameter for the shell chains and the core–shell interfacial tension.

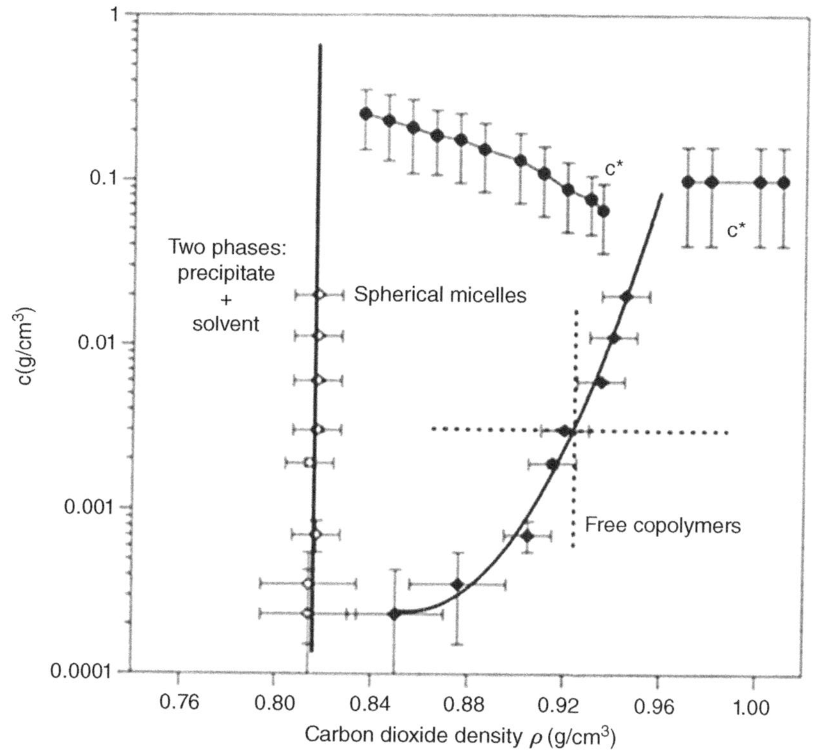

FIGURE 13.22 Polymer concentration–density phase diagram for a binary mixture of CO_2 with poly(1,1-dihydroperfluorooctyl acrylate)-*b*-poly(vinyl acetate) at 45°C ($M_n \approx$ 47,500 g/mol). Reproduced from Reference 31.

Finally, the critical micelle concentration—that is, the transition curve between spherical micelles and free copolymers—was determined to depend on density. The CMD phenomenon was also confirmed through SANS and SAXS studies [54–56]. A transition from a random coil to a polydisperse core–shell could be modeled with either increasing temperatures at 40°C or pressures at 240 bar [54].

13.3.4 Amphiphilic Graft and Comb Copolymers

Graft and comb copolymers are another fascinating class of polymers. They are made distinctive by their synthesis pathway. Graft copolymers are obtained through either the "grafting from" or "grafting onto" methods. The macromonomer route is used for the synthesis of comb copolymers. Regarding the solubility and self-assembly of graft and comb copolymers, the macromolecular parameters of study are the graft chain length, the graft chain density, and the backbone length.

An amphiphilic comb copolymer could be synthesized from the free-radical copolymerization of an 1,1-didydroperfluorooctyl acrylate monomer with a

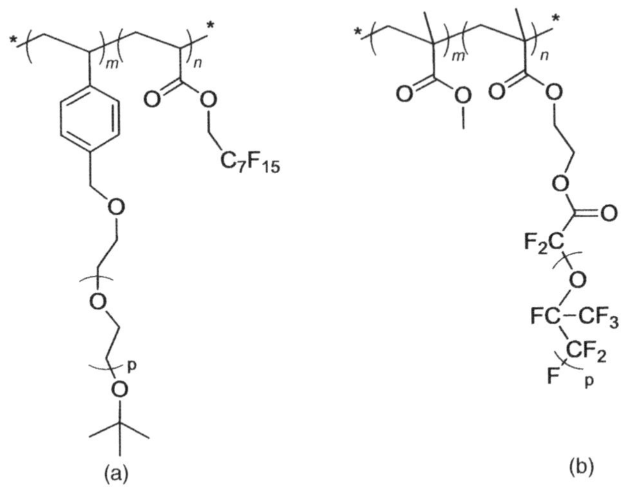

FIGURE 13.23 Molecular structures of amphiphilic comb and graft block copolymers.

poly(ethylene glycol) styrenic macromonomer (see Figure 13.23) [57]. The self-assembly of this polymer with PEO grafts having an M_n of 5000 g/mol and an overall PEG wt% of 15–18% was investigated through SANS experiments performed at 333 bar and 65°C [26]. Aggregates with an R_g of 5.6 nm and an N_{agg} of 3 were deduced from a polydisperse core–shell model. The aggregates were demonstrated to swell upon incorporation of water in binary mixtures.

Lepilleur et al. examined the phase behavior of a series of statistical perfluoropolyether-grafted poly(methyl methacrylate-*co*-hydroxyethyl methacrylate) in CO_2 (see Figure 13.21b) [58]. These copolymers were synthesized via the grafting of a carboxylic acid-terminated perfluoropolyether on a statistical poly(methyl methacrylate-*co*-hydroxyethyl methacrylate backbone [58]. For shorter polymer backbone lengths, the cloud point pressures could be decreased by increasing either graft chain lengths or graft chain density until a minimum in pressure was reached. Beyond this minimum, further increases resulted in higher cloud point pressures and the domination of molecular weight effects over lowered polymer–polymer interactions. As expected, longer polymer backbones and higher CO_2-phobic/CO_2-philic balances resulted in higher cloud point pressures. The best strategy to enhance the solubility of comb polymers was built on a high grafting density with low molecular weight grafts.

13.3.5 Amphiphilic Hyperbranched Polymers and Dendrimers

These classes of polymers are of special interest as unimolecular aggregates in sc-CO_2. The CO_2-solubility of hyperbranched polymers and dendrimers having a hydrophilic core is usually achieved via the grafting of either perfluorinated chains [59–61] or CO_2-philic polymers (see Figure 13.24) [60]. Mecking et al. thus functionalized a hyperbranched polyethyleneimine core (M_w = 5000 g/mol) by imidazole

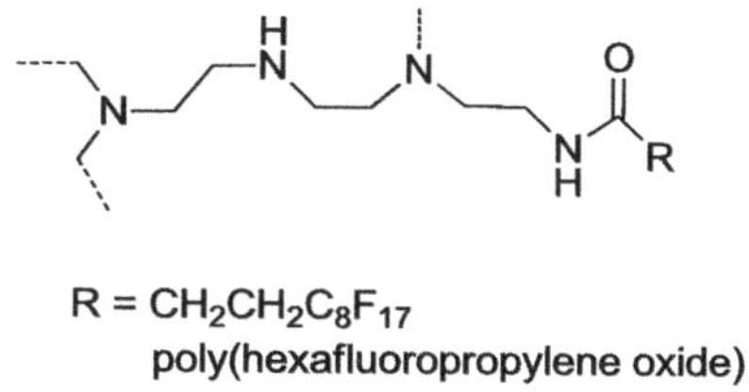

FIGURE 13.24 Molecular structure of hyperbranched polymers and dendrimers grafted with fluorinated moieties.

coupling with a long perfluorinated chain or poly(hexafluoropropylene oxide) (see Figure 13.24) [60]. At a density of 0.88 g/mL, 0.3 wt% of a perfluorinated sample with a CO_2-phobic/CO_2-philic balance of 3.2 was soluble. At a constant balance, the solubility was approximately twice as high as the perfluopolyether polymeric chain.

13.4 FLUORINATED MACROMOLECULAR SURFACTANTS FOR WATER/CARBON DIOXIDE EMULSIONS

13.4.1 Water/CO_2 Macroemulsions

Considering the design of macromolecular surfactants for the stabilization of W/C emulsions, poly(1,1-dihydroperfluorooctyl acrylate) (PFOA) and poly(1,1-dihydroperfluorooctyl methacrylate) (PFOMA) are strong candidates due to their true CO_2-philicity. Lim and coworkers explored the interfacial properties of an amphiphilic block copolymer comprising a PFOMA block (M_n = 7500 g/mol) and a PEO (M_n = 5000 g/mol) at the W/C interface as a function of salinity, temperature, and pressure (see Figure 13.25) [62]. At a concentration of 0.05 wt% and a CO_2 density of 0.9755 g/mL ($T = 25°C$, $P = 320$ bar), the interfacial tension could be lowered to 4.40 mN/m. Increasing the ionic strength with the addition of 0.15 mol/L NaCl then increased the surface tension to 5.11 mN/m as the solubility of the surfactant in water was lowered. Similarly, an increase in temperature at a constant pressure of 320 bar and a decrease in pressure to 25°C resulted in higher values of surface tension. From this, a W/C emulsion could be formed with 0.05 wt% of surfactant in a

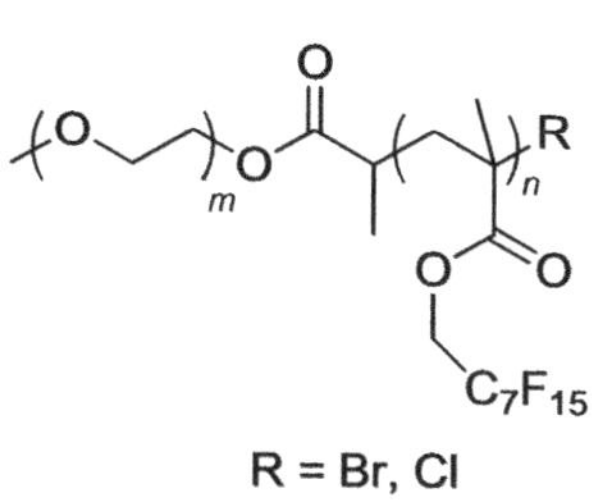

FIGURE 13.25 Molecular structure of a PEO-*b*-PFOMA block copolymer.

50:50 (w/w) water/CO_2 mixture. After stopping stirring and recirculation, creaming was observed after 30 minutes.

Following the same strategy, the formation and stability of W/C emulsions was assessed in the presence of a PEO-*b*-PFOA copolymer ($M_{n,\,PEO} = 2000$ g/mol and $M_{n,PFOA} \approx 31{,}000$ g/mol) [40]. Note that the hydrophilic/CO_2-philic balance (HCB) was lower, and the M_n of the copolymer was higher compared to the previous study. The observation of a W/C emulsion was reported from 25°C and 143 bar while flocculation or coalescence of dispersed water phase slowly occurred upon stopping of the stirrer. Notably, the formation of a W/C emulsion was reported at the cloud point pressure of the copolymer in a binary polymer/CO_2 mixture.

13.4.2 Water/CO_2 Miniemulsions

The most convincing attempt to stabilize W/C miniemulsions undoubtedly came with the use of poly(fluorinated methacrylate)-based block copolymers. Whereas creaming was observed with W/C macroemulsions stabilized with a PEO-*b*-PFOMA copolymer (see Figure 13.25), its applicability to W/C miniemulsions was successfully demonstrated [63]. Concomitantly with low values of surface tension ranging between 0.19 and 0.50 mN/m, water droplets sized between 70 and 120 nm were obtained from 1 wt% of a sample of PEO-*b*-PFOMA ($M_{n,PFOMA} = 5500$ g/mol and $M_{n,PEO} = 2000$ g/mol) in a mixture of CO_2 and 2–3 wt% of water. Variations in temperature, CO_2 density, shear rate, and HCB further allowed tuning the surface tension and the resulting average drop sizes. Remarkably, the droplet size was proportional to surface tension at the power of 2/5 while this factor equals to 3/5 for incompressible fluids. Ultimately, these W/C miniemulsions exhibited an unprecedented stability over 24 hours against flocculation.

13.4.3 Critical Flocculation Density

The concept of critical flocculation density (CFD) gives a comprehensive overview on flocculation and the relative efficiency of macromolecular surfactants. The CFD is defined as the CO_2 density below which flocculation occurs. Interestingly, the CFD of a homopolymer coincided with its UCSD, provided that the chain length was long enough to screen the attractive Van der Waals forces. This qualitatively differs from incompressible solvents where the critical flocculation temperature coincides with the theta temperature. When the attractive forces dominate over steric stabilization, flocculation then occurs at densities higher than the CFD. Regarding block copolymers, the chain of the solvophilic block must be similarly long enough to prevent flocculation above its UCSD [64].

The CFD of W/C miniemulsions stabilized by PEO-*b*-PFOMA copolymers and its dependency on molecular weight, temperature, surfactant, and water contents were investigated by Johnston and coworkers [65]. As expected, the CFD decreased with either smaller droplet size or increasing molecular weight of the CO_2-philic PFOMA block as a result of enhanced screening of the core–core attractions. The attractive forces were fully dominated when the molecular weight was higher than

30,000 g/mol. An increase in temperature also induced a decrease in the CFD since the attractive forces are inversely proportional to temperature. Inversely, when the surfactant content was decreased from 1 to 0.1 wt%, the CFD increased, which resulted from the poorer surface coverage. The W/C miniemulsions were then destabilized.

13.4.4 Water/CO_2 Microemulsions

Compared to macro- and miniemulsions, the water uptake is lower in microemulsions. By virtue of their smaller sizes and, subsequently, the more efficient packing of the surfactant at the interface, microemulsions are not subject to flocculation effects. In the field of microemulsions, the water uptake is defined as the molar ratio of water to surfactant, W_o. The corrected W_0 (called W_o^c) taking account of the solubility of water in CO_2 is defined as:

$$W_0^c = \frac{[\text{water}]_0 - [\text{water}]_s}{[\text{surfactant}]_0}$$

where $[\text{water}]_0$ is the number of moles of water in the system, $[\text{water}]_s$ the number of moles soluble in the background CO_2, and $[\text{water}]_0$ the number of moles of surfactant. W_o^c usually ranges from 10 to 100. When this ratio is lower, microemulsions are preferably referred as hydrated micelles.

While the stabilization of W/C microemulsions with Krytox-type polymers or molecular surfactants has been largely reported [66, 67], examples implying amphiphilic copolymers are scarce in literature. These examples—described below—are exclusively based on fluorinated poly((meth)acrylates) as a consequence of their lower surface tension compared to hydrocarbons and poly(siloxanes).

13.4.4.1 Amphiphilic Block Copolymers Lim and coworkers reported the formation of hydrated micelles stabilized by a poly(1,1,2,2-tetrahydroperfluorooctyl methacrylate)-based copolymers incorporating either poly(*N,N*-dimethylaminoethyl methacrylate) (PDMAEMA) (see Figure 13.26a) or poly(oligo(ethylene glycol) methacrylate) (POEGMA) hydrophilic blocks (see Figure 13.26b).

Using 0.5 wt% of a PDMAEMA-*b*-PFOMA sample ($M_{n,PDMAEMA} = 2000$ g/mol and $M_{n,PFOMA} = 10{,}000$ g/mol), the formation of hydrated aggregates could be achieved above 180 bar at 40°C [68]. However, the corrected water-to-surfactant molar ratio taking account of the intrinsic solubility of water in CO_2 (W_o^c) was only 5 and decreased to 2.4 with a higher polymer content of 1 wt%. Similarly, a W_o^c corrected ratio of 2.7 was determined with 0.25 wt% of a POEGMA-*b*-PFOMA sample ($M_{n,POEGMA} = 6500$ g/mol and $M_{n,PFOMA} = 20{,}000$ g/mol) above 140 bar at 40°C [69]. By means of comparison, a W_o up to 80 was attained in microemulsions stabilized by molecular surfactants [70].

13.4.4.2 Amphiphilic Gradient Copolymers A greater amount of water could be dispersed in sc-CO_2 using an amphiphilic gradient copolymer made of 1,1,2,2-tetrahydroperfluorodecyl acrylate and vinyl benzoic phosphonic acid (23% mol)

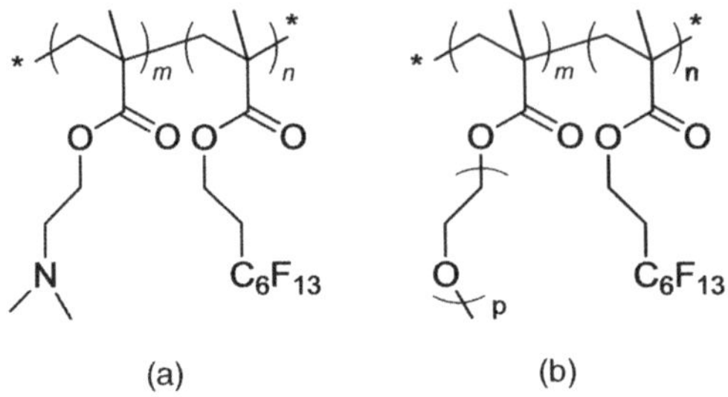

FIGURE 13.26 Molecular structures of (a) poly(*N,N*-dimethylaminoethyl methacrylate) (PDMAEMA)-*b*-poly(1,1,2,2-tetrahydroperfluorooctyl methacrylate) and (b) poly (oligo(ethylene glycol) methacrylate)-*b*-poly(1,1,2,2-tetrahydroperfluorooctyl methacrylate) block copolymers.

repeat units (see Figure 13.27) [52]. From SANS measurements performed at 40°C and 350 bar with 4.6 wt% of polymer and 0.37 wt% of D_2O, D_2O-swollen aggregates with an R_g of 2.25 nm and an N_{agg} of 8 were observed. This corresponded to a W_o^c of 13. These aggregates were stable for 8 hours as a further indication of the formation of microemulsion. The stabilization of higher water content appeared unfeasible in mild conditions of temperature ($T < 65$°C) and pressure ($P < 400$ bar).

13.4.4.3 Amphiphilic Comb Copolymers A single example of hydrated micelles stabilized by a sample of poly(1,1-dihydroperfluorooctyl acrylate)-*graft*-poly(ethylene oxide) (PFOA-*g*-PEO) was published so far (see Figure 13.23a) [57]. From SAXS experiments, the swelling of an aggregate upon the presence of 0.32 wt% of water and 0.6 wt% of the copolymer was observed at 60°C. At 300 bar, a shell radius around 12 nm and an N_{agg} of 123 were extracted from the experimental data fitted with a core–shell model. An increase of the polymer content up to 1.9 wt% resulted in an increase of this radius to 12.9 nm at 300 bar. Chillura-Martino et al. performed additional SANS experiments on this same example [26]. Using the same fitting model, they could confirm the swelling of aggregates stabilized by this graft amphiphilic copolymer. A shell radius of 12.8 nm and a slightly lower N_{agg} of 83 were determined too.

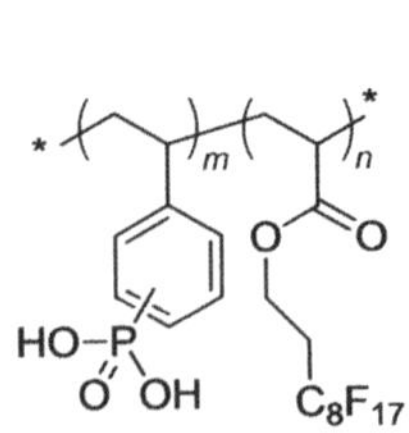

FIGURE 13.27 Molecular structure of poly(vinyl benzoic phosphonic acid)-*b*-poly(1,1,2,2-tetrahydroperfluorodecyl acrylate) gradient copolymers.

13.5 CONCLUSION

As depicted in this chapter, specific structural and electronic properties of fluoropolymers have made them the polymers of choice to be used in sc-CO_2. The study of different families of fluoropolymer demonstrated that, apart from the effect of polymer architecture and CO_2 density, the lowering of polymer–polymer interactions appeared as the main critical parameter to improve polymer solubility. Moreover, the use of those polymers as building blocks to design macromolecular surfactants allows the successful formation of micelle-like structures or the formation of emulsions in the presence of additional water. Based on this knowledge, development of new fluoropolymer families will surely pave the way for the preparation of stable w/sc-CO_2 emulsions with great promise for the development of environmental-friendly chemical processes in diverse fields of organic and inorganic synthesis and polymerization.

REFERENCES

1. Anastas, P. T.; Kirchhoff, M. M. *Acc. Chem. Res.* **2002**, *35*, 686.
2. Noyori, R. *Chem. Commun.* **2005**, *14*, 1807.
3. O'Neill, M. L.; Cao, Q.; Fang, M.; Johnston, K. P.; Wilkinson, S. P.; Smith, C. D.; Kerschner, J. L.; Jureller, S. H. *Ind. Eng. Chem. Res.* **1998**, *37*, 3067.
4. Rindfleisch, F.; DiNoia, T. P.; McHugh, M. A. *J. Phys. Chem.* **1996**, *100*, 15581.
5. Girard, E.; Tassaing, T.; Camy, S.; Condoret, J.-S.; Marty, J. D.; Destarac, M. *J. Am. Chem. Soc.* **2012**, *134*, 11920.
6. Kostov, G.; Boschet, F.; Ameduri, B. *J. Fluor. Chem.* **2009**, *130*, 1192.
7. Dardin, A.; Cain, J. B.; DeSimone, J. M.; Johnson, C. S.; Samulski, E. T. *Macromolecules* **1997**, *30*, 3593.
8. Dardin, A.; DeSimone, J. M.; Samulski, E. T. *J. Phys. Chem. B* **1998**, *102*, 1775.
9. Yonker, C. R.; Zemanian, T. S.; Wallen, S. L.; Linehan, J. C.; Franz, J. A. *J. Magn. Reson. A* **1995**, *113*, 102.
10. Diep, P.; Jordan, K. D.; Johnson, J. K.; Beckman, E. J. *J. Phys. Chem. A* **1998**, *102*, 2231.
11. Raveendran, P.; Wallen, S. L. *J. Phys. Chem. B* **2003**, *107*, 1473.
12. Raveendran, P.; Wallen, S. L. *J. Am. Chem. Soc.* **2002**, *124*, 12590.
13. Lora, M.; Lim, J. S.; McHugh, M. A. *J. Phys. Chem. B* **1999**, *103*, 2818.
14. DiNoia, T. P.; Conway, S. E.; Lim, J. S.; McHugh, M. A. *J. Polym. Sci. B Polym. Phys.* **2000**, *38*, 2832.
15. Beuermann, S.; Imran-ul-haq, M. *J. Polym. Sci. A Polym. Chem.* **2007**, *45*, 5626.
16. Bonavoglia, B.; Storti, G.; Morbidelli, M. *Macromolecules* **2005**, *38*, 5593.
17. O'Neill, M. L.; Cao, Q.; Fang, M.; Johnston, K. P.; Wilkinson, S. P.; Smith, C. D.; Kerschner, J. L.; Jureller, S. H. *Ind. Eng. Chem. Res.* **1998**, *37*, 3067.
18. McHugh, M. A.; Park, I. H.; Reisinger, J. J.; Ren, Y.; Lodge, T. P.; Hillmyer, M. A. *Macromolecules* **2002**, *35*, 4653.
19. Baradie, B.; Shoichet, M. S. *Macromolecules* **2002**, *35*, 3569.

20. Baradie, B.; Shoichet, M. S.; Shen, Z.; McHugh, M. A.; Hong, L.; Wang, Y.; Johnson, J. K.; Beckman, E. J.; Enick, R. M. *Macromolecules* **2004**, *37*, 7799.
21. Mertdogan, C. A.; DiNoia, T. P.; McHugh, M. A. *Macromolecules* **1997**, *30*, 7511.
22. Girard, E.; Marty, J. D.; Ameduri, B.; Destarac, M. *ACS Macro Lett.* **2012**, *1*, 270.
23. Hoefling, T. A.; Stofesky, D.; Reid, M.; Beckman, E. J.; Enick, R. M. *J. Supercrit. Fluids* **1992**, *5*, 237.
24. Newman, D. A.; Hoefling, T. A.; Beitle, R. R.; Beckman, E. J.; Enick, R. M. *J. Supercrit. Fluids* **1993**, *5*, 205.
25. Dupont Krytox Performance Lubricants: Product Overview; 2012. http://www2.dupont.com/Lubricants/en_US/assets/downloads/Krytox_Overview_LowRes_H-58505-5.pdf [Accessed 7th May 2012].
26. Chillura-Martino, D.; Triolo, R.; McClain, J. B.; Combes, J. R.; Betts, D. E.; Canelas, D. A.; DeSimone, J. M.; Samulski, E. T.; Cochran, H. D.; Londono, J. D.; Wignall, G. D. *J. Mol. Struct.* **1996**, *383*, 3.
27. Girard, E.; Tassaing, T.; Ladavière, C.; Marty, J. D.; Destarac, M. *Macromolecules* **2012**, *45*, 9674.
28. McHugh, M. A.; Garach-Domech, A.; Park, I. H.; Li, D.; Barbu, E.; Graham, P.; Tsibouklis, J. *Macromolecules* **2002**, *35*, 6479.
29. Kwon, S.; Bae, W.; Lee, K.; Byun, H. S.; Kim, H. *J. Chem. Eng. Data* **2006**, *52*, 89.
30. Hsiao, Y. L.; Maury, E. E.; DeSimone, J. M.; Mawson, S.; Johnston, K. P. *Macromolecules* **1995**, *28*, 8159.
31. Buhler, E.; Dobrynin, A. V.; DeSimone, J. M.; Rubinstein, M. *Macromolecules* **1998**, *31*, 7347.
32. McClain, J. B.; Londono, D.; Combes, J. R.; Romack, T. J.; Canelas, D. A.; Betts, D. E.; Wignall, G. D.; Samulski, E. T.; DeSimone, J. M. *J. Am. Chem. Soc.* **1996**, *118*, 917.
33. Blasig, A.; Shi, C.; Enick, R. M.; Thies, M. C. *Ind. Eng. Chem. Res.* **2002**, *41*, 4976.
34. Lacroix-Desmazes, P.; Andre, P.; DeSimone, J. M.; Ruzette, A. V.; Boutevin, B. *J. Polym. Sci. A Polym. Chem.* **2004**, *42*, 3537.
35. Andre, P.; Lacroix-Desmazes, P.; Taylor, D. K.; Boutevin, B. *J. Supercrit. Fluids* **2006**, *37*, 263.
36. Ribaut, T.; Lacroix-Desmazes, P.; Fournel, B.; Sarrade, S. *J. Polym. Sci. A Polym. Chem.* **2009**, *47*, 5448.
37. Ribaut, T.; Oberdisse, J.; Annighofer, B.; Fournel, B.; Sarrade, S.; Haller, H.; Lacroix-Desmazes, P. *J. Phys. Chem. B* **2011**, *115*, 836.
38. Mawson, S.; Johnston, K. P.; Combes, J. R.; DeSimone, J. M. *Macromolecules* **1995**, *28*, 3182.
39. Guo, J.; Andre, P.; Adam, M.; Panyukov, S.; Rubinstein, M.; DeSimone, J. M. *Macromolecules* **2006**, *39*, 3427.
40. Ma, Z.; Lacroix-Desmazes, P. *J. Polym. Sci. A Polym. Chem.* **2004**, *42*, 2405.
41. Guan, Z.; DeSimone, J. M. *Macromolecules* **1994**, *27*, 5527.
42. Ye, W. J.; DeSimone, J. M. *Macromolecules* **2005**, *38*, 2180.
43. Li, D.; Shen, Z.; McHugh, M. A.; Tsibouklis, J.; Barbu, E. *Ind. Eng. Chem. Res.* **2003**, *42*, 6499.

44. Wright, M. E.; Lott, K. M.; McHugh, M. A.; Shen, Z. *Macromolecules* **2003**, *36*, 2242.
45. Fink, R.; Hancu, D.; Valentine, R.; Beckman, E. J. *J. Phys. Chem. B* **1999**, *103*, 6441.
46. McClain, J. B.; Betts, D. E.; Canelas, D. A.; Samulski, E. T.; DeSimone, J. M.; Londono, J. D.; Cochran, H. D.; Wignall, G. D.; ChilluraMartino, D.; Triolo, R. *Science* **1996**, *274*, 2049.
47. Londono, J. D.; Dharmapurikar, R.; Cochran, H. D.; Wignall, G. D.; McClain, J. B.; Betts, D. E.; Canelas, D. A.; DeSimone, J. M.; Samulski, E. T.; Chillura-Martino, D.; Triolo, R. *J. Appl. Cryst.* **1997**, *30*, 690.
48. Cain, J. B.; Zhang, K.; Betts, D. E.; DeSimone, J. M.; Johnson, C. S. *J. Am. Chem. Soc.* **1998**, *120*, 9390.
49. Lo Celso, F.; Triolo, A.; Triolo, F.; Thiyagarajan, P.; Amenitsch, H.; Steinhart, M.; Kriechbaum, M.; DeSimone, J. M.; Triolo, R. *J. Appl. Cryst.* **2003**, *36*, 660.
50. Koga, T.; Zhou, S.; Chu, B. *Appl. Opt.* **2001**, *40*, 4170.
51. Edmonds, W. F.; Hillmyer, M. A.; Lodge, T. P. *Macromolecules* **2007**, *40*, 4917.
52. Ribaut, T.; Oberdisse, J.; Annighofer, B.; Stoychev, I.; Fournel, B.; Sarrade, S.; Lacroix-Desmazes, P. *Soft Matter* **2009**, *5*, 4962.
53. Zhou, S.; Chu, B. *Macromolecules* **1998**, *31*, 5300.
54. Triolo, F.; Triolo, A.; Triolo, R.; Londono, J. D.; Wignall, G. D.; McClain, J. B.; Betts, D. E.; Wells, S.; Samulski, E. T.; DeSimone, J. M. *Langmuir* **1999**, *16*, 416.
55. Triolo, A.; Triolo, F.; Lo Celso, F.; Betts, D. E.; McClain, J. B.; DeSimone, J. M.; Wignall, G. D.; Triolo, R. *Phys. Rev. E* **2000**, *62*, 5839.
56. Lo Celso, F.; Triolo, A.; Triolo, F.; Donato, D. I.; Steinhart, M.; Kriechbaum, M.; Amenitsch, H.; Triolo, R. *Eur. Phys. J. E* **2002**, *8*, 311.
57. Fulton, J. L.; Pfund, D. M.; McClain, J. B.; Romack, T. J.; Maury, E. E.; Combes, J. R.; Samulski, E. T.; DeSimone, J. M.; Capel, M. *Langmuir* **1995**, *11*, 4241.
58. Lepilleur, C.; Beckman, E. J.; Schonemann, H.; Krukonis, V. J. *Fluid Phase Eq.* **1997**, *134*, 285.
59. Cooper, A. I.; Londono, J. D.; Wignall, G.; McClain, J. B.; Samulski, E. T.; Lin, J. S.; Dobrynin, A.; Rubinstein, M.; Burke, A. L. C.; Frechet, J. M. J.; DeSimone, J. M. *Nature* **1997**, *389*, 6649.
60. Martinez, V.; Mecking, S.; Tassaing, T.; Besnard, M.; Moisan, S.; Cansell, F.; Aymonier, C. *Macromolecules* **2006**, *39*, 3978.
61. Goetheer, E. L. V.; Baars, M. W. P. L.; van den Broeke, L. J. P.; Meijer, E. W.; Keurentjes, J. T. F. *Ind. Eng. Chem. Res.* **2000**, *39*, 4634.
62. Lim, K. T.; Lee, M. Y.; Moon, M. J.; Lee, G. D.; Hong, S. S.; Dickson, J. L.; Johnston, K. P. *Polymer* **2002**, *43*, 7043.
63. Dickson, J. L.; Psathas, P. A.; Salinas, B.; Ortiz-Estrada, C.; Luna-Barcenas, G.; Hwang, H. S.; Lim, K. T.; Johnston, K. P. *Langmuir* **2003**, *19*, 4895.
64. Meredith, J. C.; Johnston, K. P. *Macromolecules* **1998**, *31*, 5518.
65. Dickson, J. L.; Ortiz-Estrada, C.; Alvarado, J. F. J.; Hwang, H. S.; Sanchez, I. C.; Luna-Barcenas, G.; Lim, K. T.; Johnston, K. P. *J. Coll. Int. Sci.* **2004**, *272*, 444.
66. Johnston, K. P.; Harrison, K. L.; Clarke, M. J.; Howdle, S. M.; Heitz, M. P.; Bright, F. V.; Carlier, C.; Randolph, T. W. *Science* **1996**, *271*, 624.

67. Sagisaka, M.; Iwama, S.; Hasegawa, S.; Yoshizawa, A.; Mohamed, A.; Cummings, S.; Rogers, S. E.; Heenan, R. K.; Eastoe, J. *Langmuir* **2011**, *27*, 5772.

68. Lim, K. T.; Hwang, H. S.; Ryoo, W.; Johnston, K. P. *Langmuir* **2004**, *20*, 2466.

69. Hwang, H. S.; Kim, H. J.; Jeong, Y. T.; Gal, Y. S.; Lim, K. T. *Macromolecules* **2004**, *37*, 9821.

70. Sagisaka, M.; Iwama, S.; Yoshizawa, A.; Mohamed, A.; Cummings, S.; Eastoe, J. *Langmuir* **2012**, *28*, 10988.

14

SEMIFLUORINATED POLYMERS FROM TRIFLUOROVINYL AROMATIC ETHER MONOMERS

JIANYONG JIN
The University of Auckland
SCOTT T. IACONO
United States Air Force Academy
DENNIS W. SMITH JR.
The University of Texas at Dallas

14.1 INTRODUCTION

In the plastics world, fluorine-containing polymers represent a rather specialized group of polymeric materials. Since the accidental invention of a bizarre white powder material (It was Teflon® as we know today) by a DuPont scientist during his ongoing research on new refrigerants in 1938 [1], the production of fluoropolymers reached an estimated world demand of 235,000 metric tons in 2011 [2]. Their many attributes include remarkable thermal and chemical attack inertness, solvent resistance, and outstanding electrical properties. These properties offset their higher cost and greater difficulty in processing than is the case for most other nonfluorinated thermoplastics. Fluoropolymers are mostly manufactured from the free radical polymerization of fluorinated olefins [3]. However, this chapter introduces a unique class of fluorinated alkenes—aryl trifluorovinyl ethers. The chemistry was developed by Dow Chemical

Handbook of Fluoropolymer Science and Technology, First Edition. Edited by Dennis W. Smith Jr., Scott T. Iacono, and Suresh S. Iyer.

Company in the early 1990s with inspiration from the perfluoroalkyl trifluorovinyl ethers and α,β,β-trifluorostyrene [4]. Compared to the perfluoroalkyl trifluorovinyl ethers, which can undergo both cyclodimerization and free radical–mediated addition polymerization, aryl trifluorovinyl ethers are solely polymerizable via cyclodimerization. Yet, with the insertion of oxygen between aromatic ring and fluorinated alkene, aryl trifluorovinyl ethers are much more stable than α,β,β-trifluorostyrene [5]. As a result, the polymerization of aryl trifluorovinyl ethers is more controllable than α,β,β-trifluorostyrene [6]. Taking advantage of these distinctive reactivities, David Babb pioneered the family of 1,2-bisaryloxy-substituted perfluorocyclobutane (PFCB) ring containing polymers [4].

Thermoplastic or thermosetting PFCB polymers are prepared by monomers containing two or multiple aryl trifluorovinyl ether ($Ar—O—CF{=}CF_2$) groups. Today, PFCB polymers represent an outstanding class of semifluorinated high performance specialty polymers with highly customizable properties including thermal, chemical, optical, surface, and ion transportation obtained through versatile copolymerization strategies [7–9]. Initially developed for aerospace and microelectronics applications at Dow Chemical, PFCB polymers find uses as materials for microphotonics [10–12], coatings [8], nanocomposite dispersing matrix [13,14], hole transport layers for light-emitting diodes [15,16], cross-linking groups in electro-optic chromophores [17,18], and more recently, PFCB polymers have emerged as promising proton exchange membrane materials for hydrogen fuel cells [19–24].

14.2 SYNTHESIS OF TRIFLUOROVINYL AROMATIC ETHER

14.2.1 Classical Synthesis for Trifluorovinyl Aromatic Ether

Trifluorovinyl aromatic ethers can be prepared by several methods. In 1966, phenyl trifluorovinyl ether was first obtained by the reaction of an alkali metal phenoxide (PhONa or PhOK) with tetrafluoroethylene in Parr bombs [25]. The yield of the reaction was mediocre because of side reactions, where reactive fluorocarbanions are trapped by the available proton donors in the reaction system to give saturated 1,1,2,2-tetrafluoroethyl phenyl ethers (Scheme 14.1).

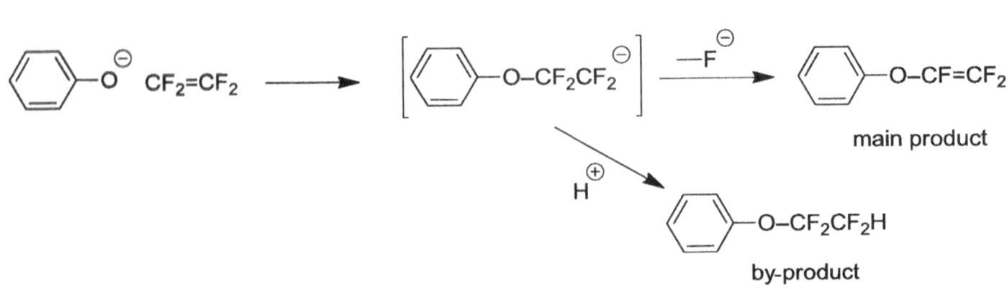

SCHEME 14.1 Synthesis of phenyl trifluorovinyl ether using tetrafluoroethylene.

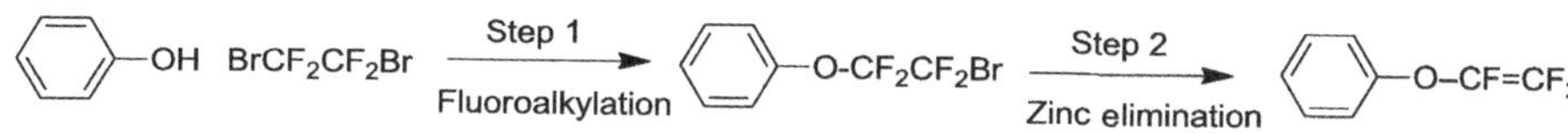

SCHEME 14.2 Preparation of aryl trifluorovinyl ethers using 1,2-dibromotetrafluoroethane.

Later on, a more efficient and mild way to prepare this compound in a two-step process (Scheme 14.2) was revealed by Babb et al. in 1993 [4]. Starting from phenolic precursors, the first step is fluoroalkylation with 1,2-dibromotetrafluoroethane ($BrCF_2CF_2Br$) to give 2-bromotetrafluoroethyl aryl ethers. In this reaction, $BrCF_2CF_2Br$ (widely used as nontoxic fire extinguishing agent) served as a fluoroalkylation agent and as a trifluorovinyl precursor.

In contrast to the well-known alkylation paths, namely, S_N1 and S_N2, the fluoroalkylation reaction was rationalized by an unusual ionic chain mechanism [26]. As shown in Scheme 14.3, the reaction was initiated with the direct attack of electron-positive bromine (δ^+) on the $BrCF_2CF_2Br$ by phenoxide. Tetrafluoroethylene ($CF_2{=}CF_2$) was generated *in situ* after loss of bromide anions. The phenoxides added to the $CF_2{=}CF_2$ to give the reactive fluorocarbanions, which were quickly terminated by bromide to form 2-bromo-tetrafluoroethyl aryl ethers. In the elimination step, zinc inserted into the C—Br bond of 2-bromo-tetrafluorethyl aryl ethers in a similar way to the preparation of Grignard reagents. Finally, the aryl trifluorovinyl ethers were obtained by the elimination of ZnBrF salt at elevated temperature.

To date, this method has become the most popular protocol to synthesize trifluorovinyl aromatic ether monomers. A variety of mono-, di-, and trifunctional trifluorovinyl ether monomers have been prepared simply by choosing the appropriate phenolic starting materials (Scheme 14.4), such as 3-trifluorovinyloxy-α,α,α-trifluorotoluene (**2**) [27], 1,3-bis(trifluorovinyloxy)benzene (**11**) [27], 4,4′-bis(trifluorovinyloxy)biphenyl (**6**), [4] and 1,1,1-tris(4-trifluorovinyloxyphenyl)ethane (**14**) [4]. Other monofunctional monomers include *p*-bromo(trifluorovinyloxy)

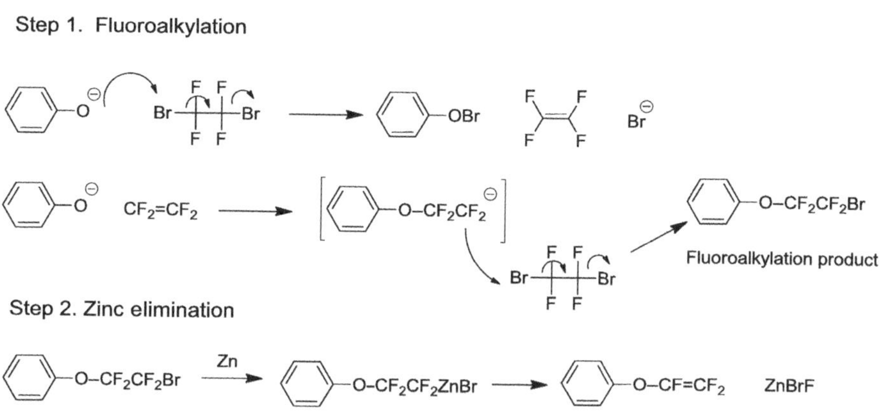

SCHEME 14.3 Postulated fluoroalkylation and zinc elimination mechanism.

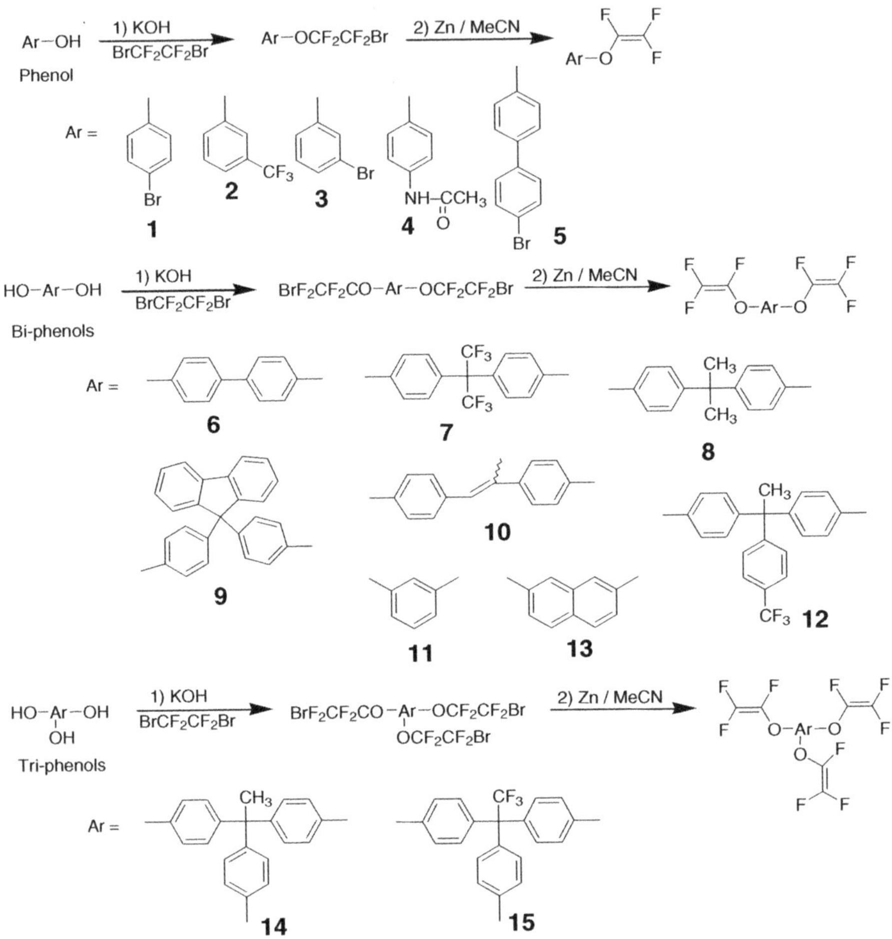

SCHEME 14.4 Mono-, di-, and trifunctional trifluorovinyl aromatic ether monomers prepared from phenolic precursors.

benzene (**1**) [28], 3-trifluorovinyloxy bromobenzene (**3**) [29], 4-(trifluorovinyloxy) acetamidobenzene (**4**) [30], and 4-bromo-4′-(trifluorovinyloxy)biphenyl (**5**) [31]. Other difunctional monomers containing hexafluoroisopropylidene (**7**) [32], bisphenol-A (**8**) [33], bisphenylfluorene (**9**) [8], α-methylstilbene (**10**) [34], resorcinol (**11**) [4], and naphthalene (**13**) [35] have also been reported. A new trifunctional trifluorovinyl ether monomer **15** [36], which is a CF_3 analog of **14**, was also prepared from this phenolic precursor route.

For the deactivated phenols, such as those that are perfluorinated or that have strong electron-withdrawing groups, the reaction gave extremely poor yield in the zinc elimination step, that is, the zinc insertion product (Ar–O–CF_2CF_2–ZnBr) exhibited unusual stability. Feiring et al. [37] at DuPont offered a new synthetic pathway

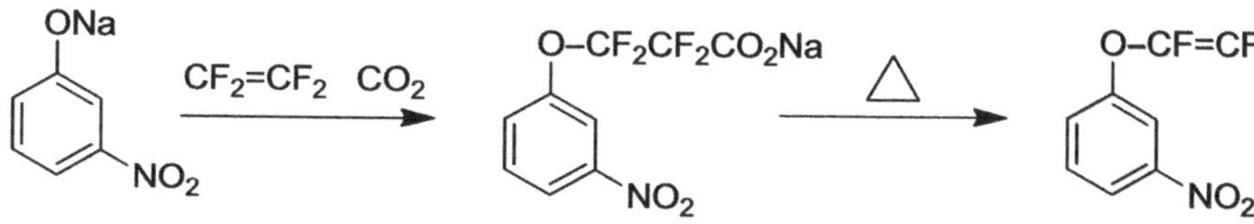

SCHEME 14.5 Nitrophenyl trifluorovinyl ether synthesized.

to access those deactivated aromatic trifluorovinyl ethers involving the reaction of nitrophenoxide with tetrafluoroethylene and carbon dioxide to give 3-nitrophenoxy-2,2,3,3-tetrafluoropropionic acid salt (Scheme 14.5). The salt was then heated to the decarboxylation temperature at 250°C to afford the nitrophenyl trifluorovinyl ether in a decent 46% yield.

14.2.2 The "Intermediate" Route to Trifluorovinyl Aromatic Ether

Although the traditional bis- and tris-phenol precursor method for the preparation of trifluorovinyl ether monomers has become general, significant problems do exist. First, both the fluoroalkylation and zinc halide elimination steps afford the undesirable tetrafluoroethyl ether ($-ArOCF_2CF_2H$) by-product [8, 27, 28]. Due to its similarity with trifluorovinyl ethers, monomer purification can be arduous as the number of functional groups increase. Compounding the problem is the step-growth polymerization requirement that high purity monomers be used in order to build high molecular weights. Furthermore, the by-product will also affect the dielectric properties of the resulting polymers [27], the near IR transparency [38–40], and the ability to exploit the telechelic nature of PFCB oligomers. Second, the zinc elimination step is sensitive to strong electron-withdrawing group substituents on the aromatic ring, resulting in dramatically decreased rates of elimination. Lastly, there are limited commercial phenolic precursors available and in some cases, the phenolic starting material must be synthesized in house.

The intermediate (*p*-bromo(trifluorovinyloxy)benzene) route was established by Smith in 1996 in pursuit of siloxane-containing aryl trifluorovinyl ether monomers [28]. *p*-Bromo(trifluorovinyloxy)benzene can be easily prepared from simple inexpensive 4-bromophenol in high yield and on the multi-kilogram scale. Surprisingly, the Grignard reagent [28] and subsequently discovered organolithium reagent [41] can be prepared smoothly while retaining the electrophilic fluorinated olefin intact. Theoretical calculations suggest that the *p*-perfluorovinylether group stabilized the *p*-carbanion and reduced the nucleophilicity of the organometallic reagents [41]. Such a discovery straightaway creates numerous opportunities to access a variety of new functional monomers. Scheme 14.6 provides a few examples of new monomers via the intermediate strategy. For example, phenyl phosphine oxide-containing monomers (**16** and **17**) were prepared by quenching the Grignard reagent of *p*-bromo(trifluorovinyloxy)benzene (**1**) with phosphorus trichloride and phenylphosphonic dichloride, respectively [31, 42–44]. The pyrimidine-containing monomer (**21**) was synthesized via a Kumada coupling reaction of Grignard reagent of **1** with

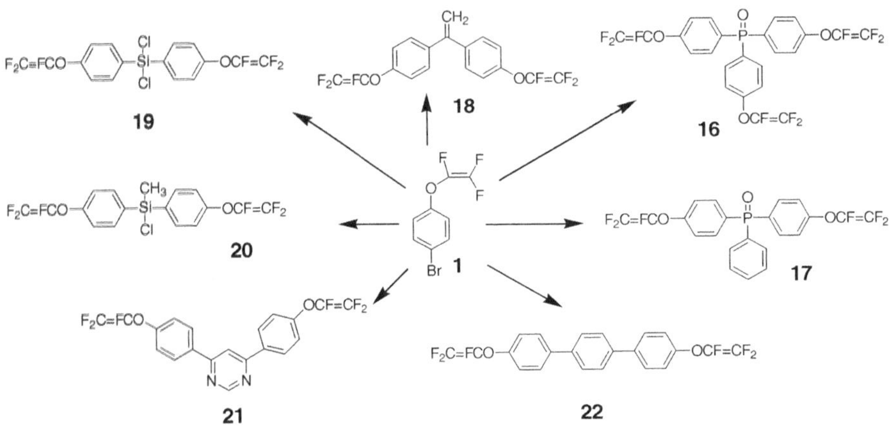

SCHEME 14.6 Trifluorovinyl ether containing monomers directly prepared from *p*-bromo-trifluorovinyloxy benzene.

4,6-dichloropyrimidine [45]. The more reactive organolithium reagent was used to prepare the vinyl (**18**) and chlorosilane monomers (**19** and **20**) [41, 46, 47]. The Suzuki cross-coupling reaction also worked well for the preparation of the terphenyl monomer (**22**) [48].

Scheme 14.7 illustrates the powerful functional group transformation capabilities of the "intermediate route" from *p*-bromo(trifluorovinyloxy)benzene to a variety of useful aryl trifluorovinyl ether derivatives. Scheme 14.8 depicts the trifluorovinylether and hexafluorocyclobutyl containing monomers and reactive resins prepared from *p*-(trifluorovinyloxy)benzene derivatives shown in Scheme 14.7. For example, condensation of silane **29** via dehydrogenative hydrolysis/methanolysis yielded the first siloxane-containing trifluorovinylether monomer (**38**) [28]. Dimerization of **29** and **30** gave the reactive perfluorocyclobutyl (PFCB) difunctional silane (**39** and **40**) [29, 49]. Fluorosilicones containing the perfluorocyclobutyl ring were synthesized by Harris et al. and their thermal properties were studied [29, 49]. Allyl(4-trifluorovinyloxy)benzene (**31**) and 4-(trifluorovinyloxy)phenyl(vinyl)silane (**32**) were synthesized and used to prepare the trifluorovinyl ether pendent polysiloxanes (**41**) via the hydrosilation reaction procedure using Karstedt's catalyst [50]. Neilson et al. prepared a series of trifluorovinylether containing Si–N–P compounds (**33**) and polymerized via the condensation reaction to yield the poly(alkyl/aryl)phosphazenes (**42**) [51]. The trifluorovinylether functionalized organosilicon reagents (**34** and **35**) were also prepared by treating the appropriate dichlorosilanes with a lithium reagent as an intermediate. These organosilicon reagents were further used to functionalize the poly(alkyl/aryl)phosphazene (**43**). Ameduri and Souzy reported the synthesis of a dialkyloxyphosphonate monomer containing the trifluorovinyl ether functionality (**36**) [5, 20, 52]. Another important derivative is the *p*-((trifluorovinyloxy)benzene)boronic acid (**37**), from which a

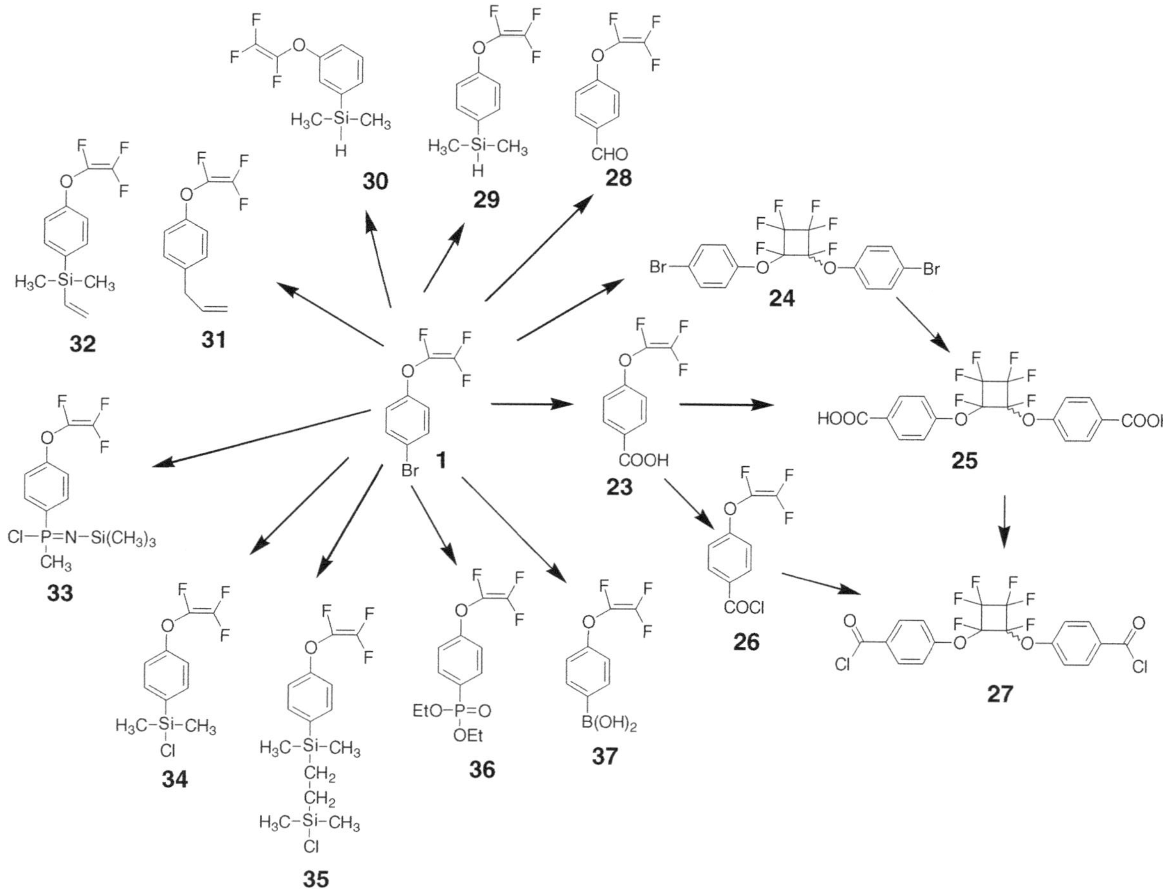

SCHEME 14.7 Other useful derivatives prepared from *p*-bromotrifluorovinyloxy benzene.

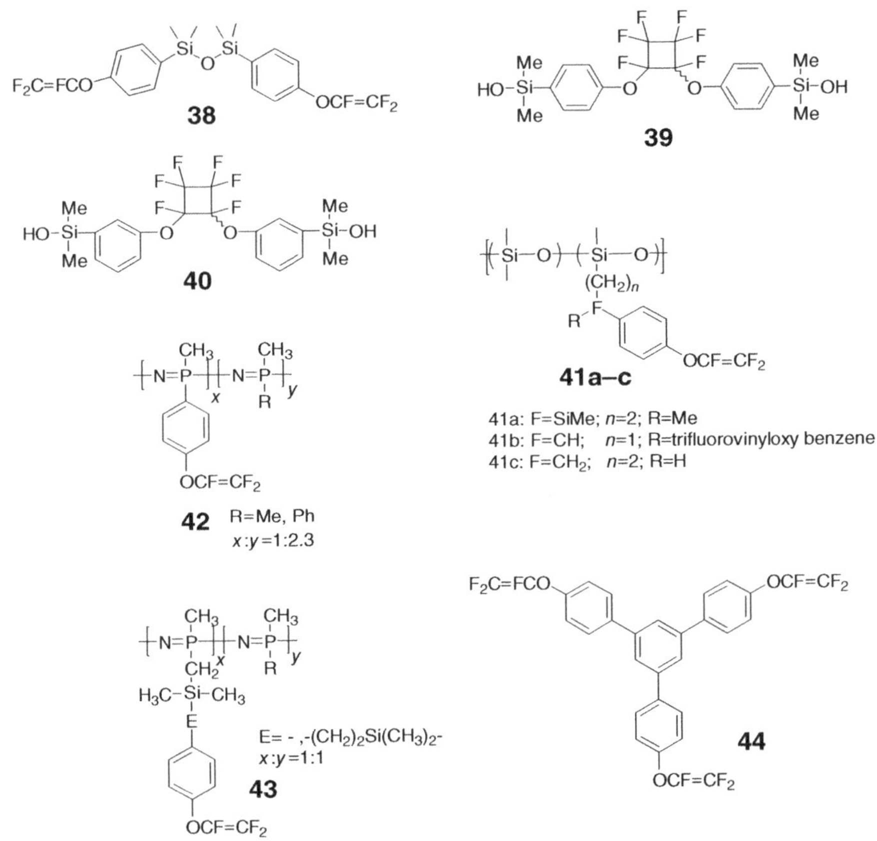

SCHEME 14.8 Selected trifluorovinylether and hexafluorocyclobutyl containing monomers and polymers prepared from *p*-trifluorovinyloxy benzene derivatives.

thermosetting PFCB monomer, 1,3,5-tris[(4-trifluorovinyloxy)phenyl]benzene (**44**) was prepared via Suzuki coupling [53].

14.3 NMR CHARACTERIZATION OF TRIFLUOROVINYL AROMATIC ETHER

The aromatic trifluorovinyl ether moiety can be best characterized by ^{19}F NMR and ^{13}C NMR. Figure 14.1 shows the typical ^{19}F NMR spectrum of trifluorovinyl aromatic ethers using 1,1-Bis[4-[(trifluorovinyl)oxy]phenyl]hexafluoroisopropylidene monomer as an example. The fluorine signals of F_1, F_2, and F_3 are widely separated due to relatively large fluorine–fluorine coupling constants compared with those between hydrogen atoms. The germinal coupling constant J_{12} is about 100 Hz. The *trans* vinylic three-bond coupling constant J_{23} gives the largest number of 110 Hz while the *cis* J_{13} is in the range of 50–60 Hz.

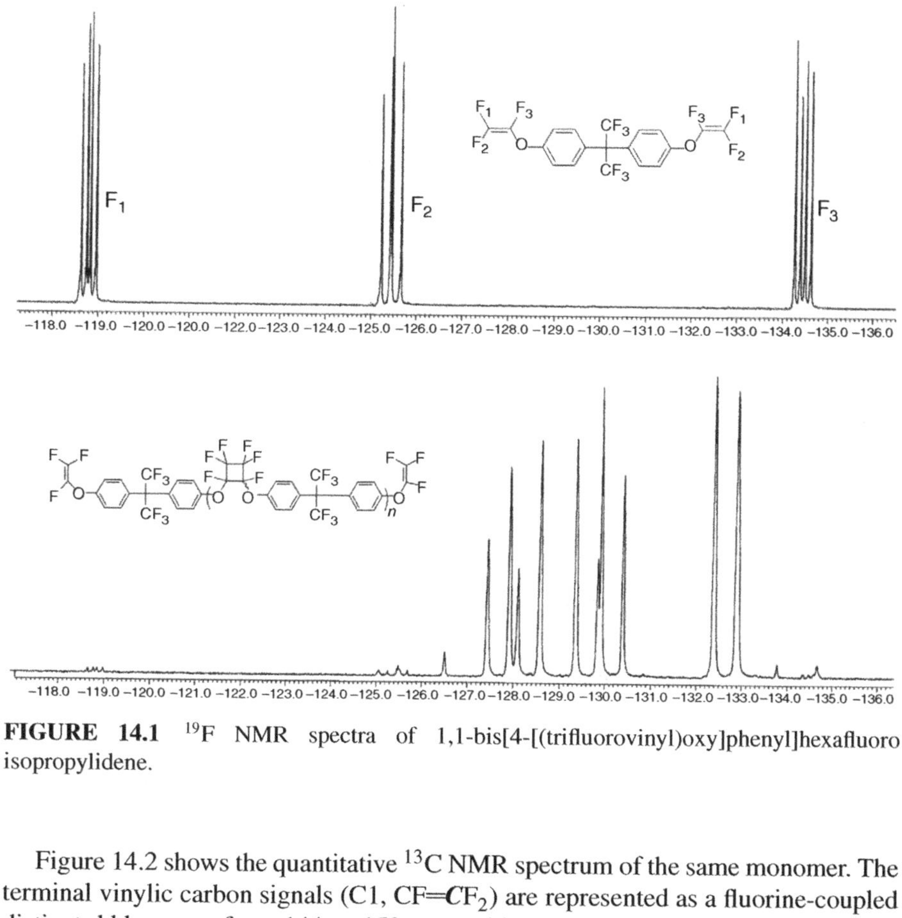

FIGURE 14.1 ^{19}F NMR spectra of 1,1-bis[4-[(trifluorovinyl)oxy]phenyl]hexafluoro isopropylidene.

Figure 14.2 shows the quantitative ^{13}C NMR spectrum of the same monomer. The terminal vinylic carbon signals (C1, CF=*C*F$_2$) are represented as a fluorine-coupled distinct ddd pattern from 144 to 150 ppm with a strong 1-bond fluorine to carbon coupling constant $^1J_{F\text{-}C}$ of 275 Hz. The internal vinylic carbon (C2, *C*F=CF$_2$) also exhibited a ddd pattern from 132 to 135 ppm, which however, appears partially overlapped with a strong aromatic carbon signal (C4).

14.4 CYCLODIMERIZATION OF TRIFLUOROVINYL AROMATIC ETHER

Cyclodimerization of fluorinated alkenes was observed as early as 1947 during the pyrolysis study of polytetrafluoroethylene (PTFE) [54]. The ability of fluorinated alkenes to dimerize with themselves has been attributed to the energy relief of fluorinated double bond strain [4]. In the same fashion, the aryl trifluorovinyl ether groups (Ar—O—CF=CF$_2$) underwent thermally activated [2+2] cycloaddition to form bisaryloxy-substituted PFCB rings. Scheme 14.9 depicts the model reaction of cycloaddition of aryl trifluorovinyl ethers. The thermal cyclodimerization of methyl

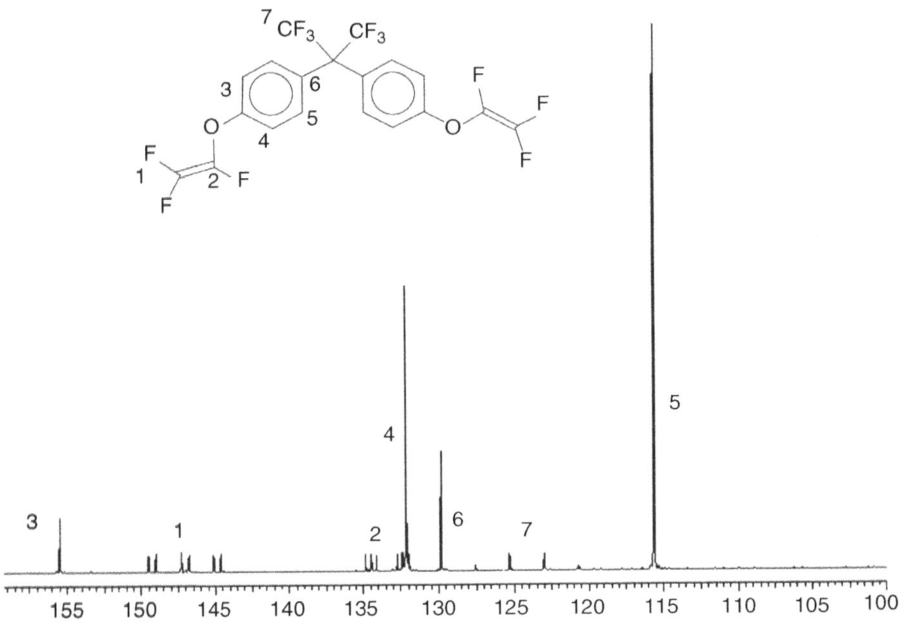

FIGURE 14.2 ^{13}C NMR spectra of 1,1-bis[4-[(trifluorovinyl)oxy]phenyl]hexafluoro isopropylidene.

4-(trifluorovinyloxy)benzoate was carried out at 180°C in the bulk for 12 hours under N_2. The diester was crystallized from methanol to give a white solid in 75% yield [55].

According to the Woodward–Hoffman rule, thermally activated [2+2] cycloaddition took place in a nonconcerted manner: the initial addition of one aryl trifluorovinyl ether to another with predominantly head-to-head regioselectivity gave a 1,4-diradical transitional state [56], followed by ring closure to form the 1,2-disubstituted PFCB ring. Although we are assuming the *trans* isomer would be of greater favorability for

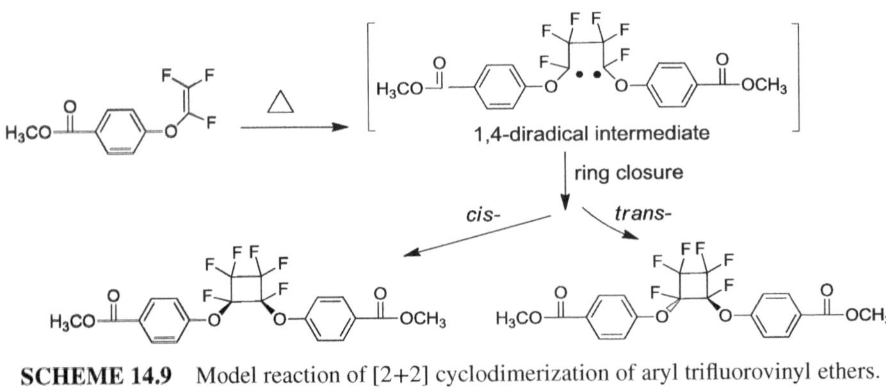

SCHEME 14.9 Model reaction of [2+2] cyclodimerization of aryl trifluorovinyl ethers.

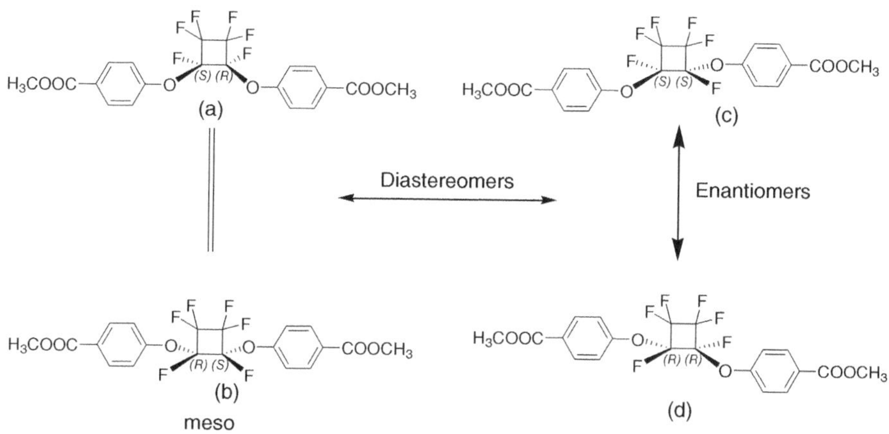

SCHEME 14.10 Stereoisomerism of 1,2-disubstituted perfluorocyclobutane (PFCB) diester model.

stability reasons, surprisingly, about equal amounts of the *cis*- and *trans*- stereoisomers were formed.

Further analysis of the products from cyclodimerization revealed that each of the *cis*- and *trans*- stereoisomers has two enantiomers. As can be seen in Scheme 14.10, the *cis*- product has two enantiomers (a) and (b), being (S, R) and (R, S) configuration, respectively, according to the Cahn–Ingold–Prelog priority rules. Both the *cis*- enantiomers are *meso* compounds and optically inactive. The *trans*- product also has two enantiomers, being (S, S) and (R, R), which exist as non-superimposable mirror images of each other. Enantiomers (c) and (d) should have identical physical and chemical properties in the absence of a chiral environment, while the *trans* and *cis* diastereomers will display differences in some physical and chemical properties, such as reactivity.

Figure 14.3 illustrates the single crystal X-ray structures of *cis* and *trans* diesters [57]. The *cis* diester has a mirror plane of symmetry, with the four carbon atoms in the PFCB ring being located in the same plane. The PFCB ring is almost a square, the bond distances between C9 and C10, C10 and C10A, C10A and C9A, C9A and C9 are 1.574, 1.563, 1.574, and 1.587 Å, respectively. The *trans* diester, on the other hand, has a twofold symmetry: the PFCB ring is not planar since a dihedral angle of 162.86° exists. The absolute configuration for this single crystal *trans* diester is (*R*, *R*) for C9 and C10.

The ^{1}H NMR and ^{19}F NMR of pure *trans* and *cis* PFCB diester are shown in Figure14.4 and 14.5, respectively, and they are distinctively different. For example, for the *trans* diester the chemical shift of the proton at *meta* position (H1) to the PFCB ring appeared about 0.15 ppm higher compared to that of the *cis* diester. The fluorine signals for pure *trans* diester appeared as a set of five peaks at −127.3, −127.8, −129.8, −130.3, and −131.9 ppm. The pure *cis* diester also exhibits a set of five peaks with different chemical shifts, which appeared at −127.8, −128.3,

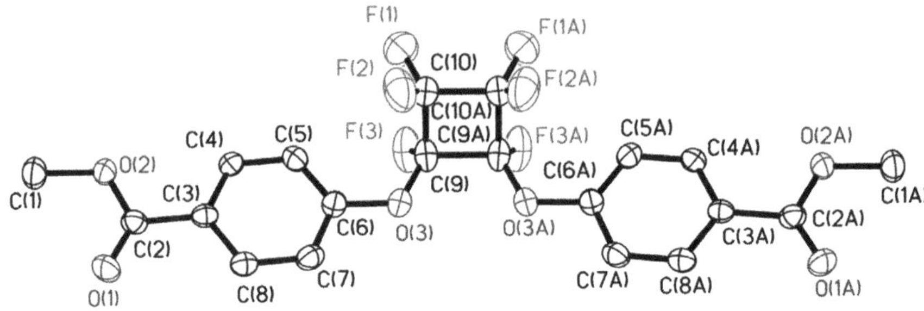

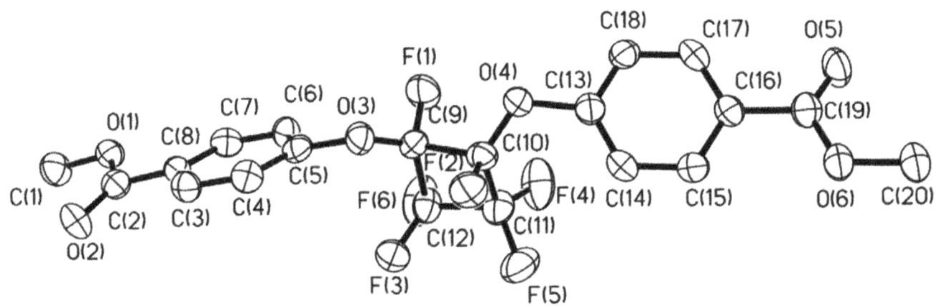

FIGURE 14.3 Perspective drawing showing 50% thermal ellipsoids of diester single crystals, *cis* stereoisomer (top), and *trans* (bottom).

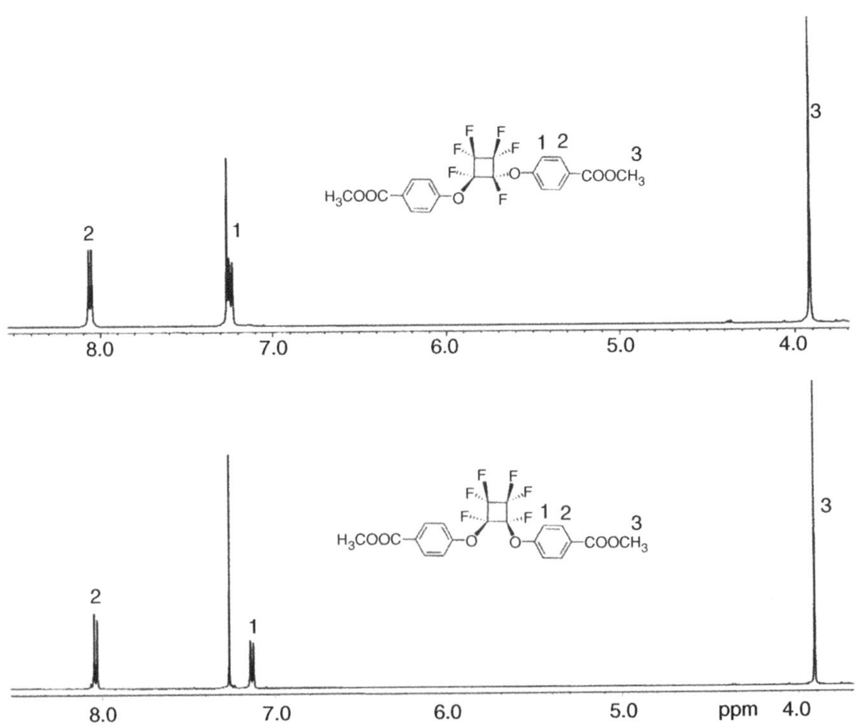

FIGURE 14.4 ^{1}H NMR spectra comparison of pure *trans* stereoisomer (top) and *cis* (bottom) diester.

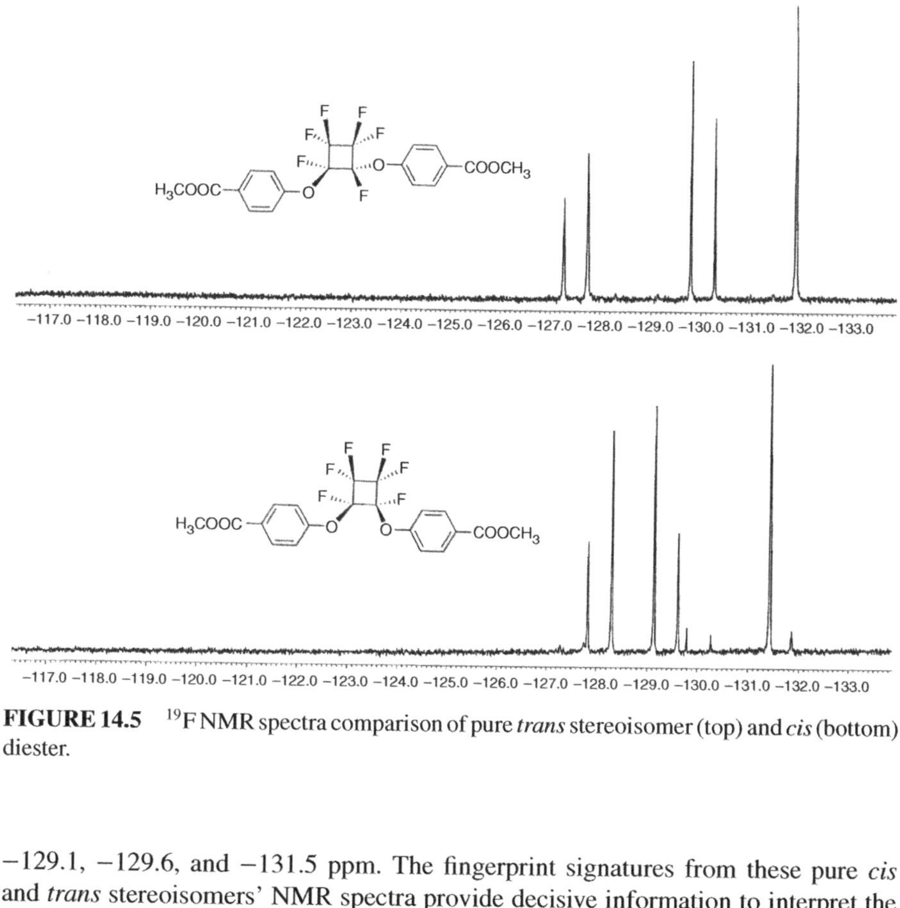

FIGURE 14.5 ^{19}F NMR spectra comparison of pure *trans* stereoisomer (top) and *cis* (bottom) diester.

−129.1, −129.6, and −131.5 ppm. The fingerprint signatures from these pure *cis* and *trans* stereoisomers' NMR spectra provide decisive information to interpret the stereoregularity of the PFCB polymer structures. For instance, Figure 14.6 is the typical ^{19}F NMR of PFCB polymer. The 10 major fluorine signals between −125 and −135 ppm represents the *cis*- and *trans*-1,2-bisaryloxyl PFCB ring. Evidently, each *cis*- and *trans*-disubstituted PFCB ring contributes five fluorine peaks to the spectrum in equal intensity. Therefore, the polymer exhibits an equal distribution of *cis*- and *trans*-1,2-disubstituted perfluorocyclobutanes.

The *cis*- and *trans*- ratio can also be identified via ^{13}C NMR spectroscopy (quantitative) as shown in Figure 14.7. At least two doublets are observed from 130 to 135 ppm. These split signals originate from the two carbons, *ortho* (4) and *ipso* (3), adjacent to the perfluorinated cyclobutane group. The corresponding carbons in the monomer are observed as singlets as shown in Figure 14.2, thus, providing strong evidence for isomer formation. Quantitative carbon integration of the *ipso* carbon signals at 131.7 (presumably *trans*) and 131.8 (*cis*) ppm gave 52% *trans* while the *ortho* carbon signals at 130 (*trans*) and 129.8 (*cis*) gave ppm 54% *trans*-1,2-disubstitution. In addition, perfluorocyclobutyl ring's carbon NMR signals appear as weak and broad multiplets in the range of 105–120 ppm (Figure 14.7 inset). Therefore, fluorine carbon decoupling needs to apply in order to obtain a perfect carbon signal from the PFCB ring.

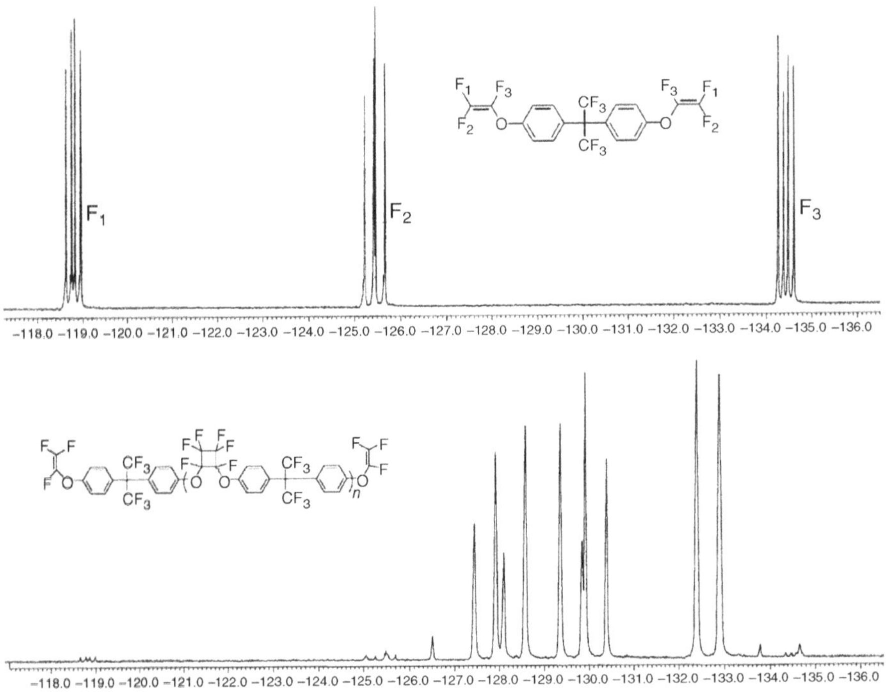

FIGURE 14.6 ^{19}F NMR spectra of poly (1,1-bis[4-[(trifluorovinyl)oxy]phenyl]hexafluor oisopropylidene).

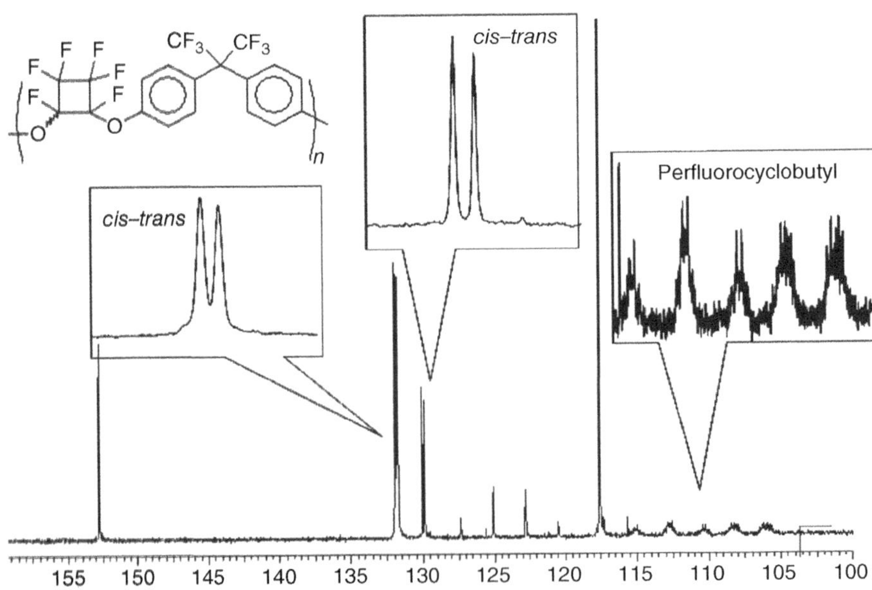

FIGURE 14.7 ^{13}C NMR spectra of poly (1,1-bis[4-[(trifluorovinyl)oxy]phenyl]hexafluor oisopropylidene).

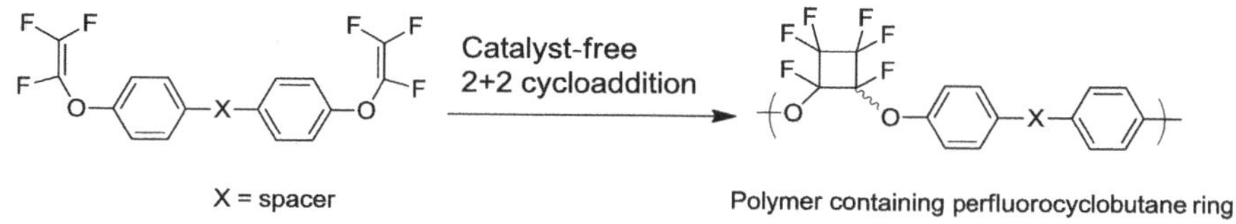

SCHEME 14.11 Synthesis of perfluorocyclobutane polymers via [2+2] cycloaddition of aryl trifluorovinyl ethers.

14.5 POLYMER SYNTHESIS VIA [2+2] CYCLOADDITION OF ARYL TRIFLUOROVINYL ETHERS

The model reaction above shows that aryl trifluorovinyl ethers have unique synthetic utilities in polymer formations. PFCB ring-containing polymers with various macromolecular architectures such as linear, branched, and cross-linked can be prepared by [2+2] cycloaddition of single molecules containing multiple aryl trifluorovinyl ether groups (Scheme 14.11). Typically, a PFCB polymer can be prepared by simply heating the aryl trifluorovinyl ether monomers in the bulk or in solution above 150°C [8]. The PFCB backbone contains equal numbers of randomly distributed *cis*- and *trans*-1,2-disubstituted hexafluorocyclobutanes. Therefore, most PFCB polymers are amorphous in nature with highly optical transparency owing to catalyst-free polymerization processes. PFCB polymers have excellent solution processability in common organic solvents and glass transition temperatures of 120–350°C. Scheme 14.12

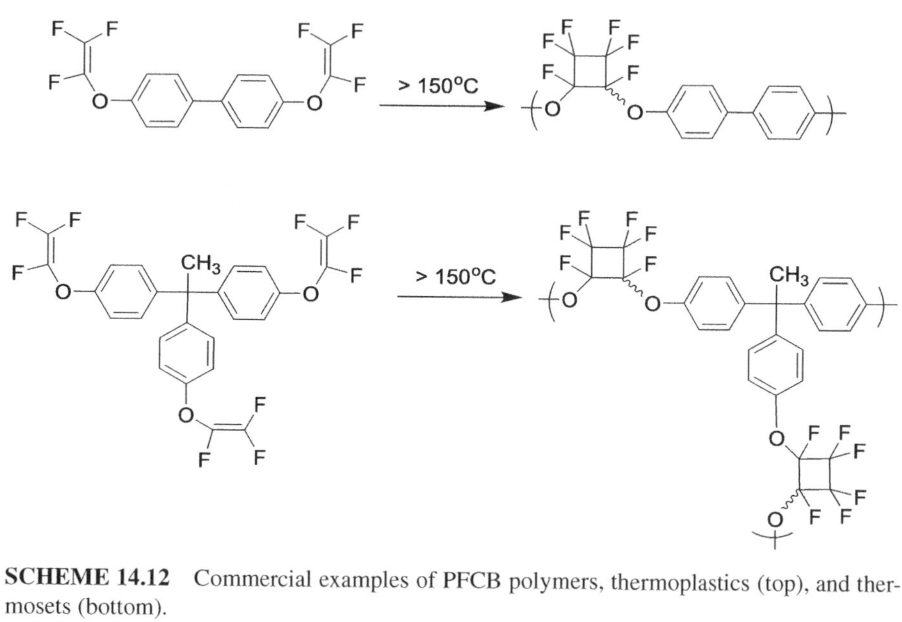

SCHEME 14.12 Commercial examples of PFCB polymers, thermoplastics (top), and thermosets (bottom).

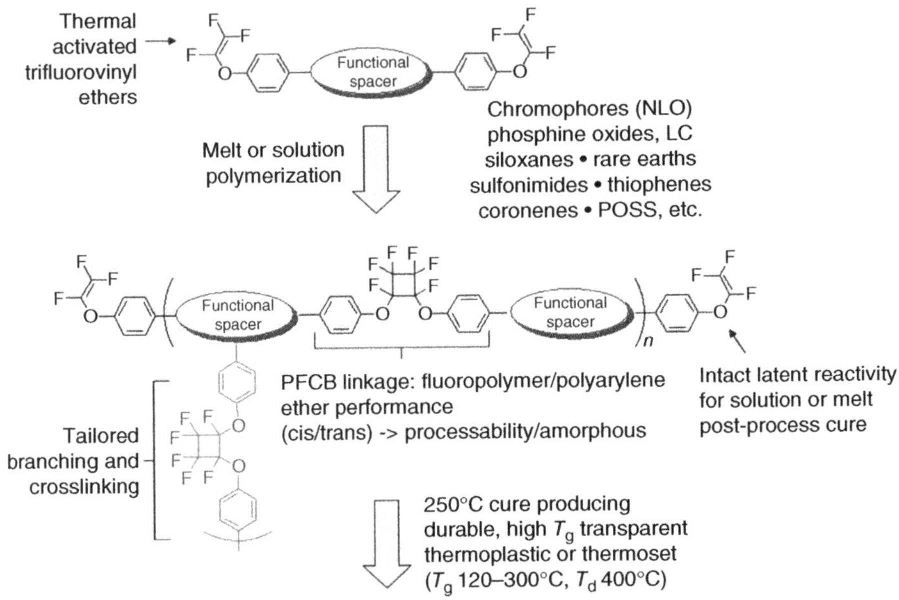

FIGURE 14.8 Design motif of high performance PFCB aryl ether polymers for processability, versatility, and performance.

shows two commercial examples of thermoplastic and thermosetting PFCB polymers [4].

The basic design scheme for functional PFCB materials is illustrated in Figure 14.8. Basically, the desired functional groups can be embedded in between trifluorovinyl aryl ether groups. These functional monomers can then be melted or solution-polymerized to optimum viscosity and molecular weight range. After processing, the intermediate stage PFCB polymer into film or other desirable geometry, final cure is applied to enhance the thermal stability and chemical durability. Enabled by the ability to produce a variety of trifluorovinyl aromatic ether monomers from classical phenolics as precursors and 4-bromo(trifluorovinyloxy)benzene via an "intermediate" route, the construction of PFCB polymers with a diverse array of functionalities is achievable.

14.6 CONCLUSIONS

In this chapter, various synthesis routes for trifluorovinyl aromatic ethers are discussed. PFCB polymers are prepared via thermally activated [2+2] cycloaddition of aryl trifluorovinyl ether monomers. The cyclodimerizaton proceeds in a stereo-random fashion giving a roughly equal distribution of *cis*- and *trans*- stereoisomers. The PFCB technology can serve as a versatile materials platform for many industrial applications. It has found applications in many different fields. The diversity of applications of PFCB polymers demonstrates the remarkable versatility of this exciting class of polymers.

ACKNOWLEDGMENTS

We thank the National Science Foundation (DMR-CAREER Award to D.S.), Defense Advanced Research Projects Agency (DARPA), Air Force Office of Scientific Research, National Textiles Center (NTC), the South Carolina NASA Space Grant Consortium, SC EPSCoR, Research Corporation and 3M Corporation for 3M Pretenured Faculty Awards (D.S.) for their financial support. We also thank Tetramer Technologies LLC (Dr Earl Wagener and Dr Chris Topping) for the donation of starting materials and monomers.

REFERENCES

1. Plunkett, R. J. Tetrafluoroethylene Polymers. US Patent 2,230,654, 1941.
2. Drobny, J. G. *Technology of Fluoropolymers*; CRC Press, 2008.
3. Scheirs, J. *Modern Fluoropolymers: High Performance Polymers for Diverse Applications*; Wiley, 1997.
4. Babb, D. A.; Ezzell, B. R.; Clement, K. S.; Richey, W. F.; Kennedy, A. P. *J. Polym. Sci. Part A: Polym. Chem.* **1993**, *31*(13), 3465–3477.
5. Souzy, R.; Ameduri, B.; Boutevin, B. *Prog. Polym. Sci.* **2004**, *29*(2), 75–106.
6. Ford, L. A.; DesMarteau, D. D. *Chem. Commun.* **2003**, (20), 2596–2597.
7. Iacono, S. T.; Budy, S. M.; Jin, J. Y.; Smith, D. W. *J. Polym. Sci. Pol. Chem.* **2007**, *45*(24), 5705–5721.
8. Smith, D. W.; Babb, D. A.; Shah, H. V.; Hoeglund, A.; Traiphol, R.; Perahia, D.; Boone, H. W.; Langhoff, C.; Radler, M. *J. Fluor. Chem.* **2000**, *104*(1), 109–117.
9. Mujkic, M.; Iacono, S. T.; Neilson, A. R.; Smith, D. W. *Macromol. Symp.* **2009**, *283–284*, 326–335.
10. Smith, D. W.; Chen, S. R.; Kumar, S. M.; Ballato, J.; Topping, C.; Shah, H. V.; Foulger, S. H. *Adv. Mater.* **2002**, *14*(21), 1585–1589.
11. Ma, H.; Luo, J. D.; Kang, S. H.; Wong, S.; Kang, J. W.; Jen, A. K. Y.; Barto, R.; Frank, C. W. *Macromol. Rapid. Commun.* **2004**, *25*(19), 1667–1673.
12. Ma, H.; Jen, A. K.; Dalton, L. R. *Adv. Mater.* **2002**, *14*(19), 1339–1365.
13. Kumar, G. A.; Chen, C. W.; Ballato, J.; Riman, R. E. *Chem. Mater.* **2007**, *19*(6), 1523–1528.
14. Gordon, J.; Ballato, J.; Smith, D. W.; Lin, J. Y. *J. Opt. Soc. Am. B* **2005**, *22*(8), 1654–1659.
15. Shao, Y.; Gong, X.; Heeger, A. J.; Liu, M.; Jen, A. K. Y. *Adv. Mater.* **2009**, *21*(19), 1972–1975.
16. Cheng, Y. J.; Liu, M. S.; Zhang, Y.; Niu, Y. H.; Huang, F.; Ka, J. W.; Yip, H. L.; Tian, Y. Q.; Jen, A. K. Y. *Chem. Mater.* **2008**, *20*(2), 413–422.

17. Ma, H.; Liu, S.; Luo, J. D.; Suresh, S.; Liu, L.; Kang, S. H.; Haller, M.; Sassa, T.; Dalton, L. R.; Jen, A. K. Y. *Adv. Funct. Mater.* **2002**, *12*(9), 565–574.
18. Jen, A. K. Y.; Ma, H.; Sassa, T.; Liu, S.; Suresh, S.; Dalton, L. R.; Haller, M. *Linear Nonlinear Opt. Org. Mater.* **2001**, *4461*, 172–179.
19. DesMarteau, D. D.; Martin, C. W.; Ford, L. A.; Xie, Y. Sulfonated Perfluorovinyl Functional Monomers. US Patent 6,384,167 B2, 2001.
20. Souzy, R.; Ameduri, B. *Prog. Polym. Sci.* **2005**, *30*(6), 644–687.
21. Kim, D. J.; Chang, B. J.; Kim, J. H.; Lee, S. B.; Joo, H. J. *J. Membr. Sci.* **2008**, *325*(1), 217–222.
22. Qian, G. Q.; Smith, D. W.; Benicewicz, B. C. *Polymer* **2009**, *50*(16), 3911–3916.
23. Lee, K. S.; Jeong, M. H.; Kim, Y. J.; Lee, S. B.; Lee, J. S. *Chem. Mater.* **2012**, *24*(8), 1443–1453.
24. Kalaw, G. J. D.; Wahome, J. A. N.; Zhu, Y. Q.; Balkus, K. J.; Musselman, I. H.; Yang, D. J.; Ferraris, J. P. *J. Membr. Sci.* **2013**, *431*, 86–95.
25. Wall, L. A. Perfluorovinylphenylethers and Their Polymers. US Patent 3,277,068 A, 1966.
26. Rico, I.; Wakselman, C. *J. Fluor. Chem.* **1982**, *20*(6), 759–764.
27. Babb, D. A.; Snelgrove, R. V.; Smith, D. W., Jr.; Mudrich, S. F. *Novel Step-Growth Polymers from the Thermal (2p +2p) Cyclopolymerization of Aryl Trifluorovinyl Ether Monomers*; ACS Symposium Series; ACS Publications, 1996; pp 431–441.
28. Smith, D. W.; Babb, D. A. *Macromolecules* **1996**, *29*(3), 852–860.
29. Rizzo, J.; Harris, F. *Polymer* **2000**, *41*(13), 5125–5136.
30. Choi, W. S.; Harris, F. W. *Polymer* **2000**, *41*(16), 6213–6221.
31. Jin, J.; Smith, D. W.; Topping, C. M.; Suresh, S.; Chen, S.; Foulger, S. H.; Rice, N.; Nebo, J.; Mojazza, B. H. *Macromolecules* **2003**, *36*(24), 9000–9004.
32. Smith, D. W.; Jin, J. Y.; Shah, H. V.; Xie, Y.; DesMarteau, D. D. *Polymer* **2004**, *45*(17), 5755–5760.
33. Liu, J.; Spraul, B. K.; Topping, C.; Smith, D. W.; McHugh, M. A. *Macromolecules* **2007**, *40*(16), 5973–5977.
34. Traiphol, R.; Shah, H.; Smith, D. W.; Perahia, D. *Macromolecules* **2001**, *34*(12), 3954–3961.
35. Ghim, J.; Shim, H. S.; Shin, B. G.; Park, J. H.; Hwang, J. T.; Chun, C.; Oh, S. H.; Kim, J. J.; Kim, D. Y. *Macromolecules* **2005**, *38*(20), 8278–8284.
36. Jin, J.; Topping, C. M.; Chen, S.; Ballato, J.; Foulger, S. H.; Smith, D. W. *J. Polym. Sci. Pol. Chem.* **2004**, *42*(20), 5292–5300.
37. Feiring, A. E.; Rozen, S.; Wonchoba, E. R. *J. Fluor. Chem.* **1998**, *89*(1), 31–34.
38. Ballato, J.; Foulger, S.; Smith, D. W., Jr. *JOSA B* **2003**, *20*(9), 1838–1843.
39. Ballato, J.; Foulger, S. H.; Smith, D. W., Jr. *JOSA B* **2004**, *21*(5), 958–967.
40. Ballato, J.; Smith, D. W.; Foulger, S. H.; Wagener, E. *P. Soc. Photo-Opt. Ins.* **2002**, *4805*, 1–8.
41. Ji, J.; Narayan-Sarathy, S.; Neilson, R. H.; Oxley, J. D.; Babb, D. A.; Rondan, N. G.; Smith, D. W. *Organometallics* **1998**, *17*(5), 783–785.
42. Babb, D. A.; Boone, H. W.; Smith, D. W.; Rudolf, P. W. *J. Appl. Polym. Sci.* **1998**, *69*(10), 2005–2012.

43. Gavrin, A.; Nebo, J.; Rice, N.; Kagumba, L.; Smith, D. W.; Jin, J.; Topping, C. M. *Space. Technol. Proc.* **2003**, *5*, 491–501.
44. Jin, J.; Topping, C. M.; Suresh, S.; Foulger, S. H.; Rice, N.; Mojazza, B. H.; Smith, D. W., Jr. *Polymer* **2005**, *46*(18), 6923–6932.
45. Qing, F. L.; Wang, R.; Li, B.; Zheng, X.; Meng, W. D. *J. Fluor. Chem.* **2003**, *120*(1), 21–24.
46. Liu, S.; Jiang, X. Z.; Ma, H.; Liu, M. S.; Jen, A. K. Y. *Macromolecules* **2000**, *33*(10), 3514–3517.
47. Ma, H.; Wu, J. Y.; Herguth, P.; Chen, B. Q.; Jen, A. K. Y. *Chem. Mater.* **2000**, *12*(5), 1187–1189.
48. Jin, J.; Smith, D. J.; Glasser, S.; Perahia, D.; Foulger, S. H.; Ballato, J.; Kang, S. W.; Kumar, S. *Macromolecules* **2006**, *39*(14), 4646–4649.
49. Rizzo, J.; Harris, F.; Meng, H. H. *Abstr. Pap. Am. Chem. S.* **1999**, *217*, U535.
50. Narayan-Sarathy, S.; Neilson, R.; Smith, D. *Polym. Prepr. Am.* **1998**, *39*, 609–610.
51. Neilson, R. H.; Ji, J. M.; Narayan-Sarathy, S.; Smith, D. W.; Babb, D. A. *Phosphorus Sulfur* **1999**, *146*, 221–224.
52. Souzy, R.; Ameduri, B.; Boutevin, B.; Virieux, D. *J. Fluor. Chem.* **2004**, *125*(9), 1317–1324.
53. Boone, H.; Smith, D.; Babb, D. *Polym. Prepr. Am.* **1998**, *39*, 812–813.
54. Lewis, E.; Naylor, M. *J. Am. Chem. Soc.* **1947**, *69*(8), 1968–1970.
55. Jin, J. Y. *Studies on Perfluorocyclobutyl (PFCB) Aryl Ether Polymers: From Fundamentals to Function*. Clemson University, 2005.
56. Mifsud, N.; Mellon, V.; Jin, J. Y.; Topping, C. M.; Echegoyen, L.; Smith, D. W. *Polym. Int.* **2007**, *56*(9), 1142–1146.
57. Spraul, B. K.; Suresh, S.; Jin, J.; Smith, D. W. *J. Am. Chem. Soc.* **2006**, *128*(21), 7055–7064.

15

COMBUSTION CHARACTERIZATION OF ENERGETIC FLUOROPOLYMER COMPOSITES

OLIVER MULAMBA AND MICHELLE PANTOYA
Texas Tech University

15.1 INTRODUCTION

Reactive composite materials that include a metallic fuel and fluoropolymer oxidizer can provide significant chemical energy for a variety of applications. Upon ignition, these composites go through an oxidation–reduction reaction that is exothermic. A subgroup of these composites is called metastable intermolecular composites (MICs) [1] or nanocomposite energetic materials and include reactants that are on the nanometer scale. Nanocomposites exhibit dramatically increased reactivity in comparison to their micron scale counterparts [2]. Due to their high energy density, high energy release, and fast reaction rate, these materials are gaining popularity for a variety of energy generation applications. For example, comparing the heats of combustion for the polymeric composition of aluminum + polytetrafluoroethylene (Al + PTFE) of 21 kJ/cm^3 to that of explosive trinitrotoluene (TNT) of 8 kJ/cm^3 shows significant potential for harnessing energy from nanocomposites [3, 4]. While explosive reactions like TNT are dominated by molecular bond–breaking chemical kinetics and produce significant power, composite reactions are limited by energy and mass diffusion between discretely separated fuel and oxidizer species. Reducing the particle size shortens the diffusion distance and offers great opportunities to enhance the power generated from these high energy density materials.

Handbook of Fluoropolymer Science and Technology, First Edition. Edited by Dennis W. Smith Jr., Scott T. Iacono, and Suresh S. Iyer.

Recent research revealed unique exothermic reactions associated with fluoropolymers and aluminum (Al) fuel particles that contribute toward their enhanced reactivity. Specifically, fluorine-based chemistry is shown to exothermically interact with the Al_2O_3 passivation shell surrounding the aluminum core. This core–shell particle structure provides new kinetics for nanocomposite reactions by enabling the passivation shell to contribute as an active exothermic participant in the reaction, rather than simply a barrier for diffusion oxidation and a heat sink in energy propagation. Exploiting chemistry of the Al_2O_3 shell is a new approach for tailoring a composite's reactivity toward a particular application.

This chapter will discuss polymeric materials that act as an oxidizer in composite energetic formulations that use aluminum as the fuel. We will focus on the synthesis and combustion characterization of these materials using various analysis methods. Fundamental combustion theories will also be explained to provide a more complete understanding of combustion in general. Many different approaches to particle and composite synthesis will be introduced as well as their impact on equilibrium reaction kinetics. The molecular structure of the fluoropolymer has recently been shown to have a significant impact on the burning performance of the composite and will be presented which includes discussion of benchmark methods for characterizing energy propagation during reaction using flame speed measurements. Ignition will also be discussed with particular focus on impact and electric initiation. By coupling discussions on chemical kinetics to ignition and energy propagation, a more complete understanding of the unique combustion behavior of fluoropolymers reacting with aluminum can be understood. The objective of this chapter is to present a broad perspective on aspects relevant to the use of polymers in energetic material combustion and provoke thoughtful inspiration for future work.

15.2 SYNTHESIS OF COMPOSITES

15.2.1 Stoichiometry

When synthesizing mixtures, the appropriate fuel to oxidizer stoichiometry must be selected. In a stoichiometric mixture, the total quantity of fuel is reacted to completion by the total quantity of oxidizer [5]. Mixtures are called fuel-lean when the amount of oxidizer is more than stoichiometric and fuel-rich when the amount of oxidizer is less than stoichiometric. An equivalence ratio, ER, distinguishes the chemical balance: stoichiometric, rich, or lean [5]. This property is defined in Equation 15.1.

$$\mathrm{ER} = \frac{(m_\mathrm{f}/m_\mathrm{o})_\mathrm{actual}}{(m_\mathrm{f}/m_\mathrm{o})_\mathrm{stoich}} \tag{15.1}$$

The variables m_f and m_o are the fuel and oxidizer masses. The subscript actual refers to the measured ratio, while the subscript stoich refers to the ratio required for stoichiometric combustion. A stoichiometric mixture has ER = 1.0; a mixture with an ER less than 1 is fuel-lean, and greater than 1 is fuel-rich. Combining reactants begins with knowledge of the chemical reaction such that the denominator in Equation 15.1 can be calculated. As an example, consider the Al + PTFE reaction:

$$4Al + 3C_2F_4 \rightarrow 4AlF_3 + 6C.$$

15.2.2 Mixing

Researchers have employed various mixing strategies to combine reactant powders. For example, in thin film energetics, metal fuel particles are combined with fluoropolymers that not only act as oxidizers in the reaction but also as a binder holding the reactants together and enabling their application as films. Energetic films have use in welding and joining as well as heat source applications [6, 7]. Energetic film deposition methods vary greatly and include electrophoretic deposition [8], sputter deposition [9], physical vapor deposition [10], and 3D macroporous structures [11]. The deposition method is extremely important because it plays a large role in particle mixing, deposition thickness, and deposition density, all of which have been shown to affect combustion behavior [12, 13]. Commonly used fluoropolymers in this application are Viton® and polyvinylidenefluoride (PVDF) [14].

Other researchers have attempted to coat aluminum particles directly with an acid, creating fluoropolymer chains that attach to the alumina shell surrounding the aluminum core particle, as illustrated in Figure 15.1 [15]. The acid shell accounts for a small percentage of the total mixture such that the mixture also requires another oxidizer to balance the reaction. This approach is referred to as surface functionalization of aluminum particles and has shown great success in controlling the reactivity of a mixture [15].

Recent work by Iacono et al. [16], employed perfluoropolyethers (PFPEs) to coat nano-aluminum particles. The nonvolatility and viscous characteristic of PFPEs make them ideal oxidizers and aid greatly in maximizing surface interactions. These coated Al particles are used in the production of an energetic Al + PFPE epoxy composite that experiences latent thermal metal-mediated oxidation. This powder

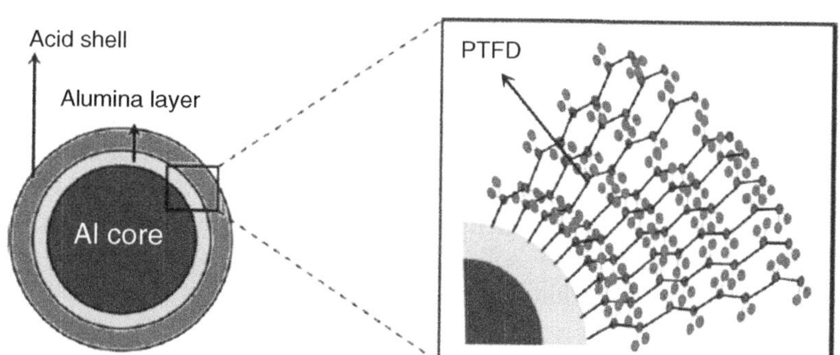

FIGURE 15.1 Functionalized fluoropolymer acid layer coating an aluminum-Al_2O_3 core–shell particle. PFTD, perfluoro tetradecanoic.

formulation which is sensitive to thermal ignition was found to have the capability of being partially cured. Therefore, moldable components can be fabricated that experience no dimensional alterations with additional curing. This work is at the forefront of epoxy-based metal + fluoropolymer energetic composites, and presents a solid foundation for further experimentation in behaviors using other metals and network polymers.

For many applications, simply combining reactant powders using a liquid-based sonification method produces suitable homogeneity within the mixture that results in reliable and repeatable combustion performance. In this approach, respective amounts of fuel and oxidizer powders are suspended in a solution of hexane or other hydrocarbon and sonicated using a sonicator (e.g., Misonix model S3000) to break up agglomerates and provide homogeneous mixing. A cyclic mixing program of 10 seconds on and off for a total of 70 seconds was found to be ideal and does not generate excessive heating within the mixture that can potentially damage particles. The mixed solution is poured into a Pyrex container and allowed to dry in a fume hood. Hexane evaporates and the remaining mixture is reclaimed using a grounded powder brush or blade. In some cases, a mesh is used to refine the powder mixture further.

15.2.3 Powders Versus Pellets

Composites can be used as powders or pressed into pellets. For example, for thermal analysis using diagnostics such as differential scanning calorimeter (DSC), powders offer better contact between the sensor and material and therefore improve the accuracy of measurement [17]. Flame speeds are typically measured for loose powder material [8]. When studying laser ignition or impact ignition, pellets are more commonly used [18–20]. When working with pellets, variations in bulk density are important and controlled primarily by the pressure applied during the pressing process. Describing bulk density is commonly done in terms of identifying a percentage of the theoretical mass density (TMD) defined in Equation 15.2.

$$\text{TMD} = \frac{1}{\sum_{i=1}^{N} \frac{M_i}{\rho_i}} \tag{15.2}$$

From Equation 15.2, (i) is an index of summation that corresponds to individual materials in the mixture and (N) is the upper bound of the summation and corresponds to the overall quantity of materials present in the mixture [21]. Also ρ and M represent the material density and its mass fraction in the mixture, respectively. A mixture that has 60% TMD is therefore 60% solids and 40% voids (air space). Loose powders are typically <20% TMD while compressed pellets typically range from 20% to 75% TMD.

For these powders, cold pressing is sufficient to achieve compacted pellets. A standard axial die used for cold pressing powders into pellets is shown in Figure 15.2. This assembly allows for specific pellet dimensions to be achieved with the use of

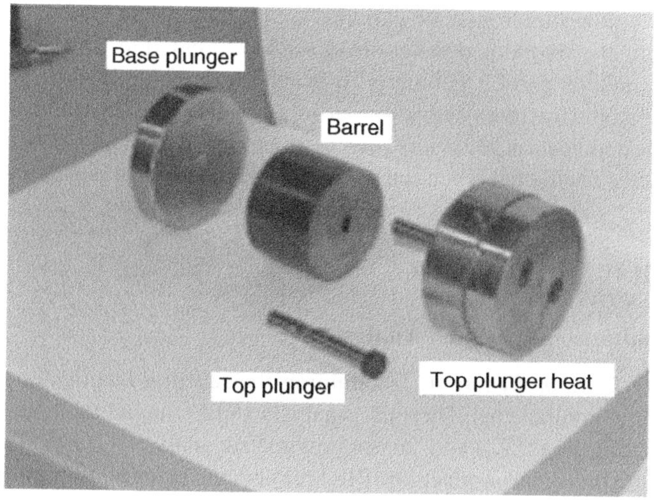

FIGURE 15.2 Cylindrical pressure die set (McMaster-Carr).

shims to vary pellet thickness. Once a %TMD is selected, the pellet volume can be calculated and the required height is applied using the shims [21].

The powder is poured into the die shaft and pressed using a hand press or a hydraulic press. A plunger can then be used to extrude the pellet. The pressures required to press pellets at different %TMDs is largely dependent on the size of the particles used. This is due to the consequences of the high pressures on the intactness of the particles [22]. Figure 15.3 shows compression pressure for micron versus nanoscale powders required to achieve a specific %TMD.

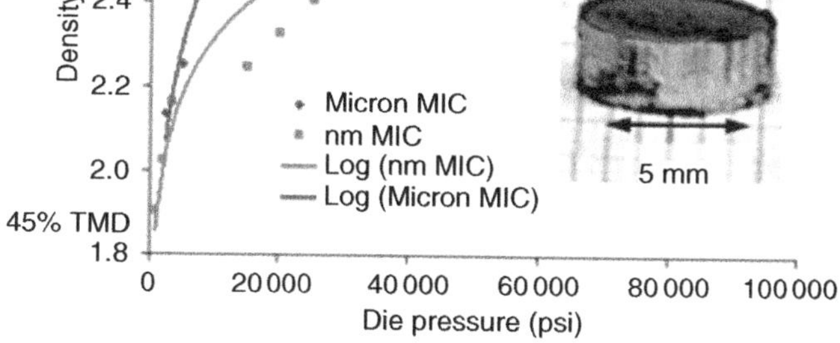

FIGURE 15.3 Pressure required to achieve specific densities, with a sample pellet illustrated.

Figure 15.3 shows micron-scale particles are easier to press than nanoscale particles and a typical consolidated pellet of Al + PTFE pressed to 50% TMD is shown. Compaction can alter particle geometry from spherical to elliptical and create imperfections, cracks, and fractures within the particle structure. Highly compact pellets can exhibit unique combustion behaviors because the reaction mechanism associated with pristine particles is affected when the particle exhibits physical deformations [22].

15.3 EQUILIBRIUM KINETICS

15.3.1 Simultaneous Thermal Analysis

15.3.1.1 Pre-ignition Reaction Examining the reaction kinetics of Al + PTFE is done using Simultaneous Thermal Analysis (STA) including both differential scanning calorimetry (DSC) and thermogravimetric analysis (TGA). These studies reveal a unique pre-ignition reaction (PIR) that occurs prior to aluminum oxidation and is illustrated in Figure 15.4 as the first exotherm in the heat flow curve (i.e., solid line) [15, 23]. This PIR is a result of aluminum oxide (Al_2O_3) reaction with fluorine ions liberated upon PTFE decomposition. Hydroxyls bonded to the Al_2O_3 shell potentially facilitate this interaction which form the catalyst aluminum (III) fluoride (AlF_3) [23]. Facilitation of Al_2O_3 fluorination due to hydroxyl groups on its surface was confirmed comparing thermal behaviors of Al_2O_3 + PTFE and calcined-Al_2O_3 + PTFE. The Al_2O_3 + PTFE habitually contained more hydroxyls than calcined-Al_2O_3 + PTFE, and the consistent decreased sizes of the PIR confirmed the assistive role of the hydroxyls [23]. The small exotherm that occurs with the PIR facilitates decomposition of PTFE and accelerates the overall reaction. Figure 15.4 shows heat flow (DSC) and mass loss (TG) curves indicating the two-stage reaction. Aluminum oxidation occurs at 550°C; therefore, the smaller exotherm observed at 450°C is described as a PIR. Partial decomposition is verified by the ~2% mass loss observed in the TG curve.

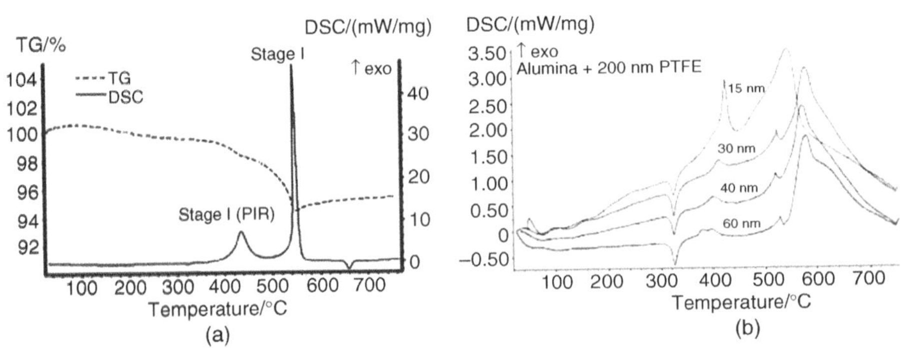

FIGURE 15.4 (a) Nano-Al particles reacting with PTFE at 20 KPM in argon, dashed line is TGA (mass loss), solid line is DSC (heat flow); (b) Al–PTFE reactions for various diameter Al particles labeled accordingly at 20 KPM in argon, only DSC curves are shown.

Comparisons between micron- and nano-sized Al particles show that the PIR is evident for both particle sizes but almost negligible with micron-sized Al particles. Based on reaction kinetics, the fluorine extraction from PTFE and the AlF_3 formation show strong dependence on Al particle size. As the particle size is reduced, the specific surface area is increased and the onset of the PIR can be manipulated. This particle size reduction causes a merging of exotherms such that the initial two on the right merge to form one as shown in Figure 15.4b [24]. In conclusion, increased surface area interactions between PTFE and Al_2O_3 allow for higher PIR exothermicity which in turn promotes the Al–PTFE reaction [24].

15.3.2 Molecular Structure

15.3.2.1 Kinetic Reactions The molecular structure of PTFE impacts the reaction kinetics and combustion behavior of Al + PTFE. With longer polymer chains there is a balanced increase in molecular weight. Thermal analysis studies show decreased surface reactions with increasing molecular weight observed through a general decrease in exothermicity of the PIR with increases in molecular weight. This behavior is illustrated in Figure 15.5; the PIR decreases with increasing PTFE chain length. Also very apparent in Figure 15.5 is the decrease in the heat of combustion with the increase in chain length, measured as the total area under the exothermic curves. With the longer chained polymers there is increased separation between successive fuel particles and an increase in the amount of PTFE that does not interact with the Al_2O_3. This increased separation may be responsible for diminished exothermicity observed by the decreased PIR.

Figure 15.6 illustrates this concept further by schematically illustrating aluminum particle interconnectivity within the polymer chain. In Figure 15.6a, there is an increase in the distance between particles and increased amount of PTFE that does not interact with Al_2O_3. In contrast, Figure 15.6b shows decreased distances between

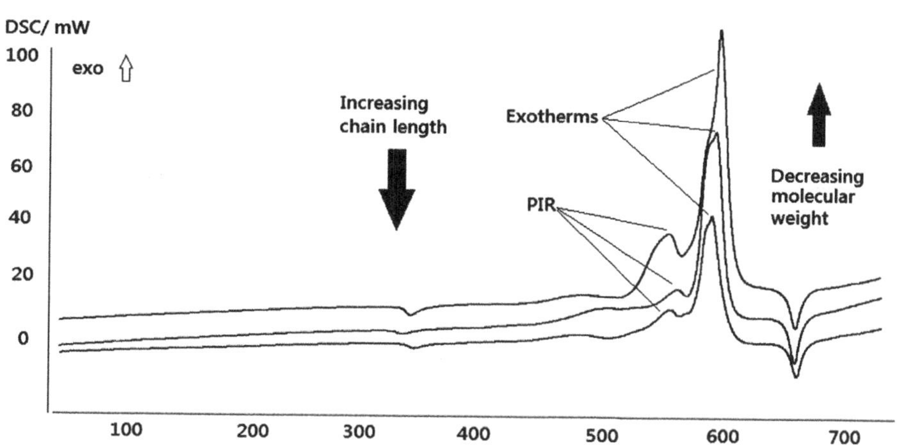

FIGURE 15.5 Reaction kinetics of PTFE with varying molecular structures.

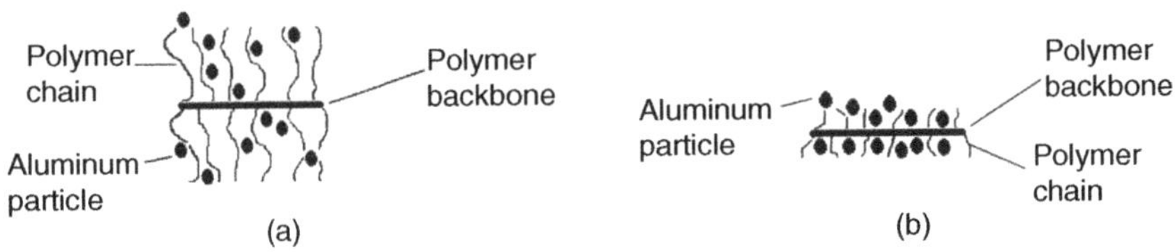

FIGURE 15.6 Effects of polymeric chain length on particle position. (a) Aluminum particles combined with longer polymer chain composite. (b) Aluminum particles combined with shorter polymer chain composite.

particles and increased amount of PTFE interacting with the Al_2O_3. The oxidation process is diffusion-limited, therefore mass transport and diffusion distances control the rate of energy release. Larger chain lengths promote greater diffusion distances and limit reactivity as observed in the decreasing PIR with increasing chain length.

15.4 NONEQUILIBRIUM COMBUSTION CHARACTERIZATIONS

15.4.1 Energy Propagation

Flame speed is typically defined for gas phase reactions as "the velocity at which the unburned gases move through the combustion wave in a direction normal to the wave surface" [25]. For example, in a flame tube, the flow rate that holds a flame stationary is directly related to the flame speed. However, in multiphase media with solid fuel and oxidizer particles within a gas medium (i.e., air), the definition of flame speed is the velocity at which the flame front propagates through the unburned particles. For non-gas generating reactions, Armstrong [26] developed well-used numerical models describing flame speed (u) based on diffusion-controlled energy propagation. Equation 15.3 presents Armstrong's relationship and shows the influence of parameters such as activation energy (E_a), thermal properties, and particle diameter (d) [26]. Also, T_f is the flame temperature, T_0 is the initial temperature of reactants, R is the universal gas constant, A is the pre-exponential factor, and α is thermal diffusivity.

$$u = 12\frac{A^{\left(\frac{-E_a}{RT_f}\right)\left(\frac{RT_f}{E_a}\right)\alpha^2}}{d^{2\frac{T_f-T_0}{T_f}}} \tag{15.3}$$

Many reactions produce gas such that Armstrong's model [26] has relatively limited application, especially when considering Al reaction with fluoropolymers. In this case, many factors contribute to propagation rate and one single model relevant to a broad class of reactions has not been developed. The physics become complicated by balancing the influences of the mode of energy propagation such as convection, compression (shock wave), diffusion, radiation as well as the reaction mechanism which may vary with aluminum particle size and reaction kinetics that may no longer be assumed first-order Arrhenius due to surface reactions with the Al_2O_3 shell.

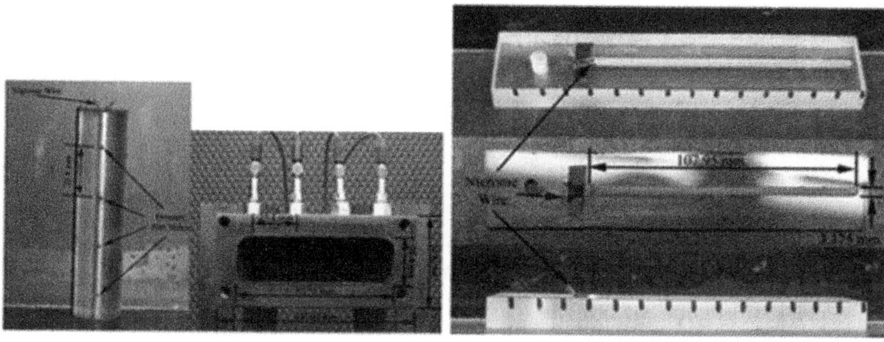

(a) (b)

FIGURE 15.7 Flame speed testing configurations: (a) closed and (b) open.

There is significant potential for advanced theories that model energy propagation in nanocomposites in the future.

Powder flame speed measurements typically fall into two categories: closed tube and open tray. For closed tube measurements, Figure 15.7a shows the experimental setup such that powder is loaded into a tube in such a way as to maintain a constant overall bulk density and ignited by a spark, hot wire, or other stimuli [27]. Optical readings can be taken using high-speed imaging diagnostics or photodiodes. High-speed cameras enable greater observation of the nature of the combustion and make accurate flame speed measurements. However, in many cases, flame speeds are so high (i.e., >1000 m/s) that a camera with the resolution and frame rate required to get reliable information may prove cost-prohibitive. Photodiodes give similar information at a much lower cost. A photodiode works by converting light into an electrical signal. An oscilloscope triggers on a certain voltage met when a threshold of light signifying combustion strikes the diode. The drawback to photodiodes is limited data from select positions so only average speed data may be inferred during analysis with little detail regarding the entrance length and steady state nature of the propagation. Pressure transducers can also be used to locate the passing flame front. Heat from the flame and gaseous products increases the pressure as the flame progresses. Bockmon et al. [27] designed an experimental setup that used both optical and pressure information by combining photodiodes and pressure transducers. In their experiments, flame speed exhibited a similar trend with pressure history [27]. Many researchers have since used this benchmark diagnostic often referred to as the "Bockmon Tube" to understand propagation of different formulations.

For open tray experiments, Figure 15.7b illustrates that loose powders are uniformly packed into a channel milled into an acrylic block. Flame speeds are typically faster than the diffusion time of energy into the acrylic such that reaction with acrylic and other obstructions associated with the acrylic base on energy propagation are negligible. A nichrome wire is used to resistively heat and initiate ignition. High-speed cameras or photodiodes are used to track the highly luminescent flame front as a function of time. Also, with an optical camera, the luminescent front is assumed to

TABLE 15.1 Flame Speeds for Al With Varying Oxidizers in Flame Tubes

Composition	Flame speed (m/s)
Al + MoO_3	960
Al + CuO	710
Al + Fe_2O_3	450
Al + NiO	270
Al + PTFE	838

travel with the flame at the same velocity. For highly convective flows, the hot product gases may be pushed ahead of the flame at a temperature capable of radiating visible light. The hot gases may increase the luminescence to a point where oversaturation leads to systemic error in flame speed calculations.

Increased flame speeds have made fluorine-based oxidizers take a front seat in combustion research. Table 15.1 provides an indication of the flame speed variations observed when Al is used with different oxidizers [28].

Observations have revealed that fluorine-based oxidizers have slower flame speeds in open configuration experiments in comparison to oxygen-based oxidizers [29]. The higher gas generation of the fluorine-based oxidizer is one factor impacting this observation. Therefore, lower gas generating mixtures show faster flame speeds in open configuration experiments. On the other hand, in closed configuration tests, fluorine-based oxidizers exhibit higher flame speed in part because high gas generation enhances convection and promotes energy propagation (Table 15.1) [29]. Also, results show that in a confined setup Al + PTFE with the shorter polymer chains were found to propagate slower, and that as the chain lengths increased so did the speeds. The shorter chain lengths showed speeds of approximately 270 m/s while the longer chains were at approximately 940 m/s. The reverse was observed for the open configuration. Though potentially contradicting, these results imply that for Al + PTFE reactions, gas generation is proportional to flame speed and as the polymer chain increases in length there is a corresponding increase in gas generation and flame speed. Therefore, tailoring energy propagation can be engineered using PTFE with controlled chains lengths (i.e., molecular weight).

15.4.2 Ignition

Electric, impact, chemical, mechanical, and thermal stimuli are a few of the initiation methods available with composite energetic mixtures [2, 20, 21, 30]. Each stimulus initiates a reaction at a different heating rate that affects the reaction mechanism and combustion behavior. For example, in ordnance applications such as small caliber munitions, a firing pin (i.e., impact) initiates the primer mixture which is composed mainly of a composite energetic material. For medium and large caliber systems, electrical means are often used. This discussion of ignition will focus largely on these two stimuli.

15.4.2.1 Impact Ignition Impact ignition is typically studied using a drop-weight apparatus schematically shown in Figure 15.8. The pellet sample is positioned on the anvil and a mass is dropped from the desired height. The energy of the drop weight is calculated using Equations 15.4 and 15.5.

$$v = (2 * g * H)^{1/2} \tag{15.4}$$

$$E = \frac{1}{2} * \frac{w}{g} * v^2 \tag{15.5}$$

In these equations, g is the acceleration due to gravity, H is the drop height, v is the velocity, and w is the weight of the drop mass. Impact is monitored using photodiodes or other imaging diagnostics to monitor light emitted indicating ignition. The measurements provide an indication of the ignition energy.

Al and PTFE are typically less impact ignition sensitive than Al combined with other oxidizers. The ductility and malleability of PTFE tend to cushion impact enough to require Al + PTFE impact ignition energies almost twice that of intermetallic or metal–metal oxide formulations. Toward this end, studies have examined the influence of carbon additives to sensitize Al + PTFE to impact ignition. Specifically, three forms of carbon additives were investigated and selected based on different physical and structural properties: spherically shaped amorphous nanoparticles of carbon,

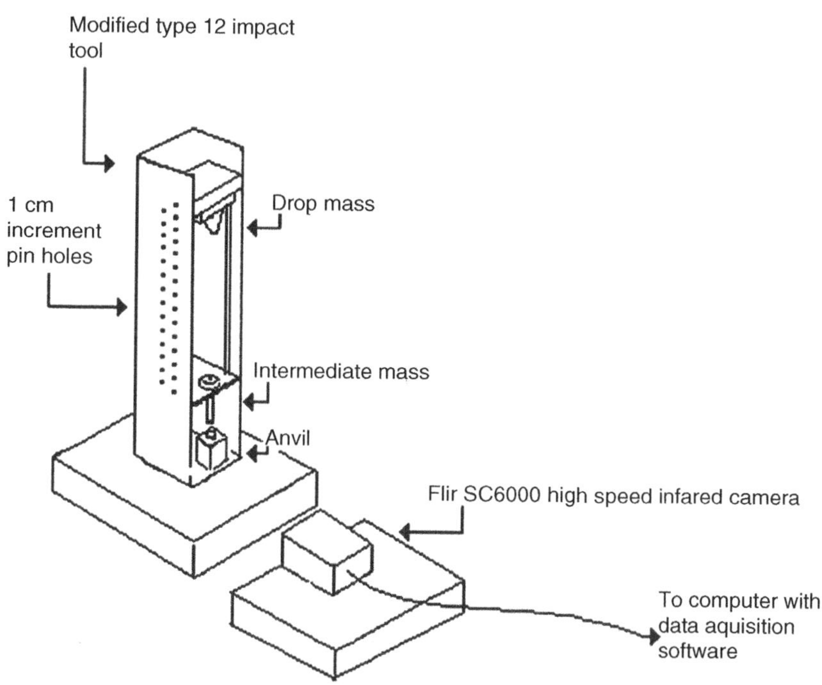

FIGURE 15.8 Drop-weight tester setup.

cylindrically shaped multiwalled carbon nanotubes (CNTs), and sheet-like graphene flakes [20]. The composites containing small concentrations of CNTs demonstrated a heightened sensitivity to impact ignition, with ignition energies on the order of 0.29 J [20]. CNTs have high Young's modulus and may promote hot spot formation spurred by friction and sheer interaction between CNTs and surrounding particles (i.e., at interfaces and grain boundaries). In contrast, graphene does not affect ignition energy or sensitivity until larger quantities are added to the composite. Graphene has a significantly higher thermal conductivity than the other additives which may promote diffusion of local thermal energy, reducing the likelihood of hot spot development. These results provide insight into factors that promote sensitivity to impact ignition thereby leading to advances in the safe handling of energetic materials. Further work is required since formation of hot spots is random on the surface being impacted and thus has not been quantified.

15.4.2.2 Electrostatic Discharge Ignition In comparison to intermetallics and aluminum–metal oxide formulations, Al + PTFE is least sensitive to electrostatic discharge (ESD) ignition [20]. Weir et al. [20] studied several Al-fueled binary composites and showed that for micron particle Al, the Al + PTFE reaction is the least sensitive to ESD ignition requiring greater than 100 mJ of energy. In fact, a study by Foley et al. used a fluoropolymer additive to decrease the ESD ignition sensitivity of aluminum with copper oxide. They showed that even small additions of the fluorocarbon, Viton A, decreased the ESD ignition sensitivity significantly [31].

Collins et al. [32] sensitized Al + PTFE with CNT and graphene additives using a similar approach to the impact studies by Kappagantula et al. that sensitized Al + PTFE to impact ignition [20]. Collins et al. [32] found that the Al particle size plays a significant role in ESD ignition sensitivity and identified a correlation between electric conductivity of the reactant mixture and ESD ignition sensitivity. The electrical conductivity of Al + PTFE was initially measured and found to be 1×10^{-7} μS/cm. Graphene nano-particles (GNP) were then added in volumetric percentages ranging from 0.5 to 4.0 vol%. Figure 15.9 shows the electrical conductivity percolation threshold for varying concentrations of GNPs [32, 33].

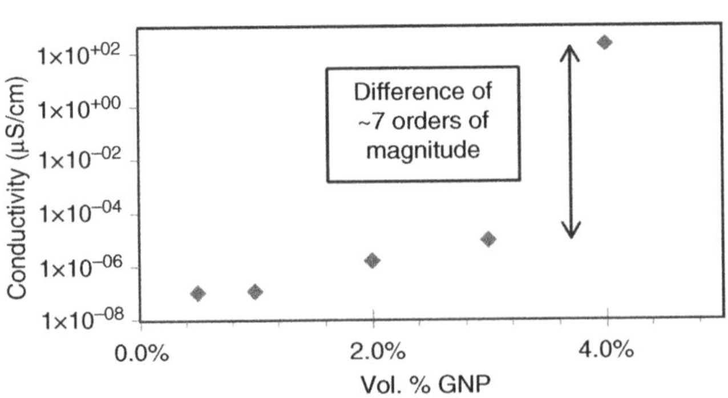

FIGURE 15.9 Electrical conductivity of Al + PTFE with added GNP.

The electrical conductivity of Al + PTFE was increased from approximately 7 orders of magnitude to 100 μS/cm, with the inclusion of 4 vol% GNPs. On the contrary, only 1 vol% CNTs showed comparable results with approximately 7 orders of magnitude increase in conductivity. The ignition sensitivity of the highly conductive mixtures was greatly decreased due to the current traveling through short circuits in the material formed by connecting nanofillers [32]. Ignition was tied to an approximate electrical conductivity of around 0.0025 μS/cm.

15.5 CONCLUSION

Fluoropolymers reacting with aluminum fuel particles show great potential for energy generation applications. This chapter focused on aluminum and polytetrafluoroethylene (Al + PTFE) composite synthesis, reaction kinetics, and combustion behavior in terms of energy propagation and ignition. Synthesis strategies include the preparation of thin energetic films, coated Al particles achieving surface functionality, and sonicating discrete particles in hydrocarbon solutions to achieve suitable mixing. Reaction kinetics of PTFE-based composites shows great dependence on Al particle size. A unique PIR exotherm increases with decreasing Al particle size. The onset of this PIR can be tailored such that the flame speed is affected. Varying the molecular structure of the PTFE provides another way of tailoring the energetic performance of these composites. Aluminum + PTFE were also shown to be sensitive to varying methods of ignition. Polymeric energetic materials such as Al + PTFE show great potential due to its tailorability and wide range of combustion behaviors.

ACKNOWLEDGMENTS

The authors are grateful for support from the Army Research Office under contract number W911NF-11-1-0439 and the encouragement of Dr Ralph Anthenien.

REFERENCES

1. Miziolek, A. W. *Adv. Mater. Manuf. Test. Info. Anal. Cent.* **2002**, *6*, 43.
2. Granier, J. J.; Pantoya, M. L. *Combust. Flame* **2004**, *138*, 373–383.
3. Yang, Y. Q.; Sun, Z. Y.; Wang, S. F.; Dlott, D. D. *J. Phys. Chem. B.* **2003**, *107*, 4485–4493.
4. Zamkov, M. A.; Conner, R. W.; Dlott, D. D. *J. Phys. Chem. C.* **2007**, *111*, 10278–10284.
5. Kuo, K. K. *Principles of Combustion*; Wiley, 1986.
6. Dai, J.; Lai, M.; LaFollette, R.; Reisner, D. Thin Film Copper Vanadium Oxide Electrodes for Thermal Batteries. In *Electrochemistry of Novel Materials for Energy Storage and Conversion*; Zaghib, K., Julien, C., Ramani, V., Eds.; Electrochemical Society (ECS Transactions): Pennington, NJ 08534-2839, USA, 2011; Vol. 33(27), pp 3–9.
7. Wang, J.; Besnoin, E.; Duckham, A.; Spey, S. J.; Reiss, M. E.; Knio, O. M.; Powers, M.; Whitener, M.; Weihs, T. P. *Appl. Phys. Lett.* **2003**, *83*, 3987–3989.

8. Sullivan, K. T.; Worsley, M. A.; Kuntz, J. D.; Gash, A. E. *Combust. Flame* **2012**, *159*, 2210–2218.
9. Zhou, X.; Shen, R.; Ye, Y.; Zhu, P.; Hu, Y.; Wu, L. *J. Appl. Phys.* **2011**, *110*.
10. Lanthony, C.; Ducere, J.M.; Esteve, A.; Rossi, C.; Djafari-Rouhani, M. *Thin Solid. Films* **2012**, *520*, 4768–4771.
11. Zhang, W.; Yin, B.; Shen, R.; Ye, J.; Thomas, J. A.; Chao, Y. *ACS. Appl. Mater. Interfaces.* **2013**, *5*, 239–242.
12. Sanders, V. E.; Asay, B. W.; Foley, T. J.; Tappan, B. C.; Pacheco, A. N.; Son, S. F. *J. Propul. Power* **2007**, *23*, 707–714.
13. Hunt, E. M.; Pantoya, M. L. *Intermetallics* **2010**, *18*, 1612–1616.
14. Bouma, R. H. B.; Meuken, D.; Verbeek, R.; Pacheco, M. M.; Katgerman, L. *Propellants. Explos. Pyrotech.* **2007**, *32*, 447–453.
15. Kappagantula, K. S.; Farley, C.; Pantoya, M. L.; Horn, J. *J. Phys. Chem. C.* **2012**, *116*, 24469–24475.
16. Miller, H. A.; Kusel, B. S.; Danielson, S. T.; Neat, J. W.; Avjian, E. K.; Pierson, S. N.; Budy, S. M.; Ball, D. W.; Iacono, S. T.; Kettwich, S. C. *J. Mater. Chem.* **2013**, *1*, 7050–7058.
17. Farley, C.; Pantoya, M. *J. Therm. Anal. Calorim.* **2010**, *102*, 609–613.
18. Comet, M.; Siegert, B.; Pichot, V.; Spitzer, D. *J. Therm. Anal. Calorim.* **2013**, *111*, 431–436.
19. Stamatis, D.; Ermoline, A.; Dreizin, E. L. *Combust. Theor. Model.* **2012**, *16*, 1011–1028.
20. Kappagantula, K.; Pantoya, M. L.; Hunt, E. M. *J. Appl. Phys.* **2012**, *112*.
21. Collins, E.; Pantoya, M.; Vijayasai, A.; Dallas, T. *Surf. Coat. Technol.* **2013**, *215*, 476–484.
22. Pantoya, M. L.; Levitas, V. I.; Granier, J. J.; Henderson, J. B. *J. Propul. Power* **2009**, *25*, 465–470.
23. Osborne, D. T.; Pantoya, M. L. *Combust. Sci. Technol.* **2007**, *179*, 1467–1480.
24. Pantoya, M. L.; Dean, S. W. *Thermochim. Acta* **2009**, *493*, 109–110.
25. Glassman, I. Y., Yetter, R. *Combustion*, 4th ed.; Academic Press, **2008**.
26. Armstrong, R. *Combust. Sci. Technol.* **1990**, *71*, 155–174.
27. Bockmon, B. S.; Pantoya, M. L.; Son, S. F.; Asay, B. W.; Mang, J. T. *J. Appl. Phys.* **2005**, *98*.
28. Dikici, B.; Pantoya, M. L.; Shaw, B. D. *Combust. Theor. Model* **2012**, *16*, 465–481.
29. Watson, K. W.; Pantoya, M. L.; Levitas, V. I. *Combust. Flame* **2008**, *155*, 619–634.
30. Kappagantula, K. S.; Clark, B.; Pantoya, M. L. *Energ. Fuel* **2011**, *25*, 640–646.
31. Foley, T.; Pacheco, A.; Malchi, J.; Yetter, R.; Higa, K. *Propell. Explos. Pyrot.* **2007**, *32*, 431–434.
32. Energetic Materials and Methods of Tailoring Electrostatic Discharge Sensitivity of Energetic Materials, Inventors: Michael A. Daniels, Ronald J. Heaps, Ronald S. Wallace, Michelle L. Pantoya and Eric Collins, filled with US Patent office October 2013. US Patent Number 2939–11946.
33. Weir, C.; Pantoya, M. L.; Ramachandran, G.; Dallas, T.; Prentice, D.; Daniels, M. *J. Electrostat.* **2013**, *71*, 77–83.

16

AMORPHOUS PERFLUOROPOLYMERS

YOSHIYUKI OKAMOTO AND FRANTIŠEK MIKEŠ
NYU Polytechnic School of Engineering
KOTARO KOIKE
NYU Polytechnic School of Engineering, Keio University
YASUHIRO KOIKE
Keio University

16.1 INTRODUCTION

Since the discovery of polytetrafluoroethylene (PTFE) in 1937, fluorinated polymers have become of increasing interest in the development of advanced materials having high thermal and oxidative stability, chemical resistance, superior electrical insulating ability, and superior optical properties [1]. PTFE and copolymers of tetrafluoroethylene (TFE) with hexafluoropropylene are crystalline and have poor optical transparency and solubility. New classes of amorphous perfluorinated polymers with extraordinary properties were developed and commercialized by DuPont (as Teflon® AF), Asahi Glass (as Cytop®), and Solvay (as Hyflon® AD) [2–8]. Teflon® AF and Hyflon® AD are a family of copolymers of 2,2-bis(trifluoromethyl)-4,5-difluoro-1,3-dioxole (**I**) and 2,2-bis(trifluoromethyl)-4-fluoro-5-trifluoromethoxy-1,3-dioxole (**II**) with TFE, respectively (see Figure 16.1). The homopolymer of dioxole (**I**) exhibits a glass transition temperature (T_g) of 330°C, which is one of the highest T_g among the amorphous polymers.

Handbook of Fluoropolymer Science and Technology, First Edition. Edited by Dennis W. Smith Jr., Scott T. Iacono, and Suresh S. Iyer.

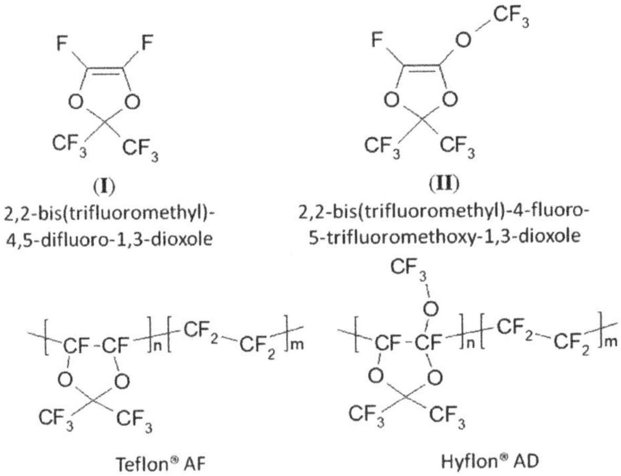

FIGURE 16.1 Chemical structures of monomers (**I**) and (**II**) and their copolymers with tetrafluoroethylene: Teflon® AF and Hyflon® AD.

16.2 EFFECT OF THE POLYMER STRUCTURE ON T_g

T_g of the dioxole polymers highly depends on the structure of the dioxole monomers. For example, when the trifluoromethyl group at the second position of (**I**) is replaced with either a fluorine atom or a long fluoroalkyl group such as pentafluoroethyl, T_g of the dioxole polymers is dramatically decreased as shown in Figure 16.2 [6].

16.3 COPOLYMERS WITH TFE–TEFLON® AF AND HYFLON® AD

The homopolymers of (**I**) and (**II**) are hard to process due to their high T_g. These monomers are copolymerized with other fluorinated olefins to improve the processability. In the case of copolymers of (**I**) and (**II**) with TFE, T_g decreases rapidly as the percentage of TFE increases. For example, T_g of commercially available Teflon® AF 1600 containing 65 mol% of dioxole (**I**) and Teflon® AF 2400 containing 87 mol% dioxole (**I**) are 160°C and 240°C, respectively. T_g of homopolymer (**II**) was 170°C, and T_g of copolymers (**II**) with TFE—the commercial Hyflon® AD 60X containing

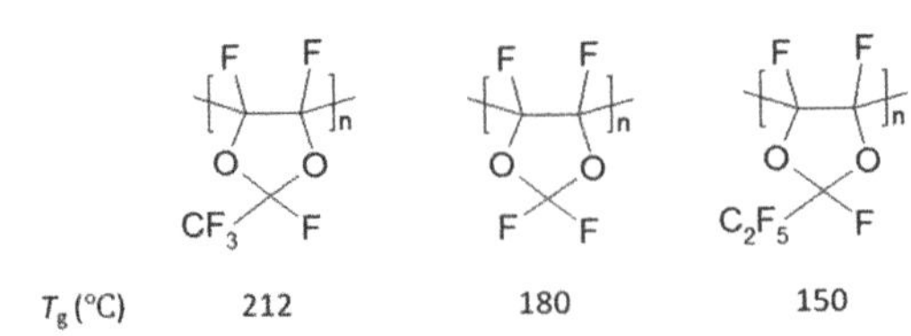

FIGURE 16.2 Effect of dioxole polymer structure on the glass transition temperature (T_g).

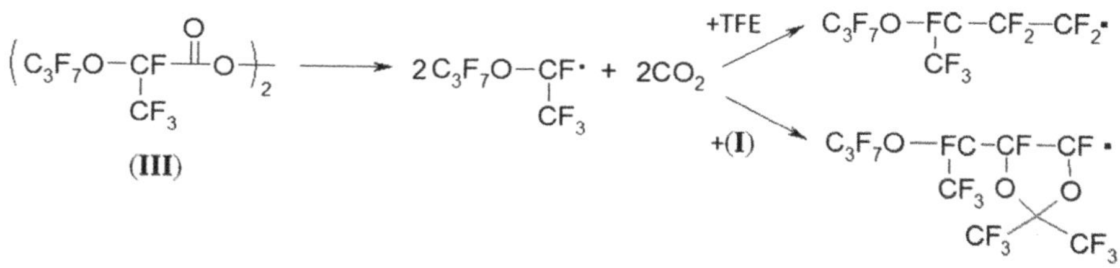

SCHEME 16.1 Thermal decomposition of bis(perfluoro-2-propoxypropionyl)peroxide and the end group of polymers.

60 mol% dioxole (**II**) and Hyflon® AD 80X containing 85 mol% dioxole (**II**)—is 110°C and 135°C, respectively [4].

TFE and monomer (**I**) were also copolymerized in supercritical carbon dioxide using a free radical initiator such as bis(perfluoro-2-propoxypropionyl)peroxide (**III**) at 35°C (the half-life time of the initiator is 40 minutes at 35°C) [9]. The decomposition of the initiator proceeds through a single-bond homolysis mechanism [10], resulting in the formation of perfluorinated end group that yields thermally stable polymers [9] (Scheme 16.1). The reaction conditions and properties of the copolymers of (**I**) and TFE obtained in supercritical carbon dioxide are shown in Table 16.1. PTFE is crystalline, so that when the amount of TFE increases in copolymers, the polymer has some microcrystalline regions. The polymers obtained in carbon dioxide have similar properties with the commercial polymers.

16.4 POLY(PERFLUORO-4-VINYLOXY-1-BUTENE)—CYTOP®

Cytop® is a homopolymer obtained by cyclopolymerization of the perfluorinated diene, perfluoro-(4-vinyloxy-1-butene) (IV). Cytop® is formed by five- and six-member rings and the five-member ring structure is predominant [3, 11] (Figure 16.3).

TABLE 16.1 Reaction Conditions and Properties of Copolymer (I) with TFE in Supercritical Carbon Dioxide

(I) in feed (mol%)	(I) in polymer (mol%)	Amount of initiator (mol%)	C^a (%)	Yield (%)	T_g (°C)
30	30	0.1	–	51	76
40	40	0.1	19	59	92
50	48	0.1	24	34	109
60	61	0.1	20	50	131
73	75	0.3	19	54	155
80	82	0.1	18	46	179
90	92	0.1	20	47	231
100	100	0.4	21	69	334

[a]Concentration of monomers (w/v), reaction time 5 h at 35°C.

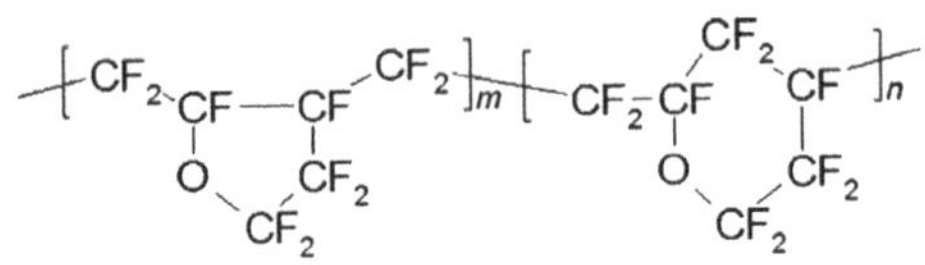

FIGURE 16.3 Chemical structure of cyclic polymer, Cytop®.

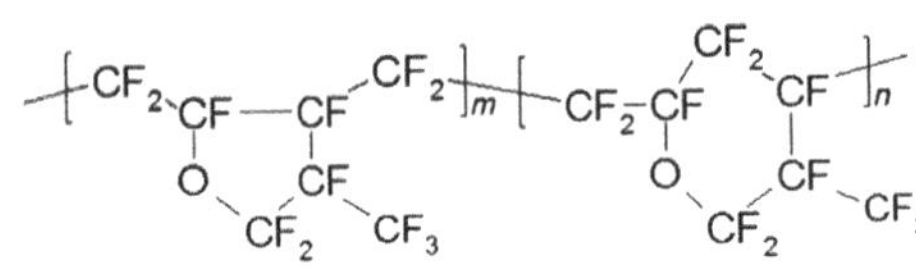

FIGURE 16.4 Chemical structure of poly(perfluoro-4-vinyloxy-3-methyl-1-butene).

T_g of Cytop® is 108°C. However, when CF_3 is substituted at the third position of the butane, perfluoro-4-vinyloxy-3-methyl-1-butene has a higher T_g value of approximately 120°C [12] (Figure 16.4).

16.5 SYNTHESES OF TEFLON®, HYFLON®, AND CYTOP® MONOMERS

Monomer (**I**) is synthesized from hexafluoroacetone and ethylene oxide in four steps as shown in Scheme 16.2 [8]. Monomer (**I**) is a colorless liquid with a boiling point of 33°C and has a high polymerization reactivity. Monomer (**II**) is synthesized by the following continuous-flow, gas-phase process (Scheme 16.3). It is also a colorless liquid with a boiling point of 30–32°C.

Cytop® is a cyclic homopolymer prepared by cyclopolymerization of perfluoro(4-vinyloxy-1-butene) (IV), which is synthesized by the following multistep process (Scheme 16.4) [3]. The monomer is a colorless liquid and its boiling point is 65°C.

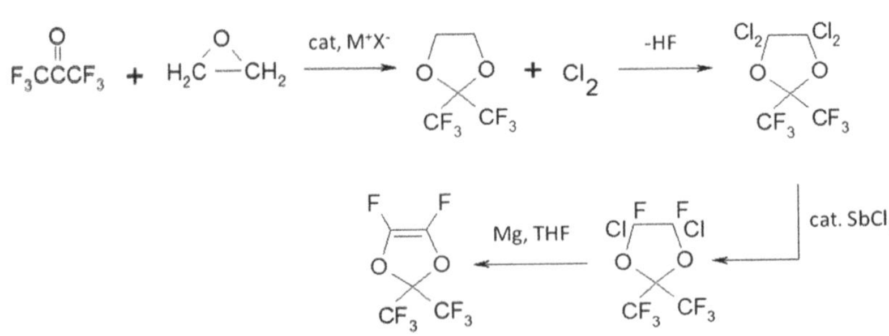

SCHEME 16.2 Synthesis of 2,2-bis(trifluoromethyl)-4,5-difluoro-1,3-dioxole.

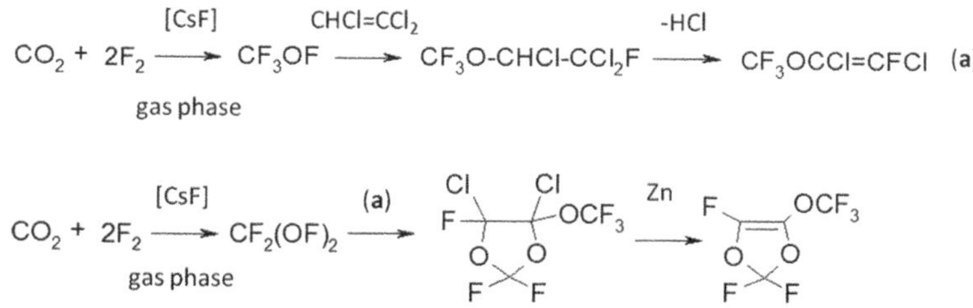

SCHEME 16.3 Synthesis of 2,2-bis(trifluoromethyl)-4-fluoro-5-trifluoromethoxy-1,3-dioxole (**II**).

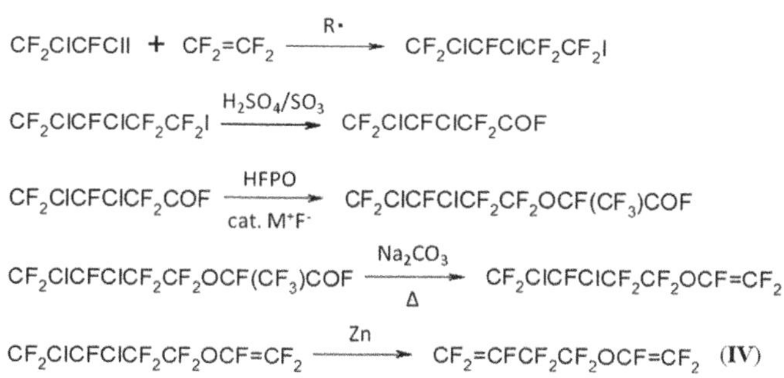

SCHEME 16.4 Synthesis of perfluoro(4-vinyloxy 1-butene).

The physical properties of these perfluoro polymers are summarized in Table 16.2. These perfluoropolymers are completely amorphous and contain no hydrogen atoms, show excellent chemical and thermal stability, and are soluble in fluorinated solvents such as hexafluorobenzene (HFB) and perfluorohexane. The dielectric constants of these fluoropolymers are considerably low, and they are almost unaffected by humidity.

TABLE 16.2 Physical Properties of Teflon® AF, Hyflon® AD, and Cytop®

	Teflon® AF		Hyflon® AD		
	1600[a]	2400[b]	60X[c]	80X[d]	Cytop®
T_g (°C)	160	240	125	140	108
Refractive index	1.31	1.29	1.33	1.32	1.34
Density (g/cm^3)	1.78	1.67	1.93	1.80	1.84
Dielectric constant ε	1.93	1.90	1.87	2.00	2.1–2.2

[a]Composition: dioxole (**I**) 65 mol% and TFE 35 mol%.
[b]Composition: dioxole (**I**) 87 mol% and TFE 13 mol%.
[c]Composition: dioxole (**II**) 60 mol% and TFE 40 mol%.
[d]Composition: dioxole (**II**) 85 mol% and TFE 15 mol%.

16.6 PERFLUORO-2-METHYLENE-1,3-DIOXOLANES

These perfluoropolymers have excellent properties in many aspects. However, the preparative methods are complex and expensive and T_g of Cytop® is relatively low. Several substituted perfluoro-2-methylene-1,3-dioxolanes (Figure 16.5) have been studied to obtain other perfluorinated polymers. The first example in the class is perfluoro-2-methylene-4-methyl-1,3-dioxolane (Figure 16.5B), which was prepared by DuPont in 1967 through perfluoropyruvyl fluoride prepared from hexafluoropropylene oxide [7]. Using a similar method, perfluoro-2-methylene-4,5-dimethyl-1,3-dioxolane (Figure 16.5D) was synthesized by Asahi Glass in 1993 [12].

Our typical preparation method for these perfluorodioxolane monomers and polymers is shown in Scheme 16.5 [13–18]. Commercially available methyl pyruvate and diols are used as the starting materials. The condensation product is then directly fluorinated in a fluorinated solvent such as Fluorinert™ FC-75 by the F_2/N_2 mixture. After flushing the system using nitrogen gas for 1 hour, fluorine gas diluted to 20% with nitrogen is blown into the reaction mixture at a flow rate of 240 L/h. The reaction is carried out at a temperature of around 25°C for over 20 hours. After the reaction is over, the mixture is neutralized with aqueous potassium hydroxide solution. The

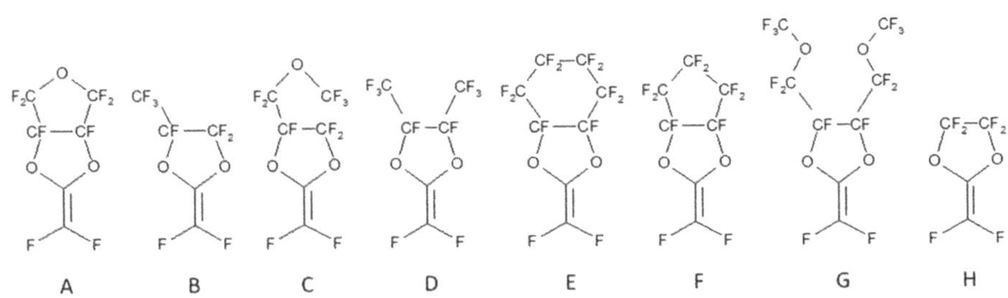

FIGURE 16.5 Chemical structures of perfluorodioxolanes.

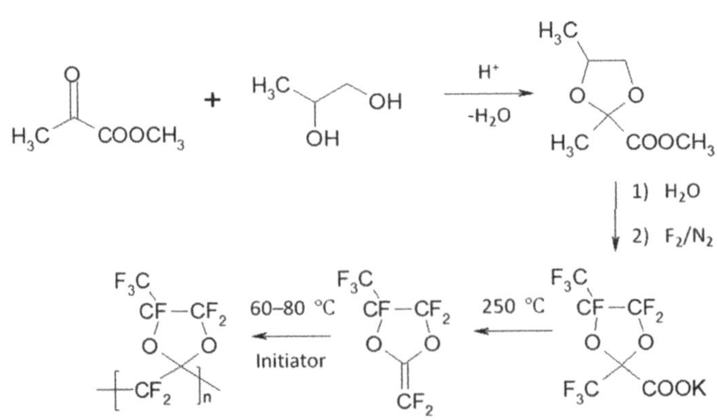

SCHEME 16.5 A typical synthetic route for a perfluorodioxolane monomer and the homopolymer.

TABLE 16.3 Effect of the Concentration of Initiator on the Properties of Poly(perfluoro-2-Methylene-4-Methyl-1,3-Dioxolane) Prepared in Bulk[a]

Initiator (wt%)	T_g (°C)	[η] (dL/g)[b]	$M \times 10^{-6}$ (g/mol)
0.14	134	0.61	3.1
0.30	133	0.52	
0.49	132	0.45	
0.80	132	0.43	2.0
3.03	131	0.23	
4.56	131	0.20	1.3

[a]Initiator: perfluorodibenzoyl peroxide, polymerization temperature: 60°C, polymerization time: 24 h, yield of polymers: 75–90%.
[b]Intrinsic viscosity in HFB at 25°C.

crude monomer is isolated upon decarboxylation of the potassium salt and then purified by fractional distillation. The fluorination process is the key step in the synthesis. The yield is as high as 75% with a careful control of the fluorination temperature, and it depends on the structure of the hydrocarbon precursor. Recently, the yield of the fluorination of similar compounds was improved to over 90% [19].

The polymerization of these monomers is generally carried out in bulk at 60–80°C using perfluorodibenzoyl peroxide or perfluoro-di-*tert*-butyl peroxide as the initiator [20] for 24–35 hours. The conversions are 70–80% and the polymers obtained are colorless and transparent. The polymers are purified by precipitation from their solutions in HFB into chloroform.

The molecular weight of the polymers can be controlled by the concentration of free radical initiator and/or chain transfer agents such as carbontetrabromide. The typical results of poly(perfluoro-2-methylene-4-methyl-1,3-dioxolane) prepared using different amounts of the initiator and chain transfer agent are shown in Tables 16.3–16.5.

TABLE 16.4 Effect of the Concentration of Initiator on the Properties of Poly(perfluoro-2-Methylene-4-Methyl-1,3-Dioxolane) Prepared in Solution[a]

Solution	Initiator (wt%)	T_g (°C)	[η] (dL/g)[b]	$M \times 10^{-6}$ (g/mol)
Perfluoro-2-hydro-2, 4-dimethyl-1,3-dioxolane	0.10	136	0.64	2.4
	0.11	135	0.45	
	0.40	134	0.31	2.1
	1.73	133	0.15	0.9
Hexafluorobenzene	0.10	134	0.30	
	0.44	135	0.20	
	1.66	134	0.15	

[a]Monomer/solvent mixture: 38/62 (w/w), initiator: perfluorobenzoyl peroxide, polymerization temperature: 60°C, polymerization time: 24 h, yield of polymers: 50–80%.
[b]Intrinsic viscosity in HFB at 25°C.

TABLE 16.5 Effect of the Concentration of the Chain Transfer Agent on the Properties of the Poly(perfluoro-2-Methylene-4-Methyl-1,3-Dioxolane)

Chain transfer agent	Initiator (wt%)	Chain transfer agent (wt%)	T_g (°C)	[η] (dL/g)[a]	$M \times 10^{-6}$ (g/mol)
Carbon tetrachloride	0.50	0	134	0.45	1.50
		0.36	132	0.36	
		1.02	131	0.32	
Carbon tetrabromide	0.538	0	133	0.57	
		0.52	125	0.048	
		1.13	120	0.039	0.17
		2.74	100	–	
Carbon tetrabromide	0.579	0	133	0.54	
		0.13	130	0.12	
		0.24	128	0.093	0.26
		0.33	127	0.074	0.25
Sulfuryl chloride	0.489	0	134	0.55	
		1.30	114	–	
		2.54	99	–	

[a]Intrinsic viscosity in HFB at 25°C.

The decomposition of perfluorodibenzoyl peroxide is performed at 60–80°C by a homolysis mechanism resulting in the formation of a pentafluorobenzyl structural unit at the polymer end (Scheme 16.6) [19, 20]. Thus, poly(perfluoro-2-methylene-1,3-dioxolanes) produced are generally thermally stable and the decomposition temperature (T_d) in the air atmosphere is >300°C.

The physical properties of these monomers and their homopolymers are given in Table 16.6. These polymers, except the polymer of monomer H, are completely

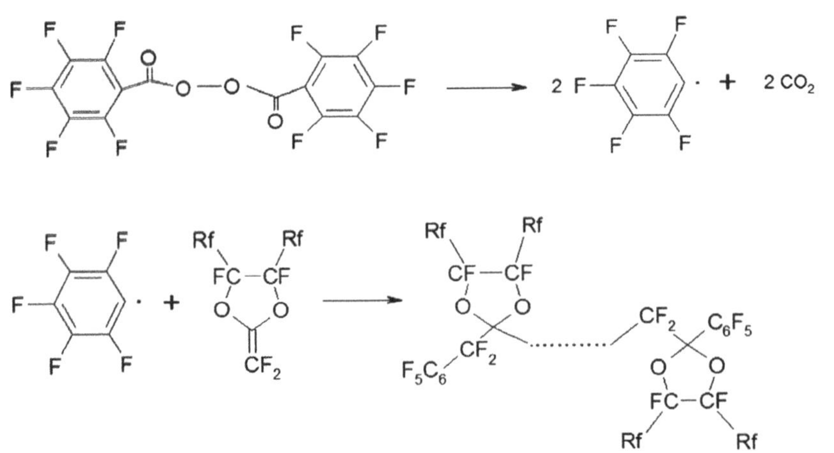

SCHEME 16.6 Decomposition of the initiator and structure of perfluoropolydioxane polymer.

TABLE 16.6 The Physical Properties of Perfluoro-2-Methylene-1,3-Dioxolane Monomers and Their Homopolymers

Monomer	A	B	C	D	E	F	G	H
Polymerization rate (×10^4 mol/L s)[a]	1.66	1.56	1.40	0.15	0.25	0.18	1.02	1.63
Polymer T_g (°C)	168	135	101	165	165	185	93	110
Polymer $RI_{632\ nm}$	1.3570	1.3310	1.3328	1.3280	1.3420	1.3460	1.3520	1.3443

[a]In FC-113, [Monomer] = 1.6 mol/L, [perfluorodibenzoylperoxide] = 0.05 mol/L, 41°C.

amorphous and exhibit extraordinary optical transmittance from the UV to the near IR region. They are chemically as well as thermally stable (T_d > 300°C). The polymers are soluble in fluorinated solvents such as HFB and Fluorinert™ FC-75. The polymer of monomer H is semicrystalline and the melting point of the crystalline part is 230°C. However, once the crystalline polymer powder is heated above the melting temperature and pressed, the prepared film becomes amorphous and flexible and does not recrystallize. The films are insoluble in any solvent including fluorinated solvents [17, 21]. The 4- and 5-dimethyl substituted 1,3-dioxolane (Figure 16.5D) has *cis* and *trans* isomers and the ratio is 27% and 73%, respectively [22]. T_g of these dioxolane polymers highly depend on the substituted group at the fourth and fifth position. T_g of dimethyl and cyclic substituted polymers (Figure 16.5A, D, E, and F) is relatively high (165–185°C) because of the sterically crowded resultant polymers. However, T_g of dialkylether substituted at the fourth and fifth positions (Figure 16.5G) is the lowest (93°C). The lower T_g is rationalized by the pendant group effect because the $-CF_2OCF_3$ group has a larger free volume than that of CF_3 and cyclic substituted compounds.

16.7 COPOLYMERS OF DIOXOLANE MONOMERS

The physical properties of perfluorodioxolane polymers are modified by copolymerizing the dioxolane monomers. The refractive indices and T_g depend on the copolymer compositions [16, 18, 23]. The copolymers can be prepared in solution and in bulk. For example, the copolymerization reactivity ratios of monomers A and C (Figure 16.6)

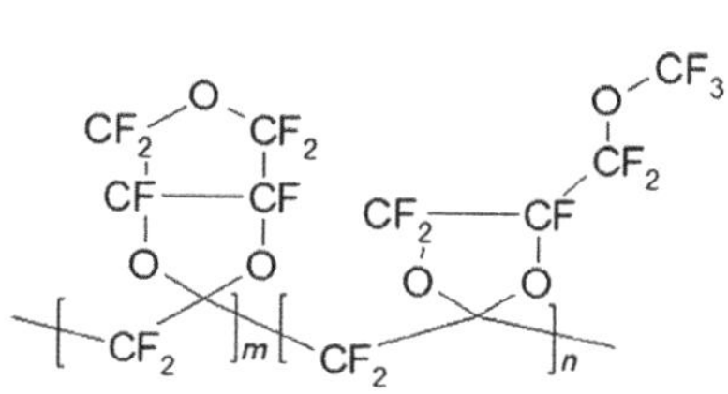

FIGURE 16.6 Chemical structure of a copolymer of perfluorodioxolanes (Figure 16.5 A and C).

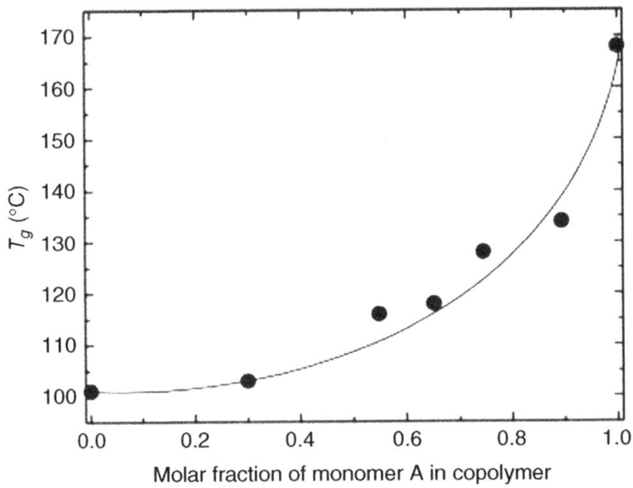

FIGURE 16.7 Dependence of T_g on the copolymer composition (monomers A and C).

are $r_A = 0.97$ and $r_C = 0.85$, respectively [18]. The obtained data show that this copolymerization yields nearly ideal random copolymers. The copolymers have only one T_g between 110°C and 165°C depending on the copolymer composition (Figure 16.7). The copolymer films prepared by casting are flexible and tough, and have high optical transparency with a low refractive index of 1.333-1.357 at 543 nm (Figure 16.8).

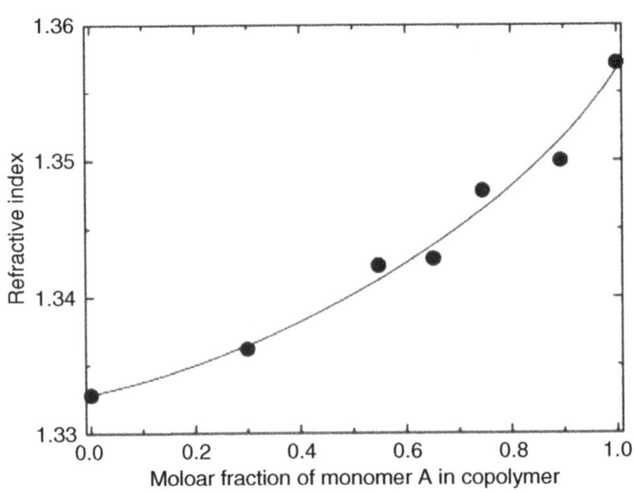

FIGURE 16.8 Dependence of the refractive index at 532 nm on the copolymer composition (monomers A and C).

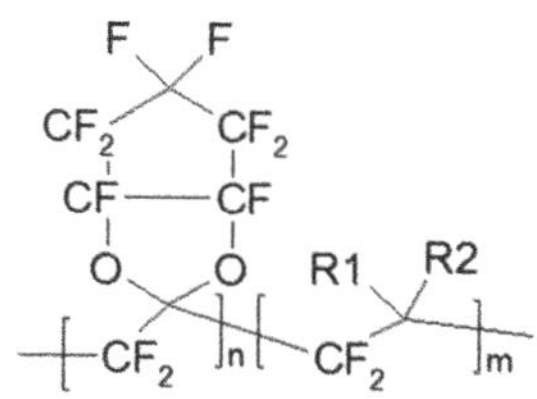

FIGURE 16.9 Structure of copolymers based on perfluoro-3-methylene-2,4-dioxabicyclo [3,3,0] octane (Figure 16.5 F) with fluorovinyl monomers.

16.8 COPOLYMERS OF PERFLUOROMETHYLENE DIOXOLANES AND FLUOROVINYL MONOMERS

Perfluoro-2-methylene-1,3-dioxolane monomers copolymerizes with various commercially available fluorovinyl monomers, too. Perfluoro-3-methylene-2,4-dioxabicyclo[3,3,0] octane (Figure 16.5F) was copolymerized with chlorotrifluoroethylene (CTFE), perfluoropropyl vinyl ether, perfluoromethyl vinyl ether, and vinylidene fluoride [23] by a free radical initiator such as perfluorodibenzoylperoxide or *tert*-butyl peroxypivalate in bulk or in solution, respectively (see Figure 16.9).

The typical results of these copolymers are shown in Table 16.7. The content of monomer F in the copolymer produced is considerably higher than that in the feed. It indicates that the reactivity of the vinyl monomers is much lower than that of the perfluorinated dioxolane monomers. The reactivity ratios of CTFE and monomer F are $r_{CTFE} = 0.74$ and $r_F = 3.64$, respectively.

The copolymers obtained are soluble in fluorinated solvents such as HFB and Fluorinert™ solvents. T_g of the copolymers is in the range of 84–145°C and are thermally stable (T_d; 320–370°C). The copolymer films are flexible and clear with a low refractive index (1.332–1.375 at 633 nm).

16.9 APPLICATIONS

These amorphous fluoropolymers are chemically as well as thermally stable, soluble in fluorinated solvents, have low dielectric constants, and the films are transparent. They have unique properties compared to traditional fluoropolymers. The amorphous polymers have high potential in many applications. The following are representative examples that are being pursued: polymer waveguides [24, 25], pellicles used in the photolithographic reproduction of semiconductor integrated circuits [26], insulators and hydrophobic surfaces for electrowetting [27, 28], polymer optical fibers [29,30], and membranes for gas separations [31–33]. Here, we describe two examples of the use of the amorphous perfluorinated polymers: optical fibers and gas separation membranes.

For optical fibers, polymethylmethacrylate (PMMA) is an excellent transparent polymer. PMMA-based plastic optical fibers (POFs) have been extensively used

TABLE 16.7 Copolymers of Monomer F (perfluoro-3-Methylene-2,4-Dioxabicyclo-[3,3,0] Octane) with Various Fluorinated Vinyl Monomers (M_2)

	Monomer in feed (mol%)		Copolymer composition (mol%, EA)[a]		Copolymer composition (mol%, NMR)[b]				
Monomer M_2	F	M_2	F	M_2	F	M_2	T_g (°C)	T_d (°C)	RI (633 nm)
$F_2C{=}CFCl$	12	88	16	84	18	82	–	–	–
$F_2C{=}CFCl$	20	80	35	65	37	63	84	322	–
$F_2C{=}CFCl$	40	60	62	38	62	39	110	330	1.3745
$F_2C{=}CFCl$	56	44	80	20	81	19	120	330	–
$F_2C{=}CFCl$	70	30	87	13	90	10	145	340	–
$F_2C{=}CFOC_3F_7$	65	35	87	13	91	9	144	342	1.3298
$F_2C{=}CFOCF_3$	21	79	51	49	60	40	154	359	1.3419
$F_2C{=}CH_2$	14	86	38	62	42	58	108	370	–
$F_2C{=}CH_2$	35	65	58	42	62	38	138	356	1.3517

[a]The composition was determined by elemental analysis.
[b]The composition was determined by NMR.

TABLE 16.8 Pure Gas Permeabilities and Selectivities of Selected Amorphous Perfluorinated Glassy Polymers at 35°C

	Teflon® AF		Hyflon® AD		
Permeability/Selectivity	2400	1600	80X	60X	Cytop®
He	–	–	430	390	170
H_2	2090	550	210	180	59
CO_2	2200	520	150	130	35
O_2	960	270	67	57	16
N_2	480	110	24	20	5.0
CH_4	390	80	12	10	2.0
He/H_2	–	–	2.0	2.1	2.8
He/CH_4	–	–	35	39	84
H_2/CH_4	5.3	6.9	18	18	30
CO_2/CH_4	5.7	6.5	13	13	18
O_2/N_2	2.0	2.4	2.8	2.9	3.2
N_2/CH_4	1.2	1.4	2.0	2.0	2.5

in short-distance data communication applications. However, high attenuation of POF based on PMMA is dominated by C–H overtone stretching and deformation vibrations. Hence, the most effective method to obtain a low loss POF is to substitute hydrogen with heavier atoms such as fluorine. Cytop® molecules consist solely of C–C, C–F, and C–O bonds. The wavelengths of the fundamental stretching vibrations of the atomic bonds are relatively long. Therefore, the vibration absorption loss of Cytop® at the region of the desired light source wavelength is negligibly small. In 2000, Asahi Glass commercialized graded-index POF based on Cytop®, Lucina™. The attenuation of the fiber is 10 dB/km at 1 μm.

The unusual solubility of gases and vapors in perfluoropolymers has several applications relevant to membrane gas separations. Perfluoropolymers have solubility selectivities that are significantly different from those of hydrocarbon-based polymers. The amorphous perfluoropolymers can be fabricated into thin, high-flux composite membranes, which possess the excellent chemical and thermal stability. Typical reported pure gas permeabilities and selectivities of these fine amorphous perfluoropolymers are shown in Table 16.8 [33].

These materials cover a broad permeability range from Teflon® AF 2400, one of the most permeable polymers, to Cytop®, the least permeable among these amorphous perfluoropolymers. However, even Cytop® is about ten times more permeable than conventional glassy polymers such as polycarbonates. The relative permeability of the perfluoropolymers is well described by their fractional free volume (FFV), a common measure of the free space in a polymer matrix available for molecular transport. For instance, both gas permeability and FFV exhibit the following order: Teflon® AF 2400 > Teflon® AF 1600 > Hyflon® AD 80X ≈ Hyflon® AD 60X > Cytop®. As the amount of bulky perfluorodioxole increases, the permeability and diffusivity increases.

REFERENCES

1. Feiring, A. E. Fluoroplastics. In *Organofluorine Chemistry: Principles and Commercial Applications*; Banks, R. E., Smart, B. E., Tatlow, J. C., Eds.; Plenum Press: New York, 1994; Chapter 15, pp 339.
2. Hung, M. H.; Resnick, P. R.; Smart, B. E.; Buck, W. H. Fluorinated Plastics, Amorphous. In *Polymeric Materials Encyclopedia*; Salamone, J. C., Ed.; CRC: Boca Raton, FL, 1996; Vol. 4; pp 2466.
3. Sugiyama, N. Perfluoropolymers obtained by cyclopolymerization and their applications. In *Modern Fluoropolymers*; Scheirs, J., Ed.; John Wiley and Sons: Chichester, 1997; Chapter 28, pp 541.
4. Milani, A.; Tommasini, M.; Castiglioni, C.; Zerbi, G.; Radice, S.; Canil, G.; Toniolo, P.; Triulzi, F.; Colaianna, P. *Polymer* **2008**, *49*, 1812.
5. Resnick, P. R. *Polym. Prepr. (Am. Chem. Soc., Div. Polym. Chem.)* **1990**, *31*(1), 312.
6. Hung, M. H. *Macromolecules* **1993**, *26*, 5829.
7. Resnick, P. R.; Buck, W. H. Teflon® AF Amorphous Fluoropolymers. In *Modern Fluoropolymers*; Scheirs, J., Ed.; John Wiley and Sons: Chichester, 1997; Chapter 22, pp 397.
8. Smart, B. E.; Feiring, A. E.; Krespan, C. G.; Yang, Z. Y.; Hung, M. H.; Resnick, P. R.; Dolbier, W. R.; Rong, X. X. *Macromol. Symp.* **1995**, *98*, 753.
9. Michel, U.; Resnick, P. R.; Kipp, B.; DeSimone, J. M. *Macromolecules* **2003**, *36*, 7107.
10. Bunyard, W. C.; Kadla, J. F.; DeSimone, J. M. *J. Am. Chem. Soc.* **2001**, *123*, 7199.
11. Yamamoto, K.; Ogawa, G. *J. Fluorine Chem.* **2005**, *126*, 1403.
12. Nakamura, M.; Sugiyama, N.; Etoh, Y.; Aosaki, K.; Endo, J. *Nippon Kagaku Kaishi* **2001**, *12*, 659.
13. Mikes, F.; Yang, Y.; Teraoka, I.; Ishigure, T.; Koike, Y.; Okamoto, Y. *Macromolecules* **2005**, *38*, 4237.
14. Liu, W.; Mikes, F.; Guo, Y.; Koike, Y.; Okamoto, Y. *J. Polym. Sci. A: Polym. Chem.* **2004**, *42*, 5180.
15. Liu, W.; Koike, Y.; Okamoto Y. *Macromolecules* **2005**, *38*, 9466.
16. Yang, Y.; Mikes, F.; Yang, L.; Liu, W.; Koike, Y.; Okamoto Y. *J. Fluorine Chem.* **2005**, *127*, 277.
17. Okamoto, Y.; Mikes, F.; Yang, Y.; Koike, Y. *J. Fluorine Chem.* **2007**, *128*, 202.
18. Yang, Y.; Mikes, F.; Yang, L.; Liu, W.; Koike, Y.; Okamoto, Y. *J. Polym. Sci. A: Polym. Chem.* **2006**, *44*, 1613.
19. Murotani, E.; Saito, S.; Sawaguchi, M.; Yamamoto, H.; Nakazima, Y.; Miyazima, T.; Okazoe, T. *J. Fluorine Chem.* **2007**, *128*, 1131.
20. Sawada, H. *Chem. Rev.* **1996**, *96*, 1779.
21. Mikes, F.; Baldrian, J.; Teng, H.; Koike, Y.; Okamoto, Y. *Polym. Adv. Technol.* **2011**, *22*, 1272.
22. Zhang, B.; Li, L.; Mikeš, F.; Koike, Y.; Okamoto, Y.; Rinaldi, P. L. *J. Fluorine Chem.* **2013**, *147*, 40.
23. Mikes, F.; Teng, H.; Kostov, G.; Ameduri, B.; Koike, Y.; Okamoto, Y. *J. Polym. Sci. A: Polym. Chem.* **2009**, *47*, 6571.

24. Zhao, Y.-G.; Lu, W.-K.; Ma, Y.; Kim, S.-S.; Ho, S. T.; Marks, T. J. *Appl. Phys. Lett.* **2000**, *77*, 2961.

25. Kuwana, Y.; Takayama, K.; Takenobu, S.; Morizawa, Y. In Perfluoropolymer Optical Waveguide Comparable to Silica in Optical Properties and Reliabilities, Proceedings of the 6th International Conference on Polymers and Adhesives in Microelectronics and Photonics, Tokyo, Japan, Jan 16–18, 2007; IEEE.

26. French, R. H.; Wheland, R. C.; Qiu, W.; Lemon, M. F.; Zhang, E.; Gordon, J.; Petrov, V. A.; Cherstkov, V. F.; Delaygina, N. I. *J. Fluorine Chem.* **2003**, *122*, 63.

27. Seyrat, E.; Hayes, R. A. *J. Appl. Phys.* **2001**, *90*, 1383.

28. Khodayari, M.; Crane, N. B.; Volinsky, A. A. Thin Solid Films 2013, 548, 632.

29. Koike, Y; Koike, K. Optical Fibers. In Polymer Science: A Comprehensive Reference; Matyjaszewski, K., Möller, M., Eds.; Elsevier, 2012; Vol. 8; Chapter 11, pp 283.

30. Koike, Y.; Koike, K. *J. Polym. Sci. B: Polym. Phys.* 2011, *49*, 2.

31. Merkel, T. C.; Bondar, V. I.; Nagai, K.; Freeman, B. D.; Yampolskii, Yu. *Macromolecules* **1999**, *32*, 8427.

32. Alentiev, A. Yu.; Shantarovich, V. P.; Merkel, T. C.; Bondar, V. I.; Freeman, B. D.; Yampolskii, Yu. P. *Macromolecules* **2002**, *35*, 9513.

33. Merkel, T. C.; Pinnau, I.; Parbhakar, R.; Freeman, B. D. Gas and Vapor Properties of Perfluoropolymers. In *Materials Science of Membranes for Gas and Vapor Separation*; Yampolskii, Yu. P., Pinnau, I., Freeman, B. D., Eds.; John Wiley & Sons: Chichester, 2006; Chapter 9, pp 251.

17

FLUOROPOLYMERS FOR SUSTAINABLE ENERGY

VINCENZO ARCELLA, LUCA MERLO, RICCARDO PIERI, PAOLO TONIOLO, FRANCESCO TRIULZI, AND MARCO APOSTOLO
Solvay Specialty Polymers

17.1 INTRODUCTION

The modern chemical industry is a source of innovative ideas that allow our society to face its modern technological challenges and durable needs. Today's chemistry is a sophisticated discipline which more and more contributes to everyone's well-being and quality of life. The development of a sustainable energy route is one of these critical societal challenges.

The United Nations General Assembly declared 2012 as the *International Year of Sustainable Energy for All* [1]. Several initiatives have been set forth thereof to face and meet energy objectives to be achieved by 2030 [2]. That witnesses the importance and breadth of this subject all over the globe, today and for the future of mankind. Indeed, according to the most recognized definition in economic and engineering, Sustainable Energy refers to "the provision of energy such that it meets the needs of the present without compromising the ability of future generations to meet their own needs [3]." Sustainable Energy development is thus based on three interlinked key components: renewable energy production, energy efficiency in use, and lowest environmental impact in terms of pollution of the overall system [4].

Handbook of Fluoropolymer Science and Technology, First Edition. Edited by Dennis W. Smith Jr., Scott T. Iacono, and Suresh S. Iyer.

In this challenging background, advanced materials are at the core of industrial innovation and enable it with better performing and sustainable products and processes. Fluoropolymers, at the top of specialty polymers selection thanks to their unique properties, are naturally part of the solution to these industrial and societal challenges.

This chapter deals with the use of such materials, with specific focus on the most spread renewable energy applications, in the three key components of Sustainable Energy: offering better performance by their use, contributing at lowering resource and energy requirements, while considering the sustainability at the end of life of the products.

17.2 THE ARENA OF RENEWABLE ENERGY TECHNOLOGIES

In order to review such wide a concept as Sustainable Energy and then identify the use of fluoropolymers therein, it is mandatory to develop a rationale to map and classify the landscape of Renewable Energy.

Several attempts are present in literature to represent the landscape of alternative energy; the focus, however, is normally directed onto a specific facet of the energy world: energy sources, production technologies, or application efficiency.

The rationale that is used at SOLVAY Specialty Polymers is meant to be more comprehensive. That is based on a pivotal object energy carrier, which is any produced and storable energy sources (secondary energy) like electricity, hydrogen, heat, around which all other technologies associated to alternative energy can be linked (Figure 17.1).

According to the scheme of Figure 17.1, in this paper the technologies are thus organized in energy production, energy conversion, and energy storage.

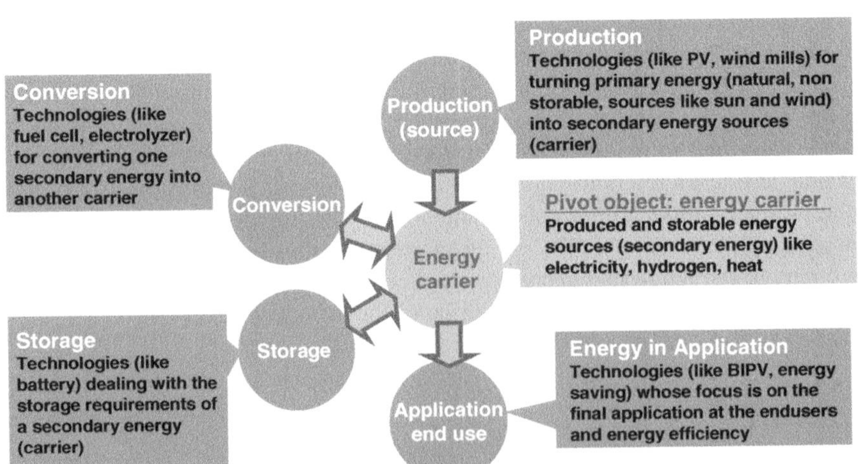

FIGURE 17.1 Map of renewable energy technologies.

TABLE 17.1 Examples of Application of the Proposed Energy Technologies Classification

Technology	IN→	Energy size	Power size	→OUT
Production				
PV	Solar light	Not controlled	Panel surface	Electricity
Fotosplitting	Solar light	Not controlled	Panel surface	Hydrogen
Windmill	Wind	Not controlled	Blade length	Electricity
Conversion				
Fuel cell	Hydrogen	Hydrogen tank	Membrane surface	Electricity
Electrolysis	Electricity	Electricity source	Membrane surface	Hydrogen
Flow battery	Chemicals	Chemicals tank	Membrane surface	Electricity
Storage				
Second battery	Electricity	Electrode/separator assembly		Electricity
Supercapacitor	Electricity	Electrode/separator assembly		Electricity
Hybrid Supercap	Electricity	Electrode/separator assembly		Electricity

To further explain the difference between conversion and storage, it is worth to note that in energy conversion technologies power and energy are independent and can be freely and autonomously modified. This is not the case for energy storage technologies where energy and power are intimately intermingled and one affects the other. Table 17.1 shows in more detail this concept and lists some technologies according to the scheme of Figure 17.1.

Goal of this paper is to discuss the usage of fluoropolymers in some selected sustainable energy technologies. Due to space limitation, after a short general introduction, one well-established technology per energy area is described in detail: photovoltaic for energy production, lithium battery for energy storage, fuel cell for energy conversion.

17.3 FLUOROPOLYMERS FOR ENERGY PRODUCTION

Main fluoromaterial applications for the established energy production technologies, PV and wind mill, are listed in Table 17.2. In the following table, only the usage of fluoropolymers in PV will be discussed.

The generation of electricity by photovoltaic devices (PV) can be divided into three main categories, based on the semiconductor used for the transformation of the light into electricity: bulk crystalline silicon, thin film technology (amorphous silicon, CIGS—Copper Indium Gallium (di)Selenide or CdTe—Cadmium Telluride), organic photovoltaic. Today, almost 90% of the market is dominated by crystalline silicon PV modules. The photovoltaic power industry is well established in many regions, with a total market for new installations of more than 26 GW in 2011.

TABLE 17.2 Fluoropolymer Usages in Energy Production

Technology	Fluoropolymer	Device Element	Main properties
PV, any generation	ECTFE ETFE	Frontsheet	Transparency Weathering resistance
	PVF PVDF ECTFE	Backsheet	Weathering resistance Electrical properties
Wind Mill	ECTFE, PTFE, PVDF, FEP	Wind turbine blades	Weathering resistance Corrosion resistance

In photovoltaic industry fluoropolymers are used in the envelope of the panel, mainly to protect the active elements of the panel from the aggression of the environment so as to grant 25 years lifetime. Figure 17.2 shows a sketch of a PV panel highlighting potential fluoropolymers usage. In the figure, ECTFE is ethylene-chlorotrifluoroethylene, ETFE is ethylene-tetrafluoroethylene, PVF is polyvinylfluoride, and PVDF is polyvinylidenefluoride.

17.3.1 Fluoropolymers in PV Panel Backsheet

The first generation of backsheet for crystalline silicon PV modules in the early 1960s was a three-layer structure: PVF (Tedlar® from DuPont)-PET (Polyethylene terephthalate)-PVF [5]. Its proven history of solid performance in demanding indoor and outdoor applications quickly pushed it to the forefront where it became the preferred choice for backsheet film with a reassuring 25-year warranty. In the three-layer assembly, the PET layer provides electrical insulation and mechanical strength while the PVF offers weatherability and opacity [6]. Other fluoroplastics can be used instead of PVF, namely PVDF, PTFE (polytetrafluoroethylene), ETFE, and ECTFE. In particular, PVDF offers equal performance to Tedlar® and its market share is increasing steadily. Indeed, during the last few years many other backsheet solutions

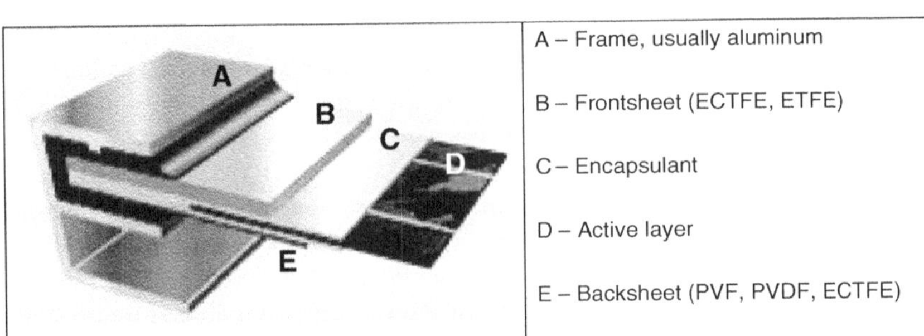

FIGURE 17.2 Cross section of a photovoltaic module.

have been proposed. As a consequence, while in 2007 more than 90% of backsheet were based on PVF, in 2011 PVF share dropped to 40%.

The goal of the backsheet is to provide UV and moisture protection, electrical insulation, and durability. To this purpose, standard international committees (such as IEC and UL) have defined a set of tests to be passed before a photovoltaic's module could be placed on the market. The IEC 61215 is one of the most used standards for qualifying a PV module [7]. The aim of these tests is to check the resistance of the PV module to all the variables that can affect and influence the life of the module itself such as electrical performance, hail and impact test, weathering, and temperature (UV and thermal cycling). One of the most important tests is the damp heat test, 1000 hours inside a climatic chamber at 85°C and 85% relative humidity. Most of the PV module producers tend to prolong well above 1000 hours of damp heat, to be sure that the module will survive after 25 years giving at least 80% of the initial power. This trend is also taken into consideration into the new version of the standard tests that is considering to have at least 2000 hours of damp heat for qualification.

In order to prevent delamination during panel lifetime, fluoropolymer films usually need a surface treatment to give adhesion during the lamination step (flame, corona, or plasma). Backside is usually nontransparent, but in some cases transparent backsheets are requested, in particular for modules that are integrated into buildings envelope (facade or roof). When transparent, the fluoropolymer film usually contains UV blocking agents.

Also, different colors are available (white, gray, black, blue, green, terracotta).

17.3.2 Fluoropolymers in PV Panel Frontsheet

Usually glass is used as frontsheet. However, the request of polymeric solutions is becoming important to make panels that are flexible and light.

Main characteristics for frontsheet are the following:

- strong, hard, tough, abrasion-resistant films retaining useful properties over a broad range of temperatures;
- excellent barrier to water vapor in a wide range of temperatures;
- very weatherable, withstanding long-term ageing outdoor exposure even in harsh conditions;
- very transparent and smooth; and
- excellent fire resistance properties.

Fluoropolymers, like Halar® ECTFE and Tefzel® ETFE are good candidates for frontsheet protection of photovoltaic modules due to their high transmittance in visible range, high stability in UV light, low permeability [5], and their excellent weathering resistance [8, 9]. Another advantage of fluoropolymers as frontsheet is their tendency to self-clean due to their low surface energy.

The high transparency of those fluoropolymers is shown in Figure 17.3 where the UV-Vis spectra of HALAR® ECTFE and TEFZEL® ETFE is compared to glass

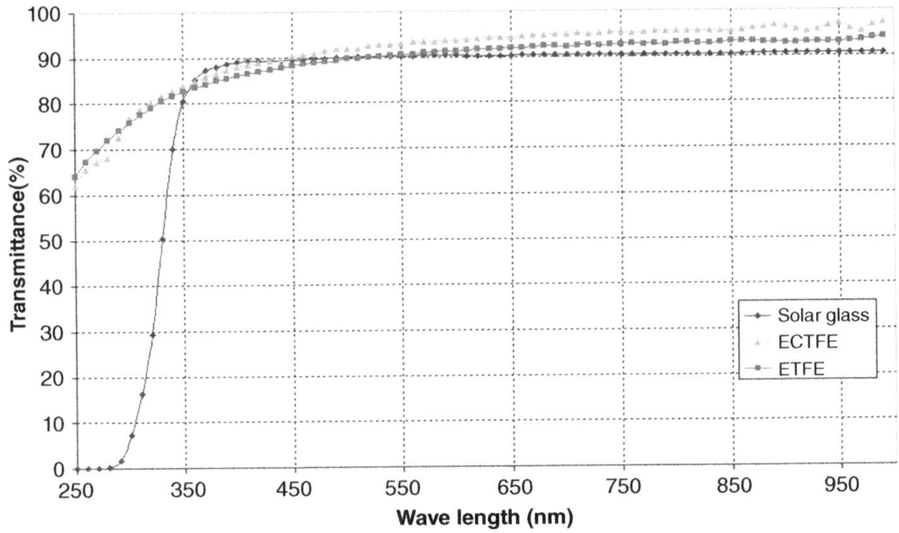

FIGURE 17.3 UV-visible transmission spectrum on 50-μm thick polymeric films (ECTFE, ETFE) compared to 3-mm Solar glass.

(Solvay internal data). Figure 17.3 shows that while glass is able to stop UV radiation up to 340 nm, the two fluoropolymers are fully transparent to UV. This is a problem for some of the materials in the inner part of the panel, for example, the ethylenevinylacetate (EVA) used as encapsulant. Indeed, while fluoropolymers are perfectly stable under UV light, many of the other materials used in a PV panel cannot withstand such energetic radiation.

To solve this issue, a new Halar® ECTFE with special UV scavenger fillers have been recently developed. To demonstrate the effectiveness of this film to protect UV-sensitive materials, the Halar® ECTFE film with UV scavengers was laminated on PET and then placed in a QUV test chamber equipped with UV-B lamp centered at 313 nm. Figure 17.4 shows the effect of UV light on PET, that becomes yellowish, while the Halar film with UV scavengers has been able to effectively protect the PET film underneath and appears almost as transparent and uncolored as the sample before UV test.

17.4 FLUOROPOLYMERS FOR ENERGY STORAGE

Main fluoromaterial applications for the current energy storage technologies are listed in Table 17.3. In the following table, only the usage of fluoropolymers in lithium batteries will be discussed.

There are two main factors behind the enormous success of commercialization of lithium batteries. One is related to the continuous evolution of mobile electronic systems such as cellular phones, laptops, and tablets, which requires new types of high

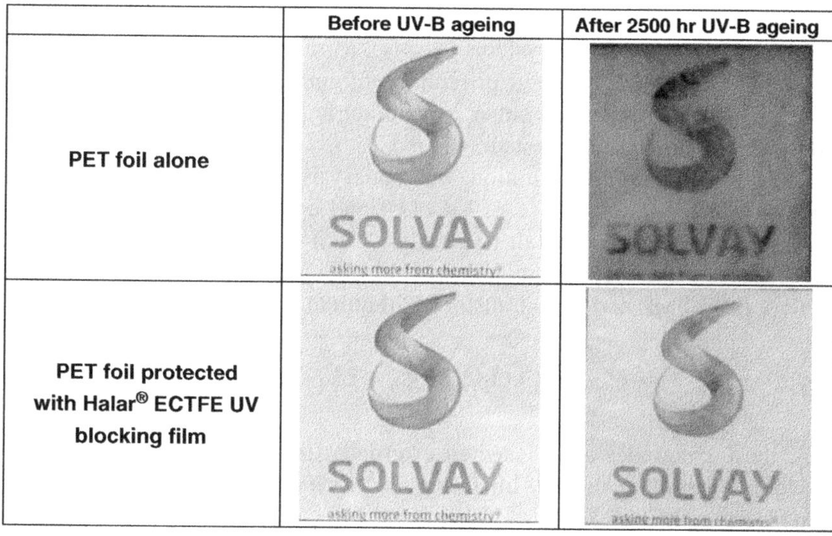

FIGURE 17.4 PET alone and PET protected by Halar® UV blocking after 2500 hours UV-B ageing (313 nm lamp).

power and high energy-density batteries. Then the continuous demand of energy and the increasing pollution due to the growing CO_2 content in the atmosphere requires the use of clean energy sources at a much higher level than today. The dream of a consistent replacement of combustion engine cars with electric or hybrid vehicles relies anyway on the availability of suitable storage systems.

Lithium-ion batteries are the best choice in both cases, because of their high energy density and design flexibility that make these system outperform the various existing technologies.

This explains why lithium batteries receive most attention at both fundamental and applied levels, and currently dominates the worldwide market sales accounting for more than 70% of the market of rechargeable batteries, with more than 4200 M cells sold in 2011 and $9.7 billion sales [10].

TABLE 17.3 Fluoropolymer Usages in Energy Storage

Technology	Fluoropolymer	Device element	Main properties
Lithium ion battery	PVDF, PTFE	Binder for electrode	Electrochemical resistance Adhesion properties
	PVDF	Separator	Electrochemical resistance Ionic conductivity
Supercap	PVDF, PTFE	Binder for electrode	Chemical resistance Adhesion properties
	PVDF	Separator	Chemical resistance Ionic conductivity

Among the different technologies in the field of rechargeable batteries, Li-based batteries outperform the other systems because, as said before, of their high energy density and flexibility that allows different configurations. The amount of energy stored in the battery, expressed either in volumetric or gravimetric unit, is directly linked to the chemistry of the system [11].

In its most common configuration, this battery is formed by a graphite anode, a lithium metal oxide cathode (e.g., $LiCoO_2$) and a porous separator soaked with a liquid solution of a lithium salt (typically $LiPF_6$) in an organic solvent mixture (ethylene carbonate–dimethylcarbonate mixture). The electrochemical mechanism of this battery is the back-and-forth transfer of lithium ions between the two electrodes:

$$6C + LiCoO_2 \, \Delta \, Li_xC_6 + Li_{(1-x)}CoO_2$$

Once these electrodes are connected externally, the chemical reactions proceed in tandem at electrodes, thereby liberating electrons and enabling the current to be trapped by the user.

Although most of the literature is dealing with developments of new active materials in order to increase battery energy and power density, focusing mainly on high voltage materials for positive electrode and silicone-based alloys for negative electrode, very few scientific papers explore the role of polymeric materials in lithium batteries.

Lithium battery is anyway a challenging application for most polymeric materials, as it demands long-term reliability as well as chemical and electrochemical resistance in the specific environment of Li cells. In case of automotive application, higher temperature performances are also required.

Many examples show the importance of fluorine chemistry in this field, where the outstanding stability of fluorinated polymers is an added value for the good performances of the battery.

This chapter summarizes the recent advances on these topics.

17.4.1 Fluoropolymer Binders

Polymers are widely used in electrode manufacturing as binders, in order to guarantee good mechanical properties to the electrode.

Table 17.4 shows the list of required characteristics for effective binders in a lithium battery. Some are related to battery's performance, while others are related to the manufacturing process itself.

The trend of battery-making companies is to increase the amount of active material filling the available restricted volume, so as to increase as much as possible energy and power density of the battery. Thus, any component other than active material, such as binder or conductive additives, should be reduced as much as possible. In this sense, the role of the binder is extremely important, since it should be able to assure good adhesion of the active material particles to the metallic electrodes and good cohesion among inorganic particles. In this way, the electrical and ionic

TABLE 17.4 Requisites for Binder

For battery performance	Good adhesion to metal collector Good cohesion of active material particles Electrochemical stability over a wide potential range Low swelling rate in nonaqueous electrolytes High melting point
For manufacturing process	Soluble in suitable solvents Slurry or solution viscosity stable over long periods Possibility to use water-based systems

conductivity is guaranteed by tight contact among particles and good homogeneity, with a beneficial impact on battery cycle life and power performances.

The polymer is usually electronically insulating, so it is very important to limit the amount of binder used in the electrode in order to avoid excessive particle coating thus limiting electrical connections. Moreover, the polymer should swell to a certain extent in the liquid electrolyte so as to assure lithium ions conductivity inside the electrode. The swelling should be limited to avoid huge volume expansion and not to lose electrical contact between active material particles.

The binder should be inactive over a wide potential range, so not to be oxidized at the positive electrode or to be reduced at the negative electrode. It is also important that the polymer has a high melting point and that the composite electrode structure remains stable even at high temperatures.

The items listed in the second section are related to the manufacturing process, as the polymer should be easily dissolved in a suitable solvent to prepare a solid dispersion (slurry) to be coated on the metal support by means of different techniques. The electrode should have enough mechanical stability in order to be machinable during the different phases of cell assembly process as pressing, slitting, winding.

A recent trend in battery industry is toward the use of water-based binders, with advantages in terms of lower environmental impact and lower processing cost since there will be no necessity of expensive sections for solvent recovery systems.

Actually polyvinylidene fluoride (PVDF) is one of the few polymers which meet most of the requirements, and it is the binder of choice for many battery makers.

Poly(vinylidene fluoride), PVDF or PVF2, is usually manufactured from radical initiated batch polymerization process in aqueous emulsion or suspension of $CH_2{=}CF_2$ monomer. PVDF is a thermoplastic that exhibits interesting properties, such as piezoelectric, pyroelectrical, and ferroelectric behaviors. PVDF has even superior dielectric permittivity arising from the strong polymerization originating from C—F bonds, and the spontaneous orientation of dipoles in the crystalline phases makes it a polar polymer with good compatibility with polar chemicals.

PVDF is inert to various solvents, oils, acids and shows low permeabilities to gases and liquids. The glass transition (T_g) and melting (T_m) temperatures are in the range of −40°C to 30°C and 160–172°C, depending on the sample (e.g., the molecular weight and the amount of defects of chaining). In fact, PVDF is typically 50–60% crystalline, where several crystalline forms have been detected, the most stable of which is the α-form [12].

Copolymerization is a general and powerful method to perform effective systematic changes in polymer properties, and is widely used in the production of commercial polymers and in basic investigations of structure–properties relations. As a result, copolymerization usually modifies the symmetry of the polymeric chain and modulates both intramolecular and intermolecular forces, so that properties such as melting point, glass transition temperature, crystallinity, stability, elasticity, permeability, and chemical reactivity may be varied within wide limits. As a matter of fact, copolymers based on VDF and chlorotrifluoroethylene (CTFE) or hexafluoropropene (HFP) represent the most important production nowadays among the thermoplastic copolymers of VDF used in lithium battery production. As the comonomer content increases, crystallinity decreases and the amorphous region is more relevant, thus chain flexibility and liquid swelling properties become more significant.

The unique features of the fluorinated nature of PVDF make it suitable for use as binder in electrode formulation.

Its intrinsically polar nature assures good adhesion and cohesion properties among active material particles.

PVDF is electrochemically stable in a wide range of voltage, which guarantees its safe use in the electrochemical environment of the lithium cell. The anodic scan toward high voltages in PVDF/carbon electrode does not show any decomposition current below 5 V versus Li^+/Li, as shown in Figure 17.5.

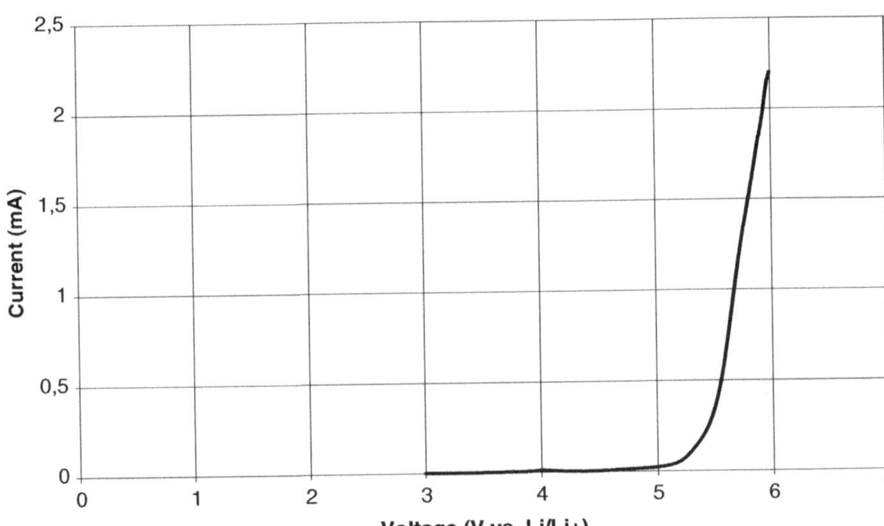

FIGURE 17.5 Anodic scan on PVDF/carbon black super-P working electrode. Scan rate: 0.1/mVs.

The role of binder crystallinity in determining adhesion was investigated measuring scratch adhesion on electrodes composed by different carbon types. Samples having higher PVDF crystallinity show higher scratch adhesion strength.

While PVDF stability at high voltages is undoubted, the question of the stability of fluorinated polymers at low voltages, as toward metallic lithium or lithiated graphite, is somehow questioned in the scientific community. The assumption is that fluorinated polymers can be attacked by lithium in a strong reductive environment, being reduced electronically to carbon and LiF.

It was shown, however, [13] that PVDF can be much more stable in anode environment compared to other fluorinated polymers such as PTFE thanks to a surface passivation that prevents further electrochemical reaction.

Moreover, it has also been shown by differential scanning calorimetry on graphitic electrodes with different binders, PVDF, PTFE, SBR–CMC (styrene butadiene–carboxymethylcellulose) [14] that Li and electrolyte are important reactants in anode thermal decomposition while the binder composition and binder level do not significantly affect the heat evolved. So VDF-based binders, incorporated into the anode films do not contribute to the thermal runaway reactions in Li-ion cells.

Among other fluoropolymer manufacturers, Solvay Specialty Polymers commercializes a wide range of VDF-based homopolymers and copolymers as binder for lithium battery under the Solef® trade name. The R&D expertise in fluorinated polymers and polymerization technology is continuously focused on the development of new tailored solutions in order to increase binder performances without losing safety requirements. Several grades are available with different molecular weight in order to combine adhesion properties and slurry processability. The new grade Solef® 5130 has been designed combining the effects of ultra-high molecular weight with the benefit of polar functional groups distributed in the polymer chain. The reinforced intermolecular interactions between polymer, active material, and metal collector result in increased performances in terms of adhesion (Figure 17.6) and chemical resistance in electrolyte. These effects are translated into higher energy density since it is possible to significantly reduce binder content, while the lower internal resistance gives access to higher power density.

17.4.2 PVDF for Lithium Battery Separators

A separator is a physical barrier between the positive and negative electrodes incorporated into most cell designs to prevent electrical shorting. Usually the separator is a microporous plastic film or other porous inert material filled with electrolyte. Separators must be permeable to the ions, electronic insulators and inert in the battery environment [15].

The materials used for the microporous polymer membranes are semi-crystalline polyolefin materials, like polyethylene (PE), polypropylene (PP), and their blends PE–PP. The preparations of the membranes can be classified into dry process and wet process [16].

While these polyolefin-based separators offer many advantages, their poor thermal stability and mechanical strength raise serious concerns regarding the electrical

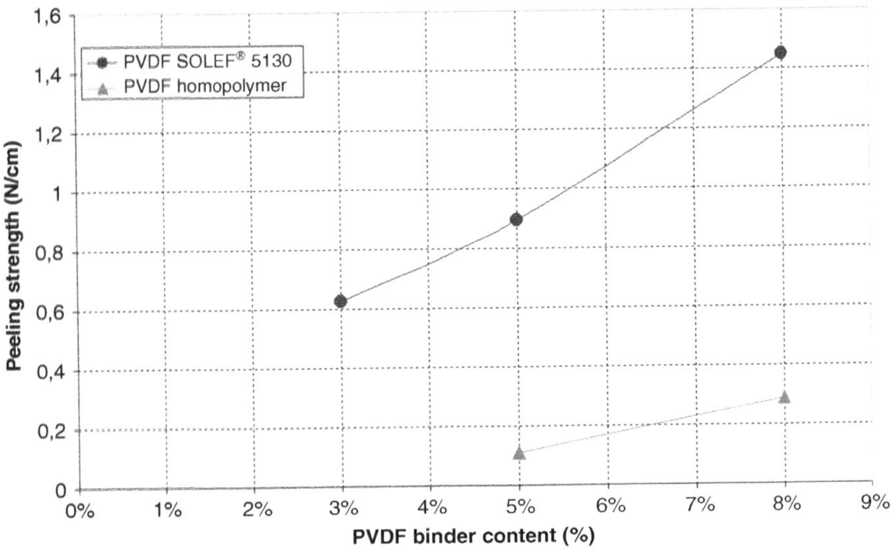

FIGURE 17.6 Adhesion at cathode with LiFePO4 active material and different PVDF grades and content.

isolation between electrodes, the failure of which could eventually lead to fire or explosion of cells.

In order to overcome these drawbacks of polyolefin-based separators, various separators have been proposed.

PE separators can be coated with PVDF-based gel electrolyte membranes, to enhance their compatibility with liquid electrolytes, reach good mechanical strengths and good thermal shutdown property [17].

Another alternative is represented by PVDF homo and copolymers porous membranes, obtained by various techniques including phase inversion [18] and electrospinning [19], while the addition of ceramic fillers such as SiO_2 [20] have also been proposed to increase the conductivity and mechanical stability of the matrix material.

Moreover, it has been shown that PVDF macroporous separators can improve the high charge rate performances of lithium-ion batteries in comparison with commercial polyolefin separators [21].

In a recent paper also poly(vinylidene fluoride–trifluorethylene) copolymers have been studied as porous membranes as separator, regarding the impact of preparation methods and porosity on the final properties of the system [22].

Another class of interesting materials for separator is composed of nonwoven films, textile products processed directly from fibers that are bonded together. The fibrous structure of nonwoven materials offers a high porosity, which is necessary for high electrolyte absorbance and low ionic resistance, excellent thermal properties, combined with cost competitiveness. However, the excessively large pore size and broad pore size distribution of conventional nonwovens, which may provoke self-discharge and internal short-circuits of cells, often hinder their successful application to lithium-ion batteries [23].

To overcome these limitations, recent works propose to use fluoropolymers to help binding the coating of ceramic powders to nonwovens [24], or through impregnation of gel-polymer electrolytes into nonwovens [25]. These composite separators may exhibit better C-rate capability and cyclability than a commercialized polyolefin separator.

17.5 FLUOROPOLYMERS FOR ENERGY CONVERSION

Main fluoromaterial applications for the current energy storage technologies are listed in Table 17.5. Most of the processes using fluoropolymers in energy transformation are related to hydrogen economy and the PFSA (perfluorosulfonic acid) ionomers.

Polymeric electrolyte membrane (PEM) fuel cells [26] are considered today as one of the most promising technologies in the field of renewable power sources and environmentally friendly energy generation, to solve the problems of oil shortage and global warming due to their high efficiency (compared to combustion cycles) and the clean exhausts [27]. Fuel cells are continuous current generators that operate a combustion process without contacting directly the fuel and the oxidant species, typically producing similar amounts of heat and electric energy.

The most commonly used reagents are air on what is called cathode side (where the reduction of oxygen takes place) and pure hydrogen on anode side (where the hydrogen oxidation takes place); combination of the two generates an operating

TABLE 17.5 Fluoropolymer Usages in Energy Storage

Technology	Fluoropolymer	Device element	Main properties
PEM FC and electrolysis	PFSA	Membrane	Electrochemical resistance Proton conductivity
		Electrode binder	Electrochemical resistance Adhesion properties
	PTFE	Membrane reinforcement	Electrochemical resistance Inertness
		Gas Diffusion layer coating	Hydrophobicity
Ph Acid FC	Fluoroelastomer	Gaskets	Chemical and thermal resistance
Ph Acid FC	PFPE	Heat exchange fluid	Chemical inertness Dielectric property
Flow battery	PFSA	Membrane	Electrochemical resistance Ionic conductivity

voltage of 0.5–0.8 Volt per cell depending on the operating current value selected. The main product of reaction consists of water that is discharged together with the exhausts, preferentially from the cathode side, where it is formed. The fuel cells are always stacked in electrical series in order to have a generator with a proper voltage; stacks are built with a sequence of 50–400 single cells, depending on the application. For some power-demanding applications, several stacks are connected with each other in order to reach a power size of some hundreds of kW.

Polymer electrolyte membrane (PEM) water electrolysis is considered the most efficient method to produce hydrogen with a high degree of purity from renewable energy resources such as wind, photovoltaic, and hydropower [28], as well as from electricity generated by conventional power sources. The process is characterized by high efficiencies and suitable current density even at low temperatures (40–80°C). PEM electrolysis systems have a number of advantages in comparison to the traditional alkaline electrolysers, such as ecological cleanliness, considerably smaller mass–volume characteristics; no corrosive electrolyte is involved, low maintenance and, essentially, a high degree of gas purity [29]. There is also the opportunity to obtain compressed gases directly from the electrolyser.

PFSA ionomers are known since the late 1960s, when the Nafion ionomers were developed by the Du Pont Company and employed as polymer electrolyte in a GE fuel cell designed for NASA spacecraft missions. Nafion polymers have found wide application especially in the chlor-alkali industry as membrane materials thanks to very high chemical inertness and low resistances to cation transport.

In more recent years, the growing interest related to cleaner energy production technologies has promoted the consideration and study of ionomers as proton-exchange membranes (PEMs) in fuel cells. PFSA are generally considered from the scientific community superior to other polymeric ionomers due to their high chemical resistance and high proton conductivity. Again, due to electrochemical stability requirements, perfluorinated materials have been preferred as the best candidates for satisfying the needs of the system. Many papers investigating properties of Nafion under many different conditions have been published [30]. Nafion is, therefore, by far the most extensively used and studied ionomer for fuel cell application.

In the mid-1980s, Ballard Power Systems showed significant improvements in fuel cell performance using ionomer membranes obtained from Dow Chemical [31]. This ionomer, commonly known as the Dow ionomer, is perfluorinated and similar in structure to Nafion, save for a shorter pendant side-chain which carries the functional ion-transporting group, wherefrom this ionomer is known as the short-side-chain (SSC) ionomer. Even though higher power-generating capability in fuel cell was demonstrated, Dow Company, after filing a series of patents, abandoned the industrial development of this interesting ionomer structure.

Probably due to the lack of commercial availability of SSC ionomer samples, the scientific literature is not very rich of works describing the characteristics of this material. This is actually quite surprising when one thinks that the first works on the SSC ionomer already put in evidence the big property differences induced by the presence of a shorter side-chain, and that the Dow patents showed possible interesting applicative advantages deriving from these differences.

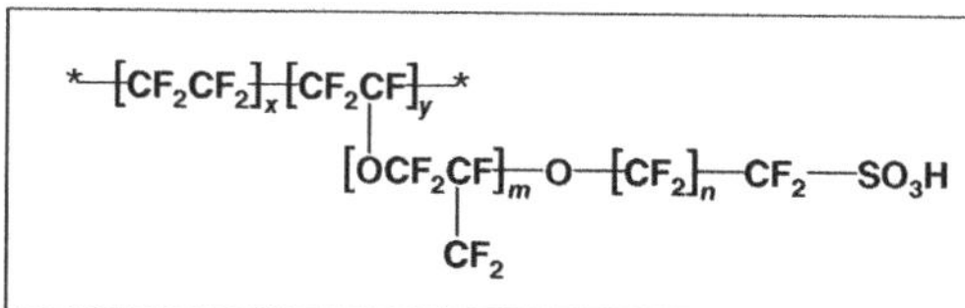

Available (or described) PFSA:

Nafion®: $m=1$ $n=1$ $x=5–13.5$
Flemion®: $m=0,1$ $n=1–5$ $x=?$
Aciplex®: $m=0,3$ $n=2–5$, $x=1,5–14$
Aquivion®: $m=0$ $n=1$ $x=3–7$

FIGURE 17.7 Structure of PFSA ionomers.

Recently, scientific literature evidenced that different companies developed perfluorosulfonic ionomers able to sustain 120°C operating conditions for long time. In some examples the relation with linear short side chain structure is evident, most of all 3M [32] and Solvay [33,34], in other cases the polymer structure is not described, like for Asahi Glass [35] or Asahi Kasey [36]. Figure 17.7 summarizes the chemical formula of the most common perfluorinated ionomers.

The same ionomer can be used as binder for both fuel cell and electrolysis electrodes. The binder purpose is to keep physical integrity of the catalyst layer and thigh value of ionic conductivity of the electrode itself.

The advantage of perfluorosulfonic materials compared to nonfluorinated competitors can be classified in:

1. Superior chemical stability: fuel cell environment is extremely oxidative (especially when reactant humidity is kept low) and peroxide species are frequently detected in both anode and cathode side [37]. C—F bond is highly resistant to radical degradation, while the C—H bond is "easily" degraded. In electrolysis application the abundance of water prevent aggressive radicals formation, but high voltage is applied during the operations thus in favor again to the most stable fluorinated materials.
2. Higher ionic conductivity: the sulfonic acid group has intrinsically a strong acid character, but, because of the electronic attraction favored by fluorine atoms in the backbone, a stronger acid character is measured when transiting from hydrogenated to fluorinated molecules. The direct consequence of a stronger acidic character is, mainly, higher ionic conductivity, an advantage that is particularly evident when dry reactants are used. This last feature is illustrated by Figure 17.8, where the conductivity of a PFSA is compared to that of an aromatic ionomer as a function of relative humidity.
3. Fluorinated materials are "softer" than hydrogenated ones. This feature had been frequently indicated as an advantage of the hydrogenated material enabling higher mechanical stability and lower gas crossover. On the other hand, the obtainment of a proper contact interface between the membrane and the electrode is necessary to guarantee low ohmic drops of the assembly and it is clearly more effectively obtained by using fluorinated materials.
4. Industrial availability of ionomeric dispersion is today limited to fluorinated ionomers one.

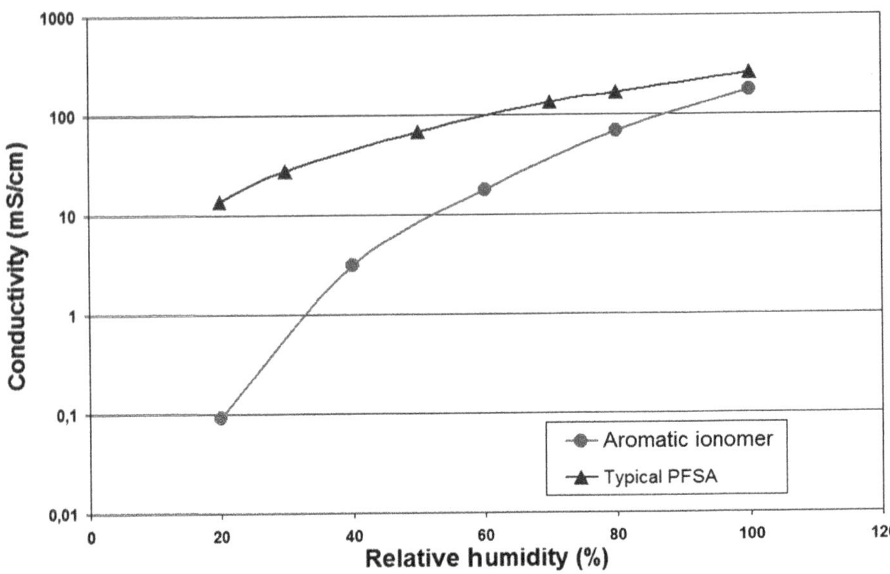

FIGURE 17.8 Comparison between typical PFSA and hydrocarbon ionomer ionic conductivity as a function of relative humidity.

Other than PFSA, different fluoropolymer are successfully used in fuel cell or electrolysis assemblies:

- PTFE is used to give hydrophobicity to the gas diffusion layers [38]. PTFE is applied in form of dispersion (deriving from polymerization latex + some additives) directly on the carbon felt, then it is sintered at high temperature (>340°C). The weight amount of PTFE used can vary from 5% to 30% depending on the operating conditions of the final product, different fluoromaterials (like perfluoropolyethers [39]) were successfully tested recently in alternative of PTFE. Thanks to the well-known hydrophobicity of fluorinated materials there are no "competitor" hydrocarbon materials successfully used for this application.
- Fluoroelastomers (FKM) can be used for the gaskets [40], especially when severe operating conditions are envisaged, this is particularly true in case of phosphoric acid fuel cell that operates at 160–200°C in presence of strong acidity given by the H_3PO_4 acid. In this case, the high performance FKMs (high chemical and thermal resistance) are quite unique and are hardly substituted with nonfluorinated rubbers. When lower temperature is envisaged (80°C), silicon or EPDM materials are often used as the cheapest alternative, even if long-term purity requirements can be questionable.
- Fluorinated fluids such as Galden® of Solvay or HFE of 3M can be used in electrochemical devices for heat exchange purpose thanks to their dielectric property and chemical inertness. One advantage of these fluids is that, in case of

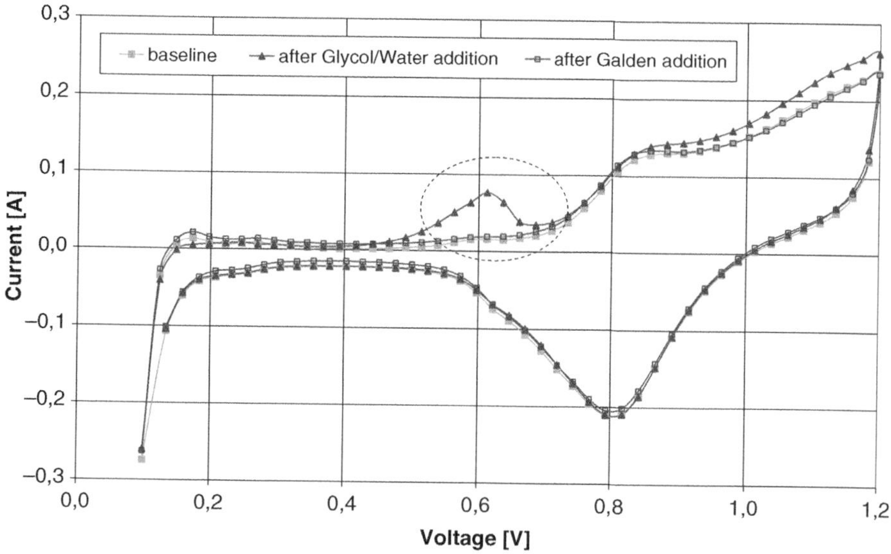

FIGURE 17.9 Cyclic voltammetry of Pt electrodes intentionally polluted with glycol or Galden cooling fluids.

accidental contact between the cooling agent and the active part of the fuel cell, no catalyst pollution will happen. On the contrary, when a water/glycol mixture is used some Pt deactivation occurs. This behavior is shown in Figure 17.9, where the dotted ellipse indicates the deactivation of Pt electrode due to water/glycol contamination.

For the flow batteries application, the introduction of fluorinated material is more complex, due to the low cost requirement that these applications need so to be competitive with other energy storage/transformation technologies, but in several cases PFSA membranes are described as the state-of-the-art technology [41].

17.6 FLUOROPOLYMERS FOR SUSTAINABILITY

The technologies discussed in this paper are known to be sustainable, meaning that they can provide an energy system based on renewable sources, independent from fossil fuels. However, in a deeper analysis, the sustainability of an energy conversion process should contemplate other aspects, like:

the lifecycle (eco-profile) of the materials associated with the process;
the impact of the required installations (volumetric, environmental); and
the nature of the energy source and the one of the produced energy.

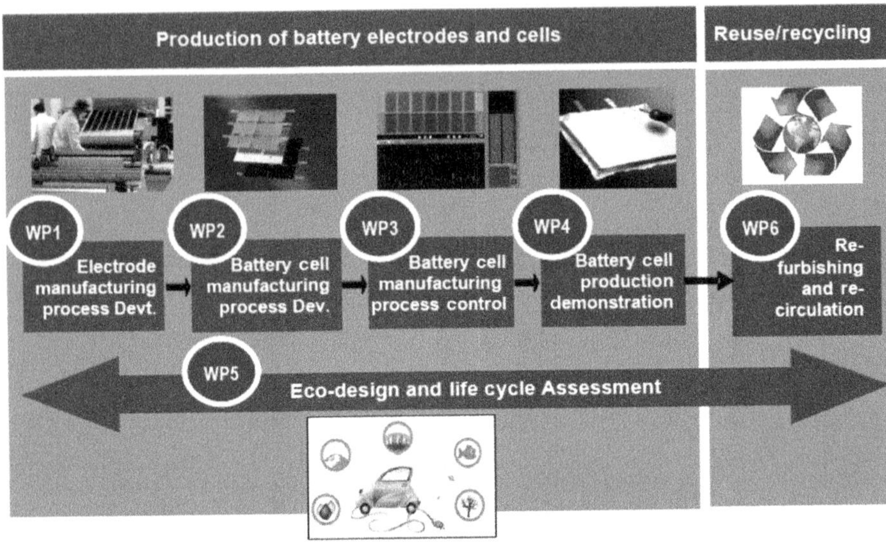

FIGURE 17.10 Work packages of ELIBAMA—FP7 funded project.

An interesting example of such integrated approach is offered by ELIBAMA, a European funded project aiming to develop innovative lithium ion batteries. ELIBAMA, that stands for European Li-Ion Battery Advanced Manufacturing, in its program contemplates the steps of reuse/recycling and eco-design/life cycle assessment (WP 5 and 6 of Figure 17.10). More and more of these elements must become part of the final package, so as to conjugate the development of innovative materials with their environmental footprint.

REFERENCES

1. General Assembly, United Nations General Assembly Resolution A/RES/65/151, February 16, 2011; pp 1–3
2. http://www.sustainableenergyforall.org/
3. Renewable Energy and Efficiency Partnership. Glossary of Terms in Sustainable Energy Regulation, 2004; p 8.
4. Prindle, B.; Eldridge, M.; Eckhardt, M.; Frederick, A. *ACEEE* **2007**, 1–53.
5. DeBergalis, M. *J. Fluorine Chem.* **2004**, *125*, 1255–1257.
6. Brandrup, J.; Immergut, E. H.; Grulke, E. A.; Abe, A.; Bloch, D. R. *Polymer Handbook*, 4th ed.; John Wiley & Sons, 2005.
7. IEC 61215—webstore.ansi.org
8. Abusleme, J.; Baert, T.; Bassi, M.; Fossati, P.; Radice, S.; Toniolo, P. 23rd EUPVSEC Proceedings: 4AV 3 52, Performances of Fluoropolymers in PV-Cells, 2008.

9. Wypych, G. *Handbook of Material Weathering*, 4th ed.; ChemTec Publishing: Toronto, 2008.
10. Pillot, C. The Worldwide Battery Market 2011–2025. BATTERIES 2012 Nice, France, October 24–26, 2012.
11. Tarascon, J. M.; Armand, M. *Nature* **2001**, *414*, 359–367.
12. Ameduri, B. *Chem. Rev.* **2009**, *109*, 6632–6686.
13. Liu, W.; Huang, X.; Guobao, L.; Wang, Z.; Huang, H.; Zhonghua, L.; Xue, R.; Chen, L. *J. Power Sources* **1997**, *68*, 344–347.
14. Roth, E. P.; Doughty, D. H.; Franklin, J. *J. Power Sources* **2004**, *134*, 222–234.
15. Arora, P.; Zhang, Z. *Chem. Rev.* **2004**, *104*, 4419–4462.
16. Zhang, S. S. *J. Power Sources* **2007**, *164*, 351–364.
17. Lee, Y. M.; Kim, J. W.; Choi, N.-S.; Lee, J. A.; Seol, W.-H.; Park, J.-K. *J. Power Sources* **2005**, *139*, 235–241.
18. Boudin, F.; Andrieu, X.; Jehoulet, C.; Olsen, I. I. *J. Power Sources* **1999**, *81–82*, 804–807.
19. Kim, J. R.; Choi, S. W.; Jo, S. M.; Lee, W. S.; Kim, B. C. *Electrochim. Acta.* **2004**, *50*, 69–75.
20. Wachtler, M.; Ostrovskii, D.; Jacobsson, P.; Scrosati, B. *Electrochim. Acta.* **2004**, *50*, 357–361.
21. Djian, D.; Alloin, F.; Martinet, S.; Lignier, H. *J. Power Sources* **2009**, *187*, 575–580.
22. Costa, C. M.; Rodrigues, L. C.; Sencadas, V.; Silva, M. M.; Rocha, J. G.; Lanceros-Méndez, S. *J. Membr. Sci.* **2012**, *407–408*, 193–201.
23. Jeong, H. S.; Choi, E. S.; Lee, S. Y. *Electrochim. Acta* **2012**, *86*, 317–322.
24. Cho, T. H.; Tanaka, M.; Ohnish, H.; Kondo, Y.; Yoshkazu, M.; Nakamura, T.; Sakai, T. *J. Power Sources* **2010**, *195*, 4272.
25. Jeong, H. S.; Kim, J. H.; Lee, S. Y. *J. Mater. Chem.* **2010**, *20*, 9180.
26. Smitha, B.; Sridhar, S.; Khan, A. A. *J. Membr. Sci.* **2005**, *259*, 10–26.
27. Larminie, S.; Dicks, A. *Fuel Cell Systems Explained*, 2nd ed.; John Wiley & Sons, **2003**.
28. Barbir, F. *Solar Energy* **2005**, *78*, 661–669.
29. Millet, P.; Mbemba, N.; Grigoriev, S. A.; Fateev, V. N.; Aukauloo, A.; Etiévant, C. *Int. J. Hydrogen Energy* **2011**, *36*(6), 4134–4142.
30. Kenneth, A.; Mauritz, B.; Moore, R. Chemical Reviews 2004-10-01, State of Understanding of Nafion.
31. Prater, K. *J. Power Sources* **1990**, *29*, 239–250.
32. Hamrock, S. HTMWG Meeting. DOE Contracts DE-FC36-02AL67621 and DE-FC36-03GO13098, 2005.
33. Ghielmi, A. *J. Power Sources* **2005**, *145*(2), 108–115.
34. Arcella, V.; Troglia C.; Ghielmi, A. *Ind. Eng. Chem. Res.* **2005**, *44*, 7646.
35. Endoh, E. Fuel Cell Seminar Abstract. Abstract No. 95, 2006.
36. Wakizoe, M. 2002 Fuel Cell Seminar Abstracts, 2002, pp 62–65.
37. Danilczuk, M.; Perkowski, A.J.; Schlick, S. *Macromolecules* **2010**, *43*(7), 3352–3358.
38. Mathias, M. F.; Roth, J.; Fleming, J.; Lehnert, W. Diffusion Media Materials and Characterisation. In *Handbook of Fuel Cells—Fundamentals, Technology and Applications*;

Vielstich, W., Gasteiger, H. A., Lamm, A., Eds.; John Wiley & Sons: Chichester, **2003**; Vol. 3, pp 517–537, ISBN 0-471-49926-9.

39. Stampino, P. G.; Molina, D.; Omati, L.; Turri, S.; Levi, M.; Cristiani, C.; Dotelli, G. *J. Power Sources* **2011**, *196*, 7645–7648.
40. Chih-Wei, L.; Chi-Hui, C.; Jinzhu, T.; Yuh-Jin, C.; Van Zeed, J. V. *Int. J. Hydrogen Energy* **2011**, *36*, 6756–6767.
41. Sukkar, T.; Skyllas-Kazacos, M. *J. Appl. Electrochem.* **2004**, *34*, 137–145.

18

EVOLUTION OF ACADEMIC BARRICADES FOR THE USE OF TETRAFLUOROETHYLENE (TFE) IN THE PREPARATION OF FLUOROPOLYMERS

DANIEL A. HERCULES, DARRYL D. DESMARTEAU, RICHARD E. FERNANDEZ, JAMES L. CLARK JR., AND JOSEPH S. THRASHER

Clemson University and Rf Consulting

18.1 INTRODUCTION AND OVERVIEW

Although tetrafluoroethylene (TFE) as a monomer is ubiquitous in the fluoropolymer industry, one rarely finds more than a few millimoles of this monomer being used at a time in academic laboratories because of the inherent safety issues with this compound (*vide infra*). For example, TFE is a known deflagrant with an energy release approximating that of black gunpowder [1]. In addition to their production facilities, most of the major fluorochemical companies have barricades in which to carry out basic and applied research with TFE. However, with the downsizing of industrial R&D budgets and the encroachment of populations on what once were remote R&D

Disclaimer. The authors make no warranties, expressed or implied, and assume no liability in connection with any use of the information presented in this Chapter. No one but persons having technical skill in this area of fluorine chemistry should attempt or repeat anything presented herein, and then at their own discretion and risk.

Handbook of Fluoropolymer Science and Technology, First Edition. Edited by Dennis W. Smith Jr., Scott T. Iacono, and Suresh S. Iyer.

facilities, the chemical industry is increasingly outsourcing more research. On the other hand, if one were to look around the academic community, not much would be found in the way of barricades for working safely with TFE. The purpose of this chapter is to document the evolution of academic TFE barricades for the purpose of preparing fluoropolymers, which to the best of our knowledge began a little over 25 years ago with a facility at Clemson University's Hunter Laboratories. As will be explained below, this initial barricade was primarily for TFE cylinder storage, from which approximately 10 g of TFE was destabilized at a time and used to prepare the initial perfluorosulfonimide polymer electrolytes in the DesMarteau laboratory [2,3]. Approaching 10 years ago, another barricade for TFE work was built as part of a new interdisciplinary science building (Shelby Hall) at the University of Alabama in Tuscaloosa, where Thrasher and coworkers carried out polymerizations with up to 100 g of TFE per batch [4,5]. More recently, with Thrasher's move to Clemson as well as the current location of the research laboratories for fluorine chemistry at the more remote Clemson University Research Park, the construction of a new TFE barricade was undertaken in which quantities of up to 1–2 kg of TFE per batch polymerization can be used. For lack of better terminology, the aforementioned TFE barricades will be called first-, second-, and third-generation facilities in the remainder of this chapter.

The task of developing a safe TFE barricade is not a trivial task especially since the fluorochemical industry will typically not share detailed plans of their barricades for obvious reasons of confidentiality and liability. On the other hand, some industrial high pressure laboratories are described in the literature, and these references will be overviewed later as they pertain to the final design and construction of the third-generation TFE facility. More important now is to describe a few of the types of conditions that one wants to avoid when using TFE, especially if not stabilized with either some type of free radical inhibitor (e.g., a terpene like D-limonene) or a diluent known to eliminate or reduce the explosive hazard of TFE (e.g., HCl, CO_2 [6], HFP [7]). As previously mentioned, TFE is a known deflagrant, which basically means that it is worse than an explosive as an explosion can backpropagate to the storage cylinder, etc. One of the most common causes of accidents when working with TFE is opening a valve too rapidly when a high pressure differential exists in the tubing/pipeline on the two sides of the valve. Due to adiabatic compression, localized heating can cause high temperatures to be reached, which can end up triggering an explosion. Such localized heating can be sufficient to initiate either the decomposition of TFE to tetrafluoromethane and carbon black or cause its dimerization to octafluorocyclobutane as shown in Equations 18.1 and 18.2, respectively [8–10]. As shown in Equation 18.3, the exothermicity of the decomposition of TFE is even worse

$$C_2F_4(g) \rightarrow CF_4(g) + C(s) \qquad \Delta H_R = -276\,\text{kJ/mol} \tag{18.1}$$

$$2\,C_2F_4(g) \rightarrow \text{c-}C_4F_8(g) \qquad \Delta H_R = -206\,\text{kJ/mol} \tag{18.2}$$

when in the presence of oxygen, and thus great care needs to be taken to eliminate oxygen by insuring the use of leak-free systems as well as the proper evacuation, purging, and degassing of equipment and reagents (e.g., solvents, initiator solutions, comonomer feeds).

$$C_2F_4(g) + O_2(g) \rightarrow 2\,COF_2(g) \qquad \Delta H_R = -628\,\text{kJ/mol} \tag{18.3}$$

18.2 FIRST-GENERATION TFE BARRICADE AT HUNTER LABORATORIES, CLEMSON UNIVERSITY

A TFE barricade was built on the top/service floor of Hunter Laboratories during its construction in 1987. From the architectural drawing shown in Figure 18.1, one can see the reinforced walls of the barricade in the upper left of the drawing (39 sq ft high pressure cell). A thick, reinforced sliding door also exists to seal off the opening to the pressure cell, and in the event of an explosion, both a window and an unreinforced ceiling serve as the "weak spot." This section of the laboratory is primarily used for TFE cylinder storage (kilogram quantities in several size 300 cylinders, where the TFE is stabilized by D-limonene), a nitrogen cylinder for purging and leak checking, and two smaller sampling cylinder for transferring and purifying small amounts of TFE (10–80 g) from the storage cylinders. The remainder of the laboratory (148 sq ft) is outfitted with two fume hoods, the gas handling system (combination glass and metal vacuum line), the polymerizer (Parr Instrument stirred reactor), and several pass-through ports between the laboratory and the barricade for the transfer of material. A schematic of the TFE gas handling system and polymerizer used in this barricade is shown in Figure 18.2. The overall design and operation of this facility has already been described in a number of publications and dissertations from the DesMarteau group [11–16], and further details on its use are also contained in another chapter by Sharif, et al. in this monograph [17].

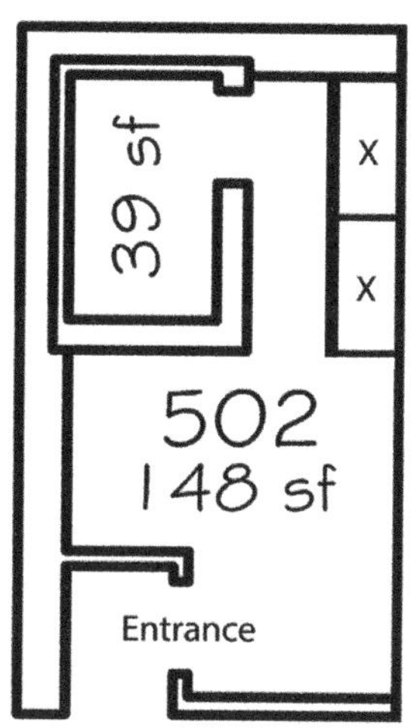

FIGURE 18.1 Architectural drawing of TFE barricade at Hunter Laboratories.

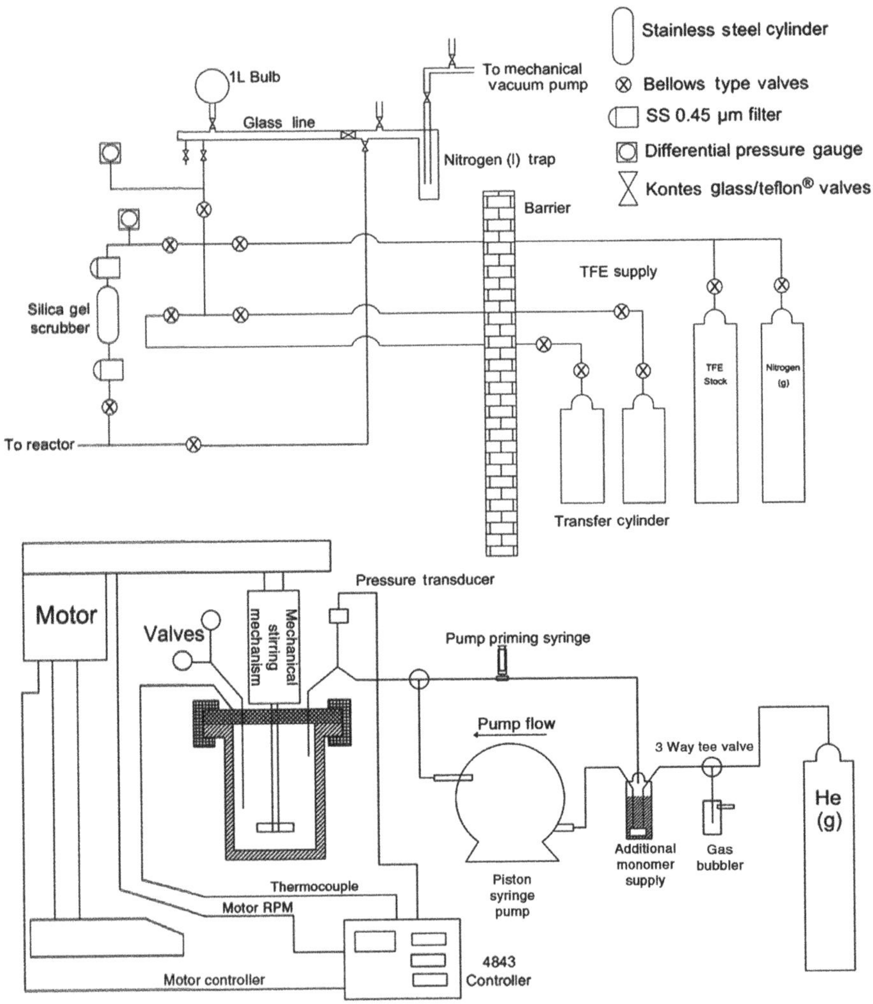

FIGURE 18.2 Schematic of TFE barricade and gas handling system located in Hunter Laboratories [13–15].

The authors feel that it is also important to briefly mention the source of the TFE cylinders in the DesMarteau facility, especially since one can no longer readily purchase TFE from the fluorochemical industry for the scale of work done or to be done in the facilities described herein. The current solution for obtaining the quantities of TFE required for ongoing research will be discussed under the section on the second-generation TFE facility, where this information is more pertinent. The initial cylinders of TFE in the first-generation TFE facility at Clemson came from the Bayonne, New Jersey plant of Imperial Chemical Industries PLC (ICI). In 1999, Asahi Glass Co., Ltd. purchased this plant, and in 2007 a decision was made to shut down this plant, thereby halting the production of fluorochemicals at the site of their American counterpart AGC Chemicals Americas, Inc. [18]. In the 1990s, DesMarteau was also able to obtain cylinders with reduced pressure of TFE from the former Peninsula Chemicals Research, Inc. (PCR).

Rather than speak more about the operation of the initial TFE facility at Clemson, several rationale for the new, third-generation TFE facility at the Clemson Research Park will be given. First, from the description given above, one can readily see that the operators are in the same laboratory when and where polymerizations with TFE are carried out, albeit on the outside of the barricade and the primary source of TFE. This is fine when only approximately 5–10 g of TFE are used per polymerization, but the desire to carry out polymerization on a significantly larger scale dictates that the operators must be put in more of a "safe haven." Furthermore, since the laboratories for fluorochemical research at Clemson have been relocated to the Clemson Research Park (eight miles from campus), the safe and appropriate transfer of chemicals and supplies between the two sites has become more problematic. Finally, the standard operating procedures (SOPs) for this TFE gas handling system call for the TFE to undergo a number of phase changes while being transferred and purified before going into the polymerizer. For example, TFE is condensed at liquid N_2 temperature into a transfer cylinder containing D-limonene, and one wonders upon warming at what temperature this inhibitor becomes effective. Although these manipulations are largely done behind the barricade, and the TFE is only destabilized immediately before use by passing though a silica gel column connected directly to one of the ports of the autoclave, one wonders if improvements can be made. The design and operation of the subsequent TFE barricades to be discussed will attempt to address these and other issues associated with scaled-up operations.

18.3 SECOND-GENERATION TFE BARRICADE AT SHELBY HALL, THE UNIVERSITY OF ALABAMA

The architectural drawing of the TFE barricade in the attic/service floor of Shelby Hall, which opened during 2004, is shown in Figure 18.3. The interior of the barricade is basically split up into three rooms and an entryway; each of the three rooms has a blowout panel in the ceiling to the outside world in the event of an explosion. The two rooms on either side of the entryway are basically lockers; one contained a vacuum system used for evacuating the gas handling system and for preparing TFE, while the other contained a high precision syringe pump (Teledyne ISCO) with which to feed the initiator solution to the polymerizer and another HPLC pump for comonomer feed. Thus, pass-through ports existed between the lockers and the backroom of the facility, which was used for TFE storage, gas handling, and polymerizations.

Although an initial request to be able to use approximately 225 g of TFE at a time was made at the design phase, the barricade was unfortunately both incorrectly engineered and built. Basically what this meant was that a safer way to store and handle TFE had to be found before this facility could be used. Fortunately, patent technology from the DuPont Company came to our rescue, as they discovered that if

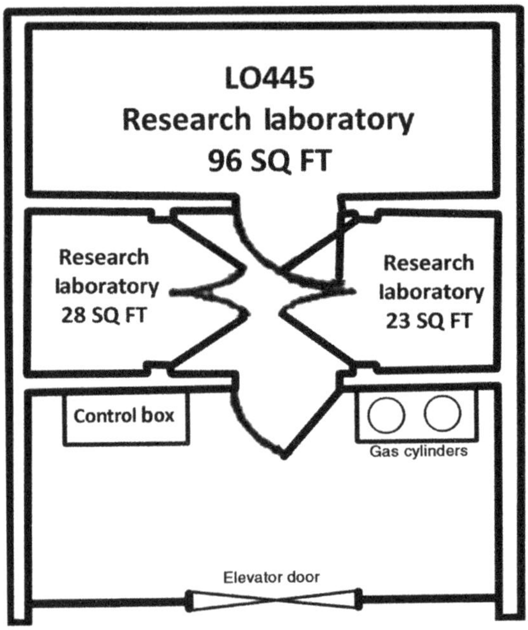

FIGURE 18.3 Architectural drawing of TFE barricade at Shelby Hall.

TFE was kept mixed with 30 mol% or more of carbon dioxide, CO_2, the TFE could not be made to explode even when fusing a Nichrome® wire (1350 °C) in a cylinder test [6]. Furthermore, DuPont only claimed the liquid composition of TFE/CO_2 in their patent [6], and since we generally operated only with gaseous mixtures of TFE and CO_2, it was felt that we were operating outside of the scope of this patent, which has recently expired.

With a safe supply of TFE conceptually in hand, one then needed to come up with a real way to prepare a mixture of TFE and CO_2. Although improved technology continues to be developed to prepare TFE from waste poly(tetrafluoroethylene) (PTFE) [19], this approach does not make sense for an academic laboratory that might not necessarily have waste PTFE for pyrolysis.

Furthermore, this approach, especially on a laboratory scale with normal equipment, yields very impure TFE, with the other primary components being hexafluoropropene (HFP), octafluorocyclobutane (OFCB), and highly toxic perfluoro-*iso*-butylene (PFIB) [20], and after purification, the TFE would still have to be mixed with CO_2. However, a search of the literature revealed that the 3M Corporation had both patented and published in the early 1950s a procedure for preparing an approximately 50:50-mol% mixture of TFE and CO_2 in 90+ percentage yield from the pyrolysis of pentafluoropropionate salts [21, 22], as shown in Equation 18.4. We have also

$$[CF_3CF_2C(O)O]^-M^+ \rightarrow C_2F_4 + CO_2 + MF, \text{ where } M = Na, K, Ca, etc. \quad (18.4)$$

benefited from the experimental procedure of Kornath and Kaufmann for this reaction, as they rediscovered this process before finding the original literature [23]. Although 3M reported a preference for using the sodium salt, we find less foaming during gas evolution with the potassium salt. Accelerating rate calorimetry (ARC) studies on these pentafluoropropionate salts also support the conclusion that gas evolution is preferential with the potassium salt; however, a small exotherm does exist in the vicinity of 175 °C for this salt, so care must be taken not to dry the salts too strongly. On the other hand, the salts must be scrupulously dried prior to pyrolysis in order to minimize the amount of pentafluoroethane, C_2F_5H, that will be formed as a by-product. This methodology has been used to prepare kilogram quantities of TFE [4,5,24].

Returning to the design of the second-generation TFE barricade and gas handling system, the latter of which is shown in the top of Figure 18.4, a number of safety features should be pointed out. First, extra shielding in the form of a 1/2-in. thick, plate steel door was added around the open sides of the southeast corner of the backroom where the TFE cylinder was stored (Figure 18.3). The control panel for the polymerization system was placed outside of the barricade so that two doors/walls existed between the polymerizer and the operators. The operators were able to both regulate monomer feed rates as well as monitor numerous readouts for temperature, pressure (vacuum and high pressure), and stirring speed from this panel. With respect to controlling the flow of TFE to the polymerization, both air-actuated valves that fail closed and a backpressure regulator (BPR) were used. Orifices were used after the main TFE cylinder valve and the two TFE air-actuated valves (labeled 1TFE and 2TFE in Figure 18.4) to prevent adiabatic compression when the valves were opened as well as any backpropagation of a flame front (flame arrestors) were an explosion to occur downstream toward the autoclave; all connections between the components of the gas handling system were made with 1/4-in. stainless steel tubing (0.035 in. wall thickness). The orifices used were basically made from tube reducers (Swagelok®) that had been welded shut and then drilled open with 0.010 in. precision holes. The orifices and BPR allowed for a very controlled flow of gases through the system. Copolymerizations with TFE, both aqueous emulsion and solution, were found to run in the presence of CO_2; however, the polymerization kinetics were slowed as one might expect [5,24]. On the other hand, if one desired to carry out polymerizations in the absence of CO_2, this was effectively accomplished by scrubbing out the CO_2 with either Ascarite® or soda lime in an in-line, water-cooled scrubber on the way to the polymerizer. Ascarite® was found to be more effective than soda lime, and in order to insure that the scrubbing was not too rapid and thereby might increase the risk of creating "hot spots," a multipoint thermocouple was used to monitor the temperature throughout the length of the Ascarite® bed [5,24]. Finally, in the instance of an "event," two "crash" buttons existed, one for closing all of the air-actuated valves at the same time and another for shutting off power to the facility.

Because of the time element, a data acquisition system was never added to this facility, and with Thrasher's move to Clemson, the entire system was decommissioned and disassembled in 2012.

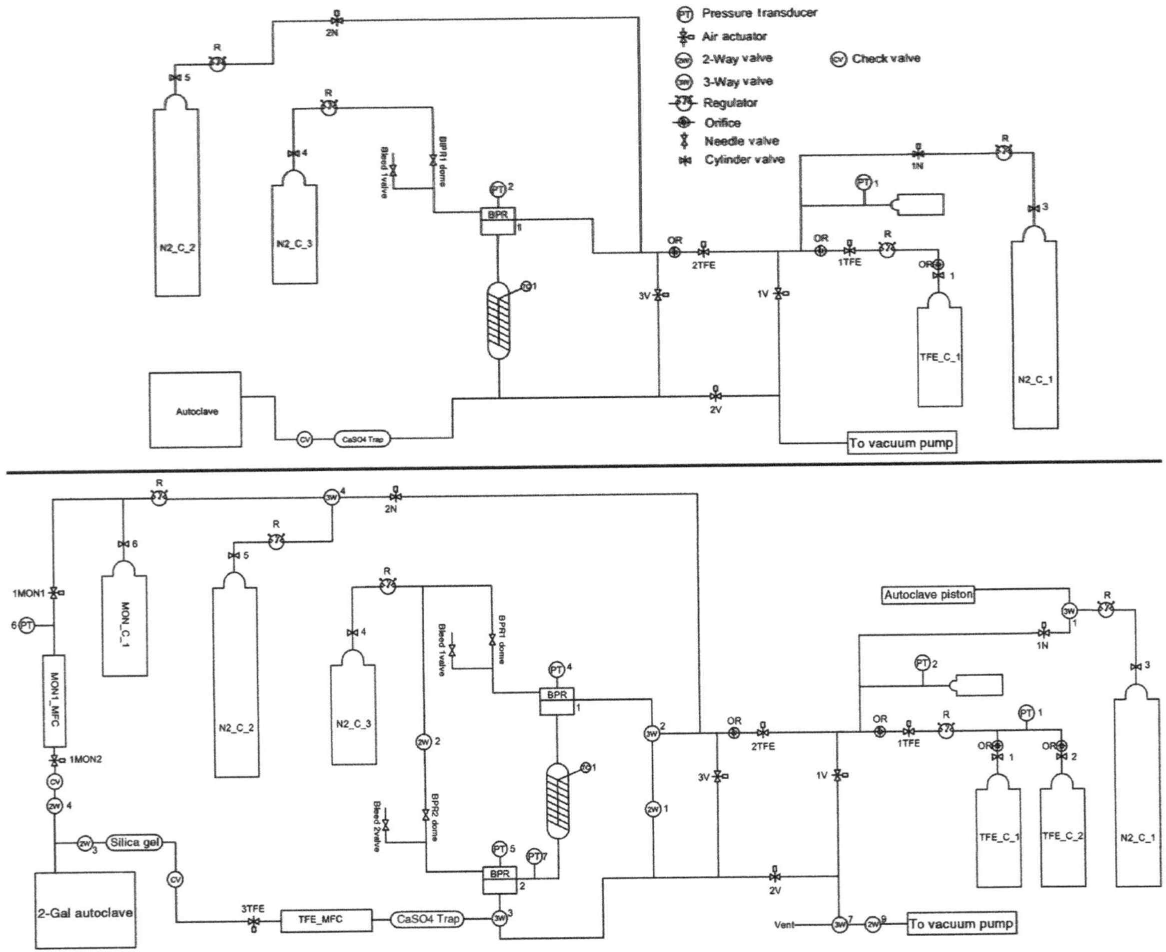

FIGURE 18.4 TFE gas handling systems. Top: system located at Shelby Hall. Bottom: system located at Clemson University.

18.4 THIRD-GENERATION TFE BARRICADE AT THE ADVANCED MATERIALS RESEARCH LABORATORY (AMRL) AND THE CLEMSON ENGINEERING TECHNOLOGY LABORATORIES (CETL), CLEMSON UNIVERSITY

The rationale for the new, third-generation TFE barricade at Clemson has already been given; however, the challenging aspects turned out to be what type of barricade to construct as well as where to locate it so that it would be in close proximity to the fluorine chemistry laboratories at AMRL. A number of designs and locations were considered before settling on the type and location of the facility, and unfortunately, we did not have the luxury of a new chemistry building being built in which to incorporate a new barricade, as was done with both Hunter Laboratories and Shelby Hall. A review of the literature did reveal a few publications giving considerable detail on the design of high pressure laboratories with high pressure cells/cubicles, where generally each cell/cubicle has a blowout wall on the same exterior wall of the laboratory facing an earthen berm [25–27]. However, other publications, some of which outlined problems with shock waves, pathway of projectiles, structural integrity, size of capital investment, and lack of flexibility for expansion with high pressure cells, suggested that the use of vented containment vessels would be more appropriate [27–30]. During our deliberations, we found that Fluitron, Inc. in Ivyland, Pennsylvania [31] markets containment vessels like those described in References 27 and 28; in fact, they appear to have both manufactured and now have for resale some of the same vessels previously used at Exxon Chemical Company (now ExxonMobil Chemical Co., a Division of ExxonMobil) high pressure laboratory in Allendale, New Jersey [32].

Unfortunately, the cost estimates for such vessels, except for perhaps the used ones, exceeded our budget, so we began to look into the cost of having an industrial fabricator build and transport such a vessel to the Clemson Research Park. At the same time, having potentially located a space within the AMRL for a hazards laboratory, we checked on the feasibility and cost that would be associated with this option. As one might be able to imagine, no building codes exist for TFE facilities; however, the potential use of a vented containment vessel dictated that the boiler code for the State of South Carolina [33] must be applied. Basically this meant that the walls of the room within the AMRL that would house this vessel had to have the same or greater strength than the walls of the vessel itself. Thus, the walls of the room would have to be strengthened, and that coupled with removing part of the roof structure in order to lift the vessel into the room with a crane and then replace the roof, meant that the costs of this option became prohibitive.

At this point in time, one of us (JLC) suggested a completely new conceptual design that from a cost point of view could take advantage of commercially available storage buildings, one built for hazardous material storage (U.S. Chemical Storage Superloc Model SL1610), which would be used as a reaction room or high pressure cell, and the other an explosive storage magazine [U.S. Explosive Storage Model ML263 (Type II/Outdoor)], which would be used as a control room. In other words, the latter building is "reverse engineered" in the sense that the operators are located

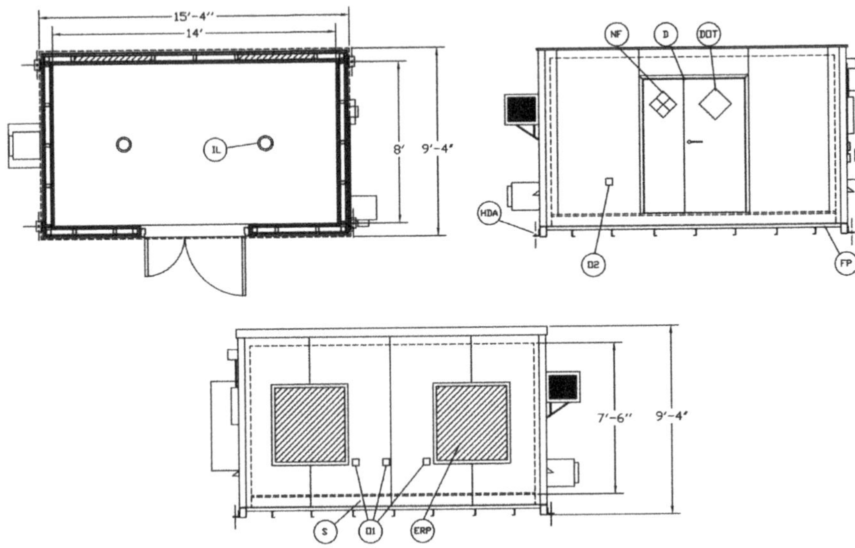

FIGURE 18.5 Reaction room building diagram. IL, internal light connector; HDA, hold down angle; O1, O2, explosion proof outlets; S, sump; EXP, explosion blowout panels; NF, NFPA sign panel; FP, forklift pockets; D, door; DOT, Department of Transportation flip chart placard (Reproduced with permission).

inside the control room while a reaction is carried in the reaction room, and in the event of an explosion, the operators remain safe in the control room. Next, a suitable location was secured for this new conceptual design of the needed TFE barricade, namely the lower parking lot adjacent to the CETL building, which was being used as a "bone yard" for old equipment from the high bay facilities along the back of the CETL building.

Each of the storage buildings is nominally a small house with its own 18,000 BTU heat pump and 200-Amp service. The nominal dimensions (both inside and out) as well as other design features of the reaction room are shown in Figure 18.5, including the two blowout panels, which release with >5 psi positive pressure, a fire suppression system, and an intake, fan, and exhaust for extra ventilation. The reaction room has a full storage capacity of 56,000 lbs, a building weight of 17,587 lbs, a floor load of 500 psf, a door load of 25 psf, a snow load of 40 psf, seismic rating of zone 4, tie-downs of anchor angle at each corner, and a wind load of 130 mph. The walls and roof are built of mineral wool insulation and multiple layers of gypsum wallboard encased between heavy-gauge galvanneal sheet steel to provide a 4-hour fire rating [34].

A diagram of the control room is shown in Figure 18.6. The exterior walls and door of this explosive storage magazine are constructed from 1/4-in. thick, ASTM A-36 prime plate steel lined with 3-in. thick hardwood, where the direction of the grain of the hardwood is rotated 90° with every 1/2 in. of overlapping thickness of

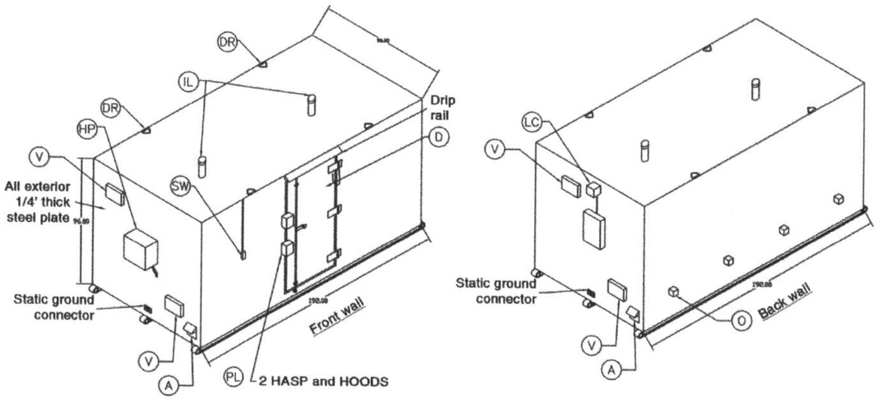

FIGURE 18.6 Diagram of the control room. A, apertures; V, vents; HP, heat pump; DR, door rings; IL, internal light connectors; D, door; SW, light switch; LC, load center (Reproduced with permission).

the hardwood. The design criteria meet or exceed ATF Type II Outdoor specifications and conform to Article 77 of the Uniform Fire Code (UFC). The control room has nominal dimensions of 16 ft (W) × 8 ft (H) × 8 ft (D), a weight of 14,376 lbs, tie downs of anchor angle at each corner, an open channel construction, and crane lifting lugs [35].

As previously mentioned, the location of the new TFE facility was to be in the former lower parking lot of the CETL building; however, a number of site preparations had to take place prior to the installation of the two modules. First, power conduits had to be brought underground from CETL to the site, and then a 20 ft × 30 ft reinforced concrete pad had to be poured, as the existing asphalt parking lot was both too sloped for and too weak to accept the weight of the new buildings. Again because of code, the reaction room was placed at least 65 ft away from the CETL building as well as 60 ft away in opposite directions from an adjacent parking lot and the closest roadway. This meant that a concrete walkway and ramp had to also be installed from the CETL building to the concrete pad, as the closest high bay space [13 ft (W) × 19 ft (L) × 25 ft (H)] in CETL was also to serve as laboratory support for the new TFE facility. Both of the modules were placed on the concrete pad facing the CETL building about 6 ft apart from each other; in other words, the door of the reaction module faces the rear of the control room, which was done for obvious safety reasons (please see Figure 18.7). Since only one side of the TFE facility was protected by an earthen berm (toward the adjacent parking lot), three Conex shipping boxes were placed in the shape of a U around the rear and sides of the reaction module. Finally, an 8-ft tall security fence was placed around the entire former parking lot with two entrance gates: a 4-ft wide gate at the concrete walkway/ramp toward the CETL building, and a 12-ft sliding gate at the rear of the yard. With the increase in size and capacity of the new facility, additional features are being introduced such as data logging capabilities, streaming video of the reaction room, mass flow controllers

FIGURE 18.7 TFE barricade located at CETL. Left: control room; right: reaction room.

(MFCs), and a new, improved gas handling system. These and other aspects of the new facility will be overviewed in the following sections.

18.4.1 Safety and Control

Our College Machine Shop fabricated a cable tray/raceway from parts from McMaster-Carr [36] that couples the control room to the reaction room. The cable tray, which is nominally 4-in. × 4-in. in diameter (please see Figure 18.8), traces power and control cables for the 2-gal and smaller autoclaves (e.g., the oven, the thermocouple, the tachometer, the pressure transducer, and the solenoid valve for

FIGURE 18.8 Cable tray/raceway connecting the reaction room with the control room.

the internal cooling), the pressure transducers in the gas handling and carbon dioxide scrubber systems, the thermocouple encoders, the MFC, the syringe pumps, the video cameras, as well as nylon tubing for control of the air-actuated valves in the gas handling system and stainless steel tubing for control of the needle valves that regulate the back pressure regulators (BPRs). Since the actuators on the remotely controlled process valves are external to the valve bodies themselves, breathing air is used in the nylon tubing to these actuators (up to about 100 psig). Furthermore, were any of the nylon tubes or the regulator to fail, we would only have breathing air emptying into the control room, which would not be a problem. On the other hand, high purity nitrogen is used for control of the BPRs (up to 700 psig) because were the diaphragm in either BPR to fail, one would only want oxygen-free nitrogen entering process lines that might be filled with TFE. Were the stainless steel tubing in the cable tray to develop a leak at one of the fittings, etc., it is possible that the control room could fill with nitrogen; however, both the control room and reaction room are equipped with oxygen deficiency monitors (for this and other reasons too). In addition, both audible and visual alarms for these monitors have been installed at the entries to both modules as well as at the entry gates to the facility.

The valve control panel shown in Figure 18.9 features all of the control valves for the air-actuated valves as well as the BPRs in the gas handling system. As with the second-generation TFE barricade, these air-actuated process valves all fail closed, and furthermore, since the air supply to all of these valves are plumbed to a master

FIGURE 18.9 Left and bottom: air-actuated valve control panel and air cylinder located in the control room; top right (from left to right): main microcontroller circuit board and high precision pressure transducer.

cutoff valve, they can all be closed with a simple 180° turn of one ball valve on the right end of the valve control panel (not in view in Figure 18.9). In addition to the control, the panel is equipped with light emitting diodes (LEDs) that indicate if each valve is currently open (green) or closed (red) via a connection to a laser setup across the shaft of each air-actuated valve. This serves as a safety feature in case for some reason a control valve gives a false indication about the actual state (open or closed) of an air-actuated valve.

In addition to the process valves, the autoclaves, MFCs, and syringe pumps (for initiator and/or comonomer feeds) are controlled remotely from various equipment in the control room, and the autoclaves are also equipped with remotely controlled, air-actuated valves for venting any excess pressure at the end of a reaction to outside the reaction room. The bottom of Figure 18.4 gives a comparison of the expanded gas handling system for the current TFE facility versus that for the previous one at Alabama (Top of Figure 18.4). A vacuum line exists inside the reaction room, which is used in a variety of ways for leak checking and evacuating components during various SOPs as well as for the preparation of TFE/CO_2 mixtures from the aforementioned pyrolysis of pentafluoropropionate salts. As before, the electrical system has a "crash" button that cuts the power to the entire facility in case of emergency. Both a fire extinguisher in the control room and the fire suppression system of the reaction room are connected to the Clemson University Fire Department in case the use of either is detected.

As with the Alabama facility, an orifice is included in the gas handling system after most process valves that handle TFE (e.g., TFE cylinder valve and 1TFE and 2TFE process valves; see Figure 18.4). As mentioned previously, these orifices afford very fine control over gas flow as well as should act as flame arrestors. The current gas handling system also has an in-line Ascarite® scrubber (with a multipoint thermocouple and closed looped, external cooling) for removing CO_2 from the gaseous TFE/CO_2 stream on its way to the polymerizer/autoclave; however, this method of CO_2 removal is only cost effective for polymerizations run in the smaller autoclaves (600 mL or less capacity). For larger autoclave runs, we modify the gas handling system into a caustic CO_2 scrubbing system and remove CO_2 from approximately 1.0 to 1.5 kg of TFE/CO_2 mixture at a time, while stabilizing the TFE with D-limonene. When ready to run a polymerization, a double-ended cylinder containing silica gel to scrub the D-limonene is connected to the entry port of the autoclave.

The reaction room is also equipped with four HD cameras capable of capturing real-time video that is recorded for later use. Furthermore, our research partners can access online video in order to observe the progress of their projects in real time. The video can be streamed with the data for a more comprehensive review.

18.4.2 Data Acquisition

The reaction setup is equipped with a data acquisition system designed in house to provide real-time measurement with high precision and data logging of all the variables. The hardware is designed using Texas Instruments microcontrollers (MSP430G2553) since they provide low power consumption, 12-bit resolution for the analog to digital converters, and the capability to emulate serial communication without any extra interface. Two sets of devices exist for the data acquisition system, namely a device for the pressure transducer analog to digital converter and the thermocouple assembly. The first one is capable of acquiring the signal of six pressure transducers, converting and transmitting the signal to the control room at a rate of 300 samples per second (see the top, right of Figure 18.9). The thermocouple device has integrated a low pass filter and a cold junction compensation filter that corrects the temperatures at the reaction room regardless of the temperature fluctuations of the room with an accuracy of 0.1°C. At the current time, there are two different sets of pressure transducer arrays: one controls the main system during polymerization reactions and the second array is used in junction with the former when scrubbing CO_2 from TFE/CO_2 mixtures.

The software is both developed using Microsoft Visual Studio .NET® and National Instruments Labview® and is capable of recording the data at various sample rates directly into an Excel file.

18.4.3 Deionized and Deoxygenated Water System

As mentioned previously, in order to work safely with TFE one needs to avoid the presence of oxygen at least down to low ppm levels. In the past, when using autoclaves of up to a 600-mL capacity, we could effectively use freeze-pump-thaw cycles to successfully degas reagents, be they aqueous- or fluorocarbon-based solutions. With the scale-up to a 2-gal autoclave and the fact that most copolymerizations with TFE are carried out via aqueous emulsion or aqueous suspension techniques, along with the use of aqueous solutions of initiators, we sought an improved method of deoxygenating the aqueous solutions required for the polymerizations. Industry often uses a steam out method, but without a source of steam at the TFE barricade, this option seemed undesirable, as the autoclave oven would have to be used as the heat source. The solution that we came up with is shown schematically in Figure 18.10. The funnel-shaped water tank (Den Hertog IBFD35 Set [37]) shown on the left side of the figure is first filled with good quality, deionized water. The water is then sparged with a stream of nitrogen (<45 psig; the large lid to the Den Hartog tank is not totally leak free, so the over pressure of nitrogen can vent here) until the dissolved oxygen level is below 1.0 ppm as measured by a dissolved oxygen meter (YSI ProODO meter [38]). Both a waterline from the bottom drain of the tank and an auxiliary nitrogen line are fed into a glove bag in which a low level oxygen atmosphere (<100 ppm) has been established (measured by a vapor oxygen sensor (Vacuum Atmospheres)). Initiator solutions as well as water/surfactant solutions can then be weighed out in the glove bag. In the case of a 600-mL capacity or smaller autoclave, the entire autoclave (bottom, head, sealing rings, and bolts) can be taken inside the glove bag for filling, while in the case of the 2-gal autoclave, the water/surfactant solution is filled into a sealable 5-L, polyethylene Jerry can for later transfer to the autoclave under an air-free environment. Initiator solutions are filled into sealable bottles for later transfer to the high precision syringe pump.

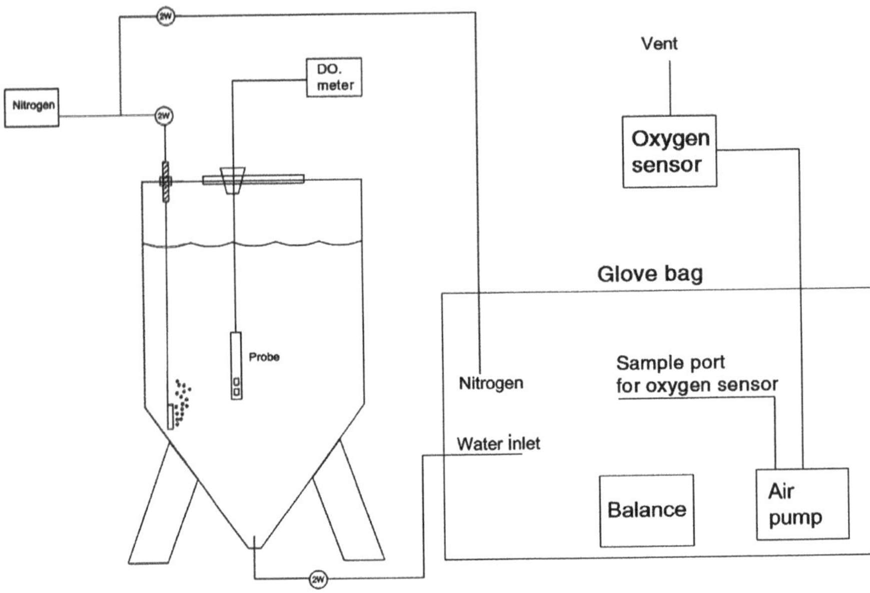

FIGURE 18.10 Deionized, deoxygenated water system, and glove bag.

18.4.4 Reactor Setup

The heart of the new TFE barricade facility is a 2-gal, floor stand autoclave (Parr Instrument [39]) shown in Figure 18.11. Interchangeable parts are also on hand where this autoclave can be converted to a 1-gal configuration. Smaller autoclaves from 100- to 600-mL capacity are also available, and the system has been designed so that most of the power and controls can be switched over ("plug-and-play") from the large autoclave to a smaller autoclave. Focusing on the 2-gal autoclave, multiple monomer co-feeds are possible in addition to the feed of the free radical initiator solution from a high precision syringe pump (Teledyne-ISCO). The desired temperature in the autoclave is maintained by balancing the external heating from an oven and the internal cooling from a temperature-controlled circulating bath. A combination of check valves and coalescence filters are used to try to prevent back streaming of any moisture from the autoclave (aqueous emulsion or suspension polymerizations) toward the MFCs. As previously mentioned, one of the feed ports is also connected to an additional remotely controlled, air-actuated valve for venting; a needle valve is connected to the exit of this valve followed by a piece of tubing that goes through the ceiling/roof of the barricade so that venting takes place outside of the reaction room. Finally, the exit end of the burst disc assembly is also hard piped to the outside world with a single piece of tubing that has nonabrupt bends in order to prevent the potential generation of hot spots/friction and rupture were the burst disc to fail and TFE be exhausted [10].

FIGURE 18.11 Front view of 2-gal reactor with accessories. Left on the table: syringe pump (Teledyne-ISCO) for initiator solution feed; middle: 2-gal floor stand reactor (Parr Instrument) with jacketed oven.

ACKNOWLEDGMENTS

The authors would like to thank Clemson University, The University of Alabama, and the various funding agencies and industries who have funded the construction, operation, and maintenance of the facilities described herein. Todd S. Sayler and Kevin T. Tice are acknowledged for their assistance with the former facility at The University of Alabama, while Cameron A. Parrish, Longyu Hu, Xiaolin Liu, Robert B. Coward, Trevyn A. Hey, Justin A. Murphy, Lauren E. Lowery, Michael E. Brady, Shane M. Williams, and Lucy M. Rummler are acknowledged for their assistance with the new facility at Clemson University. JST acknowledges the National Science Foundation (CHE-1124859) and the U.S. Department of Energy, Basic Energy Sciences (DE-FG02-05ER15718) for financial support.

REFERENCES

1. Ebnesajjad, S. *Fluoroplastics: Volume 1: Non-Melt Processible Fluoroplastics: The Definitive User's Guide and Databook*; Plastics Design Library: Norwich, NY, 2000; p 40.
2. Wainright, J. S.; Savinell, R. F.; DesMarteau, D. D.; Ma, J.-J.; Sung, K.; Zhang, L. *Proc. Electrochem. Soc.* **1994**, *94–23*, 265–274.
3. DesMarteau, D. D. Copolymers of Tetrafluoroethylene and Perfluorinated Sulfonyl Monomers and Membranes Made There From. US Patent 5,463,005, October 31, 1995.
4. Sayler, T. S.; Tice, K. T.; Beg, M. A.; Fernandez, R. E.; Thrasher, J. S. *Polym. Prepr. (Am. Chem. Soc. Div. Polym. Chem.)* **2012**, *53*, 39–40.
5. Sayler, T. S. Preparation of Perfluorinated Ionomers for Fuel Cell Applications. Ph.D. Dissertation, The University of Alabama, Tuscaloosa, AL, 2012.
6. Van Bramer, D. J.; Shiflett, M. B.; Yokozeki, A. Safe Handling of Tetrafluoroethylene. US Patent 5,345,013, September 6, 1994.
7. Van Bramer, D. J. Tetrafluoroethylene Shipping/Storage Mixtures. US Patent 5,866,727, February 2, 1999.
8. Babenko, Y. I., Lisochkin, Y. A., Poznyak, V. I. *Combust. Explos. Shock Waves* **1993**, *29*, 603–609.
9. Reza, A.; Christiansen, E. *Process Saf. Prog.* **2007**, *26*, 77–82.
10. Ferrero, F.; Meyer, R.; Kluge, M.; Schröder, V.; Spoormaker, T. *J. Loss Prevent Proc.* **2013**, *26*, 177–185.
11. Thomas, B. H.; Shafer, G.; Ma, J. J.; Tu, M.-H.; DesMarteau, D. D. *J. Fluorine Chem.* **2004**, *125*, 1231–1240.
12. Thomas, B. H.; DesMarteau, D. D. *J. Fluorine Chem.* **2005**, *126*, 1057–1064.
13. Shafer, G. J. Chemistry of Perfluoroalkylsulfonyl Methanes. Development of Vinyl Ether Monomers and Their Copolymerization with Tetrafluoroethylene. Ph.D. Dissertation, Clemson University, Clemson, SC, 2001; pp 26–28.
14. Zhou, S. Synthesis and Characterization of Perfluorinated Sulfonimide Copolymers Electrolyte Membranes. Ph.D. Dissertation, Clemson University, Clemson, SC, 2002; pp 82–84.
15. Ford, L. A. Novel Semifluorinated Polyaryl Ether Electrolytes Containing the Perfluorocyclobutane Linkage. Ph.D. Dissertation, Clemson University, Clemson, SC, 2002; pp 18–21.
16. Lu, C.; Kim, J.-H.; DesMarteau, D. D. *J. Fluorine Chem.* **2010**, *131*, 17–20.
17. Sharif, I.; Creager, S. E.; DesMarteau, D. D. *Fluorinated Ionomers and Ionomer Membranes Containing the bis[(Perfluoroalkyl)sulfonyl]imide Protecting Group,* in Handbook of Fluoropolymer Science and Technology, Smith, D. W., Jr.; Iacono, S. T.; Iyer, S. S., Eds.; John Wiley & Sons, Inc.: Hoboken, NJ, **2014**; Chap. 22, pp. 523–546.
18. Asahi Glass Co., Ltd. Press Release. Closure of Fluorochemicals Production Base in U.S. (Bayonne, NJ Plant). www.agc.com/english/**news**/2007/0928e.pdf (accessed June 28, 2013).
19. Meissner, E.; Wróblewska, A.; Milchert, E. *Polym. Degrad. Stabil.* **2004**, *83*, 163–172.
20. Molero Meneses, M.; Moreno Beltran, D.; Grau Rios, M. *Libro Actas – Congr. Nac. Med., Hig. Segur. Trab.,* 9th Conference, 1981, pp 2461–2479.

21. Hals, L. J.; Reid, T. S.; Smith, G. H. Process of Making Perfluoro Olefins. US Patent 2,668,864, February 9, 1954.
22. Hals, L. J.; Reid, T. S.; Smith, G. H., Jr. *J. Am. Chem. Soc.* **1951**, *73*, 4054.
23. Kornath, A.; Kaufmann, A. LMU München, Germany. Private Communication, 2007.
24. Thrasher, J. S. Unpublished results.
25. Craig, L. E.; Dew, J. E. *Ind. Eng. Res.* **1959**, *51*, 1249–1252.
26. Blackburn, D. W.; Reiff, H. E. *Ann. N.Y. Acad. Sci.* **1967**, *145*, 192–203.
27. Franko-Filipasic, B. R.; Michaelson, R. C. *CEP, Chem. Eng. Prog.* **1984**, *80*, 65–69.
28. Penninger, J. M. L.; Okazaki, J. K. *CEP, Chem. Eng. Prog.* **1980**, *76*, 65–71.
29. Livingston, E. H. *CEP, Chem. Eng. Prog.* **1984**, *80*, 70–75.
30. Bodurtha, F. T., Jr.; Bonifaz, C. Laboratory Barricade. US Patent 4,474,052, October 2, 1984.
31. Fluitron, Inc. Innovation in Pressure Technology: Containment Cells. http://catalog.fluitron.com/item/all-categories/containment-cells/item-1007?forward=1 (accessed September 13, 2010).
32. Folawewo, F. Fluitron, Inc. Ivyland, PA. Private communication, 2010.
33. South Carolina Department of Labor, Licensing and Regulation. Boiler Safety Program. Boiler Registration and Inspection Information. Document 200. http://www.llr.state.sc.us/POL/Boilers/Forms/Doc%20200.pdf (accessed October 8, 2010).
34. U.S. Chemical Storage. Four Hour Bi-Directional Fire Rated Buildings for Safe and Complaint Hazmat Chemical and Product Storage. http://www.uschemicalstorage.com/fire-rated-chemical-storage/superloc.html (accessed June 30, 2013).
35. U.S. Explosive Storage. Our Strength is Your Security. Outdoor Magazines. http://www.usexplosivestorage.com/index.php?option=com_content&view=article&id=93&Itemid=119 (accessed June 30, 2013).
36. McMaster-Carr. Washdown Indoor/Outdoor Stainless Steel Raceway (NEMA 4X). http://www.mcmaster.com/#standard-raceway/=nf9ufg (accessed March 14, 2012).
37. Den Hertog Industries, Inc. New 35 & 60 Gallon Full Drain Blow Molded Inductor Tanks and Poly Stand Sets. http://www.denhartogindustries.com/news/new-35-60-gallon-full-drain-blow-molded-inductor-tanks-and-poly-stand-sets/ (accessed February 7, 2013).
38. YSI: A Xylem Brand. ProODO Handheld Optical Dissolved Oxygen Meter. http://www.ysi.com/productsdetail.php?ProODO-19 (accessed February 21, 2013).
39. Parr Instrument Company. Series 4550 Floor Stand Reactors, 1 & 2 L. http://www.parrinst.com/products/stirred-reactors/series-4550-floor-stand-reactors/ (accessed March 30, 2011).

19

FLUOROPOLYMER SURFACES/INTERFACES

YUJI HIGAKI, RYOHEI ISHIGE, AND ATSUSHI TAKAHARA
Kyushu University

19.1 INTRODUCTION

Polymers with perfluoroalkyl (R_f) groups have surface characteristics that differ greatly from those of comparable hydrogenated structures, for example, excellent chemical and thermal stability [1, 2], nonadhesive properties [3], low friction coefficients [4, 5], low surface free energies [6], and antifouling behavior [7]. Fluoropolymers with R_f groups have been extensively studied in various research fields and applied as very meaningful and valuable chemicals in industry. Most of these polymers are poly(fluoroalkyl acrylate)s with long R_f groups, and these polymers have a critical surface tension (γ_c) that is much lower than that of polytetrafluoroethylene (PTFE), which is the most typical fluoropolymer. However, the water-repellent mechanism of these polymers has not been fully clarified. The authors demonstrated the water-repellent mechanism of poly(fluoroalkyl acrylate) (PFA-C_y, where y is the fluoromethylene number of the R_f groups) thin films [8–12]. PFA-C_8 is composed of rigid rod-like perfluorooctyl chains with low surface free energy. The molecular aggregation state of spin-cast PFA-C_8 thin films was investigated by surface-sensitive X-ray scattering techniques, such as grazing incidence wide-angle X-ray diffraction (GI-WAXD) measurement. The R_f groups at the side chains were oriented to the perpendicular direction to the substrate surface, forming a highly ordered lamellar structure at the air/polymer interface and resulting in a hydrophobic surface. In the

Handbook of Fluoropolymer Science and Technology, First Edition. Edited by Dennis W. Smith Jr., Scott T. Iacono, and Suresh S. Iyer.

case of PFA-C_y with long R_f groups [$y \geq 8$], the mobility of the molecular chains is low, probably because of the crystallites of the R_f groups at the surface region; therefore, the contact angle against water is stable. The authors recently tried to fabricate nanostructures for PFA-C_8 thin films by nanoimprinting lithography in which a mold with a nanostructured pattern is mechanically pressed onto a substrate coated with a polymeric material [13, 14]. The PFA-C_8 film could be nanoimprinted at room temperature, and the nanoimprinting characteristic arises from the unique properties of the R_f groups. The mobility of molecular chains is an important factor for determining the wetting behavior. However, the relationship between the mobility of molecular chains and the wetting behavior has not been evaluated directly.

Surface-tethered PFA-C_8 brush thin films represented different molecular aggregation states of the R_f groups compared with the spin-cast films [15–17]. In particular, densely grafted PFA-C_8 brush films with narrow molecular weight dispersity (MWD) ($M_w/M_n = 1.05$) prepared by surface-initiated atom transfer radical polymerization (SI-ATRP) showed a lower dynamic contact angle against water in air than that of the spin-cast film. GI-WAXD analysis of the PFA-C_8 brush film revealed that the rigid rod-like R_f groups at the outer most surface were aligned to the parallel direction to the substrate surface because the main chain of the brush was relatively stretched to the normal direction. In contrast, the R_f groups of the PFA-C_8 brush with broad MWD were oriented to the normal direction to the substrate, and the static contact angle was as high as the spin-cast film. These results indicated that water repellency of PFA-C_8 is largely governed by the aggregation structure and orientation direction of the R_f groups.

For large-scale chemical manufacturing, it has depended on organic solvents, which have given rise to serious environmental issues. The organic solvents are attempted to replace with "green" (environmentally friendly) solvents. Supercritical carbon dioxide ($scCO_2$) is one of the attractive candidates and has been utilized in a wide range of polymer processing [18–20]. $ScCO_2$ is also known to be an effective plasticization agent for many polymers and the swelling behavior of bulk polymers, such as polystyrene (PS), poly(methyl methacrylate) (PMMA), and polycarbonates has been extensively investigated over the past two decades [21]. Recently, scientific interests in the plasticization effects of $scCO_2$ have been extended to polymer thin films because of their technological importance in coating, lubrication, and adhesion, and their novel physical properties that are quite different from the bulks [22]. Intriguingly, previous results shed new light onto the excess swelling and resultant plasticization effect of polymer thin films (including polymer brushes) near the critical point of CO_2 ($T_c = 304.3$ K and $P_c = 7.38$ MPa). The excess swelling is induced by the large density fluctuations of the solvent and is rather general regardless of a choice of polymers.

In this article, we aim to introduce our recent investigation relating to the surface molecular aggregation structure and surface molecular motion of the fluoro(co)polymers with R_f side group. GI-WAXD measurement and X-ray photoelectron spectroscopy (XPS) using C_{60} ion beam were applied to the surface molecular aggregation structure analysis and the depth analysis of the fluoropolymer films, respectively. Temperature dependence of the dynamic contact angle, XPS measurement in pseudo-hydrated state, and lateral force microscopy (LFM) measurement

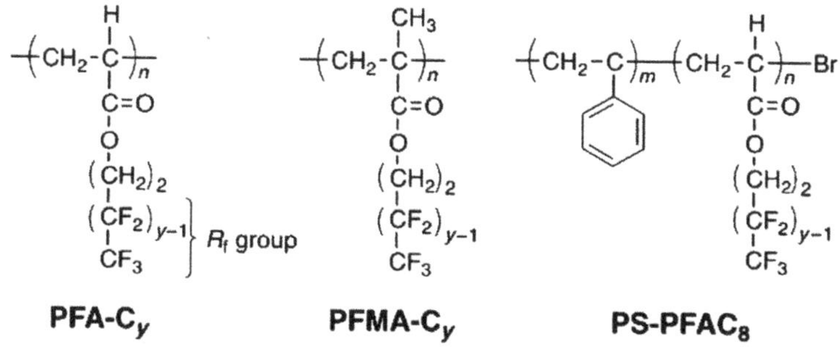

SCHEME 19.1 Chemical structures of PFA-C_y, PFMA-C_y, and PS-PFAC_8 diblock copolymer.

were applied to the investigation of the surface rearrangement of the fluoropolymers. High pressure *in situ* neutron reflectivity (NR) was applied to the investigation of the swelling state of semicrystalline CO_2-philic fluoropolymer brushes in scCO_2.

19.2 GRAZING-INCIDENCE WIDE-ANGLE X-RAY DIFFRACTION MEASUREMENT FOR SURFACE MOLECULAR AGGREGATION STRUCTURE INVESTIGATION

PFA-C_8 and PFMA-C_8 were prepared by the previously reported procedure [8,23]. The chemical structures of PFA-C_y and PFMA-C_y are shown in Scheme 19.1. Figure 19.1 shows the powder WAXD profiles of PFMA-C_y. The scattering vector (q) was defined as $(4\pi/\lambda)\sin\theta$. The PFMA-C_y with $y \leq 6$ showed no sharp peak. For

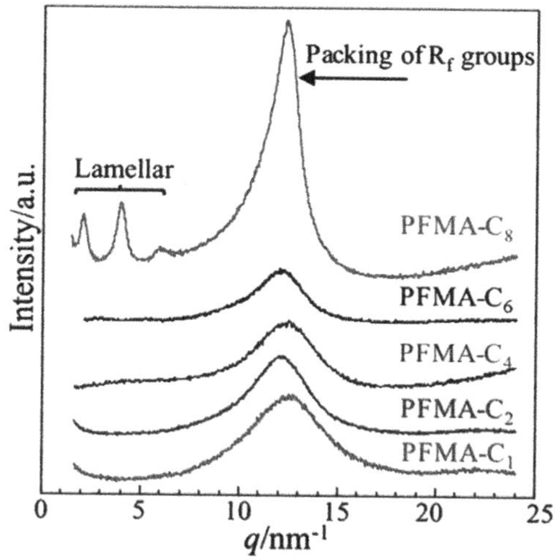

FIGURE 19.1 Powder WAXD profiles of PFMA-C_y. Reprinted with permission from Reference 23.

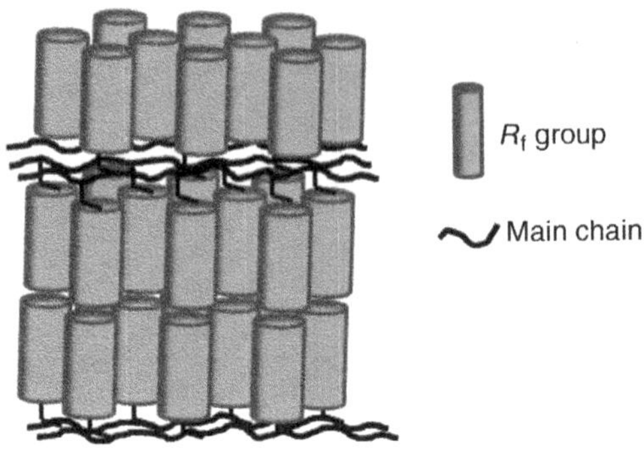

FIGURE 19.2 Schematic representation of molecular aggregation states of R_f groups of PF(M)A-C_8 film surface.

PFMA-C_8, diffraction peaks were obtained at q = 1–7 nm^{-1} and 12.3 nm^{-1}. The peaks at q = 1–7 nm^{-1} were assignable to the lamellar structure in which the R_f groups were ordered like a multilayer, and the peaks at q = 12.3 nm^{-1} were assignable to the lateral distance between R_f groups (Figure 19.2) [24–26]. Table 19.1 lists the glass transition temperature (T_g) and the melting temperature (T_m) of PFMA-C_y measured by DSC. The T_m of the R_f chain crystallites was 360 K for PFMA-C_8. The T_g were 351, 343, 310, and 300 K for the PFMA-C_y with y = 1, 2, 4, and 6, respectively, which is higher than room temperature, whereas the T_g of the PFA-C_y with $y \leq 6$ were lower than room temperature. This result indicates that the PFMA-C_y with $y \leq 6$ was in a glassy state and the mobility of the R_f groups was low at room temperature.

The fluorinated polymer (PFA-C_y, PFMA-C_y) was dissolved in HCFC-225 (concentration: 1 wt%), and a Si-wafer was coated with this solution by the spin-coating method (2000 rpm, 30 seconds). The static contact angle against water and methylene iodide, surface free energy [27], and dynamic contact angle against water for the PFMA-C_y thin films are summarized in Table 19.2. For the PFMA-C_y with $y \leq 6$,

TABLE 19.1 Glass Transition Temperature (T_g) and Melting Temperature (T_m) of PFA-C_y and PFMA-C_y

Sample	T_g/K	T_m/K
PFA-C_1	271	–
PFA-C_2	259	–
PFA-C_4	249	–
PFA-C_6	243	–
PFA-C_8	–	348
PFMA-C_1	351	–
PFMA-C_2	343	–
PFMA-C_4	310	–
PFMA-C_6	300	–
PFMA-C_8	–	360

Source: Reprinted with permission from Reference 23.

TABLE 19.2 Static Contact Angle Against Water (2 μL) and Methylene Iodide (CH_2I_2, 2 μL), Surface-Free Energy (γ_s^d: Dispersion Force Component; γ_s^p: Polar Component; γ_s: Surface-Free Energy), and Dynamic Contact Angle Against Water (50 μL; θ_a: Advancing Contact Angle; θ_r: Receding Contact Angle; θ_s: Sliding Angle) for PFA-C_y and PFMA-C_y Films

	Static contact angle (°)		Surface-free energy (mN/m)			Dynamic contact angle (°)		
Sample	Water	CH_2I_2	γ_s^d	γ_s^p	γ_s	θ_a	θ_r	θ_s
PFA-C_1	102 ± 0.9	98 ± 4.0	7.52	5.10	12.62	106 ± 5.6	45 ± 2.5	48 ± 3.0
PFA-C_2	110 ± 3.9	100 ± 5.9	7.26	2.34	9.60	129 ± 6.5	48 ± 4.5	62 ± 2.5
PFA-C_4	126 ± 4.1	101 ± 5.3	8.44	0.04	8.48	132 ± 3.6	52 ± 3.0	58 ± 1.0
PFA-C_6	123 ± 4.9	103 ± 4.4	7.35	0.25	7.60	134 ± 4.0	52 ± 2.4	62 ± 3.4
PFA-C_8	122 ± 0.9	103 ± 0.4	7.46	0.36	7.83	126 ± 0.7	98 ± 2.6	20 ± 1.1
PFA-C_8[a]	121 ± 0.5	104 ± 1.0	6.84	0.59	7.43	125 ± 0.6	100 ± 1.5	19 ± 1.0
PFMA-C_1	94 ± 0.7	73 ± 1.9	19.1	3.8	22.9	102 ± 0.7	84 ± 0.8	14 ± 0.8
PFMA-C_2	106 ± 1.6	83 ± 0.7	14.7	1.5	16.2	111 ± 0.5	90 ± 1.0	15 ± 0.7
PFMA-C_4	114 ± 0.9	102 ± 1.9	7.22	1.60	8.82	120 ± 1.9	80 ± 1.5	28 ± 1.7
PFMA-C_6	117 ± 1.2	100 ± 0.9	8.21	0.91	9.12	120 ± 1.7	76 ± 2.7	31 ± 2.8
PFMA-C_8	118 ± 0.6	103 ± 2.0	7.00	1.02	8.02	123 ± 1.5	90 ± 1.5	22 ± 3.2
PFMA-C_8[a]	120 ± 0.7	105 ± 0.9	6.42	0.80	7.22	122 ± 1.3	112 ± 2.8	7 ± 1.5

Source: Reprinted with permission from Reference 23.

[a]Annealed at 333 K for 6 h.

the static contact angle and advancing contact angle (θ_a) were high, independent of the length of the R_f groups. On the other hand, the receding contact angle (θ_r) was also large for the PFMA-C_y with $y \leq 6$, while it was small for the PFA-C_y with $y \leq 6$. The results mentioned above indicate that the surface structure of the PFMA-C_y with $y \leq 6$ is stable under water exposure. On PFA-C_y, the mobility of the molecular chains was high because of the lack in crystallinity of the short R_f groups and the low T_g to reorganize the surface chemical composition by water exposure.

The temperature dependence of the dynamic contact angle, XPS, and LFM measurements were carried out to evaluate the relationship between the surface molecular motion and wetting properties. Figure 19.3 shows the temperature dependence of the receding contact angle θ_r (left) and lateral force (right) for (a) PFMA-C_1, (b) PFMA-C_2, (c) PFMA-C_4, and (d) PFMA-C_6 thin films. θ_r showed high value at room temperature. However, with an increase in temperature, the θ_r began to decrease

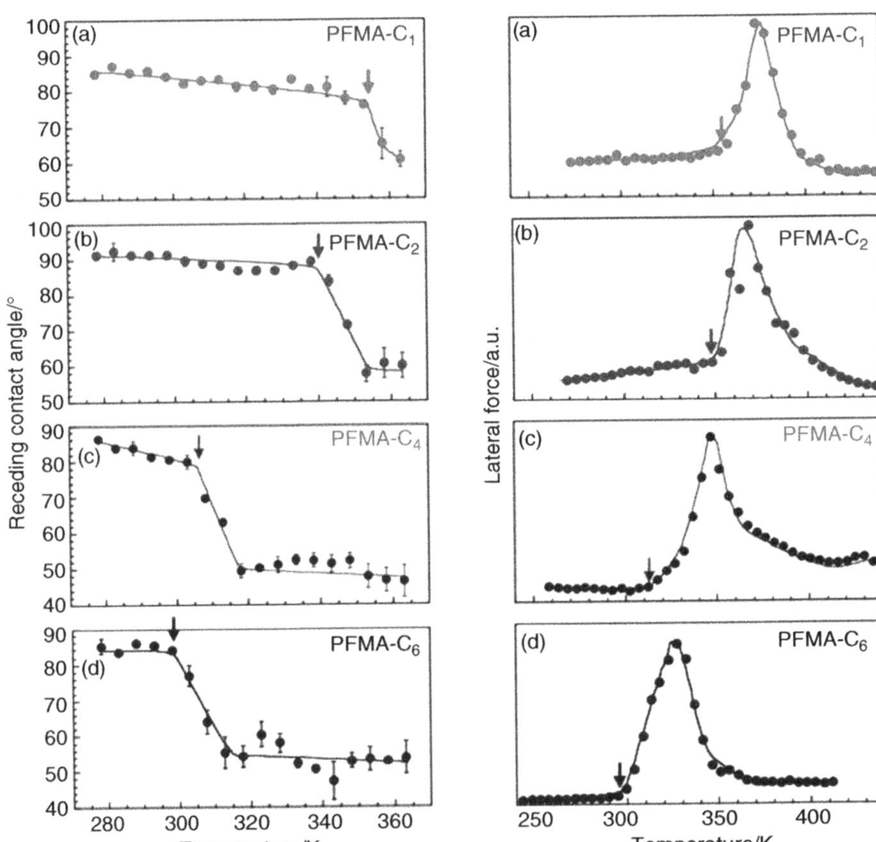

FIGURE 19.3 Temperature dependences of the receding contact angle (left) and lateral force (right) for (a) PFMA-C_1, (b) PFMA-C_2, (c) PFMA-C_4, and (d) PFMA-C_6 thin films. Arrows indicate T_g determined by DSC measurement. Reprinted with permission from Reference 23.

drastically at 353, 343, 313, and 300 K for the PFMA-C_y thin films with $y = 1, 2, 4,$ and 6, respectively. These temperatures approximately agreed with the corresponding T_g values determined by DSC (arrows). Thus, the θ_r reduction was caused by the molecular motion activated above T_g. Here, the T_g determined by DSC is the bulk properties and not surface properties. Then, the temperature dependence of the LFM measurements was conducted to evaluate the surface molecular motion of PFMA-C_y thin films. LFM measurements were carried out at various temperatures in vacuum. The lateral force is the sum of the frictional and adhesion forces acting between the sample surface and cantilever tip. Since the frictional force depends on the relaxation behavior of polymeric materials, it is possible to investigate the surface molecular motion [28–31]. Onset of curves in lateral force was observed at 353, 346, 313, and 298 K for the PFMA-C_y with $y = 1, 2, 4,$ and 6, respectively, and these temperatures could be empirically defined to be the surface T_g. These surface T_g values showed good agreement with the temperature at which θ_r began to reduce. Therefore, the surface molecular motion greatly influences the wetting behavior; below T_g, the surface molecular motion of the R_f groups is restricted by the α-methyl group to avoid the surface reorganization under water. On the other hand, above T_g, the surface molecular motion of the main chain and R_f group is activated, and the contact angle decreases by surface reorganization.

The temperature dependence of the chemical composition change in the surface region was evaluated by XPS measurements. XPS measurements were carried out with an Al-Kα X-ray source (14 kV and 200 mW), and the take-off angles were kept constant at 45°. The samples were measured in the pseudo-hydrated state to characterize the suspected surface chemical composition in water [32, 33]. The samples were pseudo-hydrated by immersion in water for 120 minutes and then dried under vacuum (60 Pa) for 120 minutes, immediately after which measurements were performed. Figure 19.4 shows the F_{1s}/C_{1s} and O_{1s}/C_{1s} values of (a) PFMA-C_1, (b) PFMA-C_2, (c) PFMA-C_4, and (d) PFMA-C_6 thin films in the dried and hydrated states. The F_{1s}/C_{1s} and O_{1s}/C_{1s} values represent the relative magnitude of F and O concentrations at the surface. In the dried state, the F_{1s}/C_{1s} and O_{1s}/C_{1s} values completely agreed with the theoretical values, which are calculated from the chemical structure of PFMA-C_y. In the pseudo-hydrated state, the F_{1s}/C_{1s} and O_{1s}/C_{1s} values did not change at below T_g. On the other hand, the F_{1s}/C_{1s} values decreased and the O_{1s}/C_{1s} values increased at above T_g. The low values of F_{1s}/C_{1s} and the high values of O_{1s}/C_{1s} are attributable to the surface reorganization and the exposure of carbonyl groups to the water interface. The results also support the relationship between the surface molecular motion and the wetting behavior of PFMA-C_y thin films.

From the powder WAXD and DSC measurements, it was revealed that PFMA-C_8 forms crystalline structures. Therefore, it is expected that the mobility of the R_f groups in PFMA-C_8 is very low and the water repellency is higher than that of the PFMA-C_y with $y \leq 6$. However, the θ_r of PFMA-C_8 was close to that of the PFMA-C_y with $y \leq 6$ before annealing, and the θ_r increased after annealing treatment (333 K for 6 hours). GI-WAXD profiles before and after annealing (at 333, 348, and 368 K for 6 hours) were compared to discuss the annealing treatment effect. GI-WAXD measurements were carried out at 300 K using a six-axis diffractometer

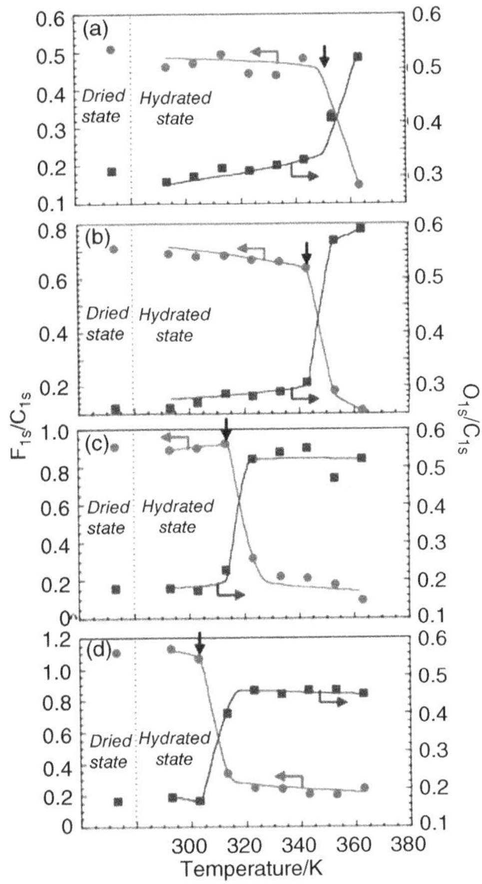

FIGURE 19.4 Temperature dependence of F_{1s}/C_{1s} and O_{1s}/C_{1s} values evaluated from XPS for PFMA-C_y thin films: (a) PFMA-C_1, (b) PFMA-C_2, (c) PFMA-C_4, and (d) PFMA-C_6 thin films in dry and pseudo-hydrated state. Reprinted with permission from Reference 23.

installed at a BL-13XU beamline in SPring-8 (Japan Synchrotron Radiation Research Institute, Hyogo, Japan) [34] for the spin-coated thin films. In the in-plane geometry, the scattering vector (q_{xy}) is parallel to the surface, and the detected profiles contain information on the crystalline structure perpendicular to the film surface. On the other hand, the information on the structure parallel to the surface is obtained from the out-of-plane geometry. Figure 19.5 shows the (a) in-plane and (b) out-of-plane GI-WAXD profiles measured at the surface region of the PFMA-C_8 thin films. In the in-plane and out-of-plane GI-WAXD profiles, peaks were observed at q_{xy} = 12.3 nm^{-1} and q_z = 1–7 nm^{-1}; these two peaks are assignable to the lateral distance between R_f groups and lamellar structures, respectively. After annealing, the peaks became sharp and the intensity increased. The R_f groups and the lamellar structures are oriented

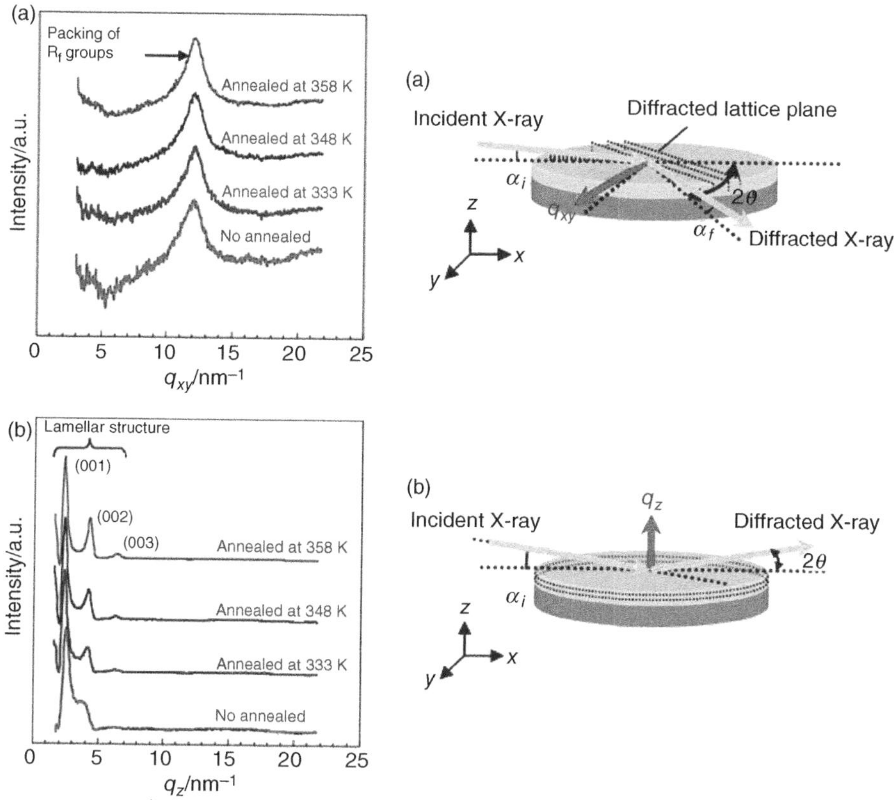

FIGURE 19.5 (a) In-plane and (b) out-of-plane GI-WAXD profiles measured at surface region of PFMA-C_8 thin films. PFMA-C_8 films were annealed at 333, 348, and 358 K for 6 hours. The wavelength of incident X-ray was 0.100 nm. Reprinted with permission from Reference 23.

perpendicular and parallel to the film surface, respectively, and both the orientation and the order of the R_f groups improved in a similar fashion to the PFA-C_y with $y \geq 8$ thin films.

19.3 XPS WITH C_{60} CLUSTER ION BEAM ETCHING FOR SURFACE DEPTH ANALYSIS OF FLUORINATED BLOCK COPOLYMER FILMS

XPS has been recognized as a powerful tool to study surface chemical composition in polymer films [35]. The depth analysis of surface structure is possible by changing take-off angle of photoelectrons within the depth range of approximately 10 nm. However, characteristic depth scales of polymer materials, such as the decay length of surface segregation for polymer blends, and outermost surface layer thickness for block copolymers are often beyond the depth limit of the angular-resolved XPS [22, 36]. Then, the XPS technique is supposed to be combined with a sputtering method. Ar ion beam has been widely used for the purpose. However, the Ar ion beam sputtering method could not be applied to the polymer films since the processing seriously damages organic materials. Recently, it has been reported that C_{60} cluster ion bombardment gently etches organic polymer surfaces. Thus, if the C_{60} cluster ion beam is used as an etching source for XPS, the depth analyses of polymer films is possible without serious sample damage. Winograd et al. have applied the technique to a PMMA thin film, and clearly presented that C_{60} ion beam is much better than the conventional Ar ion beam for the depth analysis of polymers [37]. Similar conclusion was also obtained by Sanada et al. [38]. The C_{60} cluster ion sputtering processing could be a powerful tool for the depth analysis of fluorinated block copolymers [39].

PS-PFAC$_8$ diblock copolymers with the total number-average molecular weight of 21.7k was synthesized by atom transfer radical polymerization [40]. The chemical structures of PS-PFAC$_8$ are shown in Scheme 19.1. The C_{60} ion gun was attached to a commercial XPS instrument equipped with a monochromated Al X-ray anode with 15 kV, 25 W. The photoelectron detection area was defined by the diameter of the X-ray beam, 100 μm, and the take-off angle was 45° for all the measurements employed. The incident angle of ion beam was 70° and 45° from the surface normal for 10 kV C_{60} and 500 V argon, respectively. Sputtering area on the sample was 2 mm × 2 mm. The sample current measured on the gold foil was 8 nA for C_{60} sputter and 200 nA for Ar one. The sputtering rate for 10 kV C_{60} and 500 V argon was 1 nm/min. Figure 19.6a shows C_{1s} XPS spectrum for a PS-PFAC$_8$ diblock copolymer film upon C_{60} ion beam sputtering up to 15 minutes. A C_{1s} spectrum was composed of five peaks corresponding to neutral (CCC), ether (CO), carbonyl (COO), and CF_2 and CF_3 carbons. Although the integral intensity of the whole C_{1s} peak remained unchanged even after 15 minutes, the intensity of peaks being assignable to CF_3, CF_2, and COO gradually altered with increasing time. Hence, it is plausible that C_{60} ion bombardment did not destroy internal structure of the sample. Since the sputtering rate was 1 nm/min under the current condition, the spectral change reflects structural change of the copolymer film along the direction normal to the surface down to about 15 nm. Figure 19.6b shows the time change of C_{1s} spectrum for the diblock copolymer film during the sputtering with 500 V Ar ion beam. The sputtering rate was the same as the one with C_{60} ion beam, 1 nm/min. Nevertheless, the intensity of CF_3, CF_2, and COO peaks suddenly disappeared after a few minutes. Taking into account the surface structure of the PS-PFAC$_8$ block copolymer film, the spectral change with increasing time upon Ar beam sputtering is not reasonable. It seems most likely that Ar-sputtering gives serious damages to the surface, and reliable depth profiling of polymer films may not be attained by XPS with Ar-sputtering, as well accepted in the community.

Figure 19.7 shows the depth profile for the PS-PFAC$_8$ diblock copolymer films by XPS using a C_{60} ion gun. In the panel (a), the area fraction only for CCC, CF_2, and COO peaks to the corresponding C_{1s} spectrum are presented. The striking

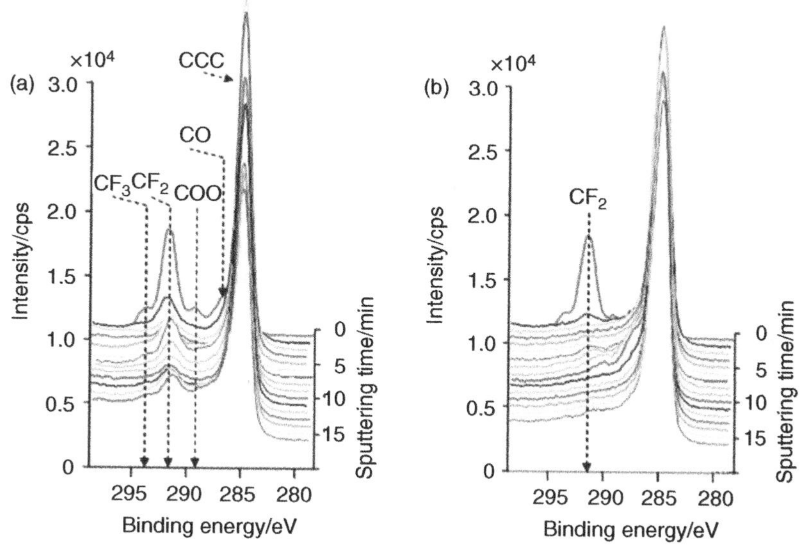

FIGURE 19.6 XPS C_{1s} spectra for PS-PFAC_8 diblock copolymer films during (a) C_{60}-sputtering and (b) Ar-sputtering for the first 15 minutes. Reprinted with permission from Reference 39.

feature of the profile is that the CF_2 and COO fractions at the outermost surface were higher than those in the interior bulk region. This means that the PFA block was preferentially segregated at the surface. The relative concentration of neutral carbons was concurrently lower at the surface than in the bulk. Interestingly, the CF_2 and COO fractions oscillated with increasing depth. Then, the depth dependence of PFA

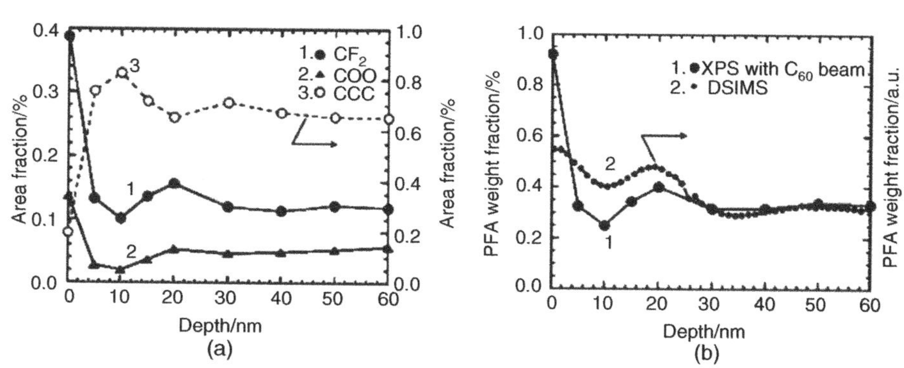

FIGURE 19.7 Depth profiles for a PS-PFAC_8 diblock copolymer film: (a) area fraction of peaks assignable to neutral, CF_2, and COO carbons to the C_{1s} spectrum by XPS with C_{60} ion bombardment and (b) PFA-C_8 weight fraction by XPS with C_{60} ion bombardment and by DSIMS. Reprinted with permission from Reference 39.

weight fraction (φ_{PFA}) at the depth of z was calculated on the basis of the following equation.

$$I_{CF2}(z)I_{CCC}(z) = (7\phi_{PFA}(z)/M_{PFA})/\left\{(2\phi_{PFA}(z)/M_{PFA}) + (8(1-\phi_{PFA}(z))/M_{PS})\right\}$$

where $I_{CF2}(z)$ and $I_{CCC}(z)$ are integrated intensity of peaks assignable to CF_2 and neutral carbons, respectively. M_{PFA} and M_{PS} are molecular weights of each repeating unit for PFA and PS. Numbers of neutral carbons included in each repeating unit of PFA and PS are 2 and 8, respectively. Similarly, seven carbons of CF_2 are included in a repeating unit of PFA. The PFA component was enriched at the surface, and was depleted beneath the segregation layer (Figure 19.7b). It is conceivable that the PS-PFAC$_8$ diblock copolymer forms lamellar structure in the film.

Molecular aggregation structure of the PS-PFAC$_8$ diblock copolymer films along the normal direction to the surface was also examined by dynamic secondary ion mass spectrometry (DSIMS), which is a well-established tool for the depth profiling of polymer materials. The PFA weight fraction at the depth of z by DSIMS was given as follows.

$$\phi_{PFA}(z) = f_{PFA} z_0 I_{F^-}(z) / \int_0^{z_0} I_{F^-}(z)dz$$

where f_{PFA}, z_0, $I_{F^-}(z)$ are overall weight fraction of PFA in the copolymer, film thickness, and secondary ion intensity of F^-, respectively. Since a matrix effect on the DSIMS profile is not included in the above equation, however, ϕ_{PFA} by DSIMS is presented in Figure 19.7b in an arbitrary unit. Acceleration voltage of primary O_2^+ ions at 45° off normal incidence was 4 keV, and current density was approximately 80 nA over the area of 0.16 mm^2. The film surface was covered by a buffer layer of deuterated polystyrene (dPSt) and then was sputter-coated by a thin gold layer with the thickness of 20 nm to avoid charging up during the measurement. The depth dependence of ϕ_{PFA} was not deconvoluted by the instrumental function either, meaning that the real depth profile should be sharper than the presented one [41,42]. The XPS depth profile with C_{60} ion sputtering was in good accordance with the DSIMS result. Thus, it can be claimed that XPS with a C_{60} gun enables one to gain access to the depth analysis even for soft materials such as polymers. The striking difference between XPS with C_{60} ion bombardment and DSIMS could be seen at the outermost surface of $Z = 0$. In the case of DSIMS, the structure even at the outermost surface was damaged by ion beam. On the other hand, that is not the case for XPS with C_{60} ion beam, namely the outermost surface remained intact. If only the outermost surface region and the underneath layer are focused, the XPS with C_{60} ion beam would be superior to the DSIMS due to gentle etching. In addition, the quantitativity for the measurement should be better for XPS with C_{60} ion beam than for DSIMS. However, it should be reminded that XPS signal at a depth attained after the gentle etching by C_{60} ion beam is always integrated over the depth range of three times inelastic mean-free path of photoelectrons. Of course, the angular-resolved XPS is also effective in the depth region down to 10 nm for polymers. For the last case, complicated mathematical treatments, like the inverse Laplace transform, are necessary to extract the real depth profile from the XPS spectra obtained at various take-off-angles of photoelectrons.

19.4 NEUTRON REFLECTIVITY MEASUREMENTS FOR THE SWELLING STRUCTURE INVESTIGATION OF PFA-C_8 BRUSHES BY SUPERCRITICAL CARBON DIOXIDE

PFA-C_8 brush thin films were prepared by SI-ATRP in the presence of Ethyl 2-bromoisobutylate (EB) as a free initiator using Copper(I) bromide / 4,4'-Dinonyl-2,2'bipyridyl (C_9bpy) catalyst, as shown in Scheme 19.2 [15, 17]. Molecular weight and MWD of the PFA-C_8 were determined by SEC of unbound polymers produced in the same reaction batch to be $M_n = 217{,}000$ and $M_w/M_n = 2.20$, respectively. The brush films had the flat and smooth surfaces with root-mean-square roughness of about 0.84 nm in a dry state based on AFM observation.

The swelling behavior of the PFA-C_8 brush thin film with 60 nm thickness in high pressure CO_2 was investigated by *in situ* NR measurements. Figure 19.8a shows the schematic view of the NR set-up. The specular NR measurements were performed at the NG-7 reflectometry, the National Institute of Standards and Technology, Center for Neutron Research (NCNR). The wavelength (λ_N) of neutron beams was 0.47 nm with $\Delta\lambda_N/\lambda_N = 2.5\%$. The details of the high pressure NR experiments and high pressure cell has been described elsewhere [44]. The scattering length density (SLD) of the dry PFA-C_8 brush thin film was calculated to be 3.20×10^{-4} nm^{-2} with the bulk density of 1.60 g/cm^3 so that deuterium labeling was not necessary. The SLD of CO_2, which varies from 0.0004×10^{-4} to 2.0×10^{-4} nm^{-2} in the pressure range of $0.1 < P < 13.8$ MPa at $T = 309$ K, was calculated based on the bulk density obtained by the equation of CO_2 state given by Huang et al. [45]. The *in situ* NR experiments were conducted under the isothermal condition ($T = 309$ K) with elevated pressures up to $P = 13.8$ MPa. The PFA-C_8 brush thin film was exposed to CO_2 for up to 4 hours prior to data acquisition in order to ensure that equilibrium swelling was achieved. The NR curves of the PFA-C_8 in scCO_2 are plotted as a function of pressure in Figure 19.8b. The solid lines in Figure 19.8b correspond to the best-fit curves based on the volume fraction profiles of the polymer chains in Figure 19.8c. As the pressure increases, the intervals of the fringes become smaller, indicating that the brush is expanded at higher pressures. We found that the brush height increased from 60 nm

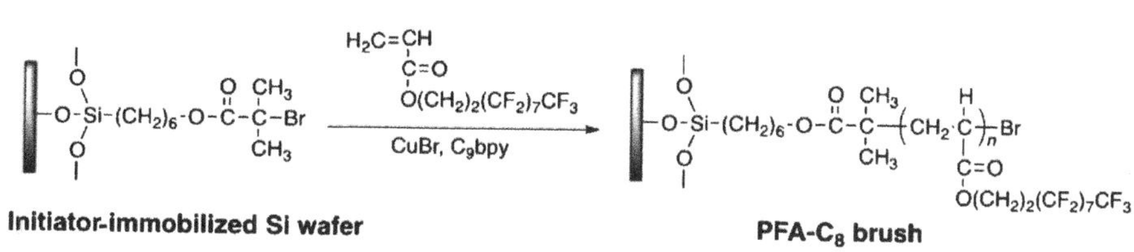

SCHEME 19.2 Preparation for PFA-C_8 brush thin films on the flat silicon wafer by atom transfer radical polymerization.

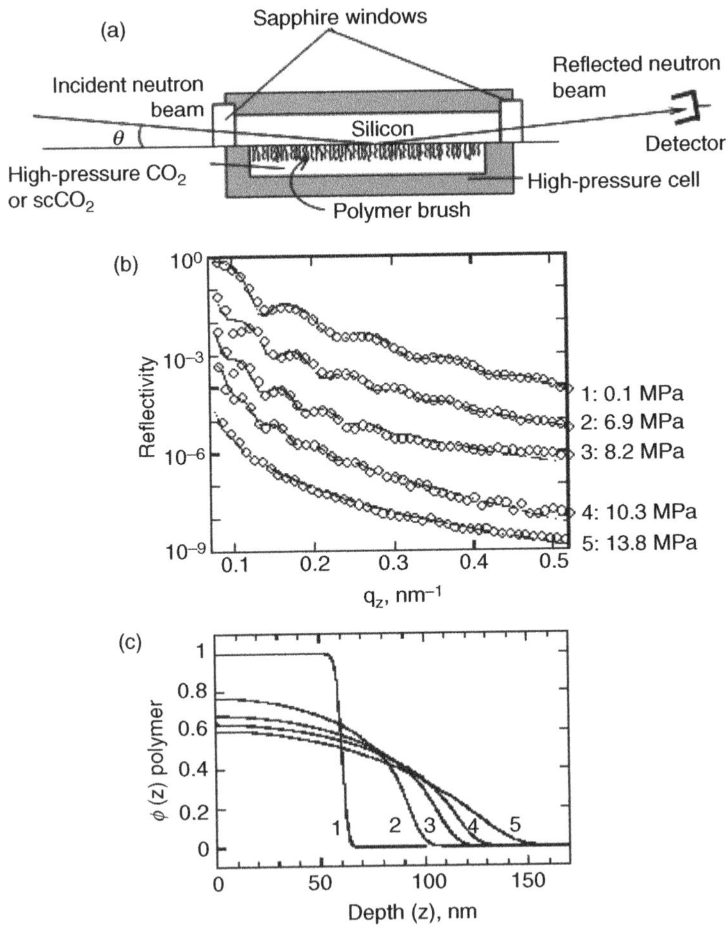

FIGURE 19.8 (a) The schematic geometry of the NR set-up. (b) NR profiles (circles) for the PFA-C_8 brush thin film at different CO_2 pressures. Consecutive reflectivity curves have been offset from each other for clarity. The solid lines correspond to the calculated reflectivity curves based on the volume fraction profiles of the polymer chains in (c). Reprinted with permission from Reference 43.

in the dry state to 120 nm at $P = 8.2$ MPa, that is, the ridge condition, and further reached to 150 nm at $P = 13.8$ MPa. When the pressure was increased to 13.8 MPa, the fringes disappeared because of the parabolic interface structure of the swollen brush, indicating the solvent quality of CO_2 at $P = 13.8$ MPa is good for the PFA-C_8 brush. This is in good agreement with the fact that α is close to 1 at $P = 13.8$ MPa, as will be discussed later.

Figure 19.9a shows the linear extension (S_f) of the brush calculated by the following equation.

$$S_f = (L - L_0)/L_0$$

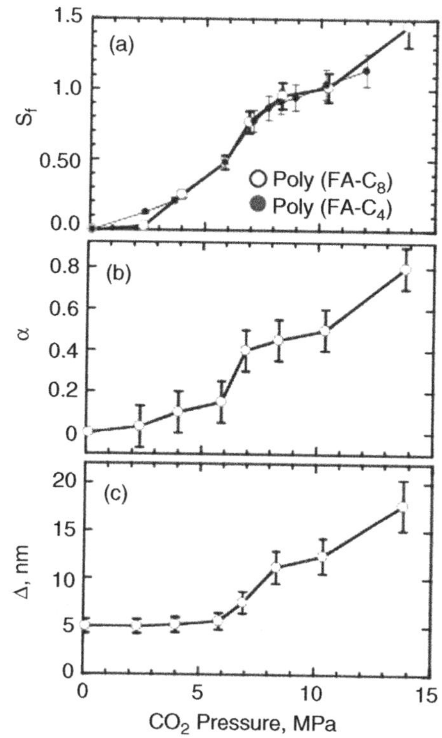

FIGURE 19.9 Pressure dependence of (a) the linear extension (S_f), (b) profile exponent (α), and (c) interfacial width (Δ) of PFA-C_8 brush (open circle) and PFA-C_4 brush (filled circle) thin film at $T = 309$ K in CO_2. Reprinted with permission from Reference 43

where L and L_0 are the measured heights of the swollen and unswollen polymer brush, respectively. The S_f values of the PFA-C_8 brush at 309 K increased with CO_2 pressure up to 13.8 MPa. It is also noted a noticeable excess peak near the "density fluctuation ridge" ($P_{ridge} = 8.2$ MPa at $T = 309$ K) where the excess swelling of various spin-cast (nongrafted) CO_2-phobic polymer thin films as well as CO_2-philic polymer thin films takes place [46]. *In situ* swelling behavior of an amorphous PFA-C_4 (M_w = 304,000, $M_w/M_n = 1.3$, the dry thickness of 58 nm) brush was also determined by NR technique [47]. Interestingly, the swelling behavior of the semicrystalline PFA-C_8 brush is in good agreement with that of the amorphous PFA-C_4 brush, except for the low pressure region ($P < 4.1$ MPa). It is known that the solubility of CO_2 to semicrystalline polymers is poor due to the rigid crystalline structures that prevent CO_2 molecules from penetrating into a polymer matrix. With this in mind, it is reasonable to suppose that the large sorption of CO_2 molecules induces the melting of the semicrystalline PFA-C_8 brush at $P > 4.1$ MPa.

Figure 19.9b shows the CO_2 pressure dependence of α [48]. In this study, α represents the affinity of the polymer against CO_2. In the limit of infinitely long chains, the values, $\alpha = 1$ (a parabolic shape) for a good solvent, $\alpha = 0.5$ for a θ solvent, and $\alpha = 0$ (a step function) for an extremely poor/nonsolvent, are predicted. We found that the α values increase from 0 at atmospheric pressure (i.e., a poor solvent) to ~0.5 at around $P = 8$ MPa, and 0.8 ± 0.1 at $P = 13.8$ MPa, indicating that the solvent quality of CO_2 for the PFA-C_8 brush is improved significantly as pressure increases. This is further confirmed by the variation of the interfacial (tail) width (Δ) (Figure 19.9c). The width between the polymer and CO_2 layers, which was initially 5 nm at atmospheric pressure, increased up to almost 20 nm at $P = 13.8$ MPa.

19.5 SUMMARY

Investigation of surface and interface analyses of fluoro(co)polymers was introduced. GI-WAXD measurement is a powerful tool for the investigation of the surface molecular aggregation structure of the fluoropolymers. Temperature dependence of the dynamic contact angle, XPS measurement in pseudo-hydrated state, and LFM measurement can be applied to the investigation of the surface rearrangement of the fluoropolymers. The surface T_g of the fluoropolymers clearly relate with the dynamic wetting properties. XPS measurement with C_{60} ion beam bombardment is available for the depth analysis of the block copolymer films with fluoropolymer component because of the soft etching characteristics. The Ar-sputtering gives serious damages to the surface, whereas the C_{60} ion sputtering inhibits the sample damage to give reliable depth profiling of polymer films. *In situ* NR measurement affords detail insights of the swelling state of semicrystalline CO_2-philic fluoropolymer brushes in supercritical carbon dioxide. The significant swelling of the fluoropolymer brush with increasing CO_2 pressure was clearly observed, and the large sorption of CO_2 molecules induces the melting of the semicrystalline PFA-C_8 brush.

ACKNOWLEDGMENTS

We acknowledge Prof. T. Koga (Stony Brook University) for helpful discussion and technical support on the NR experiment. We are extremely grateful to Prof. K. Tanaka (Kyushu University) for discussion and his insight in the surface analysis. We are also grateful to M. Hikita, Dr K. Honda, Dr H. Yamaguchi for their assistance with the work presented here. We gratefully acknowledge Dr M. Morita (Daikin Industries Ltd.) for supplying the FA-C_y and FMA-C_y monomers. The synchrotron radiation GI-WAXD measurements were performed at the BL13XU in the SPring-8 with the approval of the Japan Synchrotron Radiation Research Institute (JASRI) (Proposal No. 2008B1498).

REFERENCES

1. Lousenberg, R. D.; Shoishet, M. S. *J. Polym. Sci. A: Polym. Chem.* **2000**, *38*, 1344.
2. Forsythe, J. S.; Hill, D. J. T. *Prog. Polym. Sci.* **2000**, *25*, 101.
3. Zisman, W. A. *Ind. Eng. Chem.* **1963**, *50*, 18.

4. Blanchet, T. A.; Kennedy, F. E. *Wear* **1992**, *153*, 229.
5. Blanchet, T. A. Wear of Polytetrafluoroethylne and PTFE Composite. In *Polymer Tribology*; Briscoe, B. J., Ed.; Imperial College Press: London, UK, 2009; pp 347–374.
6. Pittman, A. G. Surface Properties of Fluorocarbon Polymers. In *Fluoropolymers*; Wall, L. A., Ed.; Wiley-Interscience: New York, 1972; pp 419–449.
7. Schmidt, D. L.; Coburn, C. E.; DeKoven, B. M.; Potter, G. E.; Meyers, G. F.; Fisher, D. A. *Nature* **1994**, *368*, 39.
8. Honda, K.; Morita, M.; Otsuka, H.; Takahara, A. *Macromolecules* **2005**, *38*, 5699.
9. Honda, K.; Yakabe, T.; Koga, T.; Sasaki, S.; Sakata, O.; Otsuka, H.; Takahara, A. *Chem. Lett.* **2005**, *34*, 1024.
10. Honda, K.; Morita, M.; Sasaki, S.; Sakata, O.; Takahara, A. *Trans. Mater. Res. Soc. Jpn.* **2006**, *32*, 239.
11. Honda, K.; Morita, M.; Takahara, A. *Kobunshi Ronbunshu* **2007**, *64*, 181.
12. Honda, K.; Yamaguchi, H.; Kobayashi, M.; Morita, M.; Takahara, A. *J. Phys. Conf. Ser.* **2008**, *100*, 012035.
13. Honda, K.; Morita, M.; Takahara, A. *Soft Matter* **2008**, *4*, 1400.
14. Honda, K.; Morita, M.; Masunaga, H.; Sasaki, S.; Takata, M.; Takahara, A. *Soft Matter* **2010**, *6*, 870.
15. Yamaguchi, H.; Honda, K.; Kobayashi, M.; Morita, M.; Masunaga, H.; Sakata, O.; Sasaki, S.; Takahara, A. *Polym. J.* **2008**, *40*, 854.
16. Yamaguchi, H.; Honda, K.; Kobayashi, M.; Morita, M.; Masunaga, H.; Sakata, O.; Sasaki, S.; Takata, M.; Takahara, A. *J. Phys. Conf. Ser.* **2009**, *184*, 012009.
17. Yamaguchi, H.; Kikuchi, M.; Kobayashi, M.; Ogawa, H.; Masunaga, H.; Sakata, O.; Takahara, A. *Macromolecules* **2012**, *45*, 1509.
18. DeSimone, J. M. *Science* **2002**, *297*, 799.
19. Kendall, J. L.; Canelas, D. A.; Young, J. L.; DeSimone, J. M. *Chem. Rev.* **1999**, *99*, 543.
20. Weibel, G. L.; Ober, C. K. *Microelectron Eng.* **2003**, *65*, 145.
21. Wissinger, R. G.; Paulaitis, M. E. *J. Polym. Sci. Polym. Phys.* **1987**, *25*, 2497.
22. Sánchez, I. C. In *Physics of Polymer Surfaces and Interfaces*; Sanchez, I. C., Ed.; Butterworth Heinemann: Boston, **1992**. CHAPTER 3. Mechanical Properties of Polymer Interfaces.
23. Honda, K.; Morita, M.; Sakata, O.; Sasaki, S.; Takahara, A. *Macromolecules* **2010**, *43*, 454.
24. Takahara, A.; Morotomi, N.; Hiraoka, S.; Higashi, N.; Kunitake, T.; Kajiyama, T. *Macromolecules* **1989**, *22*, 617.
25. Volkov, V. V.; Plate, N. A.; Takahara, A.; Kajiyama, T.; Amaya, N.; Murata, Y. *Polymer* **1992**, *33*, 1316.
26. Bunn, C. W.; Howells, E. R. *Nature* **1954**, *174*, 549.
27. Owens, D. K.; Wendt, R. C. *J. Appl. Polym. Sci.* **1969**, *13*, 1741.
28. Johnson, J. L.; Kendall, K.; Roberts, A. D. *Proc. R. Soc. London A* **1971**, *324*, 301.
29. Kajiyama, T.; Tanaka, T.; Takahara, A. *Macromolecules* **1995**, *28*, 3482.
30. Hammerschmidt, J. A.; Gladfelter, W. L.; Haugstad, G. *Macromolecules* **1999**, *32*, 3260.
31. Tanaka, K.; Takahara, A.; Kajiyama, T. *Macromolecules* **2000**, *33*, 7588.
32. Takahara, A.; Jo, N. J.; Kajiyama, T. *J. Biomater. Sci. Polym. Ed.* **1989**, *1*, 17.

33. Takahara, A.; Takahashi, K.; Kajiyama, T. *J. Biomater. Sci. Polym. Ed.* **1993**, *5*, 183.

34. Sakata, O.; Furukawa, Y.; Goto, S.; Mochizuki, T.; Uruga, T.; Takeshita, K.; Ohashi, H.; Matsushita, T.; Takahashi, S.; Tajiri, H.; Ishikawa, T.; Nakamura, M.; Ito, M.; Sumitani, K.; Takashi, T.; Simura, T.; Saito, A.; Takahashi, M. *Surf. Rev. Lett.* **2003**, *10*, 543.

35. Briggs, D. In *Practical Surface Analysis*; Seah, M. P., Ed.; Wiley: Chichester, **1990**. CHAPTER 2. X-Ray Photoelectron and Auger Electron Spectroscopy.

36. Jones, R. A. L.; Richards, R. W. *Polymers at Surfaces and Interfaces*; Cambridge University Press: Cambridge, 1999.

37. Szakal, C.; Sun, S.; Wucher, A.; Winograd, N. *Appl. Surf. Sci.* **2004**, *231*, 183.

38. Sanada, N.; Yamamoto, A.; Oiwa, R.; Ohashi, Y. *Surf. Interface Anal.* **2004**, *36*, 280.

39. Tanaka, K.; Sanada, N.; Hikita, M.; Nakamura, T.; Kajiyama, T.; Takahara, A. *Appl. Surf. Sci.* **2008**, *254*, 5435.

40. Hikita, M.; Tanaka, K.; Nakamura, T.; Kajiyama, T.; Takahara, A. *Langmuir* **2004**, *20*, 5304.

41. Hall, P. M.; Morabito, J. M.; Panousis, N. T. *Thin Solid Films* **1977**, *41*, 341.

42. Kawaguchi, D.; Tanaka, K.; Takahara, A.; Kajiyama, T. *Macromolecules* **2001**, *34*, 6164.

43. Yamaguchi, H.; Gin, P.; Arita, H.; Kobayashi, M.; Bennett, S.; Satija, S. K.; Asada, M.; Koga, T.; Takahara, A. *RSC Adv.* **2013**, *3*, 4778–4785.

44. Koga, T.; Seo, Y. S.; Shin, K.; Zhang, Y.; Rafailovich, M. H.; Sokolov, J. C.; Chu, B.; Satija, S. K. *Macromolecules* **2003**, *36*, 5236–5243.

45. Huang, F. H.; Li, M. H.; Lee, L. L.; Starling, K. E.; Chung, F. T. H. *J. Chem. Eng. Jpn.* **1985**, *18*, 490–496.

46. Koga, T.; Seo, Y. S.; Zhang, Y.; Shin, K.; Kusano, K.; Nishikawa, K.; Rafailovich, M. H.; Sokolov, J. C.; Chu, B.; Peiffer, D.; Occhiogrosso, R.; Satija, S. K. *Phys. Rev. Lett.* **2002**, *89*, 125506–1.

47. Koga, T.; Gin, P.; Yamaguchi, H.; Endoh, M. K.; Asada, M.; Sendogdular, L.; Kobayashi, M.; Takahara, A.; Akgun, B.; Satija, S. K.; Sumi, T. *Polymer* **2011**, *52*, 4331–4336.

48. Karim, A.; Satija, S. K.; Douglas, J. F.; Ankner, J. F.; Fetters, L. J. *Phys. Rev. Lett.* **1994**, *73*, 3407.

20

FLUOROPOLYMER DIELECTRICS

MUSTAPHA RAIHANE
Faculty of Sciences and Techniques, Marrakech, Morocco
BRUNO AMEDURI
Institute Charles Gerhardt of Montpellier, France

20.1 INTRODUCTION

Fluorinated polymers [1–7] are attractive niche macromolecules because of their remarkable properties. Thus, these materials have found applications in various high technology fields. Two families of fluorinated polymers can be considered. The first one corresponds to those where the fluorinated groups are located in the polymer backbone, which are chemically inert, thermally stable, and have low refractive index and low dielectric constant. The polymers or copolymers having crystalline structures with large dipole moments can exhibit high piezoelectricity when their main chain exhibits an all-trans conformation. Well-known examples are the poly(vinylidene fluoride) (PVDF) [8] especially β-type conformation, the crystalline copolymers of vinylidene fluoride and trifluoroethylene [2,4,9], and the terpolymers of vinylidene fluoride, trifluoroethylene, and chlorotrifluoroethylene [5,6]. Novel ferroelectric polymers based on PVDF and its copolymers with hexafluoropropene proposed for high electric energy storage and low loss performance have been studied by Zhu et al. [10–13] The second family concerns the polymers containing a fluorinated dangling group that enhances the surface properties, it exhibits low dielectric constant. In addition, several fluoroalkylacrylate and methacrylate polymers were developed and their physical properties were studied [14–19]. Dielectric relaxation studies were

Handbook of Fluoropolymer Science and Technology, First Edition. Edited by Dennis W. Smith Jr., Scott T. Iacono, and Suresh S. Iyer.

achieved on a series of poly(fluoroalkylmethacrylate)s. The α relaxations above T_g and γ relaxations below T_g were observed and assigned to reorientation of segments and local molecular motion of fluoroalkyl side groups [20].

Indeed, high dielectric constant polymers are becoming crucial components in advanced electronic devices, such as memory and gate dielectrics for integrated circuits, stationary power generation, and miniature capacitors for telecommunication [21]. Polymers containing cyano groups directly attached to the main chain are of potential interest for these developments owing to the high dipole moments that can be induced along the macromolecular chain. Vinylidene cyanide (or 1,1-dicyanoethylene, VCN) has often been employed in the preparation of high polar polymers as a useful starting monomer [22, 23]. Electric applications of VCN were pioneered by Miyata et al. [24, 25] or from the NASA [26], and these authors observed piezoelectric properties of amorphous poly(VCN-*alt*-VAc) copolymers (where VAc stands for vinyl acetate). Therefore, studies that include the copolymerization reactions of VCN with various monomers have been of growing relevance in the last few years.

Other polymers based on commercially available cyano monomer such as acrylonitrile (AN) have been used. The dielectric behavior of poly(acrylonitrile), PAN, in its pristine and cyclized states has been studied [27, 28]. A soluble organic PAN material was investigated for its feasibility of gate dielectric applications. Functional transistor characteristics were achieved in air for implementation of PAN in organic thin film transistors using poly(3-hexylthiophene) as a semiconductor material [29]. Poly(acrylonitrile-*ter*-methyl methacrylate-*ter*-Indene) terpolymers have been prepared and thoroughly investigated by isochronal dielectric tests showing that their glass–rubber transition values (T_g) are highly dependent on the composition [30]. Cyano monomers such as methacrylonitrile, acrylonitrile, and methyl β,β-dicyanoacrylate copolymerized with vinyl acetate or isopropenyl acetate and vinyl formate [31]. The pyroelectric constants of these copolymers were measured and gave a good approximation for a piezoelectric behavior [32]. Copolymers based on AN with captodative monomers such as α-cyanovinyl acetate [33] and methyl α-acetoxyacrylate [34] were synthesized by radical copolymerization, and their molecular dynamics were studied by dielectric spectroscopy [35, 36].

Advances in portable electronic devices, stationary power systems, and hybrid electric vehicles create demand for low cost, compact, and high performance electrical energy storage devices [37, 38]. The development of dielectric polymers (capacitor systems) with high energy density is governed by the dielectric constant of the materials that separate the opposite static charges. The most promising and exciting avenues for the development of dielectric materials with high energy density are: (i) study of "structure–property" centered on the polymeric chain by increasing its polarity from the incorporation of the monomers with a strong dipole moment (Theory of Fröhlich) [39] and (ii) the introduction of inorganic nanoparticles with high permittivity into polymer matrices to form dielectric polymer nanocomposites [40].

The present chapter outlooks the synthesis and characterization of polymers bearing C–CN and C–F groups. In fact, introduction of additional dipolar and structure-orienting groups, such as cyano comonomers, may result in improved dielectric properties of fluorinated polymers and therefore their electric energy storage.

20.2 MONOMERS

20.2.1 Fluorinated Monomers

Three kinds of fluorinated monomers have been used: (meth)acrylates, styrenes bearing fluorinated atom or groups, and vinylidene fluoride (VDF) (Scheme 20.1).

2,2,2-Trifluoroethyl methacrylate (MATRIF) and 2,2,2-trifluoroethyl acrylate (ATRIF) monomers are commercially available products, but they can be synthesized from methacryloyl or acryloyle chloride and 2,2,2-trifluoroethanol in the presence of triethylamine as a base [41]. The purification of ATRIF and MATRIF monomers was carried out by distillation (at 59°C/100 mmHg and 46°C/125 mmHg, respectively).

4-Fluorostyrene (4FS) monomer is also commercially available product and can be distilled before used (bp. 67°C/50 mmHg).

VDF is a colorless, flammable, and nearly odorless gas that boils at −82°C; its freezing point is −144°C [4]. VDF exhibits several advantages: it is a nontoxic, environmentally friendly gas in contrast to chlorotrifluoroethylene (CTFE), nonexplosive contrarily to tetrafluoroethylene (TFE) or trifluoroethylene (TrFE). Various syntheses to prepare VDF are reported in a review [5]. In fact, the chlorination of chlorofluorocarbon (CFC) 152 (i.e., 1,1-difluoroethane) to 1-chloro-1,1-difluoroethane, followed

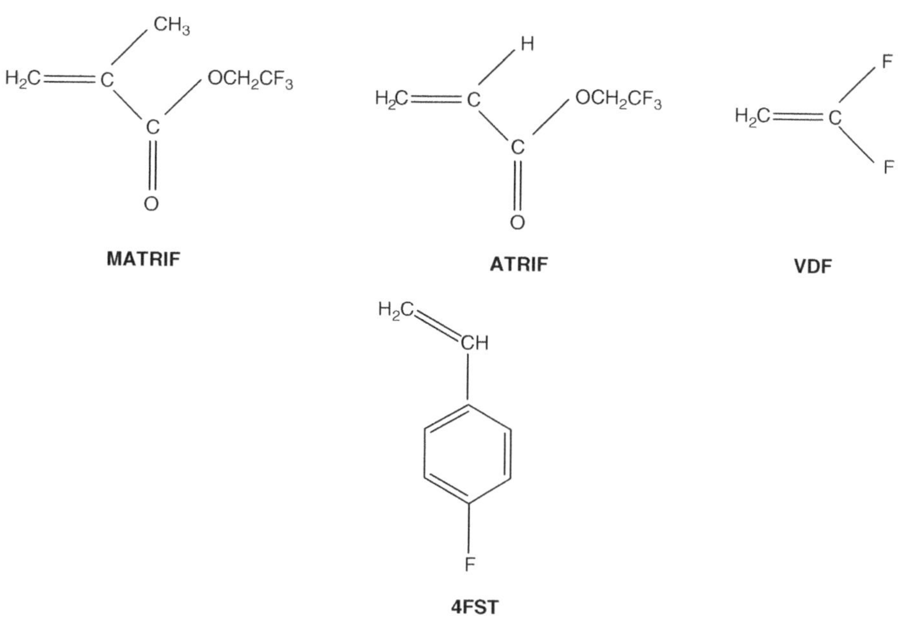

SCHEME 20.1 Structures of fluorinated monomers.

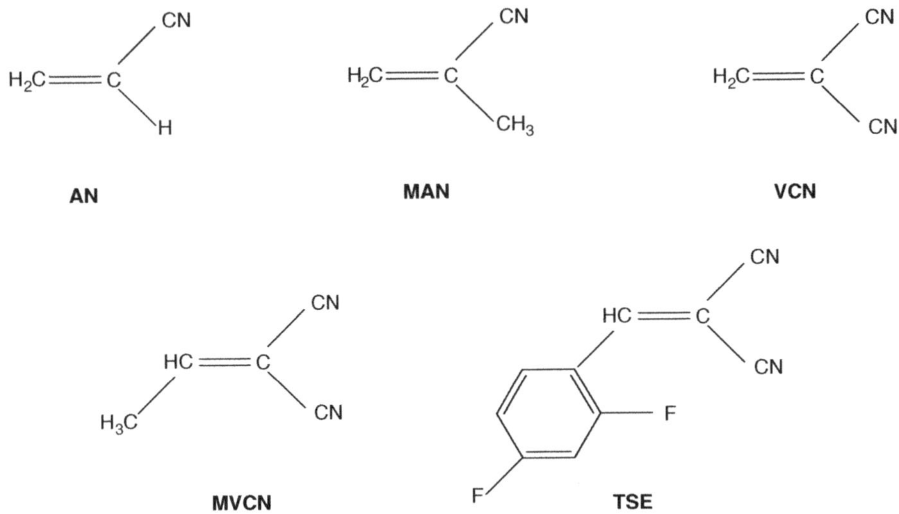

SCHEME 20.2 Structures of cyano monomers.

by a dehydrochlorination at about 700–900°C in the gas phase, is the main industrial process to VDF, developed by Arkema, Kureka, Daikin, DuPont, Solvay, and others.

20.2.2 Cyano Comonomers

Cyano monomers are olefin compounds bearing one or two cyano groups (Scheme 20.2).

Acrylonitrile (AN) and methacrylonitrile (MAN) are commercially available products. They are colorless liquids and they may turn yellow in the presence of impurities. Most industrial AN and MAN are produced according to the Sohio process [42, 43] corresponding to ammoxidation of propylene or isobutylene phase in the presence of a catalyst at 400–510°C.

Vinylidene cyanide (VCN), methyl vinylidene cyanide (MVCN), and 2-(2,4-difluorophenyl)-1,1-dicyanoethylene (TSE) are noncommercially available monomers.

VCN has been synthesized according to a modification of the procedure described by Bell et al. [44], by pyrolytic decomposition of 1,1,3,3-tetracyanopropane (TCP) performed at a temperature ranging between 180°C and 250°C (Scheme 20.3).

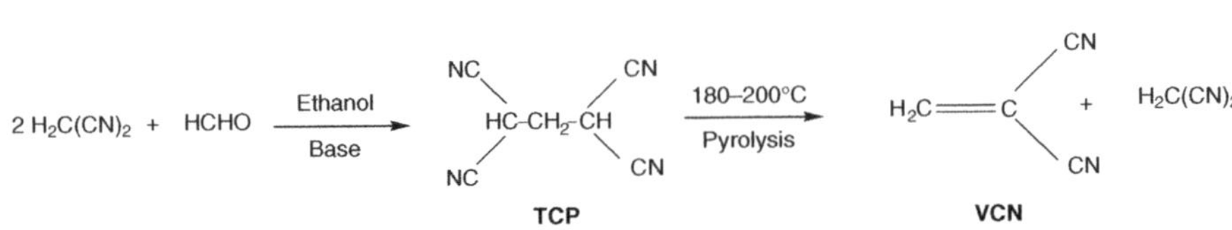

SCHEME 20.3 Synthesis of vinylidene cyanide (VCN) [44].

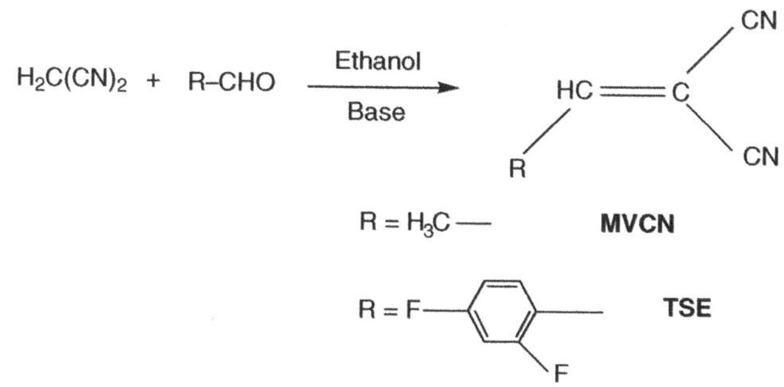

SCHEME 20.4 Synthesis route of methyl vinylidene cyanide (MVCN) and 2-(2,4-difluorophenyl)-1,1-dicyanoethylene (TSE) [45–48].

1,1,3,3-Tetracyanopropane was prepared in the first step from formaldehyde and malononitrile as a compound possessing mobile hydrogen, in the presence of β-alanin as a base. Vinylidene cyanide (VCN) can be stored over P_2O_5. VCN is difficult to prepare and is highly reactive since a bit moisture enables its homopolymerization [23].

However, methyl vinylidene cyanide (MVCN) and 2-(2,4-difluorophenyl)-1,1-dicyanoethylene (TSE) were synthesized in one-step Knoevenagel condensation by Carson and Stoughton [45] (Scheme 20.4) from appropriate aldehyde and malononitrile as a compound possessing mobile hydrogen in the presence of base (β-alanin or piperidine) [47–49]. MVCN and TSE monomers can be purified by distillation (bp. 65°C/1 mmHg) and by recrystallization from 2-propanol (Melting point = 148°C), respectively.

20.3 HOMOPOLYMERIZATION AND PROPERTIES

The highly electrophilic vinylidene cyanide (VCN) readily undergoes homopolymerization by anionic propagation, whereas its homopropagation efficiency in the presence of radical initiators is very poor [49]. Furthermore, the homopolymer slowly undergoes carbon chain scission in the presence of moisture, a process catalyzed by bases [50] yielding a hard, white, nonmeltable, and hydrolytically unstable polymer [51]. VCN homopolymer starts to degrade from about 160°C by a depolymerization mechanism (or unzipping), as suggested by Grassie and McGuchan [52]. On the other hand, several copolymers of VCN have been obtained by free-radical polymerization, and alternating structures with electron-donating comonomers can result in materials with interesting properties (section 20.4.2).

As VCN is difficult to synthesize, other polymers based on commercially available cyano monomer such as acrylonitrile (AN) have also been used.

Poly(acrylonitrile), PAN, is easily produced by free radical polymerization using bulk, solution, emulsion, and by dispersion. Seventy to eighty percent of PAN has been commercially used in producing carbon fibers because of its high softening point [53]. PAN contains a high level of permanent dipole with strong interchain interactions between nitrile groups, with the presence of a stable amorphous phase in 90–150°C and the frequency range of 100 Hz–100 kHz [54]. These properties make PAN a potential candidate as dielectric material.

As expected, under free radical conditions, MVCN and TSE monomers are unable to homopolymerize because of steric difficulties encountered in homopolymerization of 1,1 and 1,2-disubstituted ethylene [55]. When $(C_6H_5)_3P$, NaCN or 1,4-diazabicyclo[2,2,2]octane (DABCO) were used as anionic initiators, the bulk polymerization of MVCN slowly showed an increase of viscosity and turned glassy (mainly trimer) [56]. However, very thin film of poly(methyl vinylidene cyanide), PMVC, was formed by Langmuir–Blodgett (LB) monolayer deposition from a water subphase [57]. X-ray diffraction (XRD) studies in a water vapor ambient and in vacuum showed well-ordered films of PMVC, obtained by LB deposition. The electronic structure of PMVCN with averaged molecular weight (M_w = 2780 g/mol) displayed a strong dipole ordering [58]. Investigation of ferroelectricity in PMVCN has been studied [59].

Poly(vinylidene fluoride), PVDF, is usually manufactured by a radical initiated polymerization process in aqueous emulsion or suspension (involving pressures of 10–300 atm) at temperatures of 10–130°C, regarded as heterogeneous reactions requiring a fluorinated surfactant [60]. Other polymerization procedures of VDF have been reviewed by Ameduri et al. [4, 5] PVDF is typically 50–70% crystalline with five distinct crystal polymorphs named α, β, γ, δ, and ε. These different structures (especially α, β forms) enable PVDF to exhibit dielectric properties. Piezoelectric transducers were developed at the beginning of the twentieth century, notably by Langevin, and immediately found application in detecting the sound waves emanating from submerged submarines. Nowadays, piezoelectric PVDF materials are ubiquitous and used in all sorts of applications, including crystal oscillators, transducers in telephone speakers, headphones, sonar arrays, mechanical actuators, etc.

Fluorinated poly(methacrylates) or poly(acrylates), rich in trifluoromethyl groups, exhibit superior performance of chemical inertness, excellent weatherability, low refractive index, lower dielectric constant, and special surface properties [14, 61]. Poly(2,2,2-trifluoroethyl methacrylate), poly(MATRIF), is an important class of such materials. It has been extensively used in high performance coatings [17], photoelectric communications, and microelectronics [62]. Poly(MATRIF) is easily produced by free radical polymerization using bulk, solution, and emulsion polymerization methods [63]. Structural characterization of ^{13}C NMR of poly(MATRIF) prepared by radical and anionic polymerization has been studied. Syndiotactic structure was obtained by radical initiator in contrast to an isotactic structure achieved by anionic polymerization [64].

In contrast to MATRIF monomer, 2,2,2-Trifluoroethyl acrylate (ATRIF) and its (co)polymers have not been extensively studied. Free-radical homopolymerization of ATRIF and other fluorinated acrylates was carried out under various conditions and the stereostructures of the obtained polymers were investigated [65]. ATRIF monomer seemed promising to form thin films that have high hydrophobicity and good barrier properties [66]. In 1985, Narita et al. [67] reported the radical copolymerization of ATRIF with styrene, and the assessed reactivity ratios. To study the incorporation of cyano monomers into the poly(ATRIF) and poly(MATRIF) units, the radical homopolymerization of MATRIF and ATRIF was performed as a reference using AIBN as a free radical initiator in dioxan [64] and acetonitrile [68] as the solvents, respectively.

20.4 RADICAL COPOLYMERIZATION

20.4.1 Random Copolymers by Conventional Radical Copolymerization

Random copolymers based on acrylonitrile and methacrylonitrile with ethyl α-acetoxyacrylate [69] were synthesized by radical copolymerization. The microstructure of the resulting copolymers was characterized by ^{13}C NMR spectroscopy, and some reactivity ratios could be assessed. In addition, copolymers based on methyl vinylidene cyanide (MVCN) and vinyl acetate [70], styrene [71], or substituted styrene also exhibit interesting dielectric behavior [72].

20.4.1.1 Poly(M_C-co-MATRIF) Copolymers (M_C Stands for AN, MAN, or MVCN) Poly(M_C-*co*-MATRIF) copolymers were synthesized from equimolar amounts of cyano monomers and MATRIF initiated by AIBN (Scheme 20.5) [73]. The yields of the copolymerizations were ranging between 40% and 60%.

These copolymers were characterized by ^{1}H, ^{13}C, ^{19}F NMR, and IR spectroscopy, while their molecular weights (M_n) were ranging between 13000 and 24000 g/mol. The compositions of the copolymers were calculated by measuring integrals of some characteristic signals in the ^{1}H NMR spectra and were confirmed by elemental analyses. Table 20.1 shows the molar percentages of incorporated cyano monomer calculated from elemental analyses and assessed by ^{1}H NMR in the copolymers of MATRIF. The copolymer compositions of acrylonitrile with nonfluorinated acrylic monomers such as ethyl acrylate [74] and methyl methacrylate [75] were calculated using elemental analyses. The incorporations of AN were close to 41% and 53%, respectively, and showed that the lateral CF_3 group slightly affects the reactivity of AN.

The same conclusion was proposed for the copolymer based on MAN. Table 20.1 also displays the molar percentage of MVCN in the copolymer that is very low (5%).

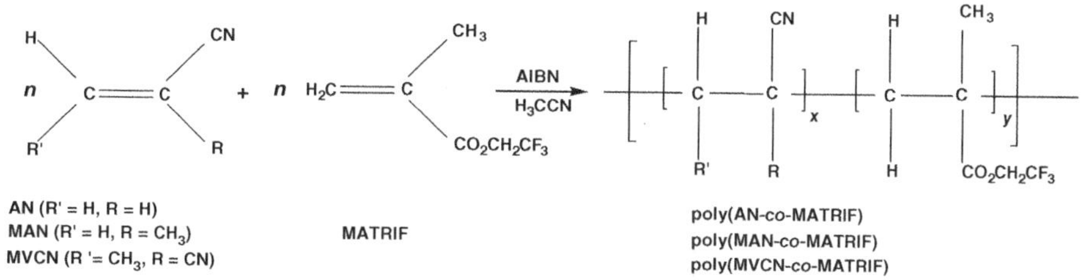

SCHEME 20.5 Synthetic routes of copolymers based on 2,2,2-trifluoroethyl methacrylate (MATRIF) [73].

TABLE 20.1 Percentages of MATRIF Incorporated in Various Poly(M_C-*co*-MATRIF) Copolymers Calculated from Elemental Analyses or ^{1}H NMR Spectroscopy [73]

	Elemental analyses				*^{1}H NMR*
	% of atom			%molar[a]	%molar
Copolymers	%N	%O	%F	(cyano monomer)	(cyano monomer)
Poly(AN-*co*-MATRIF)	5.1	15.4	27.5	43.3	44.0
Poly(MAN-*co*-MATRIF)	5.4	46.8	46.9	46.8	45.0
Poly(MVCN-*co*-MATRIF)	0.9	18.5	33	5.1	4.6

[a]Values of the percentages of incorporated cyano monomer calculated from arithmetic mean of nitrogen, oxygen, and fluorine percentages.

MVCN monomer is less reactive toward the growing macroradical terminated by a MATRIF unit.

The results of the Alfrey–Price Q–e parameters [76] (where Q and e take into account the stabilization by resonance and the polar effects of the monomer) are commonly used to predict the monomer reactivity ratios. The Alfrey–Price parameters are known for AN, MAN, MVCN, and MATRIF [68]. The calculated values of the reactivity ratios, r_1 and r_2, and the product ($r_1 \times r_2$) are summarized in Table 20.2.

As evidenced by the values of the ($r_1 \times r_2$) product for poly(AN-*co*-MATRIF) and poly(MAN-*co*-MATRIF) copolymers close to 1, statistical structures have been obtained. Furthermore, the reactivity ratio of MVCN close to zero means that MVCN does not homopolymerize. The reactivity ratio of MATRIF is higher ($r_2 = 32.5$) indicating that the probability of homopolymerization of MATRIF is quite high. This result explains the low percentage of incorporation of MVCN in the poly(MVCN-*co*-MATRIF) copolymer (5%) calculated from elemental analyses and NMR spectroscopy.

Differential scanning calorimetry (DSC) analyses showed that the copolymers were amorphous, T_g ranging between 77°C and 86°C (Table 20.3). The thermal degradation of poly(AN-*co*-MATRIF), poly(MAN-*co*-MATRIF), and poly(MVCN-*co*-MATRIF) copolymers was studied by means of thermogravimetry. Table 20.3

TABLE 20.2 Reactivity Ratios of Cyano Monomers (r_1) and MATRIF (r_2) and the Values of the Product ($r_1 \times r_2$) Calculated from the Alfrey–Price Parameters ($T = 80°$C) [73]

Copolymers	r_1	r_2	$r_1 \times r_2$
Poly(AN-*co*-MATRIF)	0.31	3.04	0.93
Poly(MAN-*co*-MATRIF)	0.87	1.10	0.96
Poly(MVCN-*co*-MATRIF)	0	32.52	0

r_i represents the ratio of the homopropagation rate to the crosspropagation rate.

TABLE 20.3 Glass Transition and Decomposition Temperatures of Poly(M-*co*-MATRIF) Copolymers [73]

Copolymer	T_g (°C)	T_d (°C)[a]	$T_{d'}$ (°C)[a]
Poly(AN-*co*-MATRIF)	77	240	345
Poly(MAN-*co*-MATRIF)	86	230	264
Poly(MVCN-*co*-MATRIF)	69	200	221

[a]Copolymers heated under nitrogen at 10°C/min; T_d, temperature of beginning of degradation; $T_{d'}$, temperature of degradation at 10% of loss weight.

lists their temperatures when the copolymers start to decompose and their degradation temperature corresponding to 10% of weight loss. Thermogravimetric analyses under inert atmosphere indicated that thermal decomposition occurred from 200°C, 230°C, and 240°C for copolymers containing MVCN, MAN, and AN, respectively (Figure 20.1).

20.4.1.2 Poly(AN-co-ATRIF) Copolymers Radical copolymerization based on acrylonitrile (AN) and 2,2,2-trifluoroethyl acrylate (ATRIF) initiated by AIBN was investigated in acetonitrile [77]. The resulting poly(AN-*co*-ATRIF) copolymers were characterized by ^{1}H, ^{13}C and ^{19}F NMR and IR spectroscopy, and size exclusion chromatography (SEC). The ^{1}H NMR spectra enabled to determine the composition of the resulting poly(AN-*co*-ATRIF) copolymers. The copolymer compositions were assessed by using integrals of the signals of ester methylene protons (CH_2–CF_3) assigned to ATRIF in the 4.60–4.90 ppm range, and also the methylene protons signal of the backbone (CH_2) of both units in the 1.70–2.45 ppm. The kinetics of radical copolymerization of AN with ATRIF was investigated from several experiments achieved at 70°C from initial $[AN]_0/[ATRIF]_0$ molar ratios ranging between 20/80 and 80/20 and enabled to determine the reactivity ratios of both comonomers (Table 20.4).

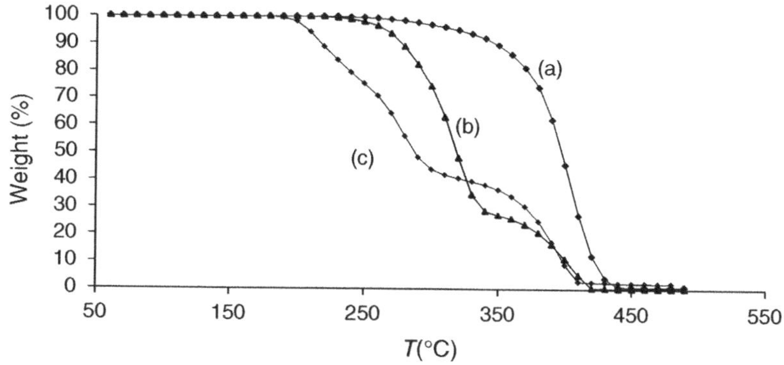

FIGURE 20.1 TGA thermograms under nitrogen of: (a) poly(AN-*co*-MATRIF), (b) poly(MAN-*co*-MATRIF), and (c) poly(MVCN-*co*-MATRIF) copolymers [73].

TABLE 20.4 Results of the Radical Copolymerization of Acrylonitrile (AN) and 2,2,2-Trifluoroethyl acrylate (ATRIF) Initiated by AIBN (1 wt%) at 70°C for 15 min [77]

		AN composition (mol%)		SEC[b]		DSC[c]	TGA[d]
Exp.	Yield (wt%)	Feed (f_{AN})	Copolymer[a] (F_{AN})	M_n (g/mol)	PDI	T_g (°C)	T_d (°C)
1	12	20	23	195,000	2.05	17.0	358
2	10	40	47	84,000	2.55	25.5	362
3	12	50	56	64,000	2.25	33.0	340
4	11	60	62	98,000	1.83	36.0	368
5	12	80	84	n.d.[e]	n.d.[e]	61.0	350

[a]Calculated from ^{1}H NMR spectroscopy.
[b]SEC in THF using polystyrene standards.
[c]T_g midpoint under nitrogen, 10°C/min.
[d]T_d (onset) under nitrogen, 10°C/min.
[e]Not determined.

Three different laws were used to assess the reactivity ratios r_1 of AN (1) and r_2 of ATRIF (2): Fineman and Ross' method [78], Kelen and Tüdos law [79], and the revised patterns scheme [80]. From the monomer–polymer copolymerization curve, the Fineman–Ross and Kelen–Tüdos laws (Figure 20.2) enabled to assess the reactivity ratios ($r_{AN} = r_1 = 1.25 \pm 0.04$ and $r_{ATRIF} = r_2 = 0.93 \pm 0.05$ at 70°C) while the revised patterns scheme led to $r_{12} = r_{AN} = 1.03$, and $r_{21} = r_{ATRIF} = 0.78$ at 70°C. In all cases, $r_{AN} \times r_{ATRIF}$ product was close to unity, which indicates that poly(AN-*co*-ATRIF) copolymers exhibit a random structure.

The Igarashi's theory [81] clearly indicated that the copolymers obtained from the radical copolymerization of AN and ATRIF exhibit a statistic structure as the content of heterodyads (M_{ATRIF}–M_{AN}) in the copolymers varied from 25 to 37 mol% when the molar percentage of ATRIF in the copolymer decreased from 77% to 38% (Figure 20.3).

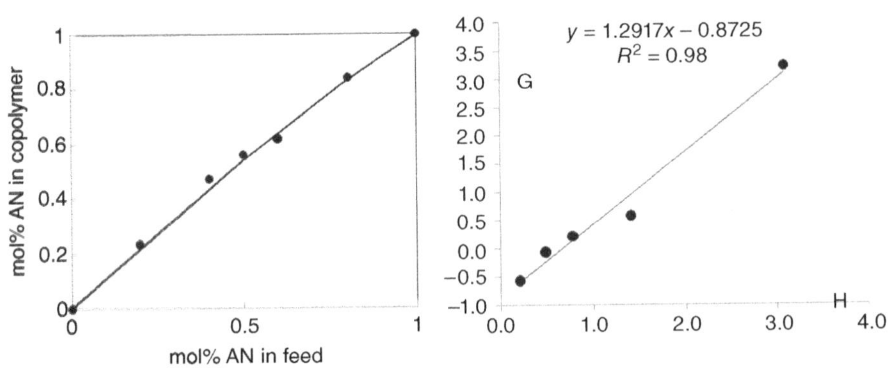

FIGURE 20.2 Fineman–Ross graphic method G versus H (right) and AN monomer–polymer composition curve, calculated from ^{1}H NMR spectroscopy (left) [77].

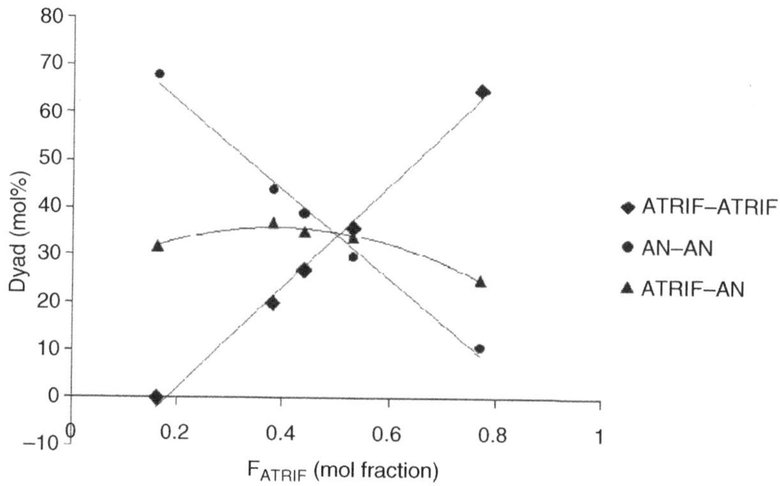

FIGURE 20.3 Dyad monomer sequence fraction versus the ATRIF mol fraction in the poly(AN-*co*-ATRIF) copolymers [77].

The thermal properties of poly(AN-*co*-ATRIF) copolymers were investigated by DSC and by thermogravimetry. The glass transition temperature, T_g, increased from 17°C to 61°C as the mole percentage of ATRIF decreased from 77% to 16% in the copolymer (the molar% F also decreased from 34% to 13% in the copolymer) (Table 20.4).

This may arise from CF_3 group of ATRIF, which could have a plasticizing effect in the copolymers. On the other hand, the incorporation of ATRIF in AN units reduces the T_g and hence should improve the film forming property. TGA thermograms, recorded under nitrogen (Figure 20.4), feature the evolution of the thermostability of poly(ATRIF) homopolymer and all poly(AN-*co*-ATRIF) copolymers with the molar composition (from 20% to 80%). Their onset decomposition temperatures are listed in Table 20.4 [77].

The poly(AN-*co*-ATRIF) copolymers were degraded at higher temperatures than the poly(ATRIF) homopolymer. On the other hand, the thermal stabilities were improved by the incorporation of AN comonomer. In addition, the TGA thermograms exhibit two steps of thermal degradation of these copolymers.

20.4.2 Alternating Copolymers by Conventional Radical Copolymerization

VCN copolymerizes with various comonomers such as methyl methacrylate [82], isopropenyl acetate [83], vinyl benzoate [82], styrene [84], substituted styrenes [71], and vinyl esters of fatty acids [85]. All these copolymerizations led to alternating copolymers, and their microstructures were characterized by ^{13}C NMR spectroscopy.

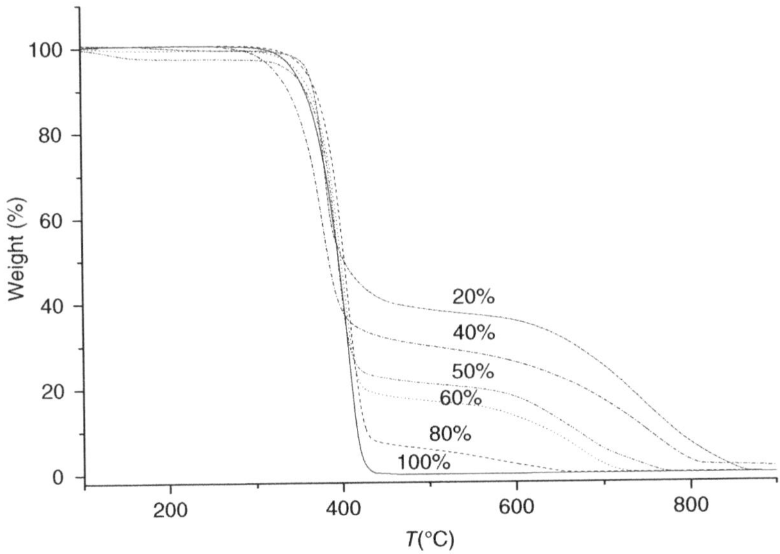

FIGURE 20.4 TGA thermograms of poly(AN-*co*-ATRIF) copolymers under N_2 at 10°C/minute (*x*% represents the percentage of ATRIF in the copolymer) [77].

20.4.2.1 Poly(VCN-alt-MATRIF) Copolymer A novel copolymer of vinylidene cyanide (VCN) and 2,2,2-trifluoroethyl methacrylate (MATRIF) was synthesized by a free radical process in bulk in a 52% yield from an equimolar comonomer feed (Scheme 20.6) [86]. The copolymer's composition and microstructure were analyzed by FT-IR, ^{1}H and ^{13}C NMR spectroscopy, SEC, and elemental analysis. Elemental analysis led to C = 46.62%, H = 3.29%, N = 9.00% (%F and %O complement to 100%) [86, 87].

The incorporation of VCN in the copolymer was 42 mol% (25 wt%) according to the nitrogen content (9.0%) determined by elemental analysis. However, it was slightly higher (39 mol%) when calculated from the ^{1}H NMR spectrum (Figure 20.5).

This later assessment was deduced from the integrals of the resonances from the side methylene chain of MATRIF in the 4.5–4.8 ppm range and the main chain methylene resonances of both co-units in the 1.8–2.8 ppm range.

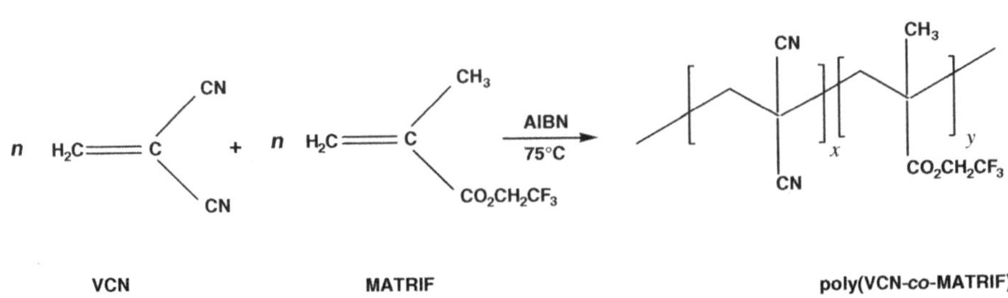

SCHEME 20.6 Synthesis of poly(VCN-*co*-MATRIF) copolymer [86, 87].

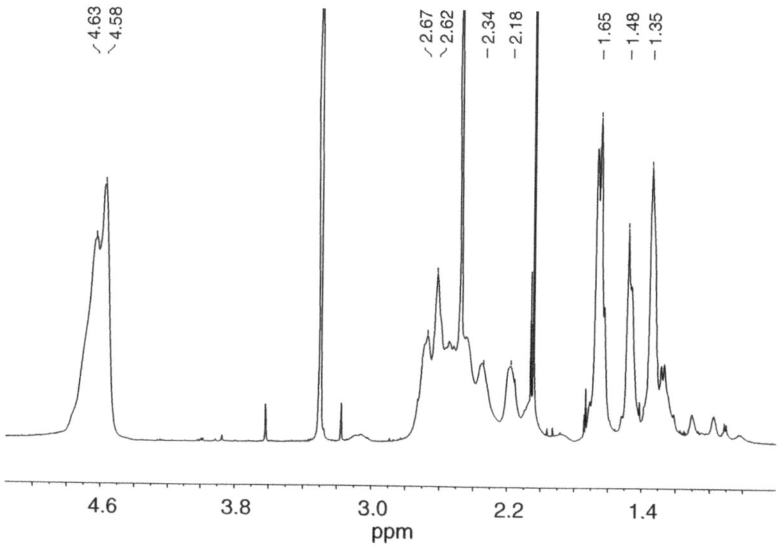

FIGURE 20.5 ^{1}H NMR (600 MHz) spectrum of poly(VCN-*alt*-MATRIF) copolymer recorded in DMSO-d$_6$ [86].

SEC analysis gave $\overline{M}_n$ = 136 kg/mol (in polystyrene standards) and $\overline{M}_w/\overline{M}_n$ = 1.52, suggesting a free radical polymerization mechanism with negligible contribution from chain transfer to the polymer [55].

Monomer composition and comonomer sequence distribution were determined by high resolution ^{1}H and ^{13}C NMR spectroscopy. In the ^{1}H NMR spectrum, MATRIF α-CH_3 resonance pattern is observed in 0.82–1.67 ppm (Figure 20.6), three signals were observed and can be associated to the (M) centered VMV, MMV, and MMM triads (where V and M stand for VCN and MATRIF, respectively) [86]. Integration of the three sets of resonances yields the following ratio VMV/MMV/MMM = 37/55/8.

Deeper NMR characterization showed that three well-resolved carbonyl resonances in the ^{13}C NMR spectrum were assigned to the MATRIF-centered triads VMV, VMM, and MMM, respectively. The presence of VCN dyads (e.g., in VVM and VVV sequences) was shown to be marginal or even absent. Hence, the microstructure of this poly(VCN-*co*-MATRIF) copolymer did not significantly differ from that of other copolymers of VCN, with a predominantly alternating and stereoirregular structure in which the presence of VCN homosequences is rare. The low reactivity ratios calculated for the radical polymerization of both comonomers according to the revised pattern scheme of Jenkins model [80] led to $r_{VCN} = 0.1$ and $r_{MATRIF} = 0.3$ at 75°C and their product $r_{VCN} \times r_{MATRIF} = 3 \times 10^{-2}$, suggesting an alternating tendency of that radical copolymerization.

Incorporation of VCN induced a sharp increase in the glass transition temperature of the amorphous poly(VCN-*alt*-MATRIF) copolymer (T_g = 100°C) with respect to that of the poly(MATRIF) parent homopolymer (T_g = 68°C). On the other hand, the presence of main chain carbons substituted with two nitrile groups induces a sharp

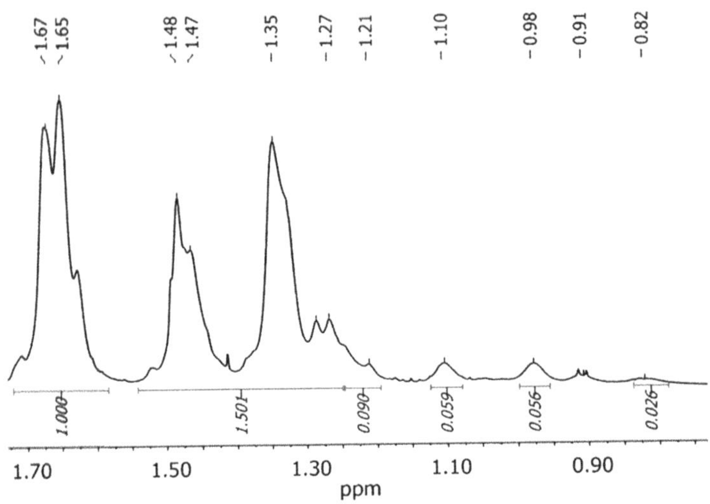

FIGURE 20.6 Expanded α-CH_3 region of the ^1H NMR 600 MHz spectrum of poly(VCN-*alt*-MATRIF) copolymer [86].

increase of T_g, as a result of chain stiffening caused by hindered rotation about main chain C–C bonds and by strong intermolecular dipolar interactions.

The TGA thermogram of poly(VCN-*alt*-MATRIF) copolymer indicates that both the main thermal degradation process, with onset at 368°C, and a minor weight loss at around 222°C are mainly associated with random main-chain scission (Figure 20.7) [87]. The thermal stability of this copolymer is thus greatly improved with respect to both parent homopolymers. In agreement with the NMR observations, the high thermal stability of the copolymer is also indicative of a low fraction of VCN–VCN

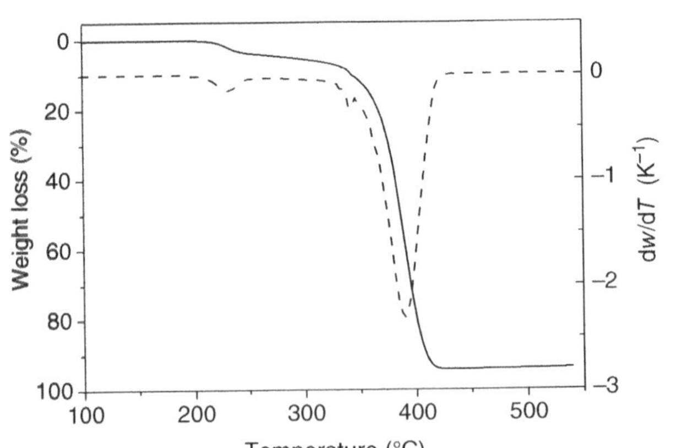

FIGURE 20.7 Weight loss curve (solid line) and its derivative (dashed line) from the TG analysis of poly(VCN-*alt*-MATRIF) run at 10°C/min under nitrogen purge [87].

dyad, which would promote the onset of decomposition at much lower temperatures around 160°C [52].

The thermal decomposition under nonoxidative conditions of a poly(VCN-*alt*-MATRIF) copolymer was investigated by thermogravimetry (TG) and Pyrolysis-GC-MS [87]. The main kinetic parameters evaluated by isothermal (only for the main degradation step) and nonisothermal methods (Friedman, Flynne Walle Ozawa, and Kissinger, respectively) indicate a simple first order decomposition mechanism with a dominant random scission depolymerization reaction. The high activation energies (between 231 and 295 kJ/mol) assessed according to the different procedures for the main degradation process are also consistent with a depolymerization mechanism involving the scission of C–C bonds in main chain.

20.4.2.2 Poly(TSE-co-4FS) Copolymers Copolymerization of fluorine ring substituted 2-phenyl-1,1-dicyanoethene (TSE) with 4-fluorostyrene (4FST) was prepared from an equimolar monomer feed composition in the presence of a 1,1′-Azobis(cyclohexanecarbonitrile) (ABCN) as the radical initiator [88]. The yield was close to 59% (Scheme 20.7). Homopolymerization of 4FST was conducted under identical conditions as those in the copolymerization (Yield = 76%).

The structure of this copolymer was characterized by IR, ^{1}H and ^{13}C NMR, GPC, DSC, and TGA. The composition of the poly(TSE-*co*-4FST) copolymer was calculated from nitrogen analysis (%N = 7.22%), the molar percentage of cyano monomer (TSE) was closed to 38%. The molecular weight calculated from SEC using polystyrenes as standards was around 8000 g/mol.

High T_g of poly(TSE-*co*-4FST) copolymer (T_g = 156°C), in comparison with that of 4-fluorostyrene homopolymer (P4FST) (T_g = 104°C) indicates a substantial decrease in chain mobility of the copolymer due to the high dipolar character of the trisubstituted ethylene monomer units.

The thermogravimetric analysis showed that the thermal decomposition of poly(TSE-*co*-4FST) copolymer occurred from 260°C (T_d *onset*) (Figure 20.8) [88]. The copolymer decomposed in two stages with a rapid decomposition of 93% of the

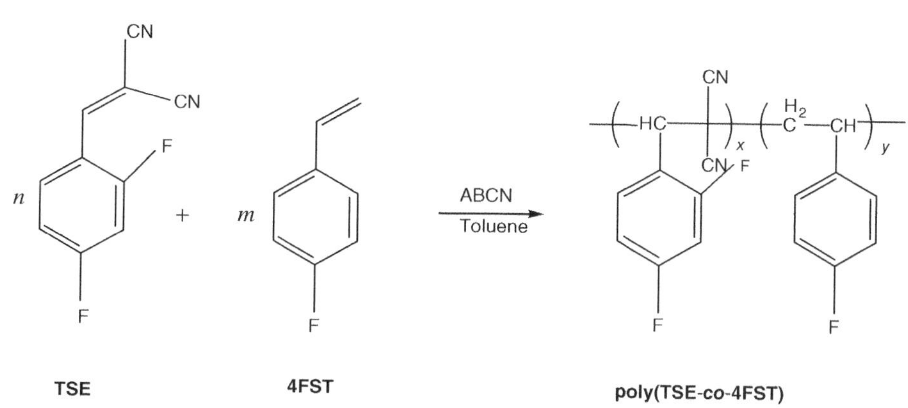

SCHEME 20.7 Synthesis route of poly(TSE-*co*-4FST) copolymer [88].

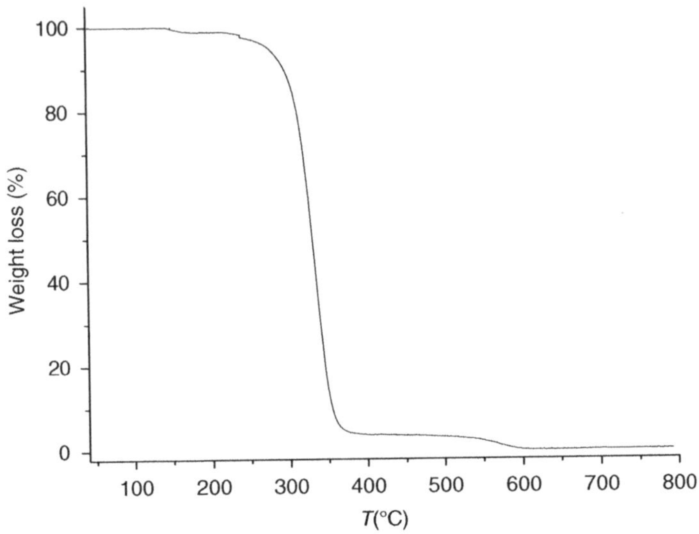

FIGURE 20.8 TGA thermogram of poly(TSE-*co*-4FS) copolymer under nitrogen [88].

sample in the 275–390°C range followed by a slower second stage decomposition in 480–650°C range. According to the % of residual copolymer, the depolymerization reaction could be suggested as the mechanism of its thermal decomposition.

The kinetics of radical copolymerization of 4FST (M_1) monomer with TSE (M_2) comonomer was investigated from five experiments achieved at 70°C from initial $[M_1]_0/[M_2]_0$ molar ratios ranging between 30/70 and 70/30 to achieve a conversion less than 20%. The copolymers were characterized by elemental analysis which enabled to assess the mol% of both comonomers in the copolymers (Table 20.5).

The relative reactivity of M_1 and M_2 can be estimated by Fineman–Ross law (G *versus* H). In the absence of the self-propagation of TSE monomer ($k_{22} = 0$, $r_2 = 0$), Figure 20.9a shows that reactivity of 4FST radical toward TSE (r_1) is around 0.6. The product $r_1 \times r_2$ is close to zero attesting a tendency to alternating structure of poly(TSE-*co*-4FS) copolymer. The experimental curve (i.e., the molar ratio of 4FST (M_1) units in the copolymer (F_1) *versus* the molar ratio of (M_1) units in the feed (f_1)

TABLE 20.5 Radical Copolymerization of 4FST (M_1) and TSE (M_2) [88]

f_1 of M_1 in the feed (mol%)	Yield (%)	Nitrogen in copolymer (wt%)	F_1 of M_1 in copolymer (mol%)
30	18	7.56	60
40	14	7.08	63
50	17	6.81	64
60	15	6.39	67
70	13	5.45	70

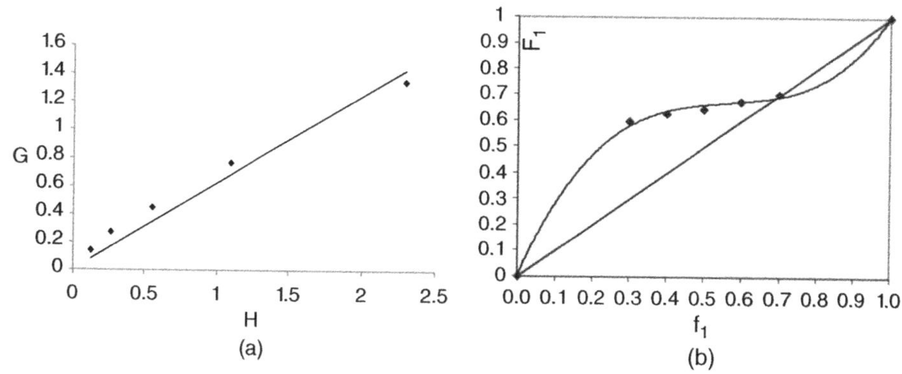

FIGURE 20.9 (a) Fineman–Ross graphic method G versus H for the assessment of the reactivity ratios of 4FST and TSE (M_2) at 70°C and (b) composition curve F_1 (Molar percentage of M_1 in the copolymer) versus f_1 (Molar percentage of M_1 in the feed) [88].

is also shown in Figure 20.9b. Because $r_1 < 1$ and $r_2 = 0 < 1$ for this copolymer, the azeotropic composition was found to be 71 mol% of 4FST. As the molar% of 4FST is more than 50% (around 65%, Figure 20.9b), the structure of the copolymer has an alternating structure trend with homosequence of 4FST.

20.4.3 PVDF-*b*-PM$_{CN}$ Block Copolymers Prepared by Controlled Radical Polymerization (M$_{CN}$: AN, MAN, and VCN)

Controlled radical polymerization (CRP) [89, 90] of VDF has been scarcely investigated and successful attempts concern iodine transfer polymerization (ITP) [91–94]. Indeed, ITP of VDF was pioneered in the late 1970s at the Daikin Company [95] and currently led to commercially available thermoplastic elastomers.

To develop dielectric polymers with C–CN and C–F groups, the syntheses of poly(vinylidene fluoride)-*b*-poly(AN, MAN, VCN) (PVDF-*b*-PM$_{CN}$) block copolymers (Scheme 20.8), using the iodine transfer polymerization (ITP) of acrylonitrile (AN), methacrylonitrile (MAN), and vinylidene cyanide (VCN), in the presence of PVDF-I, was reported. In a first step, the ITP of vinylidene fluoride (or 1,1-difluoroethylene, VDF) with $C_6F_{13}I$ initiated by *tert*-butyl peroxypivalate (TBPPI)

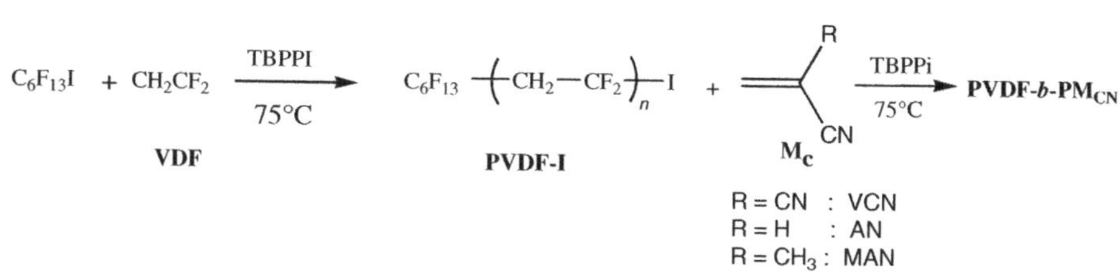

SCHEME 20.8 Synthesis of PVDF-*b*-PM$_{CN}$ block copolymers by ITP of cyano monomers M$_{CN}$ [89].

was achieved [91,96] (first step of Scheme 20.8). In the further step, those PVDF-I blocks were subsequently used as original macromolecular chain transfer agents to achieve the ITP with cyano monomers (M_{CN}) (second step of Scheme 20.8) for the preparation of PVDF-*b*-PM_{CN} block copolymers [89].

Perfluoroalkyl iodides are known to exhibit a weak CF_2–I bond dissociation energy (45 kJ/mol) [97] that enables the C–I bond to be easily broken in various ways: thermally, photochemically, biochemically, electrochemically, and with various initiators or catalysts. That is the reason why C_6F_{13}–I has been chosen as a chain transfer agent (CTA) for this ITP. The full details of the procedure of this ITP are supplied in references [91,96,98]. The synthesis of PVDF-*b*-PM_{CN} block copolymers was carried out at 75°C, initiated by TBPPI. The ITP process was been reviewed and described in references [91,98].

The VDF blocks were characterized by ^{1}H and ^{19}F NMR spectroscopies and show the presence of both end-groups,–CH_2CF_2–I and –CH_2CF_2–I, hence confirming a previous study [96–98]. As –CH_2I bond is less reactive than –CF_2I, PVDF blocks bearing –CH_2I end-group were inactive, while the blocks that contained a –CF_2I extremity were able to be reactivated. The final PVDF-*b*-PM_{CN} products could be purified by washing with acetone, which enabled to solubilize the PVDF blocks and to get off the nonreactive PVDF blocks.

As expected, as the refractive indexes of fluorinated polymers are very low, SEC analysis displayed negative signal assigned to PVDF block response, whereas, the final PVDF-*b*-PM_{CN} block copolymers were identified by positive signals. The values of M_n (Table 20.6) show that the reactivity of cyano monomers toward the VDF macroradicals can be classified in the following decreasing order: AN > MAN > VCN.

The TGA shows that these macromolecules are stable up to at least 270°C except the block copolymer based on VCN. The nonthermal stability of this copolymer arises from the fact that PVCN is known for its low degradation temperature [52].

Figure 20.10 displays the DSC thermograms. It is observed a clear change of baseline for all analyzed copolymers. This is attributed to T_g, ranging between 62°C and 86°C (Table 20.6). However, the T_g of PVDF block is not evidenced by DSC

TABLE 20.6 Molecular Weights and Thermal Properties of PVDF and PVDF-*b*-PM_{CN} Block Copolymers [89]

Block Copolymers	SEC[a] M_n (g/mol)	DSC[b] T_g	T_m (°C)	TGA[c] T_d (°C)
PVDF	4200	–	157	271
PVDF-*b*-PAN	51,100	76	155	287
PVDF-*b*-PMAN	27,600	86	150	263
PVDF-*b*-PVCN	6400	62	–	105

[a]SEC or GPC: using DMF as the solvent.
[b]For DSC: the samples were heated at 10°C/min; the values of the glass transition temperature (T_g) reported herein are taken at the half-height of the capacity jump on second run. T_m represent the melting point.
[c]T_d, temperature of beginning of degradation (2 wt%) under air.

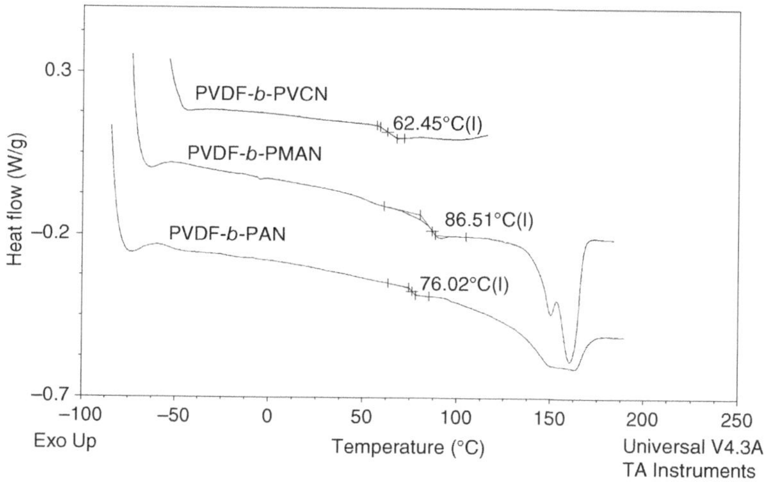

FIGURE 20.10 DSC thermograms of PVDF-*b*-PMAN, PVDF-*b*-PVCN, and PVDF-*b*-PAN block copolymers under nitrogen [89].

because of the high crystallinity of PVDF block. The lower values of these T_gs compared to those of the literature could be due to the small size (DPn) of the polymeric chain [99]. If the structural state of PAN is compared to that of PMAN, the only difference is the replacement of H by CH_3 in the main chain. Its effect on the T_g is interpreted by rigid, bulky, and inflexible substituents on carbon chains, which prevents the rotation about single bonds in the main chain. No cyanopolymers have clear melting points noted by DSC because of high coercive forces and low thermal stability. Several endothermic transitions in the temperature range of 20–380°C have been observed for PAN and a quite high heating rate (40°C/min) is necessary to evidence the melting points of cyanoblocks [100]. The melting points given in Figure 20.10 and in Table 20.6 correspond to the melting points of PVDF blocks. X-ray diffraction patterns of PVDF-*b*-PM_{CN} copolymers show evidence of crystallinity, which was confirmed by their melting points. It was noted that the crystallinity of PVDF decreased when the cyano blocks were inserted. The large interactions between cyano dipoles may contribute to the aggregation between chain molecules. From DSC measurements, using the enthalpy of theoretical melting (104 J/g [101]), the crystallinity rates obtained were estimated to 59% for the PVDF block, 14% for the PVDF-*b*-PAN, and 18% for PVDF-*b*-PMAN block copolymers.

20.5 DIELECTRIC PROPERTIES

20.5.1 Introduction

Different from a conductor or a semiconductor, a dielectric is an electric insulator, whose band gap is so large that no electric charges can freely flow through the material. Instead, dielectric polarization is induced by the applied electric field, resulting in accumulations of positive charges on one side of the dielectric toward the field and negative charges on the opposite side [102]. These charge accumulations enabled a direct application, namely, electric charge storage in capacitors. Since the energy density (U) of a dielectric capacitor polymer is proportional to its permittivity (ε_r) [40] and the square of the electric field E ($U = 0.5\ \varepsilon_r \varepsilon_0\ E^2$, where ε_0 is the vacuum permittivity), many efforts have been devoted to the enhancement of the material's permittivity and/or dielectric breakdown strength [103]. The combination of high permittivity nanoparticles and a high breakdown strength polymer matrix provides a potential to achieve high permittivity nanocomposite materials for high energy density capacitor [104]. In addition, according to the Theory of Fröhlich [95], polar polymers obtained by polymerization of the monomers with a strong dipole moment are good candidates to the development of dielectric polymers (capacitors systems) with high energy density. Cyanofluoropolymers synthesized above and bearing strong dipole moments such as C–CN could be consider as polar polymers. Thus, their dielectric properties have been performed by dielectric relaxation spectroscopy (DRS) measurements in a frequency and temperature range. The complex dielectric function is [105]:

$$\varepsilon^*(f) = \varepsilon'(f) - i\varepsilon''(f) \qquad \text{(Equation 20.1)}$$

where f, ε', and ε'' represent the frequency, the real part (dielectric or permittivity constant (capacity)), and the imaginary part (dielectric loss (conductivity)), respectively, where $\tan\delta = \varepsilon''/\varepsilon'$ as the dissipation factor.

The dielectric relaxation data obtained were deconvoluted using a sum of the model function introduced by Havriliak–Negami [106]:

$$\varepsilon^*(\omega) = \varepsilon_\infty + \sum_j \frac{\Delta\varepsilon_j}{\left[1 + (i\omega\tau_{\mathrm{HN}_j})^{\alpha_{\mathrm{HN}_j}}\right]^{\beta_{\mathrm{HN}_j}}} \qquad \text{(Equation 20.2)}$$

where j and $\Delta\varepsilon = \varepsilon_s - \varepsilon_\infty$ stand for the number of relaxation processes the dielectric strength $\Delta\varepsilon$ (i.e., the difference between the real permittivity values at the low and high frequency limits), respectively. $\tau \approx (2\pi f_{max})^{-1}$ is the characteristic relaxation time, while α_{HN} and β_{HN} are the shape parameters ($0 < \alpha_{\mathrm{HN}} < 1, 0 < \alpha_{\mathrm{HN}}\beta_{\mathrm{HN}} < 1$). When the data are influenced by low frequency conductivity contribution, an additional term $i\sigma/\omega^c\varepsilon_0$ was added to the dielectric loss, where ε_0 is the vacuum permittivity, σ and c are fitting parameters: σ is related to the dc conductivity of the sample and the parameter c ($0 < c \leq 1$) reflects bulk conductivity of ions for $c = 1$, while $c < 1$ reflects interfacial polarizations.

20.5.2 Dielectric Properties of Poly(M_C-*stat*-MATRIF) Copolymers (M_C Stands for AN, MAN, or MVCN)

Dielectric properties of poly(AN-*stat*-MATRIF), poly(MAN-*stat*-MATRIF), poly(MVCN-*stat*-MATRIF) copolymers as well as poly(MATRIF) homopolymer have been studied [107]. The complex dielectric was measured isothermally in the

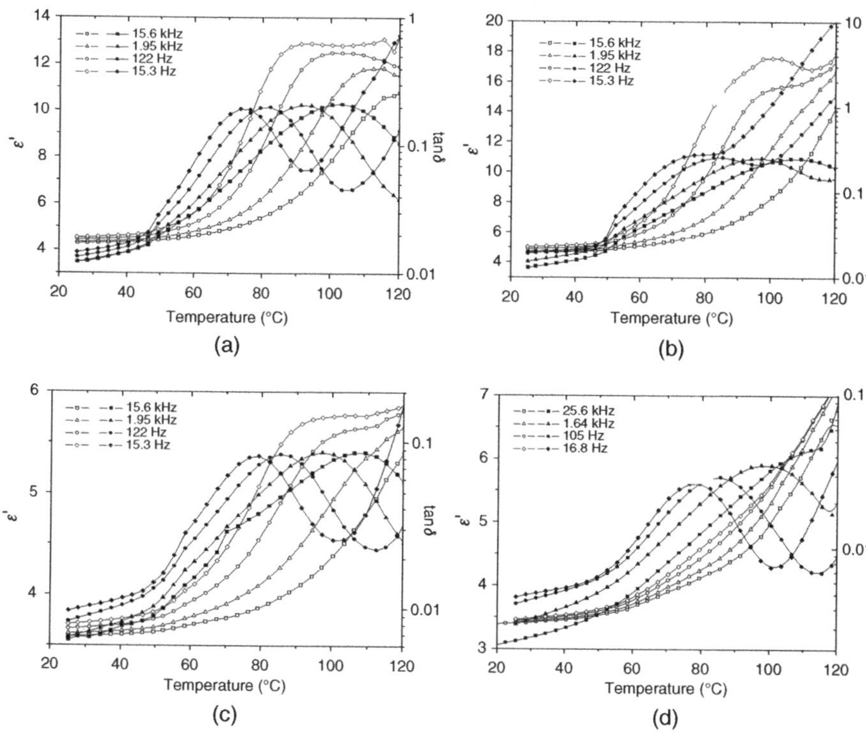

FIGURE 20.11 Dielectric spectra of the variation of permittivity (ε′) (open symbols) and tanδ (closed symbols) versus temperature at different frequencies for (a) poly(AN-*stat*-MATRIF); (b) poly(AN-*stat*-MATRIF); (c) poly(MVCN-*stat*-MATRIF) copolymers, and (d) poly(MATRIF) homopolymer [107].

frequency range from 10^{-1} to 10^{6} Hz. The temperature of the sample was varied from 25°C to 120°C in steps of 3 K.

Figure 20.11 exhibits the plots of the permittivity (ε′) and the dielectric loss (tanδ) versus the temperature at different frequencies of such (co)polymers.

These figures have been recorded only up to 120°C because of the strong influence of ionic conductivity for temperatures above.

From the assessment of ε′ and tanδ, it can be pointed out that below T_g (i.e., below 50°C), the complex permittivity is low and almost independent of temperature indicating that the motion of cyano dipoles is completely frozen in the copolymers of MATRIF [108]. The values of ε′ (Table 20.7) for poly(AN-*stat*-MATRIF), poly(MAN-*stat*-MATRIF), and poly(MVCN-*stat*-MATRIF) copolymers were 4.5, 5.0, and 3.8, respectively. They arise from the polarity of the copolymer chains due to the contribution at lower temperature of the local polarization of the polar moieties to ε′. These values are high compared to that of the poly(MATRIF) homopolymer which is ε′ = 3.5. The low value obtained for MVCN is in agreement with the low level of incorporation found by elemental analysis.

TABLE 20.7 α-Process Dielectric Increments, for the Copolymers and Homopolymer of MATRIF [107]

Sample	ε'_{∞}	$\Delta\varepsilon$ (100°C)
Poly(AN-*stat*-MATRIF)	4.5	8.5
Poly(MAN-*stat*-MATRIF)	5.0	11.0
Poly(MVCN-*stat*-MATRIF)	3.8	2.2
Poly(MATRIF)	3.5	1.1

At 50°C, associated to the glass transition temperature, T_g, a relaxation phenomenon assigned to the α-relaxation is observed for all the copolymers. This relaxation is related to the cooperative reorientation motion of large segments containing backbone (CN) groups, as suggested by Furukawa et al. [108], and this was confirmed by the large apparent dielectric relaxation strength. A similar behavior was observed in common amorphous cyano polymers based on vinylidene cyanide (VCN) [109], methylvinylidene cyanide (MVCN) [72], AN, and MAN [35, 36].

Above 80°C, these α-relaxations are sometimes overlapped at low frequencies and high temperatures by dc conduction due to ionic impurities [108].

Figure 20.12 shows a comparison of the permittivity values (ε′) and the dielectric losses (tanδ) of these materials (copolymers and homopolymer) *versus* the temperature at 120 Hz.

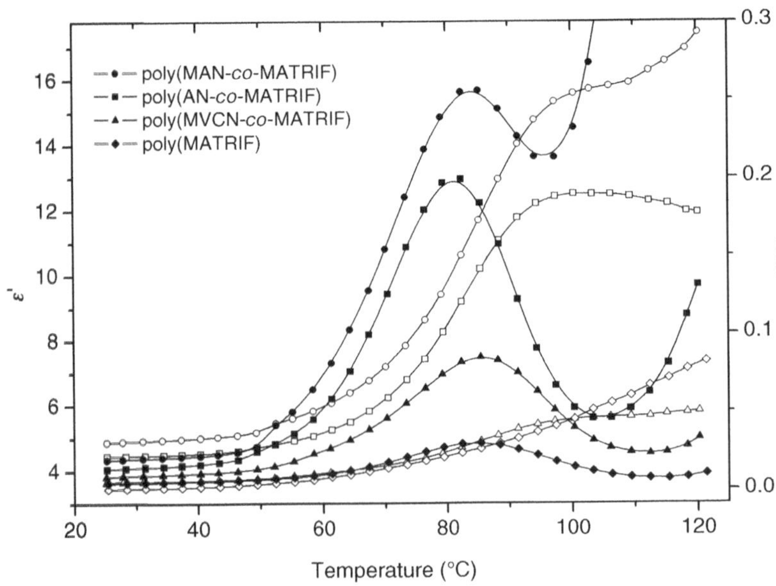

FIGURE 20.12 Dielectric spectra regarding the variation of permittivity ε′ (open symbols) and tanδ (closed symbols) versus temperature at 122 Hz for MATRIF copolymers compared to the homopolymer [107].

The dielectric losses of poly(MVCN-*stat*-MATRIF) copolymer are half an order of magnitude weaker than those shown for poly(AN-*stat*-MATRIF) and poly(MAN-*stat*-MATRIF) copolymers because of its lower polarity which is similar to that of poly(MATRIF) homopolymer.

The influence of the α-relaxation process on ε' value taking into account the dielectric strength $\Delta\varepsilon$ ($\Delta\varepsilon = \varepsilon'_0 - \varepsilon'_\infty$) was studied. This parameter is proportional to the number of dipoles oriented in the applied electric field. $\Delta\varepsilon$ values which are assessed from the curves of ε' versus the temperature and from the Havriliak–Negami (HN) equation (Equation 20.2) are listed in Table 20.7. It is observed that the dielectric relaxation strength is not comparable to those of most amorphous polymers for which $\Delta\varepsilon$ is generally comprised between 2 and 5 [110]. This behavior is not linked to the presence of fluorine atoms in the side chain of MATRIF because they do not contribute to the polarization significantly (indeed $\Delta\varepsilon = 1.1$ for poly(MATRIF)). The increase in polarity of poly(AN or MAN-*stat*-MATRIF) copolymers can be attributed to the high dipole moment of AN and MAN monomers and to the increase of mobility due to an increase of spacing between the chains, that is, a decrease of inter-chain interactions induced by the lateral substituents. This hypothesis was confirmed by a calculation of dipole moments for propane nitrile and isobutyronitrile which gave 2.97 D and 3.01 D, respectively. Concerning the poly(MVCN-*stat*-MATRIF) copolymer (the dipole moment of 1,1-dicyanoethane was found 3.18 D), the $\Delta\varepsilon$ value is not significant since the percentage of incorporated MVCN is very low (5%). As expected, only a small increase in dielectric relaxation strength ($\Delta\varepsilon' = 2.2$) was registered. The value of the dielectric constant in the glassy state (room temperature) is in agreement with the contribution of local polarization. Hence, the introduction of polar comonomers such as AN, MAN, MVCN induces one of the highest values of permittivity for polymers, under T_g.

20.5.3 Dielectric Properties of Poly(AN-*stat*-ATRIF) Copolymer

Both poly(AN-*stat*-ATRIF) copolymer and poly(ATRIF) homopolymer were dielectrically characterized over a frequency range from 10^{-1} to 10^6 Hz, and covering a temperature range from 223 to 393 K [68].

The real and imaginary components of the complex dielectric permittivity, $\varepsilon^*(f)$, are supplied in Figure 20.13 for poly(ATRIF) homopolymer (a and b) and for poly(AN-*stat*-ATRIF) copolymer (c and d), for some representative temperatures above the glass transition temperature, T_g. A dominant relaxation process is visible being associated with the dynamic glass transition, α-relaxation. As expected, the maximum of each peak is shifted toward higher frequencies as the temperature increases due to an enhancement of dipolar mobility in the origin of cooperative motions above T_g. At high temperatures, and low frequencies, the dielectric spectra are dominated by charge transport that will be analyzed further on.

The real permittivity, at any temperature above T_g of each polymer, as well as the intensity of the dielectric loss peaks observed for the main α-relaxation has been reported to be always greater for poly(AN-*co*-ATRIF) copolymer than for

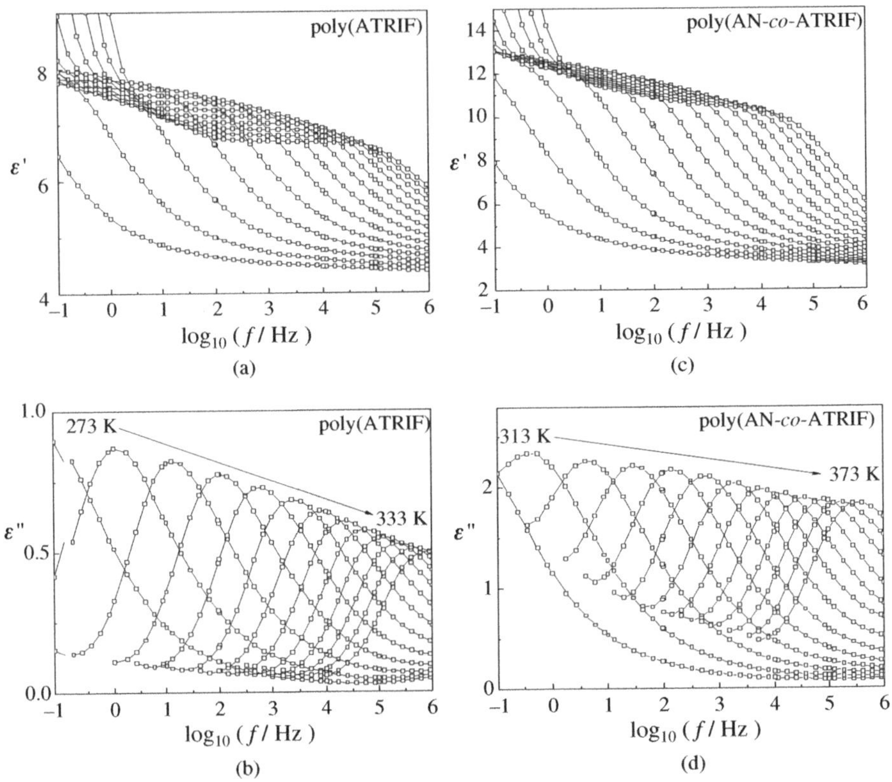

FIGURE 20.13 Real permittivity, ε′(*f*), and dielectric loss spectra, ε″(*f*), collected isothermally for: poly(ATRIF) homopolymer between 273 and 333 K (a) and (b); poly(AN-*co*-ATRIF) copolymer between 313 and 373 K, (c) and (d); all isotherms presented in steps of 5 K [68].

poly(ATRIF) homopolymer. In Figure 20.13, the respective isochronal dielectric plots are taken from the isothermal data of ε′(*T*) and tanδ(*T*) at 1 kHz.

Both ε′(*T*) curves (Figure 20.14a) for poly(ATRIF) homopolymer and poly(AN-*stat*-ATRIF) copolymer present the normal step rising which is the signature of the α-relaxation associated to the dynamical glass transition, occurring at temperatures close to the glass transition determined by calorimetry ($T_{\mathrm{g,homopolymer}}$ 273 K and $T_{\mathrm{g,copolymer}}$ 312 K) [68]. The decrease observed at temperatures above the ε′ step is the expected behavior for the real permittivity as predicted by the Fröhlich–Kirkwood equation [34, 111]. The difference in the temperature location of the abrupt ε′ steps is a further evidence of the significant shift to higher temperatures of the copolymer glass transition relatively to the homopolymer, confirmed also by the deviation of the tanδ–peak, as presented in Figure 20.14b.

As the dielectric technique probes molecular mobility, through the reorientation of dipoles, the observed shift to higher temperatures means that the molecular dynamics associated with the glass transition, becomes more hindered in the copolymer. The

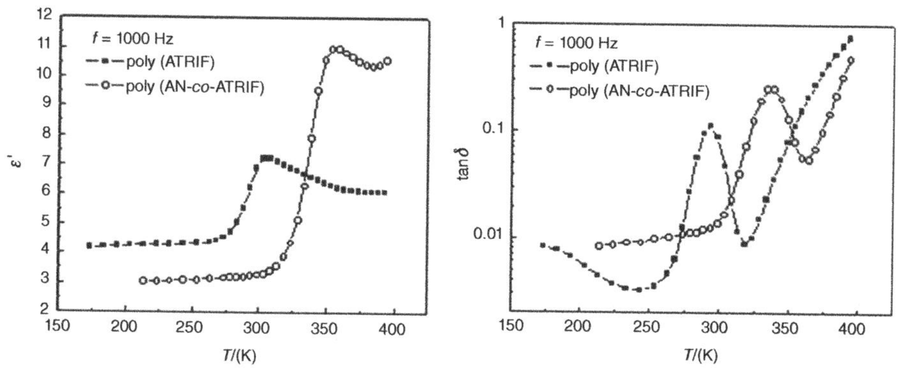

FIGURE 20.14 Temperature dependency of (a) the real part of dielectric permittivity ε′ (*T*) and (b) the dielectric loss tangent $\tan\delta(T) = \varepsilon''(T)/\varepsilon(T)$, for poly(ATRIF) homopolymer (black squares) and poly(AN-*stat*-ATRIF) copolymer (gray circles) [68].

slow behavior of this cooperative motion arises from the substitution of part of the bulky $CO_2CH_2CF_3$ side-groups by the more rigid CN groups, which, as mentioned above, increases the stiffness of the polymeric main-chain. From the magnitude of the step in the ε′ (*T*) traces, it can be concluded that a higher dielectric strength is associated with the α-relaxation of poly(AN-*co*-ATRIF) copolymer relatively to the homopolymer, indicating that a higher dipole moment is relaxing in the copolymer as it is expected by the inclusion of the highly polar cyano (CN) moieties rigidly attached to the main-chain that relax coupled to the segmental motion in the origin of the α-process.

The characteristic relaxation times obtained by the HN fit to the data, after conversion to τ_{max} (Equation 20.3), are plotted *versus* the inverse of temperature in Figure 20.15 for poly(ATRIF) homopolymer and poly(AN-*co*-ATRIF) copolymer.

$$\tau(T) = \tau_\infty \exp\left(\frac{B}{T - T_0}\right) = \tau_\infty \exp\left(\frac{DT_0}{T - T_0}\right) \quad \text{(Equation 20.3)}$$

where τ_∞, B and T_0, the so-called Vogel temperature, are empirical parameters characteristic of the material.

The parameters obtained from the VFTH fit (Figure 20.15) of the plot of $\ln\tau(T)$ versus allow to estimate the glass transition temperature at $\tau = 100$ s [112], as 269.5 K and 307.1 K for poly(ATRIF) homopolymer and poly(AN-*stat*-ATRIF) copolymer, respectively. These data were in reasonable agreement with DSC. Another parameter resulting from the HN fit (Equation 20.2) is the dielectric strength, Δε presented in the upper inset of Figure 20.15. As can be seen, Δε generally decreases with the temperature as predicted by the Fröhlich–Kirkwood equation [34, 111]. This is also evident from the evolution of the peaks height in the raw data (Figure 20.15) where the same trend is observed for both poly(ATRIF) and poly(AN-*stat*-ATRIF) polymers.

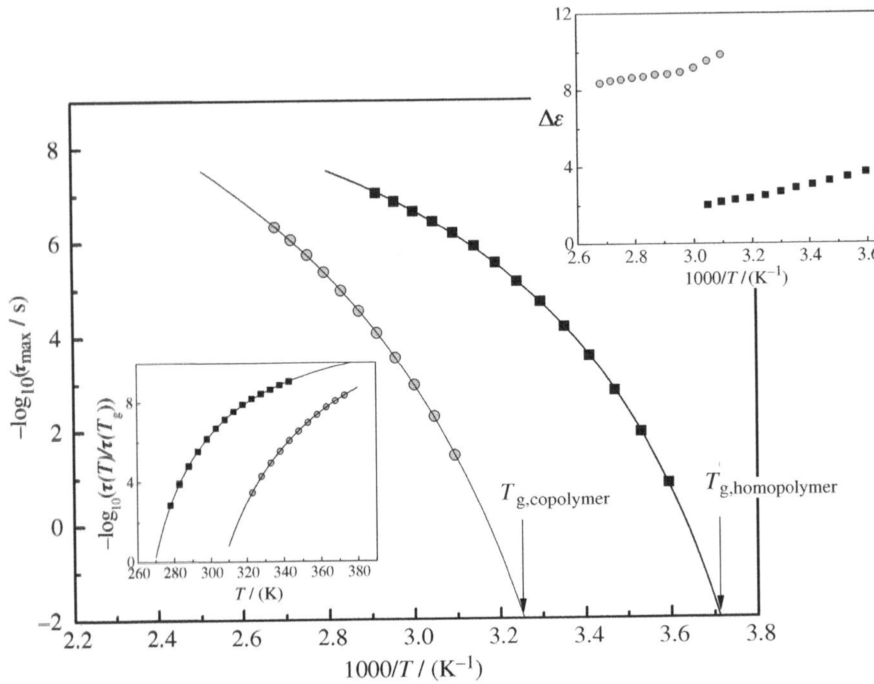

FIGURE 20.15 Temperature dependence of the relaxation times of both poly(ATRIF) homopolymer (squares) and poly(AN-*stat*-ATRIF) copolymer (circles); the arrows indicate the glass transition temperatures estimated from the VFTH equations at $\tau = 100$ s [68].

The estimated Δε values confirm the higher dielectric response of the copolymer, attaining a factor of ~2.6 higher than that of the homopolymer. As previously mentioned, this increase of the relaxation strength in the copolymer comes from an increase in the dipolar moment, more precisely, in the average square value of the dipole moment, per monomer unit. The high value of the dipole moment of the cyano group, ~4 D in several aliphatic compounds [113, 114], explains the higher value in the dielectric strength in the copolymer. The dipole moment of the $CO_2CH_2CF_3$ moiety is around 1.34 D.

The existence of secondary relaxations at temperatures below T_g is clearly obvious for poly(ATRIF) homopolymer from the tanδ(*T*) plots at the fixed frequency (1 kHz) shown in Figure 20.14b. The same plot for poly(AN-*stat*-ATRIF) copolymer shows only a nearly constant plateau. On the other hand, the existence of subglass mobility for the copolymer is noted, but much less intense than that of the homopolymer. Because the nitrile groups are bounded directly to the polymer backbone, its movement is highly correlated to the motion of the polymeric backbone and therefore their contribution to the secondary process is rather small. This is the reason why the dielectric response in the copolymer is much smaller compared to that of the homopolymer. The isothermal dielectric loss spectra, ε″(*f*) confirm the latter observations for the poly(ATRIF), where at least two secondary relaxations (β and γ) can be identified.

20.5.4 Dielectric Properties of Poly(TSE-*alt*-4FS) Copolymer (TSE Stands for 2-(2,4-difluorophenyl)-1,1-dicyanoethylene)

Broadband dielectric spectroscopy were performed to characterize molecular motion in P4FST homopolymer and copolymer based on 2-(2,4-difluorophenyl)-1,1-dicyanoethylene (TSE) with 4-fluostyrene [88]. The goal of this study is to explore the incorporation of fluorinated TSE units in the dielectric properties of these copolymers and to establish the structure–properties relationship. The temperature is ranging from −150°C to 300°C and a frequency interval from 0.001 to 10 Mhz.

Figure 20.16 displays isochronal scans temperature at fixed frequencies (103 Hz, 1.09 kHz, 11.6 kHz, and 123 kHz) of the real permittivity ε' (Figure 20.16a) and the dissipation factor tanδ (Figure 20.16b) for P4FST homopolymer.

Two successive parts can be pointed out from the analysis of ε' and tanδ according to the temperature zone: the first one from 0°C to 95°C while the second one ranges from 95°C to 160°C.

In the glassy state of P4FST in the region between 0°C to 95°C (below T_g = 104°C), the ε' and tanδ had low dependence on frequency, no dielectric relaxation was observed in tanδ curves (Figure 20.16a), the dipolar orientation is mainly blocked [110]. As a consequence, the real part of permittivity is almost constant ($\varepsilon' = 2.1$) corresponding to typical value for the atomic and electronic polarization [35]. This value is similar to the insulating commercially available polystyrene (i.e., $2.5 < \varepsilon' < 2.65$ at 50 Hz and 20°C) [115].

In 95–160°C region (above T_g), molecular dipoles are able to orient under the external field. An additional effect of polarization was observed which increases the permittivity of the P4FST (i.e., $\varepsilon' = 5.35$ at 140°C and 1.09 kHz). Through tanδ (Figure 20.19b), it is possible to note a relaxation process known under α-relaxation which is related to the cooperative reorientation motion of large segments above T_g. The dielectric loss is lower around 10^{-2} (i.e., 0.06 at 1.09 kHz). Similar behavior was observed in the atactic polystyrene [116].

Figure 20.17 shows the dielectric parameters ε' (Figure 20.17a) and tanδ (Figure 20.17b), respectively, versus the temperature for the following frequencies: 6.27 Hz, 18.8 Hz, 1.52 kHz, 123 kHz, and 1 MHz for poly(TSE-*alt*-4FST) copolymer.

The temperature range can be divided into two regions: below 150°C and above 150°C (i.e., above T_g = 156°C).

In the zone ranging from 40°C to 150°C, it can be pointed out from the assessment of ε' and tanδ that below T_g, the complex permittivity is low and almost independent of temperature and frequency. The value of ε' was 2.41 and it arises from the polarity of the copolymer chains comparatively to the P4FST homopolymer ($\varepsilon' = 2.13$).

In the glass transition temperature domain, a relaxation phenomenon assigned to the α-relaxation was observed as the cooperative reorientation motions of large segments containing backbone (CN) groups, as suggested by Furukawa et al. [108]

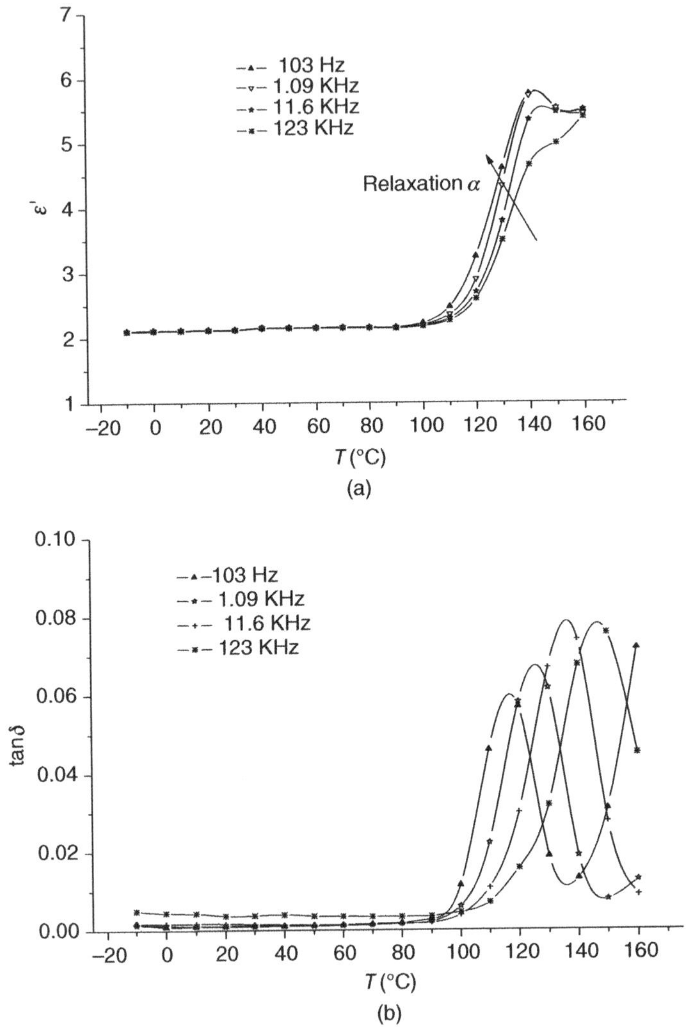

FIGURE 20.16 Dielectric permittivity ε′ (a) and dissipation factor tanδ (b) versus temperature of P4FST homopolymer at various frequencies [88].

In order to determine the relaxation rate of ε′ relaxation, the activation energy was evaluated using the Arrhenius relation (Equation 20.4):

$$f = f_0 \exp\left(\frac{E_\mathrm{a}}{k_\mathrm{B}T}\right) \tag{20.4}$$

where $f_0 = 1/2\pi\tau_0$, f, E_a, and k_B stand for the frequency at which M'' (or tanδ) maxima occurs, the activation energy, and the Boltzmann constant, respectively. According to the range of frequency of the present analysis, a linear variation of the log f

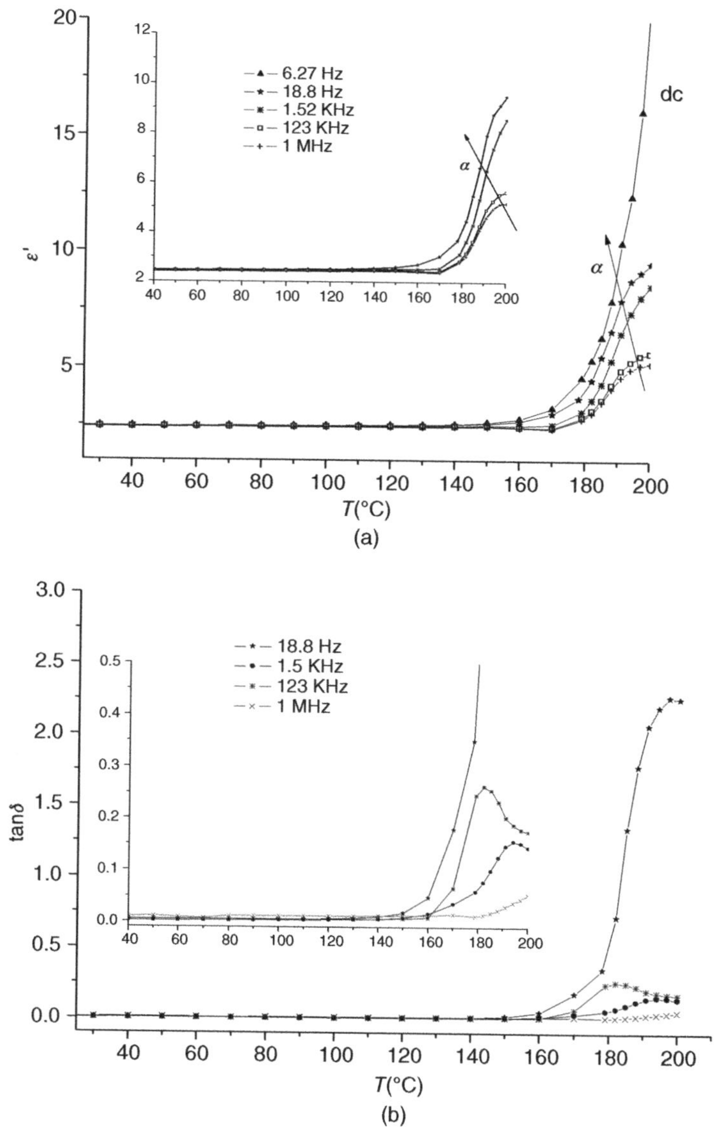

FIGURE 20.17 The dielectric permittivity ε′ (a) and the dissipation factor tanδ (b) versus temperature of copolymer based on TSE (2-(2,4-difluorophenyl)-1,1-dicyanoethylene) and 4FST at various frequencies [88].

versus 1/*T* was used to calculate the apparent activation energy. The obtained values are 396 kJ/mol and 497 kJ/mol for P4FST homopolymer and poly(TSE-*alt*-4FST), respectively [88]. These energies, around a few hundreds of kJ/mol, confirm the presence of a primary relaxation phenomenon associated with the glass transition of the copolymers and are similar to those of copolymers based on cyanide monomers such

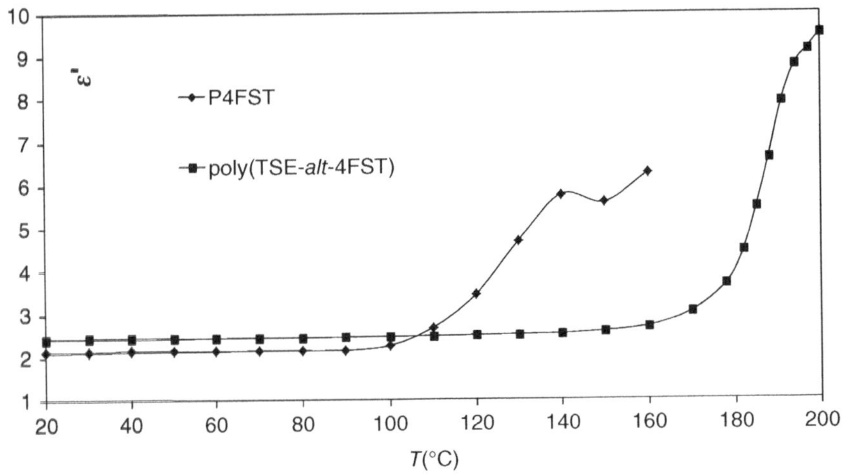

FIGURE 20.18 Dielectric spectra of the variation of permittivity ε' versus temperature at 18.8 Hz for P4FST homopolymer and at 12 Hz for poly(TSE-*alt*-4FST) copolymer [88].

as copolymer based on 2-methyl-1,1-dicyanoethylene with 4FST ($E_a = 445.8$ kJ/mol) [72].

Figure 20.18 exhibits a comparison of the permittivity values of homopolymer and copolymer versus the temperature at 18.8 Hz and 12 Hz, respectively.

The values of $\Delta\varepsilon$ can be accessed from the curves of ε' versus the temperature (Figure 20.18). The calculated values of $\Delta\varepsilon$ ($\Delta\varepsilon = \varepsilon'_s - \varepsilon_\infty$) are 3.6 at 140°C and 6.7 at 197°C for P4FST homopolymer and poly(TSE-*alt*-4FST) copolymer, respectively.

The value of the dielectric relaxation strength $\Delta\varepsilon$ of poly(TSE-*alt*-4FST) copolymer above T_g ($\Delta\varepsilon = 6.7$) is lower than those of copolymers based on methyl vinylidene cyanide (MVCN) and 4FST ($\Delta\varepsilon = 10.9$) [72]. In fact, the substituent on position 2 of TSE comonomer (2,4-diflurostyrene group) has bulkier steric effect more than the methyl group in the MVCN comonomer. This low flexibility of the poly(TSE-*alt*-4FST) copolymer chain limits the reorientation of the CN groups in the direction of the applied electric field, thus reducing the value of the dielectric relaxation strength $\Delta\varepsilon$.

20.5.5 Dielectric Properties of PVDF-*b*-PM$_{CN}$ Block Copolymers (M$_{CN}$: AN, MAN, and VCN)

The dielectric spectra were obtained using the samples with thicknesses ranging between 80 and 150 μm, and in the temperature −130°C to 150°C range, except for the PVDF-*b*-PVCN block copolymer (−130°C to 100°C) [89]. The values of the dielectric permittivities versus temperature (Figure 20.19) show the existence of two relaxations. At room temperature (100 Hz), the values for PVDF homopolymer, PVDF-*b*-PAN, PVDF-*b*-MAN, and PVDF-*b*-PVCN block copolymers are 6.3, 3.9, 5.0, and 5.5 respectively.

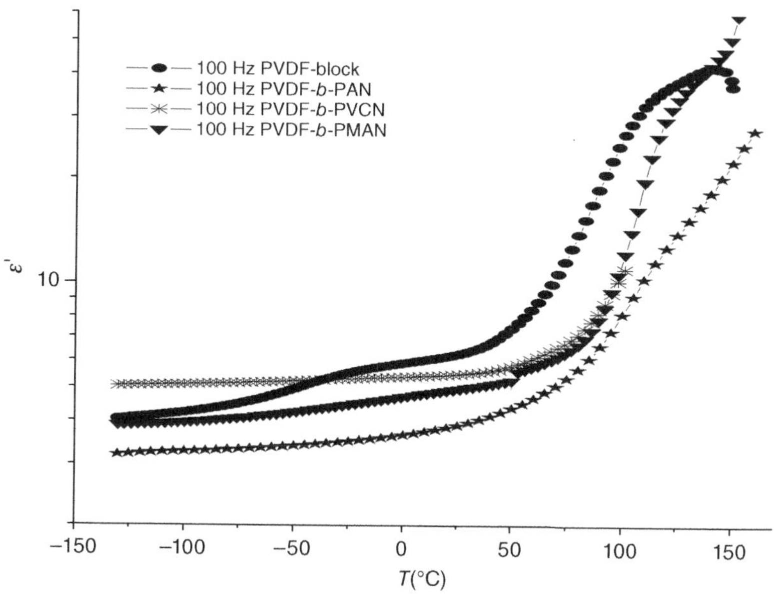

FIGURE 20.19 Dielectric permittivities of various block copolymers and PVDF versus the temperature (for 100 Hz) [89].

To minimize the effects of conductivity (electrode polarization), the formalism of electric modulus using M^* is introduced. In analogy to mechanical relaxation, the electric modulus M^* is defined by the following [Equation 20.5; 117]:

$$M^* = \frac{1}{\varepsilon^*} = \frac{1}{\varepsilon' - j\varepsilon''} = \frac{\varepsilon'}{\varepsilon'^2 + \varepsilon''^2} + j\frac{\varepsilon''}{\varepsilon'^2 + \varepsilon''^2} = M' + jM'' \qquad \text{(Equation 20.5)}$$

where M' and M'' are the real and the imaginary electric modulus, respectively. Figure 20.20 shows the plots of the electric modulus (M'') versus frequency at different temperatures for PVDF block and for the PVDF-*b*-PAN, PVDF-*b*-PMAN, PVDF-*b*-PVCN block copolymers. Figures 20.20b, 20.20c and 20.20d show that the PVDF-*b*-PAN, PVDF-*b*-PMAN, and PVDF-*b*-PVCN block copolymers exhibit at least two wide relaxations.

Figure 20.19 shows that in the glassy state (below −80°C/100 Hz) the value of ε' for PVDF-*b*-PVCN ($\varepsilon' = 5$) is higher than that of PVDF block ($\varepsilon' = 4.1$). This arises from the small amount and high polarity of PVCN incorporated in this block copolymer. A different behavior was observed, when both other polycyano blocks were incorporated. In agreement to to their lower polarity and their high proportion, their ε' values were lower.

PVDF exhibits four relaxations [118, 119], when temperature and frequency are changed. For example, at 1 kHz it is possible to evidence γ relaxation (−80°C) originated from the chain rotation in the amorphous phases, β relaxation (−20°C)

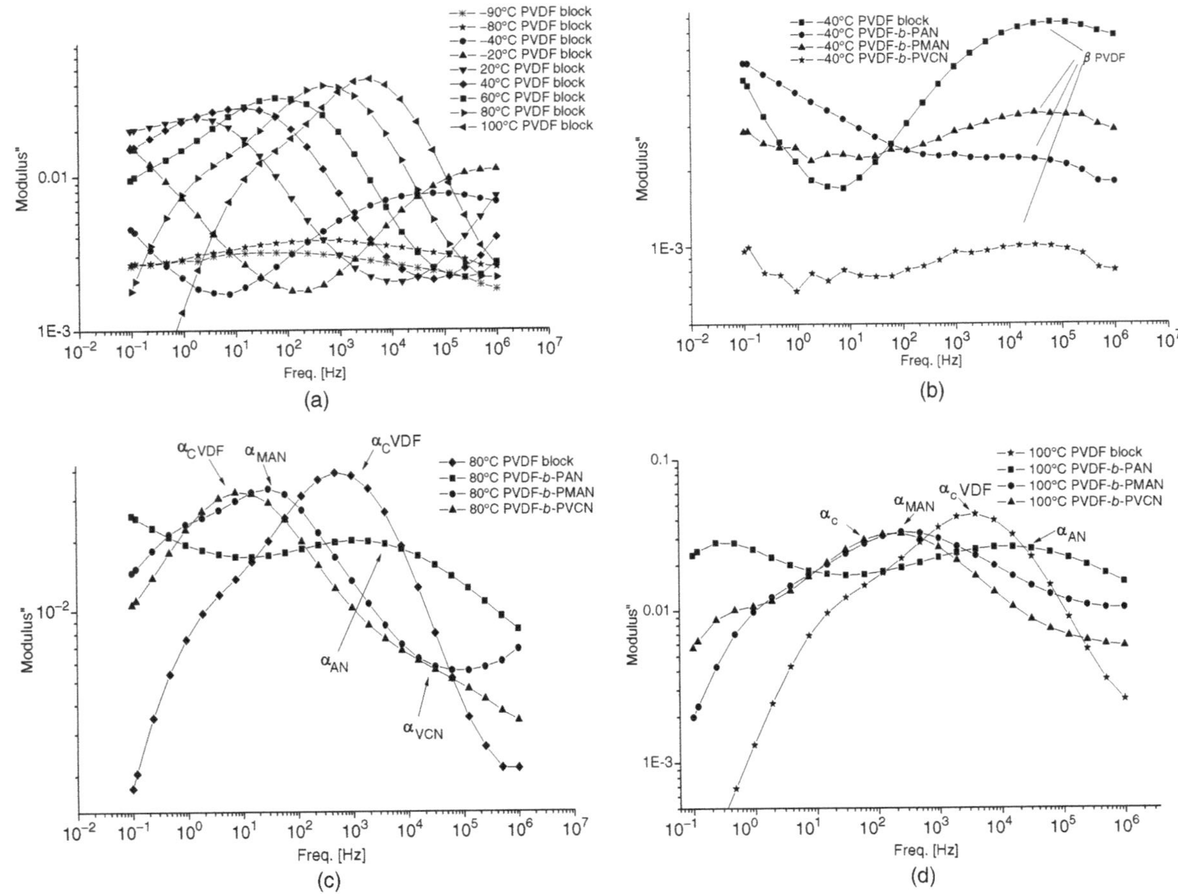

FIGURE 20.20 Imaginary part of the dielectric modulus (M'') versus frequency for PVDF block at different temperatures (a) for the PVDF blocks and its block copolymers: (b) at −40°C, (c) at 80°C, and (d) at 100°C [89].

due to the micro-brownian motion in the amorphous phase (linked with the glass transition since the relaxation seems to exhibit "Vogel–Fulcher–Tammann behavior"), β' relaxation (50°C), which is attributed to fold motions in the amorphous phases and the ac relaxation (80°C) assigned to the defects in the crystals.

For PVDF blocks, the variation of imaginary part of the dielectric modulus (M'') as a function of frequency (Figure 20.20a) shows almost the same behavior for the β and α_c relaxations. However, these relaxations are shifted toward lower temperatures while γ and β' cannot be distinguished for the PVDF block. At −40°C (at 104 Hz), which corresponds to the glass transition region (β), a relaxation due to the local conformational motion is noted. The large apparent dielectric relaxation strength confirms this hypothesis. Another relaxation (α_c) occurred at temperatures above 40°C (10 Hz), and this is due to the motion in the crystalline phase, which has a very high ratio as evidenced by DSC and X-ray diffraction measurements [89]. Above 80°C, these relaxations are overlapped at low frequencies by dc conduction due to ionic impurities [72]. The intensities of both relaxations increase with the increasing temperature especially for the relaxation associated to the glass transition.

In Figures 20.20b,c,d, the first relaxation at low temperatures (−40°C/104 Hz), which is not quite clear (Figure 20.20b) for the block copolymers, can be assigned to β relaxation, which can be found in PVDF blocks at the vicinity of its glass temperatures. This hypothesis is in good agreement with what has been noted for the PVDF block and from the literature [120, 121]. At such low temperatures, the poly(cyano) blocks were frozen [73] and they did not induce any relaxation. This behavior is different from that observed for the PVDF block at its T_g vicinity. This can be explained by the presence of the cyano blocks. The same phenomenon was observed by Hahn et al. [120] on PVDF/PMMA blends known to display a high miscibility. By changing the percentages of both polymers, these authors concluded that this micro-brownian relaxation of PVDF decreases until it disappears for 20 wt% of PVDF in the mixture.

With regard to the high temperature relaxation, it is known that for the cyano-containing copolymers, a dielectric relaxation arising from the glass transition was observed at around 60°C, and no subglass dielectric relaxations were noted. Cyano groups have cooperative energy, and hence only a macroscopic conformational change occurs. The dielectric behavior shows that the CN dipole rotations occur even just below T_g. The same behavior was observed for the copolymer based on VCN and vinyl acetate [24, 25]. This allows to conclude that recorded relaxations are mainly originating from poly(cyano) blocks at their glass transition temperatures as assessed by DSC (Table 20.6). These relaxations are highlighted using modulus plots. Figure 20.20d shows that this wide α relaxation is located in the vicinity of the T_gs of PAN, PMAN, and PVCN blocks. Relaxations can also be found, which are attributed to α_c (relaxation occurring in the crystalline phases of PVDF blocks). In addition, α_c is observed for the PVDF-*b*-PVCN and PVDF-*b*-PMAN block copolymers but not for PVDF-*b*-PAN (Figure 20.20c). This possibly arises from the high molecular weight of PAN with reference to PVDF block molecular weight, which does not allow the relaxation of the crystalline phase to appear.

20.5.6 Dielectric Properties of Poly(VCN-*alt*-MATRIF) Copolymer

The dielectric properties were investigated in the 120–433 K temperature range, so well below and above the glass transition temperature and at different frequencies in the range 0.4 mHz–10 MHz. The real components of the complex dielectric permittivity, ε' and $\tan\delta$ are listed in Figure 20.21 for poly(MATRIF) homopolymer ((a) and (b)) and the poly(VCN-*alt*-ATRIF) copolymer ((c) and (d)), at fixed frequencies 10 Hz, 117 Hz and 10 kHz [121].

Poly(MATRIF) homopolymer (Figures 20.21a, b) exhibits a dynamic scenario with three relaxation processes, similar to what is observed by the corresponding nonfluorinated homopolymer based on ethyl methacrylate PEMA: (i) α-process related to calorimetric glass transition; (ii) β-process (visible above and below T_g), due to carbonyl group rotation involving the rocking motion of segmental unit, as demonstrated by the group of Spiess [123] for poly(methyl methacrylate), PMMA and poly(ethyl methacrylate, PEMA); and (iii) γ-process (active below T_g): analogous found in poly(fluoroalkyl methacrylate)s, attributed to an internal motion of perfluoroalkyl groups in the side chain [20].

Figures 20.21c and d show that poly(VCN-*alt*-MATRIF) copolymer exhibits a dynamic scenario with four relaxation processes: two relaxations are above T_g, merging at low temperatures, while two others are below T_g. The addition of VCN unit increases the dielectric constant as well as the value of calorimetric glass transition T_g: (i) α_1-relaxation (Debye-like and strong, i.e., low fragility index $m = 44$ [68]); (ii) α_2-relaxation (broad and fragile, high fragility index $m = 101$ [68]), more separated from β-one: effect of rigidity of VCN segment; (iii) β-relaxation (faster), due to a less efficient packing: similar effect for random copolymers based on butyl methacrylate (*n*-BMA) with styrene (ST) named poly(*n*-BMA-*stat*-ST) copolymer [123].

In both cases, additionally the dc conductivity contribution (due to the drift of ionic impurities) is responsible for the upturn of $\tan\delta$ at high temperature and the electrode polarization for the downward trend at very high temperatures for $f = 1$ Hz.

Figures 20.21 (a, c) show that the permittivity below T_g is close to 3.0 and 3.2 for poly(MATRIF) homopolymer and poly(VCN-*alt*-MATRIF) copolymer, respectively. Above T_g, the α relaxation phenomena occur with an important increasing of ε' for the poly(VCN-*alt*-MATRIF) copolymer ($\varepsilon' = 20$ at 425 K) comparatively to poly(MATRIF) homopolymer ($\varepsilon' = 4$ at 370 K). This result arises from the increasing polarity of poly(MATRIF) by the incorporation of cyano groups (VCN).

20.6 APPLICATIONS

The applications of dielectric polymers are numerous and a growing interest has recently been shown on fluorinated materials that exhibit such properties [10–13, 123–127]. In fact, advanced pulsed power and power conditioning technologies for power generation/regulation and all electric vehicles/ships require a capacitor material to achieve a high discharged electric energy density as well as low loss performance [127, 128]. On the other hand, advanced dielectric applications,

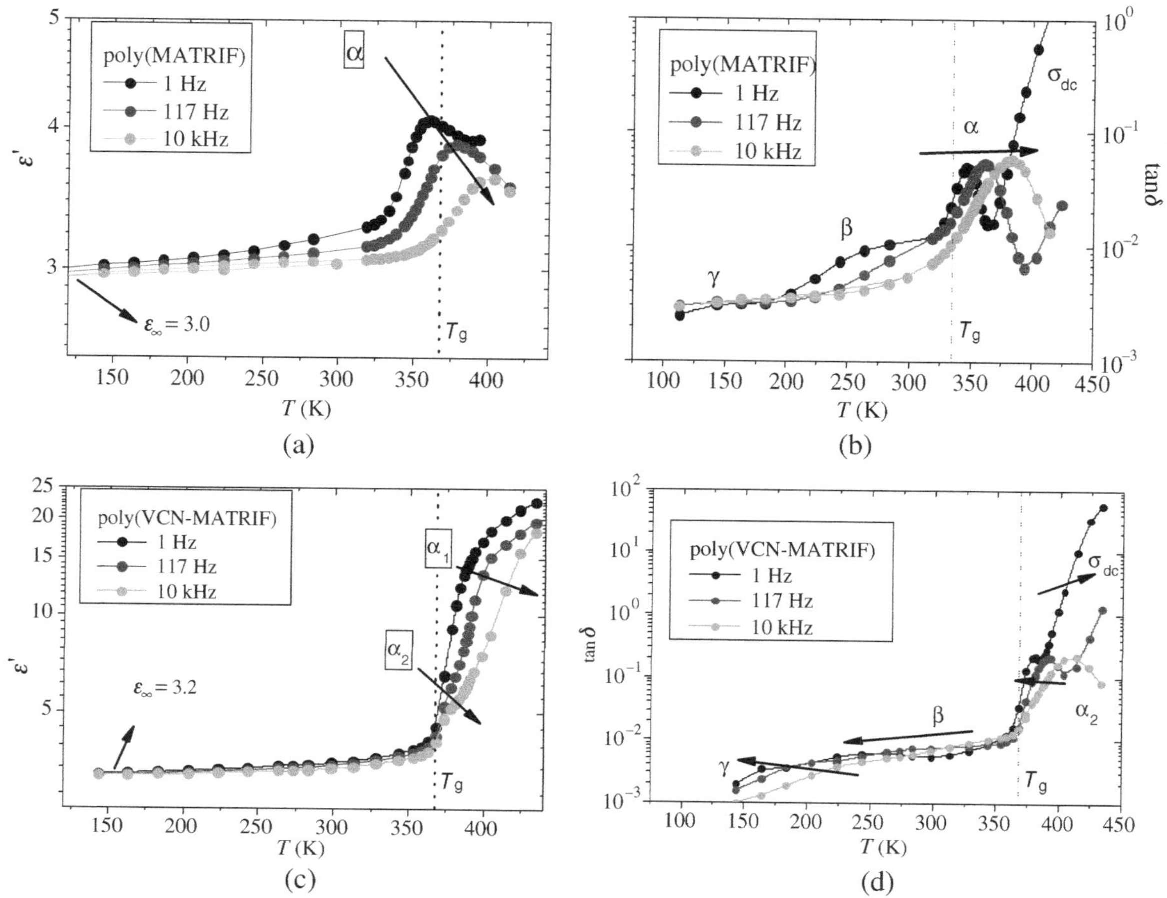

FIGURE 20.21 Temperature dependency of real permittivity $\varepsilon'(T)$ and $\tan\delta$ collected for: poly(MATRIF) homopolymer (a) and (b); poly(VCN-*alt*-MATRIF) copolymer (c) and (d) [121].

such as power conditioning, electric launch platform, and all electrical or hybrid vehicles, need a polar material, due to their large effective dielectric constants [129–131] and in turn high electrical energy storage [132]. Among commercially available polymer capacitor films, PVDF stands out because of its high dielectric constant and high dc breakdown strength (~700 MV/m for ~10 μm capacitor grade films) [131]. Electrical and electronic applications and future trends of fluoropolymers based on vinylidene fluoride such as jacketing of cables for plenum areas including signaling, communication, and power lines, insulating compounds for jacketing wires and cable assemblies, cathodic protection, industrial power control systems, high temperature wiring, and wire and cable sheathing have been reviewed by Ameduri [5]. Fluorinated materials prepared from different polymorphisms and crystallite sizes were obtained in PVDF and P(VDF-*stat*-HFP) (HPF: hexafluoropropene) films and their practical application of ferroelectric polymers for advanced high energy density dielectric film capacitors were reported by Zhu et al. [12, 13]. New polar dielectric polymers for high electric energy storage and low loss applications such poly(VDF-*co*-TrFE-*co*-CTFE)-*g*-PS (14%) graft copolymer was synthesized. A significantly reduced ferroelectric loss and relatively high discharged energy density were observed because of this antiferroelectric-like behavior even at high poling fields [10]. Recently, Zhu et al. [126] reported multilayer films for high energy density storage. The 65-layer poly(carbonate)/poly(methyl methacrylate)/Poly(VDF-*co*-HFP) (PC/PMMA/ Poly(VDF-*co*-HFP)) system exhibited a 25% enhanced dielectric breakdown strength, 50% higher energy density, 40% decreased hysteresis loss energy density, and slower ion migration behavior. It was determined that the improved dielectric properties were attributed to the high interactivity between PMMA layers and both PC and poly(VDF-*co*-HFP) layers. Furthermore, cyano polymers bearing CN groups are interesting as dielectric materials. This is exemplified by the well-known piezoelectricity of the poly(VCN-*alt*-VA) copolymer [24, 25], the discovery of which stimulated investigations on the copolymerization of cyano-containing monomers with various vinyl and diene comonomers [23, 26, 71, 82–85]. Electrical properties of copolymers based on acrylonitrile have also been reported [35, 36].

The above sections mentioned polar polymers containing C–F and C–CN (alternating copolymers of VCN, block copolymers of PVDF, and random copolymers of AN) are a potential candidate as dielectric materials and find many applications in the field above.

20.7 CONCLUSION

This chapter summarizes (i) the synthesis of fluoro and cyano monomers, (ii) the preparation and characterization of their respective homopolymer and copolymers, and (iii) the study of the dielectric behavior of the resulting cyanofluoropolymers.

Cyano monomers such as MVCN, VCN, and TSE were synthesized. Copolymers based on MATRIF were produced by radical copolymerization with AN, MAN, MVCN, and VCN. Kinetics of radical copolymerization of AN with ATRIF has also been studied. Copolymer based on 2-(2,4-difluorophenyl)-1,1-dicyanoethylene (TSE), and 4-fluorostyrene (4FST) was investigated.

The resulting copolymers were characterized by ^{1}H, ^{13}C, and ^{19}F NMR and IR spectroscopy, and SEC. Their compositions were assessed by ^{1}H NMR and elemental analyses. The structure of the obtained copolymers was evidenced by the assessment of the reactivity ratio and NMR spectroscopy. From monomer–polymer composition curve and the Fineman–Ross, Kelen–Tüdos and Jenkins laws, we can conclude that the copolymers based on AN, MAN exhibit random structures, while the copolymers based on VCN with MATRIF had alternating structures. In addition, poly(TSE-*co*-4FST) copolymer presents an alternating structure terminated with homosequence of 4FST.

The syntheses of poly(vinylidene fluoride)-*b*-poly(AN, MAN, or VCN) (PVDF-*b*-PM_{CN}) block copolymers achieved by the iodine transfer polymerization (ITP) of acrylonitrile (AN), methacrylonitrile (MAN), and vinylidene cyanide (VCN), in the presence of PVDF-I, are also original materials with interesting dielectric properties.

These fluorocyanopolymers were dielectrically characterized in a wide range of frequencies and temperatures. The dominating relaxation process detected in these materials is the α-relaxation, associated with the dynamic glass transition. A VFTH temperature dependence of the relaxation times was found for these fluorocyanopolymers. The polarity–dielectric constant relationship has been established. Actually, the inclusion of CN group into fluorinated units enhances the dielectric increment and makes them potential candidates for film capacitors.

It can be expected that much work should be carried out to find out new fluorocyanopolymers bearing even better properties, that should attract the interest of academic and industrial researchers.

ACKNOWLEDGMENTS

All coworkers (from France (University Claude Bernard-Lyon), Italy (University of Pisa), Sweden (KTH, Stockholm), Tunisia (University of Sfax), USA (DePaul University of Chicago), Portugal (University of Lisbon)) quoted in the references below are acknowledged for their efficient contributions. The authors also thank all the financial supports for different programs: from the CNRT (Morocco)-CNRS (France); CNRST (Morocc)-NSF (USA) collaboration; Program of collaboration VR-MENA-Sweden; PROTARS III (Morocco), Moroccan-Tunisia program, and Collaboration Projects Italy-Morocco.

NOMENCLATURE

AN	Acrylonitrile
MAN	Methacrylonitrile
MVCN	Methyl Vinylidene Cyanide
VCN	Vinylidene Cyanide

TSE	*2-(2,4-difluorophenyl)-1,1-dicyanoethylene*
VDF	Vinylidene Fluoride
MATRIF	2,2,2-Trifluoroethyl methacrylate
ATRIF	2,2,2-Trifluoroethyl acrylate
4FST	4-fluorostyrene
AIBN	Azobis-isobutyronitrile
ABCN	1,1'-Azobis(cyclohexanecarbonitrile)
TBPPI	*tert*-butyl peroxypivalate

REFERENES

1. (a) Wall, L. A. *Fluoropolymers*; Wiley Interscience: New York, 1972. (b) Feiring, A. E. Fluoroplastics. In *Organofluorine Chemistry: Principles and Commercial Applications*; Banks, R. E., Smart, B. E., Tatlow, J. C., Eds.; Plenum Press: New York, 1994; Chapter 15, pp 339–372. (c) Willoughby, B. G.; Banks, R. E. Fluoropolymers. In *Encylopedia of Advanced Materials*; Bloor, B. G., Brook, R. J., Flemings, M. C., Mahajan, S., Cahn, R. W., Eds.; Pergamon: Oxford, UK, 1994; pp 887–895.
2. Seilers, R. Polyvinylidene fluoride in the chemical industries. In *Modern Fluoropolymers*; Scheirs, J., Ed.; Wiley: New York, 1997; Chapter 25, pp 487–506.
3. Hougham, G.; Cassidy, P. E.; Johns, K.; Davidson, T. *Fluoropolymers: Synthesis and Applications*; Plenum Press: New York, 1999; Vol. 1–2.
4. Ameduri, B.; Boutevin, B. *Well-Architectured Fluoropolymers: Synthesis, Properties and Applications*; Elsevier: Amsterdam, The Netherlands, 2004.
5. Ameduri, B. *Chem. Rev.* **2009**, *109*(12), 6632.
6. Boschet, F.; Ameduri, B. *Chem. Rev.* **2014**, *114*, 927.
7. Patil, Y.; Ameduri, B. *Prog. Polym. Sci.* **2013**, *38*, 703.
8. Lovinger, A. J.; Davis, G. T.; Furukawa, T.; Broadhurst, M. G. *Macromolecules* **1982**, *15*, 323.
9. Uchidari, M.; Iwamoto, T.; Iwata, K.; Tamur, M. *Rep. Prog. Phys. Jpn.* **1979**, *22*, 345.
10. Guan, F.; Wang, J.; Yang, L.; Tseng, J.-K.; Han, K.; Wang, Q.; Zhu, L. *Macromolecules* **2011**, *44*, 2190.
11. Guan, F.; Pan, J.; Wang, J.; Wang, Q.; Zhu, L. *Macromolecules* **2010**, *43*, 384.
12. Guan, F.; Wang, J.; Pan, J.; Wang, Q.; Zhu, L. *Macromolecules* **2010**, *43*, 6739.
13. Zhu, L.; Wang, Q. *Macromolecules* **2012**, *45*, 2937.
14. Boutevin, B.; Pietrasanta, Y. *Les Acrylates et Polyacrylates Fluorés*; Erec: Paris, France, 1989.
15. (a) Chiantore, O.; Lazzari, M.; Aglietto, M.; Castelvetro, V.; Ciardelli, F. *Polym. Degrad. Stab.* **2000**, *67*, 461. (b) Alessandrini, G.; Aglietto, M.; Castelvetro, V.; Ciardelli, F.; Peruzzi, R.; Toniolo, L. *J. Appl. Polym. Sci.* **2000**, *76*, 962. (c) Lazzari, M.; Chiantore, O.; Castelvetro, V. *Polym. Int.* **2001**, 50, 863.
16. Xiong, S. D.; Guo, X. L.; Li, L.; Wu, S.; Chu, P. K.; Xu, Z. *J. Fluorine Chem.* **2010**, *131*, 417.
17. Obata, M.; Matsuura, N.; Mitsuo, K.; Nagai,H.; Asai, K.; Harada, M.; Hirohara, S.; Tanihara, Ma.; Yano, S. *J. Polym. Sci. A: Polym. Chem.* **2010**, *48*, 663.

18. (a) Princi, E.; Vicini, S.; Pedemonte, E.; Arrighi, V.; McEwen, I. J. *J. Appl. Polym. Sci.* **2007**, *103*, 90. (b) Alyamac, E.; Soucek, M. D. *Prog. Org. Coat.* **2011**, *71*, 213.
19. (a) He, G.; Zhang, G.; Hu, J.; Sun, J.; Hu, S.; Li, Y.; Liu, F.; Xiao, D.; Zou,H.; Liu, G. *J. Fluorine Chem.* **2011**, *132*, 562. (b) Zhang, S.; Zhao, J.; Chu, G.; Zhang, L.; Xu, A.; Liu, G. *J. Fluorine Chem.* **2011**, 132, 915.
20. Koizumi, S.; Tadano, K.; Tanaka, Y.; Shimidzu, T.; Kutsumizu, S.; Yano, S. *Macromolecules* **1992**, *95*, 6563.
21. Gerhard-Multhaupt, R. *Ferroelectrics* **1987**, *75*, 385.
22. Gilbert, H.; Miller, F. F.; Averill, S. J.; Schmidt, R. F.; Stewart, F. D.; Trumbull, H. L. *J. Amer. Chem. Soc.* **1954**, *76*, 1074.
23. Conciatore, A. B.; Trapasso, L. E.; Stackman, R. W. Vinylidene Cyanide Polymers. In *Encyclopedia of Polymer Science and Technology*, 1st ed.; John Wiley & Sons, Inc.: New York, 1971; Vol. 14, p 581.
24. Miyata, S.; Yoshikawa, M.; Tasaka, S.; Ko, M. *Polym. J.* **1980**, *12*, 875.
25. Tasaka, S.; Miyasato, K.; Yoshikawa, M.; Miyata, S.; Ko, M. *Ferroelectrics* **1984**, *57*, 267.
26. Harrison, J. S.; Ounaies, Z. Piezoelectric Polymers; NASA/Cr2001–21142. ICASE, 2001; Report No. 43.
27. Thünemann, A. F. *Macromolecules* **2000**, *33*, 1790.
28. Kaji, H.; Miura, N.; Shmidt-Rohr, K. *Macromolecules* **2003**, *36*, 6100.
29. Hui-Lin, H.; Wei-Chang, Y.; Ya-Lien, L.; Tri-Rung, Y. *Appl. Phys. Lett.* **2007**, *91*, 23501.
30. Mohammady, S. Z.; Elkholy, S. S.; Elsabee, M. Z. *Polym. Inter.* **2007**, *56*, 7.
31. Way, T. F.; Hall, H. K., Jr. *Polym. Bull.* **1990**, *24*, 151.
32. (a) Padias, A. B.; Chu, G.; Kalnin, I.; Hall, H. K., Jr. *Polym. Prep.* **1991**, *32*(1), 537. (b) Hall, H. K., Jr.; Padias, A. B.; Chu, G.; Lee, H. Y.; Kalnin, I.; Sansone, M.; Breckenridge, G. *J. Poly. Sci. A: Poly. Chem.* **1992**, *30*(11), 2341.
33. Montheard, J. P.; Mesli, A.; Belfkira, A.; Raihane, M. *Macromol. Rep.* **1994**, *A31*, 1.
34. Montheard, J. P.; Zerroukhi, A.; Ouillon, I.; Raihane, M.; Pham, Q. T. *J. Macromol. Sci. A: Pure. App. Chem.* **1997**, *34*, 291.
35. Arous, M.; Kallel, A.; Kaddami, H.; Lahcini, M.; Belfkira, A.; Raihane, M. *J. Appl. Polym. Sci.* **2009**, *114*, 1094.
36. Raihane, M.; Zerroukhi, A.; Kaddami, H.; Lahcini, M.; Boiteux, G. *Polym. Inter.* **2013**, *62*(4), 684.
37. Office of Basic Energy Science, US Department of Energy: Washington, DC; 2007.
38. Whittingham, M. S. *MRS. Bull.* **2008**, *33*, 411.
39. Fröhlich, H. *Theory of Dielectrics*; Clarendon Press: Oxford, 1958.
40. Wang, Q.; Zhu, L. *J. Polym. Sci. B: Polym. Phys.* **2011**, *49*, 1421.
41. Lazzari, M.; Chiantore, O.; Castelvetro, V. *Polym. Int.* **2001**, *50*, 863.
42. *The Sohio Acrylonitrile Process*; American Chemical Society National Historic Chemical Landmarks: Washington, DC, November 17, 2007.
43. Aoki, K.; Nakamura, K. S. Process of the Preparation of Acrylonitrile and Methacrylonitrile. US Patent 4,600, 541, 1986.
44. Bell, R. A.; Brown, B. E.; Duarte, M.; Howard-Lock, H. E.; Lock, C. J. L. *Can. J. Chem.* **1987**, *65*, 261.

45. Carson, B. B.; Stoughton, R. W. *J. Am. Chem. Soc.* **1928**, *50*, 2825.
46. Montheard, J. P.; Boinon, B.; Rodriguez, F.; Pham, Q. T.; Waton, H. *Makrom. Chem.* **1991**, *192*, 341.
47. Atlas, S.; Raihane, M.; Kharas, G. B.; Hanawa, E.; Hill, B. L.; Davi, J. D.; Mueller, A. *J. Macromol. Sci. A: App. Polym. Sci.* **2010**, *47*(6), 491.
48. Kharas, G. B.; Hanawa, E.; Hill, B. L.; Flynn, K. T.; Bajaj, N. H.; Bittar, Y.; Dow, M. R.; Moloney, S.; Nguyen, T. A.; Pitts, S. A.; Serret, C. A.; Valadez, J.; Vallejo, A. C.; Atlas, S.; Raihane, M. *J. Macromol. Sci. A: App. Polym. Sci.* **2010**, *47*(10), 973.
49. Gilbert, H.; Miller, F. F.; Averill, S. J.; Carlson, E. J.; Folt, V. L.; Heller, H. J.; Stewart, F. D.; Schmidt, R. F.; Trumbull, H. L. *J. Am. Chem. Soc.* **1956**, *78*, 1669.
50. Gilbert, H.; Miller, F. F.; Averill, S. J.; Schmidt, R. F.; Stewart, F. D.; Trumbull, H. L. *J. Am. Chem. Soc.* **1954**, *76*, 1074.
51. Spraque, B. S.; Green, H. E.; Reuter, L. F.; Smith, R. D. *Angew.Chem.* **1962**, *1*, 425.
52. Grassie, N.; McGuchan, R. *Eur. Polym. Mater.* **1972**, *8*, 245.
53. Clingerman, M. L. Development and Modelling of Electrically Conductivité Composites Materials. Ph.D. Dissertation, Michigan Technological University, **2001**.
54. Gupta, A. K.; Chand, N. *J. Polym. Sci. B: Polym. Phys.* **1980**, *18*, 1125.
55. Odian, G. *Principles of Polymerization*, 4th ed.; Wiley: New York, **2004**.
56. Way, T.F.; Hall, H. K., Jr. *Polym. Bull.* **1990**, *24*, 151.
57. Ilie, C. C.; Rosa, L. G.; Poulsen, M.; Takacs, J. M. *Integr. Ferroelectr.* **2011**, *125*, 98.
58. Xiao, J.; Rosa, L. G.; Poulsen, M.; Qfeng, D.; Sahadeva Reddy, D.; Takacs, J. M.; Cai, L.; Zhang, J.; Ducharme, S.; Adowben, P. *J. Phys. Condens. Matter.* **2006**, *18*, L155.
59. Poulsen, M.; Ducharme, S.; Sorokin, A. V.; Reddy, S.; Takas, J. M.; Wen, Y.; Kim, J.; Adenwalla, S. *Ferroelectrics Lett.* **2005**, *32*, 91.
60. (a) Blaise, J.; Grimaud, E. (Elf Atochem S.A.) US Patent 4,025,709, 1977. (b) Bacque, X.; Lasson, P. (Solvay S.A). Eur. Pat. Appl. 387938, 1990.
61. (a) Sangermano, M.; Bongiovanni, R.; Malucelli, G.; Priola, A.; Thomas, R. R.; Kausch, C. M.; Kim, Y. *J. Polm. Sci. A: Polym. Chem.* **2006**, *44*, 6943. (b) Kwon, S.; Bae, W.; Lee, K.; Byun, H. S.; Kim, H. J. *J. Chem. Eng. Data.* **2007**, *52*, 89.
62. (a) Choi, D.; Yeom, E. H.; Park, M.; Kim, J. K.; Kim, B. C. *J. Appl. Polym. Sci.* **2004**, *93*, 2082. (b) Vasilopoulou, M.; Douvas, A. M.; Palilis, L. C.; Bayiati, P.; Alexandropoulos, D.; Stathopoulos, N. A.; Argitis, P. *Microelectron. Eng.* **2009**, *86*, 1142.
63. (a) Ha, J. W.; Park, I. J.; Lee, S. B. *Macromolecules* **2005**, *38*, 736. (b) Liu, Z. T.; Chen, J. G.; Liu, Z. W.; Lu, J. *Macromolecules* **2008**, *41*, 6987. (c) Liang, J. Y.; He, L.; Zheng, Y. S. *J. Appl. Polym. Sci.* **2009**, *112*, 1615.
64. Passaglia, E.; Aglietto, M.; Ciardelli, F.; Mendez, B. *Polym. J.* **1994**, *26*, 1118.
65. Liu, W.; Nakano, T.; Okamoto, Y. *J. Polym. Sci. A: Poly. Chem.* **2000**, *38*, 1024.
66. Le, D. V.; Kendrick, M. M.; O'Rear, E. A. *Langmuir* **2004**, *20*, 7802.
67. Narita, T.; Hagiwara, T.; Hamana, H. *Makromol. Chem. Rapid. Commun.* **1985**, *6*, 5.
68. Merino, E. G.; Atlas, S.; Raihane, M.; Belfkira, A.; Lahcini, M.; Hult, A.; Dionísio, M.; Correia. N. *Eur. Poly. J.* **2011**, *47*, 1429.
69. Tanaka, H.; Okazaki, T.; Tezuka, Y.; Hongo, T.; Takahashi, Y. *Polymer* **2002**, *43*, 1189.
70. Montheard, J. P.; Boinon, B.; Raihane, M.; Pham, Q. T. *Polym. Commun.* **1991**, *32*, 567.

71. Monthéard, J. P.; Boinon, B.; Belfkira, A.; Raihane, M.; Pham Q. T. *Makromol. Chem.* **1993**, *194*(10), 2839.

72. Raihane, M.; Montheard, J. P.; Boiteux, G. *Macromol. Chem. Phys.* **2000**, *201*, 2365.

73. Raihane, M.; Ameduri, B. *J. Fluorine Chem.* **2006**, *127*, 391.

74. Brar, A. S.; Sunita. *Eur. Polym. J.* **1992**, *7*, 803.

75. Brar, A. S.; Sunita. *Polymer* **1993**, *34*, 3391.

76. Alfrey, T.; Price, C. C. *J. Polym. Sci.* **1974**, *2*, 101.

77. Atlas, S.; Raihane, M.; Lahcini, M.; Hult, A.; Malkoch, M.; Ameduri, B. *J. Polym. Sci. A: Poly. Chem.* **2013**, *51*, 3856–3866. doi:10.1002/pola.26783

78. Fineman, M.; Ross, S. D. *J. Polym. Sci.* **1950**, *5*, 259.

79. Kelen, T.; Tüdos, F. *J. Macromol. Sci. Chem.* **1975**, *A9*, 1.

80. (a) Jenkins, A. D. *Eur. Polym. J.* **1989**, *25*, 721. (b) Jenkins, A. D. *J. Macromol. Symp.* **1996**, *111*, 159. (c) Jenkins, A. D. *J. Polym.Sci. A: Polym. Chem.* **1996**, *34*, 3495. (d) Jenkins, A. D. *J. Polym. Sci. A: Polym. Chem.* **1999**, *37*, 113. (e) Jenkins, A. D. *Macromol. Symp.* **2001**, *111*, 159.

81. Igarashi, S. *J. Polym. Sci. Polym. Lett. Ed.* **1963**, *1*, 359.

82. Maruyama, Y.; Jo, Y. S.; Inoue, Y.; Chûjô, R.; Tasaka, S.; Miyata, S. *Polymer* **1987**, *28*(7), 1087.

83. Inoue, Y.; Kawaguchi, K.; Maruyama, Y.; Jo, Y. S.; Chûjô, R.; Seo, I.; Kishimoto, M. *Polymer* **1989**, *30*(4), 698.

84. Inoue, Y.; Kashiwazaki, A.; Maruyama, Y.; Jo, Y. S.; Chûjô, R.; Seo, I.; Kishimoto, M. *Polymer* **1988**, *29*(1), 144.

85. Ohta, Y.; Inoue, Y.; Chûjô, R.; Kishimoto, M.; Seo, I. *Polymer* **1990**, *31*(8), 1581.

86. Raihane, M.; Castelvetro, V.; Bianchi, S.; Atlas, S.; Ameduri, B. *J. Polym. Sci. A: Polym. Chem.* **2010**, *48*(21), 4900.

87. Castelvetro, V.; Raihane, M.; Bianchi, S.; Atlas, S.; Bonaduce, I. *Polym. Degr. Stab.* **2011**, *96*, 204.

88. Atlas, S.; Raihane, M.; Kharas, G. B.; Kaddami, H.; Arous, H.; Kallel, A. *J. Macromol. Sci. A: Pure App. Chem.* **2012**, *49*(12), 1.

89. Meskini, A.; Raihane, M.; Stevenson-Royaud, I.; Boiteux, G.; Seytre, G.; Ameduri, B. *J. Non-Cryst. Solids* **2010**, *356*, 688.

90. (a) Matyjaszewski, K.; Davis, T. P. *Handbook of Radical Polymerization*; Wiley-Interscience: New York, **2002**. (b) Matyjaszewski, K., Ed. *Controlled/Living Radical Polymerization: From Synthesis to Materials*; ACS Symposium Series 944; American Chemical Society: Washington, DC, 2006. (c) Matyjaszewski, K.; Gnanou, Y.; Liebler, L. *Macromolecular Engineering: Synthetic Aspects*; Wiley VCH: Weiheim, Germany, 2007; Vol. 1, pp 171–204.

91. Boyer, C.; Valade, D.; Ameduri, B.; Lacroix Desmazes, P.; Boutevin, B. *J. Polym. Sci. A: Polym. Chem.* **2006**, *44*, 5763.

92. Kostov, G.; Boschet, F.; Buller, J.; Badache, L.; Brandstadter, S.; Ameduri, B. *Macromolecules* **2011**, *44*, 1841.

93. Girard, E.; Marty, J. D.; Ameduri, B.; Destarac, M. *ACS Macro.Lett.* **2012**, *1*, 270.

94. Patil, Y.; Ameduri, B. *Polym. Chem.* **2013**, *4*, 2783.

95. (a) Oka, M.; Tatemoto, M. Vinylidene Fluoride-hexafluoropropylene Copolymer Having Terminal Iodines. In *Contemporary Topics in Chemical Reviews*; Bailey, W. J., Tsuruta, T., Eds.; Plenum Press: NewYork, **1984**; Vol. 4, pp 763–781. (b) Tatemoto, M. *Int. Polym. Sci. Technol.* **1985**, *12*, 85; Translated in English from Nippon GomuKyokaishi, **1985**, *57*, 761. (c) Tatemoto, M.; Shimizu, T. Thermoplastic Elastomers. In *Modern Fluoropolymers*; Scheirs, J., Ed.; Wiley: New York, 1997; Chapter 30, pp 565–576.
96. Boyer, C.; Valade, D.; Sauguet, L.; Ameduri, B.; Boutevin, B. *Macromolecules* **2005**, *38*, 10353.
97. Luo, Y.R. *Comprehensive Handbook of Chemical Bond Energies*; CRC Press: Boca Raton, FL, **2007**; pp 9–75.
98. Boutevin, B.; David, G.; Boyer, C.; Tonnar, J. F.; Lacroix-Desmazes, P.; Ameduri, B.; *Chem. Rev.* **2006**, *106*, 3936.
99. Fox, T. G.; Flory, P. J. *J. Appl. Phys.* **1950**, *21*, 581.
100. Andray, A. L. *Polyacrylonitrile Polymer Data Handbook*; Oxford University Press Inc., **1999**; p 254.
101. Nakagawa, K.; Ishida, Y. *J. Polym. Sci. B: Polym. Phys.* **1973**, *11*, 2153.
102. Kao, K. C. *Dielectric Phenomena in Solids: With Emphasis on Physical Concepts of Electronic Processes*; Elsevier, Academic Press: Boston, MA, 2004.
103. Ducharme, S. *ACS Nano.* **2009**, *3*, 2447.
104. Li, J.; Seok, S. I.; Chu, B.; Dogan, F.; Zhang, Q.; Wang, Q. *Adv. Mater.* **2009**, *21*, 217.
105. Kremer, F.; Schönhals A. *Broadband Dielectric Spectroscopy*; Springer-Verlag: Berlin, Germany, 2003.
106. (a) Havriliak, S.; Negami, S. *Polymer* **1967**, *8*(4), 161. (b) Havrilia, S.; Negami, S. *J. Polym. Sci. C: Polym. Symp.* **1966**, *14*, 99.
107. Meskini, A.; Raihane, M.; Ameduri, B.; Hakme, C.; Sage, D.; Stevenson, I.; Boiteux, G.; Seytre, G.; Kaddami, H. *Eur. Polym. J.* **2009**, *45*(3), 804.
108. Furukawa, T.; Nakajima, K.; Koizumi, N.; Date, M. *Jpn. J. App. Phys.* **1987**, *26*, 1039.
109. Zou, D.; Iwasaki, S.; Tsutsui, T.; Saito, S.; Kishimoto, M.; Seo, I. *Polymer* **1990**, *321*, 1888.
110. Hedvig, P. *Dielectric Spectroscopy of Polymers*; Adam Hilger: Bristol, UK, 1977.
111. Schönhals, A.; Kremer, F. Analysis of dielectric spectra. In *Broadband Dielectric Spectroscopy*; Schönhals, A.; Kremer, F. Eds.; Springer-Verlag: Berlin, Germany, **2003**; Chapter 3.
112. Richert, R.; Blumen, A. *Disorder Effects on Relaxational Processes: Glasses, Polymers, Proteins*; Springer: Berlin, Germany, 1994.
113. Blythe, A. R. *Electrical Properties of Polymers*; Cambridge University Press: Cambridge, UK, 1979.
114. Schwarz, M.; Kitchman, L. A.; Tucker, R. W.; Nelson, E. R. *J. Chem. Eng. Data.* **1970**, *15*(2), 341.
115. Dubois, J. C. Technique de l'ingénieur, traité Plastiques et composites, Référence AM3140, Date de publication: 10 avr. 2001.
116. Boyer, R. F.; Turley, S. G. In *Molecular Motion in Polystyrene*; Meier, D. J.; Gordon & Breach Science Publisher: London, UK, 1978; Vol 4, p 333.
117. (a) Macedo, P. B.; Moynihan, C. T.; Bose, R. *Phys. Chem. Glasses* **1972**, *13*, 171. (b) Tsangaris, G. M.; Psarras, G. C.; Kouloumbi, N. *J. Mater. Sci.* **1998**, *33*, 2027.

118. Yano, S. *J. Polym. Sci. A: Polym. Chem.* **1970**, *28*, 1057.
119. Kakutani, H. *J. Polym. Sci. A: Polym. Chem.* **1970**, *28*, 1177.
120. Hahn, B.; Wendorff, J.; Yoon, D. Y. *Macromolecules* **1985**, *18*(4), 718.
121. Capaccioli, S.; Castelvetro, V.; Rolla, P. A.; Raihane, M.; Atlas, S.; Meskini, A. α-and β-Dielectric Relaxations in Fluorinated Methacrylate Polymers and Copolymers of Vinylidene Cyanide. *Results Presented at the 6th International Conference on Broadband Dielectric Spectroscopy and Its Application*, Madrid (Spain), September 7–10, 2010; Paper 23, p 201.
122. Kuebler, S. C.; Schaefer, D. J.; Boeffel, C.; Pawelzik, U.; Spiess, H. W. *Macromolecules* **1997**, *30*, 6597.
123. Kahle, S.; Korus, J.; Hempel, E.; Unger, R.; Höring, S.; Schröter, S.; Donth, E. *Macromolecules* **1997**, *30*, 7214.
124. Zhou, Z.; Carr, J.; Mackey, M.; Yin, K.; Schuele, D.; Zhu, L.; Baer, E. *J. Polym. Sci. B: Polym. Phys.* **2013**, *51*, 978.
125. Carr, J. M.; Mackey, M.; Flandin, L.; Schuele, D. E.; Zhu, L.; Baer, E. *J. Polym. Sci. B: Polym. Phys.* **2013**, *51*, 882.
126. Yang, L.; Li, X.; Allahyarov, E.; Taylor, P. L.; Zhang, Q. M.; Zhu, L. *Polymer* **2013**, *54*, 1709.
127. Sarjeant, W. J.; Zirnheld, J.; MacDougall, F. W. *IEEE Trans. Plasm. Sci.* **1998**, *26*, 1368.
128. Tan, Q.; Irwin, P.; Cao, Y. *IEEJ Trans. Fund. Mate.* **2006**, *126*, 1153.
129. Lovinger, A. J. *Science* **1983**, *220*, 1115.
130. Lu, Y.; Claude, J.; Zhang, Q. M.; Wang, Q. *Macromolecules* **2006**, *39*, 6962.
131. Karawasa, N.; Goddard, W. A. *Macromolecules* **1992**, *25*, 7268.
132. Chu, B.; Zhou, X.; Ren, K.; Neese, B.; Lin, M.; Wang, Q.; Bauer, F.; Zhang, Q. M. *Science* **2006**, *313*, 1887.

21

FLUOROPOLYMERS—ENVIRONMENTAL ASPECTS

Klaus Hintzer and Werner Schwertfeger
3M/Dyneon GmbH

21.1 INTRODUCTION

In the early 1930s, researchers of IG-Farbenindustrie in Hoechst/Frankfurt (Germany) studied systematically the first polymerizations of fluoroethenes; the Hoechst researchers had already prepared polychlorotrifluoroethylene (PCTFE) and polytetrafluoroethylene (PTFE) including copolymers, recognizing the outstanding properties of these polymers [1]. The first patent application for a fluoropolymer was filed in October 1934 by Schloffer/Scherer [2]. PTFE was also discovered independently by chance in 1938 by Plunkett from DuPont (USA) while investigating fluorinated refrigerants. The unique properties of PTFE were recognized during the Manhattan Project; where there was an urgent need for a material withstanding the highly corrosive environment during the process of separating the isotopes of UF_6 for the first atomic bomb. PTFE apparently fulfilled all the needs, spurring the development of processing and production methods for these unique polymers. In 1946, PTFE was commercialized by DuPont under the trademark Teflon® [3].

Fluoropolymers are defined as polymers consisting of only carbon and fluorine in the polymer backbone. For this reason, fluorosilicones and fluoroacrylate polymers are quite often not referred to as fluoropolymers.

The unique properties of fluoropolymers are due to the fact that the polymer backbone is formed by the strong carbon–carbon bond (C–C ~340 kJ/mol) and the

Handbook of Fluoropolymer Science and Technology, First Edition. Edited by Dennis W. Smith Jr., Scott T. Iacono, and Suresh S. Iyer.

extremely stable carbon–fluorine bond (C–F ~490 kJ/mol; for comparison: C–H ~420 kJ/mol). Substitution of fluorine for hydrogen in a polymer improves three key physical properties:

- increased service temperatures and reduced flammability;
- the low surface energy provides nonstick properties/anti-adhesiveness, low coefficient of friction, self-lubricating effects, and lower solubility in hydrocarbons;
- excellent electrical and optical properties in low–high frequency loss rates and low refractive indices.

PTFE, PCTFE, and all other fluoropolymers (Figure 21.1) gained immediate acceptance during commercialization in the various markets. During the ensuing decades, many fluoropolymers, including fluorothermoplastics and fluoroelastomers, were developed. The entire sales volume for fluoropolymers is today more than 230,000 tons per year (world consumption for fluoroplastics in 2012: 216,000 tons [4]; world consumption for fluoroelastomers in 2009: ~20,000 tons [5]). The total market value is more than US$ 4 billion.

Due to the outstanding product properties, fluoropolymers are indispensable materials. Fluoropolymers are serving highly demanding applications in a diverse range of industries (Table 21.1) which no other polymer class can provide.

As with all matured polymers, but in particular due to the persistent nature of fluoropolymers, all individual steps of the whole life cycle (Figure 21.2) are scrutinized under environmental and sustainability aspects.

21.2 MONOMER SYNTHESIS, POLYMERIZATION/ WORKUP PROCESSES

All commercial large-scale synthesis for the five major fluoromonomers are based on a chlorination/fluorination processes of C_1/C_2 hydrocarbons, mostly including a dehydrohalogenation step at higher temperatures [7]. See, for example, the synthesis of the most important monomers TFE/HFP in Scheme 21.1. These manufacturing processes are quite energy consuming (e.g., the making of 1 ton TFE requires almost 5 tons of chlorine and the energy consumption is >10,000 kWh/ton TFE).

$$CH_4 + Cl_2 \longrightarrow CHCl_3 + 3\,HCl$$

$$CHCl_3 + HF \longrightarrow CHClF_2\ (\text{R 22}) + 2\,HCl$$

$$2\,CHClF_2 \xrightarrow{\Delta} F_2C{=}CF_2,\ CF_3CF{=}CF_2 + 2\,HCl$$

SCHEME 21.1

During the 1980s, it was recognized that chlorofluorocarbons (CFCs) and chlorofluorohydrocarbons have a significant ozone depleting potential; with the Montreal Protocol (1987) [8] all critical CFC's were banned for dispersive applications and

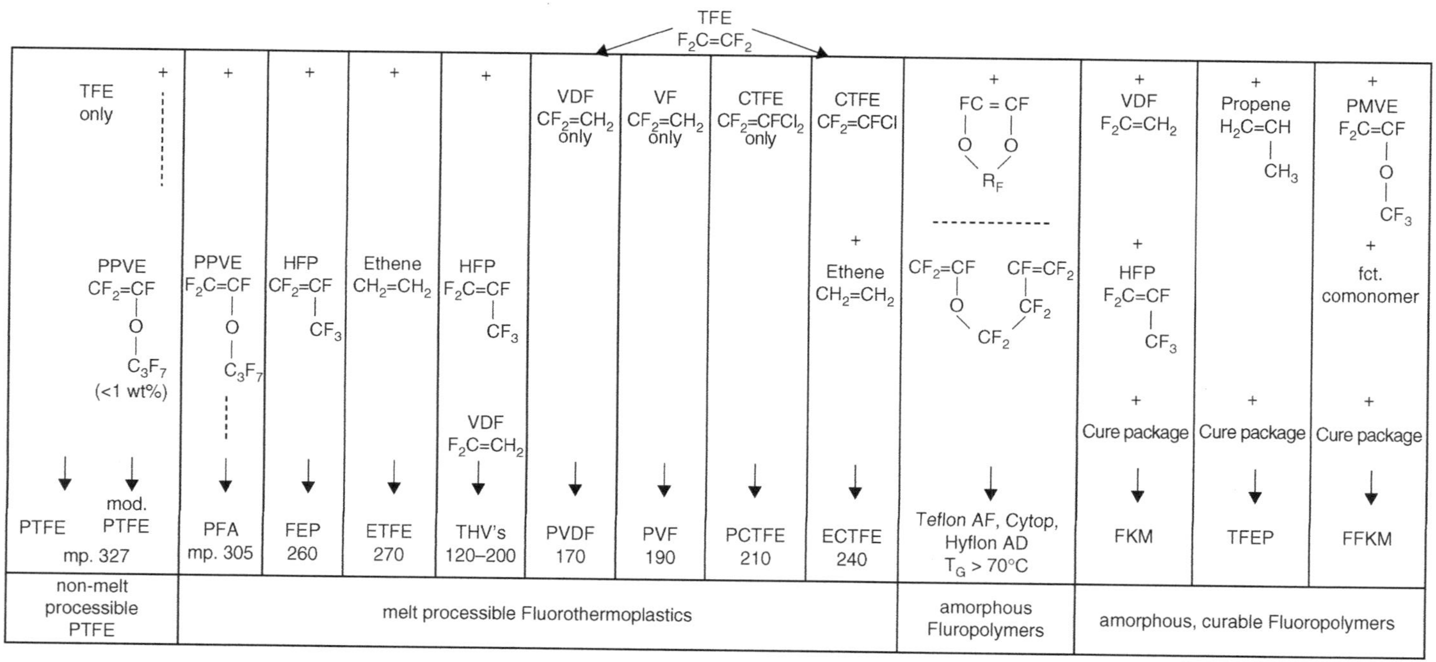

FIGURE 21.1 Fluoropolymer technology base.

TABLE 21.1 Major Applications of Fluoropolymers [6].

Industry/ application area	Key properties	Typical uses
Chemical processing	Chemical resistance Good mechanical properties Thermal stability Cryogenic properties	Gaskets, vessel liners, valve and pipe liners, tubing, coatings
Electrical and communications	Low dielectric constant High volume/surface resistivity High dielectric breakdown voltage Flame resistance, thermal stability Low refractive indices	Wire and cable insulation, connectors, optical fibers
Automotive and office equipment	Low coefficient of friction Good mechanical properties Cryogenic properties Chemical resistance Low permeation properties	Seals and rings in automotive power steering, transmission, and air-conditioning. Copier roller and food processing equipment covering. Fuel management systems
Houseware	Thermal stability Low surface energy Chemical resistance	Cookware coatings
Medical	Low surface energy Stability Excellent mechanical properties Chemical resistance	Cardiovascular grafts, heart patches, ligament replacement
Architectural fabrics/films	Excellent weatherability Flame resistance Low surface energy	Coated fabrics and films for buildings and roofs, front/backside films for solar applications
Polymer additives	Low coefficient of friction Flame resistance	Polyolefin processing to avoid surface defects Anti-dripping agents
Semiconductor industry	Chemical resistance High purity Thermal stability	Process surfaces Wafer carriers Tubing, valves, pumps, and fittings

as refrigerants, with ample transition times provided to the industry to phase out their use. The making of fluoromonomers based on chlorine/fluorine containing precursors was further allowed, but with significant limitations on gaseous emissions. The German-based Hoechst AG established during the 1980s proactively closed loop systems and thermal oxidizers in order to incinerate off-gas streams from the overall monomer synthesis (also the exposure limits of some fluoromonomers (e.g., HFP) are low [9]), as well as any kind of off-gases emissions from polymerization and workup processes. The challenge in building up such thermal oxidizers is to cope with the extreme corrosion in the combustion units as well as in the HF-absorbers and to recycle the CaF_2 back into the process.

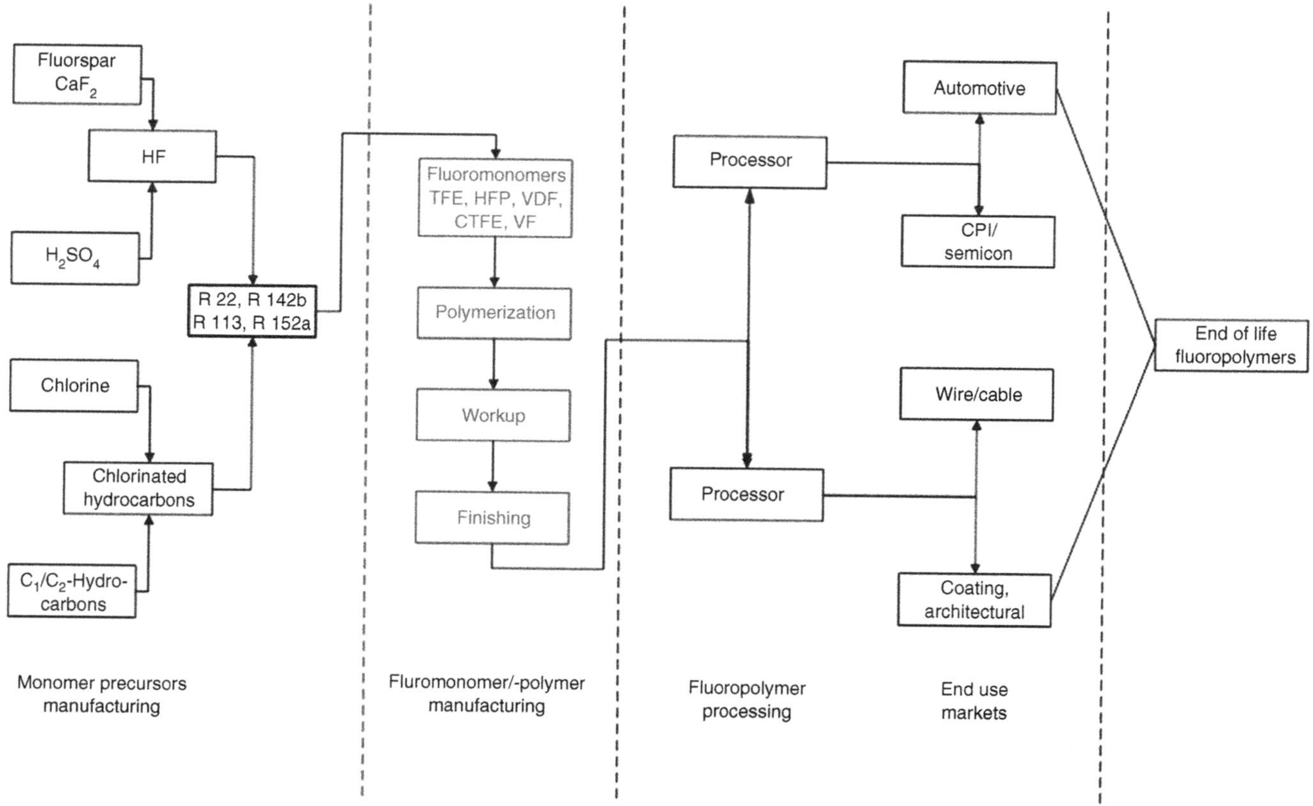

FIGURE 21.2 Fluoropolymer value chain.

The principal method for synthesizing fluoropolymers is free radical polymerization, as the other typical methods, for example, cationic polymerization proves ineffective, due to the electrophilic nature of fluoroolefins. Fluoroolefins can be polymerized by anionic catalysts, but termination by fluoride ion elimination prevents formation of high molecular weights. Coordination catalysts do not effect polymerization of fluoroolefins.

The free radical polymerizations are mostly water based, either as aqueous suspension polymerization (mostly applied for PTFE-polymers) or as aqueous emulsion polymerization in the presence of emulsifiers, most preferably in the presence of fluorinated emulsifiers. The use of such fluorinated emulsifiers, however, provides their own challenges (see chapter 21.3).

In previous times, radical polymerization of fluorinated olefins in chlorinated fluorocarbons (e.g., in R113—CF_2Cl-$CFCl_2$) for fluorinated copolymers was widespread; also many workup processes (e.g., agglomeration steps) used chlorinated fluorocarbons. Due to the high emissions of these ozone-depleting solvents and due to the Montreal Protocol, these processes had to be changed to either environmentally friendly solvents (e.g., to CF_3-$(CF_2)_4$-CF_2H for ETFE polymerizations [10]) or completely to water-based systems.

The polymerization in supercritical media (e.g., in Sc CO_2)—originally introduced as the alternative "green" polymerization technology—did not find widespread applicability [3, 11]. Other free radical polymerization processes, for example, in the gas phase or in ionic liquids are still of academic interest, but are not widely adopted in industrial manufacturing processes.

Tetrafluoroethene has high tendency to self-decomposition into carbon and carbon tetrafluoride:

$$F_2C = CF_2 \rightarrow C + CF_4 \quad \Delta H\ 320\ \mathrm{kJ/mol}$$

The generated power of this decomposition is close to explosives and can be initiated by so-called "hotspots" formed during the PTFE-polymerization (primarily during suspension PTFE-polymerizations). Due to the high heat of polymerization (~170 kJ/mol) and the excellent insulating properties of PTFE, "hotspots" can develop if the mixing/agitation and cooling of the aqueous suspension in the kettle are not adequate or if—in the case of aqueous emulsion polymerization—coagulation occurs during the polymerization. The hotspots float to the water surface and ignite the TFE-gas phase. One of the major challenges for PTFE manufacturers was to avoid these violent explosions; for example, by polymerizing PTFE at low temperatures and TFE pressures [12]. It was also recommended to run PTFE-suspension polymerizations under inert gas blankets [13] and PTFE-emulsion polymerizations in the presence of "tamers" which become active at higher temperatures [14]. The risk of violent explosions requires that the entire polymerization equipment system, including transfer lines of TFE-monomers, is protected by rupture disks and flame arrestors. Fluoropolymer producers are sharing their findings and experiences to prevent these types of incidents and to reduce risks in handling fluoropolymers [15–17].

21.3 FLUORINATED EMULSIFIERS AND ALTERNATIVES

21.3.1 General Aspects

The emulsion (co)polymerization of fluorinated, especially perfluorinated fluoromonomers requires an emulsifier to stabilize the polymer particles in the liquid phase [18]. Besides the ability to stabilize the particles and to prevent coagulation, the emulsifier needs to be inert under the polymerization conditions. In particular, chain transfer reactions should be avoided under any circumstances, as they tend to decrease the molecular weight of a polymer, whereas the desired molecular weight is usually high for fluoropolymers [19]. Thus, from the very early days of emulsion polymerization highly or preferably perfluorinated emulsifiers have been used. These include perfluorinated carboxylic acids as well as sulfonic acids in the form of their salts, especially ammonium salts. Although the variety of salts fulfilling the requirements is high and quite a few of them are claimed in the patent literature, only a limited number of substances were actually used in industrial applications. These include the ammonium salts of perfluorooctylsulfonic acid (PFOS) and perfluorooctanoic acid (PFOA), respectively. Application of PFOS and its ammonium salt as polymerization emulsifier was minimized over time in favor of the PFOA ammonium salt (APFO); PFOS was mainly used for the production of impregnation materials. Starting in the last decade, PFOA and similar substances were replaced by new substances.

21.3.1.1 Production and Fate of Perfluorinated Emulsifiers in the Past Industrial production of PFOS and PFOA is predominantly based on electrochemical fluorination (ECF) of hydrocarbon precursors followed by hydrolysis of the acid fluoride and salt formation [7, 20] (Scheme 21.2).

$$C_8H_{17}\text{-}SO_2Cl + HF \longrightarrow C_8F_{17}\text{-}SO_2F \longrightarrow C_8F_{17}\text{-}SO_3H \longrightarrow [NH_4][C_8F_{17}\text{-}SO_3]$$

$$C_7H_{15}\text{-}COCl + HF \longrightarrow C_7F_{15}\text{-}COF \longrightarrow C_7F_{15}\text{-}COOH \longrightarrow [NH_4][C_7F_{15}\text{-}CO_2]$$

SCHEME 21.2

According to their synthetic pathways, PFOS and PFOA were rarely used as pure substances but consisted of homologs and isomers that were separated from the ECF product mix. APFO was produced by an alternative pathway as well, which resulted in the pure straight chain substance. The telomerization of perfluoroethyl iodide with tetrafluoroethene (TFE) results in homologs [21], but the perfluorooctyl iodide can be separated and purified by fractional distillation. The iodide is converted to the carboxylic acid fluoride using fuming sulfuric acid and finally hydrolyzed with water to give PFOA [22] (Scheme 21.3).

$$CF_2{=}CF_2 + IF_5 + I_2 \longrightarrow CF_3\text{-}CF_2\text{-}I + CF_2{=}CF_2 \longrightarrow CF_3\text{-}CF_2\text{-}(CF_2\text{-}CF_2)_3\text{-}I \rightarrow \rightarrow$$

$$[NH_4][CF_3\text{-}CF_2\text{-}(CF_2\text{-}CF_2)_2\text{-}CF_2\text{-}CO_2]$$

SCHEME 21.3

An alternative route to nonbranched perfluorinated carboxylic acids starts from trifluoromethyliodide. After telomerization with TFE, a mixture of odd-numbered telomer iodides is obtained which can be separated by fractional distillation [21]. Predominantly the C7–C9 cut ($n = 3$ and 4) was used to produce the corresponding acid fluorides which are converted to the ammonium salt via standard procedures (Scheme 21.4). Another route to straight chain PFOA is a recrystallization process starting from a mixture of straight chain and branched PFOA. Trichloromethane is used as solvent in this purification procedure [23].

$$CF_3I + CF_2{=}CF_2 \longrightarrow CF_3\text{-}(CF_2\text{-}CF_2)_n\text{-}I \longrightarrow CF_3\text{-}(CF_2\text{-}CF_2)_{n-1}\text{-}COF \longrightarrow [NH_4][CF_3\text{-}(CF_2\text{-}CF_2)_{n-1}\text{-}CO_2]$$

SCHEME 21.4

Although fluorinated emulsifiers are high-priced chemicals, no serious attempts were made for many years to recover and recycle these substances. Depending on the workup and application, they remained on the fluoropolymer and were destroyed during further processing, especially during thermal procedures like sintering and extrusion. But precipitation of the fluoropolymers from the polymerization latices and drying processes at low temperatures resulted in the release of emulsifiers to the environment, either airborne or by aqueous media. More than 80% of the emulsifier is found in the aqueous process effluent while the rest might remain on the polymer [24].

Industrial wastewater treatment plants can only remove a part of the emulsifiers from the aqueous effluent by absorption to the sludge as they are practically not biodegradable and cannot be precipitated as Ca-salts. As PFOA and PFOS are environmentally persistent, a wide range of chemical processes have been developed to treat (highly) diluted aqueous solutions of fluorocarbon acids [25].

Early attempts to remove and recycle PFOA are described by Kuhls and Weiss [26]. Up-concentration of fluoropolymer dispersions via ultrafiltration gave a permeate, which showed a high portion of the emulsifier originally contained in the untreated dispersion. The permeate was further treated with a weakly basic anion exchanger resulting in an adsorption of the fluorinated emulsifier reaching a recovery rate of 80% and more. The loaded ion exchange resin was treated with a regeneration mixture of aqueous ammonia to elute the PFOA. Another regeneration mixture is described by Kuhls [27], which consists of water, a water-miscible organic solvent, and a mineral acid and can remove up to 100% of the loaded PFOA from the anion exchange resin.

21.3.1.2 Recovery/Removal and Recycling of Perfluorinated Acids from Fluoropolymers and Industrial Waste Streams In the last decade of the twentieth century, PFOS and PFOA as well as other fluorinated substances came under scrutiny because they are environmentally persistent and potentially bioaccumulative. Analytical developments resulted in the detection of these substances widespread in the environment [28]. Triggered by these environmental concerns, as well as by the high price of the fluorinated emulsifiers, methods were developed which allowed effective recovery and recycling of fluorinated emulsifiers from these industrial production processes.

21.3.1.2.1 From Off-Gases Off-gases, especially from polymer drying processes, often contain substantial amounts of PFOA or its salts. Scrubbing such gases with aqueous NaOH solution [29] or K_2CO_3 solution [30] results in the absorption of the PFOA. In the case of K_2CO_3 as scrubber fluid, the PFOA K-salt is formed which has a low solubility in the scrubber solution and can be removed from the system as floating solid. Recycling of the PFOA is achieved by acidification of the mixture followed by esterification, distillation, and finally conversion of the pure ester to the ammonium salt APFO [31].

21.3.1.2.2 From Wastewater Streams Coagulation of fluoropolymer dispersions and subsequent washing of the precipitated polymers with water results in large volumes of wastewater streams with relatively low concentration of perfluorinated emulsifier. In addition, it contains other anions which are part of the polymerization process or are formed during the polymerization. Depending on the polymerization recipe, oxalic acid, sulfate, peroxodisulfate, fluoride, and other anions can be present. Anion exchange resins themselves may contain hydroxyl, chloride, and sulfate anions. Except for peroxodisulfate, all other anions can be replaced from the anion exchange resin by the perfluorinated acids during the loading process. This allows high loading of the resin and results in late breakthrough of the perfluorinated acid.

Application of the ion exchange process described by Kuhls and Weiss [26] to separate the perfluorinated emulsifier from the wastewater stream results in rapid clogging of the ion exchange column. The reason for this behavior is due to the fact that the wastewater contains small amounts of colloidal fluoropolymer particles which are not visible to the naked eye. These particles are kept in solution by the presence of the emulsifier which is absorbed on the particle surface. Removing the absorbed emulsifier by binding it chemically to the anion exchange resin results in the flocculation of the residual polymer which stays in the ion exchange resin bed and ultimately clogs the bed. Pretreatment of the wastewater to remove the polymer particles using chemical and physical methods (e.g., flocculation, filtration of flocculated particles) was claimed [32], but ultimately a robust process was developed by the addition of small amounts of a nonionic emulsifiers to the wastewater stream [33]. This stabilizes the polymer particles after removing the fluorinated emulsifier and results in a fluidized ion exchange bed which can be reused after regeneration without decrease of activity. Any type of basic anion exchange resin is suitable for this process, but the best results were achieved with strongly basic resins. Regeneration of the loaded resin can be carried out using various eluting systems, for example, with bases like NaOH, NH_3, or quarternary ammonium hydroxide in an aqueous solvent mixture [34], ammonium salts in aqueous solvent [35], or mineral acid in a water–solvent mixture [27]. Medium strength anion exchange resins can be regenerated by using a mixture of aqueous ammonia and methanol [24].

21.3.1.2.3 Recovery/Removal via Granulated Active Carbon Bed Granulated active carbon (GAC) beds are used to remove undesired fluorinated low molecular weight substances from manufacturing plant effluents. The first application of this method to perfluorinated emulsifiers was described by Russian authors [36]. Although the active carbon bed eliminates most of the perfluorinated acids from the aqueous effluent, the applied regeneration is not efficient and the regenerated active carbon shows a decreased activity which limits the number of reuses. This was overcome by the addition of small amounts of a nonionic emulsifier to the wastewater stream [37]. No clogging of the active carbon bed was observed. Regeneration of the carbon bed, loaded with PFOA, was carried out by esterifying the PFOA by heating the carbon with a mixture of sulfuric acid, water, and methanol. The PFOA was removed as the methyl ester. The regenerated active carbon can be reused without loss in efficiency.

21.3.1.2.4 Emulsifier Recovery/Removal from Fluoropolymer Dispersions Fluoropolymer dispersions contain all of the fluorinated emulsifier used during the polymerization step unless a part of it is lost during workup (e.g., up-concentration) and stabilization with nonionic nonfluorinated emulsifiers. In any event, some fluorinated emulsifier always remains in the dispersion, as all of these steps cannot remove the fluorinated emulsifier completely, even if they are applied several times, for example, by up-concentration via ultrafiltration with repeated dilution of the up-concentrated dispersion [26]. Although further processing of fluoropolymer dispersions usually destroys the bulk of the fluorinated emulsifier, it is desirable to recover and reuse the expensive polymerization aid completely while keeping or even improving the quality of the dispersions.

Fluoropolymer dispersions may be stabilized by the addition of nonionic surfactants like Triton X 100® or Genapol X 080®. Contacting such a stabilized fluoropolymer dispersion with a strong basic anion exchanger like Amberlite IRA 402® in an ion exchange process removes the fluorinated emulsifier completely from the dispersion without clogging of the ion exchange resin bed or affecting the properties of the dispersion [38, 39]. Instead of using the anion exchange resin in the form of a fixed bed, which is passed by the fluoropolymer dispersion, the anion exchange resin can be added as such to the dispersion and is removed after a given contact time by filtration [40]. In the following workup steps, the treated dispersion is usually up-concentrated and adjusted with further additives for the desired application. After rinsing the loaded anion exchange resin with water, it is exposed to a regeneration liquid to remove the fluorinated emulsifier. The regenerated resin is employed again in the ion exchange process without capacity loss.

21.3.1.2.5 Workup of Regeneration Liquids All of the mentioned ion exchange processes as well as the active carbon bed process and the scrubbing of off-gases yield aqueous (regeneration) liquids containing the fluorinated emulsifier. Depending on the composition of the aqueous liquids, the following recycling process is carried out after acidification using a strong mineral acid.

The preferred process consists of mixing sulfuric acid, methanol, and the fluorinated emulsifier liquid resulting in a strongly acidic mixture containing the methyl ester of the perfluorinated emulsifier acid, that is, methyl perfluorooctanoate (Me-PFOA). Distillation of Me-PFOA yields a ternary mixture containing Me-PFOA along with water and methanol. Condensation of the vapor results in two phases. The lower phase is more or less pure ester while the upper phase consists of water and methanol. The lower phase is separated for further workup and the upper phase is returned to the reaction mixture. The esterification process continues as long as there is fluorinated emulsifier left in the distillation sump and the volatile ester is distilled off. As soon as the formation of the lower phase has ended, a methanol–water mixture is removed by further distillation until no more methanol is left in the distillation sump. The condensed methanol–water mixture can be reused in the distillation process, ensuring that traces of Me-PFOA which may be left in the process liquids, after the phase separation has ended, are returned to the regeneration process. Once the distillation sump is free of the fluorinated emulsifier or its derivatives, it can be further treated by conventional wastewater processes [31]. The crude ester is purified by fractional distillation followed by conversion to the ammonium salt APFO.

21.3.1.2.6 Direct Workup of Loaded Absorbents (Anion Exchange Resins or Granulated Active Carbon) Not all absorption processes using either anion exchange resins or GAC are equipped with a regeneration unit. In this case, the absorbent is removed from the absorption column and replaced by fresh absorbent. Workup of the loaded absorbent can be achieved via direct regeneration. The absorbent is transferred to a distillation reactor with a suitable mixing device avoiding milling of the absorbent particles. After addition of water, methanol, and sulfuric acid, the mixture is heated to start the distillation, which is carried out in the same way as described in section 21.3.1.2.5. The regenerated absorbents are separated from the liquid phase and treated for further use, for example, by washing with water [41]. This process is suitable as well for absorbents which need to be replaced and still contain residual fluorinated emulsifiers. After such treatments the remaining amount of fluorinated emulsifier usually is quite low, which allows standard incineration of the spent absorbents.

21.3.1.2.7 PFOA Purification by Oxidation Perfluorocarboxylic acids used as emulsifiers in polymerization processes require a high purity. Impurities can suppress the polymerization initiation or lead to chain transfer processes during polymerization. Such impurities can result from synthesis (e.g., via perfluoroalkyl iodides) or recycling (e.g., via anion exchange processes) of the perfluorinated acids. This resulted in the development of a purification process which destroys such impurities using strong oxidizers [42]. Redox systems with a standard electrode potential of more than +1.30 V can be used. Suitable oxidizers include persulfates and dichromates. Because of the strength of the oxidizers used, the process requires perfluorinated acids with low water content. This can be achieved by employing concentrated sulfuric acid or oleum. After addition of the oxidizing agent, the reaction mixture is heated to destroy the impurities. Finally, the purified perfluorinated acid is separated by vacuum distillation.

21.3.1.2.8 Emulsifier Removal by Other Methods One approach to reduce the emulsifier content of fluoropolymer dispersions is based on the thermal degradation of the emulsifiers at sufficiently high temperatures [43]. To avoid adverse effects for the dispersion, a nonionic stabilizer is added. The stabilized dispersion is heated to about 100°C for several hours. During this time, the emulsifier content drops. Depending on the heating time, the resulting residual emulsifier content can be negligible. This procedure is predominantly useful for α-branched carboxylic acid salts, which are converted to hydrogen-containing ethers according to Scheme 21.5.

SCHEME 21.5

In another approach, removal of perfluorinated acids from aqueous solutions can be achieved by the addition of an aluminum salt solution followed by the addition of a strong mineral acid. After heating the mixture, an organic layer forms, that is then separated and washed with a mineral acid solution. Distillation of the organic layer results in the recovered perfluorinated acid [44].

Alternatively, aqueous solutions of fluorinated emulsifiers can be treated in a flow reactor using a catalyst bed consisting of metal oxides, for example, a combination of aluminum and nickel oxides. Using elevated temperatures and a sufficiently long contact time, emulsifiers like APFO or APFDO (see below) can be reduced from 800 mass ppm to about 10 mass ppm or less. The destruction of fluorinated emulsifier can be detected by testing the catalyst for absorbed emulsifier. This includes treatment of the catalyst with a mixture of hydrochloric acid, methanol, and water. The absence of detectible emulsifier leads to the conclusion that the fluorinated acids had been destroyed [45].

Furthermore, PFOA can be separated from wastewater streams generated by the coagulation of fluoropolymers by an up-concentration process. After flocculation of residual polymer particles, the wastewater is set to pH 10 followed by a distillation in vacuum. This can be done using one or more distillation procedures. Up-concentration is continued until the emulsifier concentration reaches at least 5% of the aqueous solution. Then the pH is set to 1, and the free emulsifier acid is removed from the water by extraction with a suitable solvent and worked up further [46].

Up-concentration of fluoropolymer dispersions is often carried out by the addition of nonionic surfactants followed by heating. The mixture separates in two phases. The upper, clear phase is separated by decantation. It contains a part of the fluorinated emulsifier [47,48].

Ultrafiltration of fluoropolymer dispersions using oscillating membranes results in a permeate containing a part of the fluorinated emulsifier. After dilution of the up-concentrated dispersion with water and nonionic surfactant, the ultrafiltration is repeated. This process may be carried out several times, resulting in an up-concentrated fluoropolymer dispersion with low fluorosurfactant content [49].

If recycling methods cannot be applied, for example, for technical, environmental, or economic reasons, PFOA or similar substances can be removed by degradation technologies. A compilation of possible pathways is listed by Vecitis et al. [25]. These technologies include incineration of sludge containing absorbed fluorinated substances, photolysis in the presence of persulfate, direct UV photolysis or photocatalysis in the presence of tungsten compounds or TiO_2. The efficiency and the application to industrial scale are sometimes quite limited.

A quite advanced technology to degrade fluorinated acids in water is the electrochemical oxidation with hydroxyl radicals; commercially available boron-doped diamond electrodes [50] or PbO_2 electrodes [51, 52] can generate $^{\bullet}OH$ hydroxyl radicals, which can degrade completely any kind of fluorinated acid.

21.3.2 PFOA Phase Out and Replacement

Over the last 25 years, an increasing number of scientific papers have been published dealing with findings of PFOS and PFOA widely spread in the environment and in the blood of general population (the half-time in blood is in the range of a couple of years) [53]. Accordingly the persistence, potential toxicity, and bioaccumulation issues of PFOS and PFOA came under scrutiny, with the consequence that the topic was discussed by the US Environmental Protection Agency (EPA). In May of 2000, 3M, a manufacturer of PFOS and PFOA decided to cease the production of PFOS and PFOA [54]. The phase out of PFOA was a challenge for the fluoropolymer industry, as there were no other suitable emulsifiers readily available which could replace PFOA or its homologs.

Fluoropolymer production is certainly not the only source for PFOA found in the environment. Another source is related to impregnating and stain repellent agents as well as similar products manufactured on the basis of perfluorooctyl iodide. Usually ethene is reacted with perfluorooctyl iodide (or its homologs), followed by hydrolysis of the iodide to the hydroxyl group. This results in the so-called fluorotelomer alcohols (Scheme 21.6) which are the basis for impregnating and stain repellent agents. Although no PFOA as such is used for these products, the degradation of the perfluorooctyl derivatives leads to PFOA or its homologs [55] and contributes to the overall environmental occurrence of perfluorinated carboxylic acids.

$$R_f\text{-}CF_2\text{-}I + CH_2{=}CH_2 \longrightarrow R_f\text{-}CF_2\text{-}CH_2\text{-}CH_2\text{-}I \longrightarrow R_f\text{-}CF_2\text{-}CH_2\text{-}CH_2\text{-}OH \rightarrow \rightarrow R_f\text{-}CO_2H$$

R_f = e.g. C_7F_{15}

SCHEME 21.6

In 2006, due to various findings with PFOA, the US-EPA invited eight companies—including fluoropolymer manufacturers—to participate in a voluntary program to reduce global emissions of PFOA by 95% by 2010 and to work toward the elimination of PFOA by 2015 [56].

Meanwhile PFOS and perfluorinated carboxylic acids with C_{11}–C_{14} chain length are listed in Europe as "substances of very high concern" (SVHC); and are regulated in many countries [57]; recently, it was also recommended to list PFOA as SVHC-compound by the European Chemicals Agency (ECHA) [58].

21.3.2.1 Containment Strategy The German-based 3M Company, Dyneon GmbH (previously owned by Hoechst AG) was already pioneering potential options

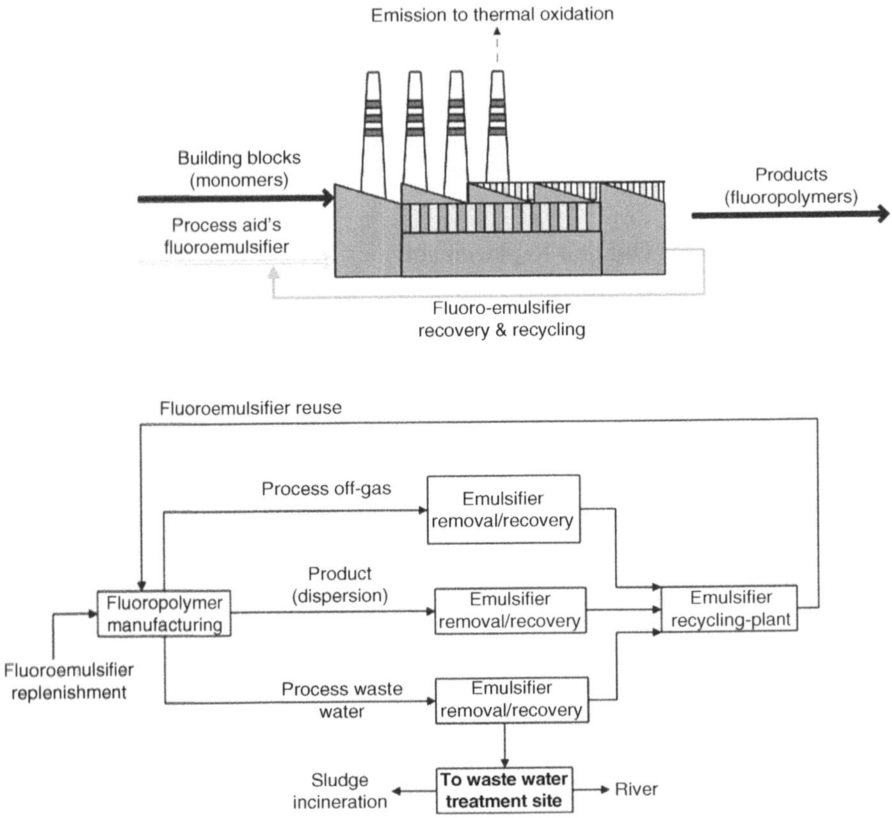

FIGURE 21.3 Fluoroemulsifier containment strategy.

to recover/remove and to recycle fluorinated emulsifiers—including PFOA—from off-gases, wastewater streams, aqueous dispersions, and products since the very beginning of the 1990s. By the mid 1990s, a large scale facility to recover PFOA from off-gas streams was implemented by using scrubbing systems. During the following years, large-scale units to recover PFOA from aqueous wastewater streams and aqueous fluoropolymer dispersions—using anion-exchange technologies—were implemented; a recycling facility to reuse the recovered PFOA was installed in parallel.

With the so-called "containment strategy" (Figure 21.3), new state-of-the-art technology was established, with a proven record of robustness and large-scale applicability. These recovery/recycling technologies can be used for APFO and a variety of fluoroemulsifiers and are nowadays installed in a number of fluoropolymer manufacturing facilities around the world. The overall recapture rates for fluoroemulsifiers are approximately 98%.

21.3.2.2 Emulsifier Replacement The US EPA Stewardship Program calls for zero emission during production and zero product content of APFO and related substances by 2015, but the initiative also suggests that manufacturers replace these substances in their manufacturing processes with chemicals having an improved toxicity and ecotoxicity profile or, even better, to develop polymerization techniques which require less or no fluorinated emulsifier.

21.3.2.3 Dyneon's Approach to Replace PFOA Although 3M, through its Dyneon GmbH subsidiary, had focused all PFOA-based polymerizations into one manufacturing site and had implemented containment technologies allowing the recovery/recycling of more than 95% PFOA by 2007, there were still some minor losses either through wastes which were incinerated or where APFO was destroyed by other means, for example, by thermal or chemical degradation. On the other hand, the increasing concerns regarding PFOA led the company to find a replacement emulsifier technology. In 2005, an R&D program was started to develop a substance addressing the following requirements:

- Improved toxicity profile compared to APFO.
- Reduced toxicological or ecotoxicological impacts.
- Potentially not persistent in the environment.
- Production based on available starting materials.
- Fulfills all requirements for the polymerization process with minor process adjustments.
- Recovery/recycling in already existing equipment with high recycling rates.

The failure of a replacement emulsifier candidate on any one of these criteria ruled it out as an acceptable APFO replacement.

Following these selection criteria, a large number of potential emulsifiers was synthesized and tested. Finally, a monohydrofluorocarboxylic acid ammonium salt, internally referred to as "ADONA," was selected. ADONA is based on a readily available perfluorinated vinyl ether. Several different production routes for ADONA are available. The first synthesis [59] starts with the addition of methanol to the vinyl ether under radical conditions. The resulting alcohol is oxidized to the carboxylic acid [59, 60]. After reaction with methanol, the resulting ester is separated and purified. Finally, the ester is converted to ADONA (Scheme 21.7).

$$CF_3\text{-}O\text{-}CF_2\text{-}CF_2\text{-}CF_2\text{-}O\text{-}CF{=}CF_2 + CH_3OH \longrightarrow CF_3\text{-}O\text{-}CF_2\text{-}CF_2\text{-}CF_2\text{-}O\text{-}CHF\text{-}CF_2\text{-}CH_2OH$$

$$CF_3\text{-}O\text{-}CF_2\text{-}CF_2\text{-}CF_2\text{-}O\text{-}CHF\text{-}CF_2\text{-}CH_2OH + KMnO_4 \longrightarrow CF_3\text{-}O\text{-}CF_2\text{-}CF_2\text{-}CF_2\text{-}O\text{-}CHF\text{-}CF_2\text{-}COOH$$

$$CF_3\text{-}O\text{-}CF_2\text{-}CF_2\text{-}CF_2\text{-}O\text{-}CHF\text{-}CF_2\text{-}COOH + CH_3OH \longrightarrow CF_3\text{-}O\text{-}CF_2\text{-}CF_2\text{-}CF_2\text{-}O\text{-}CHF\text{-}CF_2\text{-}COOCH_3$$

$$CF_3\text{-}O\text{-}CF_2\text{-}CF_2\text{-}CF_2\text{-}O\text{-}CHF\text{-}CF_2\text{-}COOCH_3 + NH_3 \longrightarrow [NH_4][CF_3\text{-}O\text{-}CF_2\text{-}CF_2\text{-}CF_2\text{-}O\text{-}CHF\text{-}CF_2\text{-}CO_2]$$

(ADONA)

SCHEME 21.7

An alternative reaction sequence starts from MV31 as well. Here, MV31 is converted with sodium cyanide in an aqueous reaction mixture resulting directly in the carboxylic acid salt. After acidification, the fluorinated acid can be separated by distillation. Conversion with ammonia yields ADONA [61] (Scheme 21.8).

$$CF_3\text{-}O\text{-}CF_2\text{-}CF_2\text{-}CF_2\text{-}O\text{-}CF{=}CF_2 + NaCN + H_2O \longrightarrow Na[CF_3\text{-}O\text{-}CF_2\text{-}CF_2\text{-}CF_2\text{-}O\text{-}CHF\text{-}CF_2\text{-}CO_2]$$

$$Na[CF_3\text{-}O\text{-}CF_2\text{-}CF_2\text{-}CF_2\text{-}O\text{-}CHF\text{-}CF_2\text{-}CO_2] + \text{Acid} \longrightarrow CF_3\text{-}O\text{-}CF_2\text{-}CF_2\text{-}CF_2\text{-}O\text{-}CHF\text{-}CF_2\text{-}COOH$$

$$CF_3\text{-}O\text{-}CF_2\text{-}CF_2\text{-}CF_2\text{-}O\text{-}CHF\text{-}CF_2\text{-}COOH + NH_3 \longrightarrow [NH_4][CF_3\text{-}O\text{-}CF_2\text{-}CF_2\text{-}CF_2\text{-}O\text{-}CHF\text{-}CF_2\text{-}CO_2]$$

(ADONA)

SCHEME 21.8

Another synthetic pathway employs formic acid t-butyl ester which is added to MV31 by a radical reaction. Conversion of the addition product to the free acid and neutralization with ammonia results in ADONA [62].

ADONA was introduced by Dyneon in the polymerization and recovery/recycling processes starting 2008, and by the beginning of 2009 Dyneon had completely replaced APFO with ADONA [63].

ADONA was extensively tested regarding toxicology and ecotoxicology. No significant adverse effects have been identified [64].

21.3.2.4 APFO Replacements by Other Fluoropolymer Manufacturers Other fluoropolymer producers have pursued their own APFO replacements. According to literature, different PFOA replacements seem to be selected by the following companies (Scheme 21.9):

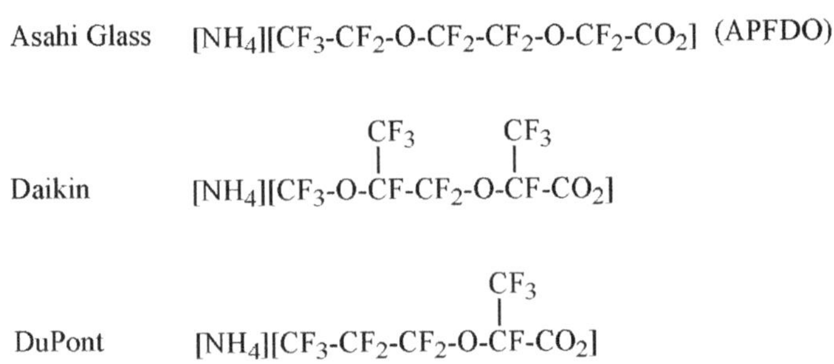

SCHEME 21.9

APFDO/EEA [65] can be made, for instance, by fluorination of nonfluorinated or partially fluorinated precursors with subsequent conversion of the acid fluoride to the ammonium salt (Scheme 21.10).

e.g. $CH_3\text{-}CH_2\text{-}O\text{-}CH_2\text{-}CH_2\text{-}O\text{-}CH_2\text{-}CO_2R + F_2 \longrightarrow CF_3\text{-}CF_2\text{-}O\text{-}CF_2\text{-}CF_2\text{-}O\text{-}CF_2\text{-}COF \rightarrow \rightarrow$

$[NH_4][CF_3\text{-}CF_2\text{-}O\text{-}CF_2\text{-}CF_2\text{-}O\text{-}CF_2\text{-}CO_2]$

SCHEME 21.10

The Daikin product may be produced by the addition of 1,2-epoxy hexafluoropropane (HFPO) to carbonylfluoride [66] and conversion of resulting acid fluoride to the ammonium salt [67] (Scheme 21.11). Alternatively partially/perfluorinated emulsifiers based on tetrafluorooxetane were described [68].

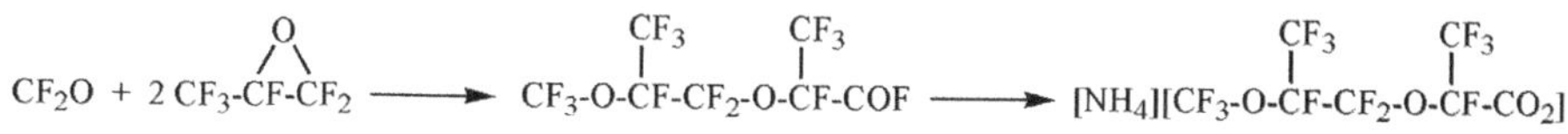

SCHEME 21.11

Daikin announced in December 2007 that it intended to cease the use of PFOA by the end of 2012. According to a paper released by Daikin in December 2012, the use of PFOA was completely stopped. But it was still a minor by-product in some products, so the release of PFOA was reduced to less than 1% compared to the year 2000 [69].

The DuPont emulsifier may be prepared by dimerization of HFPO [66] and conversion of the acid fluoride to the ammonium salt (Scheme 21.12). Some toxicological data for DuPont's HFPO-dimer-based emulsifier are published in EPA 8e submissions [70].

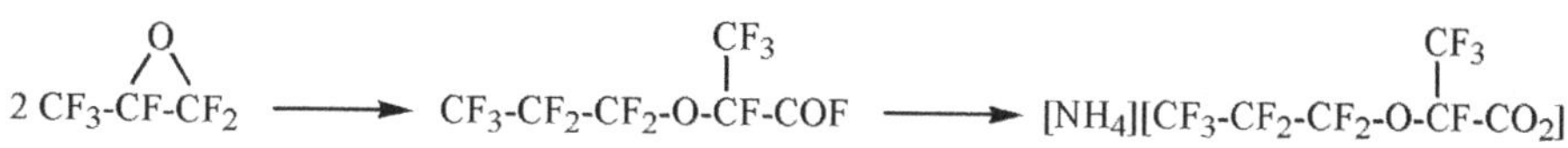

SCHEME 21.12

*21.3.2.5 **Fluoroemulsifier-Free Polymerizations*** The various fluoropolymer manufactures are exploring polymerization approaches to eliminate the use of highly fluorinated emulsifiers during the aqueous emulsion polymerizations. In cases where lower molecular weight polymers are generated and in the presence of chain-transfer agents, a couple of new options are feasible:

- amorphous standard co/terpolymers comprising TFE, HFP, VDF—suited for elastomer applications—can be polymerized without any emulsifier [71,72]
- for VDF-containing polymers, a variety of nonfluorinated compounds have been evaluated to stabilize the fluoropolymer particles during emulsion polymerization; these compounds include: polyvinyl/acrylic acids [73], derivatives of polyethylene/propylene glycols [74], alkylphosphate esters [75], vinyl acids [76], siloxanes [77], silanes [78], long-chain hydrocarbon acids [79,80], and derivatives of sugars [81].
- so-called "surfmers," which are surfactants that can also act as monomers in the polymerization reaction. Surfmers include compounds with a polymerizable double bond of the general formula:

$$CX_2 = CX - (O)_n - R - Z \quad X = H, F; n = 0, 1, Z = COO^-, SO3^-,$$
$$R = \text{fluorinated alkyl group } C_1 - C_{10}$$

can be used to generate specific classes of fluoropolymers [82, 83]. The polar surfmers will be incorporated into the fluoropolymer backbone and stabilize the polymer particles like a classical fluoroemulsifier.

Ideally, at the end of polymerization, the surfmers are completely incorporated and the final polymer and polymer latices are essentially free of unpolymerized fluorinated surfmers.

The changing landscape with regard to regulation of APFO and related fluorosurfactants have led fluoropolymer manufacturers to re-evaluate and in some cases introduce polymerization approaches, which in previous times were considered not economically feasible.

21.4 FLUOROPOLYMER RECYCLING

There are quite a variety of different sources and different types of fluoropolymers for recycling purposes. Consequently, one has to consider these issues from various angles:

21.4.1 Scrap from Fluoropolymer Manufacturers and Processors

21.4.1.1 PTFE Materials The amounts of scrap, wet waste materials, or off-specification materials of unfilled PTFE from manufacturers are usually in the lower percentage range. In contrast, the amount of scrap resins from processors in making semifinished parts or end use articles are usually in the range of 10–30% [84], and in some areas it is above 50%. The high prevalence of scrap materials are due to the very specific processing technologies for PTFE (moulding, sintering, machining, and cutting). For unfilled PTFE-scrap resin, today there exist three established recycling paths:

1. Sintered, unfilled PTFE-scrap resin is cleaned from all contaminant and milled into certain particle size classes, which can be reused, for example, in Ram extrusion applications. This so-called Repro-PTFE can also be mixed with virgin PTFE to a certain content; such Repro-PTFE materials get specific designations [85] and are typically used for less demanding end-use applications.
2. Clean, unfilled PTFE can be thermally degraded into low molecular weight PTFE. Such processes are commercially established. The thermal degradation of high molecular weight scrap PTFE occurs at temperatures of about 500°C in ovens, kneaders, or preferably in extruders [86–88]; the obtained low molecular weight PTFE-materials are further milled into very small particle sizes of a few microns—such materials are often called PTFE micropowders.
3. Alternatively, clean and unfilled PTFE can be degraded by high energy radiation such as X-ray, gamma ray, or electron beam [3, 89]. The degradation of high molecular scrap PTFE by electron beam irradiation is commercially widely used, and in practice continuous processes are used to improve the economics. After irradiation, the material is milled to the desired smaller particles sizes.

The thermo/radiation degraded PTFE-micropowders are mostly used as additives to plastics, inks, oils, lubricants, and coatings to import fluoropolymer-like properties such as reduced wear rates and friction.

21.4.1.2 PTFE Compounds In contrast to clean, unfilled PTFE, where some recycling opportunities are well established and the recycling rates reach a significant level, there are no large-scale recycling technologies for PTFE compounds. This is primarily due to the presence of the large variety of different fillers (e.g., glass fibers, graphites, carbons, metals, ceramics, organic fillers, pigments) and to the variable amounts of fillers in the PTFE compounds. For the significant amount of scrap streams (at least 10–30%), considering that landfilling is not a sustainable option and in particular that landfilling is getting more and more regulated in Europe, PTFE compound scrap is prone to be recycled into TFE/HFP-monomers (see below).

21.4.1.3 Perfluorinated/Partially Fluorinated Thermoplasts/Elastomers Clean, unfilled, unpigmented, and uncured scrap materials from manufacturing and processing of perfluorinated/partially fluorinated thermoplasts or elastomers are generated in the low percent range. Nearly all of these materials are recycled back into the different processes; the end use properties are almost unaffected.

21.4.2 End of Life (EOL)—Fluoropolymers

In some cases, used perfluorinated fluoropolymers (e.g., PTFE, PFA) are recycled by special cleaning processes and are ending up in the Repro-PTFE or micropowder-PTFE-market. Perfluorinated thermoplasts (e.g., PFA) are reused in applications where the quality requirements (e.g., lot traceability) are much lower. Overall, the lion's share of used fluoropolymers is, however, ending up in landfills, in incineration plants, or in blast furnaces. Communal waste incinerators can tolerate only very limited amounts of fluoropolymers due to the high corrosion due to hydrofluoric acid formed in the process.

21.4.3 Alternative Recycling Concepts

21.4.3.1 Perfluoropolymers The manufacturing of TFE/HFP monomers consumes large amounts of energy (>10,000 kWh/ton TFE). In the past, there were many efforts to recover perfluorinated monomers from scrap materials. All of the approaches were based on the findings that perfluorinated polymers can be pyrolyzed under high temperature conditions into TFE/HFP [90–95] (Scheme 21.13). The pyrolysis back to monomers of perfluorinated polymers is a rare property in the materials industry; there are only a few other polymer classes [e.g., Polystyrene, Poly(methylmethacrylate)] which can be converted into monomers upon heating.

$$\sim CF_2\text{-}CF_2\text{-}CF_2\text{-}CF_2\sim \xrightarrow{\text{"}CF_2\text{"}} CF_2{=}CF_2,\ CF_2{=}CF\text{-}CF_3$$

SCHEME 21.13

This process is a convenient route to prepare small quantities of TFE on lab scale; but so far as the authors are aware, no industrial scale facility has been installed. This may, however, soon be changing, as new energy-sound and robust processes for the high temperature conversion (HTC) of perfluoropolymers have been developed, which are also suited for large-scale HTC units [94–96]. The new HTC process designs are capable of pyrolyzing all kinds of perfluoropolymers back into TFE/HFP with yields of more than 80% [95]. The process is particularly well suited to convert filled PTFE compounds into monomers—so far there has been no viable technology for recycling PTFE compounds. The HTC-pyrolysis can recycle all kinds of filled fluoropolymer compounds (containing, e.g., carbons, graphites, glass fibers, metals incl. sulfides/oxides) and unfilled materials from manufacturing, processing and end-of-life materials. Even perfluorinated polymers with functional groups can be subject to these processes (Scheme 21.14) which are used in fuel cell applications or for membranes in electrochemical processes.

$$\sim(CF_2\text{-}CF_2)_n\text{-}(\underset{\displaystyle |\atop \displaystyle O\text{-}R_F\text{-}X}{CF}\text{-}CF_2)_m\sim \quad [X = SO_3^-, CO_2^-]$$

SCHEME 21.14

The HTC process is complementary with existing recycling methods for unfilled PTFE (e.g., Repro-PTFE, micropowder). These processes may finally close the loop for perfluorinated polymers, see Figure 21.4 (i.e., by avoiding landfilling and combustion). Currently, 3M/Dyneon are planning an industrial scale HTC-pilot unit in Germany to demonstrate feasibility and robustness. The environmental benefit in pursuing this recycling path is showing quite significant reductions in raw material usage, waste streams, and in the overall CO_2 balance (see Figure 21.5).

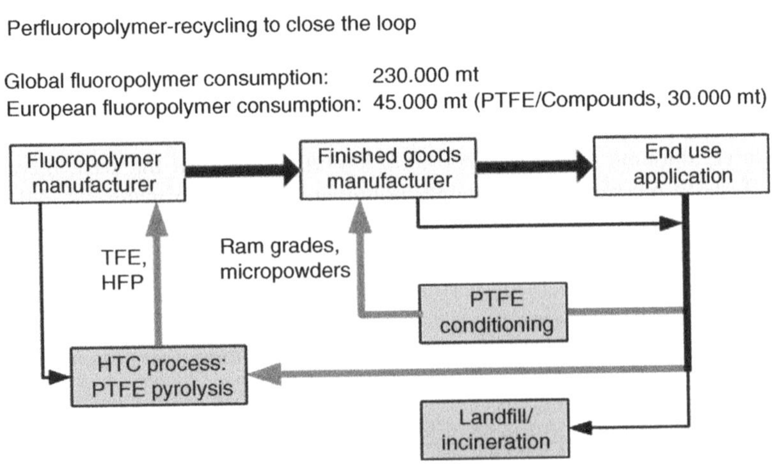

FIGURE 21.4 Perfluoropolymer recycling options.

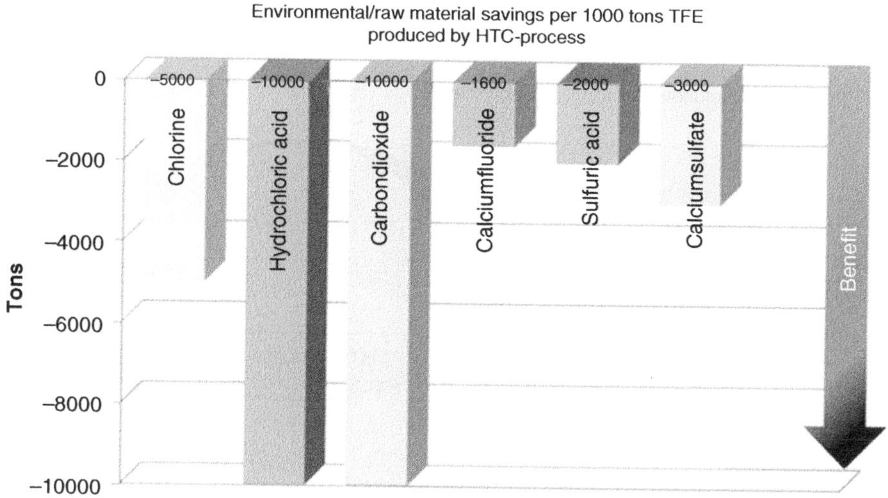

FIGURE 21.5 Environmental benefit pyrolyzing perfluoropolymers back to TFE.

21.4.3.2 Partially Fluorinated Polymers Partially fluorinated polymers (e.g., THV, ETFE, PVDF, and the like) as well as perfluorinated polymers in combination with larger amounts of hydrocarbon-containing polymers (e.g., blends, laminates) can be recycled by HTC processes into monomers only with low/moderate yields [97]; usually these EOL-polymers are landfilled or incinerated in small portions due to the corrosive HF.

In Europe and in particular in Germany, CaF_2 is considered as one of the most important 14 strategic raw material [98]. This classification has spurred investigations to optimize the incineration/combustion of fluoropolymer-containing materials to recover HF and/or CaF_2 from the off-gases. Such processes are known in the literature [99]; the challenge is to establish robust and energy sound processes and to apply the know-how from thermal oxidizers of fluorocarbon gases in order to recover HF and/or CaF_2 in an efficient manner for reuse. The implementation of such processes would finally close the remaining loop (see Figure 21.6).

21.5 SUMMARY

The fluoropolymer industry has already mastered successfully many of the important challenges in respect to the environmental impact of fluoropolymer manufacturing (e.g., ban of chlorine-containing fluorocarbons, explosion risks, PFOA-phase out). Life cycle assessments, including ecobalances (e.g., according to ISO 14040/14025) have been initiated also to demonstrate the value of specific fluoropolymers (recently such a study about ETFE was published [100]). While a comprehensive view of the impact of fluoropolymer manufacturing can be tedious to assemble, as the worldwide existing data from the various databases (e.g., ProBas [101], EcoInvent [102]) are not

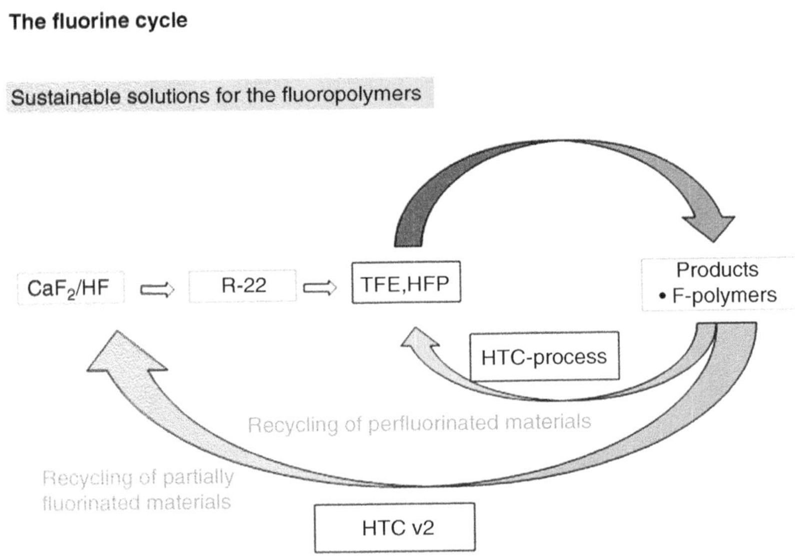

FIGURE 21.6 The closed fluoropolymer fluorine cycle.

always consistent, these data will surely be consolidated, as the understanding of the environmental impact of fluoropolymer manufacture remains a priority for national and international observers.

Based on all these achievements, it is foreseeable that sooner or later the industry will establish closed manufacturing loops throughout the whole value chain (with close to zero emissions). A closed fluorine cycle (see Figure 21.6) will be one such target, with optimized raw material streams and sound energy balances and reduced environmental burdens.

The realization of this vision will also ensure further future growth opportunities and it will stimulate the development of new fluoropolymers with unique properties.

NOMENCLATURE

Chemical Abbreviations

Monomers/Fluorochemicals:

TFE	Tetrafluoroethene
HFP	Hexafluoropropene
VDF, VF2	Vinylidene fluoride
R22	Chlorodifluoromethane
R113	1,1,2-Trichlorotrifluoroethane
R142b	1-Chloro-1,1-difluoroethane
R152b	1,1-Difluoroethane
PFOA	Perfluorooctanoic acid

PFOS	Perfluorooctylsulfonic acid
APFO	Ammonium salt of perfluorooctanoic acid
Me-PFOA	Methyl-perfluorooctanoate
ADONA	Ammonium 3-H-perfluoro-4,8-dioxanonanoate
APFDO	Ammonium perfluoro-3,6-dioxaoctanoate
PPVE	Perfluoropropylvinylether
VF	Vinylfluoride
CTFE	Chlorotrifluoroethene
	Polymers:
PTFE	Polytetrafluoroethylene
PFA	Perfluoroalkoxy-copolymer
FEP	Flourinated ethene propene-copolymer
ETFE / ET	Ethene tetrafluoroethene-copolymer
PVDF	Polyvinylidene fluoride
THV™ FP	Tetrafluoroethene, hexafluoropropene and vinylidene fluoride-terpolymer
PCTFE	Polychlorotrifluoroethylene
ECTFE	Ethene chlorotrifluoroethene-copolymer
PVF	Polyvinyl fluoride
FKM, FFKM	Fluoroelastomers, perfluoroelastomers
TFEP	Tetrafluoroethene propene-copolymer

REFERENCES

1. Wetzel, W. *NTM Z. Gesch. Wissen. Tech. Med.* **2005**, *13*, 79.
2. Schloffer, F.; Scherer, O. (IG-Farbenindustrie), DP677 091 (1934).
3. Ebnesajjad, S. Fluoroplastics 1 and 2, Plastic Design Library (2000).
4. Kaelin, Th.; Will, R. K.; Yamaguchi, Y. CEH Marketing Research Report: Fluoropolymers (2012).
5. Löchner, U.; Inoguchi, Y. DEH Marketing Research Report: Fluoroelastomers (2010).
6. Ebnesajjad, S.; Khaldkar, P. R. Fluoropolymer Applications in Chemical Processing Industries, Plastic Design Library (2005).
7. Siegemund, G.; Schwertfeger, W.; Feiring, A.; Smart, B.; Behr, F.; Vogel, H.; McKusick, B. *Fluorine Compounds, Organic, in Ullmann's Encyclopedia of Industrial Chemistry*; Wiley-VCH: Weinheim; doi:10.1002/14356007.a11 349, accessed 5 February 2014.
8. Montreal Protocol. http://www.unep.org/ozonaction, accessed 5 February 2014.
9. TLVs and BEIs, *Threshold Limit values*, ACGIH, Signature Publications, 2011.
10. Fukunaga, S.; Funaki, A. (Asahi Glass), EP2239282 (2009).
11. DeYoung, J. P.; Romack, T. J.; DeSimone, J. M. In *Topics in Applied Chemistry, Fluoropolymers 1*; Kluwer Academic/Plenum Publishers, 1999; ISBN 0-306-46060-2.
12. Hartwimmer, R. (Farbwerke Hoechst), DE1720801 (1968).
13. Felix, B.; Löhr, G.; Hofmeister, W.; Hengel, R. (Hoechst), DE4022405 (1990).
14. Hintzer, K.; Loehr, G.; Killich, A.; Staudt, H.-J. (3M), EP1353965 (2001).
15. Epsch, R.; Goldmann, G.; Killich, A.; Löhr G.; Staudt, H.-J. *Chem. Eng. Technol.* **2002**, *25*, 277.

16. Epsch, R.; Goldmann, G.; Killich, A.; Löhr G.; Staudt, H.-J. *Chem. Eng. Technol.*, **2002**, *25*, 393.
17. *Guide for the Safe Handling of Fluoropolymer Resins*; Plastics Europe, Brussels, 2012.
18. Berry, K. L. (DuPont), US2559752 (1951).
19. Scheirs, J., Ed. Structure/Property Considerations for Fluoropolymers and Fluoroelastomers to Avoid In-service Failure. Modern Fluoropolymers; John Wiley & Sons, Chichester, 1997; ISBN 0-471-97055-7.
20. Simons, J. H.; Harland, W. J. *J. Electrochem. Soc.* **1949**, *95*, 47.
21. Brace, N. O. *J. Fluorine Chem.* **1999**, *93*, 1.
22. von Werner, K.; Gisser, A. (Hoechst), US4400325 (1983).
23. Kimura, K. (Shin Akita Kasei), JP01/117840 (1989), Chem. Abstr. 111 (1989) 173619x.
24. Führer, S.; Hintzer, K.; Löhr, G.; Schwertfeger, W. (3M), US6642415 (2003).
25. Vecitis, C. D.; Park, H.; Cheng, J.; Mader, B. T.; Hoffmann, M. R. *Front. Envir. Sci. Engin. China*, **2009**, *3*, 129.
26. Kuhls, J.; Weiss, E. (Hoechst), US4369266 (1983).
27. Kuhls, J. (Hoechst), US4282162 (1981).
28. Houde, M.; Martin, J. W.; Letcher, R. J.; Solomon, K. R.; Muir, D. C. G. *Environ. Sci. Technol.* **2006**, *40*, 3463.
29. Sulzbach, R. A.; Kowatsch, W.; Steidl, D. (Dyneon), US5990330 (1999).
30. Sulzbach, R. A.; Grasberger, R.; Brandenburg, R. A. (Dyneon, DuPont), US6245923 (2001).
31. Obermeier, R.; Stefaniak, G. (Hoechst), EP632009 (1995).
32. Felix, B.; Sulzbach, R. A.; Fuehrer, S.; Kaiser, T.; Kniep, H.; Budesheim, A. (3M), US6613941 (2003).
33. Felix, B.; Zipplies, T.; Fuehrer, S.; Kaiser, T.; Budesheim, A. (Dyneon, Axiva), US6518442 (2003).
34. Fuehrer, S.; Löhr, G.; Schwertfeger, W. (Dyneon), US6436244 (2002).
35. Maurer, A.; Hintzer, K.; Schwertfeger, W.; Obermaier, E.; Weber, A. (3M), US20060205828 (2006).
36. Bistrov, G. A.; Kuznetzova, E. M.; Kaboshina, T. A. *Plasticheskie Massy* **1990**, 75.
37. Hintzer, K.; Obermaier, E.; Schwertfeger, W. (3M), US7018541 (2006).
38. Hintzer, K.; Löhr, G.; Schwertfeger, W.; Sulzbach, R. A. (Dyneon), EP1273597 (2003).
39. Blädel, H.; Hintzer, K.; Löhr, G.; Schwertfeger, W.; Sulzbach, R. A. (3M), US6833403 (2004).
40. Combes, J. R.; Johnson, D. W.; Breske, S. T. (DuPont), US20100093894 (2010).
41. Führer, S.; Hintzer, K.; Obermaier, E.; Schwertfeger, W. (3M), EP1561742 (2005).
42. Mayer, L.; Löhr, G. (Hoechst), EP566974 (1993).
43. Johnson, D. W.; Teter, K. S. (DuPont), WO2010129842 (2010).
44. Schultz, J. A. (DuPont), US6437159 (2002).
45. Nomura, J.; Matsuoka, Y. (Asahi Glass), US20100324352 (2010).
46. Funaki, H.; Yanase, K.; Kamiya, H.; Uehara, M.; Nagatomo, K.; Nishimura, Y.; Ohno, S.; Hirano, S. (Asahi Glass, Sasakura Engineering), US20050150833 (2005).
47. Marks, B. M.; Whipple, G. H. (DuPont), US3037953 (1962).

48. Miriglioni, A.; Deregibus, E.; Kapeliouchko V. (Solvay Solexis), EP 1714986 (2006).
49. Britnell, A.; Simpson, M.; Conheady, J.; Hosokawa, K. (Daikin America), US20060241214 (2006).
50. Zhuo, Q.; Deng, S.; Yang, B.; Huang, J.; Wang, B.; Zhang, T.; Yu, G. *Electrochim. Acta*, **2012**, *77*, 17.
51. Fath, A. (Hansgrohe), DE102009013380 (2010).
52. Wabner, D.; Grambow, C. *J. Electroanal. Chem. Interf. Chem.* **1985**, *195*, 95.
53. Report of an OECD Workshop on Perfluorocarboxylic Acids (PFCAs) and Precursors, Stockholm, Sweden, November 20–22, 2006; ENV-JM-MONO(2007)11-2.pdf
54. http://solutions.3m.com/wps/portal/3M/en_US/PFOS/PFOA/information/phase-out-technologies/, accessed 5 February 2014.
55. Ellis, D. A.; Martin, J. W.; De Silva, A. O.; Mabury, S. A.; Hurley, M. D.; Sulbaek Andersen, M. B.; Wallington, T. J. *Environ. Sci. Technol.* **2004**, *38*, 3316.
56. 3M/Dyneon Progress Report 2009. www.epa.gov/oppt/pfoa/pubs/stewardship/index.html, accessed 5 February 2014.
57. Zushi,Y.; Hogarth, J. N.; Masunaga, S. *Clean Tech. Environ. Policy* **2012**, *14*, 9–20.
58. http://echa.europa.eu/de/registry-of-submitted-svhc-intentions, accessed 5 February 2014.
59. Hintzer, K.; Juergens, M.; Kaspar, H.; Koenigsmann, H.; Maurer, A.; Schwertfeger, W.; Zipplies, T. (3M), EP1963247 (2005).
60. Hintzer, K.; Vogel, D. E.; Guerra, M. A.; Ignatowska, J.; Roeschenthaler, G.-V.; Shyshkov, O.; Vogel, K. M.; Zipplies, T. (3M), WO2011050131 (2011).
61. Zipplies, T.; Hintzer, K.; Shyshkov, O. (3M), WO2010071730 (2010).
62. Hintzer, K.; Zipplies, T.; Shyshkov, O. (3M), WO2011066156 (2011).
63. http://www.epa.gov/opptintr/pfoa/pubs/stewardship/preports4.html; 2010, accessed 5 February 2014.
64. Gordon, S.C. *Regul. Toxicol. Pharmacol.* **2011**, *59*, 64.
65. Funaki, H.; Seki, R.; Oharu, K.; Kamiya, H. (Asahi Glass), EP1698662 (2005).
66. Millauer, H.; Schwertfeger, W.; Siegemund, G. *Angew. Chem. Int. Ed. Engl.* **1985**, *24*, 161.
67. Fu, T.-W.; Swearingen, S. H.; Ichida, T.; Yoshii, S. (Daikin, DuPont), US7253315 (2007).
68. Morita, S.; Tanaka, Y.; Washino, K.; Tsuda, N.; Kishine, M. (Daikin), US7589234 (2009).
69. www.daikin.com/chm/pfoa/ (released December 21, 2012).
70. DuPont 8e Submissions 8EHQ-0113-16436P and 8EHQ-0113-16478Q. www.epa.gov/opptintr/tsca8e/pubs/8monthlyreports/2013/8ejan2013.html, accessed 5 February 2014.
71. Kaspar, H.; Hintzer, K.; Dewitte, G.; Schwertfeger, W. (3M), EP1392744 (2002).
72. Kaspar, H.; Scott, P. J.; Hintzer, K.; Löhr, G. (3M), EP1556426 (2009).
73. Durali, M.; Hedhli, L. (Arkema), WO2007018783 (2006).
74. Amin-Sanayei, R.; Durali, M. (Arkema), WO2008073686 (2007).
75. Lyons, D. F. (DuPont), US20080262177 (2008).
76. Durali, M.; Hedhli, L. (Arkema), WO2012030784 (2011).
77. Wille, R. A.; Durali, M.; Hedhli, L.; Antoun, S. Y. (Atofina), EP1462461 (2004).

78. Bissinger, P.; Dadalas, M.; Hintzer, K.; Mayer, L.; Schwertfeger, W.; Zipplies, T. (3M), EP2069407 (2006).
79. Wille, R. A.; Durali, M.; Hedhli, L.; Amin-Sanayei, R.; Schmidhauser, J. (Arkema), EP1591461 (2004).
80. Hintzer, K.; Kaspar, H.; Traunspurger, H.; Zipplies, T. (3M), WO2007120348 (2006).
81. Zipplies, T.; Hintzer, K.; Dadalas, M.; Frey, O.; Lochhaas, H. (3M) WO 2011/014715 (2010).
82. Tsuda, N.; Tanaka, Y.; Morita, S.; Sawada, Y.; Shimizu, T. (Daikin), EP1739100 (2005).
83. Tsuda, N.; Nanba, Y.; Sawada, Y.; Shimizu, T. (Daikin), EP1681308 (2004).
84. http://www.mikro-technik.eu/html/body-reproflon-ptfe.html, accessed 13 February 2014.
85. ISO 12086 Plastics-Fluoropolymer dispersions, and moulding and extrusion materials; Part 1, Designation system and bases for specifications (ISO 12086-1: 2006; CEN, Bruxelles).
86. Hartwimmer, R.; Weiss, E. (Hoechst), US3813449 (1974).
87. Hartwimmer, R. (Hoechst), US4076760 (1978).
88. Straßer, H.; Felix, B.; Reiff, U.; Richter, H.-B.; Zipplies, T. (Hoechst), DE4437994 (1994).
89. Hoffmann, T.; Heller, M.; Engelhardt, T.; Taeger, A.; Marks, H.; Lehmann, D. *GAK – Gummi, Fasern, Kunststoffe* **2012**, *65*, 784.
90. Meissner, E.; Wroblewska, A.; Milchert, E. *Polym. Degrad. Stabil.* **2004**, *83*, 163.
91. Van der Walt, I. J.; Bruinsma, O. S. L. *J. Appl. Polym. Sci.* **2006**, *102*, 2752.
92. Arkles, B. C.; Bonnett, R. N. (DuPont), US3832411 (1974).
93. Schöttle, T.; Hintzer, K.; Staudt, H.-J.; Weber, H. (Hoechst), US5432259 (1995).
94. Ichida, T.; Homoto, Y. (Daikin), WO2003074456 (2009).
95. Hintzer, K.; Kämpf, G.; Kolbeck, T.; Zipplies, T.; Willert-Porada, M.; Gerdes,T.; Schmidt-Rodenkirchen, A. (3M), EP2346801 (2009).
96. Van der Walt, I. J.; Grunenberg, A. T.; Nel, J. T. (South African Nuclear Energy Corp.), WO2007063462 (2007).
97. Van der Walt, I. J.; Hintzer, K.; Löhr, G. (Nuclear Energy Corporation of South Africa, 3M), WO2001058840 (2001).
98. Enterprise and Industry Directorate General http://ec.europa.eu/enterpise/policies/raw-materials/files/docs/repnt-b_en.pdf
99. Abe, R.; Morita, H.; Hirai, H.; Uehara, M. (Nittetsu Kakoki, Asahi Glass), JP2005231984 (2005).
100. Institut Bauen und Umwelt/Vector Foiltec, Nowofol, Dyneon; Declaration-number EPD-VND-2011111-D (2011).
101. http://www.probas.umweltbundesamt.de, accessed 13 February 2014.
102. Frischknecht, R.; Jungbluth, N.; Althaus, H.-J.; Doka, G.; Dones, R.; Heck, T.; Hellweg, S.; Hischier, R.; Nemecek, T.; Rebitzer, G. *Int. J. Life Cycle Ass.* **2005**, *10*, 3; http://dx.doi.org/10.1065//ca2004.10.181.1

22

FLUORINATED IONOMERS AND IONOMER MEMBRANES CONTAINING THE BIS[(PERFLUOROALKYL) SULFONYL]IMIDE PROTOGENIC GROUP

IQBAL SHARIF, STEPHEN CREAGER, AND DARRYL D. DESMARTEAU
Clemson University

22.1 INTRODUCTION AND OVERVIEW

Polymer electrolyte membrane (PEM) fuel cells convert chemical energy into electrical energy using hydrogen gas or methanol (usually) as fuel, giving water as the only by-product when hydrogen is the fuel. The high efficiency and low pollution of fuel cells make them an attractive green energy option. Among the many issues affecting the performance of a PEM fuel cell, the proton exchange membrane plays a crucial role. The development of high performance polymers for use as solid polymer electrolytes (SPEs) in PEM fuel cells has been under intense investigation. A schematic diagram for a PEM fuel cell is shown in Figure 22.1 [1]. In a typical single cell, two electrodes (usually consisting of dispersed platinum supported on carbon) are separated by an SPE membrane that facilitates transport of hydrated protons and water but does not allow the flow of electrons. At the anode, hydrogen fuel dissociates into electrons that flow through the conductive electrode and out of the cell, and positive

Handbook of Fluoropolymer Science and Technology, First Edition. Edited by Dennis W. Smith Jr., Scott T. Iacono, and Suresh S. Iyer.

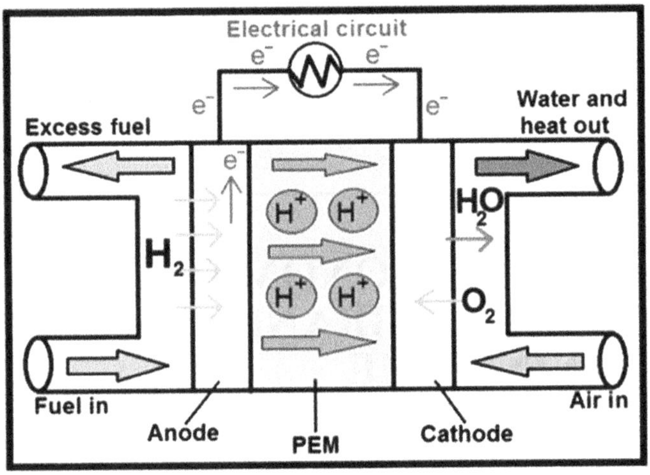

FIGURE 22.1 A proton exchange membrane fuel cell.

hydrogen ions migrate through the membrane electrolyte to the cathode where they combine with oxygen and the returning electrons to form water.

Many different polymers have been explored as PEMs including sulfonated polystyrenes, poly-α,β,β-trifluorostyrenes, polysulfones, poly(aryl ether ketones) (PAEK), poly(aryl ether ether ketones) (PEEK), partially and fully fluorinated fluoroalkylsulfonic acid polymers, and others [2–11]. Many hydrocarbon polymers suffer from poor chemical resistance [8]; therefore, fully or partially fluorinated polymers such as DuPont's Nafion®, Dow's XUS PFSA, and Ballard's BAM3G01 are considered to be promising [8–11]. Due to its favorable properties and widespread commercial availability, Nafion® is one of the polymers of choice for much PEM fuel cell research and development. Nafion®, a perfluorosulfonic acid (PFSA) ionomer, was introduced in the 1960s by E.I. du Pont de Nemours and Co. and has extraordinary chemical, mechanical, thermal, and electrochemical properties [10–12]. Dow Chemical Company later introduced their short-side-chain version of a PFSA known as XUS [9]. The latter is now available from Solvay Specialty Polymers as Aquivion® PFSA. Both Nafion® and Aquivion® PFSA are copolymers of tetrafluoroethylene (TFE) with a trifluorovinyl ether (TFVE) comonomer containing a sulfonyl fluoride group. The latter group is hydrolyzed to the sulfonic acid group after copolymerization (Scheme 22.1). These materials show good performance as thin membranes in PEM fuel cells operating below 100°C.

Great interest exists in a PEM fuel cell that could operate above 120°C. At these temperatures, issues of many systems would be improved, including reduced CO poisoning of the platinum catalyst [8] and diminished need for stack cooling. Carbon monoxide is often present as an impurity in the hydrogen fuel that is introduced into the fuel cell so strategies for mitigating CO poisoning are needed. Unfortunately, PFSA ionomers lose hydration above 80°C and proton conductivity decreases, and

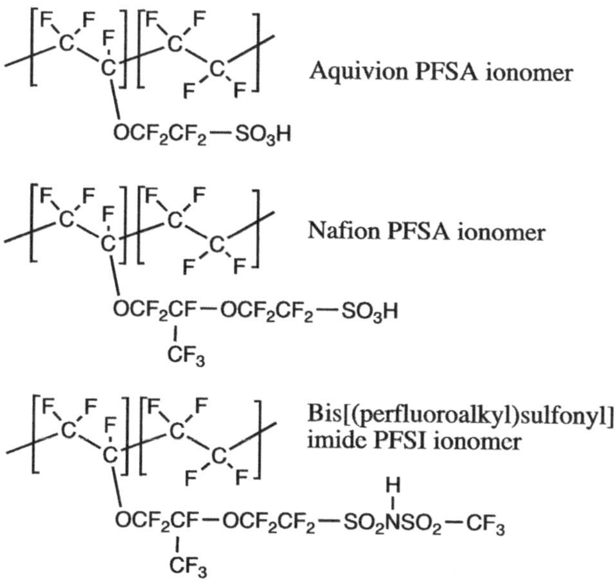

SCHEME 22.1 Ionomers.

at still higher temperatures, sulfonic acids can further dehydrate forming an anhydride ($-SO_2OSO_2-$) which leads to further loss of protonic conductivity [13]. There exists a need for a polymer that would overcome the deficiencies of the PFSA-based polymers. A different, strongly acidic functional group, capable of operating in a PEM at higher temperatures without significant decomposition while maintaining good proton conductivity, is needed. However, the number of suitable acid functions is very limited.

We embarked on a program to produce a PEM that would operate at higher temperatures (>110°C) and that had equal or better proton-conducting properties than the PFSA polymers [14]. The polymers were based on the bis[(perfluoroalkyl)-sulfonyl]imide (PFSI) acid functional group incorporated into a perfluorinated ionomer. The PFSI ionomer (Scheme 22.1) evolved from small-molecule work on perfluoroalkylsulfonyl derivatives to form carbon and nitrogen xenon bonds [15, 16]. Beginning in the late 1970s, numerous Brφnsted nitrogen superacids were synthesized that are known as PFSIs [17]. The parent member of this family is $(CF_3SO_2)_2NH$, which has a pK_a value of 7.8 in acetic acid ($CF_3SO_2H = 4.7$) [18] and was shown in gas-phase acidity measurements to have a much greater acidity than CF_3SO_3H and HI [19]. The relatively high acidity associated with $(R_fSO_2)_2NH$ and the high oxygen solubility in perfluorocarbon compounds led to excellent oxygen reduction kinetics on platinum when $(R_fSO_2)_2NH$ was used as an additive in a typical phosphoric acid test fuel cell or as the bulk aqueous electrolyte [14, 20]. It was decided to initiate a program to produce the first perfluorinated PEM containing the sulfonimide acid functional group.

$$\underset{1}{(CF_3SO_2)_2)} + NH_3 \longrightarrow \underset{2}{CF_3SO_2NH_2}$$

$$CF_3SO_2NH_2 + NaOH \longrightarrow \underset{3}{CF_3SO_2NHNa}$$

$$CF_3SO_2NHNa + HMDS \xrightarrow[CH_3CN]{} \underset{4}{CF_3SO_2NNa{-}SiMe_3}$$

SCHEME 22.2 Synthesis of sodium salt of trimethylsilyltrifluoromethylsulfonimide **4**.

Herein we review details on synthesis of the TFVE monomer $CF_2{=}CFOCF_2CF(CF_3)OCF_2CF_2{-}SO_2N(Na)SO_2CF_3$ **9**, its copolymers with TFE **10**, and some physical properties of these polymers. Electrochemical characterization of earlier versions of these polymers has been reported previously [21–24], as well as some preliminary reports on their preparation and properties [14, 25–30].

22.2 MONOMER SYNTHESIS

The synthesis of TFVE monomer **9** is shown in Schemes 22.2–22.4. The perfluorinated sulfonyl fluoride monomer **5**, which is used in the production of DuPont's polymer Nafion®, offers an excellent starting material for the production of 3,6-dioxa-Δ^7-4-trifluoromethyl perfluorooctyl trifluoromethyl sulfonimide monomer **9**. Previous work has shown that the reaction of sulfonyl fluorides with silylated sulfonamides yields the corresponding sulfonimide [17, 18, 31, 32]. This reaction offers a viable route to the synthesis of the perfluorinated sulfonimide monomer **9** in high yield from **5**, $CF_2{=}CFOCF_2CF(CF_3)OCF_2CF_2SO_2F$. Since **5** was not readily available, it had to be synthesized before coupling with sodium trimethylsilyl trifluoromethyl sulfonamide **4**. Although the synthesis of **5** is well known from patents, to the best of our knowledge, the laboratory synthesis of this monomer has not been published.

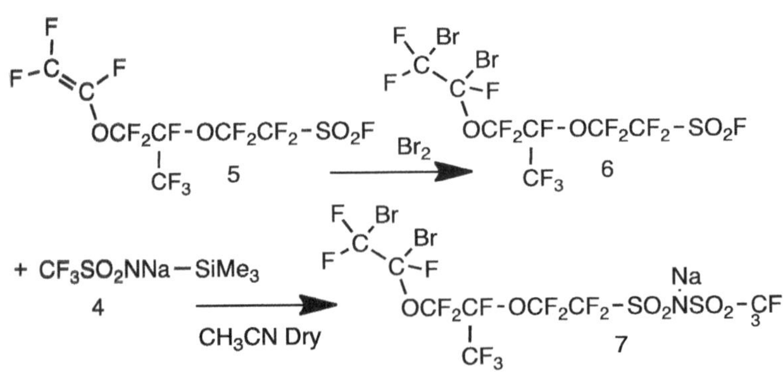

SCHEME 22.3 Synthesis of protected dibromosulfonimide.

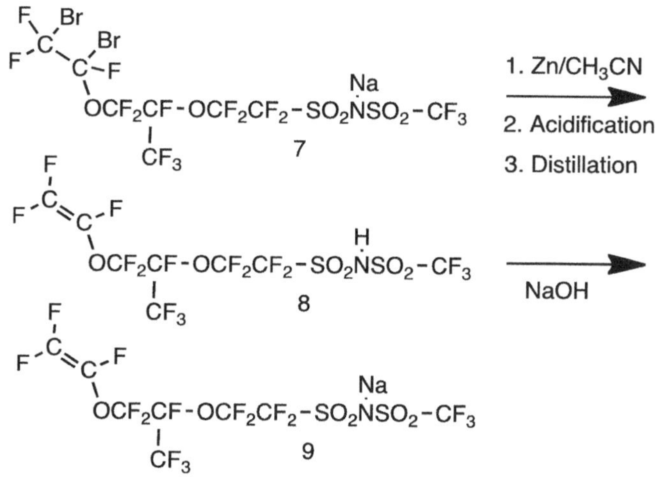

SCHEME 22.4 Synthesis of **9** sodium 3,6-dioxa-Δ^7-4-trifluoromethyl perfluorooctyl trifluoromethyl sulfonamide.

The synthesis of **5** involves a series of steps each forming several products, described in our previous publication [33].

Preparation of monomer sodium salt of sodium 3,6-dioxa-Δ^7-4-trifluoromethyl perfluorooctyl trifluoromethyl sulfonimide 9.

Trifluoromethane sulfonic anhydride **1** was ammoniated to give sulfonamide **2** [34, 35]. The amide **2** is then neutralized to **3** with NaOH (aq) and then silylated using hexamethyldisilazane (HMDS) to give **4** (Scheme 22.2). Attempts at direct reaction of the silylated material **4** with monomer **5** resulted in vinyl attack and formed a complex mixture; therefore, the vinyl group of monomer **5** must be protected with bromine. The bromine-protected sulfonyl fluoride **6** is coupled with the silylated sulfonimide **4** to give protected sulfonimide **7** (Scheme 22.3). The coupled product **7** is then dehalogenated with activated zinc and acidified to give the acid form of the monomer **8** which is purified by distillation. Because an aqueous emulsion polymerization is used, the acidic monomer **8** must be converted to its salt form, sodium 3,6-dioxa-Δ^7-4-trifluoromethyl perfluorooctyl trifluoromethyl sulfonimide **9** using sodium hydroxide [33] (Scheme 22.4).

22.3 COPOLYMERIZATION

TFE and the perfluorosulfonimide monomer **9** readily undergo a free-radical polymerization in an aqueous emulsion system (Scheme 22.5) to produce copolymer **10**. The copolymerization of the perfluorosulfonimide monomer **9** with TFE does not behave as a typical emulsion polymerization. Typical emulsion polymerizations

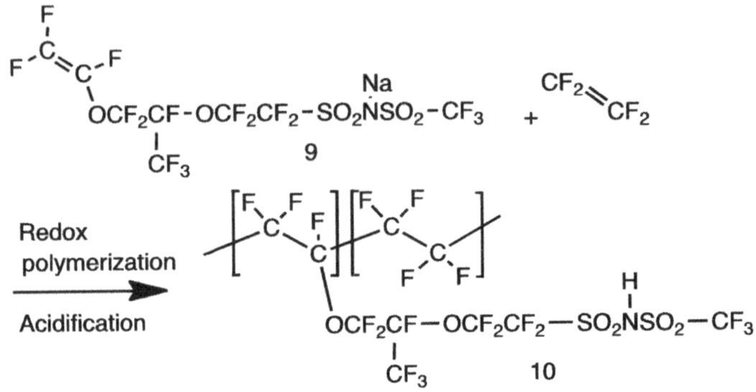

SCHEME 22.5 Copolymerization of PFSI monomer **9** with TFE.

use a micelle-forming surfactant with a hydrophobic monomer with some of the monomer solubilized in the micelle interior [36]. The majority of the monomer will be dispersed throughout and stabilized by the emulsifying agent. In the case of the perfluorosulfonimide monomer **9**, the monomer itself can behave as an emulsifying agent with both a hydrophobic region and a hydrophilic region. Therefore, the monomer readily dissolves in the aqueous reaction medium, contrary to typical emulsion polymerizations.

Scheme 22.6 shows the redox reaction involved in the emulsion polymerization. The semicontinuous copolymerization of the perfluorosulfonimide monomer **9** with TFE was initiated by $K_2S_2O_8$/$NaHSO_3$/catalytic iron (Scheme 22.6), which produces both the sulfate radical anion and the hydrogen sulfite radical at room temperature. Both radical species can initiate polymerization. Ammonium perfluoro-octanoate and Na_2HPO_4/NaH_2PO_4 are present as the emulsifier and pH buffer, respectively. The source of catalytic iron in our polymerizations is thought to be from the buffer salts as well as the 316 stainless steel autoclave. Polymerization of the perfluorosulfonimide monomer **9** with TFE employed a semibatch reactor system as shown in Figures 22.2 and 22.3. Early work used a batch reactor and produced blocky copolymers due to the changing ratio of the two monomers during polymerization. The continuous semibatch reactor system used a high pressure metering syringe pump for continuous addition of aqueous perfluorosulfonimide monomer **9** to the autoclave at a rate

$$K_2S_2O_8 \longrightarrow 2K^+ + S_2O_8^{-2}$$

$$NaHSO_3 \longrightarrow Na^+ + HSO_3^-$$

$$S_2O_8^{-2} + Fe^{+2} \longrightarrow SO_4^{-2} + SO_4^{\bullet -} + Fe^{+3}$$

$$HSO_3^- + Fe^{+3} \longrightarrow Fe^{+2} + HSO_3^{\bullet}$$

SCHEME 22.6 The persulfate–bisulfite–iron system.

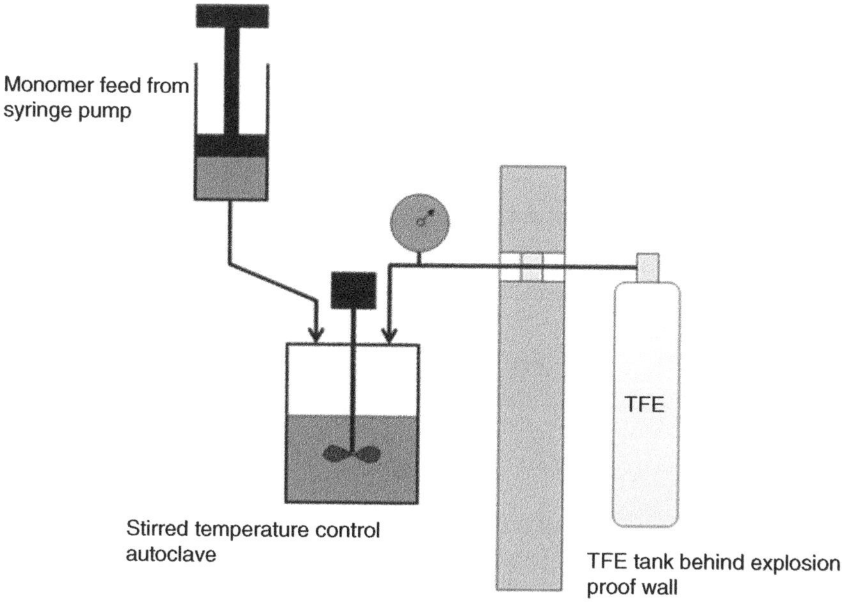

FIGURE 22.2 Aqueous emulsion copolymerization of TFVESI monomers with TFE.

approximating the uptake of **9** into polymer. After the initial charge of TFE to the reactor, the pressure was maintained at 145–150 psi by incremental (every few minutes) addition of TFE. This system allowed the concentration of the two monomers to remain fairly constant throughout the copolymerization. The TFE consumption was monitored by pressure drops throughout the reaction. The polymerization was stopped when the desired amount of TFE was consumed. Unlike typical emulsion polymerizations, our emulsions are extremely viscous at room temperature at the end of the polymerization. The polymer can be coagulated from the thick latex emulsion using 70 wt% HCl to produce a white, fibrous polymer when dried. All polymerizations based on consumed TFE had approximately a 50% conversion of the total perfluorosulfonimide monomer **9** added into the polymer, and the remaining monomer **9** may be recovered. The amount of TFE consumed in the reaction was estimated using the ideal gas equation based on the headspace of the reactor and the total pressure drop of the TFE. By plotting the number of moles of TFE consumed versus the time of the polymerization, it was possible to monitor the polymerization process and achieve very reproducible results.

Figure 22.3 shows a schematic diagram of the layout of the TFE barricade room at Hunter Laboratory in the Clemson Chemistry Department where these polymerization reactions were conducted. TFE gas is provided from a pressurized gas cylinder that is housed behind a protective wall. TFE is highly reactive and can dimerize into perfluorocyclobutane in a highly exothermic reaction, so limonene is placed in the TFE gas cylinder as an inhibitor. During polymerization, TFE is passed through a

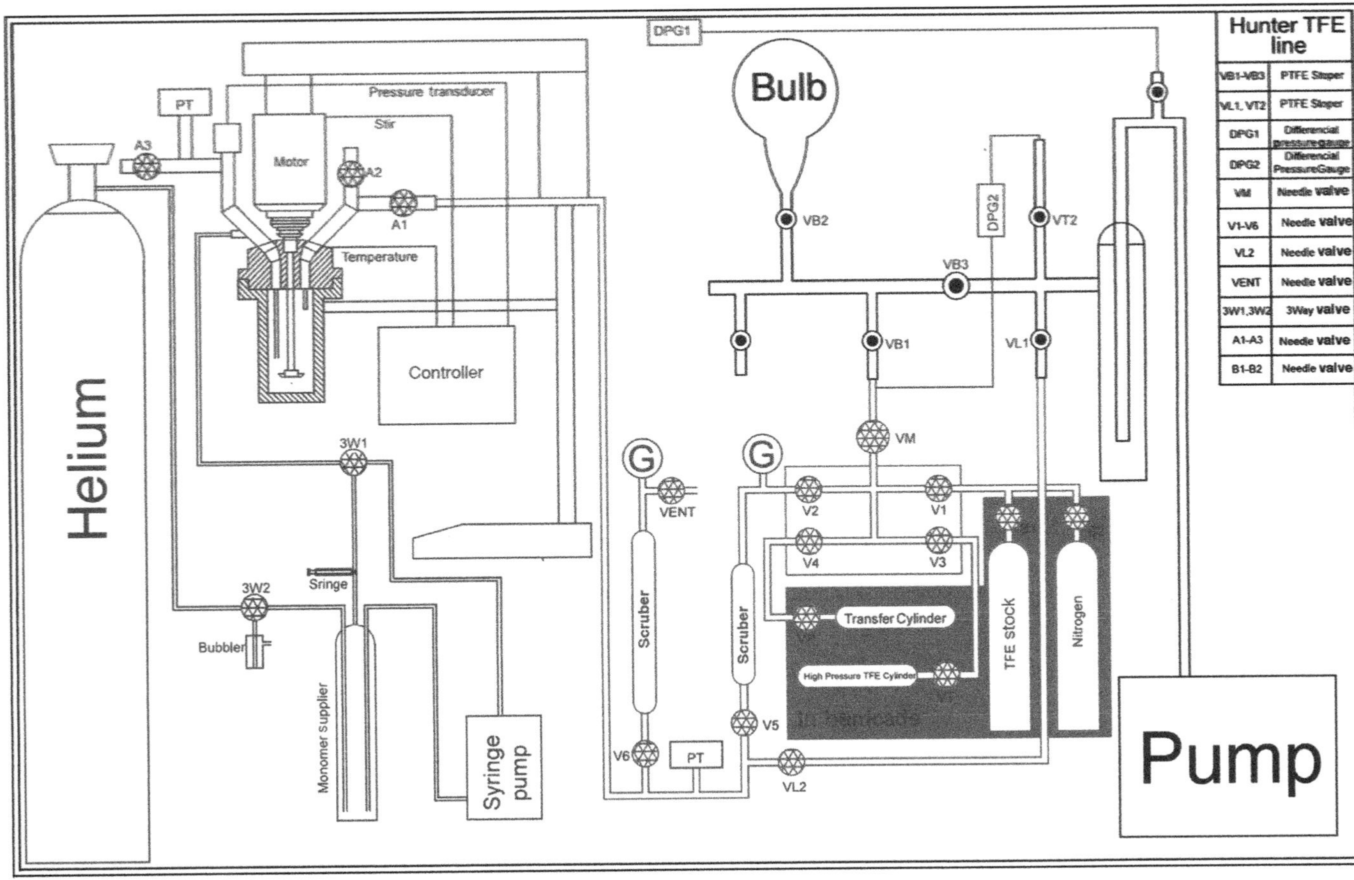

FIGURE 22.3 A schematic diagram of TFE facility in the bomb room at Hunter Chemistry Department.

silica scrubber to scavenge limonene. TFE is transferred from a large cylinder at a pressure of approximately 120 psi into a high pressure TFE cylinder of volume 500 cm^3 through a transfer cylinder. First, the TFE is transferred to a transfer cylinder that contains limonene using a glass bulb at a pressure of about 760 mmHg. The transfer cylinder is cooled to liquid nitrogen temperature (−196°C) during TFE transfer. After transferring about 80 g of a mixture of TFE and perfluorocyclobutane, the transfer cylinder is warmed to approximately −80°C to −100°C, and then the TFE is transferred into a high pressure cylinder.

Perflurocyclobutane has a boiling point of −6°C so it stays in the transfer cylinder, whereas TFE (about 80 g) having a boiling point of −76°C is transferred to the high pressure cylinder. Usually, about 2–4% of perflurocyclobutane is present in TFE. After transfer, the high pressure TFE cylinder at −196°C is evacuated under vacuum to remove any volatiles. The pressure in the high pressure TFE cylinder after TFE transfer is usually 500–550 psi. During polymerization, the TFE pressure is kept at 120–140 psi in the autoclave reaction vessel. The autoclave (450 cm^3) is evacuated and leak-tested with the vacuum line and then monomer **9**, buffers, surfactant, and initiator in deionized water are introduced through valve A2 into the reactor. TFE is introduced through A1 at a pressure of about 120–140 psi. The addition solution containing the monomer and buffers is added to a cylinder, which is purged with helium gas and introduced into the autoclave through syringe pump at about 0.15 mL/min. TFE is added to autoclave as it is consumed, as a specified pressure is maintained.

Table 22.1 shows some typical copolymerizations involving monomer **9** with TFE to produce ionomers having different equivalent weights (EWs) or ion-exchange capacities. The polymerizations were reproducible within experimental error, and the ion-exchange capacity (IEC) of the copolymer can be changed by varying either the stirring rate, the pressure of TFE in the reactor, the time of polymerization, or the amount of monomer in the reactor.

22.4 THERMAL CHARACTERIZATION OF SULFONIMIDE POLYMERS

Perfluorosulfonimide copolymers **10** have excellent thermal stability. The polymer's typical onset temperature for thermal decomposition is around 375–400°C in the acid form and around 480°C in the salt form. Figure 22.4 shows thermographs of three bulk copolymer materials at different EWs in the acid form, acquired under a nitrogen atmosphere. The first step in the weight loss at around 375°C corresponds to loss of the perfluorosulfonimide functional group. The next step in the weight loss corresponds to the poly(tetrafluoroethylene) (PTFE)-like backbone at 400–580°C. Figure 22.5 shows a typical differential scanning calorimetry (DSC) scan of three polymer **10** copolymer samples with different EWs. A broad endotherm at 100–170°C is indicative of coordinated water. The second endotherm at 320–350°C occurs at nearly the same temperature as the melting temperature of virgin PTFE [37]; therefore, this peak is assigned to the melting (T_m) of PTFE crystallites in the copolymers. The

TABLE 22.1 Typical Reaction Conditions for the Copolymerization of Sulfonimide Monomer 9 with TFE Which Gives Polymer 10

Polymers **10**	173	174	184	185	186	187	192	193	194	195	221
Reactor (mL)	450	450	450	450	450	450	450	450	450	450	450
Monomer autoclave (g)	2.04	2.04	2.04	1.36	2.04	2.04	2.04	1.36	2.04	2.04	2.04
Surfactant $C_7F_{15}CO_2NH_4$ (g)	2.06	2.06	2.06	2.06	2.06	2.06	2.06	2.06	2.06	2.06	2.06
$Na_2HPO_4 \cdot 7H_2O$ (g)	2.81	2.81	2.81	2.81	2.81	2.81	2.81	2.81	2.81	2.81	2.81
NaH_2PO_4 (g)	1.41	1.41	1.41	1.41	1.41	1.41	1.41	1.41	1.41	1.41	1.41
$K_2S_2O_8$ (g)	0.37	0.37	0.37	0.37	0.37	0.37	0.37	0.37	0.37	0.37	0.37
$NaHSO_3$ (g)	0.28	0.28	0.28	0.28	0.28	0.28	0.28	0.28	0.28	0.28	0.28
H_2O D.I. (mL)	260	260	260	260	290	290	290	260	290	290	230
Pressure of TFE (psi)	145–150	145–150	140–150	130–136	140–150	140–150	135–140	150–160	150–160	135–140	100–108
Stirring speed (rpm)	650	650	650	650	650	650	650	650	550–650	510	650
Temperature (°C)	10	10	10	10	10	10	10	10	10	10	10
Reaction time (h)	7	6	5.5	5.5	4	6	4	6	4	2.5	4
Total TFE drop (psi)	52	44	41	40	31	55	36	49	20	25	25
Average TFE drop (psi/h)	7.4	7.3	7.45	7.27	7.75	9.01	9	8.16	5	10	6.25
TFE consumed (g)	2.85	2.41	2.24	2.19	1.69	3.01	2	2.68	1.09	1.37	1.37
Added monomer (g)	2.17[a]	2.17	2.17	1.44	2.17	2.17	2.17	1.44	2.17	2.17	2.17
$Na_2HPO_4 \cdot 7H_2O$ in addition	0.32	0.32	0.32	0.32	0.32	0.32	0.32	0.32	0.32	0.32	0.32
NaH_2PO_4 in addition (g)	0.16	0.16	0.16	0.16	0.16	0.16	0.16	0.16	0.16	0.16	0.16
Addition volume (mL)	29	29	29	29	29	29	29	29	29	29	29
PTFE recovered (g)	0.79	0.65	0.26	0.8	0.2	0.525	0.19	0.9	0	0	0
Polymer weight(g)	4.76	4.87	2.31	4.92	2.66	4.03	3.2	5.23	1.29	1.23	3.72
IEC(mgH/g)	0.82	0.979	0.759	0.616	0.862	0.81	0.857	0.688	0.936	1.04	0.968
Equivalent weight (EW)	1225	1021	1317	1624	1160	1235	1167	1454	1067	958	1033
TFE in polymer(g)	2.52	2.13	1.3	3.18	1.34	2.15	1.62	3.16	0.6	0.49	1.65
Monomer in ionomer (g)	2.24	2.74	1.01	1.74	1.32	1.88	1.58	2.07	0.69	0.74	2.07
Monomer:TFE	1:6.5	1:4.5	1:7	1:10.5	1:5.9	1:6.6	1:6	1:9	1:5	1:4	1:4.6
Water uptake (%)	61	74	45	35	63	55	64	40	65	69	62

[a]TFVE monomer could only be added after 3 h of polymerization.

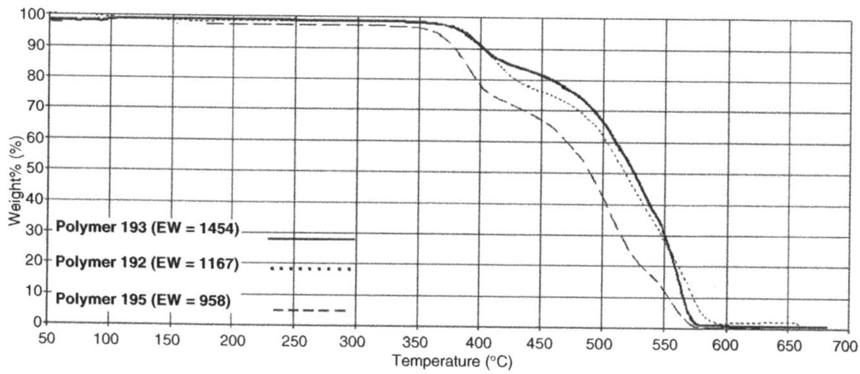

FIGURE 22.4 thermogravimetric analysis (TGA) of perfluorosulfonimide copolymer of **10**, polymers 192, 193, and 195.

heat of fusion (ΔH_{fusion}) can be calculated for the second endotherm and is related to the extent of PTFE-like crystallinity in the polymer [38]. As seen in Figure 22.5, as the ion-exchange capacity of the polymers increases, the extent of PTFE-like crystallinity decreases; this finding is due to the introduction of more bulky side chains in the copolymer which reduce the crystalline order and reduce the crystallization of the PTFE-like macromolecular chains.

22.5 PFSI IONOMER DISSOLUTION AND MEMBRANE FORMATION

Membranes of typical PFSA materials such as Nafion® and Aquivion® are often melt-processed in the precursor sulfonyl fluoride form ($-SO_2F$) and then hydrolyzed to the sulfonic acid using caustic media followed by acidification [39,40]. In the case of the sulfonimide polymer **10**, membranes cannot be formed by melt-processing since the polymer exhibits no melting point; however, membranes of the polymer can be prepared by solution-casting. A considerable amount of work has been published on preparing PFSA polymer membranes from the solution used to repair pinholes in chlor-alkali membranes and as coating materials for modified electrodes [39–43]. For the PFSA materials, solution-cast membranes can be formed using a variety of solvents including dimethylformamide (DMF), dimethylsulfoxide (DMSO), and ethylene glycol (EG).

Although membranes are obtained using a variety of solvents, we find that the preferred solvent for film formation of a PFSI polymer **10** is a mixture of DMF, methanol, and deionized water heated at 80–90°C in the ratio of 200:19:10 mL. The acid form of the polymer is blended with DMF:methanol:water to form a viscous liquid and this process converts the acid function into the dimethylammonium salt form. This procedure gives a dispersion in which the polymer is not completely dissolved. The dispersion can be filtered using a nylon mesh to remove large particulates and the solvent can be removed under a slight vacuum around 90°C to obtain what Gebel describes as a "compact nearly dried solution" [41]. The remaining clear film

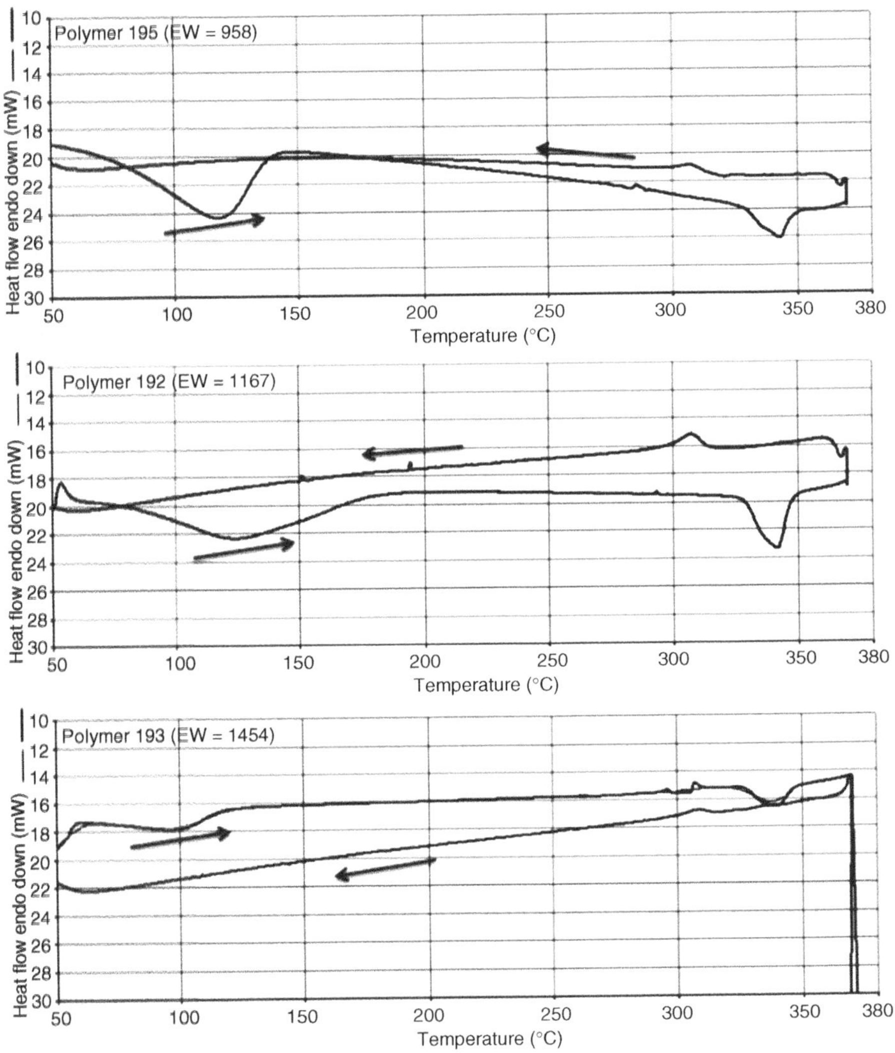

FIGURE 22.5 DSC of perfluorosulfonimide copolymer of **10**, polymers 192, 193, and 195.

is brittle and will readily dissolve in the casting solvent; this finding indicates that only small reentaglement and rearrangement of the polymer chains has occurred. The PFSI solution-cast membranes were found to exhibit better properties when annealed between 200°C and 250°C [41]. The poor physical properties of the unannealed cast membranes have been attributed to the supermolecular features of the polymer, that is, the PTFE crystallites and the ionic aggregates [42]. When casting, the polymer chains become immobile and crystallization cannot occur. Thermal excitation provides the necessary mobility which allows the polymer to reorient and form the

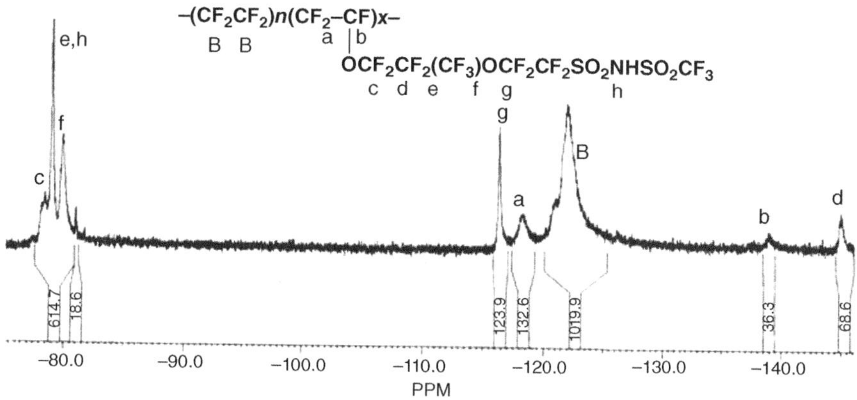

FIGURE 22.6 ^{19}F NMR of a sulfonimide:TFE copolymer **10** in the acid form.

PTFE crystallinity and the ionic aggregation needed to form a high quality membrane [39–42]. Sulfonimide polymer **10** membranes that are annealed between 225°C and 250°C also exhibit some discoloration due to organic impurities in the film-casting process. Subsequent soaking in hot nitric acid restores a colorless appearance and fully converts the polymer to the protonated form.

The ^{19}F NMR spectrum (Figure 22.6) of the dissolved PFSI polymer **10**, which was prepared by dissolution in a mixture of *n*-propanol, methanol, and deionized (D.I.) water as solvent, is consistent with the expected copolymer structure. The vinyl fluorine atoms (chemical shifts $\delta = -112.4, -120.8, -135.6$ ppm) observed in the monomer were not observed in the copolymer. Likewise, a new broad unresolved resonance at $\delta = -120.4$ ppm is indicative of aliphatic fluorine atoms associated with the PTFE backbone of the copolymer. Resonances for the other fluorine atoms in the structure were broad and unresolved, but similar in chemical shift to that of the monomer.

22.6 EQUIVALENT WEIGHTS OF PFSI IONOMERS AND MEMBRANES

Equivalent weights for PFSI ionomers immediately following precipitation from the reaction medium and for membranes formed as described in section 22.5 were measured by acid–base titration. In brief, the procedure is as follows. A known amount of dried polymer (in a typical measurement, approximately 0.0670 g of polymer) and 0.10 g of NaCl were dissolved in 30 mL of deionized water, and the resulting solution was refluxed in a 150-mL conical-shaped flask at 80°C overnight with constant stirring. The flask was covered with a latex balloon and then filled with nitrogen gas through a septum. The polymer swelled up in water. A few drops of phenolphthalein were added to the solution, which was then titrated with a 0.0100 M

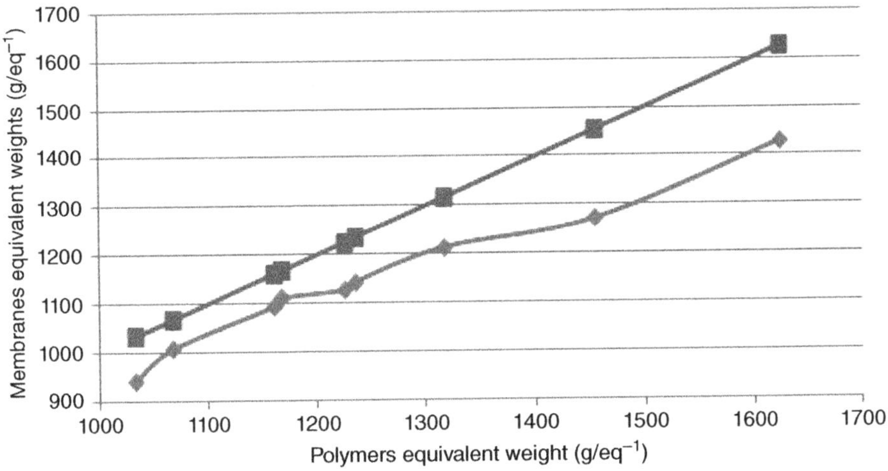

FIGURE 22.7 Equivalent weights (EWs) of polymers versus EWs of their membranes for polymers 173, 184, 185, 186, 187, 192, 193, 194, 221. Diamonds, data for membranes; squares, trend line assuming equal EW for polymers and membranes.

solution of standardized NaOH. The solution was remained in an 80°C hot oil bath during the titration. After obtaining a pink color, an excess (2 mL) of standardized NaOH solution was added, and the solution was left stirring overnight at 80°C under a nitrogen atmosphere. The pink-colored solution was then back titrated with a standardized HCl solution, 0.0125 M. The same procedure was applied to nearly all the polymers in Table 22.1, in both the as-precipitated form and in membrane form.

Figure 22.7 is a plot of EWs of several of the ionomers from Table 22.1, with the membrane EW on the *y*-axis and the as-precipitated polymer EW on the *x*-axis. If the ionomer samples are pure and the process of polymer dissolution and film casting occurs with the complete transfer of all ionomers, it is expected that the EWs of the membranes and the as-precipitated ionomers should be the same. For this reason, a guideline is included in the plot for the case of EWs for as-precipitated ionomers and membranes being equal. A close look at the plot shows that for all samples, the membrane EW is less than the EW of the as-precipitated ionomer. EW discrepancies are typically between 75 and 200 g/eq, and are greater for the higher EW samples. The most likely cause of these discrepancies is that ionomer dissolution into the DMF/methanol/water solvent occurs with some fractionation, such that very high EW fractions are not dissolved and are transferred to the membranes. Consistent with this possibility, we note that a small portion of the ionomer is usually collected on the nylon mesh filter following ionomer dissolution. The very high EW ionomer fractions are very PTFE-like in character and are likely to have low solubility in all solvents. It is important to recognize that this fractionation occurs. Among other things, it tells us that in studies of membrane proton conductivity as a function of ionomer EW, the membrane EW must be used, not the EW of the ionomer used to form the membrane.

22.7 WATER UPTAKE BY MEMBRANES

Ionomer membranes are typically very good water absorbers which is fortunate because water is important in helping to promote rapid proton transport. On the other hand, water uptake causes swelling and can negatively impact membrane mechanical properties. In designing the ideal proton exchange membrane for fuel-cell applications, it is important to achieve just the right balance in water uptake, providing just enough water to keep proton conductivity high but not so much as to cause too much swelling or degradation of mechanical properties.

Water uptake in PFSI membranes was studied using two procedures, as follows. In the first procedure, a membrane sample is reflexed in water overnight at 80°C followed by decanting of the water, membrane cleaning, and removal of adherent liquid water with absorbent paper. The membrane sample is immediately weighed on an analytical balance as it loses water quickly. The membrane is then thoroughly dried under vacuum at 80°C overnight and weighed again without delay as it absorbs moisture rapidly. The second procedure differs from the first only in that the membrane is heated in a beaker of water at 60°C for 1–2 hours prior to the initial weighing. The water uptake of Nafion® 112, (28%) using the second procedure was nearly the same as reported in the literature [44] which serves to validate the technique.

Table 22.2 and Figure 22.8 present data on water uptake by PFVSI:TFE copolymer membranes from Table 22.1. Water uptake is always higher by the first method involving overnight treatment with refluxing water; this is not unexpected since this method is more aggressive and would be more likely to soften the membrane and allow for greater water penetration. A clear trend of higher water uptake by membranes having lower EW is also seen. This finding is sensible; the lower EW ionomers have more proton and ionic group content and can absorb more water to solvate those ions, and they have less PTFE-like character which makes them softer and more able to swell to absorb even larger amounts of water.

TABLE 22.2 Water Uptake of PFVSI:TFE Membranes 10 and Their Blends

Polymer identity	EW of polymer (g/eq)	EW of membrane (g/eq)	Water uptake refluxed in water overnight (wt%)	Water uptake at 60°C for 2 h % (wt%)
174	1021	977	74	38
184	1317	1213	45	23
185	1624	1426	35	20
186	1160	1093	63	32
187	1235	1142	55	30
193	1454	1270	40	21
194	1067	1007	65	37
221	1033	941	75	40
Nafion® 112	1100	1059	60	28

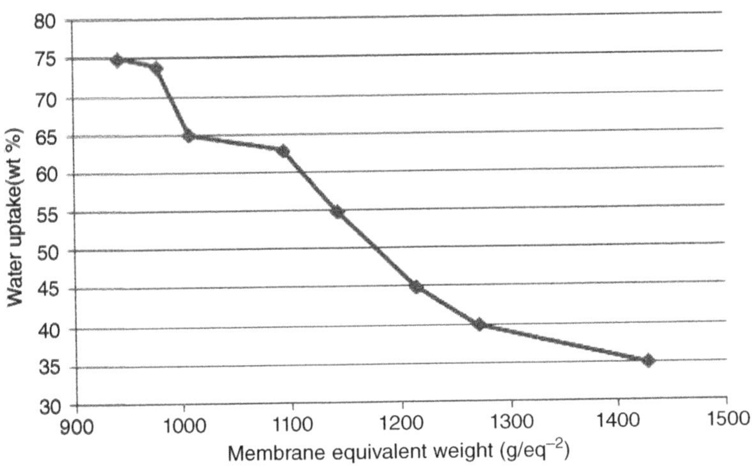

FIGURE 22.8 Water uptake of PFSI membranes 174, 184, 185, 186, 187, 193, 194, and 221 versus their EWs by method 1 involving overnight refluxing in water.

22.8 PROTON CONDUCTIVITY OF PFSI MEMBRANES

The proton conductivity of ionomer membranes was measured under variable relative humidity (RH) and temperature conditions using a Scribner model 850C compact fuel cell test system, coupled with a BekkTech model BT 112 4-point-probe conductivity test cell fixture (Figure 22.9). Before measuring conductivity, the desired RH was achieved in the test fixture by changing the dew point of a humidified gas stream that flows through the anode side of the test fixture. All data were acquired after membrane samples reached their equilibrium state at a fixed humidity and temperature. A typical procedure is as follows. First, the anode temperature (dew point) was set to 71°C and the cell (test fixture) temperature to 80°C to give 69% RH in the test fixture. Then, the membrane was allowed to equilibrate for about 2 hours to reach an equilibrium state. Next, the anode temperature (dew point) was changed to provide the appropriate RH in the test fixture, and the fixture was allowed to stand for 15 minutes each time to allow the membrane to reach equilibrium before resistance (impedance) readings were taken. To ensure reliable data acquisition, we compared the proton conductivity of Nafion® 112 in the literature [42] with the one measured in this study under similar conditions, and obtained a good agreement.

Figure 22.10 (*y*-axis) shows the conductivity of membrane 194 (made from a sample of polymer 194, EW of polymer = 1067 g/eq, for which the membrane formed by solvent-casting has an EW of 1007 g/eq) compared to Nafion® 112 over 20–100% RH at 80°C. The conductivity of membrane 194 is higher at 20–30% RH and at 90–100% RH compared to Nafion® 112 and is similar to Nafion in the 30–90% RH range. In both cases, the conductivity drops by approximately 25× when decreasing RH from 100% to 20%. This behavior is commonly observed for both PFSA and PFSI ionomers, and is thought to reflect many factors including changes in

FIGURE 22.9 Conductivity cell BT 212 connected to fuel cell test system 850C.

ionomer morphology/domain structure and also proton hydration as RH is changed. The fact that the PFSA and PFSI ionomers show similar decreases in conductivity with decreasing RH suggests that they have similar morphologies, with PTFE-like domains providing structural integrity and ionic domains providing pathways for proton transport.

Figure 22.11 shows conductivity versus RH for membrane 221 which has a relatively low EW of 941 g/eq. Conductivity of this membrane is approximately 3× higher than that of Nafion 112 at all RH values tested. This finding underscores the importance of EW in determining ionic conductivity, that is, even relatively modest changes in EW can have large changes in conductivity. This fact suggests that changes in EW involve more than just simple changes in the concentration of charge

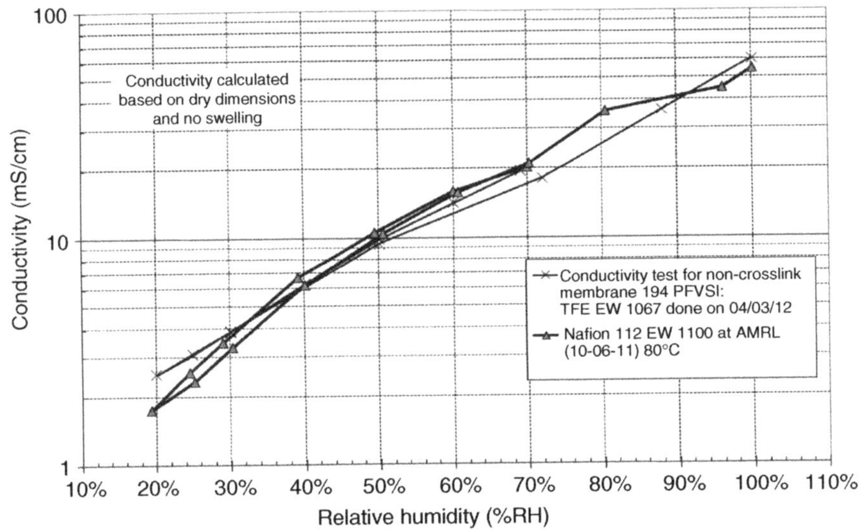

FIGURE 22.10 Conductivity of membrane 194 over a range of RH values, compared to Nafion 112 at 80°C.

carriers, they probably also involve changes in ionomer morphology and connectivity between/among the continuous ion-transporting pathways in the membrane.

Ionic conductivity for ionomer membranes is also expected to be a strong function of the EW of the ionomer comprising the membrane. A convenient way of comparing ionomers having different EWs is via their conductivity under water-saturating conditions, at 100% RH. Table 22.3 presents ionic conductivity data for a series of

TABLE 22.3 Ionic Conductivity of PFSI Ionomer Membranes

Polymer identity	EW of polymer (g/eq)	EW of membrane (g/eq)	Thickness (microns)	Conductivity at 80°C at 100% RH (S/cm)
173	1225	1126	38	0.0767
174	1021	977	30	0.0918
184	1317	1213	54	0.0676
185	1624	1426	26	0.0472
186	1033	993	34	0.0963
186	1033	993	25	0.0828
187	1235	1130	28	0.086
192	1167	1111	48	0.1055
193	1454	1270	26	0.0622
194	1067	1007	21	0.1058
221	1033	941	40	0.145
Nafion® 112	1100	1059	55	0.055

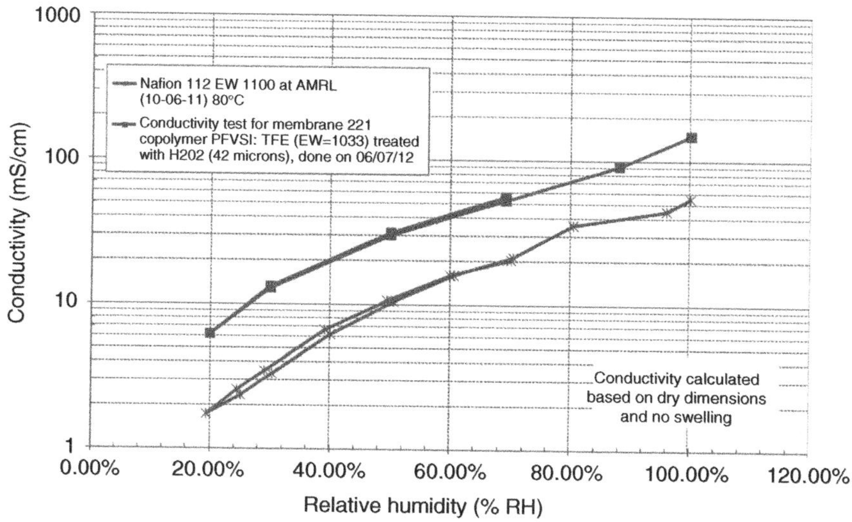

FIGURE 22.11 Conductivity of membrane 221 over a range of RH values, compared with Nafion 112 at 80°C.

membranes under these conditions, and Figure 22.12 presents a plot of the conductivity at 80°C and 100% RH for a series of ionomers for which the membrane EW varies between approximately 950 and 1450 g/eq. A clear trend of diminished ionic conductivity with increasing ionomer EW is observed. This is obviously an expected result since higher EW means fewer protonic charge carriers and fewer ionic groups to comprise a continuous ionic domain to support bulk ion transport.

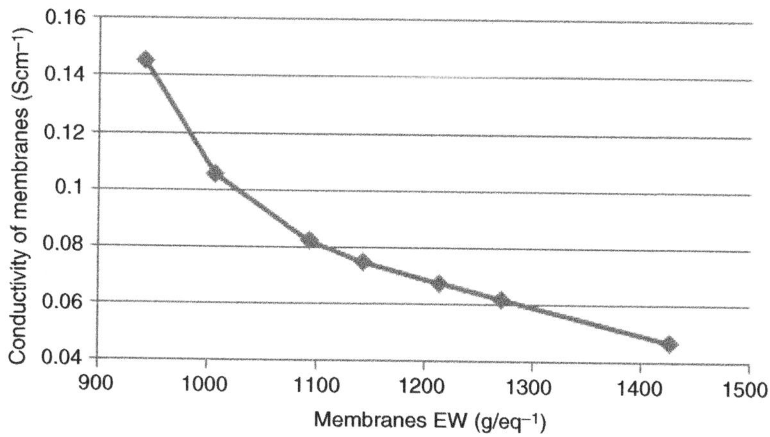

FIGURE 22.12 Conductivity of membrane 184, 185, 186, 187, 193, 194, 221 versus their EWs at 80°C and 100% RH.

22.9 SUMMARY

This chapter has summarized the synthesis and basic characterization of perfluorinated ionomers prepared by copolymerization of TFE with TFVE monomers functionalized with PFSI functional groups. The resulting ionomers have good thermal and chemical properties and good dimensional stability and are good proton conductors over a wide range of relative humidities. A modest fractionation in EW is noted when preparing membranes from precipitated ionomer samples, reflecting loss of some higher EW fractions during the dissolution/evaporation process. Clear trends in water uptake and ionic conductivity with variation in ionomer EW are noted and rationalized in terms of an increase in ionic character and a decrease in PTFE-like crystallinity in membranes as the ionomer EW decreases. Conductivity values are similar to those of PFSA ionomers such as Nafion, for PFSI samples that are still at a relatively early stage of development. Prospects for further improvement of ionomer and membrane properties by continued materials development are excellent.

ACKNOWLEDGMENTS

The authors gratefully acknowledge the contributions of many coworkers, too numerous to mention individually, over the years who have contributed in countless ways to the development of this class of ionomer materials. Each person contributed in his or her own way, and all contributions are necessary to move the materials development forward. The authors gratefully acknowledge the Division of Chemical Sciences, Geosciences, and Biosciences, Office of Basic Energy Sciences of the U.S. Department of Energy through Grant DE-FG02-05ER15718 for financial support of the work on the synthesis and characterization of fluoropolymer electrolytes.

APPENDIX: DETAILED REPRESENTATIVE PROCEDURES FOR PFSI IONOMER AND MEMBRANE

Typical Copolymerization of Monomer 9

$F_2C{=}CFOCF2CF(CF_3)OCF_2CF_2SO_2N(Na)SO_2CF_3$ With TFE to Give Polymer 10 194 Monomer **9** 2.04 g (3.41 mmol), surfactant $C_7F_{15}COONH_4$, 2.06 g, (4.78 mmol), buffers Na_2HPO_4 = 2.81 g (22 mmol), NaH_2PO_4 = 1.41 g (13 mmol), and $NaHSO_3$ 0.28 g (2.69 mmol) were added to 260 mL of deionized, deoxygenated water at 10°C in a 500-mL separatory flask and purged with nitrogen gas. Also, $K_2S_2O_8$ (initiator) was dissolved in 30 mL of deionized, deoxygenated water at 10°C in a 50-mL flask and purged with nitrogen gas. The addition solution was prepared by dissolving 0.32 g Na_2HPO_4, 0.16 g NaH_2PO_4, and 2.17 g monomer in 30 mL of deionized, deoxygenated water and purged with nitrogen gas. Autoclave 450 mL was connected to vacuum line through valve A1 in the bomb room as shown in Figure 22.3. 1/16th in. tubing was connected through a metering pump to deliver addition solution during polymerization. The addition metering pump main selector valve was set to the bypass position. The autoclave was purged with nitrogen twice and evacuated. Five minutes before polymerization started, $K_2S_2O_8$ solution was injected to the separatory flask containing monomer solution. Then this solution was sucked through A2 valve into the autoclave (Figure 22.3; already set to 10°C temperature with cryostat using ethanol/water), followed by 30 mL of D.I. deoxygenated water injected into the separatory flask to wash away the reaction mixture. Then TFE was added to a pressure of 150–160 psi and this pressure was maintained during polymerization. The addition solution was already added to the metering pump reservoir and was degassed with helium for 20 minutes. The addition metering pump was primed and the main selector valve was turned to reactor and micrometer was set to deliver liquid at a rate of 0.15 mL/min (estimated time to deliver 30 mL of solution is three and a half hours). The polymerization was carried out for 4 hours, at which point 2 mL of addition monomer remained in the cylinder. The pump was then switched off and the main selector valve was turned to the bypass position and the reaction was stopped. In total, 20 psi (10.9 mmol) of TFE was consumed. The unreacted TFE was vented into the hood. The autoclave was opened; there was no insoluble polymer obtained. The emulsion solution (310 mL) was divided into two portions, 110 and 200 mL. The portion of 110 mL was used for subsequent studies and membrane fabrication, and the portion of 200 mL was left for blending.

To the 110 mL portion, 50 mL of concentrated HCl (37%) was added slowly and stirred vigorously overnight to precipitate the polymer, which was then filtered under suction and washed with deionized water until the eluant was neutral. This procedure gave gel-like polymer (5.35 g). To this polymer, 200 mL of deionized water was added and stirred at 70–80°C overnight to remove surfactant, buffer, and monomer. The polymer was filtered and washed with D.I. water until no more foam appeared in filtrate. This gave 4.81 g of gel-like polymer. This polymer was refluxed in 50 mL of HNO_3 (70%) overnight twice to convert completely into the –NH form, then filtered and washed with D.I. water (2 L) until the eluant was neutral. This polymer was again heated at 80°C in 100 mL of D.I. water overnight, filtered, and washed with D.I. water until the eluant was neutral. This procedure gave 2.02 g of white wet polymer. It was dried under dynamic vacuum overnight at 100°C to give 0.46 g of white polymer.

Preparation of Membrane 194

An amount of 0.2590 g of polymer 194 was dissolved into 60 mL of DMF, 5.4 mL of methanol, and 3 mL of deionized water in a 100-mL round-bottomed flask equipped with a water condenser. The solution was heated at 80°C with vigorous stirring overnight. The solution was allowed to cool to room temperature. The polymer mixture was blended into a solution for 2 minutes using a stainless steel blender (speed 4) and filtered on a nylon cloth (stockings). The resulting viscous solution was poured into an 80-mL stainless steel dish precleaned with HNO_3 (70%), and then the solution was degassed under vacuum in an oven. The contents of the oven were then heated to 80°C under vacuum pump with a slow flow of nitrogen gas into the oven at 25 in. of Hg for 14 hours. The remaining DMF was removed at 180°C at full vacuum for 12 hours; the membrane was annealed at 250°C for 4 hours under vacuum. The oven was brought to 210°C and hot, deionized water was added to the dish in order to peel off the membrane. The membrane was then boiled in 50 mL of 70% HNO_3 overnight twice to convert the dimethyl ammonium salt back into the protonated form of the PFSI group. The dimethyl ammonium salt is formed by the reaction of –NH of polymer with DMF. Nitric acid was decanted and the membrane was washed with water until the eluant was neutral. The membrane was then boiled in water twice overnight. The membrane surface was cleaned with kimwipes and quickly weighed (0.3740 g). After 5 minutes, its weight decreases to 0.2950 g (loss of 21% of weight in 5 minutes). It loses weight when the wet membrane is left in open air. The membrane was dried in vacuum oven at 100°C overnight to give 0.2130 g of dry membrane. The dry membrane also gains weight when left in open air (after 1 minute, the weight of dry membrane was 0.2230 g, a 5% increase in weight in 1 minute) and the weight is steady after 1 minute. The thickness of membrane was found to be 23 microns.

REFERENCES

1. Kordesch, K.; Simader, G. *Fuel Cells and Their Applications*; VCH: New York, 1996; Chapter 4.
2. Ehrenberg, S. G.; Serpico, T.; Wnek, G. E.; Rider, J. N. Fuel cell incorporating novel ion conducting membrane. US Patent 5,468,574, 1995.
3. Wei, J.; Stone, C.; Steck, A. Trifluorostyrene and substituted trifluorostyrene copolymeric compositions and ion exchange membrane formed therefrom. US Patent 5,422,411, 1995.
4. Andreola, C.; Chlanda, F. P.; Huang, J. P. US Patent 5,643,968, 1997.
5. Brack, H. P.; Bonorand, L.; Buhrer, H. G.; Scherer, G. G. *Polym. Prepr. Am. Chem. Soc. Div. Polym. Chem.* **1998**, *39*, 897.
6. Lehtinen, T.; Sundholm, G.; Holmberg, S.; Sundholm, F.; Bjornbom, P.; Bursell, M. *Electrochem. Acta.* **1998**, *43*, 1881.
7. Feiring, A. E.; Rozen, S. Fluorinated ion exchange polymers and intermediates therefor. US Patent 5,668,233, 1997.
8. Kordesch, K.; Simader, G. *Fuel Cells and Their Applications*, VCH: New York, **1996**, Chapter 2.
9. Ezzell, B. R.; Carl, W. P.; Mod, W. A. Preparation of vinyl ethers. US Patent 4,358,412, 1980.
10. Resnick, P. R. Ion exchange polymer film, consisting of fluorinated polymer with N-monosubstituted sulfonamido groups method and apparatus for electrolysis of alkali or alkaline earth metal halide. US Patent 4,085,071 A19780418, 1978.
11. Connoly, C.; Gresham, D. J. Fluorocarbon vinyl ether polymers. US Patent 3,282,875, 1966.
12. Zaluski, C; Xu, G. *Macromolecules* **1994**, *27*, 6750.
13. Hassani, M. O.; Germain, A.; Brunel, D. A. *Tetrahedron Lett.* **1981**, *22* (1), 65–68.

14. DesMarteau, D. D. *J. Fluor. Chem.* **1995**, *72*, 203.
15. Lebland, R. D.; DesMarteau, D. D. *J. Chem. Soc. Chem. Commun.* **1974**, *14*, 555.
16. Foropoulos, J. F., Jr.; DesMarteau, D. D. *J. Am. Chem. Soc.* **1982**, *104*, 4260.
17. DesMarteau, D. D.; Zhu, S.; Pennington, W. T.; Gotoh, Y.; Witz, M.; Zuberi, S. *J. Fluor. Chem.* **1989**, *45*, 24.
18. Foropoulos, J. F., Jr.; DesMarteau, D. D. *Inorg. Chem.* **1984**, *23*, 3720.
19. Koppel, I. A.; Taft, R. W.; Anvia, F.; Zhu, S. -Z.; Hu, L. -Q.; Sung, K.; DesMarteau, D. D.; Yagupolski, L. M.; Yagupolski, Y. L.; Ignat'ev, N. V. et al. *J. Am. Chem. Soc.* **1994**, *116*, 304.
20. Razaq, M.; Razaq, A.; Yeager, E.; DesMarteau, D. D.; Singh, S. *J. Electrochem. Soc.* **1989**, *136*, 385.
21. Creager, S. E.; Sumner, J. J.; Ma, J. J.; DesMarteau, D. D. *J. Electrochem. Soc.* **1998**, *145*, 107.
22. Creager, S. E.; Sumner, J. J.; Bailey, R. D.; Ma, J. -J.; Pennington, W. T.; DesMarteau, D. D. *Electrochem. Solid-State Lett.* **1999**, *2*, 434.
23. Savett, S. C.; Atkins, J. R.; Sides, C. R.; Harris, J. L.; Thomas, B. H.; Creager, S. E.; Pennington, W. T.; DesMarteau, D. D. *J. Electrochem. Soc.* **2002**, *149*, A1527.
24. Atkins, J. R.; Sides, C. R.; Creager, S. E.; Harris, J. L.; Pennington, W. T.; Thomas, B. H.; DesMarteau, D. D. *J. New Mat. Elec. Syst.* **2003**, *6* (1), 9.
25. DesMarteau, D. D. Copolymers of tetrafluoroethylene and perfluorinated sulfonyl monomers and membranes made therefrom. US Patent 5,463,005, 1995.
26. Rollins, H. W.; Lin, F.; Johnson, J.; Ma, J. -J.; Liu, J. -T.; Tu, M. -H., DesMarteau, D. D.; Sun, Y. P. *Langmuir* **2000**, *16*, 8031.
27. Rollins, H. W.; Whiteside, T.; Shafer, G. J.; Ma, J. -J.; Tu, M. -H.; Liu, J. -T.; DesMarteau, D. D.; Sun, Y. P. *J. Mater. Chem.* **2000**, *10*, 2081.
28. DesMarteau, D. D.; Creager, S. E.; Savett, S.; Thomas, B. *Polym. Mat. Sci. Eng.* **1999**, *80*, 600.
29. DesMarteau, D. D.; Ma, J. -J.; Tu, M. -H.;. Liu, M. -T; Thomas, B. *J. McClellan Polym. Mat. Sci. Eng.* **1999**, *80*, 598.
30. Ma, B.; Bunker, C. E.; Simons, K.; Liu, J. -T.; Ma, J. -J.; Martin, C. W.; DesMarteau, D. D.; Sun, Y. -P. *J. Electroanal. Chem.* **1998**, *459*, 15.
31. Meußdoerffer, Von J.; Niederprüm, H. *Chemiker Ztg.* **1972**, *96*, 10, 582.
32. DesMarteau, D. D.; Witz, M. *J. Fluor. Chem.* **1991**, *52*, 7.
33. Thomas, B. H.; Shafer, G; Ma, J. J.; Tu, M.; DesMarteau, D. D. *J. Fluor. Chem.* **2004**, *125*, 1231–1240.
34. Zhang, J.; DesMarteau, D. D.; Zuberi, S.; Ma, J. -J.; Xue, L.; Gillette, S. M.; Blau, H., Gerhardt, R. J. *J. Fluor. Chem.* **2002**, *116*, 45.
35. Zhang, J.; DesMarteau, D. D. *J. Fluor. Chem.* **2001**, *111*, 253.
36. Utracki, L. A.; Weiss, R. A., Eds. *Multiphase Polymers: Blends and Ionomers*; ACS Symp. Ser. No. 395; Am. Chem. Soc.: Washington, DC, **1989**; Chapter 15.
37. Moore, R. B.; Martin, C. R. *Macromolecules* **1989**, *22*, 3594.
38. Bilmeyer, F. W. *Textbook of Polymer Science*, 3rd ed.; John Wiley & Sons, Inc.: New York, 1984.
39. Moore, R. B.; Cable, K. M; Croley, T. L. *J. Memb. Sci.* **1992**, *75*, 7.

40. McCain, G. H.; Covitch, M. J. *J. Electrochem. Soc.* **1984**, *131*, 1350.
41. Gebel, G.; Aldebert, P.; Pineri, M. *Macromolecules* **1997**, *20*, 1425.
42. Slade, S.; Campbell, S. A.; Ralph, T. R.; Walsh, F. C. *J. Electrochem. Soc* **2002**, *149*(12), A1556
43. Moore, R. B.; Martin, C. R. *Anal. Chem.* **1986**, *58*, 2570.
44. Li, Q; He, R; Berg, R. W.; Hjuler, H. A.; Bjerrum, N. J. *Solid State Ionics* **2004**, *168*, 177–185

23

FLUORINATED SILSESQUIOXANES

SEAN RAMIREZ AND JOSEPH MABRY
Edwards Air Force Base

23.1 INTRODUCTION AND OVERVIEW

Polyhedral oligomeric silsesquioxanes (POSS) are thermally robust chemical structures possessing an inorganic core [$SiO_{1.5}$] and an organic periphery, either aliphatic or aromatic [1–3]. The properties of these molecules can be finely tuned via chemical manipulation of their organic periphery, allowing for easy polymer blending, covalent attachment of small molecules, and incorporation of POSS along a polymer backbone. The solubility of POSS compounds is highly dependent on their organic periphery [1–5]. The impact of chemical functionality on the periphery of POSS on solubility has been well documented [5–7]. The incorporation of POSS into polymers often produces nanocomposites with improved properties, such as higher glass transition temperature (T_g), mechanical strength, thermal and chemical resistance, and ease of processing [8–10]. These properties are used in the development of high performance materials for commercial applications in the aerospace and medical industries [8–13]. Various POSS analogues have been reported with applications related to semiconducting materials [14], atomic oxygen resistant coatings [9], and low surface energy coatings (discussed in this chapter). Due to these useful attributes, attempts to develop alternative silsesquioxane functionalization methods have been undertaken by many research groups.

The recent development of fluoroalkyl chains attached to POSS has led to low surface energy fluoroalkyl polyhedral oligomeric silsesquioxanes (F-POSS) (Figure 23.1) [15–18]. These robust nanofillers are used for low surface energy applications,

Handbook of Fluoropolymer Science and Technology, First Edition. Edited by Dennis W. Smith Jr., Scott T. Iacono, and Suresh S. Iyer.

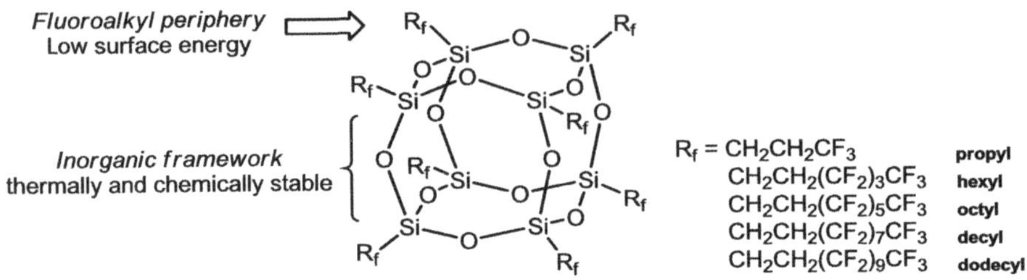

FIGURE 23.1 F-POSS general structure; with a typical T_8 core.

particularly in nonwetting coatings and oil–water separation [19–24]. The functionality surrounding the silsesquioxane core is chemically inert due to nonpolarizable fluoroalkyl groups. As the length of the pendant fluoroalkyl chain increases, from propyl to decyl, their solubility decreases in common organic solvents (tetrahydrofuran, chloroform, etc.), which limits their solubility to fluorinated solvents (hexafluorobenzene, AK-225) [15]. F-POSS structures typically are crystalline solids with rigid fluoroalkyl chains due to steric and electronic repulsion. The crystallinity and low surface energy values of long-chain F-POSS compounds ($CF_2 \geq 5$) give these structures their unique superhydrophobic and oleophobic properties. The fluorodecyl F-POSS (**4**) compound possesses one of the lowest surface energy values known (γ_{sv} = 9.3 mN/m) for any crystalline solid [25]. When incorporated into polymers, these compounds impart their nonwetting properties on their host substrates [26–28]. With proper surface texture, polymers blended with F-POSS, such as poly(dimethylsiloxane) (PDMS), have exhibited omniphobic behavior, that is, the surfaces repel nearly every liquid imaginable [29]. These spectacular attributes have led to a large amount of research into exploiting these properties by incorporation of F-POSS into polymers.

Although physical blending of F-POSS into polymers imparts excellent properties, these materials can be prone to surface abrasion, phase separation, low solubility in common organic solvents, and poor mechanical robustness. To circumvent these potential issues, the development of an incompletely condensed F-POSS was highly desired. With modification, such a POSS compound can be covalently attached to a host polymer. Examples in the literature have reported incompletely condensed silsesquioxanes with short-chain functionality that demonstrated a synthetic feasibility to produce reactive F-POSS compounds. For example, (3,3,3-trifluoropropyl)$_7$-$Si_7O_9(ONa)_3$, has been derivatized with long-chain fluorinated trichlorosilanes to produce low surface energy materials [16]. However, shorter propyl chains do not possess the same magnitude of liquid repellency as longer fluoroalkyl chains. Recently, a synthetic approach developed by Ramirez et al. opens up the closed cage F-POSS into an incompletely condensed silsesquioxane with functional disilanols [16, 30]. This new method follows similar synthetic strategies to those developed by Feher and coworkers [31–33]. Feher's work consisted of opening completely condensed POSS cages through multistep syntheses to create silanols via either an acidic or a basic route. Ramirez's modified version of the acidic route

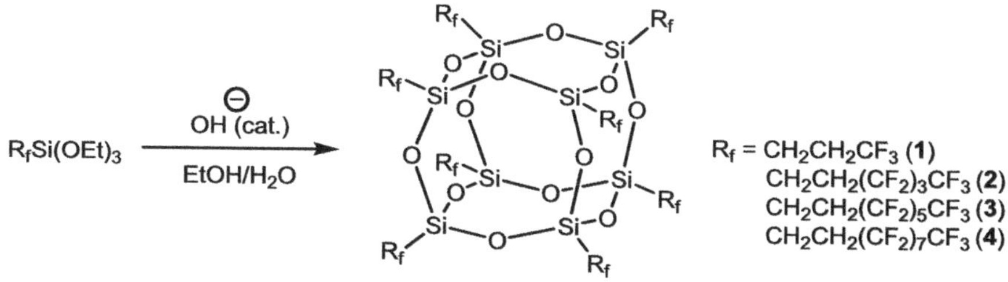

FIGURE 23.2 Direct synthesis of octahedral fluorinated POSS compounds.

for F-POSS produced a disilanol F-POSS capable of further functionalization with a variety of organic molecules. This chapter will discuss recent developments with the synthesis, functionalization, incorporation into polymer matrices, and applications of fluorinated silsesquioxanes.

23.2 SYNTHESIS OF FLUORINATED SILSESQUIOXANES

Fluorinated POSS compounds fluorodecyl (**FD,4**), fluorooctyl (**FO,3**), and fluorohexyl (**FH,2**) were produced by the base-catalyzed hydrolysis of trialkoxy silanes (Figure 23.2) [15]. These compounds tend to condense into T_8 cages, rather than cage mixtures, as has been previously observed in the base-catalyzed synthesis. Although other cage structures including the T_{10} and T_{12} have been observed during their synthesis, the yields of T_8 for these reactions are nearly quantitative. This is significant because the usual method to produce T_8 cages is the acid-catalyzed hydrolysis of trichlorosilanes, which produces an undesirable acidic by-product, lowering the overall yield of the reaction. Other methods to produce fluorinated silsesquioxanes have been examined as well, including the thiol-click chemistry of acylate long-chain fluoroalkyls and octathiol POSS [34, 35]. Theses structures were primarily used for self-wrinkling patterned surfaces, through the surface migration of low surface energy fluoroalkyl chains [34]. The synthesis of these structures focused on modifying already closed POSS cages.

Corner capping with commercially available functionalized fluoroalkyltrichlorosilanes afforded diverse architectures with linear fluoroalkyl chains (**6–10**), a branched hexafluoroiso-propyl (**9**), and a heptafluoroisopropoxypropyl ether (**10**) (Figure 23.3) [16]. Hepta(3,3,3-trifluoropropyl)tricycloheptasiloxane trisodium silanolate **4** was used as an intermediate for the preparation of fluorinated POSS compounds **6–12** by “corner capping” with fluoroalkyltrichlorosilanes. The intermediate salt (**5**) is stable in dry air but decomposes to silsesquioxane resin upon exposure to moisture. The intermediate adduct is the result of a pathway to fully condensed cage structures by simply controlling feedstock stoichiometry.

Unfortunately, the synthesis of other fluorinated trisodium POSS salts seems to be limited to the trifluoropropyl derivative and is not a general synthetic route.

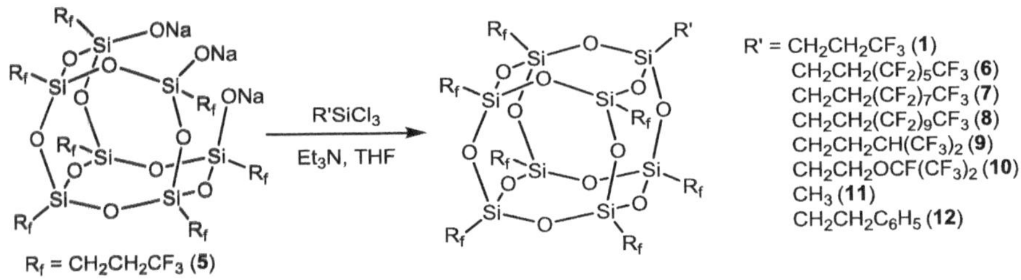

FIGURE 23.3 Synthesis of corner capped FP compounds.

In order to successfully open the chemically resistant long-chain F-POSS, a three-step synthetic procedure was developed following a similar method first employed by Feher and coworkers., as discussed earlier [33, 36]. The first step of this synthesis involves opening the closed caged F-POSS (**4**) with a trifluoromethanesulfonic acid (HOTf) to form a triflate intermediate compound (Figure 23.4) [31]. An important part of this reaction is the chemical equilibrium between the open- and closed-cage silsesquioxane frameworks. The ^{29}Si resonances at −62.6, −65.0, and −66.7 ppm (Si ratio 2:2:4) are associated with the triflate intermediate. Attempts to isolate this derivative have resulted in a reversion of this mixture to compound **4**, as the ditriflate intermediate is highly unstable. Although this makes characterization of the triflate intermediate difficult, it allows the open-caged structure to be effectively trapped with a bridged sulfate compound, $NBut_4HSO_4$. Starting material **4** is also observed with a strong peak at −66.3 ppm. The bridging sulfate moiety stabilizes the electron-withdrawing F-POSS cage framework, leading to another NMR observed intermediate with ^{29}Si resonances at −64.5, −65.7, and −67.3 ppm (Si ratio 2:2:4). A viable isolation procedure for intermediate bridged sulfate has not been attained with most attempts leading back to compound **4**. The bridged sulfate reaction mixture is subsequently added to a polar fluorinated solvent/water mixture to convert bridged sulfate to disilanol F-POSS (**13**). This solvent mixture allows for the introduction of water into a highly fluorinated environment. The main side product produced during each step of this synthesis was compound **4**, which can be subsequently recycled for further reactions.

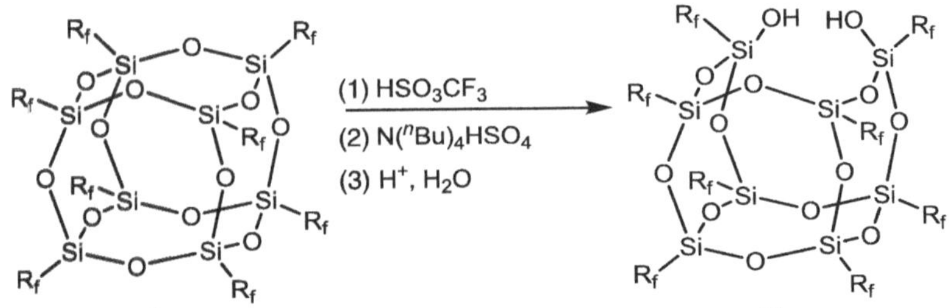

FIGURE 23.4 Synthesis of incompletely condensed long-chain fluoroalkyl silsesquioxane.

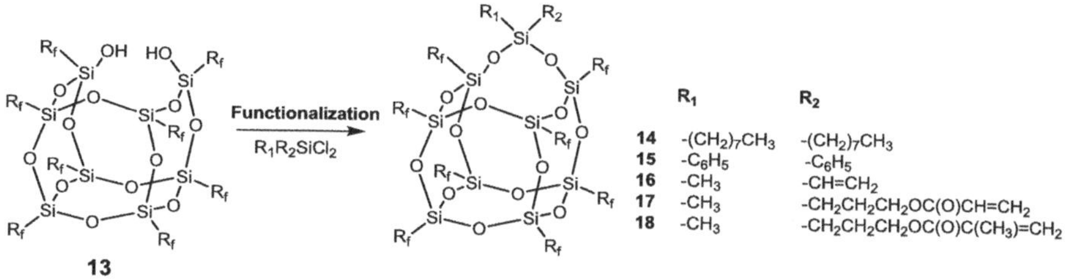

FIGURE 23.5 Edge-capping reactions of open cage disilanol.

Once synthesized, the reactive disilanol is modified by simple chemical condensation reactions with chlorosilanes to produce functional F-POSS compounds (Figure 23.5). Several functional dichlorosilanes have been reacted with compound **13** to produce a small library of compounds. An example condensation of **13** with (3-acryloxypropyl) methyldichlorosilane in the presence of triethylamine produced compound **17** with the loss of HCl [31]. The primary side product during these reactions is the initial starting material, compound **4**. Characterization of these reactions proves to be simple due to a new Si resonance observed in all the ^{29}Si NMRs of the synthesized F-POSS structures. Interestingly, the functionalization of the POSS cage influences the immediate Si–$\underline{C}H_2$ group, providing another diagnostic signal in ^{13}C NMR. Once modified, the properties of fluorinated silsesquioxanes can be further modified through the choice of reactive dichlorosilane. For example, longer chain alkyl functionalities change the wetting properties of surfaces coated with the F-POSS from hydrophobic and oleophobic to hydrophobic and oleophilic due to the interaction of the alkyl chain from the surface of F-POSS.

23.3 CRYSTAL STRUCTURE AND MELTING POINT

Single crystal X-ray structures of a variety of fluoroPOSS compounds have assisted in the elucidation of properties for these molecules (Table 23.1). The structure of the hexyl (**2**) and decyl (**4**) F-POSS were solved (Figure 23.6) [14]. These structures each contain substantial amounts of disorder due to the long fluoroalkyl chains. The disorder from the fluorinated chains was due to the thermal motion of these segments. As fluoroalkyl chain length increases from propyl to decyl, the influence of the chains on the overall crystal structure increases on the crystal packing. Short-chain fluoropropyl F-POSS (**1**) exhibited an intermolecular dimeric Si–F close contact with neighboring structures in the ORTEP structures (Figure 23.7) [37]. These contacts were attributed to increasing the overall melting point of the F-POSS structure (~230°C) for the short-chain F-POSS compared with longer chain decyl (~140°C). Crystallographic data revealed as fluoroalkyl chain length increased, the periphery became more helical, and the Si–F interaction was not as strong, thereby lowering the overall melting point. The near-parallel arrangement of the fluoroalkyl chains, resulting from the intramolecular interactions between silicon and fluorine atoms leads to low surface

TABLE 23.1 Crystallographic Data for F-POSS Compounds [14, 16, 31]

Compound	Space group	Unit cell dimensions	Unit cell volume (Å^3)	Z, density (g/cm^3)	R-factor
11 MethylFluoropropyl T_8	Monoclinic P 2(1)/c	a = 13.3164(16) Å b = 17.129(2) Å c = 20.726(3) Å α = 90° β = 105.007(2)° γ = 90°	4566.2(10)	4, 1.616	4.80
1 Fluoropropyl T_8	Tetragonal I 41/a	a = 21.1129(7) Å b = 21.1129(7) Å c = 11.1505(7) Å α = 90° β = 90° γ = 90°	4970(4)	1, 1.686	3.05
2 Fluorohexyl T_8	Triclinic P1	a = 11.806(5) Å b = 12.393(5) Å c = 15.729(6) Å α = 72.073(6)° β = 76.024(6)° γ = 66.151(5)°	2009.0(14)	1, 1.987	3.89
4 Fluorodecyl T_8	Triclinic P1	a = 10.352(1) Å b = 21.984(2) Å c = 28.653(3) Å α = 102.082(1)° β = 95.702(1)° γ = 90.907(1)°	6340.2(1)	2, 2.092	6.76
13 Disilanol	Monoclinic P 2(1)/c	a = 11.833(2) Å b = 57.141(11) Å c = 19.069(4) Å α = 90° β = 92.20(3)° γ = 90°	12,884(4)	2, 2.068	9.15

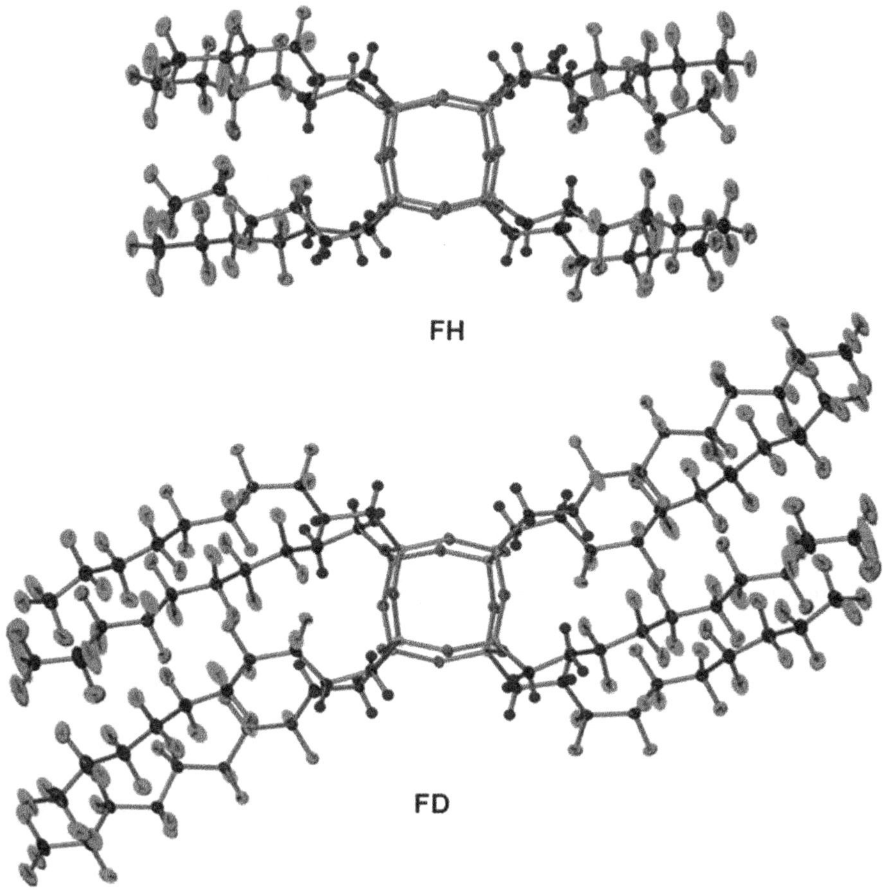

FIGURE 23.6 ORTEP representation of FH (**2**) and FD (**4**) POSS at 103 K, with thermal ellipsoids set at 50% probability. Reproduced with permission from Reference 14. Copyright 2008.

energy properties of these materials. The low surface energy values of neat films of these compounds are attributed to the high crystallinity of the fluoroalkyl chains. During cooling of F-POSS structures a phase change occurs, often causing these crystals to crack.

The crystal structure of the incomplete silsesquioxane (**13**) was recently solved and displayed the same near parallel arrangement of fluoroalkyl chains observed for compounds **3** and **4** (Figure 23.8) [31]. The highly disordered crystal structure revealed a dimerization of silsesquioxanes due to an eight-member hydrogen bonded ring between the disilanol groups. The dimerization apparently helped stabilize this molecule, because these structures can easily revert back to starting completely condensed F-POSS **4** via dehydration. The combination of the flouroalkyl chain

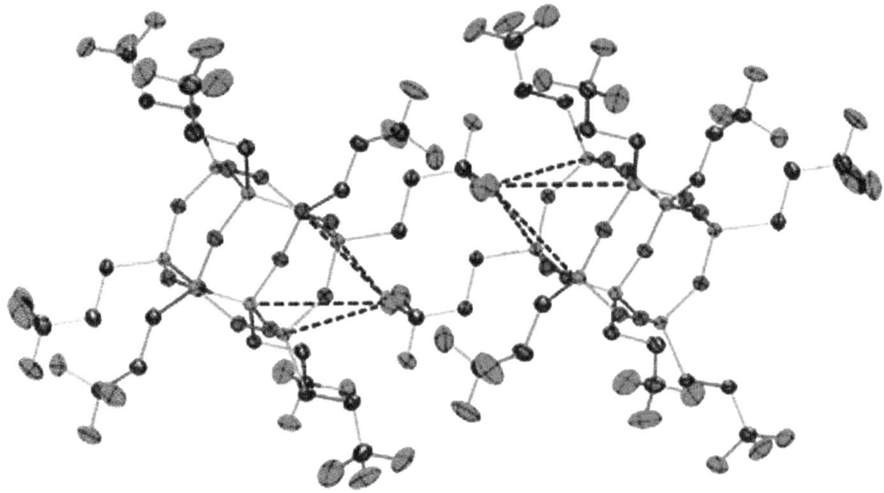

FIGURE 23.7 Dimeric Si–F close contact ORTEP at 193 K with displacement ellipsoids shown at 30% probability. Hydrogen atoms and THF molecules omitted for clarity. Reproduced with permission from Reference 38. Copyright 2012.

interaction and hydrogen bonding are the driving forces for the crystallization of this structure. Hydrogen bonding between silanols has been observed for nonfluoroalkyl POSS, typically between trisilanol structures. Dimerization for aliphatic and aromatic POSS structures has been observed but never for an F-POSS compound. A melting point for this compound was not determined because the structure reverts back to closed cage FD when heated.

When **13** is functionalized with various organic moieties, the melting point can be tuned and is dependant on the functionality attached. Intensive studies by Iacono and coworkers showed the melting point for functionalized **1** was tunable with functionality as well. The melting points for F-POSS-related compounds is summarized in Table 23.2. From this table, it is clear that the intermolecular interactions of cage–cage interactions can be highly influential on the overall melting points. Specifically, Iacono's work illustrated the suppression of melting point, as longer chain fluoroalkyl groups are added to fluoropropyl POSS corner. Interestingly, a substantial spike in melting point values was observed for short-chain propyl functionalities. The authors attributed the higher melting point values to a strong intermolecular Si–F interaction. Other work reported by Ramirez and coworkers. demonstrated the ability of organic substitutes to break up various Van der Waals interactions between fluoroalkyl chains and lower the melting point of functional F-POSS compounds. All of these compounds have relatively close molecular weights due to the long-chain fluoroalkyl periphery on the POSS structure. For example, compound **14** possesses two long aliphatic octyl chains protruding from the POSS cage edge, which lowers the melting point substantially from 141°C (for **4**) to ~65°C. When a more rigid functional group such as a diphenyl is added the melting point is raised back to 120°C. From the table it

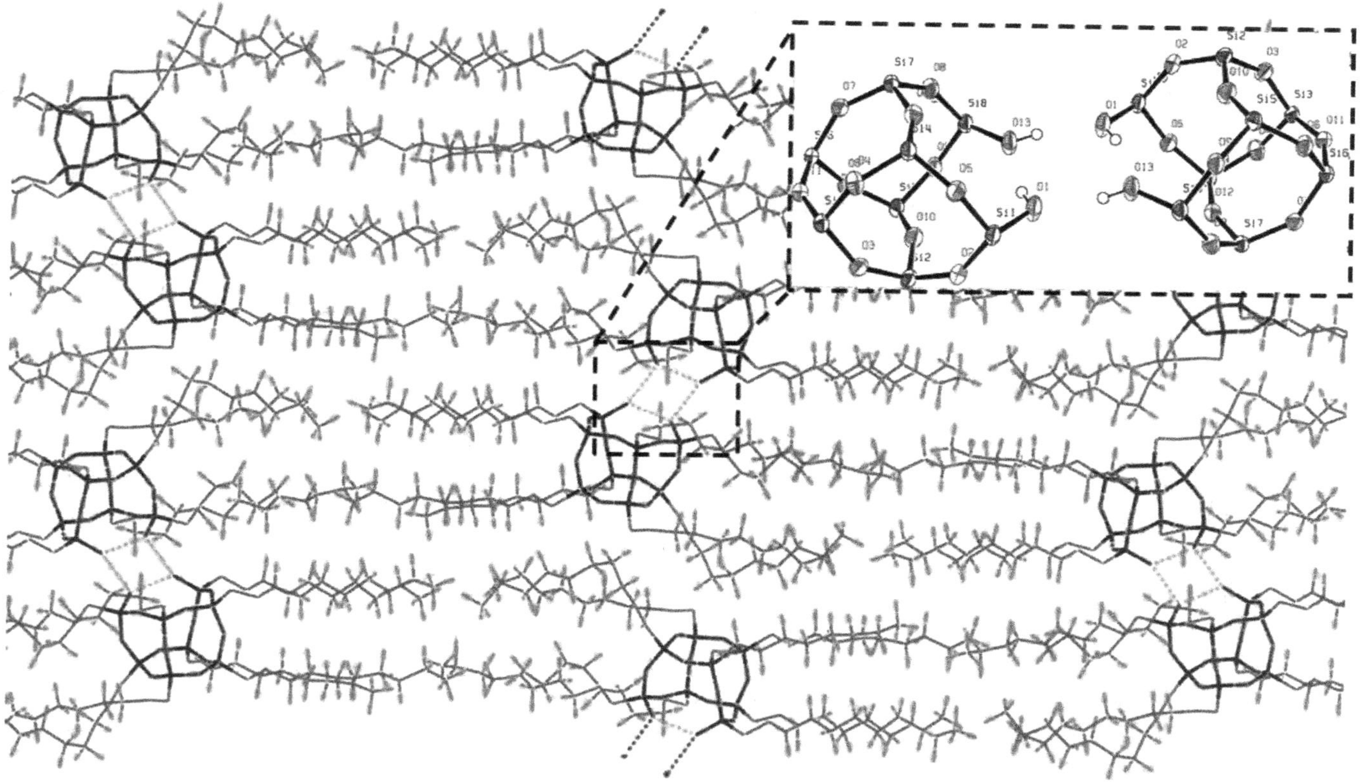

FIGURE 23.8 ORTEP representation of compound **13** at 102 K with ellipsoids set at 50% probability. Green F, white C, red O, blue Si. Hydrogen atoms and disorder not shown for clarity (top). View of lattice structure illustrating intermolecular hydrogen bonding between silanols. Insert zoom in on disilanol hydrogen bonding interaction (bottom). Reproduced with permission from Reference 32. Copyright 2011.

TABLE 23.2 Melting Points and MW Data for F-POSS Compounds

F-POSS type	Melting point (°C)	Molecular weight (g/mol)	Reference
F-POSS T_8 (**1**) propyl	243–237	1193	[14]
F-POSS T_8 (**3**) Octyl	120	3193	[14]
F-POSS T_8 (**4**) Decyl	150	3993	[14]
$-CH_2CH_2(CF_2)_5CF_3$ (**6**)	104–106	1443.9	[16]
$-CH_2CH_2(CF_2)_7CF_3$ (**7**)	88–89	1543	[16]
$-CH_2CH_2(CF_2)_9CF_3$ (**8**)	105–107	1643	[16]
$-CH_2CH_2CH(CF_3)_2$ (**9**)	234–236	1275	[16]
$-CH_2CH_2O(CF_3)_2$ (**10**)	70–71	1849.35	[16]
$-CH_3$ (**11**)	168–170	1111	[16]
$-CH_2CH_2C_6H_5$ (**12**)	109–111	1201	[16]
$Si-(OH)_2$ (**13**)	Compound reverts to T_8	4011	[31]
$Si-((CH_2)_7CH_3)_2$ (**14**)	64.8–66.9	4264	[31]
$Si-(C_6H_5)_2$ (**15**)	120.2–122.8	4192	[31]
$Si-(CH_3)(CH{=}CH_2)$ (**16**)	83.8–84.8	4080	[31]
$Si((CH_3)((CH_2)_3OC(O)CCH{=}CH_2)$ (**17**)	76.2–78.2	4166	[31]
$Si-(CH_3)((CH_2)_3OC(O)C(CH_3){=}CH_2)$ (**18**)	73.2–75.9	4180	[31]

is evident that more flexible functionality (compounds **10**, **14**, **16**, **17**) lowers melting points substantially due to breaking up the factors that influence crystalline packing of POSS structures. Less substantial influences arise from rigid (**15**) or functionalities on F-POSS cage structures.

23.4 PROPERTIES

The properties of fluoroPOSS compounds are unique and have been studied intensely since their initial synthesis. From these studies, it was found that the longer chain F-POSS compounds possess the lowest surface energy values. Contact angle measurements are typically used to measure the surface properties of these compounds. Solubility of F-POSS becomes more limited as fluoroalkyl chain length increases. Long-chain F-POSS compounds are limited to fluorinated solvents. However, the solubility of F-POSS can be manipulated through surface functionalization of the cage. Only two methods currently achieve this through functionalization of short-chain (propyl) trisilanol, or long-chain (decyl) disilanol. Propyl F-POSS (**1**) was modified with a variety of fluoroalkyl chains (Figure 23.3) via reaction with trichlorosilanes. Disilanol F-POSS (**13**) was modified with a variety of organic functionalities via reaction with dichlorosilane. These moieties range from unreactive alkyl chains to vinyl-based monomers.

Contact angle measurements found that these functionalities influence the surface properties of F-POSS [31]. The solubility was also expanded to nonfluorinated solvents when longer organic groups were attached. These groups are thought to both help break up the crystallinity of the cage and increase miscibility with solvent. Contact angle measurements (θ_{adv} and θ_{rec}) for F-POSS compounds are summarized in Table 23.3. These measurements for water and hexadecane help demonstrate the nonwetting behavior of F-POSS structures and the influence of functionality on the POSS cage. Contact angle measurements were performed on smooth surfaces to limit the influence of surface roughness and geometry, both of which can be significant factors. From this data, it becomes apparent that longer fluoroalkyl chain length is more desirable for nonwetting behavior for both water and hexadecane. Lower contact angles values and higher hysteresis ($\theta_{adv} - \theta_{rec}$) were observed for shorter fluoroalkyl chains. The highest values were observed for fluorodecyl chain-type compounds. Similar to melting point values, wetting behavior for F-POSS compounds can be influenced by chemical functionalization to the cage structure. Adding functionality to F-POSS can influence the overall wetting behavior of F-POSS when the functionality is substantial enough. This becomes most apparent when comparing hexadecane wetting on compounds **4** versus **14**. A large contact angle hysteresis is observed for compound **14** due to the aliphatic octyl chains interaction with the aliphatic hexadecane droplet. The organic functionality influence decreases rapidly when either the chain length of the nonfluorinated functionality is shortened (**16**, **17**, **18**) or the organic moiety is more rigid (**15**).

23.5 F-POSS COMPOSITES

The additions of F-POSS to polymers have led to the production of a variety of composites with surface energies that can be tailored for their desired applications [15–25]. These properties rely on the low surface energy provided from F-POSS and their host polymer matrix. Depending on that matrix, nanocomposites have been produced that are superhydrophobic, superoleophobic, or superomniphobic. Initial composites were found to be superhydrophobic and oleophobic with a mixture of F-POSS and poly(methylmethacrylate) (PMMA) (Figure 23.9) [15]. The lotus leaf is hydrophobic but has no oil resistance while the F-POSS coated leaf is omniphobic. Electrospun fibers of these materials were spun cast on lotus leaf, metal grids, and fabrics to produce a textured surface morphology. These surfaces were found to be omniphobic with unprecedented surface properties. This simple addition of F-POSS to different polymers leads to a rapid decrease in the overall surface energy of the synthesized composites and provides a facile route to systematically tune the surface energy of the produced composite over a very wide range.

Superhydrophobic surfaces display water contact angles greater than 150° and low contact angle hysteresis ($\theta_{adv} - \theta_{rec}$) [17, 26]. Through theoretical studies and experimentation, it was demonstrated that the combination of surface chemistry, roughness, and geometry were all factors for nonwetting properties [15–17, 23–26, 38–40]. The addition of a liquid droplet to a surface leads to either the Wenzel

TABLE 23.3 Contact Angle Measurements for F-POSS Compounds

	Water		Hexadecane	
F-POSS type	(θ_{adv})	(θ_{rec})	(θ_{adv})	(θ_{rec})
F-POSS T_8 (**1**) propyl [25]	$111 \pm 1°$	$95 \pm 2°$	$48 \pm 2°$	$39 \pm 3°$
F-POSS T_8 (**2**) hexyl [25]	$120 \pm 1°$	$103 \pm 4°$	$73 \pm 1°$	$39 \pm 2°$
F-POSS T_8 (**3**) octyl [25]	$122 \pm 2°$	$97 \pm 2°$	$76 \pm 4°$	$45 \pm 2°$
F-POSS T_8 (**4**) decyl [25]	$122 \pm 2°$	$116 \pm 2°$	$80 \pm 1°$	$61 \pm 3°$
F-POSS T_8 (**19**) isobutyl [25]	$102 \pm 2°$	$61 \pm 2°$	$52 \pm 2°$	$37 \pm 3°$
$Si{-}(OH)_2$ (**13**) [31]	$116.8 \pm 0.4°$	$111 \pm 0.6°$	$77.4 \pm 0.4°$	$74.4 \pm 0.8°$
$Si{-}((CH_2)_7CH_3)_2$ (**14**) [31]	$117.9 \pm 0.5°$	$95.5 \pm 0.4°$	$69.1 \pm 1.2°$	$23.1 \pm 1.2°$
$Si{-}(C_6H_5)_2$ (**15**) [31]	$116.2 \pm 0.4°$	$110.5 \pm 0.5°$	$76.0 \pm 0.8°$	$73.2 \pm 0.4°$
$Si{-}(CH_3)(CH{=}CH_2)$ (**16**) [31]	$116.2 \pm 0.4°$	$100.6 \pm 0.8°$	$78.4 \pm 0.3°$	$70.6 \pm 2.3°$
$Si((CH_3)((CH_2)_3OC(O)CCH{=}CH_2)$ (**17**) [31]	$118.2 \pm 1.0°$	$90.6 \pm 1.0°$	$76.8 \pm 0.3°$	$64.8 \pm 1.0°$
$Si{-}(CH_3)((CH_2)_3OC(O)C(CH_3){=}CH_2)$ (**18**) [31]	$117.1 \pm 0.6°$	$93.8 \pm 1.5°$	$78.1 \pm 0.4°$	$63.0 \pm 1.2°$

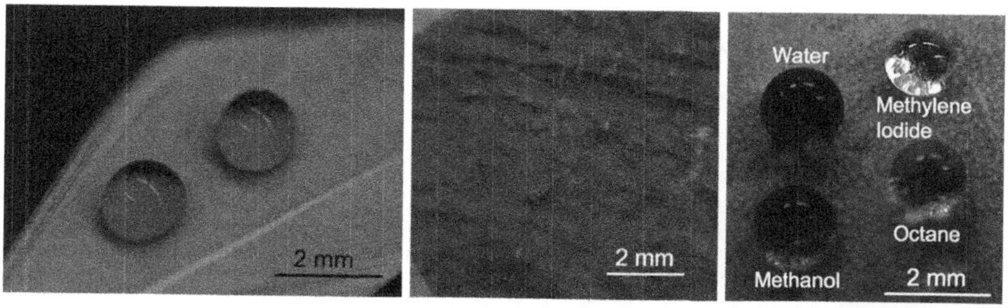

FIGURE 23.9 Duck feather coated with F-POSS (**4**) with a rapeseed oil droplet (γ_{lv} = 35.7 mN/m) colored with oil red O (left). Droplet of octane (γ_{lv} = 21.7 mN/m) on a lotus leaf coated with **4** (center). Droplets of water (γ_{lv} = 72.1 mN/m), methylene iodine (γ_{lv} = 50.1 mN/m), methanol (γ_{lv} = 22.7 mN/m), and octane (γ_{lv} = 21.7 mN/m) on a lotus leaf coated with PMMA and F-POSS (44% wt.) electrospun fibers, indicating the presence of microscopic pockets of air and the formation of a composite interface. Reproduced with permission from Reference 15. Copyright 2008.

state [41] (fully wetted) or Cassie–Baxter [42] state (solid–liquid–air interface). Numerous publications go into the full details on the importance of surface roughness and geometry for nonwetting surfaces [15–17, 24, 38–40]. Overall, an engineered textured surface with the appropriate surface roughness and re-entrant curvature can lead to the establishment of a metastable Cassie–Baxter state for nonwetting surfaces. These publications also illustrate the importance of multiscale roughness for designing nonwetting surfaces. The combination of these surfaces with F-POSS structures has led to the development of omniphobic surfaces capable of repelling low surface tension liquids (γl_v < 25 mN/m) [25].

For example, the composites formed by blending F-POSS molecules with a relatively hydrophilic polymer, PMMA yield superhydrophobic surfaces. The addition of F-POSS molecules allows the surface energy to be systematically tuned with composites within the range γ_{SV} = 9–35 mN/m [15, 17]. Significant surface segregation (or blooming) of F-POSS (**4**) molecules toward the air interface was observed for these samples due to their extremely low surface energy. As a result, only a small amount of **4** (~10 wt.%) was needed to sufficiently cover the surface of the spin-coated blend. For **4**, concentrations greater than ~10 wt.%, both the advancing and receding contact angles reach a plateau at $\sim\theta_{adv} = \theta_{rec} = 123°$. These studies illustrated the need for only a small amount of F-POSS to achieve the desired surface nonwetting properties. It is clear that the surface of the electrospun blended fibers that contain greater than ~10 wt.% POSS are superhydrophobic, displaying both θ_{adv}, $\theta_{rec} > 150°$.

Recognizing the presence of re-entrant surface features in commercial fabrics, the dip-coating process was used to deliver a coating of fluorodecyl POSS molecules onto the fabric surface, bestowing superoleophobicity [19]. Technoflon, a fluoroelastomer binder, was added to the dip-coating solution to inhibit the formation of fluorodecyl POSS crystallites and yield a more conformal and elastomeric coating. Superomniphobic surfaces were recently produced from electrospun coatings of cross-linked PDMS and F-POSS on top of steel wire meshes, producing a

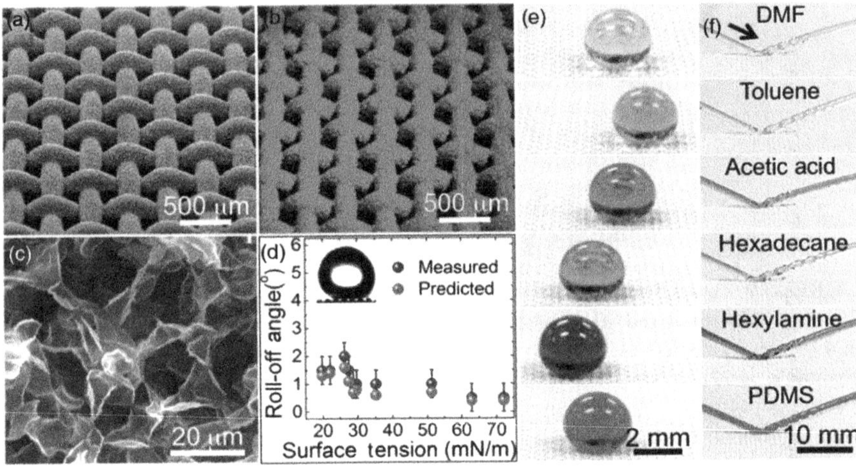

FIGURE 23.10 (a) SEM image of the hierarchically structured surface illustrating the electrospun coating of cross-linked PDMS + 50 wt% compound **4** on a stainless steel wire mesh 70. (b) Elemental mapping of fluorine. (c) SEM image illustrating the re-entrant curvature of electrospun texture. (d) Roll-off angles for Newtonian liquids on the surfaces shown in (a). The inset shows an ethanol droplet rolling on the surface at a roll-off angle $\omega = 2°$. (e) Different liquid droplets on surface. (f) Bouncing liquids on surface shown on (a). Reproduced with permission from Reference 29. Copyright 2013.

hierarchically structured surface (Figure 23.10) [29]. These surfaces were resistant to virtually all liquids including organics, acids, bases, solvents, and polymer solutions, thus serving as a chemical shield against organic/inorganic, polar/nonpolar, and Newtonian/non-Newtonian liquids.

23.6 COVALENT BINDING OF F-POSS TO POLYMERS

There is only one example of covalently bound F-POSS to polymers from Ramirez et al. [43]. In that work, reverse-addition chain fragmentation termination (RAFT) polymerization of F-POSS PMMA copolymers was achieved in a fluorinated solvent and F-POSS-MA was determined to be an active monomer in RAFT polymerization. Molecular weights were determined with SEC-MALLS using fluorinated solvent as the mobile phase. The work demonstrated the difficulty characterizing an F-POSS functionalized copolymer. As expected, increased F-POSS composition in PMMA copolymers enhanced nonwetting properties, while coating surfaces and fabrics with F-POSS copolymers were found to produce nonwetting materials in the Cassie–Baxter state due to added surface structure of the cotton fabrics. These copolymers represented the first examples of long-chain F-POSS being covalently bound to polymers chains. Contact angles for both water and hexadecane increased relative to neat PMMA. Hexadecane was found to wet neat PMMA and 1 wt.% F-POSS composition films. However, the surfaces became oleophobic with low contact angle hysteresis once F-POSS compositions reached 5 wt.%. The contact angle hysteresis was lower for the highest copolymer composition (25 wt.% F-POSS) at 8°, compared with pure monomer (**17**) at 15°. The difference in contact angle was attributed to the increased fluorine content provided by F-POSS along the polymer backbone and on the surface. Hydrophobicity for the coated surfaces increased with F-POSS composition. From these measurements only 5 wt.% F-POSS was sufficient to obtain low surface energy properties. Using RAFT polymerization with F-POSS-MA will potentially enable the production of block copolymers with superhydrophobic/oleophobic regions. The authors are currently expanding upon this work by synthesizing polymers with both hydrophilic and hydrophobic qualities. This work potentially solves a major surface robustness problem that can plague F-POSS composites. The phase separation of F-POSS can be beneficial because F-POSS will bloom to the surface, where it is generally desired, but this phase separation can also be detrimental for compatibilization of F-POSS with polymers. Producing F-POSS copolymers was found to solve this compatibilization problem because the F-POSS copolymers can be mixed directly with nonfluorinated polymers to produce homogenous mixtures.

23.7 APPLICATIONS

Fluorinated silsesquioxanes have many potential applications as drop-in modifiers for wetting-resistant surfaces [19, 23]. Initial work demonstrated that, combined with surface roughness and geometry, F-POSS composites were capable of producing superhydrophobic (contact angles > 150°) and superoleophobic surfaces, as well as iceophobic coatings [44, 45]. The initial work performed by Cohen and McKinley (and others) used surfaces with re-entrant texture and F-POSS to produce robust superoleophobic surfaces [15, 17]. These surfaces were capable of wetting resistance to low surface tension liquids, such as decane and octane. The combination of textured surfaces with high $-CF_2$, CF_3 groups results in a very low surface energy material. This work demonstrated that F-POSS surface migrates due to the low surface energy of the compounds. Electrospun mats of PMMA/F-POSS composites were also examined and produced wetting-resistant surfaces. Other examples of applications include switchable liquid wettable surfaces (Figure 23.11) [38]. PEMA–POSS (polyethylmethacrylate-POSS) composites were found to possess a wetting switchability dependant on their annealing parameter due to the combined effect of low T_g PEMA (ensured mobility of polymer chains) and low surface energy F-POSS (provides nonwetting properties).

The wetting switchability was expanded upon recently to produce hygro-responsive membranes that are both superhydrophilic and superoleophobic, in air and water for separating oil–water mixtures [46]. This gravity-driven technology relies on the combination for cross-linked poly(ethylene glycol) and F-POSS to enable this methodology to succeed. The hygro-responsive membranes can be both superhydrophobic and superoleophobic. The membranes are oleophobic in air and underwater to be capable of separating oil–water emulsions, with efficiency ≥99.9% (Figure 23.12) [46]. The membranes were found to be scalable to separate large

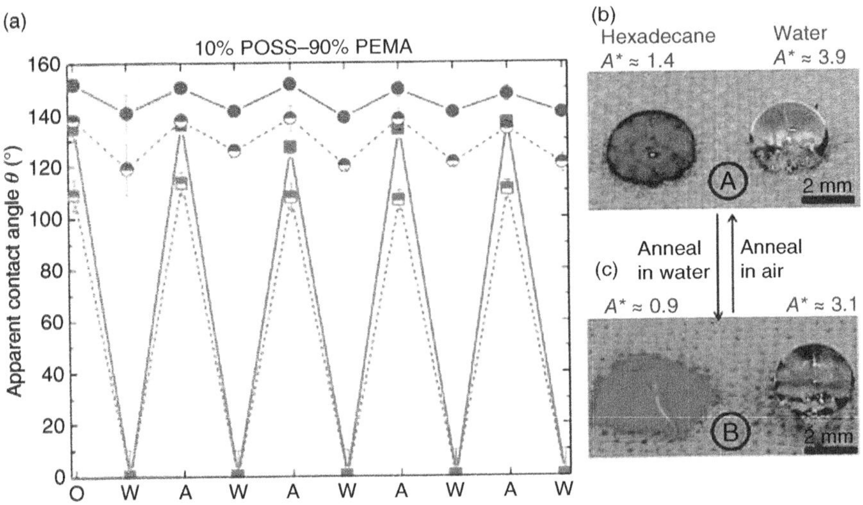

FIGURE 23.11 Switchable liquid wettability on a 10 wt.% 4-coated polyester fabric surface. (a) Apparent advancing (filled symbols) and receding (half-filled symbols) contact angle data with water (blue circles) and hexadecane (red squares) on a 10 wt.% 4-coated polyester fabric surface in the original as made (O), water-annealed (W), and air-annealed (A) states. (b) Small droplets of water and hexadecane forming robust composite interfaces on the dip-coated fabric in the air-annealed state (A). (c) Fully wetted hexadecane droplet along with water droplet forming a robust composite interface on the dip-coated fabric in the water-annealed state (W). Reproduced with permission from Reference 39. Copyright 2009.

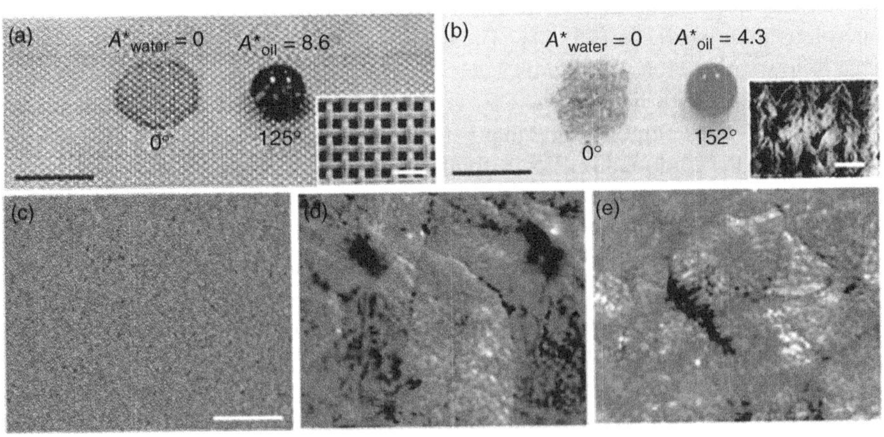

FIGURE 23.12 (a, b) Droplets of water (dyed blue) and rapeseed oil (dyed red) on stainless steel mesh 100 (a) and polyester fabric (b). Both surfaces have been dip coated with 20 wt.% 4 + x-PEGDA blend. Scale bars, 5 mm. Insets: morphologies of the respective dip-coated mesh and fabric surfaces. Scale bars, 500 μm. (c–e) AFM phase images of surfaces coated with x-PEGDA (c), a 10 wt.% 4 + x-PEGDA blend (d) and 20 wt.% 4 + x-PEGDA blend €. The phase angle ranges are 0–115° (c), 0–25° (d), and 0–21° (e). Reproduced with permission from Reference 47. Copyright 2012.

volumes of water–oil emulsions at high efficiency ($Q_{water} \approx 43{,}200\ L/m^2/h$). These membranes were found to operate at continuous separation for up to 100 hours at a time, leading to applications in oil–water cleanup, wastewater treatment, and emulsion separation.

SUMMARY

This chapter covered the recent development in fluorinated silsesquioxane synthesis, functionalization, properties, and applications. The synthesis of these structures follows the traditional methodology of producing POSS cages. The chain length of the fluoroalkyl periphery was shown to be the strongest influence on overall surface properties and morphology. Functionalization of these structures has proven to be extremely difficult with success stemming from modification of incompletely condensed silsesquioxane silanols. These structures were found to be reactive toward chlorosilane compounds. Functionality at the cage edge or corner was influential toward the wetting and melting properties of F-POSS compounds. Monomer derivatized F-POSS compounds were capable of producing various copolymers with a retention of their nonwetting properties. Due to the low surface energy properties of F-POSS, research efforts have focused on exploiting these properties for applications related to ice-phobicity, antiwetting coatings, and oil–water separation capabilities.

ACKNOWLEDGMENTS

We gratefully acknowledge the Air Force Office of Scientific Research and the Air Force Research Laboratory, Aerospace Systems Directorate for financial support.

REFERENCES

1. Cordes, D. B.; Lickiss, P. D.; Rataboul, F. *Chem. Rev.* **2010**, *110* (4), 2081–2173.
2. Lickiss, P. D.; Rataboul, F. *Adv. Organomet. Chem.* **2008**, *57*, 1.
3. Provatas, A.; Matisons, J. G. *Trends Polym. Sci.* **1997**, *5*, 327.
4. Moore, B. M.; Ramirez, S. M.; Yandek, G. R.; Haddad, T. S.; Mabry, J. M. *J. Organomet. Chem.* **2011**, *696*.
5. Yandek, G. R.; Moore, B. M.; Ramirez, S. M.; Mabry, J. M. *J. Phys. Chem. C.* **2012**, *116* (31), 16755–16765.
6. Milliman, H.W.; Boris, D.; Schiraldi, D.A. *Macromol.* **2012**, *45*, 1931–1936.
7. Misra, R. A.; Alp, H.; Jarrett, W. L.; Morgan, S. E. *Polymer* **2009**, *50*, 2906–2918.
8. Joshi, M. B., Bhupendra, S. *J. Macro. Sci. Polym. Rev.* **2004**, *C44* (4), 389–410.
9. Phillips, S. H. H.; Timothy, S.; Tomczak, S. J. *Cur. Opin. Solid State Mater. Sci.* **2004**, *8* (1), 21–29.
10. DeArmitt, C. W.; Wheeler, P. *Plast. Addit. Compd.* **2008**, *10* (4), 36–39.

11. Ghanbari, H. C., Brian, G.; Seifalian, A. M. *Macromol. Rapid Comm.* **2011**, *32* (14), 1032–1046.
12. Tanaka, K. A.; Adachi, S.; Chujo, Y. *J. Polym. Sci. A: Polym. Chem.* **2009**, *47*, 5690–5697.
13. Nguyen, T. P.; Lee, C. W.; Hassen, S.; Le, H. C. *Solid State Sci.* **2009**, *11* (10), 1810–1814.
14. Mabry, J. M.; Vij, A.; Iacono, S. T.; Viers, B. D. *Angew. Chem., Int. Ed.* **2008**, *47*, 4137–4140.
15. Tuteja, A.; Choi, W.; Mabry, J. M.; McKinely, G. H.; Cohen, R. E. *Proc. Natl. Acad. Sci. USA.* **2008**, *105*, 18200–18205.
16. Iacono, S. T.; Vij, A.; Grabow, W.; Smith, D. W., Jr.; Mabry, J. M. *Chem. Commun.* **2007**, *47*, 4992–4994.
17. Tuteja, A.; Choi, W.; Ma, M.; Mabry, J. M.; Mazzella, S. A.; Rutledge, G. C.; McKinley, G. H.; Cohen, R. E. *Science* **2007**, *318*, 1618–1622.
18. Iacono, S. T.; Budy, S. M.; Smith, D. W., Jr.; Mabry, J. M. *J. Mater. Chem.* **2010**, *20*, 2979–2984.
19. Kota, A. K.; Mabry, J. M.; Tuteja, A. *Surf. Innov.* **2013**, *1*, 71–83.
20. Tuteja, A.; Mabry, J. M. *Adv. Silicon Sci.* **2012**, *4*, 179–193.
21. Kota, A. K.; Li, Y.; Mabry, J. M.; Tuteja, A. *Adv. Mater.* **2012**, *24*, 5838–5843.
22. Kota, A. K.; Kwon, G.; Choi, W.; Mabry, J. M.; Tuteja, A. *Nat. Commun.* **2012**, *3*, 2027/1–2027/8.
23. Meuler, A. J.; Chhatre, S. S.; Nieves, A. R.; Mabry, J. M.; Cohen, R. E.; McKinley, G. H. *Soft Matter* **2011**, *7*, 10122–10134.
24. Kwon, G.; Kota, A. K.; Li, Y.; Sohani, A.; Mabry, J. M.; Tuteja, A. *Adv. Mater.* **2012**, *24*, 3666–3671.
25. Chhatre, S. S.; Guardado, J. O.; Moore, B. M.; Haddad, T. S.; Mabry, J. M.; McKinley, G. H.; Cohen, R. E., *ACS Appl. Mater. Interfaces* **2010**, *2*, 3544–3554.
26. Choi, W.; Tuteja, A.; Chhatre, S.; Mabry, J. M.; Cohen, R. E.; McKinley, G. H. *Adv. Mater.* **2009**, *21*, 2190–2195.
27. Iacono, S. T.; Budy, S. M.; Smith, D. W., Jr.; Mabry, J. M. *J. Mater. Chem.* **2010**, *20*, 2979–2984.
28. Kobaku, S. P. R.; Kota, A. K.; Lee, D. H.; Mabry, J. M.; Tuteja, A. *Angew. Chem. Int. Ed.* **2012**, *51*, 10109–10113.
29. Pan, S.; Kota, A. K.; Mabry, J. M.; Tuteja, A. *J. Am. Chem. Soc.* **2013**, *135*, 578–581.
30. Ramirez, S. M.; Diaz, Y. J.; Campos, R.; Haddad, T. S.; Mabry, J. M. *ACS Symp. Ser.* **2012**, *1106*, 95–109.
31. Ramirez, S. M.; Diaz, Y. J.; Campos, R.; Stone, R. L.; Haddad, T. S.; Mabry, J. M. *J. Am. Chem. Soc.* **2011**, *133* (50), 20084–20087.
32. Feher, F. J.; Soulivong, D.; Nguyen, F. *Chem. Commun.* **1998**, *12*, 1279–1280.
33. Feher, F. J.; Soulivong, D.; Nguyen, F.; Ziller, J. W. *Angew. Chem., Int. Ed.* **1998**, *37* (19), 2663–2667.
34. Gan, Y.; Jiang, X.; Jin, J. *Macromolecules* **2012**, *45*, 7520–7526.
35. Dutkiewicz, M.; Maciejewski, H.; Bogdan, M.; Karasiewicz, J. *Organometallics* **2011**, *30*, 2419–2153.
36. Feher, F. J.; Nguyen, F.; Soulivong, D.; Ziller, J. W. *Chem. Commun.* **1999**, *17*, 1705–1706.

37. Kettwich, S. C.; Pierson, S. N.; Peloquin, A. J.; Mabry, J. M.; Iacono, S. T. *New J. Chem.* **2012**, *36*, 941–946.
38. Chhatre, S. S.; Tuteja, A.; Choi, W.; Revaux, A.; Smith, D.; Mabry, J. M.; McKinley, G. H.; Cohen, R. E. *Langmuir* **2009**, *25*, 13625–13632.
39. Leng, B.; Shao, Z.; de With, G.; Ming, W. *Langmuir* **2009**, *25* (4), 2456–2460.
40. Marmur, A. *Langmuir* **2008**, *24* (14), 7573–7579.
41. Wenzel, R.N. *Ind. Eng. Chem.* **1936**, *28*, 988–994.
42. Cassie, A. B. D.; Baxter, S. *Trans. Faraday Soc.* **1944**, *40*, 546–551.
43. Ramirez, S. M.; Diaz, Y. J.; Sahagun, C. M.; Duff, M. W.; Lawal, O. B.; Iacono, S. T.; Mabry, J. M. *Polym. Chem.* **2013**, *4*, 2230–2234.
44. Meuler, A. J.; McKinley, G. H.; Cohen, R. E. *ACS Nano* **2010**, *4*, 7048–7052.
45. Meuler, A. J.; Smith, J. D.; Varanasi, K. K.; Mabry, J. M.; McKinley, G. H.; Cohen, R. E. *ACS Appl. Mater. Interfaces* **2010**, *2*, 3100–3110.
46. Kota, A. K.; Kwon, G.; Choi, W.; Mabry, J. M.; Tuteja, A. *Nat. Commun.* **2012**, *3*, 2027/1–2027/8.

24

MULTIDIMENSIONAL NMR OF FLUOROPOLYMERS

XIAOHONG LI

Soochow University

JESSI BAUGHMAN, CHUN GAO, LINLIN LI, PETER L. RINALDI, AND ERIC B. TWUM

University of Akron

ELIZABETH F. MCCORD

E. I. du Pont de Nemours and Co.

FAITH J. WYZGOSKI

The Ohio State University—Mansfield

24.1 INTRODUCTION

Fluoropolymers can take on an amazing variety of characteristics depending on the details of their molecular structures. Modern methods of polymer synthesis have been adapted to provide tremendous flexibility in designing fluoropolymer structures so that materials can be prepared for a variety of applications. Understanding the microstructures of these polymers is essential to probe their structure–property relationships and to improve the overall performance of fluoropolymer materials. Modern spectroscopic tools are sorely needed to keep up with the requirements to characterize fluoropolymers so that proof of preparation of the desired structures is obtained and the quantities of the desired structural elements can be measured.

Handbook of Fluoropolymer Science and Technology, First Edition. Edited by Dennis W. Smith Jr., Scott T. Iacono, and Suresh S. Iyer.

Modern NMR instruments have enormous flexibility to control pulse timing events, so that an elaborate array of multidimensional NMR experiments can be implemented for complex structure determination. Implementation of these methods is particularly relevant for fluoropolymers, which often have three NMR active nuclei (^{1}H, ^{19}F, and ^{13}C) that are ideally suited for creating useful multidimensional NMR experiments.

In this chapter, we describe a variety of methodologies for applying multidimensional NMR (mostly 2D- and some 3D-NMR) for the characterization of fluoropolymers. Space limitations preclude a comprehensive survey of the literature. Instead, a few of the primary methodologies are described involving combined use of multidimensional NMR methods for structure elucidation. Then, a selected group of papers were reviewed to illustrate the applications of these methodologies to the characterization of some of the most common classes of fluoropolymers, including homo- and copolymers with poly(vinylidene fluoride), fluorinated polyethers, fluorinated ionomers, poly(vinyl fluoride) and its copolymers, and polytetrafluoroethylene (PTFE) and its copolymers.

The reader is assumed to be familiar with basic 1D- and 2D-NMR techniques for organic and polymeric structure studies, which might be achieved by reading a text such as that by Silverstein et al. [1] and a recent review on polymer NMR spectroscopy [2].

24.2 METHODOLOGY

24.2.1 NMR of Fluorine-Containing Materials

The basic NMR characteristics of ^{1}H, ^{19}F, and ^{13}C can be described as in Table 24.1. In fluoropolymers, these three NMR active nuclei often all present which provides three separate probes for studying their structures in simple 1D-NMR experiments. When purchasing instruments for these applications, the manufacturer must pay careful attention to probe design in order to simultaneously minimize probe background signals in the ^{1}H, ^{19}F, and ^{13}C spectra. Many probes designed for ^{1}H detection have components made of fluoropolymers to eliminate ^{1}H background signals which would then produce broad humps in the ^{19}F baseline that cause integration problems in 1D and produce ridges in 2D spectra. If the probe is to be used for studying fluoropolymers, these components can be replaced with ceramics based on inorganic materials.

TABLE 24.1 Basic NMR Characteristics of ^{1}H, ^{19}F, and ^{13}C

Nuclei	Spin, I	Natural abundance %	Frequency (MHz, at 11.74T)	Receptivity relative to ^{13}C	Reference compound	Shift range (ppm)
^{1}H	1/2	99.9885	500	5.87×10^{3}	Me_4Si	10
^{19}F	1/2	100	470.2	4.90×10^{3}	Me_4Si	200
^{13}C	1/2	1.07	125.7	1	CCl_3F	400

TABLE 24.2 Ranges of Typical $^nJ_{XY}$ Couplings in Fluoropolymers

n	$^nJ_{CH}$ (Hz)	$^nJ_{CF}$ (Hz)	$^nJ_{HF}$ (Hz)	$^nJ_{FF}$(Hz)	$^nJ_{HH}$(Hz)
1	125–175	250–350	–	–	–
2	0–5	30–50	30–50	200–300	10–20
3	3–8	5–10	8–15	5–15	3–8
4	–	–	–	10–90	–
5	–	–	–	2–30	–
6–7	–	–	–	1–7	–

The fluorocarbon-based polymers have some unique NMR characteristics because of the ^{19}F nuclei which produce special homo- and heteronuclear J-couplings. Table 24.2 summarizes typical J-couplings found in the organic polymers discussed in this review. Although not specifically dealing with fluoropolymers, a useful and comprehensive compilation of J-couplings in fluorinated compounds can be found in a publication by Foris [3].

In normal hydrocarbon-based polymers, only ^{1}H–^{13}C double resonance experiments are needed for ^{13}C-detected NMR and single resonance for ^{1}H experiments. With fluoropolymers triple resonance experiments are desirable for ^{13}C NMR, and double resonance for ^{1}H NMR, otherwise couplings arising from fluorines would introduce enormous complexity to the spectra. As an example Figure 24.1 shows ^{13}C 1D-NMR spectra acquired with various permutations of ^{1}H and ^{19}F decoupling from a neat sample of Krytox® dimer. All three spectra were obtained under the same conditions with the same parameters except for the decoupler settings. In Figure 24.1a, both ^{1}H and ^{19}F were decoupled simultaneously, which provided the best results, with all peaks simplified to singlets having maximum signal intensities. In comparison, the spectrum in Figure 24.1b obtained with only ^{19}F decoupling, and the spectrum in Figure 24.1c obtained with only ^{1}H decoupling, have poor signal to noise and are more complicated due to the remaining J_{CH} and J_{CF} couplings, respectively. In particular, the Z_1-type-CHF peaks were simplified and improved resolution and signal-to-noise level, revealing resonances of isomeric structures in Figure 24.1a.

Another example in Figure 24.2 shows an expansion from the ^{1}H 1D-NMR spectrum of poly(vinylidene fluoride-*co*-tetrafluoroethylene) P(VDF-co-TFE) containing signals from —$CH_2CF_2\underline{H}$ chain-end groups [4]. The small unresolved splittings in the coupled spectrum (Figure 24.2a) is $^3J_{HH}$ coupling to the vicinal CH_2 protons. The large triplet splitting is $^2J_{HF} \approx 50$ Hz to the two geminal fluorines; this large splitting is removed in the spectrum obtained with ^{19}F decoupling (Figure 24.2b). In this case, decoupling provides spectral simplification and signal-to-noise improvements so that additional weak chain-end signals from monomer-sequence effects are revealed.

The similar NMR properties of ^{19}F and ^{1}H as described in Table 24.1 means that most NMR experiments used for ^{1}H would also work to degree for ^{19}F. With modern technologies, a large variety of multidimensional NMR experiments are feasible to probe the correlation within these three nuclei of fluoropolymers, of which the most basic are homonuclear 2D-NMR experiments discussed in Chapter 24.2.2. When

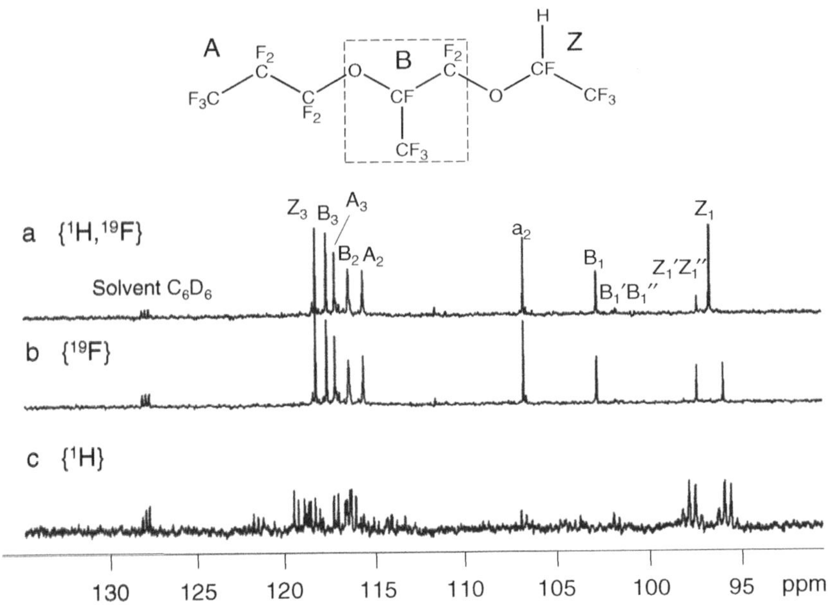

FIGURE 24.1 Downfield (CF_n) regions from the 125 MHz ^{13}C 1D-NMR spectra of a neat Krytox® dimer sample, obtained with (a) both ^1H and ^{19}F decoupling; (b) only ^{19}F decoupling; and (c) only ^1H decoupling. Reprinted with permission from University of Akron.

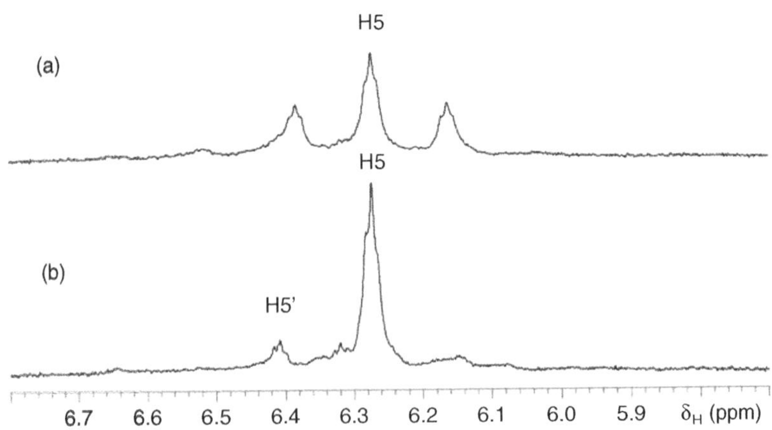

FIGURE 24.2 Expansion from the 500 MHz ^1H 1D-NMR spectra of P(VDF-co-TFE): (a) ^1H NMR spectrum without ^{19}F decoupling and (b) ^1H NMR spectrum with broadband ^{19}F decoupling. Reprinted with permission from Reference [4].

protons and fluorines are present in the polymer, both ^{1}H–^{1}H and ^{19}F–^{19}F 2D-NMR experiments are possible. Heteronuclear correlated 2D-NMR experiments between these nuclei like ^{19}F–^{1}H HETeronuclear COrrelation (HETCOR) are also possible. When sufficient sample is available, both $^{1}H\{^{13}C\}$ and $^{19}F\{^{13}C\}$ heteronuclear correlated 2D-NMR experiments such as Heteronuclear Single Quantum Correlation (HSQC) and Heteronuclear Multiple Bond Correlation (HMBC) experiments can be performed.

All the 2D-NMR experiments have the same basic format which can be divided into four units: preparation, evolution, mixing, and detection periods. For an A-B correlated 2D-NMR experiment, the preparation period usually starts with a delay during which the spins are allowed to return to equilibrium; this is followed by a pulse or a cluster of pulses and delays that transfers the magnetization from B to A nuclei, called coherence transfer, just before the evolution period. The evolution period (t_1) is a variable time delay, increased in a stepwise manner from an initial value of zero to a final value $ni \times 1/sw$ (related with spectral window $sw1$ in the indirectly detected dimension of nucleus A), provides the key to the generation of the second dimension. The mixing period serves to transfer the magnetization back from A to B nuclei. Then the detection (t_2) period is needed to collect the FID (Free Induction Decay) data. It produces a spectrum similar to the one obtained from a 1D experiment of the detected nucleus. Each t_1 value in the 2D sequence is repeated *nt* times and *np* data points (FID's) are stored (*np* is related to the acquisition time *at* and *sw* is the direct detection dimension spectral window of nucleus B).

During the multidimensional NMR experiments, effective coherence transfer is important, which can be accomplished via accurate J-modulation (e.g., Insensitive Nuclei Enhanced by Polarization Transfer, INEPT) or dipole–dipole (Nuclear Overhauser Effect, NOE) interaction. As seen in the table, the couplings with ^{19}F are quite different from those with ^{1}H and ^{13}C. The large and varied ^{19}F couplings create opportunities and permit a larger variety of experiments, while at the same time, it also create complicated spectra with complex cross-peak patterns. The $^{1}J_{CF}$ and $^{2}J_{CF}$ in Table 24.2 is sufficient for the $^{19}F\{^{13}C\}$ heteronuclear 2D-NMR experiments detecting one-bond and two-bond ^{19}F–^{13}C connectivity, while the $^{3}J_{FF}$ and $^{4}J_{FF}$ long-range couplings can be utilized to study the polymer structure through ^{19}F–^{19}F interaction. Some of the possible J-couplings can serve as the basis for 2D-NMR correlations [2]. The possible atomic connectivity information acquired are illustrated in Scheme 24.1.

Another unique NMR property for fluoropolymers comes from the large spectral dispersion of ^{19}F as shown in Table 24.1. The general ^{19}F spectral window is 2 times that of ^{13}C and 40 times that of ^{1}H which requires many more t_1 increments to be collected in order to obtain sufficient digital resolution in the indirectly detected ^{19}F dimension. Adequate digital resolution in a ^{19}F–^{19}F COrrelation SpectroscopY (COSY) experiment might require 10 or more times as many increments as the corresponding ^{1}H–^{1}H COSY experiment. Failure to obtain adequate digital resolution in the f_1 dimension might produce inaccurate cross-peak positions or lack of distinction between the positions of cross-peaks due to closely spaced multiplets. More importantly, very often both large and small couplings are present in the ^{19}F multiplets.

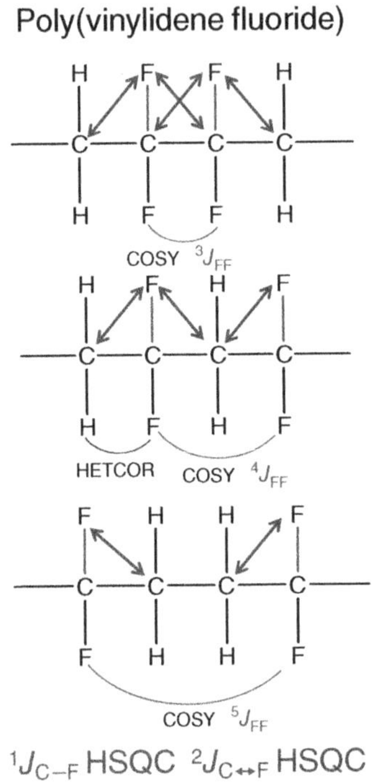

SCHEME 24.1 J-couplings utilized to study the possible atomic connectivity in PVDF.

Since the cross-peak components have both positive and negative phases (depending on the relative signs of couplings) failure to adequately digitize the signals in the f_1 dimension leads to inadequate resolution and cancellation of antiphase cross-peak components. Thus, cross-peaks which should easily be detected in the COSY spectrum based on the presence of couplings and the concentration of the species might disappear from the 2D spectrum. Another unique aspect of ^{19}F–^{19}F COSY spectra is the predominance of cross-peaks from $^2J_{FF}$ and $^4J_{FF}$ couplings compared to the relatively weak (or absence of) cross-peaks $^3J_{FF}$ and $^5J_{FF}$ couplings. The broad spectral window also makes it very difficult to excite the ^{19}F magnetization along the whole window equally. Selective excitation experiments are designed using different shaped pulses to treat each type of ^{19}F (CF, CF_2, and CF_3/CF_2O) as if they were different nuclei [5].

24.2.2 Homonuclear Experiments

Homonuclear experiments fall into two categories, chemical-shift correlated experiments and 2DJ spectroscopy. The former permits identification of chemical shift correlations via J-coupling between atoms (e.g., COSY) or via spatial proximity of two atoms to produce dipole–dipole relaxation (NOESY, Nuclear Overhauser

Enhancement SpectroscopY). The 2DJ spectroscopy produces 2D-NMR spectra that separate J-coupling and chemical shift into separate dimensions.

COSY is the most commonly utilized 2D-NMR experiment for fluoropolymer analysis. In routine applications, the setup of ^{19}F–^{19}F COSY experiments and use of the resulting spectra are very similar to those of ^{1}H–^{1}H COSY experiments which are routinely used in the study of hydrocarbon-based structures. This simplest of 2D-NMR experiments is usually composed of two 90° pulses flanking the t_1 evolution period, which minimizes the difficulties from nonuniform excitation of the wide ^{19}F spectral bandwidth (compared to most other 2D-NMR experiments which include 180° pulses having twice the length and about half the excitation bandwidth). The experiment is relatively insensitive to miscalibrated pulses; in fact, the COSY-45 sequence (second pulse is 45° rather than 90°) is often preferred since it produces weaker diagonal signals, which do not contain useful information and only mask weak cross-peaks falling near the diagonal.

Several complicating factors are noteworthy in the application of ^{19}F–^{19}F COSY experiments. Cross-peaks in ^{1}H–^{1}H COSY 2D-NMR spectra primarily result from $^{2}J_{HH}$ and $^{3}J_{HH}$ couplings. However, in ^{19}F–^{19}F COSY 2D-NMR spectra cross-peaks can be produced by $^{2}J_{FF,}$ $^{3}J_{FF}$, $^{4}J_{FF}$, and even $^{5}J_{FF}$ couplings. In fact, the cross-peaks from four-bond couplings are often more prominent than those from three-bond couplings (the latter are sometimes absent from the spectrum) [2, 6]. Second, COSY cross-peak patterns between resonances of atoms A and B consist of an $n \times m$ pattern of antiphase cross-peaks, where n and m represent multiplicities of resonances A and B. If the digital resolution in the 2D-NMR spectrum is insufficient to resolve some of the smaller couplings in these multiplets, then the antiphase multiplet components will cancel one another and eradicate the cross-peak even though the coupling and the signal strength should result in a strong cross-peak pattern.

Additional complications arise due to the apparent differences in the mechanisms producing $^{n}J_{FF}$ couplings compared to $^{n}J_{HH}$ couplings. The $^{3}J_{HH}$ are generally determined by the Karplus relationship with dihedral angle. However, the exact relationships between conformation and $^{n}J_{FF}$ are not completely understood except that the proximity of nonbonding orbitals of two F atoms which are close in space usually produce large J_{FF} values; $^{4}J_{FF}$ (10 Hz and larger) and even $^{5}J_{FF}$ values are often larger than $^{3}J_{FF.}$ Finally, $^{2}J_{FF}$ couplings are often 200–300 Hz, resulting in very broad multiplet patterns [6, 7]. The combination of broad cross-peak patterns and the many small couplings present can sometimes create problems with resolution of overlapping cross-peak patterns when many overlapping resonances exist in the spectrum from many slightly different structures, as is often the case with polymer NMR spectra. As a consequence, structural analysis and the significance associated with the presence of cross-peaks are more challenging using ^{19}F–^{19}F COSY.

A few alternative COSY methods have been developed to improve the quality of spectra with large numbers of overlapping correlations or with large spectral windows. Selective COSY utilizes long-shaped pulses to excite a small window within the field of correlations in a spectrum. The smaller spectral window of the selective COSY experiment enhances digital resolution and reduces experiment time due to the reduced number of 2D increments required to obtain high digital resolution. The improved digital resolution reduces the cancellation of antiphase multiplet components from overlapping cross-peaks that can occur in the traditional COSY experiment [5]. The optional addition of selective inversion pulses in the middle of the t_1 evolution period eliminates couplings in the f_1 dimension and produces spectra with simplified cross-peak patterns and improved signal intensity. Figure 24.3 shows a comparison of standard COSY data with selective COSY for regions containing the CF—CF_2 correlations from the structure at the top of the figure. The top plot shows a selected region from the standard phase-sensitive COSY spectrum; two sets of cross-peaks are seen (highlighted by the boxes) from four-bond couplings across the ether linkages joining fragments A-B and B-C. The middle plot is the corresponding region of a phase-sensitive selective COSY spectrum. The standard COSY spectrum shows broad, poorly resolved cross-peaks from the poor digital resolution, while the selective COSY spectrum exhibits an enormous amount of detail from the numerous J_{FF} couplings. The bottom spectrum was obtained with a ^{19}F inversion pulse in the middle of t_1 to remove ^{19}F homonuclear couplings in the f_1 dimension, resulting in single lines in this dimension. Four sets of cross-peaks are resolved in each of the regions, from the four stereoisomers (and their mirror images). The cross-peak patterns in the selective COSY spectrum have been simplified by the removal of many of the nearby CF and CF_2 couplings.

Total correlation spectroscopy (TOCSY) experiments produce a complete set of cross-peaks among all nuclei in the same spin system. This can be helpful in identifying nuclei of a ring or sections of a polymer chain. ^{19}F TOCSY spectra can be difficult to interpret in polymers due to overlapping resonances. Use of ^{19}F–^{19}F TOCSY spectra is uncommon due to the difficulties with producing an adequate spin-lock pulse over the large ^{19}F spectral window [8].

The NOESY experiment is helpful in the identification of cross-peaks between resonances of vicinal fluorines, often missing from COSY spectra. Cross peaks resemble those in COSY spectra, but are formed only for nuclei that have sufficient dipole–dipole interactions (usually for ^{19}F atoms that are within 0.4–0.5 nm of one another). The number of intervening bonds or the presence of J-coupling does not influence cross-peak formation in NOESY [8].

The 2D J-resolved (2DJ) spectroscopy technique is used to measure scalar coupling constants. It is most useful in spectra with severe spectral overlap. The 2DJ spectrum, after data processing, results in a plot of chemical shift along the f_2 axis and homonuclear J-coupling along the f_1 axis. This effectively isolates multiplets due to homonuclear scalar couplings at the corresponding chemical shift of the multiplet. Homonuclear 2DJ spectroscopy offers a convenient method for distinguishing between J_{FF} homonuclear and J_{HF} heteronuclear couplings (and between J_{HH} and J_{HF} couplings) as the splittings from the latter will remain along the f_2 dimension. This is especially useful on instruments that do not have the capability of H/F double resonance. Furthermore, the pulse sequence used for this experiment removes the line-broadening effect of magnetic field inhomogeneity in the f_1 dimension, resulting in narrow lines and the ability to resolve very small homonuclear couplings.

A study by Bruch [9] is a classic example of the use of ^{19}F homonuclear 2DJ spectroscopy for the study of fluoropolymers. Scalar couplings of poly(vinyl fluoride)

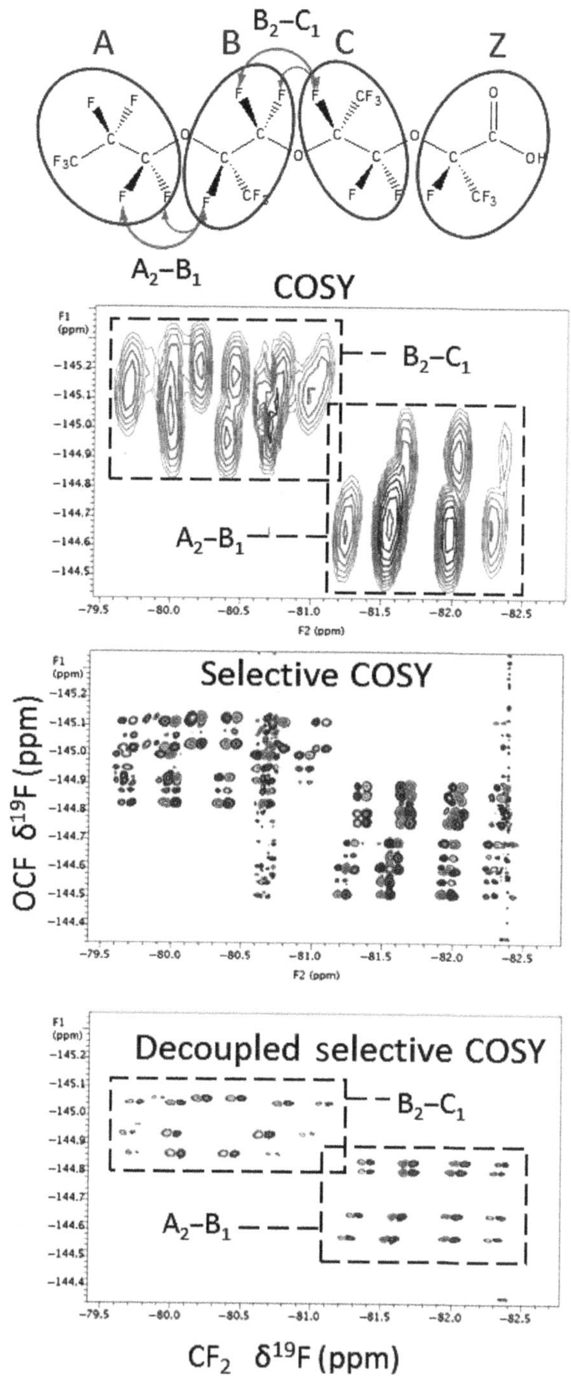

FIGURE 24.3 Perfluorinated polyether oligomer of Krytox® fluoropolymer: (top) standard COSY, (middle) selective COSY, and (bottom) selective COSY with f_1 decoupling. Reproduced with permission from Reference 5.

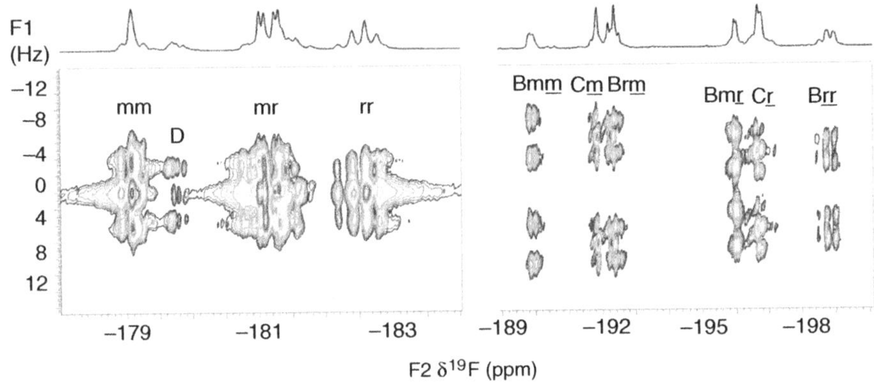

FIGURE 24.4 Peak-containing regions from the 470 MHz ^{19}F homonuclear 2DJ spectrum of poly(vinyl fluoride) showing the (a) low field and (b) high field.

(PVF) were too hard to measure in COSY spectra due to the extensive cross-peak overlap. A recently acquired ^{19}F homonuclear 2DJ spectrum of PVF intended to replicate the originally published data is shown in Figure 24.4. The data were collected with continuous ^{1}H decoupling to remove heteronuclear J_{HF} couplings in both the f_1 and f_2 dimensions. Without ^{1}H decoupling, the multiplets will be split in the f_2 dimension by the many $^{n}J_{HF}$ heteronuclear couplings. In the spectra of this polymer with so many hydrogen atoms, this is very advantageous.

In Figure 24.4a, the triplets for the mm (−178.9 ppm), mr (−180.9 ppm), and rr (−182.3) type A fluorines of head-to-tail backbone structure result from $^{4}J_{FF}$ couplings to fluorines on adjoining monomer units. The measured $^{4}J_{FF} \approx 4$ Hz for type A fluorines in r dyads is less than that of type A fluorines in m dyads ($^{4}J_{FF} \approx 5$ Hz), the magnitude of the scalar coupling is dependent on the stereochemistry. The two doublets from type D fluorines near −179.5 and −181.8 ppm have a $^{4}J_{FF} \approx 7$ Hz and are assigned to A″/D couplings; the fluorine in the other direction along the backbone is five bonds away and does not produce resolvable couplings.

Figure 24.4b shows a series of doublet of doublets for the resonances from fluorines B and C of inverted monomer units. A doublet of doublets is observed for type B fluorines, indicating unequal couplings to fluorines on adjoining monomer units. Based on the nomenclature in the paper, m and r represent the relative configurations of fluorines separated by four bonds and m̲ and r̲ represent the relative configurations of fluorines separated by three bonds. It was found that $^{4}J_{FF}$ couplings between fluorines in m and r dyads are 5 and 4 Hz, respectively, and that $^{3}J_{FF}$ couplings between fluorines in m̲ and r̲ dyads are 13 and 9 Hz, respectively. The m̲ type C fluorines at −191.5 ppm produce a doublet of triplets from the 13.1 Hz coupling (doublet) to vicinal m̲ type B fluorines and the 3.0 Hz coupling (triplet) to both type A′ and type D fluorines five bonds away. The couplings at −191.5 ppm confirm the assignment of type C fluorines, which had been previously debated. The overlap in this region made assignments difficult based on previously published COSY spectra and also hindered the determination of scalar coupling constants for the type C fluorines at −192.0 ppm. The homonuclear 2DJ spectra provided beautiful resolution of all the complex multiplets in the ^{19}F NMR spectrum of this polymer.

24.2.3 Heteronuclear Shift Correlation Experiments

The presence of three NMR active nuclei in many fluoropolymers provides the possibility of performing a number of permutations of heteronuclear correlated 2D-NMR experiments. ^{1}H–^{19}F heteronuclear correlated experiments (with either ^{1}H or ^{19}F detection) provide enormous sensitivity as both nuclei have high γ and essentially 100% natural abundance. When sufficient sample is available, ^{1}H–^{13}C and ^{19}F–^{13}C heteronuclear correlated experiments are possible; however, these experiments are 100-fold less sensitive as only ^{1}H/^{19}F correlations to ^{13}C (1.1% abundance) are detected.

24.2.3.1 ^{1}H–^{19}F Heteronuclear Correlation ^{1}H–^{19}F heteronuclear experiments are ideal for analyzing the structures of fluoropolymers containing both hydrogen and fluorine atoms. Typical ^{1}H–^{19}F heteronuclear experiments include ^{1}H–^{19}F HETCOR, ^{1}H–^{19}F COSY, and ^{1}H–^{19}F gradient heteronuclear multiple quantum coherence (gHMQC). Since the abundance of ^{19}F and ^{1}H nuclei are both 100%, ^{1}H–^{19}F heteronuclear experiments can provide excellent sensitivity, so that the weak resonances from end-groups can be observed. Note that there are some requirements for the hardware to perform this kind of experiment. A special probe is needed that can be optimally tuned to the ^{1}H and ^{19}F resonance frequencies simultaneously. In addition, the console must be able to simultaneously generate two high band frequencies.

In ^{1}H–^{19}F HETCOR experiments, either ^{1}H or ^{19}F can be used in the directly detected f_2 dimension. Due to the large spectral window of ^{19}F resonances, direct observation of ^{19}F resonances is preferred to obtain sufficient digital resolution of the ^{19}F resonances [6]. Otherwise, a large number of t_1 increments in the f_1 dimension would be required to get comparable resolution (doubling the digital resolution in the f_1 dimension would require more than double the experiment time) [10]. One of the disadvantages of ^{1}H–^{19}F heteronuclear experiments is that cross-peaks identification is not as clear as in ^{1}H–^{13}C heteronuclear correlated 2D spectra. The ^{1}H–^{13}C heteronuclear 2D-NMR experiments can be devised to provide one-bond or multi-bond (typically two- or three-bond) correlations, based on specific needs. This is due to the large difference between one- and two-/three-bond ^{1}H–^{13}C coupling constants ($125 < {}^{1}J_{CH} < 175$ Hz and $0 < {}^{2}J_{CH} < 5$ Hz) which are used to optimize polarization transfer delays in the pulse sequences [7]. In comparison, the two and three bonds ^{1}H–^{19}F couplings are similar (see Table 24.2). As a result, two-bond and three-bond correlations are indistinguishable in the ^{1}H–^{19}F heteronuclear NMR spectra. In ^{19}F–^{1}H HMQC and HMBC experiments, the delays $\times = 1/(2 \times {}^{n}J_{HF})$ must be optimized based on the n-bond couplings to improve cross-peak intensities, and it is hard to see correlations if the n-bond couplings are small [10]. In comparison, it is not necessary to optimize the delays based on the n-bond couplings in the ^{1}H–^{19}F COSY experiments. Macheteau's work [10] suggests that ^{1}H–^{19}F COSY could provide

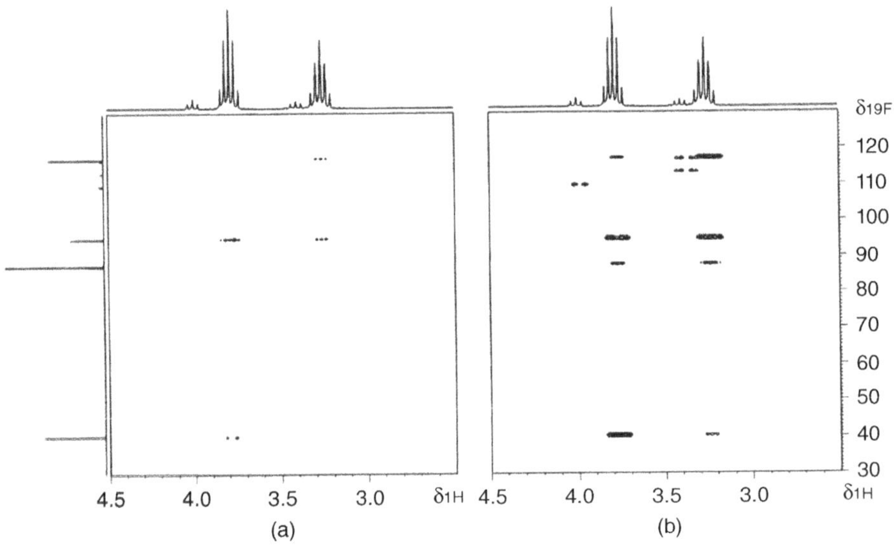

FIGURE 24.5 Comparison of ^1H–^{19}F correlation 2D-NMR experiments of $C_2F_5(VDF)_2I$ in acetone-d_6 at 25°C:^1H–^{19}F HMQC NMR spectrum and (b) ^1H–^{19}F hetero-COSY NMR spectrum. Reproduced with permission from Reference [10].

more information on multi-bond correlations between proton and fluorine atoms than ^1H–^{19}F gHMQC and gHMBC spectra. Figure 24.5 shows the ^1H–^{19}F HMQC (a) and ^1H–^{19}F hetero-COSY (b) of a model for vinylidenefluoride (VDF) homopolymer, with ^{19}F and ^1H 1D spectra plotted along the sides. Based on their results, ^1H–^{19}F hetero-COSY shows strong three-bond ^1H–^{19}F correlations and weak five-bond correlations. In comparison, only three bonds ^1H and ^{19}F correlations appear on the ^1H–^{19}F HMQC spectrum in this case. In addition, ^{19}F NMR requires a shorter pulse width than ^1H NMR in order to cover a larger spectral width compared to the narrow spectral window for resonances of ^1H nuclei. The resonances far away from the transmitter may not receive full excitation. To solve this problem, a selective ^1H–^{19}F heteronuclei COSY experiment has been used.

24.2.3.2 ^1H–^{13}C and ^{19}F–^{13}C Heteronuclear Correlation Experiments ^1H–^{13}C correlation 2D-NMR experiments are routinely done in most laboratories. The early ^1H–^{13}C correlation experiments involved ^{13}C detection. These ^{13}C-detected experiments, such as the HETCOR experiment, used to identify 1-bond ^{13}C–^1H correlations [11–13] and COLOC (correlation spectroscopy via long range coupling), used to identify multiple bond correlations [14], have relatively low sensitivity because they involve detection of the low γ ^{13}C nuclei. They have been replaced by ^1H-detected experiments due to the higher sensitivity detection of the high γ nucleus. The ^1H-detected HMQC and HSQC experiments are now preferred to the HETCOR experiment. Likewise, the ^{1}H-detected HMBC and heteronuclear 2-bond correlation (H2BC) [15] experiments have replaced the ^{13}C-detected experiments such as COLOC. Despite their lower sensitivity, two advantages of the ^{13}C-detected experiments (relative to ^{1}H detection) are first that fewer problems with solvent resonances are encountered, and secondly that enhanced digital resolution in the ^{13}C dimension is obtained with very little increase in the total NMR experiment time by simply increasing the acquisition time.

Most modern dual-channel NMR spectrometers are capable of performing ^{1}H-detected heteronuclear 2D-NMR experiments. There are several variants of these heteronuclear experiments; those that use pulsed magnetic field gradients (PFGs) for suppression of undesired signal components are most useful for obtaining clean, artifact-free spectra. In addition, incorporation of multiplicity editing, adiabatic pulses, and selective pulses can be used to enhance the information content and quality of the resulting spectra. Detailed information on these ^{1}H{^{13}C} correlation experiments and their application to organic, inorganic, and hydrocarbon-based polymers can be found in most NMR text books.

Like their ^{1}H{^{13}C} counterparts, inverse-detected ^{19}F{^{13}C} experiments are preferred to direct-detected ^{13}C{^{19}F} experiments. In principle, ^{1}H{^{13}C} correlated experiments can be readily adapted to collect ^{19}F{^{13}C} chemical shift correlated 2D-NMR spectra by taking into account the differences between the C—F and C—H couplings when setting the coherence transfer delays and the different resonance frequencies of ^{19}F and ^{1}H. However, their application to the study of fluoropolymers is not widespread. In practice, many other factors must be considered in order to obtain good quality ^{19}F{^{13}C} correlated 2D-NMR spectra from fluoropolymers. The high frequency channel of the ^{1}H detection probe must be tunable to the ^{19}F frequency. If the fluoropolymer contains hydrogen atoms, coupling to the third NMR active nucleus can introduce complications in the NMR spectra. When such materials are studied, it is useful to have a three-channel instrument and probe so that various permutations of the NMR experiments involving pulses and decoupling of the ^{1}H, ^{19}F, and ^{13}C frequencies can be performed.

24.2.3.3 Heteronuclear ^{19}F–^{13}C Experiments for 1-Bond Fluorine–Carbon Correlations The first important consideration in the ^{19}F–^{13}C NMR experiment is the selection of optimized pulse sequence delays, which are dependent on the J-couplings. In typical organic structures, $^1J_{\rm CH}$ values span a range of 125–165 Hz (i.e., 145 Hz ± 10%). In ^{1}H{^{13}C} correlation experiments, if an average $^1J_{CH}$ $\approx$ 145 Hz is assumed, optimized delays of 1/4 J $\approx$ 1.7 ms usually give decent correlations for all CH groups, as a delay which is misset by 10% does not drastically diminish the intensities of cross-peaks. In organofluorine compounds, $^1J_{\rm CF}$ coupling constants typically span a range from 150 to 350 Hz and are dependent upon the electronegativities of the substituents on carbon; thus the $^1J_{\rm CF}$ values for CF, CF_2, and CF_3 groups are approximately 150, 250, and 350 Hz, respectively. Optimizing the delays for any one coupling produces large errors in the delays of the other groups, dramatically diminishing their cross-peak intensities. It is therefore more appropriate to run separate experiments, one with the ${}^1J_{CF}$ coupling optimized for CF groups and another with the ${}^1J_{CF}$ coupling optimized for CF_2 and CF_3 fluorine resonances.

Rather than using HMQC, it is preferable to use HSQC-based pulse sequences. In the former, cross-peak patterns are split by J_{FF} homonuclear couplings in both the f_1 and f_2 dimensions (${}^2J_{FF}$ = 100–300 Hz, ${}^3J_{FF}$ = 5–15 Hz, ${}^4J_{FF}$ = 10–90 Hz, and ${}^5J_{FF}$ = 2–30 Hz, [16,17]). These complex cross-peak patterns greatly diminish signal intensities and lead to broad cross-peak patterns which become difficult to resolve from one another. In HSQC-based experiments, the splittings from J_{FF} couplings are eliminated from the f_1 dimension leading to sharper and more intense cross-peak patterns.

The simple HSQC pulse sequences are less desirable than HMQC pulse sequences because they usually contain a larger number of 180° pulses which create problems from nonuniform excitation of the large ${}^{19}F$ spectral windows, however, new variations of the HSQC sequence incorporate adiabatic pulses [18]. These adiabatic pulses compensate for resonance offset effects and misset 180° pulses and are able to uniformly excite the large spectral windows encountered in the ${}^{19}F$ chemical shift dimension [19]. Another approach is to use selective ${}^{19}F$ excitation pulses [20]. In this approach, shaped pulses are used to selectively excite regions of interest in the ${}^{19}F$ chemical shift dimension (e.g., CF, CF_2, or CF_3 regions). This latter approach is facilitated by the large spectral dispersion of the resonances in ${}^{19}F$ NMR spectra. Furthermore, when separate experiments are performed for the three regions, delays can be optimized for the J_{CF} couplings of the fluorines in each region.

Of course for hydrogen-containing fluoropolymers, it is often useful to do both ${}^{19}F\{{}^{13}C\}$ and ${}^1H\{{}^{13}C\}$ HSQC experiments but with continuous ${}^1H$ decoupling in the former and with ${}^{19}F$ decoupling in the latter. A detailed discussion of the complications associated with ${}^{19}F$ decoupling in these experiments has been presented in the literature [5]. For a more detailed discussion of 1-bond ${}^{19}F\{{}^{13}C\}$ correlation experiments, readers are referred to a review by Battiste and Newmark [6].

24.2.3.4 Heteronuclear ${}^{19}F$–${}^{13}C$ Experiments for Multiple-Bond Fluorine–Carbon Correlations Multiple bond ${}^{19}F$–${}^{13}C$ correlation experiments provide the additional spectral data needed to determine atomic connectivities and to obtain unambiguous assignments of ${}^{19}F$ and ${}^{13}C$ resonances from fluoropolymers. Results from ${}^1J_{CF}$ and ${}^2J_{CF}$ correlation experiments (and also those from ${}^3J_{HF}$ HETCOR) can be used to provide information to distinguish between the different ${}^{19}F$–${}^{19}F$ couplings (${}^3J_{FF}$, ${}^4J_{FF}$, and ${}^5J_{FF}$) present in fluoropolymers [21–23]. HMBC 2D-NMR spectra are commonly used to provide C–H and C–F multiple bond correlations for structure determination [21, 24, 25]. In the ${}^1H\{{}^{13}C\}$ HMBC experiments, 2- and 3-bond correlations are usually observed, however, in the case of ${}^{19}F\{{}^{13}C\}$ HMBC, usually only 2-bond correlations are observed. This is due to the diminishing magnitude of ${}^nJ_{CF}$ as the number of bonds separating C and F increases. Due to the relatively small ${}^3J_{CF}$ coupling constants in fluoropolymers (typically less than 10 Hz), which results in large coherence transfer delays, and the short T_2 for ${}^{19}F$ (compared to ${}^1H$), signal losses due to T_2 relaxation makes the detection of 3-bond correlations difficult.

Even for small molecules, where the ^{19}F T_2 relaxation is relatively long, it is often difficult to detect $^2J_{CF}$ and $^3J_{CF}$ correlations in a single experiment because of the large difference between $^2J_{CF}$ and $^3J_{CF}$. As with HMQC experiments, the numerous $^nJ_{FF}$ couplings produce splittings in the f_1 dimension, broadening and diminishing the intensities of the cross-peaks. Furthermore, HMBC experiments are generally run without decoupling of ^{13}C during the acquisition time (f_2 dimension) to prevent cancellation of antiphase signal components, leading to additional reduction of cross-peak intensities and line broadening in the f_2 dimension.

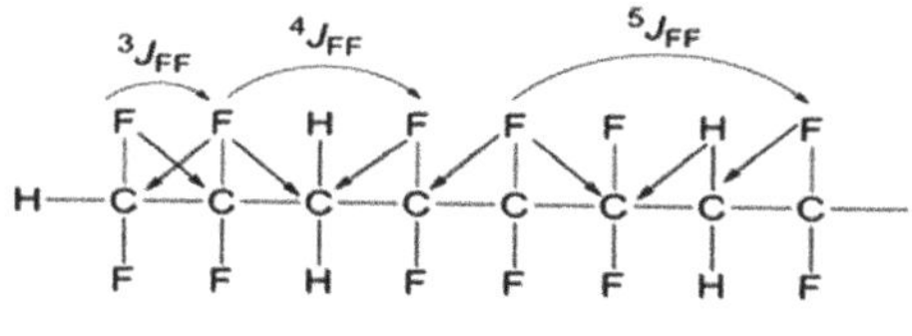

1

Schorn et al. [22] used the standard HMQC pulse sequence with delays optimized for $^2J_{CF}$ to provide the needed two-bond C—F correlations used to assign the resonances in a bis(n-perfluoroalkyl)zinc complexes. To be able to distinguish correlations due to $^1J_{CF}$ leakage from $^2J_{CF}$ correlations in the $^2J_{CF}$ experiment, no ^{13}C decoupling was applied during the acquisition period. Cross-peaks from $^1J_{CF}$ correlations were easily identified from their large doublet splitting in the ^{19}F chemical shift dimension.

HSQC-type pulse sequences have also been used to provide $^2J_{CF}$ correlations in fluoropolymers. In the work by Rinaldi et al. [5, 26], the standard HSQC pulse sequence was modified such that it could be tuned to run either $^1J_{CF}$ or $^2J_{CF}$ optimized experiments. In order to suppress $^1J_{CF}$ correlations in the two-bond experiment, a TANGO (testing for adjacent nuclei by gyration operator) pulse cluster was incorporated at the beginning of the pulse sequence. The phase of the 135° pulse in the TANGO pulse cluster could be set such that $^1J_{CF}$ correlations are destroyed in the two-bond experiment and *vice versa*. This modified HSQC pulse sequence can be used to perform band-selective experiments; problems with nonuniform bandwidth excitation associated with the use of $^{19}F\{^{13}C\}$ HSQC are therefore eliminated. The sequence can also be used with both ^{13}C and ^1H decoupling, providing enhanced sensitivity.

Figure 24.6 shows an example of the use of $^1J_{CF}$ (a) and $^2J_{CF}$ (b) $^{19}F\{^{13}C\}$ HSQC spectra to the study of P(VDF-*co*-TFE). In this figure, 0 and 2 represent backbone CH_2 and CF_2 groups (from carbons with 0 and 2 attached fluorines). The CF_2 fluorines in 020 3-carbon sequences contain adjoining CH_2 groups, therefore the fluorine resonances show 2-bond C—F correlations to the CH_2 carbon resonances (left regions in Figure 24.6b). The CF_2 fluorines in 022/220 3-carbon sequences contain adjoining CH_2 and CF_2 groups, therefore the fluorine resonances show two-bond C—F correlations to both the CH_2 carbon resonances (middle regions in Figure 24.6b). The CF_2 fluorines in 222 3-carbon sequences contain adjoining CF_2 groups, therefore the fluorine resonances show two-bond C—F correlations to the CF_2 carbon resonances (right regions in Figure 24.6b). The combined data from the two experiments permit

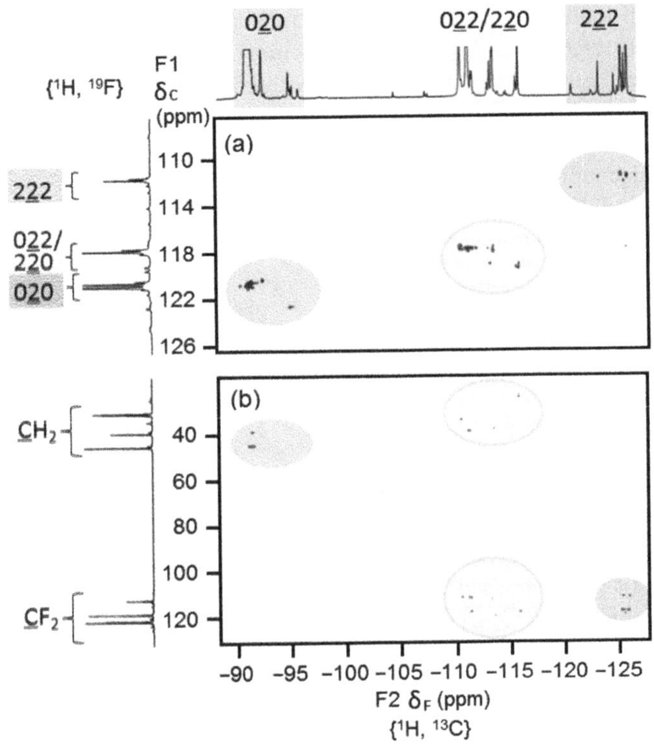

FIGURE 24.6 The $^{19}F\{^{13}C\}$ 2D-NMR spectra of P(VDF-*co*-TFE) (a) one bond ($^1J_{FC}$) (b) two bond ($^2J_{FC}$). Reprinted with permission from Reference [4]

identification of the structures and resonance assignments for 3-carbon sequences (020, 022/220, and 222) in poly(vinylidene fluoride) (PVDF).

24.2.4 Identification of Chain and Branch Ends

It is known that the end-groups in fluoropolymers, including both chain ends and chain branches, have an important influence on their properties [27–31]. In addition, the end-groups, which are typically produced during the polymerization process, can reveal information about the mechanisms of initiation, chain transfer, and termination [26, 30]. Therefore, a good understanding of the structures of end-groups is important to explore the structure–property relationships in fluoropolymers. According to studies by Imran-ul-haq et al. [29] and Herman et al. [31], the end-groups in PVDF can affect the polymer morphology, crystal form, and crystallinity. Madorskaya et al. [28] prepared PVDF with different initiators and investigated the influences of end-groups on the thermal stability and degradation mechanism of the polymer. In the study by Hedhli et al. [27], the influence of long chain branches (LCBs) in PVDF on the rheological properties of the polymer were discussed and the mechanism of forming these LCBs was studied. In this section, the use of T_1 relaxation and Diffusion-Ordered Spectroscopy (DOSY) NMR experiments to identify the chain ends and branch ends is presented.

Unlike small molecules, whose spin–lattice relaxation times are typically controlled by their tumbling motions, the relaxation times of polymers are governed by the local segmental motions of groups of atoms. The atoms of the chain ends and the branch ends, typically those attached to the last few carbons, are more mobile and exhibit longer relaxation times. In comparison, the motions of the atoms in the polymer backbone are more restricted and therefore have shorter T_1's. In this way, the spin–lattice relaxation experiment provides a way to distinguish the resonances of chain ends/branch ends from those of the backbone.

DOSY is a useful tool to analyze mixtures by providing valuable information about molecular size and shape [32]. In a typical DOSY experiment, a series of spectra are collected at different gradient pulse amplitudes; various processing methods are then used to produce a 2D spectrum with chemical shift in the f_2 dimension and diffusion coefficients in the f_1 dimension. Generally, the larger the molecular size, the smaller the diffusion coefficient. It is known that synthetic polymers consist of polymer chains with a distribution of molecular weights (MWs). The use of DOSY to determine the MW distributions of polydispersed polymers has been reported by Chen et al. [33]. Therefore, the DOSY NMR experiment has the potential to distinguish between the resonances of the chain ends and branch ends in polymers.

In the DOSY experiment, a weighted average of the diffusion coefficients of the atoms contributing to a resonance is usually measured. The diffusion coefficients of the atoms in the polymer are controlled by the overall translational motion of the entire molecule. Therefore, the measured diffusion coefficient for atoms in polymers is the average of the atoms from both high and low MW polymer molecules. Atoms in the polymer backbone and branches have diffusion coefficients that are highly weighted by the contribution from the high MW polymer molecules that are characterized by slow diffusion. Therefore, they should have smaller diffusion coefficients. In comparison, the atoms from chain ends, which are more highly represented in the low MW molecules, will have larger diffusion coefficients.

Twum et al. reported the use of ^{19}F T_1 and DOSY to identify the chain ends and branch ends of PVDF [26]. Figure 24.7 shows the ^{19}F DOSY NMR spectrum of PVDF. In this figure, the groups whose resonances are labeled with B have smaller diffusion coefficients, and were therefore assigned to the fluorine atoms in the polymer backbone. The groups with resonances I″ and J″ are attributed to the fluorine in bold in the structure $-CH_2CF_2CH_2\mathbf{CF}_2H$. Due to the larger diffusion coefficient of J″ than that of I″, the resonances of J″ and I″ were proposed to be resonances of polymer chain ends and short chain branch ends, respectively. In addition, according to the T_1 relaxation study by Twum et al., the T_1s of the resonances from the polymer backbone range from 0.33 to 0.60 seconds. The larger T_1 value of the resonance J″, which is 1.96 seconds, confirms that it is from the polymer chain end. As for I″, its T_1 value is between that of polymer backbone and chain end. This supports assignment of I″ to a short chain branch end, whose mobility is more restricted than that of a polymer chain end.

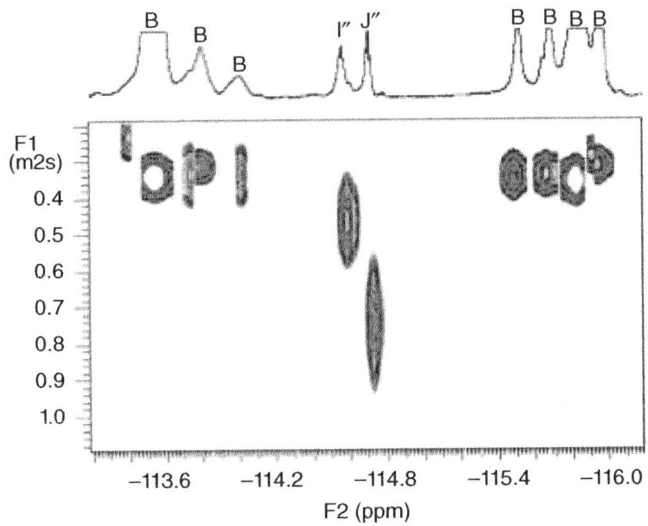

FIGURE 24.7 Expansion from the ^{19}F DOSY 2D-NMR spectrum of PVDF. Reprinted with permission from Reference 26.

24.2.5 3D-NMR

When 2D-NMR does not provide sufficient spectral dispersion, 3D-NMR experiments are useful; they can also occasionally provide structure information that is not available from separate 2D-NMR experiments. 3D-NMR experiments can be implemented by combining two 2D-NMR experiments or by incorporating pulse sequence elements that produce stepwise magnetization transfer between nuclei as is done in elements of the HSQC experiment. The latter is applicable when three NMR active nuclei with $I = 1/2$ are present in the molecule and is commonly used in NMR characterization of proteins uniformly labeled with ^{13}C and ^{15}N, where $^{1}J_{HN}$ and $^{1}J_{CN}$ values are used to produce 3D-NMR spectra which correlate the chemical shifts of ^{1}H, ^{13}C, and ^{15}N having specific relationships in the protein structure (e.g., among the resonances of –NH–C(O)– amide fragments or among –NH–CH(R)– amino acid fragments of a protein) [34]. The overlapping fragments from two or more experiments such as these can provide complete structures and resonance assignments of protein backbones. The sensitivity of such 3D-NMR experiments is governed by the γ of the nuclei which serve (typically ^{1}H) as a source of initial NMR magnetization, the γ of the detected nuclei (typically ^{1}H), and the product of the fractions of the $I = 1/2$ isotope of the three nuclei whose shifts are correlated. In order to make ^{1}H/^{13}C/^{15}N correlated 3D-NMR experiments feasible, biochemists increase the abundances of ^{13}C and ^{15}N (natural abundance 1.1% and 0.37%, respectively) by uniform isotopic labeling.

In fluoropolymers, the presence of three nuclei with $I = 1/2$ (^{1}H, ^{19}F, and ^{13}C) in many cases makes it possible to do chemical-shift correlated 3D-NMR experiments.

The high abundances of both ^{1}H and ^{19}F make it possible to do these experiments without resorting to expensive and difficult isotopic labeling. Either ^{1}H- or ^{19}F-detected experiments are feasible and both can be performed with similar detection limits. Although many classes of 3D-NMR experiments are possible, two are obvious. (1) If chemical-shift correlated experiments are performed using coherence transfer delays optimized for $^1J_{CF}$ and $^1J_{CH}$ couplings, the ^{1}H/^{19}F/^{13}C chemical shift correlations defining structure elements highlighted by the red bonds in **2** can be identified. (2) If chemical-shift correlated experiments are performed using coherence transfer delays optimized for $^2J_{CF}$ and $^1J_{CH}$ couplings the ^{1}H/^{19}F/^{13}C chemical shift correlations defining structure elements connected by the blue lines in **2** can be identified. Combined data from the two experiments can give complete structure information and resonance assignments, and in many cases, only the latter experiment is necessary [35, 36]. The detection limits for these experiments are similar to those of ^{1}H{^{13}C}- and ^{19}F{^{13}C}-HSQC experiments which can be routinely performed on fluoropolymers.

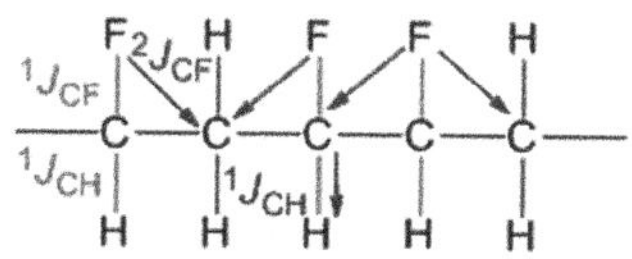

2

24.3 APPLICATIONS

The following sections describe selected applications of various methodologies involving combined use of the multidimensional NMR experiments described in section 24.2. The sections are divided based on the major classes of fluoropolymers which have commercial and/or scientific significance. These include PVDFs, perfluorinated polyethers, poly(chlorotrifluoroethylene)s, fluorinated ionomeric polymers, and PTFEs. The applications serve as examples of how multidimensional NMR experiments can be combined to solve structure problems in each of these fluoropolymer classes.

24.3.1 Polyvinylidene Fluoride

The copolymerizations of VDF with a variety of monomers, as well as their applications, were reviewed in Ameduri's paper [37]. The structural variations present in these polymers include monomer inversions, monomer/stereo-sequence effects (in copolymers containing monomers with branch-forming structure elements), chain ends, and short/LCBs. The characterization of these polymers in early studies predominantly depended on 1D-NMR and empirical chemical shift additivity rules [38, 39]. With the improvement in both NMR hardware and pulse sequence development, 2D-NMR characterization of these polymers has become more popular.

For VDF-containing polymers, the NMR spectra are typically extremely complicated, not only because of the structural variations in the polymers but also due to the complicated splitting patterns caused by the abundance of couplings among and between ^{1}H and ^{19}F. In Katoh's study, the ^{13}C NMR spectrum of PVDF was obtained with simultaneous ^{1}H and ^{19}F decoupling; in both, the resolution and sensitivity were improved considerably by elimination of all J-coupling interactions [40]. To study the monomer sequence effects in PVDF, they also collected ^{19}F–^{13}C HETCOR and ^{19}F–^{19}F COSY 2D-NMR spectra, which produced resonance assignments completely based on NMR. It should be mentioned that when sufficient sample is available, the abundance of couplings with multiplets resolved can provide an amazing wealth of information about polymer structure.

Hedhli et al. [27] used 1D-NMR and ^{1}H–^{13}C gHSQC 2D-NMR experiments to confirm the presence of branches in PVDF, and the mechanisms of forming these branches were also proposed. In Wormald's work, the detailed structures in VDF telomeres were fully characterized by ^{1}H–^{19}F HETCOR and COSY, enabling them to calculate the degree of polymerization and the percentage of reverse units [41].

In work by Li et al. [42], a combination of 2D-NMR experiments were used to study the backbone microstructure of P(VDF-*co*-TFE). The carbon sequences were assigned systematically from lower level 3-carbon sequences to higher level 7- and 9-carbon sequences. First, one- and two-bond ^{19}F{^{13}C} gHSQC were used to assign the 3-carbon sequences as in Figure 24.6. Compared with the ^{19}F–^{13}C HETCOR experiment in Katoh's work, the ^{19}F{^{13}C} gHSQC experiment, which is based on single quantum coherence, provided better sensitivity and cleaner spectra. In Figure 24.6b, the resonances between −110 and −115 ppm show correlations with both CH_2 carbon and CF_2 carbons, and therefore they are assigned to 022 sequences, in which 0 indicates CH_2 groups without attached fluorine and 2 indicates CF_2 groups with two attached fluorines. Similarly, the resonances of the fluorines in 020 and 222 sequences are easily assigned since they are only correlated with CH_2 or CF_2 carbons. In addition, the ^{13}C resonances of 3-carbon sequences in the ^{13}C dimension can be assigned based on the one-bond ^{19}F{^{13}C} gHSQC spectrum (Figure 24.6a). Once 3-carbon sequences are assigned, they can be used as the basis to assign 5-carbon sequences with the help of the ^{19}F–^{1}H gHETCOR spectrum and the two-bond ^{19}F{^{13}C} gHSQC spectrum. Finally, the ^{19}F–^{19}F gDQCOSY (Double Quantum filtered COSY) spectrum, provides 3-, 4- and 5-bond F—F correlations, but does not identify which is which. The one- and two-bond ^{19}F{^{13}C} gHSQC spectra are used for that purpose. Referring to structure **1**, if two fluorines are separated by three bonds, the fluorines' resonances will be correlated to each other's carbon resonances; if two fluorines are separated by four bonds, their resonances will be correlated to the resonance of same intervening carbon; if two fluorines are separated by five bonds their resonances will be correlated to the resonances of two different intervening carbons. Thus, combining the data from these two separate types of experiments permitted the identification of central C—F group in 5-, 7-, and 9-carbon sequences along the polymer backbone.

In addition to identifying the microstructure in the backbone, it is also important to know the detailed structure of chain ends and branches. One of the greatest difficulties in characterizing the structures of end-groups is their low concentration. Therefore, sensitivity is a first priority when choosing from a variety of NMR experiments. In Twum's work, ^{19}F T_1 and DOSY experiments were carried out to identify the resonances of chain ends and branch ends in PVDF [26]. Then, combined data from ^{1}H–^{19}F gHETCOR and ^{19}F–^{19}F selective COSY 2D-NMR experiments were used to determine the structures of these end-groups, as well as the penultimate monomers. In ^{1}H–^{19}F gHETCOR and ^{19}F–^{19}F selective COSY experiments, both ^{1}H and ^{19}F nuclei are high γ nuclei with essentially 100% natural abundance. Compared with ^{1}H{^{13}C} and ^{19}F{^{13}C} gHSQC experiments, the sensitivity of these experiments are 100-fold better, as the natural abundance of ^{13}C is only 1.1%. In addition, by selectively exciting the spins of interest in selective COSY experiments, both resolution and sensitivity were enhanced, making it possible to observe the weak correlations from low occurrence structures. Based on these experiments, these authors were able to determine the structures of chain ends, short chain branch ends, and LCBs.

24.3.2 Perfluorinated Polyethers

Perfluoropolyethers (PFPEs) are a family of liquid fluoropolymers that have unique properties such as high thermal stability, low volatility, low *Tg*, chemical inertness, low friction coefficient. PFPEs have the properties of a good lubricant (lower surface energy, low volatility, and good oxidative stability); they are therefore used as lubricants in many applications such as in the aerospace industry [43]. The properties of PFPEs are influenced by their structural characteristics such as MW, MW distribution, type of end-groups, and the structure of the chain [44]. NMR spectroscopy is an ideal technique to study these features of the polymer.

Howell et al. [45] prepared a series of poly(hexafluoropropylene oxide) (PHFPO)-halides (poly-HFPO-CF_2X, where X = I, Br, Cl, and F) and used ^{19}F 1D NMR to determine the chemical shift and also the geminal coupling constant of the CF_2X end-group. In the case of the CF_2I end-group, the expected AB patterns were well resolved and the geminal coupling was measured as 200 Hz. For the chloro analogs, the two expected doublets overlapped, although the geminal coupling (203 Hz) could still be measured. However, for the bromo analogs, a broad peak with unresolved couplings was seen.

The mechanism of anionic ring-opening polymerization of HFPO with alkali metal fluorides as catalyst was studied by Kostjuk et al. [46] using ^{19}F and ^{1}H 1D-NMR, and MALDI-TOF-MS spectroscopy. ^{19}F NMR resonance assignments were made by comparing the changes in signal intensities for two different MWs and also by inspection of the chemical shifts for the different resonances.

Takasaki et al. [47] used ^{19}F{^{13}C) HSQC, ^{19}F{^{13}C} constant time inverse-detected gradient accordion rescaled long-range HMBC (CIGAR-HMBC), and ^{19}F–^{19}F COSY to study two model compounds: perfluoro(2,5,8-trimethyl-3,6,9-trioxadodecanoyl) fluoride (PFTTF) (3) and perfluoro(2-ethoxyethane)sulfonic acid (PFESA) (4). These low MW model compounds have structures (–OCF_2–, $CF(CF_3)O$–, –$OCF_2CF_2SO_3H$) similar to those found in HFPO and Nafion® perfluoro ionomer precursor; resonance assignments from these model compounds were used to provide assignment for the ^{13}C and ^{19}F resonances in Nafion® perfluoro ionomer precursor.

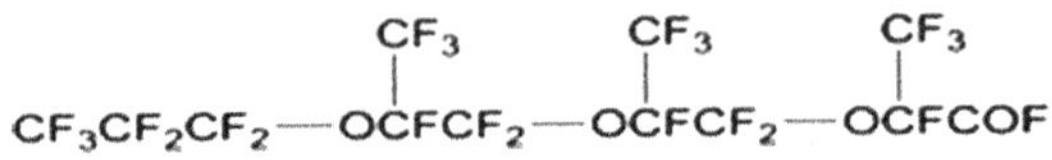

3

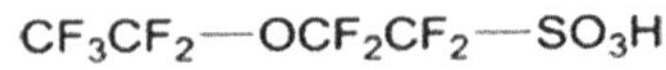

4

The structure of PFTTF (3) was separated into four substructures (5–8), $^{19}F\{^{13}C\}$ HSQC and $^{19}F\{^{13}C\}$ CIGAR-HMBC were used to provide $^{1}J_{CF}$ and $^{2}J_{CF}$ correlations, respectively. Thus ^{13}C and ^{19}F assignments for the resonances in the four substructures were obtained. Using correlations obtained from the COSY spectrum, the connections between the different substructures were established. The COSY spectrum of 3 provided $^{2}J_{FF}$, $^{3}J_{FF}$ and $^{4}J_{FF}$ correlations between fluorine resonances in the different substructure, hence the atomic connectivity of the four substructures through the ether linkages was determined. The different COSY correlations ($^{2}J_{FF}$, $^{3}J_{FF}$ and $^{4}J_{FF}$) were confirmed by using the HSQC and CIGAR-HMBC. Based on the chemical shifts of the assigned ^{19}F and ^{13}C resonances in 3 and 4, and on ^{19}F coupled and decoupled 1D ^{13}C spectra of Nafion® perfluoro ionomer precursor, the ^{19}F and ^{13}C resonances of Nafion® perfluoro ionomer precursor were determined.

5

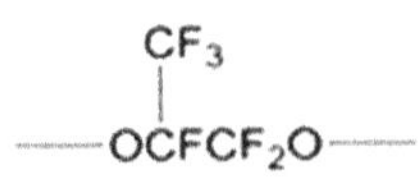

6

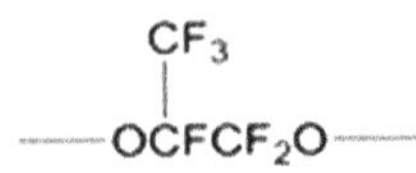

7

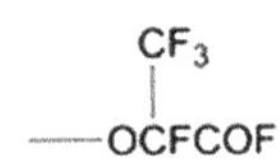

8

Li et al. [5] used a combination of 2D-NMR methods to study a similar perfluoropolyether. They examined carboxyl-terminated low MW oligomers of PHFPO which served as a model for Dupont's Krytox® fluoropolymers. They used spectra from selective one- and two-bond $^{19}F\{^{13}C\}$ HSQC experiments, similar to the data shown in Figure 24.8, to identify the atomic connectivities within each polyether unit. Note that the correlations to ^{19}F resonances A_3 and A_1 occur at slightly different chemical shifts in the one-bond experiment (top spectra) compared to those in the two-bond spectra (bottom spectra). This is due to a one-bond ^{13}C isotope effect which shifts resonances of attached fluorines up-field by approximately 0.1 ppm compared to the resonances of fluorines attached to ^{12}C. The two-bond isotope effect is 10-fold smaller and is too small to notice in these 2D-NMR spectra. Similar sets of correlations were observed for each monomer unit in the structure.

Once the resonances of each unit were identified, the attachments were determined by identifying $^{4}J_{FF}$ correlations in selective COSY spectra like those shown in Figure 24.3. In the f_1-decoupled selective COSY spectra sufficient chemical shift dispersion and spectral resolution was achieved to identify most of the stereo-sequences present in the oligomers. A similar approach enabled them to identify the structures of all the chain ends in these oligomers and to assign their resonances [48].

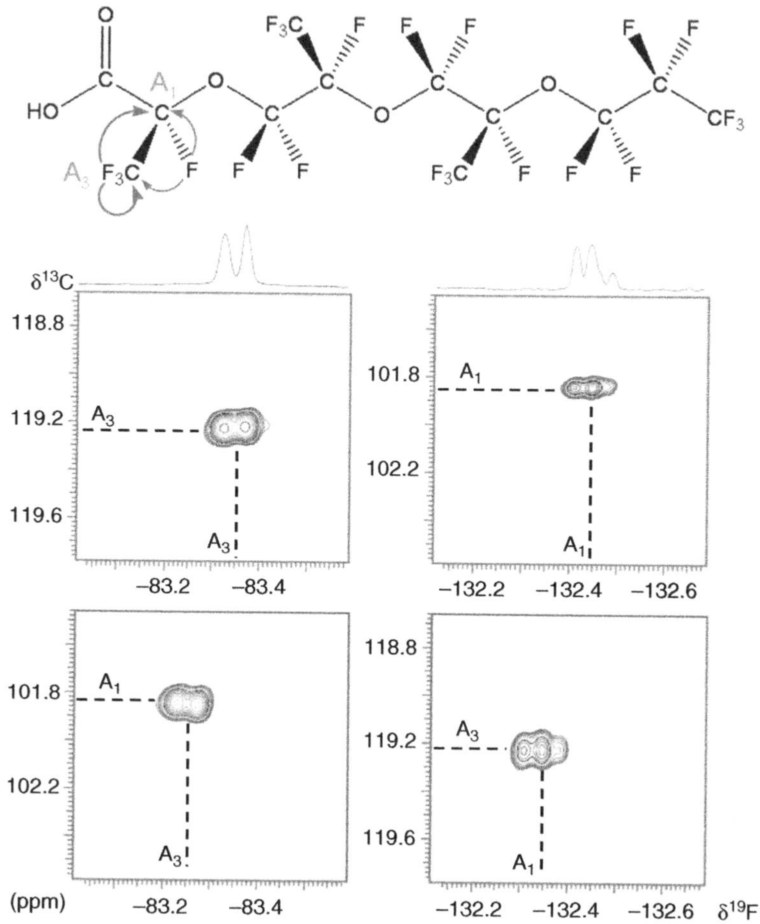

FIGURE 24.8 Selected regions from the $^{19}F\{^{13}C\}$ HSQC spectra of PHFPO oligomer: data in red show $^{1}J_{CF}$ correlations and data in blue show $^{1}J_{CF}$ correlations as illustrated on the structure at the top of the figure. Reproduced with permission from Reference 5.

24.3.3 Poly(chlorotrifluoroethylene)

Poly(chlorotrifluoroethylene) (PCTFE, Kel-F® fluoropolymer) is a thermoplastic semicrystalline polymer with (–CF_2CClF–) repeating units [49]. Typically, PCTFE is synthesized via a free radical polymerization using bulk, suspension, or emulsion techniques [50]. Compared with PTFE (Teflon® fluoropolymer), which has similar chemical properties, there is asymmetry in the CTFE monomer unit due to the presence of the chlorine atom in CFCl groups. Consequently, the NMR features of PCTFE are more complicated due to the possible monomer- and stereo-sequence variations. Since PCTFE has various distinguished properties, including thermal, chemical, and radiation resistance, low vapor permeability, and electrical insulation [51,52], it has been employed in a wide range of applications, such as lubricating oils, industrial grease, cable jacketing, protective coatings for sensitive electronics, other types of coatings, and other industrial uses [53]. Most importantly, PCTFE polymers maintain their excellent performance over an extensive temperature range, from −240 to 200°C [51]. There is an increasing interest in the copolymerization of PCTFE with other monomers, such as vinyl ethers, vinyl acetates, and propylene [54–56]. As a result, the NMR characterization of these polymers is attracting the attention of a growing number of polymer chemists.

Solubility is one of the biggest hurdles to characterizing the structures of PCTFE by solution NMR. In the work by Tiers and Bovey [57], the sample was dissolved in 3,3′-bis-trifluoromethylbiphenyl at 140°C and the ^{19}F 1D-NMR spectrum was collected at the temperature of 150°C. By comparing the ^{19}F spectra of PCTFE and model compounds (*meso* and *racemic* $CF_2ClCFClCF_2CFClCF_2Cl$), they were able to identify the resonances of *meso* and *racemic* CF_2 fluorines. To obtain better resolution, Ishigure et al. [54] utilized ^{19}F homonuclear decoupling experiments to study the stereo-sequences in an alternating copolymer, poly(chlorotrifluoroethylene-alt-propylene) (PCTFE-alt-PR). With the necessity of characterizing fluoropolymers with more complicated structures, the application of multidimensional NMR techniques is becoming essential.

In the work by Carnevale et al. [56], a combination of 2D-NMR experiments were used to study the stereo-sequences in the alternating copolymer of poly(chlorotrifluoroethlene-alt-ethyl vinyl ether) (PCTFE-EVE). Structures **9** and **10** show two possible diastereoisomeric backbone dyad units. Their mirror images are not shown here as the enantiomers cannot be distinguished from each other based on NMR data obtained in achiral media. In each structure, there are two stereogenic centers, and the terms meso and racemic are used loosely here since the substituents on each stereo-center are different. They used a combination of ^1H–^1H COSY (with ^{19}F decoupling), $^1H\{^{19}F\}$ HMQC, ^{19}F J-resolved, and $^1H\{^{13}C\}$ HSQC 2D-NMR experiments to identify two sets of resonances from two distinguishable isomers.

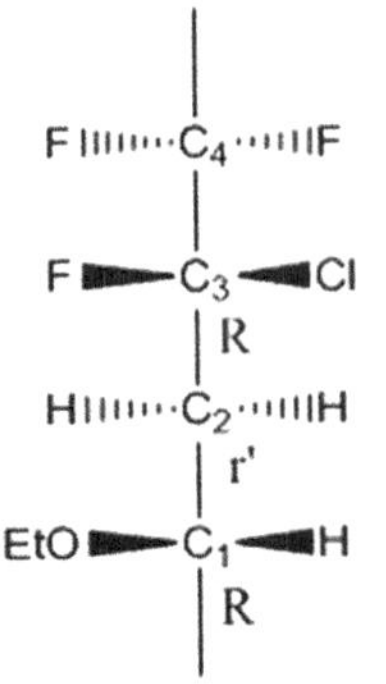

9

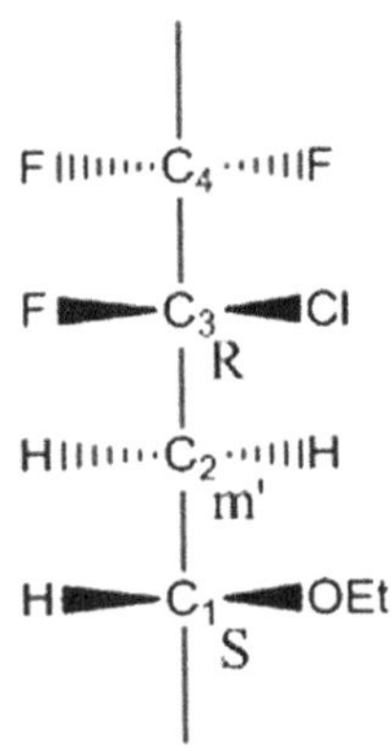

10

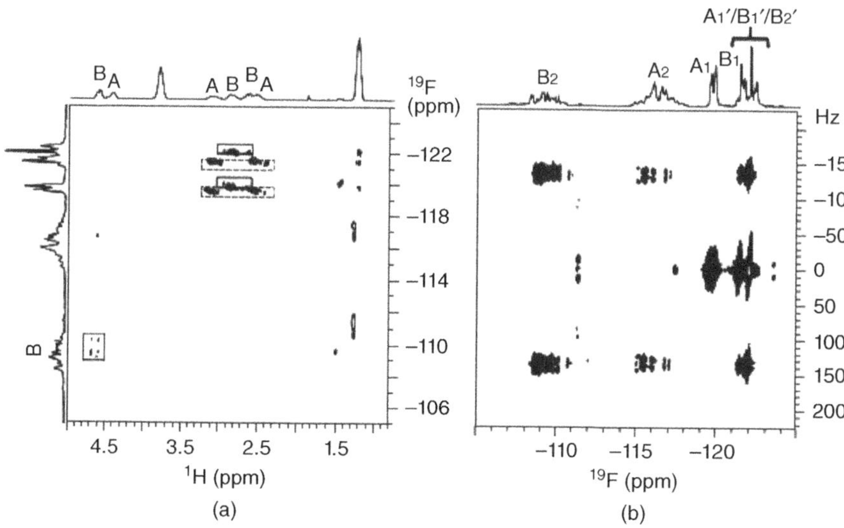

FIGURE 24.9 400 MHz NMR spectra of P(CTFE-*co*-EVE) in $CDCl_3$: (a) ^1H–^{19}F HMQC spectrum and (b) J-resolved ^{19}F NMR spectrum. Reproduced with permission from Reference [56].

Figure 24.9a shows the ^1H–^{19}F HMQC spectrum; the resonances between −119 and −123 ppm were attributed to the fluorine atoms in CFCl groups since they correlate with the ^1H resonances of $C(2)H_2$ groups. In addition, a ^{19}F J-resolved 2D-NMR spectrum was collected to identify the ^{19}F resonances from CF_2 groups due to the difference between $^2J_{FF}$ geminal couplings(≈250 Hz) and $^3J_{FF}$ vicinal couplings (≈10 Hz). Figure 24.2b shows the ^{19}F J-resolved 2D-NMR spectrum; the resonances with large doublet splittings (A2/B2) are attributed to CF_2 groups and the resonances with unresolved couplings near $f_1 = 0$ (A1/B1) are attributed to CF groups. The resonances at −109.4 and −116.8 ppm only show a coupling of about 260–280 Hz, which is consistent with the geminal $^2J_{FF}$ in CF_2 groups. It is noteworthy that the resonances at −124.4 and −122 ppm have both $^2J_{FF}$ and $^3J_{FF}$ coupling.

The data suggest that both CF_2 and CFCl fluorines contribute to the resonances between −121.4 and −122.0 ppm. The $^2J_{FF}$ couplings prove the existence of CF_2 fluorines in this region. The presence of CFCl fluorines can be confirmed by the ^1H–^{19}F HMQC data. Finally, Carnevale et al. used density functional theory (DFT) calculations to predict the relative chemical shifts of **9** and **10**. As a consequence, they were able to assign the two sets of resonances to their respective stereoisomers.

24.3.4 Ionomeric Polymers

Fluorinated ionomers are known for their use as ion-permeable electrolyte membranes most commonly in chloro-alkali electrolysis, electrochemical devices, and H_2/O_2 fuel cells. The polymer membrane facilitates the transport of protons between the electrodes and acts as a barrier for mixing of fuels. In order to achieve optimal performance, this polymer should have high proton conductivity, thermal stability, and prolonged resistivity toward radical deformation over the lifetime of the fuel cell. Common polymers that meet these requirements currently in production are the aliphatic perfluorosulfonic acid (PFSA) membranes, Nafion® perfluoro ionomer (Dupont), Aciplex® (Asahi Chemicals Co.), and Dow® (Dow Chemicals; also known as Hyflon® ion) fluoropolymers (**11**). The Nafion® perfluoropolymer membrane comonomer, otherwise known as perfluoro sulfonylfluoride ethyl propyl vinyl ether (PSEPVE), remains the most studied of the few groups that have examined PFSA polymers with multidimensional NMR. However, very little work has been published on multidimensional solution NMR of these materials due to the difficulties of preparing solutions of these polymers.

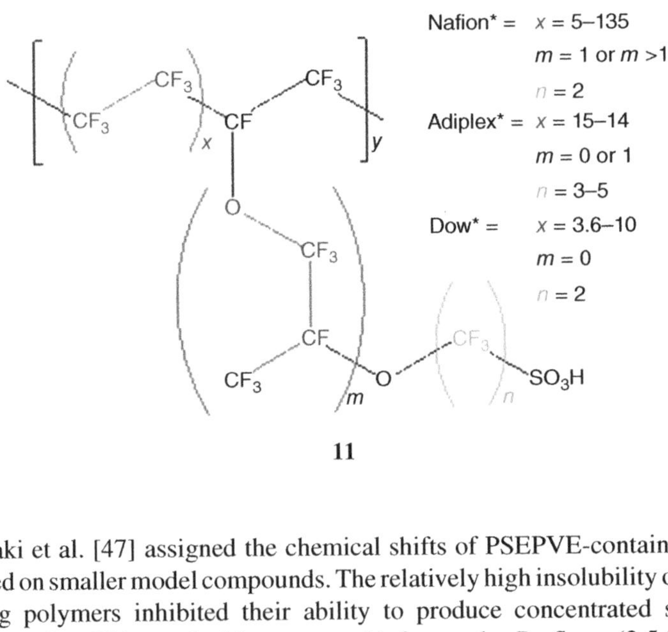

11

Takasaki et al. [47] assigned the chemical shifts of PSEPVE-containing copolymers based on smaller model compounds. The relatively high insolubility of PSEPVE-containing polymers inhibited their ability to produce concentrated samples for 2D-NMR studies. This resulted in spectra with few peaks. Perfluoro(2,5,8-trimethyl-3,6,9-trioxadodecanoyl)fluoride and perfluoro(2-ethoxyethane) sulfonic acid were chosen to represent two halves of PSEPVE. These compounds were then studied with $^{19}F\{^{13}C\}$ HSQC, ^{19}F–^{19}F COSY, and $^{19}F\{^{13}C\}$ CIGAR-HMBC.

Baiagern et al. [8] assigned hydrolyzed poly(TFE-co-PSEPVE) chemical shifts through characterization of the prepolymer, the sulfonyl fluoride form of Nafion® perfluoropolymer membrane (**12**). It was assumed that the chemical shifts of the prepolymer would closely match those of the final polymer since the structures are nearly identical. The perfluorinated Nafion® prepolymer was characterized with ^{19}F–^{19}F selective COSY and $^{19}F\{^{13}C\}$ gradient crisis^2-HSQC (gc2HSQC).

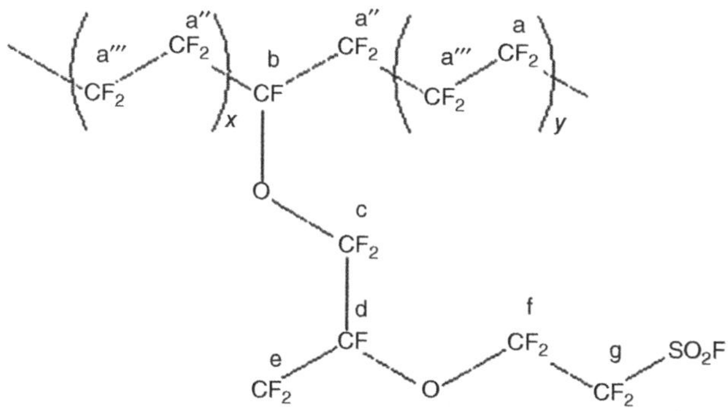

12

The ^{19}F–^{19}F COSY produced spectra with poorly resolved, artifact-laced spectra that made it difficult to identify correlations. Large peaks along the diagonal further complicated assignments, as smaller, nearby cross-peaks were masked. ^{19}F–^{19}F selective COSY using E-BURP excitation pulses resolved these issues by narrowing the excitation window to select areas, resulting in improved digital resolution. An additional 180° inversion pulse was used in the middle of the t_1 evolution period to simplify multiplet patterns through multiple frequency decoupling. A comparison of the COSY and selective COSY spectra is shown in Figure 24.3. Through use of ^{19}F–^{19}F selective COSY, all correlations from the main chain of the perfluorinated Nafion® prepolymer were identified. Most notably, correlations up to $^{6}J_{FF}$ allowed the assignment of two previously unassigned —OCF_2 groups, (c) at −77.5 ppm and (f) at −78.0 and −79.4 ppm, shown in Figure 24.10.

The remaining one- and two-bond correlations were identified through the use of gc2HSQC. The backbone CF_2 groups (a), (a′), and (a″) were uniquely identified through the use of $^{1}J_{CF}$ and $^{2}J_{CF}$ experimental delays. All correlations of the perfluorinated Nafion® prepolymer were identified from $^{1}J_{CF}$ and $^{2}J_{CF}$ couplings. A few cross-peaks were left unidentified in the gc2HSQC spectrum.

24.3.5 Poly(tetrafluoroethylene)

PTFE has often been characterized by solid-state NMR methods; but publications regarding 2D-NMR experiments in solution are rare because the homopolymer lacks solubility in typical solvents. However, characterization of tetrafluoroethylene (TFE) copolymers has been accomplished in solution using both ^{1}H–^{19}F hetero-COSY and its selective version, while employing pulsed field gradients. Macheteau et al. [10] demonstrated the utility of these 2D-NMR methods for obtaining resonance assignments that could not be resolved by ^{1}H–^{19}F HMQC and HMBC experiments alone. Their studies of the fluorinated terpolymer designated as XC2000S that contained VDF, TFE, and allyloxypropandiol (AOPD) (CH_2=$CHCH_2OCH_2CHOHCH_2OH$) in a 65%:35%:13% ratio showed the usefulness of COSY experiments to identify copolymerization sites and to characterize chain-end structures. The authors stated

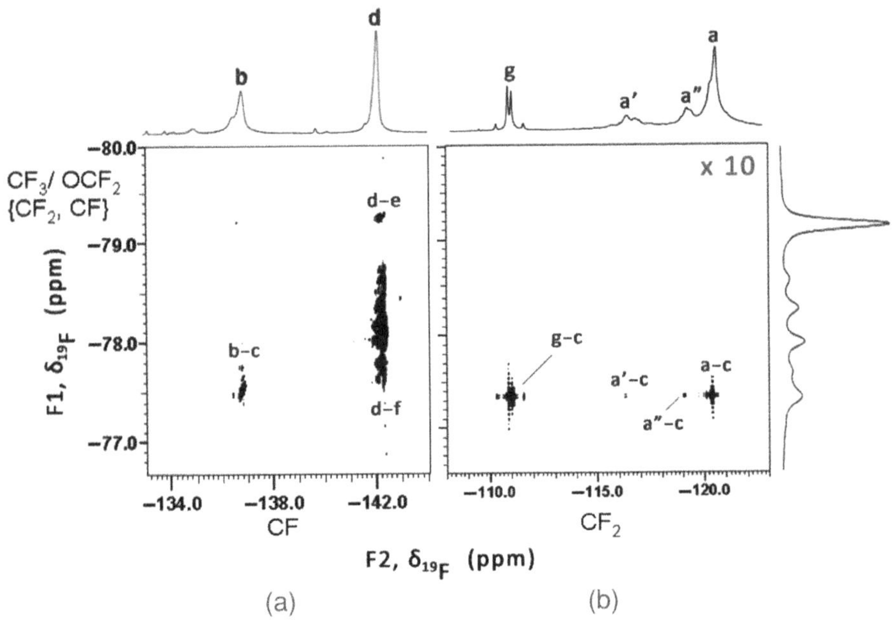

FIGURE 24.10 The ^{19}F–^{19}F selective COSY spectrum of perfluorinated Nafion® prepolymer acquired using selective excitation of the —OCF2, CF3 (f_1) and the (a) CF, and (b) CF2 ($f2$) regions. Reproduced with permission from University of Akron.

that the major advantage of their ^1H–^{19}F hetero-COSY pulse sequence was that there was no delay dependent on *n*-bond coupling, so the experiment was not tuned to reject correlations from all but a narrow band of couplings as is usually the case in HSQC-type experiments Pulsed field gradients were also employed to eliminate undesired signals and to select N- or P-type peaks.

The pulse sequence used for these COSY experiments placed ^{19}F chemical shift (with its large spectral width) in the indirectly detected dimension to minimize the evolution period (t_1). Because fluoropolymers have rapid T_2 relaxation times (~15 ms); the authors also chose ^1H detection to optimize signal detection. Their selective ^1H–^{19}F hetero-COSY experiment utilized the larger spectral width for ^{19}F (~80 ppm) versus ^1H (~8 ppm) to better separate and select the ^{19}F signals. In the selective version, the first 90° pulse in the sequence was replaced by a selective excitation using a shaped pulse.

For the ^1H–^{19}F hetero-COSY spectrum shown in Figure 24.11, strong correlations were observed for small chain-end structures such as CF_2CH_3 (−107.7, 1.8), CH_2CF_2H (−115, 6.4), and CF_2CF_2H (−138.4, 6.92) with the values in parenthesis representing $\delta = {}^{19}F$ and ^1H, respectively. The authors also demonstrated the effectiveness of ^1H–^{19}F selective hetero-COSY techniques for fluoropolymer analysis by selectively exciting a small ^{19}F signal (−114.38 ppm) that was affected by the presence of AOPD. This signal was evident in an overlapping region of the nonselective ^1H–^{19}F hetero-COSY spectrum of the terpolymer (Figure 24.12), but was not

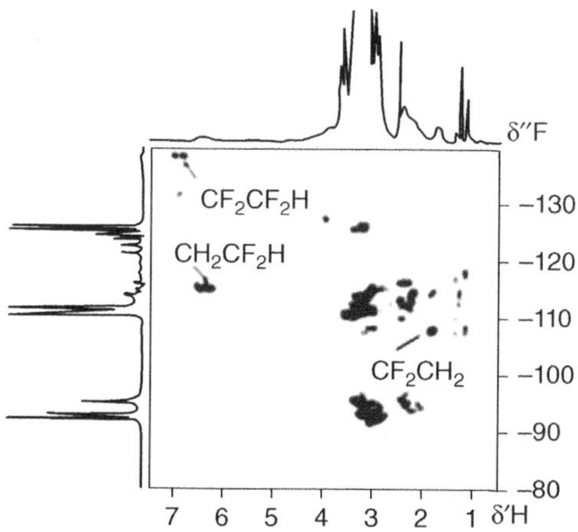

FIGURE 24.11 $^1H-^{19}F$ Hetero-COSY correlation map of the VDF/TFE/AOPD terpolymer in DMSO-d_6, 47°C, with the 1D ^{19}F and ^1H along the side and top. Reproduced with permission from Reference [10].

observed in a $^1H-^{19}F$ hetero-COSY spectrum of VDF/TFE copolymer. The value of combining $^1H-^{19}F$ selective hetero-COSY and its selective versions with other 2D-NMR methods permitted better interpretation of complex fluoropolymer signals.

2D-NMR methods have also been used to determine coupling constants and resonance assignments for an alternating copolymer of TFE and propylene **13**. Research on an Aflas® fluoroelastomer sample dissolved in THF-d_8 was reported by Kuroki

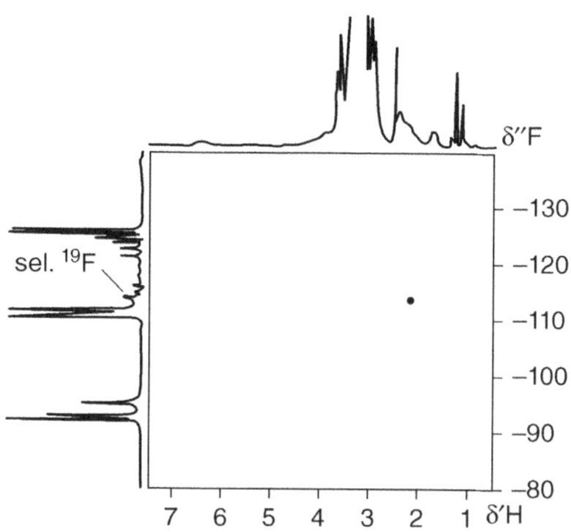

FIGURE 24.12 $^1H-^{19}F$ selective hetero-COSY correlation map of the VDF/TFE/AOPD terpolymer in DMSO-d_6, 47°C, with the selected ^{19}F signal at −114.38 ppm; ^{19}F and ^1H 1D-NMR spectra are placed along the side and top. Reproduced with permission from Reference [10].

[58]. According to the author, analysis of the CH_3 region of the 1D ^{1}H NMR spectrum showed that the sample was highly alternating.

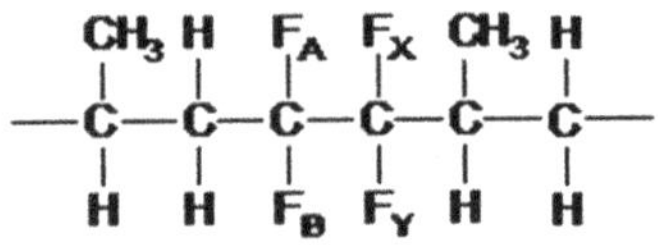

P TFE P

13

Figure 24.13a shows the ^{19}F 1D-NMR spectra of the poly(tetrafluoroethylene-*co*-propylene) poly(TFE-P); the complex peak patterns were mainly attributed to $^2J_{FF}$ couplings which made peak assignment difficult. They obtained a ^{19}F J-resolved (Figure 24.14) and ^{19}F COSY (Figure 24.15) 2D-NMR spectra which helped with the assignments. In the J-resolved 2D-NMR spectrum, if the data are skewed about the $f_1 = 0$ axis by 45° in frequency space, a spectrum with pure ^{19}F chemical shift in the f_2 dimension and pure J_{FF} coupling in the f_1 dimension is obtained. A projection of this 2D-NMR spectrum onto the f_2 axis produces a pure chemical shift spectrum (with homonuclear couplings removed). This "broadband homonuclear decoupled spectrum" is shown across the top of the 2D-NMR plot in Figure 24.14, and is reproduced in Figure 24.13b below the "normal" ^{19}F 1D-NMR spectrum. The ^{19}F broadband homonuclear decoupled spectrum shows eight resolved peaks numbered from 1 to 8 (−110.3(1), −112.6(2), −113.1(3), −115.2(4), −116.0(5),

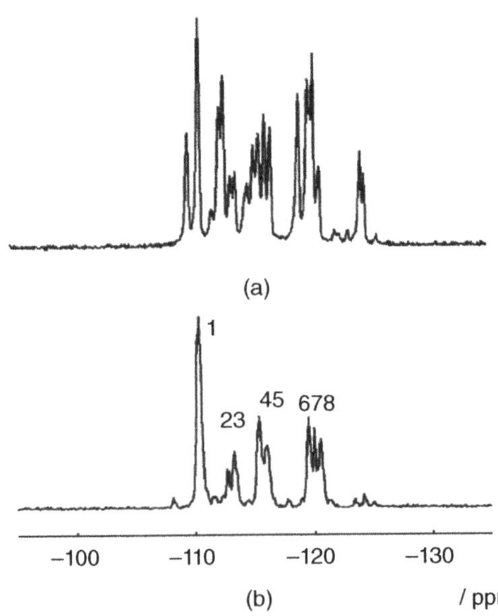

FIGURE 24.13 (a) ^{19}F solution state NMR spectrum of alternating TFE-P copolymer in THF-d_8 solution at room temperature by a simple single pulse experiment and (b) 1D J-resolved ^{19}F solution state NMR spectrum of alternating TFE-P copolymer at room temperature.

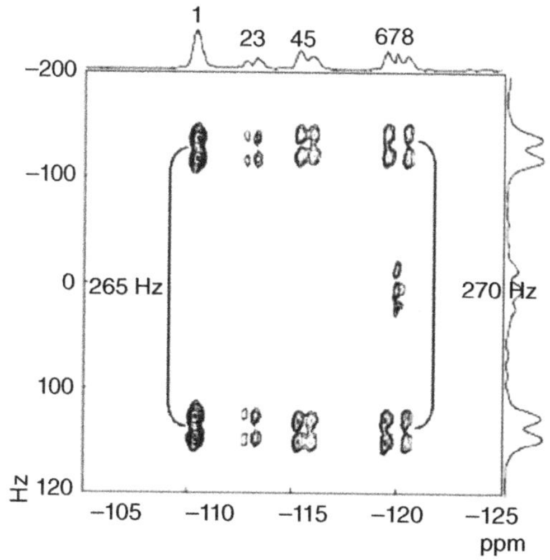

FIGURE 24.14 ^{19}F homonuclear J-resolved 2D-NMR spectrum at room temperature.

−119.5(6), −119.9(7), and −120.4(8) ppm) The author labeled geminal fluorine nuclei on the carbon atom adjacent to CH_2 groups as AB and those on carbon atoms adjacent to $CHCH_3$ groups as XY. According to Figure 24.14, the J splittings were assigned as $^2J_{1,2,3} = 271$ Hz to AB fluorines and $^2J_{4,5,6,8} = 270$ Hz to XY fluorines. Peak 7 had a very small coupling and was assigned to fluorines of TFE—TFE sequences.

To assign AB—XY connections, a ^{19}F–^{19}F COSY experiment was used (Figure 24.15). Large cross-peaks for AB pairs (1 and 2) and (1 and 3) were observed in

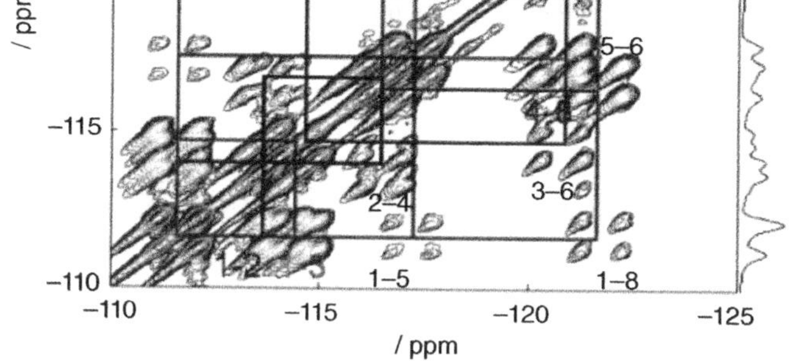

FIGURE 24.15 ^{19}F COSY NMR spectrum of TFE-P alternating copolymer.

addition to those for XY pairs (4 and 8) and (5 and 6) which were attributed to $^2J_{AB}$ and $^2J_{XY}$ couplings. Small cross-peaks arising from ^{3}J-couplings between AB and XY were also observed for (1 and 5), (1 and 8), (2 and 4), and (3 and 6). Two (AB–XY) pairs (1, 2–4, 8) and (1, 3–5, 6) were assigned respectively as *meso* and *racemic* configurations.

24.4 CONCLUSIONS

NMR spectra of fluoropolymers provide abundant structural information from the many NMR interactions presented in these systems. For the many possibilities of performing multidimensional/multipulse/multiple-resonance NMR experiments, we have only presented here the tip of the iceberg. Taking care of all the spin interactions to properly implement multipulse NMR techniques can be extremely challenging yet very necessary. The need for the information provided from advanced NMR experiments makes the efforts worthwhile.

The insolubility of some polymers limits the applications of solution NMR studies, this is also a frequent problem for fluoropolymers. Recent development of ultra-fast magic-angle spinning (MAS, >40 kHz) probes has made it possible to obtain high resolution spectra approaching the quality of solution spectra. At these high spinning speeds, both chemical shift anisotropy and dipolar couplings can be removed from the spectra. Liquids-like spectra from solid samples can be obtained using simple one-pulse experiments and low power decoupling methods [2]. Tremendous advances in the characterization of insoluble fluoropolymers can be expected as many simple solution NMR pulse sequences (with minor modifications) are applied to the study of solid materials [59]. Development of NMR methodology for characterization of the important materials produced in the fluoropolymer area should be a rewarding field for future research.

REFERENCES

1. Silverstein, R. M.; Webster, F. X.; Kiemle, D. J. *Spectrometric Identification of Organic Compounds*, 7th ed.; John Wiley & Sons: Somerset, NJ, 2005.
2. Rinaldi, P. L.; Li, X.; Li, L.; Paudel, L.; Twum, E. B. *Polymer Science: A Comprehensive Reference*, 2nd ed.; Elsevier, 2012; Vol. 2, p 111.
3. Foris, A. *Magn. Reson. Chem.* **2001**, *39*, 386.
4. Li, L.; Twum, E. B.; Li, X.; McCord, E. F.; Fox, P. A.; Lyons, D. F.; Rinaldi P. L. *Macromolecules* **2012**, *45*, 9682.
5. Li, X.; McCord, E. F.; Baiagern, S.; Fox, P. A.; Howell, J. L.; Sahoo, S.; Rinaldi, P. L. *Magn. Reson. Chem.* **2011**, *49*, 413.
6. Battiste, J.; Newmark, R. A. *Prog. Nucl. Magn. Reson. Spectrosc.* **2006**, *48*, 1.
7. Rinaldi, P. L.; Baughman, J.; Li, L.; Paudel, L.; Twum, E. B.; Zhang, B.; Li, X.; McCord, E. F.; Wyzgoski, J. F. *eMagRes.* **2013**, *2*, 109.

8. Baiagern, S. *Multidimensional NMR Studies of the Polymers*; University of Akron, 2008; p 82.
9. Bruch, M. D. *Macromolecules* **1989**, *22*, 151.
10. Macheteau, J. P.; Oulyadi, H.; Hemelryck, B.; Bourdonneau, M.; Davoust, D. *J. Fluor. Chem.* **2000**, *104*, 140.
11. Bodenhausen, G.; Freeman, R. *J. Magn. Reson.* **1977**, *28*, 471.
12. Freeman, R.; Morris, G. A. *J. Chem. Soc., Chem. Commun.* **1978**, *16*, 684.
13. Bax, A.; Morris, G. A. *J. Magn. Reson.* **1981**, *42*, 501.
14. Kessler, H.; Griesinger, C.; Zarbock, J.; Loosli, H. R. *J. Magn. Reson.* **1984**, *57*, 331.
15. Nyberg, N. R.; Duus, J. O.; Sorensen, O. W. *J. Am. Chem. Soc.* **2005**, *127*, 6154.
16. Dolbier, W. R. *Guide to Fluorine NMR for Organic Chemists*; Wiley: Hoboken, NJ, 2009; p 18.
17. Twum, E. B.; McCord, E. F.; Fox, P. A.; Lyons, D. F.; Rinaldi, P. L. *Macromolecules* **2013**, *46*, 4892.
18. Kupce, E.; Freeman, R. *J. Magn. Reson.* **1995**, *A115*, 273.
19. Newmark, R. A.; Webb, R. J. *J. Fluor. Chem.* **2005**, *126*, 355.
20. Cheatham, S.; Groce, J. *J. Fluor. Chem.* **2004**, *125*, 1111.
21. Ribeiro, A. A. *J. Fluor. Chem.* **1997**, *83*, 61.
22. Schorn, C.; Naumann, D.; Scherer, H.; Hahn, J. *J. Fluor. Chem.* **2001**, *107*, 159.
23. Twum, E. B.; Li, X.; McCord, E. F.; Fox, P. A.; Lyons, D. F.; Rinaldi, P. L. Fluorine Containing Polymers: Advances in Fluorine-Containing Polymers. In *ACS Symposium Series*; Smith, D., Iacono, S., Kettwich, C., Boday, D., Eds.; ACS, 2012; Vol. 1106, Chapter 12, p 187.
24. Bax, A.; Summers, M. F. *J. Am. Chem. Soc.* **1986**, *108*, 2094.
25. Summers, M. F.; Marzilli, L. G.; Bax, A. *J. Am. Chem. Soc.* **1986**, *108*, 4285.
26. Twum, E. B.; Gao, C.; Li, X.; McCord, E. F.; Fox, P. A.; Lyons, D. F.; Rinaldi, P. L. *Macromolecules* **2012**, *45*, 5501.
27. Hedhli, L.; Mekhilef, N.; Moyses, S.; Lewis, R. H. *Macromolecules* **2008**, *41*, 2011.
28. Madorskaya, L. Y.; Loginova, N. N.; Panshin, Y. A.; Lobanov, A. M. *Polym. Sci. U.S.S.R.* **1983**, *25*, 2490.
29. Imran-ul-haq, M.; Tiersch, B.; Beuermann, S. *Macromolecules* **2008**, *41*, 7453.
30. Pianca, M.; Barchiesi, E.; Esposto, G.; Radice, S. *J. Fluor. Chem.* **1999**, *95*, 71.
31. Herman, Uno, T.; Kubono, A.; Umemoto, S.; Kikutani, T.; Okui, N. *Polymer* **1997**, *38*, 1677.
32. Johnson, C. S. *Prog. Nucl. Magn. Reson. Spectrosc.* **1999**, *34*, 203.
33. Chen, A.; Wu, D.; Johnson C. S. *J. Am. Chem. Soc.* **1995**, *117*, 7965.
34. Schwalbe, H. *Chem. Bio. Chem.* **2003**, *4*, 135.
35. Li, L.; Zhang, B.; Wyzgoski, F.; Li, X.; McCord, E. F.; Rinaldi, P. L. *ACS Macro Lett.* **2013**, *2*, 141.
36. Li, L.; Rinaldi, P. L. *Macromolecules* **1997**, *30*, 520.
37. Ameduri, B. *Chem. Rev.* **2009**, *109*, 6632.
38. Cais, R. E.; Kometani, J. M. *Anal. Chem. Acta* **1986**, *189*, 101.

39. Murasheva, Y. M.; Shashkov, A. S.; Dontsov, A. A. *Polym. Sci. U.S.S.R.* **1981**, *23*, 711.
40. Katoh, E.; Ogura, K.; Ando, I. *Polym. J.* **1994**, *26*, 1352.
41. Wormald, P.; Ameduri, B.; Harris, R. K.; Hazendonk, P. *Polymer* **2008**, *49*, 3629.
42. Li, L.; Twum, E. B.; Li, X.; McCord, E. F.; Fox, P. A.; Lyons, D. F.; Rinaldi, P. L. *Macromolecules* **2012**, *45*, 9682.
43. Kasai, P. H. *Macromolecules* **1992**, *25*, 6791.
44. Karis, T. E.; Marchon, B.; Hopper, D. A.; Siemens, R. L. *J. Fluor. Chem.* **2002**, *118*, 81.
45. Howell, J. L.; Lu, N.; Perez, E. W.; Friesen, C. M.; Novak, I.; Waterfeld, A.; Thrasher, J. S. *J. Fluor. Chem.* **2004**, *125*, 1513.
46. Kostjuk, S. V.; Ortega, E.; Ganachaud, F.; Ameduri, B.; Boutevin, B. *Macromolecules* **2009**, *42*, 612.
47. Takasaki, M.; Kimura, K.; Kawaguchi, K.; Abe, A.; Katagiri, G. *Macromolecules* **2005**, *38*, 6031.
48. Li, X.; McCord, E. F.; Fox, P. A.; Howell, J. L.; Rinaldi, P. L. *Int. J. Polym. Anal. Ch.* **2012**, *17*, 161.
49. Brown, E. N.; Rae, J. P.; Orler, E. B. *Polymer* **2006**, *47*, 7506.
50. Brasure, D. E.; Ebnesajjad, S. *Encyclopedia of Polymer Science and Engineering*; John Wiley & Sons, Inc.: New York, **1989**; p 468.
51. Millet, G. H.; Kosmala, J. L. *Kirk-Othmer Encyclopedia of Chemical Technology*; John Wiley& Sons, Inc.: New York, 2000.
52. Louis, F.; Ronald, E. *Synthetics, Mineral Oils, and Bio-Based Lubricants*; CRC Press, 2005.
53. Peter, E. N. *Plastics: Thermoplastics, Thermosets, and Elastomers. In Handbook of Materials Selection*; John Wiley & Sons, Inc.: New York, 2007; p 349.
54. Ishigure, K.; Tabata, Y.; Oshima, K. *Macromolecules* **1975**, *8*, 177.
55. Baradie, B.; Shoichet, M. S. *Macromolecules* **2002**, *35*, 3569.
56. Carnevale, D.; Wormald, P.; Amerduri, B.; Tayouo, R.; Ashbrook, S. E. *Macromolecules* **2009**, *42*, 5652.
57. Tiers, G. V. D.; Bovey, F. A. *J. Polym. Sci. Pol. Chem.* **1963**, *1*, 833.
58. Kuroki, S. *Polym. J.* **2009**, *41*, 449.
59. Zhou, D. H. Fast Magic Angle Spinning for Protein Solid-State NMR Spectroscopy. In *Solid-State NMR Studies of Biopolymers*; McDermott, A. E., Polenova, T., Eds.; John Wiley & Sons, Ltd., 2010; Chapter 18.

25

MELT PROCESSIBLE PERFLUOROPLASTICS FOR DEMANDING APPLICATIONS

Paul Brothers, Gregory Chapman, Kimberly Farnsworth, and Richard Morgan

E.I. du Pont de Nemours and Co.

25.1 INTRODUCTION AND OVERVIEW

Polymeric plastics, which are made by the free radical polymerization of tetrafluoroethylene (TFE) and, optionally, other perfluorinated monomers, can be found in some of the most demanding applications. By comparison to their hydrocarbon counterparts, the density of perfluoroplastics is significantly higher and the cost per unit weight is generally greater. In addition, equipment used in the melt processing of perfluoroplastics generally requires more exotic materials of construction. As a result, fabricated items incorporating perfluoroplastics are produced with significant financial investment. It is, therefore, not surprising that these polymers are often the polymers of last resort. The attendant value of perfluoroplastics is found in their unique combination of properties. This chapter highlights a few, among thousands, of the applications in which melt processible perfluoroplastics have found utility.

Perfluoroplastics have a multitude of useful properties. Any one of these properties may possibly be found in certain hydrocarbon-containing polymers, metals, or other materials. However, it is in the applications that require a combination of two or more of these properties where the perfluoroplastic finds unique utility. The unique properties and performance of perfluoroplastics are highly related to their structure.

Handbook of Fluoropolymer Science and Technology, First Edition. Edited by Dennis W. Smith Jr., Scott T. Iacono, and Suresh S. Iyer.

25.1.1 Chemical Resistance

The size of the fluorine atom is ideal for producing a very uniform and almost complete cover over the polymer backbone. In addition, the extremely high electronegativity of fluorine results in exceptional bond strength with the carbon–fluorine bond being one of the strongest in organic chemistry [1]. The net result is that carbon–carbon bonds of the perfluoroplastic backbone are protected by the surrounding sheath of fluorine which is held tightly by the extremely strong carbon–fluorine bonds. It is the strength of the carbon–fluorine bond, the degree of coverage and the lack of molecular interactions that give rise to the excellent chemical resistance of perfluoroplastics.

25.1.2 Electrical Properties

Perfluoroplastics are nonpolar with a symmetrical arrangement that centers the charge balance in the middle of the polymer chain. The lack of polarity leads to exceptional electrical performance with a low dielectric constant and dissipation factor. Perfluoroplastics further benefit from the rigid, linear chain which maintains consistent dielectric properties across a wide temperature range [2].

25.1.3 Low Coefficient of Friction

The van der Waals interactions in perfluoroplastics are extremely low. As a result, polytetrafluoroethylene (PTFE) has the lowest coefficient of friction of any known polymer. This property is due to the chemistry and structure described above which also give rise to the excellent electrical properties, thermal stability and chemical inertness.

25.1.4 High Temperature Stability

The high degree of crystallinity, high molecular weight (MW) and rigid helical chain structure of the PTFE combine to produce a very high melting point. The "as polymerized", high MW PTFE has a melting point on first heating, that is in excess of 335°C, and a melting point of 327°C for subsequent melts. The polymer has an upper continuous use temperature of 260°C. The inclusion of perfluorinated comonomers with side chains in the polymer backbone reduces the crystallinity, reduces the melting point, and creates entanglements that allow good physical properties with an MW that is one to two orders of magnitude lower than PTFE and a viscosity that is six to seven orders of magnitude lower than PTFE. Perfluoroalkoxy (PFA) resins contain approximately 1.5 mole percent perfluoropropyl vinyl ether (PPVE) or approximately 3.5 mole percent perfluoroethyl vinyl ether (PEVE) as the comonomer. The comonomer reduces the melting point to 285–315°C but maintains the PTFE continuous use temperature of 260°C [3]. Fluorinated ethylene propylene (FEP) resins utilize hexafluoropropylene (HFP) comonomer to introduce a bulky perfluoromethyl group. If used alone to produce FEP, HFP must be incorporated at about eight mole percent [4]. This reduces the melting point to around 260°C and allows the retention of most of the properties of PTFE except for the upper continuous use temperature which drops to 200°C due to the greater steric effect at the branch point on the tertiary carbon.

25.1.5 Low Temperature Performance

Perfluorinated polymers exhibit high strength, toughness, and self-lubrication even at low temperatures, thereby allowing them to be routinely used at cryogenic temperatures. PTFE resins are useful from −268°C and are highly flexible from −79°C [5]. PFA performs well in cryogenic applications at −196°C [6] and FEP maintains reasonable tensile properties down to −251°C [7].

25.1.6 Long-Term Environmental Stability

In addition to the thermal stability of fluoroplastics that insure minimal changes in the polymer after long-term exposure at extreme temperatures, the strength of the carbon–fluorine bond also contributes to excellent resistance to sunlight (i.e., ultraviolet (UV) light), salt spray, and other components of weathering and environmental exposure. Perfluoroplastics are comparatively transparent to UV light. Many other polymers require stabilizers, plasticizer, or fillers to achieve the desired level of processing or physical properties. Examples are plasticizers in polyvinyl chloride (PVC) or UV stabilizers in polyethylene (PE). For most applications, perfluoroplastics are inherently usable without additives.

25.1.7 Fire Performance

Flammability, or combustion, can be viewed as the exothermic reaction of a fuel with oxygen. Hydrocarbon compounds react exothermically with oxygen to form carbon dioxide and water by the reaction:

$$\sim CH_2 \sim +3/2\,O_2 \rightarrow CO_2 + H_2O$$

The enthalpy of reaction (ΔH_{Rxn}) is −146.9 kcal/g mol indicating that the oxidation of a hydrocarbon is thermodynamically a highly exothermic process that releases significant energy. Since fluorine is more electronegative than oxygen, it is a much stronger oxidizer than oxygen itself; therefore, the oxidation of a fluorine bond is actually an endothermic (energy-consuming) reaction. Perfluorinated polymers contain numerous fluorine bonds, so they do not easily combust. In fact, the complete oxidation of a fluorocarbon backbone is +8.4 kcal/g mol, which thermodynamically indicates that this is a slightly endothermic process:

$$\sim CF_2 \sim +3/2\,O_2 \rightarrow CO_2 + OF_2$$

The oxidation of the C–C bonds in a fluoropolymer to form carbonyl fluoride is a more realistic reaction in a fire scenario:

$$\sim CF_2 \sim +1/2\,O_2 \rightarrow COF_2$$

Although this is exothermic with an enthalpy of reaction of −54.7 kcal/g mol, it still releases only a little more than one-third of the energy that occurs from the oxidation of a hydrocarbon [8].

25.1.8 High Purity

Since perfluoroplastics may be used without additives, which may degrade and/or leach out of the polymer with time, they are useful in high purity applications. For polymers used in especially high purity applications, particular attention is brought to the materials of construction of equipment used in the manufacture of the polymer, the quality of starting ingredients, and the isolation and packaging of the "as polymerized" polymer. For ultrahigh purity applications, the perfluoroplastic may be treated with elemental fluorine to eliminate end groups and other moieties which may degrade in the presence of harsh chemicals or on exposure to high temperatures or have the potential for leaching and release of cations and anions [9]. These moieties, although extremely small in number, may result from the organic and inorganic initiators used for the free radical polymerization, the chain transfer agents used for control of polymer MW or other components used in the manufacturing process.

25.2 PERFLUOROPLASTICS IN WIRE AND CABLE APPLICATIONS

From some of the original applications of PTFE in military and aerospace applications to new grades of FEP that provide the ultimate combination of electrical performance and fire safety for today's ever more demanding data communications markets, fluoroplastics have historically had a strong connection with the wire and cable industry. Their unique combination of exceptional properties allow them to provide superior performance in many demanding applications, including wire and cable insulation for communications, military, aerospace, manufacturing and many other industries. A few of the most notable properties that are critical for success in these demanding applications include excellent thermal stability, exceptional chemical resistance, low surface energy, wide operating range, and high purity with the ability to use in virgin form without fillers and additives. In addition, the requirement for outstanding dielectric properties across a wide temperature range is paramount. The ever-increasing demand for bandwidth has pushed requirements towards insulations with low dielectric constant and dissipation factor. These demands are being addressed through focus on the base polymer, improved stability in the wire coating process, and renewed emphasis on foamed insulating materials. Improvements in the dielectric performance of the base polymer are normally achieved by reducing or eliminating polarity in the polymer chain, which pushes towards the limit of PE or PTFE. For instance, the use of ethylene chlorotrifluoroethylene (ECTFE) as a plenum primary coating has rapidly diminished as electrical requirements have tightened. Finally, as will be discussed in the next section, superior fire performance is the key in one of the largest wire and cable applications. Perfluorinated polymers have inherently outstanding flame performance as shown by very low fuel loads, a limiting oxygen index (LOI) of >95%, and lack of sustained combustion once the ignition source is removed. Although the performance decreases significantly with the addition of other atoms, even partially fluorinated polymers are much better than their hydrocarbon counterparts.

25.2.1 Plenum Cables

Fluoroplastics have been used as wire coatings in military and aerospace applications since their discovery. They have also found their way into wire and cable applications that required extreme performance based on thermal extremes (cryogenic and high temperature applications), chemical resistance, etc., but the major market expansion in fluoroplastic use in the wire and cable industry occurred in North America when the National Electric Code (NEC) allowed polymers with low smoke-generation and flame-spread to be installed in building plenums without metal conduits.

A plenum is an open space within a building that is used as part of a Heating, Ventilation, and Air Conditioning (HVAC) system. This includes both the ducts used to supply the air and the generally open air returns that exist above a suspended ceiling or under a raised floor. This open space is a convenient place to run cables since it is a continuous area that is not visible to building occupants. However, the lack of fire blocks and free air flow in the plenum allows the potential for rapid fire spread between rooms. This has resulted in the development of special fire code ratings designed to insure that the cables in the plenum do not easily burn if exposed to the heat from a fire, minimize flame spread, and limit the amount of smoke generation. In the United States, plastics used in the construction of plenum cable are regulated under the National Fire Protection Association (NFPA) standards 90A (Standard for the Installation of Air Conditioning and Ventilating Systems) and 262 (Standard Method of Test for Flame Travel and Smoke of Wires and Cables for Use in Air Handling Spaces).

The plenum application is dominated by FEP due to its combination of outstanding dielectric properties, melt processibility and the low flammability of perfluoroplastics. The electrical performance is critical to insure high speed data transmission rates at the ever-increasing bandwidth required by the modern information age. The low flammability is essential to meet the flame spread and smoke generation requirements as codified in NFPA 262 and Underwriters Laboratory UL-910 Steiner Tunnel Test. In the UL-910 test, 24–foot-long cables are supported by a one foot wide cable rack. The rack is filled with a single layer of cables. The cables are ignited by an 88 KW (300,000 BTU/h) methane flame. Flame spread is aided by a 240 ft/min draft. During the 20-minute test, flame spread is observed through small windows spaced one foot apart. Smoke is measured by a photocell installed in the exhaust duct. To qualify, cables must have a flame spread of less than 5 feet beyond the end of the $4\frac{1}{2}$-foot ignition flame, a peak optical density of 0.5 maximum (33% light transmission), and a maximum average optical density of 0.15 (70% light transmission).

Consistent melt processibility is required to economically manufacture primaries to the tight tolerances required to meet the stringent electrical requirements of the codes and standards organizations such as the Institute of Electrical and Electronics

Engineers (IEEE) P802.3ba 40Gb/s and 100Gb/s Ethernet Task Force., the Telecommunications Industry Association (TIA) Category 6 and Category 6a or Augmented Category 6 (ANSI/TIA/EIA-568-B.2-10), or the international standard ISO/IEC 11801 Class F and Class F_A.

The strong demand for plenum cables led to the need for higher melt extrusion rates, which were accomplished through the incorporation of perfluoroalkyl vinyl ether (PAVE) termonomers that allow good stress crack resistance at much higher melt flow rate [10]. This provides for the opportunity to improve line speed, yield, and product uniformity during the manufacturing process [11].

25.2.2 Limited Combustible Cables

The plenum concept has been so successful that many plenums contain abandoned cables where each cable can individually pass the smoke and flame spread requirements, but when added together, they can have a measurable fuel value. The next logical evolution in fire safety is to limit the overall fuel load in the plenum, which has led to the definition of a limited combustible cable [12]. These cables have a maximum potential heat value of 3500 BTU/lb (NFPA 259) in addition to the smoke and flame spread requirements common to other plenum cables. As described earlier, the best way to minimize fuel load with a polymer is to use a perfluorinated compound, and in this case, both primary insulation and jacket benefit from this attribute [13].

25.2.3 High Performance Data Cables

The uniformity of the dielectric material and consistency in spacing of the conductors in the final construction becomes increasingly evident as the constructions begin to push the dielectric limitations of the materials. These efforts focus on the engineering design of the cable and the quality systems during manufacturing.

Foamed fluoroplastics have been used in coaxial cables since the 1960s, but they are receiving increased interest for twisted-pair applications since the dielectric performance of all solids can be improved by the uniform addition of voids. Current efforts focus on smaller and more uniform voids that can be incorporated into thin-walled products such as micro coaxial cables and twisted pair cables. The use of advanced foamed perfluoroplastic resins provides the cable designer with options for improving cable performance and/or reducing cable weight [14, 15].

25.2.4 Military and Aerospace

Military and aerospace applications often require the highest level of reliability for systems where failure can have catastrophic consequences. Fluoroplastics fill critical needs for insulation in terrestrial, aviation, and space applications. Many of these applications are defined by Military Specifications (MIL SPEC) with two of the most important being MIL-C-17 for critical coaxial cable applications and MIL-W-22759 for fluoropolymer-insulated single conductor electrical wires primarily used as aircraft wire.

High MW PTFE tends to be favored in applications where the potential for extreme temperatures, harsh chemicals, and other extremely aggressive environments exist. This is particularly true when combined with high frequency requirements such as coaxial cable applications where the excellent dielectric properties are also required. Unfortunately, such PTFE cannot be processed by conventional melt processing techniques. With the recent developments in foamed perfluoroplastic products, economical cables with properties very near to those of expanded PTFE are possible.

The melt processible fluoroplastics are often desired due to the cost benefits of melt extrusion over paste extrusion. FEP, PFA and specially formulated melt processible perfluoroplastics are used in many of these applications; however, in some of these applications, perfluoroplastics may not be the ideal choice. In cases where high cut-through resistance and better tensile properties are required, it is often desirable to employ a partially fluorinated polymer such as ETFE (ethylene-tetrafluoroethylene). ETFE is the copolymer of ethylene and TFE [16] that normally includes an additional termonomer to increase the flexibility required in commercial applications [17]. The increased physical and electronic interactions of the ETFE polymer chain are responsible for the comparatively enhanced physical properties. Additionally, the partially fluorinated polymers may be cross-linked to further improve physical properties. These benefits, however, are obtained at the expense of the unique properties of perfluoroplastics discussed in the Introduction and Overview.

25.2.5 Electric Heat Tracing Cable

The high continuous use temperature, chemical resistance, and good weathering characteristics of fluoroplastics make them a logical choice for electric heat tracing with the grade determined by the temperature rating of the heat tracing and the expected environment. This is one of the wire and cable applications where the importance of temperature ratings is sufficient to justify the higher price of PFA. Fluoroplastics are commonly used as the jacketing material for all types of heat tracing. Semiconductive PFA is also used in self-regulating cable where the carbon-filled PFA is extruded between two parallel wires. The carbon particles provide a resistive pathway for current to flow between the two wires and generate heat. As the fluoroplastic element heats up, it expands and increases the distance between carbon particles, thereby reducing the current and controlling the temperature.

25.2.6 Other Wire and Cable Applications

Fluoroplastics are routinely used as wire coatings in cryogenic and liquid oxygen service due to their low temperature flexibility and good dielectric properties. Extremely thin layers of insulations can be applied on the conductor by coating the fluoroplastic directly from the dispersion form. This is especially useful for high temperature magnet wire where the product benefits from the electrical insulating properties and high end use temperature rating.

Excellent chemical resistance combined with the broad temperature rating makes perfluoroplastics a natural fit for wiring in the chemical process industry (CPI). The grade tends to be based on the required temperature rating with PFA utilized at the highest temperatures and FEP used in slightly lower temperature applications.

Fluoroplastics are also used in a number of other niche wire applications such as automotive, appliance wire, and connectors. The environment in the engines and engine compartments of automobiles are constantly seeing increasing temperature and harsher chemicals that require the performance of fluoroplastics. Appliance and electronic hookup wire benefits from both the high temperature performance that allows use with high soldering temperatures and the good dielectric properties. PTFE is often used in connectors for high frequency applications.

25.2.7 Compounded Products

Many polymers are compounded to impart other properties and/or to lower cost. Some polymers, like PVC, require plasticizers or additives for virtually all applications. The quantity and variety of industrial applications of fluoroplastics that use neat resin without any additives is actually a tribute to the robustness of their inherent properties, but it should not conceal the potential for compounded products.

The wire and cable industry has used fluoroproducts compounded with pigments and nucleants for decades. Semiconductive grades were introduced in the 1990s, but they tend to be limited to specialty applications like self-regulating heater cables described above. New areas of focus include jacket compounds for limited combustible cables and nanocomposites.

Compounding can assist with the limited combustible requirements mentioned above. Although jackets made with perfluoroplastic have been around since the advent of the plenum specifications, their market share has declined due to low productivity and relatively high cost. New products that combine compounding technology with the inherently low fuel load of perfluorinated polymers promise to open new windows for jacket compounds that possess the low flammability of perfluorinated resins combined with improved processibility.

The incorporation of nanoparticles into polymers has generated considerable excitement and activity since the interfacial effects at that scale can impart significant physical property improvements [18]. The low surface energy of perfluoroplastics increases the difficulty of producing fully exfoliated nanocomposites. The incorporation is easier with partially fluorinated compounds such as polyvinylidene fluoride (PVDF) or fluoroelastomers [19], but even PTFE compounds have been produced for wire and cable applications [20].

25.2.8 Life Cycle

The lack of additives could also be advantageous throughout the cable's entire life cycle. Many of the materials currently used in wire and cable tend to be highly filled, often with components that degrade or leach out. This can lead to reduction of physical, electrical, or flammability properties over time, especially for compounds that require plasticizers. The fillers can also make the materials difficult or impossible to recycle. Perfluorinated polymers do not suffer from aging or weathering, so cables could theoretically be used for generations without degradation. If the cables become obsolete due to technological advancements, the fluorinated copolymers are conducive to easy recycling. The outstanding thermal stability permits multiple extrusions without degradation and the high density of the polymer aids in separation from other polymers.

25.3 PERFLUOROPLASTICS IN SEMICONDUCTOR INDUSTRY APPLICATIONS

High purity PFA fluorocarbon resins have been used by the semiconductor industry for more than 20 years. These resins are uniquely designed to meet the demanding needs in these high performance applications [21, 22]. In addition to good manufacturing practice, the chemical composition and structure of PFA are responsible for these unique and desirable properties that combine to meet the numerous critical parameters for semiconductor industry applications including outstanding chemical resistance, permeation resistance, excellent thermal stability, extremely high purity (as defined by the Semiconductor Equipment and Materials Industry Standard SEMI F-57: Specification for Polymer Components used in Ultrapure Water and Liquid Chemical Distribution Systems), surface smoothness of finished parts, and ease of processibility.

The properties and processibility of PFA have allowed for the development of a wide variety of components that are used in critical wet chemical applications in the semiconductor industry including high purity bulk chemical systems, wet etching systems, stripping systems, cleaning systems, chemical mechanical planarization (CMP) system components, analytical equipment, and high purity chemical manufacturing.

25.3.1 High Purity Bulk Chemical Systems

Bulk chemical systems involve the handling and transport of large amounts of chemicals throughout a production facility. The systems often include automated equipment for diluting and dispensing chemicals. PFA is widely used for the piping, tanks, lining of fluid handling components such as pumps, pressure regulators and valves, and filter housings that compose the bulk chemical handling systems. The use of PFA also protects the purity of the chemicals that are being handled.

25.3.2 Wet Etching, Stripping, and Cleaning Systems

In wet processing, silicon wafers are cleaned and etched, and then cleaned of photoresist and residuals from the etching process. Components made from PFA are used in tools and fluid delivery systems to help prevent contamination. System components which are commonly made from PFA or lined with PFA include piping, valves, fitting, pumps, baths, sinks, tanks, dip tubes for sensors, filter housings, and wafer carriers.

25.3.3 Chemical Mechanical Planarization

Chemical mechanical planarization is used to level the wafer surfaces before vapor deposition and subsequent processing steps. The working fluid is a slurry of fine abrasive particles, chemicals, and water. Components made with PFA help guard against contamination of process fluids and slurries that come in contact with the wafers. For slurries and chemically aggressive deionized water used for dilution and washing, piping of highly pure PFA help prevent introduction of contaminants.

25.3.4 Analytical

Analyses of process fluid composition, determination of particulate and bacterial levels and other laboratory procedures are the critical steps in semiconductor manufacture. Fluid chemistries must be analyzed and kept on specification to maintain yields. For accuracy and repeatability, labware and equipment must not react with or contaminate the samples being studied. Reagents for analytical work as well as the process fluids are held in containers of PFA for protection against contamination. Because PFA is unaffected by virtually all chemicals and solvents, containers may be reused multiple times. Apparatus for wet analytical analyses may be interconnected with tubing made from PFA [23]. Laboratory apparatus including beakers and other lab equipment may also be made from PFA to help maintain chemical purity and prevent inaccurate analyses.

25.3.5 High Purity Chemical Manufacturing

High purity chemicals and solvents used in the semiconductor industry are manufactured using equipment that resists corrosion and leaching. It is often more cost effective to make products that are inherently pure rather than trying to achieve high purity through refining steps. To guard purity, the CPI uses corrosion-resistant fluid-handling components when producing chemicals for use in the semiconductor industry. PFA vessels and linings may be used in process vessels and columns for reactions, distillations, absorptions, and other processes. PFA-lined pipes and fittings are used for transporting ultrapure fluids and PFA tubing is used when a smaller diameter is sufficient. A wide range of lined valves, pumps storage containers, and other fluid handling components are available with PFA linings. Containers for shipping high purity chemicals are also lined with PFA and because the linings are unaffected by contact with virtually all chemicals, they can be reused.

25.3.6 Processing/Fabrication

The melt processibility of PFA allows for a wide variety of conventional melt-processing techniques to be used with this resin. Common processing techniques used to create parts for use in the semiconductor industry include [24, 25]

- injection molding;
- extrusion;
- rotomolding/rotolining; and
- transfer molding.

25.3.6.1 ***Injection Molding*** Injection molding involves slow injection of melted resin into mold cavities. The molding equipment must be constructed of corrosion-resistant metal in order to minimize corrosion and contamination. Parts commonly made via this technique include fittings, valves, and wafer carriers.

25.3.6.2 ***Extrusion*** Standard extrusion technology may be used to produce tubing from PFA resins. Special corrosion resistant metal alloys must be used for the equipment in order to prevent metal contamination and equipment damage. The resin is melted in the barrel of the extruder, pushed through a heated die of specific dimensions, and then quenched. PFA tubing and pipe linings are made via extrusion.

25.3.6.3 ***Rotomolding/Rotolining*** Rotomolding/Rotolining is particularly useful for making free-standing, hollow parts and seamless linings of metal vessels. This is a slow, nonshear processing technique in which the PFA resin (typically in small bead form) is heated to 350–360°C and then simultaneously rotated on two perpendicular axes inside of a mold.

25.3.6.4 ***Transfer Molding*** Flow meter linings, valve and fitting liners, and tees can be made via a transfer molding process with the valve or fitting serving as the mold. Melted PFA resin is forced into the fitting at a temperature above its melt point. After the resin transfer is complete, the fitting is cooled under pressure.

25.4 APPLICATIONS FOR LOW MW PTFE (MICROPOWDER OR FLUOROADDITIVE)

A very high MW, in the millions, is required to provide good mechanical properties to PTFE. However, there are some applications of PTFE that do not require the mechanical properties borne of high MW. High-MW PTFE has a tendency to form a fibrous network when handled, thereby making it difficult to use as an additive to other materials. The low MW PTFE counterpart does not exhibit this tendency and is employed as an additive to nonfluorinated materials to provide them with some of the properties of perfluorinated plastics. These low MW PTFE products are usually referred to as "fluoroadditives" or "micropowders". All major fluoropolymer manufacturers produce "micropowders". In addition, other companies purchase scrap and off-specification high MW PTFE and convert it into micropowders. Although these applications are not well known, millions of pounds of PTFE micropowders are commercially sold each year. The low surface energy and good lubricity of PTFE are retained even with a low MW. By the addition of PTFE micropowders, these properties can be imparted to a variety of other materials.

The production and use of low MW PTFE as an additive to other materials was first proposed in the early 1960s [26]. Addition of low MW PTFE to thicken grease was proposed in 1970 [27] and use as an additive to other materials grew through the late 1970s and 1980s. It was found by the late 1970s that low MW PTFE "wax" provided better properties to coatings and inks at lower addition levels than could be accomplished with corresponding hydrocarbon wax. High-MW PTFE was used as an additive to plastics by the early 1980s but most such addition has now been converted to low MW PTFE.

25.4.1 Micropowder Production

Polymer chemistry and architecture are key to the properties for which PTFE micropowders are used and the production process dictates much of the architecture. The PTFE micropowders are produced by three general processes: (1) direct polymerization of low MW PTFE [28–32], (2) irradiation and grinding of high MW PTFE [33–35], and (3) thermal degradation of high MW PTFE [36, 37].

Polymerization of PTFE may be carried out by either suspension or dispersion polymerization. There are major differences between suspension and dispersion polymerization. Suspension polymerization employs only a very small amount of surfactant, if any, and utilizes high agitation rates. During the polymerization, polymer separates from the water medium as stringy particles of several hundred microns. This polymer, normally referred to as PTFE granular resin, is then put through a cutter to reduce size before further processing. The cut granular polymer has a surface area of 1–4 m^2/g. By contrast, dispersion polymerization employs enough surfactant that the polymerization produces stable submicron PTFE particles. The polymer is removed from the reactor as an aqueous emulsion. Separation of polymer from the emulsion is a separate process step, usually called coagulation. The polymer particles from the coagulation of PTFE dispersion are agglomerates of sub-micron (50–250 nm) primary particles and can have a surface area of 8–30 m^2/g. The most typical dispersion-type PTFE has a primary particle size of 200–250 nm and a surface area of 8–12 m^2/g.

Particle size and surface area are important properties for many micropowder applications and each micropowder application has its own requirements. Due to their small primary particle size, the powders from dispersion polymerization are the more easily suspended in low viscosity media. However, some applications require a larger particle size. Particulate additives can increase the viscosity of fluids or even polymer melts and the effect is a function of the additive surface area. This can be a positive or negative factor depending on the application.

Suspension and dispersion polymerization processes for production of high MW PTFE have been widely used since the 1940s but process modification for direct polymerization of low MW PTFE has been more recent. Dispersion polymerization of low MW PTFE is currently most commonly employed. The major difference in polymerization between high and low MW PTFE is the addition of a chain-transfer-agent (CTA). Since the addition of a CTA reduces polymerization rate, higher initiator levels and other additives may be needed. Several CTAs have been employed but hydrocarbons such as ethane or methane are most commonly used. The amounts of CTA and initiator are used to control MW. Direct polymerization provides PTFE with a narrow MW distribution and polymer end-group types that are dependent on the CTA and the initiator types. Low MW PTFE has been polymerized in organic liquids but water is the more usual polymerization medium. All of the major fluoropolymer manufacturers include directly polymerized low MW PTFE in their product lines.

As noted above, surface area is an important property for many micropowder applications. The only directly polymerized low MW PTFE available commercially is based on dispersion polymerization (high surface area) so either irradiated granular or thermally degraded PTFE is employed when a low surface area is preferred for the application.

The largest micropowder production volumes are produced via irradiation/grinding of high MW PTFE and there are many patent and journal references for this technique in addition to those cited above. The irradiation is usually carried out with a high energy electron beam and the irradiation dose is controlled to achieve the desired MW reduction (MW drops with increasing irradiation dose). The PTFE feedstock may be high MW off specification or scrap resin, first or marginal quality PTFE powder, or aqueous PTFE dispersion. The feedstock may be from either suspension ("granular") or dispersion polymerization but granular feedstock is more common. The polymer may be spread out thinly in trays or placed in an agitated vessel during irradiation. Water may be added to the polymer during irradiation for process cooling and polymer end-group control.

Grinding of the irradiated PTFE is generally accomplished with an air-jet mill since the desired particle size is usually quite small (<30 μm). Granular PTFE and scrap may be passed through a cutting step before jet-milling. The polymer becomes easier to grind with an increased irradiation dose, so high doses are used when the objective is a smaller particle size. The biggest advantages of the irradiation process are that it can employ the lowest cost feedstock and it is relatively easy to change the desired particle size and product MW. The process disadvantages are that MW distribution is generally broad and the resulting molecular end-groups are carboxylic acid and acid fluorides that may cause product problems for some applications. Product surface area is usually dependent on the feedstock type (dispersion or suspension polymerized polymer).

25.4.2 Addition to Printing Inks

The largest single use of low MW PTFE is as an additive to printing ink. Before low MW PTFE "wax" was developed, hydrocarbon waxes were used as a component in inks and coatings. The addition of waxes to printing ink protects the ink pigments from abrasion and inhibits printed pages from sticking to one another (typically called "blocking") [38, 39]. Even though low MW PTFE is typically more expensive than hydrocarbon wax, it has been found to be much more effective. Thus, much less

(~1/3) low MW PTFE is required to achieve a specified abrasion resistance or block resistance. Any type of wax tends to reduce print gloss and the lower level of required PTFE wax affords prints higher gloss. The particulate nature of PTFE micropowder appears to be important in improving abrasion resistance. A particle size that is slightly larger than the print thickness is believed to protrude slightly above the print surface and provides the contact point with other surfaces. This obviously means that particle size control is critical for this application. Particles that are smaller than the print thickness may precipitate to the print bottom and afford no benefit so a narrow particle size distribution is preferred. Most low MW PTFEs used in printing ink is based on suspension polymerization or scrap PTFE to minimize the possibility of excessively small particles. The ideal particle size depends on the printing process and print thickness but a typical size is 3–4 μm [40]. Other print properties improved by addition of low MW PTFE are a lower surface energy, resistance to soiling, and a silkier feel. The use of mixtures of hydrocarbons and PTFE waxes in printing ink is also widely practiced.

25.4.3 Addition to Coatings

There are two general types of coatings that may contain PTFE micropowders. Coatings may be either permanent or temporary. Permanent coatings are intended to last for some extended time whereas temporary coatings typically have a very limited lifetime. Permanent coatings of thermoplastic or thermosetting resins that contain PTFE particles are used on bakeware, iron soles, and chemical vessels. Such coatings benefit from improved abrasion resistance, lower surface energies, and easier cleaning. The permanent coating formulations typically have high viscosities that allow the use of fairly large PTFE particles. The maximum size for "smooth" coatings may be 20 μm but coatings with a rough or "matte" finish can be provided by even larger PTFE particles. A high concentration of PTFE particles has been used to make coatings opaque that would otherwise be clear.

Temporary coatings are usually produced from PTFE suspended in a low viscosity liquid. They are used as lubricating coatings on items such as mechanical fasteners or moving parts and as release coatings that are applied to molds used to produce plastic or elastomer parts.

25.4.4 Addition to Plastics

It has been found that the addition of fluoroplastics to a wide variety of thermoplastics can afford compounds/alloys superior properties, especially wear and friction properties. High-MW PTFE was first used for this purpose but processing problems were experienced that limited its possible addition level to less than about 5%. Improvements in wear performance are largely a function of the PTFE addition level. It was eventually found that much higher loadings (up to 20%) of low MW PTFE afforded good wear and lubricity without causing processing problems. Addition to plastics is the second largest application of PTFE micropowder use. Plastic resin producers have product lines that include products which contain PTFE micropowders but much of these compounded products are produced by specialty compounders.

The degree of wear/friction improvement depends on the particular plastic to which the PTFE is being added [41]. As examples, the above referenced study showed that the addition of 15% low MW PTFE to a 6/6 polyamide, a polyacetal resin, and a polycarbonate reduced the specific wear rates by 97%, 46%, and 97%, respectively. The dynamic friction coefficients were reduced by 76%, 48% and 97%, respectively. Although these are reasonable comparisons, wear and friction performance can depend on the exact contact conditions such as velocity, pressure, and countersurface composition. Commercial compounds of thermoplastics with PTFE will vary depending on the application requirements but an addition level of 15–20% is usually considered optimum for the best wear performance.

The preferred average particle size for use as an additive to plastics is 10–20 μm and the preferred PTFE source is from suspension polymerization [42]. One concern with dispersion-polymerized PTFE is that their sub-micron primary particles may be easily "picked" from a plastic surface during wear. Larger particles may be imbedded in the plastic surface, thereby inhibiting removal. An equivalent weight of small particles also means more particles that might increase nucleation during crystallization of some plastics. An excessively large particle might appear as a visible part defect or act as a stress point to cause mechanical failure of a plastic part.

There have been some reports of the addition of PTFE to plastic resins to improve their soil and stain repellency and/or barrier properties; however, the addition of noncrystalline fluorochemicals is more common for this purpose [43].

High MW PTFE is added to some commercial thermoplastics to enhance resistance to fire. The PTFE forms a char on the surface of the plastic preventing dripping of polymer melt and inhibiting the fire. The usual explanation for its effect has been that a spider web structure of the fibrillated high MW PTFE supports the char. Although high MW PTFE is commonly used for this purpose, there have been reports of using low MW PTFE for this purpose as well.

25.4.5 Addition to Lubricants

The primary reason to use lubricants is to reduce friction and wear between two interacting surfaces. Hydrocarbon oils have the proper friction properties to meet these requirements but their low viscosity may cause them to be forced out of the contact region between interfaces. Powders of low MW PTFE may be added to liquid lubricants to provide "reserve" lubrication in case the liquid phase is forced out. Low-MW PTFE that is used this way is sometimes called an "extreme pressure" or "boundary" additive. The type of PTFE used in lubricants may be from either suspension or dispersion polymerization but the small particle size of dispersion-type PTFE is usually preferred to help maintain dispersion in the oil. Many Journal articles and patents have been published that report the performance of lubricating oils with and without the addition of PTFE. For example, Rico et al. [44] provided the results of an extreme pressure wear study of steel balls (Shell four-ball test) with several mineral oils containing four different percentages (1–10%) of PTFE.

Wear scars became smaller and the load/speed at which welding of metal took place increased significantly as the PTFE level was raised. The PTFE used for this testing was produced via dispersion polymerization.

Greases are liquid hydrocarbons (or fluorocarbons) that have been thickened with additives to a higher viscosity. The most common thickeners are clays and fatty acid soaps but low MW PTFE may also be used, especially with fluorocarbon oils. As with low viscosity oil lubricants, PTFE is used in hydrocarbon grease as a reserve to provide lubrication under more extreme conditions. However, PTFE is also used because of its white color and superior temperature and chemical resistance. Some hydrocarbon greases are thickened with a mixture of PTFE and clay or soap thickener. Fluorocarbon greases almost always utilize low MW PTFE as their thickener. Although PTFE can act as a lubricant in grease, the primary lubricant is the oil so the objective is usually to use the minimum amount of PTFE that is necessary to achieve the desired grease viscosity. Grease viscosity is a function of the particle size and surface area of the thickener, so small size, high surface area PTFE micropowders are desired for such products. The thickening of fluorinated oils with low MW PTFE having a small particle size and high surface area is discussed by G. Bell [45] and Maccone et al. [46]. Bell evaluated the thickening efficiency of PTFE powders with surface areas from 2.3 to 20 m^2/g and found the required amount of PTFE to attain their goal viscosity was reduced from 34.3 to 20.0%. The emphasis of Maccone et al. was the development of a PTFE with a primary particle size of 50 nm and a surface area of 30 m^2/g. About 50% less of this PTFE was required compared to a control (230 nm particle size and 8 m^2/g) to produce grease with a specific viscosity.

25.4.6 Addition to Elastomers (Rubbers)

Elastomers or rubbers are typically cross-linked amorphous polymers that are used above their glass transition temperature. There are many variations of the basic elastomer types and some benefit from the addition of low MW PTFE more than others. The elastomer variations include basic rubber stocks, crosslinking agents, and additives. The potential advantages of the addition of low MW PTFE are improved abrasion resistance, reduced friction, improved chemical resistance, and less "stickiness" [47]. The abrasion resistance of elastomers is highly dependent on their mechanical strength so PTFE addition is more beneficial for the weaker elastomers such as silicones and fluoroelastomers. Reduced friction is dependent on the PTFE being at the surface of elastomer parts and will depend on the manufacturing process of the parts and whether any wearing of the initial surface has taken place. Improved oil resistance has been claimed via addition of low MW PTFE to a variety of elastomers [48,49].

25.5 APPLICATIONS FOR AQUEOUS PERFLUOROPLASTIC DISPERSIONS

The melt processible perfluoroplastics highlighted in the above sections are often polymerized in totally aqueous media and are therefore available as aqueous dispersions of perfluoroplastic particles with a mean volumetric particle size of 150 to

250 nm. These "raw" dispersions are most often concentrated to higher solids content and stabilized with nonionic surfactants. The same is true for aqueous dispersions of high MW PTFE. Detailed discussion of applications of the isolated resins of high MW PTFE (i.e., PTFE fine powder and granular resins) have been omitted in this chapter since these resins cannot be considered conventional melt processible polymers. However, the melt processing techniques used for producing coatings from dispersions of the melt processible polymers and high MW PTFE are often the same. In many applications, the coating formulation may include mixtures of PTFE and PFA or FEP. The formulated coatings may be applied by any number of techniques such as spray coating (electrostatic and conventional), dip coating, spin coating, coil coating, roller coating and curtain coating. For details on formulation and processing of coatings, the reader is referred to an excellent reference by McKeen [50].

25.5.1 Coatings on Metal Cookware

Probably the most widely recognized application of perfluoroplastic coatings is in consumer cookware. The cooking process can be a somewhat harsh chemical environment with acids, such as present in tomato based products, held at elevated temperatures. In addition, many food substances can be extremely sticky. Normally during the cooking process, utensils are used which also imparts some mechanical abuse to the cooking surface. Key properties for this application are high thermal stability, good chemical resistance and excellent release. For the consumer, this translates into a durable, easily cleaned surface. The perfluoroplastic is not inherently resistant to extreme mechanical abuse at high temperatures, nor is it likely to adhere directly to the metal substrate. The coatings are therefore formulated with other additives in order to strike the correct balance of properties. Generally the coating for consumer cookware is applied in multiple layers with each layer having different formulation and functionality.

25.5.2 Coatings on Woven Substrates

Among the aqueous perfluoroplastic dispersions, high MW PTFE dispersions are used in the highest volume. For coatings on woven glass cloth, it is normally the only perfluoroplastic component in the formulation. A huge amount of coated glass cloth is produced each year. It is normally manufactured by a dip coating process where the woven glass fabric is passed through a reservoir of formulated dispersion, dried and then sintered. In the sintering step, the temperature is raised to above the melting point of the PTFE in order to cause the particles to coalesce. Multiple passes through the coater is normally required to build the desired coating thickness.

The applications for PTFE-coated glass cloth are numerous. As an architectural fabric, the translucent fabric allows light to be transmitted. This affords protection from the environment while keeping a level of indirect sunlight inside the structure. The flexible coated fabric is unaffected by UV light and the PTFE provides resistance to acid rain and other environmental conditions. The surface is, in effect, self-cleaning because the nonstick properties allow rain to wash off deposits. Nonflammability and stain resistance of the PTFE fabric is also advantageous.

Another use for PTFE-coated glass fabric is in industrial cooking. The flexible fabric may be used as a conveyor belt through the cooking zone of industrial equipment where the flexibility, thermal and chemical resistance, and the release properties of the polymer are key. Because PTFE exhibits excellent performance at a very wide range of temperatures, the fabric may also be employed as the conveyor belting in coolers and freezers.

In some applications, the PTFE coating is not a continuous coating and the polymer is not sintered. Most notable is where an open weave glass cloth is utilized as a filter fabric. For this application, the cloth is dip-coated and the dispersion is dried. While in service, the filter fabrics often flex due to the force of the media being filtered. Also, pressure pulses may be employed in order to dislodge filtered particulate. In this application, the main property of value is the lubricity of the PTFE. Since glass is inherently abrasive, the presence of PTFE particles provides lubrication during movement of the glass fibers and thereby extends the life of the fabric. Equally important is the benefit of chemical resistance and improvement in nonstick properties.

25.5.3 Coatings for Industrial Applications

Perfluoroplastic dispersions which are formulated into paints find utility in many different, and sometimes unexpected, industrial applications. A few such examples are coatings on molds, coatings on automotive paint racks, coatings on air conditioner pistons, coatings on equipment and fasteners in the marine environment, coatings for down-hole oil and gas applications, and coatings inside metered dose inhalers.

Rubber compounds such as those used in the manufacture of running shoe soles are injection or compression molded to form the final shape. These compounds are prone to sticking to the metal mold. In such cases, perfluoroplastic coatings are very useful in providing a durable, nonstick surface on the molds. The key properties of value are the thermal stability and release properties of the perfluoroplastic.

Automotive paint racks are used to hold parts while they are being painted and these racks must be used with several different colors. The perfluoroplastic coatings when applied to automotive paint racks allow the user to easily and quickly remove paint on the support rack with a simple high pressure water lance. In this application, both the nonstick property and the chemical resistance of the perfluoroplastic coating to the automotive paint formulation are critical.

Aluminum pistons which are used to compress the refrigerant in automotive air conditioners are often coated with a perfluoroplastic formulation. These coatings must be resistant to attack by the refrigerant used in the system and must be able to operate at the elevated temperatures that are produced upon compression of the refrigerant gas. However, the main utility of the perfluoroplastic coating in this application lies in its lubricity. By minimizing wear of this moving part, the lifetime of the compressor is greatly increased.

The down-hole oil and gas environment is especially challenging. Conditions vary from well to well but usually involve a corrosive, high temperature environment with a multitude of organic and possibly inorganic components in the fluid being removed from the well. During operation of the well, various substances may precipitate on the well wall, thereby reducing production rates. Depending on the well, the fouling or scaling can be from asphaltenes, paraffins, waxes, inorganic compounds, etc. and combinations thereof. A properly formulated perfluoroplastic coating which is tightly bonded to the wall of the well provides a surface that is smooth, chemically resistant and possesses low surface energy. These coatings are always multiple layer systems which are formulated to maximize adhesion to metal, to maximize permeation resistance to small and/or corrosive molecules and to maximize the nonstick properties of the coating system. Use of these coatings can greatly reduce fouling rates which, for the end user, translates to reduced cleaning and maintenance cost, reduced production losses due to shutdowns, and reduced production rate losses between cleanups.

Perfluoroplastic coatings may be found on the inside of many metered dose inhalers. In this application, the objective is to keep all active ingredients dispersed in the fluid media. The thin perfluorinated surface provides a very nonreactive and impermeable barrier. As a result, the organic active ingredients are prevented from partitioning to the container wall and thereby changing the concentration of the drug to be delivered.

Not all perfluoroplastic coatings need be applied from liquid formulations. The aqueous dispersions can be isolated, dried and the resulting powder processed into an optimal particle size for powder coating applications. Some industrial vessels, equipment, and ductwork utilize powder-coated perfluoroplastic for thin layers of protection. One of the largest applications of perfluoroplastic powder coatings is in industrial bakeware. This type of cookware is normally used in automated systems and cycled through the baking process thousands of times. After each trip through the baking process, the final product is expected to be released from the cooking surface by gravity. In these applications a liquid primer is normally applied to provide adhesion to the substrate and then a topcoat of perfluoroplastic is applied via a powder coating process. Such coatings are formulated for durability and to utilize the nonstick property of perfluoroplastics to maximize long-term release.

Another large application of powder-coated perfluoroplastic is in automotive weld nuts and bolts. In the assembly of automobiles, nuts and bolts are often welded onto sheet metal parts for subsequent final assembly. In the normally automated spot welding process, sparks of molten metal may land on the threads of the nut or bolt. In the subsequent dipping of the part into a paint bath, the threads may be coated with paint. In these applications, the threads of the nuts or bolts are powder coated with high temperature perfluoroplastic such as PFA. The coating is melted to form a continuous layer. This perfluoroplastic coating is not formulated to bond tightly to the metal substrate. During the welding process, splattered molten metal will not attach to the metal thread because a barrier of perfluoroplastic is in place. Furthermore, the perfluoroplastic does not degrade in the short-term high temperature exposure of the welding process. In addition, the subsequent dipping or spraying of the automotive part with paint results in the paint being placed on top of the perfluoroplastic surface. During assembly by automated torque wrenches, the free bolt being screwed into a welded nut or a free nut being screwed onto a welded bolt will simply peel off the loosely attached perfluoroplastic on the welded fastener along with any metal spatter or dried paint. With no obstructions, the fasteners seat properly with predetermined torque. In this process, the threads of the fastener are protected from both molten metal and paint. The properties that are key to this application are the nonstick characteristics and high temperature performance of the perfluoroplastic.

There are some applications in which clear coatings are desired, but probably the most unique is the coating of light bulbs. There are some uses for light bulbs where even the remote possibility of broken glass falling to surfaces below is not acceptable. Examples are in food preparation or serving areas and in hospital operating rooms. In these cases, the high temperature melt processible perfluoroplastic, PFA, may be found. PFA powder coatings may be applied to glass light bulbs and melted to form a continuous encapsulating layer. If the bulb is broken, the PFA coating prevents the glass fragments from dispersing into the environment. The key properties of use in this application are transparency and high temperature stability.

25.5.4 Compounded Products

A convenient way to disperse additives in perfluoroplastics is by blending with aqueous dispersions. Generally, such mixing is done with "raw" or "as polymerized" dispersions that have not been concentrated or stabilized. After mixing, the dispersion is coagulated and the compounded product is dried. The resulting highly dispersed mixture can be formed into shapes for further processing or use. A typical use for such compounded products would be as bearing surfaces. In this application, the perfluoroplastic provides lubricity and chemical resistance while the filler provides other properties such as wear resistance or electrical conductivity. In gasket or packing applications, the filler may also impart creep resistance to the perfluoroplastic structure.

25.6 CONCLUSION

In this chapter, we have highlighted a few of the many applications of melt processible perfluoroplastics and attempted to show the combination of properties that make such plastics an ideal fit for the application. Out of necessity, many applications have been omitted. Examples include extruded films which can be used from such simple applications as release sheets and vacuum bags in composite molding to more complex, integral parts of laminated structures such as interlayer dielectrics in circuit boards. All of the applications found in the semiconductor industry may also be found in the larger CPI where all of the properties, with the possible exception of ultra high purity, come to bear. In general, the use of perfluoroplastics in demanding applications is only limited by need and imagination.

Except for a small introduction in the section on aqueous dispersions, high MW PTFE is one perfluoroplastic which is not covered in this chapter. This is because, in the strictest sense, resins of high MW PTFE are not melt processible. Granular PTFE resins are processed by sintered metal technology techniques. PTFE fine powders are processed by paste extrusion techniques. The structures created with the use of these resins can be quite similar to melt processible perfluoroplastic parts or can be very different such as structures with controlled porosity. The applications of high MW PTFE are as numerous and as varied as those of melt processible perfluoroplastics. However, one common characteristic is that all of the applications utilize cleaver processing techniques to exploit combinations of the unique properties which are inherent in the particular polymer.

REFERENCES

1. O'Hagan, D. *Chem. Soc. Rev.* **2008**, *37*(2), 308–319.
2. Fernandes, T. *BICSI Newsletter* **2001**, *11*, 5–8.
3. Harris, J.; McCane, D. Polymers of Perfluoroalkoxy Perfluorovinyl Ethers. US Patent 3,132,123, May 5, 1964.
4. Bro, M. I.; Sandt, B. W. Novel Perfluorocarbon Polymers. US Patent 2,946,763, July 26, 1960.
5. DuPont. *Teflon® PTFE Fluoropolymer Resin Handbook*; Bulletin H-37051-4; DuPont Fluoroproducts: Wilmington, DE, 2004.
6. DuPont. *Teflon® PFA Fluoropolymer Resins Properties Handbook*; Bulletin H-37052-3; DuPont Fluoroproducts: Wilmington, DE, 1998.
7. DuPont. *Teflon® FEP Fluoropolymer Resins Product and Properties Handbook*; Bulletin H92068-1; DuPont Fluoroproducts: Wilmington, DE, 2003.
8. Chapman, G. A. *The Past and Future of Fluoropolymers in Wire and Cable*. IWCS, Session 6-1; 2004.
9. Imbalzano, J. F.; Kerbow, D. L. Stable Tetrafluoroethylene Copolymers. US Patent 4,743,658, May 10, 1988.
10. Carlson, D. P. Tetrafluoroethylene Terpolymers. US Patent 4,029,868, June 14, 1977.
11. Earnest, T. R.; Faverau, D. A.; McKee, N. D.; Tooley, P. A. High Melt Flow Fluoropolymer. US Patent 7,122,609, October 17, 2006.
12. Clarke, F. B., III; Gewain, R. G. *International Limited Combustible Plenum Cable Fire Test Project*; The Fire Protection Research Foundation, March 2001.
13. Sampson, R. *Cabling Bus. Mag.* **2003**, *13*, 20.
14. Thuot, G. G.; Xiao, L. *Wire Cable Technol. Int.* **2012**, *40*(3), 66.
15. Young, R. T.; Thuot, G. G. *Antenna Syst. Technol.* **2011**, 9–12.
16. Hanford, W. E.; Roland, J. R. Tetrafluoroethylene Copolymers. US Patent 2,468,664, April 26, 1949.
17. Carlson, D. P. Copolymers of Ethylene/Tetrafluoroethylene and of Ethylene/Chlorotrifluoroethylene. US Patent 3,624,250, November 30, 1971.
18. Michalczyk, M. J.; Sharp, K. G.; Stewart, C. W. Fluoropolymer Nanocomposites. US Patent 5,726,247, May 10, 1998.
19. Apostolo, M.; Triulzi, F. *J. Fluorine Chem.* **2004**, *125*, 303.
20. Polastri, F.; Ligabue, O.; Kapeliouchko, V.; Garrison, D.; Marchese, E. *PTFE Nanocomposites in Tubes and W&C*; IWCS, Session 6-2; 2007.
21. Carlson, D. P.; Schmiegel, W. Organic Fluoropolymers, In *Ullmann's Encyclopedia of Industrial Chemistry*, 5th Edition; Gerhartz, W., Ed.; Wiley-VCH: Weinheim, **1988**; Vol. A11, p 408.

22. DuPont. *Meeting Semiconductor Needs for Cleanliness and Reliability*; DuPont Teflon Fluoropolymer Resin Product Information; DuPont Fluoroproducts: Wilmington, DE, 2003.
23. DuPont. *Teflon® PFA HP Safeguards Purity in Analytical Systems*; Application Profile; DuPont Fluoroproducts: Wilmington, DE, 2001.
24. Ebnesajjad, S. *Fluoroplastics: Melt Processible Fluoropolymers: The Definitive Users Guide and Databook*; Plastics Design Library Series; William Andrew Publishing Inc.: Norwich, NY, 2003; Vol. *2*.
25. Gangal, S. V.; Brothers, P. D. Perfluorinated Polymers, Tetrafluoroethylene–Perfluorovinyl Ether Copolymers. In *Encyclopedia of Polymer Science and Technology*, 4th Edition, John Wiley & Sons, Hoboken, NY, 2014; Vol. 9, pp. 547–559.
26. Brady, J. D. Telomers of Tetrafluoroethylene. US Patent 3,067,262, December 4, 1962.
27. Skehan, J. T. Grease Composition. US Patent 3,505,229, April 7, 1970.
28. Green, L. Q.; Moses, R. W. Aqueous Fluorocarbon Telomer Dispersion and Process of Preparing Same. US Patent 3,105,824, October 1, 1963.
29. Kuhls, J.; Fitz, H.; Haasemann, P. Fluorocarbon Waxes and Process for Producing Them. US Patent 3,956,000, May 11, 1976.
30. Morgan, R. A.; D'Haenens, L. G. P. J. Non-Chalking Release/Wear Coating. US Patent 6,060,167, May 9, 2000.
31. Mayer, L.; Tonnes, K.; Bladel, H. Polytetrafluoroethylene Micropowders, Their Preparation and Use. US Patent 5,641,571, June 24, 1997.
32. Morgan, R. A. Directly Polymerized Low Molecular Weight Granular Polytetrafluoroethylene. US Patent 7,176,265, February 13, 2007.
33. Dillon, J. A. Process for Non-Destructive Radiation Degradation of Polytetrafluoroethylene. US Patent 3,766,031, October 16, 1973.
34. Derbyshire, R. L. Treatment of Sintered Poly-Tetrafluoroethylene with Irradiation and Heat to Produce a Grindable Material. US Patent 4,220,511, September 2, 1980.
35. Neuberg, W. B.; Luniewski, R. Apparatus and Method for Radiation Processing of Materials. US Patent 4,777,192, Oct 11, 1988.
36. Hartwimmer, R.; Weiss, E. The Process for Preparation of Fluorocarbon Waxes. US Patent 3,813,449, May 28, 1974.
37. Hinds, C. Tetrafluoroethylene Products with Enhanced Crystallinity and Processes for Producing the Same. US Patent 6,555,648 B1, April 29, 2003.
38. Carlick, D. J.; Bassemir, R. W.; Krishnan, R.; Durand, R. R. Low Rub Printing Ink. US Patent 5,158,606, October 27, 1992.
39. Carroll, J. R.; Bradley, R. M.; Kalmikoff, A. I. *Mod. Paint Coat.* 1993.
40. Ebnesajjad, S.; Morgan, R. A. *Fluoropolymer Additives*; Plastics Design Library; Elsevier: Waltham, MA, 2012; p 152.
41. Mens, J. W.; de Gee, M. A. W. J. *Wear* **1991**, *149*, 255–268.
42. Ebnesajjad, S.; Morgan, R. A. *Fluoropolymer Additives*; Plastics Design Library; Elsevier: Waltham, MA, 2012; p 116.
43. Sterling, R. E.; Goldberg, E. P. Thermoplastic Polymers with Dispersed Fluorocarbon Additives. US Patent 6,541,558, April 1, 2003.
44. Rico, E. F.; Minondo, I.; Cuervo, D. G. *Wear* **2007**, *262*, 1399–1406.

45. Bell, G. A. The Effect of PTFE Thickener Particle Characteristics on Grease Formulation and Performance, In *7th ELGI Annual General Meeting*, 1995.

46. Maccone, P.; Palamone, G.; Kapeliouchko, V. *Eurogrease* **2004**, May/June, 17–21.

47. Ebnesajjad, S.; Morgan, R. A. *Fluoropolymer Additives*; Plastics Design Library; Elsevier: Waltham, MA, 2012; pp 182–184.

48. Fukunaga, N. Oil-Repellent Fluororubber Composition. US Patent 5,428,096, June 27, 1995.

49. Park, E. H. Reinforced Elastomeric Seal. US Patent 7,658,387, February 9, 2010.

50. McKeen, L. W. *Fluorinated Coatings and Finishes Handbook: The Definitive User's Guide and Handbook*; Ebnesajjad, S., Ed.; Plastics Design Library Series; William Andrew Publishing Inc.: Norwich, NY, 2006.

INDEX

Handbook of Fluoropolymer Science and Technology, First Edition. Edited by Dennis W. Smith Jr., Scott T. Iacono, and Suresh S. Iyer.

Printed and bound by CPI Group (UK) Ltd, Croydon, CR0 4YY
18/09/2026
14964971-0002